MW01013301

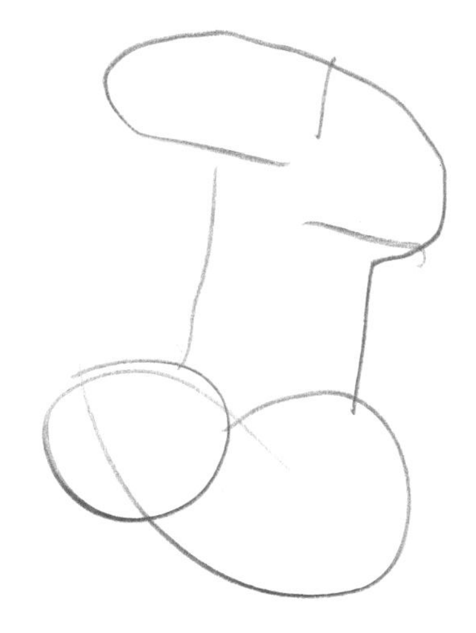

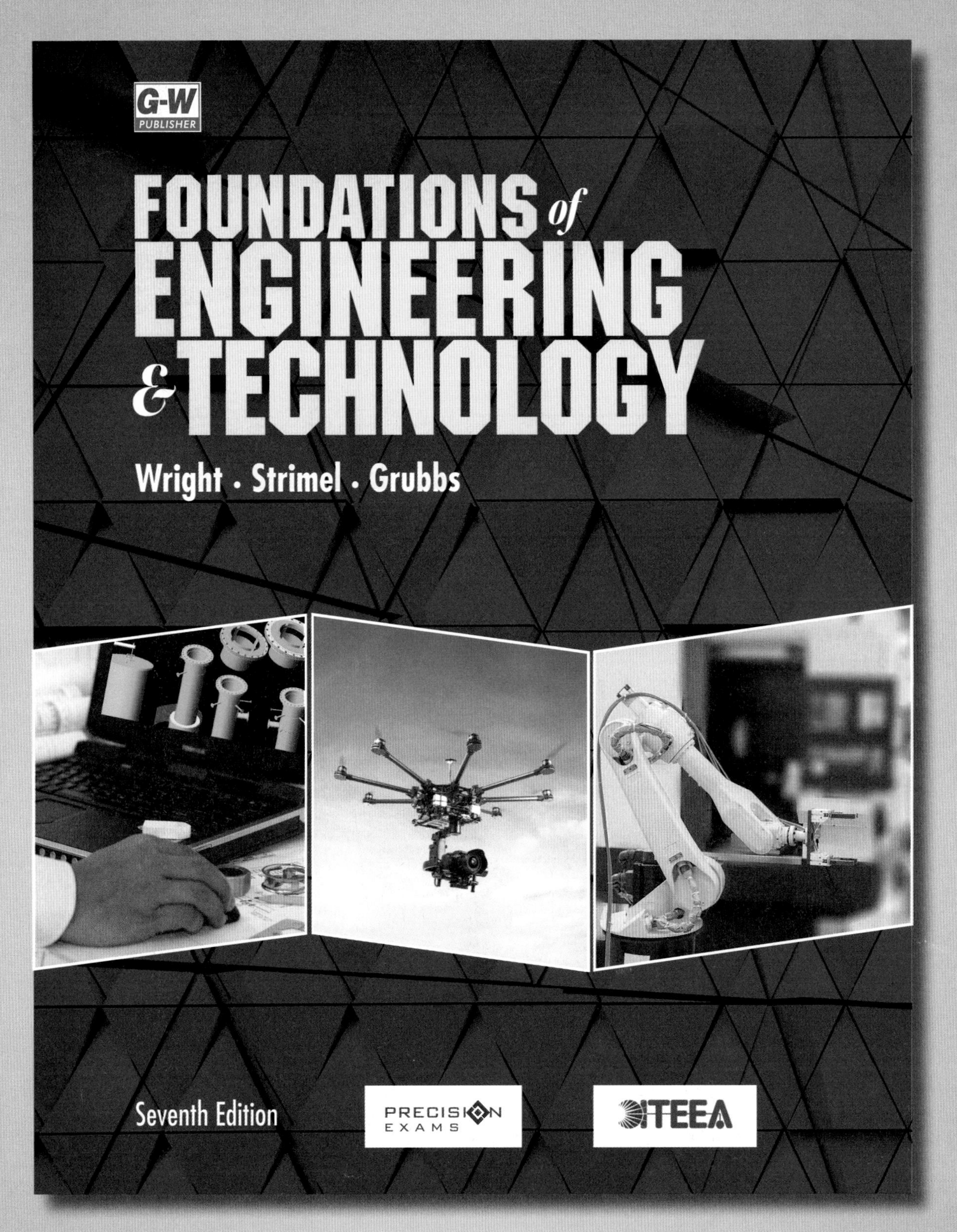

A comprehensive study of technology and of the engineering needed to create technology.

Guided Tour

Learning Objectives clearly identify the knowledge and skills to be obtained when the chapter is completed.

Technical Terms list the key terms to be learned in the chapter.

G-W Learning Companion Website Activity Icon identifies related content available on the G-W Learning companion website.

Close Reading features provide students with questions to think about while reading the chapter.

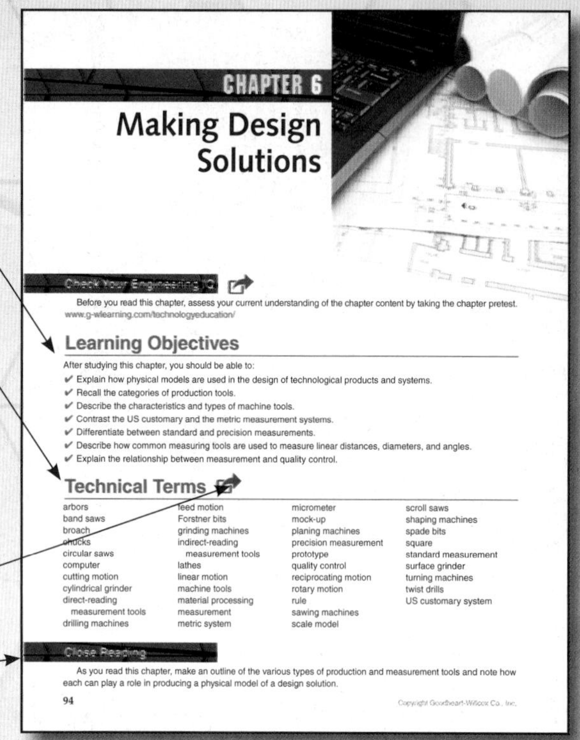
CHAPTER 6

Making Design Solutions

Check Your Engineering IQ

Before you read this chapter, assess your current understanding of the chapter content by taking the chapter pretest.
www.g-wlearning.com/technologyeducation/

Learning Objectives

After studying this chapter, you should be able to:
- Explain how physical models are used in the design of technological products and systems.
- Recall the categories of production tools.
- Describe the characteristics and types of machine tools.
- Contrast the US customary and the metric measurement systems.
- Differentiate between standard and precision measurements.
- Describe how common measuring tools are used to measure linear distances, diameters, and angles.
- Explain the relationship between measurement and quality control.

Technical Terms

arbors	feed motion	micrometer	scroll saws
band saws	Forstner bits	mock-up	shaping machines
broach	grinding machines	planing machines	spade bits
chucks	indirect-reading measurement tools	precision measurement	square
circular saws	lathes	prototype	standard measurement
computer	linear motion	quality control	surface grinder
cutting motion	machine tools	reciprocating motion	turning machines
cylindrical grinder	material processing	rotary motion	twist drills
direct-reading measurement tools	measurement	rule	US customary system
drilling machines	metric system	sawing machines	
		scale model	

Close Reading

As you read this chapter, make an outline of the various types of production and measurement tools and note how each can play a role in producing a physical model of a design solution.

94

Copyright Goodheart-Willcox Co., Inc.

Tomorrow's Technology Today features are found at the beginning of each section. They highlight emerging technologies.

Nanotechnology

Many of today's technological products seem to be getting smaller and smaller—from machines to computer chips. It is estimated that, within the next 50 years, these products will continue to decrease in size so much that thousands of machines will be able to fit on the point of a pencil! In the coming decades, nanotechnology, or "the science of the small," will use these tiny machines for many amazing applications. One nanometer equals one-millionth of a millimeter. We are talking about extremely tiny machines.

All matter in the universe is made up of atoms. Cells are nature's tiny machines. We still have a lot to learn about constructing materials on such a small scale. With new developments in nanotechnology, however, we are getting closer. Many scientists believe we will be able to manipulate each individual atom of an object in the near future.

Nanotechnology combines engineering with chemistry. The goal of nanotechnology is to manipulate atoms and place them in a specific pattern. These patterns will produce a certain kind of structure. Depending on the atoms used and the patterns created, nanotechnology can be applied in several different areas. The most promising uses for this technology are in medicine, construction, computers, and the environment.

This branch of technology is expected to greatly impact the medical industry. Nanotechnology devices will be programmed to attack cancer cells and viruses and rearrange their molecular structures to make them harmless. The tiny robots might also be able to slow the aging process. This might increase life expectancy. Since these machines will be able to perform work a thousand times more precisely than current methods, they can also be programmed to perform delicate surgeries.

The construction industry will benefit from nanotechnology as well. The tiny machines will be used to make much stronger fibers. There is hope that, one day, we will be able to replicate anything, from diamonds to food.

Computer parts continue to get smaller all the time. They will, however, soon reach their limit. Nanotechnology will be used to make much smaller storage devices with the capability to hold much more information.

Finally, nanotechnology can be used to better the environment. We are currently running out of many natural resources. With nanotechnology, we will be able to construct more of them. Manufacturing with nanotechnology produces much less pollution than traditional manufacturing. In addition, these devices can potentially be programmed to rebuild the ozone layer and clean up oil spills.

Nanotechnology has many exciting potential uses. This technology, however, also has its challenges. Similar to all new technology, it raises questions about its impacts on people, society, and the environment.

Technology Explained features briefly explain how technological devices and systems work.

Illustrations have been designed to clearly and simply communicate the specific topic. Illustrations have been completely replaced and updated for this edition.

Technical Terms appear in bold italics where they are defined.

STEM Connections provide information on topics relevant to chapter materials that connects the content to science, technology, engineering, or math.

20 Foundations of Engineering & Technology

Hammers are also used to remove nails. This requires the exertion of heavy pressure. Using their knowledge of mathematics and science, early engineers looked to mechanical advantage to develop a way to remove nails. ***Mechanical advantage*** is a measure of the ability to amplify the amount of effort exerted using some type of mechanical device. Using mathematical practices to calculate mechanical advantage, engineers solved the problem by adding a claw to the back of the hammer, **Figure 2-3.** The claw is used as a lever, amplifying the ability to pry a nail out of the material without exerting a great deal of effort. This process of developing and improving technologies using knowledge of mathematics and science is engineering.

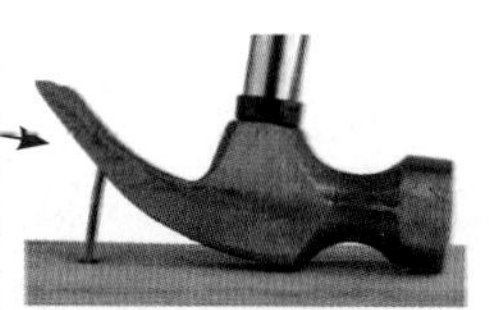
OlegSam/Shutterstock.com

Figure 2-3. An example of mechanical advantage is using the claw on a hammer to remove a nail.

STEM Connection: Science and Mathematics — Calculating Mechanical Advantage

Tools and machines are developed to extend human capabilities by achieving mechanical advantage. Recall that mechanical advantage is a measure of the ability to amplify the effort exerted using some type of mechanical device. When greater mechanical advantage is achieved, greater tasks can be accomplished. Determining the best ways to achieve mechanical advantage when using machines requires scientific knowledge of materials and forces and mathematical abilities to understand and calculate the level of mechanical advantage achieved.

Tools and machines are classified as one or more of six simple machines. Simple machines are the most basic mechanisms used to change a force being exerted to achieve some type of task. The six simple machines are the lever, wheel and axle, inclined plane, wedge, pulley, and screw. The claw found on a hammer or an adze is used to remove nails. It is considered a lever because it provides leverage to remove nails. A lever is comprised of a rigid bar that rests on a fulcrum, or a pivot point, to make moving objects easier.

Claws achieve mechanical advantage by converting a small input force into a larger output force. Mechanical advantage is calculated as a ratio of output force to the input force. Therefore, if a claw produces an output force 10 times greater than the force applied, it has a mechanical advantage of 10. To calculate the estimated mechanical advantage of a lever, use this formula:

$$\text{mechanical advantage (MA)} = \frac{\text{effort arm length}}{\text{load arm length}}$$

The effort arm length is the distance between the spot where effort is applied and the fulcrum. The load arm length is the distance between the load that will be moved and the fulcrum. See **Figure A.** The effort arm length is 5" and the load arm length is 1/2", providing a mechanical advantage of 10. This means that the claw multiplies the input force exerted 10 times.

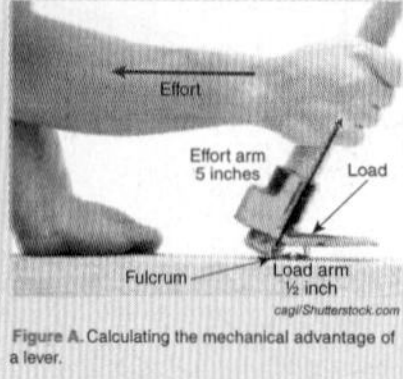

cagi/Shutterstock.com

Figure A. Calculating the mechanical advantage of a lever.

Copyright Goodheart-Willcox Co., Inc.

Technology Explained — 3-D Printer

3-D printer: A device that produces three-dimensional solid objects from binary data on a digital file created using computer-aided design software.

3-D printers are changing the way products are produced. 3-D printing is an additive manufacturing process used to produce almost any three-dimensional product by adding layer by layer of material to create the product.

To print a three-dimensional product, a three-dimensional blueprint of the product is first made using computer-aided design (CAD) software. This blueprint can then be sent to the 3-D printer to print the object. The most common material used to print three-dimensional objects is plastic. However, materials such as glass, wax, metal, and even paper can be used.

Some 3-D printers have a spool of plastic material in a string-like form.

iStock/Thinkstock.com/Russell Bonam

Figure A. A three-dimensional sphere being printed.

Academic Connection: History — Ancient Architecture

Civilizations throughout history are often defined by their structures. This is because a civilization's cultural traditions and religions, elements and resources of certain geographical regions, and technological development influenced the practices and processes for planning and producing their structures. These influences shaped the way structures look and the methods in which they were built during many different time periods at diverse locations around the world. For this reason, an ancient Egyptian structure, **Figure A**, is very different from an ancient Grecian-Roman structure, **Figure B.**

Early Egyptian societies believed in many gods and carried out a variety of rituals in respect to these deities. The design of structures was often under the direction of a priest who would plan the shape and form of structures to facilitate the rituals and to honor their gods. On the other hand, early Greek and Roman architecture followed their societies' focus on civic life. They built with newly discovered materials such as marble and concrete. The structures were planned to accommodate municipal and governmental activities. The use of different materials between the two societies makes it possible to distinguish between their architecture.

Nestor Noci/Shutterstock.com

Figure A. Ancient Egyptian structure.

sFaseevo/Shutterstock.com

Figure B. Ancient Grecian-Roman structure.

Academic Connections provide information on topics relevant to chapter materials that connects the content to communication, history, literacy, and essay writing.

Think Green Product Life Cycle and Reduction

In the manufacturing industry, *product life cycle* refers to the stages a product goes through from its design through its production to its eventual disposal. Examining this entire process helps manufacturers determine ways to increase the efficiency of the production while minimizing wastes that are costly and harmful to the environment. Finding ways to minimize waste throughout a product's life cycle is called *reduction*. One example of reduction is when manufacturers change the size and shape of a product design to use mostly materials in their standard sizes. This practice reduces the processing of the materials, minimizing waste and reducing waste disposal costs. For instance, a standard sheet of plywood is typically 4′ × 8′. Therefore, if a product is designed to use 2 full sheets of plywood in its production versus using only partial sheets, the amount of energy for separating or cutting the sheets is minimized and the amount of waste to be disposed of is reduced.

Now think about ways you can reduce waste in common activities. Examples are making two-sided photocopies, using cloth instead of paper items, and turning off lights when you are not using them. What other examples can you think of?

You may be surprised at how small adjustments in your everyday life can help the environment.

Think Green notes highlight key items related to sustainability, energy efficiency, and environmental issues.

Career Connection Automotive Technician

Automotive technicians use skill and knowledge to inspect, maintain, and repair automobiles and light trucks. They review the description of a problem, use a diagnostic approach to locate the problem, repair the vehicle, and test the repair. Automotive technicians must be able to use a variety of testing equipment. This equipment includes onboard and stationary diagnostic equipment and computers.

High school students interested in becoming automotive technicians can take courses such as technology and engineering education, automotive repair, electronics, and computers. Interested students will need to complete some vocational or postsecondary education programs in automotive service for entry-level positions. These programs usually last six months to a year. Many are hands-on programs, centered on topics such as brake maintenance or engine performance.

PATIWIT HONGSANG/Shutterstock.com

Individuals who are organized, have strong customer service skills, enjoy troubleshooting, and are hands-on workers will perform well in this field. It is also important for automotive technicians to possess the necessary licenses, certifications, and registrations. Automotive technicians may work as automotive body and glass repairers, diesel service technicians, heavy vehicle and mobile equipment service technicians, and small engine mechanics. One of the largest professional organizations for automotive technicians is the National Institute for Automotive Service Excellence (ASE).

Employment in this area is expected to grow. Automotive technicians may have the following potential projects:

- Identify automotive problems and diagnose using computer software.
- Explain automotive issues to customers.
- Repair and replace worn and damaged parts.
- Perform routine care and maintenance.

Career Connection features can provide a path for career success.

Engineering Design Challenges provide laboratory activities requiring creativity and critical thinking skills.

TSA Modular Activity features provide additional activities intended to develop skills used in TSA competitive events.

Engineering Design Challenge

Vertical Structure Challenge

Background

You have learned about the forces that structures must be able to support and withstand. An assortment of structural members of different shapes and sizes is used to support the structure and withstand these forces. Determining the best design for the structure requires an understanding of structural analysis concepts in order to compute the load capacity of structural components. Using this knowledge, your team will design and build a model of an efficient vertical load-bearing structure, using limited materials.

Situation

Your team will use their civil engineering knowledge and skills to build the tallest, strongest, and most efficient tower using a maximum of 24 sheets of standard copy paper. The paper will be used to create the structural members of the tower. You may use 3/4″ cellophane tape to join the paper members. The tower must be at least 50 cm tall and span a 10 cm gap, **Figure A**. Your tower must support at least 2.5 kg and withstand an earthquake test, using an earthquake-testing platform (a movable platform connected to a rigid frame by rubber bands or springs). Once the tower passes the earthquake testing, place it over the 10 cm gap as shown in **Figure A**. The tower cannot be attached to the platform. To test the load-bearing capacity of the structure, place a bucket on top of the structure, then slowly add sand until the structure fails.

Desired Outcomes

Your team must design and construct a tower that meets the following requirements:

- Stands at least 50 cm tall
- Spans a 10 cm gap
- Composed of no more than 24 sheets of paper
- Supports at least 2.5 kg during earthquake testing

Your team must also develop a technical report that describes how the tower was designed using

364

TSA Modular Activity

Structural Engineering

Activity Overview

In this activity, you will create a balsa-wood bridge and determine its failure weight (the load at which the bridge breaks).

Materials

- Grid paper
- 20′ of 1/8″ × 1/8″ balsa wood
- 3″ × 5″ note card
- Glue

Background Information

General. There are several types of bridges. The length of the span and available materials generally determine the type of bridge used in a particular situation. For this activity, a truss design is considered the most efficient.

Truss bridges. The truss bridge design is based on the assumption that the structural members carry loads along their axes in compression or tension. The members along the bottom of the bridge carry a tensile load. Those along the top of the truss carry a compressive load. The members connecting the top and bottom chords (members) can be in tension or compression.

Gussets. Gussets are plates connected to members at joints to add strength. The structural steel members are welded or bolted to the gusset. When designing your bridge, include a gusset at each joint, if possible.

Wood properties. Due to its molecular structure, wood can normally carry a greater load in tension than it can in compression. Also, a shorter member can carry a greater compressive load than a longer member.

Guidelines

- Create a scale sketch of the bridge before building.
- Two pieces of balsa wood can be glued together along lengthwise surfaces. No more than two pieces of balsa can be glued together. Do not use excessive amounts of glue.
- Gussets cut from the 3″ × 5″ card can be no larger than the diameter of a US quarter. One gusset cannot touch another gusset. This plate cannot be sandwiched between two pieces of balsa wood.
- The bridge design must take into account the loading device. Your instructor will provide specific guidelines for bridge length, bridge width, and the required details for attachment of a loading device.
- Your bridge will be weighed before being loaded.

Evaluation Criteria

Your project will be evaluated using the following criteria:

- Accuracy of sketch, compared to completed bridge.
- Conformance to guidelines.
- Efficiency (failure weight ÷ bridge weight).

366

292 Foundations of Engineering & Technology

Summary

- Secondary manufacturing involves the actions used to change industrial materials into products.
- Casting and molding processes give materials shape by introducing a liquid material into a mold.
- Forming processes apply force through a forming device to cause the material to change shape.
- Separating processes remove excess material to make an object of the correct size and shape. Machining and shearing are two types of separating processes.
- Conditioning processes, which change the internal properties of a material, include mechanical and chemical conditioning.

Check Your Engineering IQ

Now that you have finished this chapter, see what you learned by taking the chapter posttest.
www.g-wlearning.com/technologyeducation/

Test Your Knowledge

Answer the following end-of-chapter questions using the information provided in this chapter.

1. Secondary manufacturing processes change industrial materials into _____.
2. Casting and molding processes give materials shape by introducing a(n) _____ material into a mold.
3. In forming processes, the force applied is in a range that is _____.
   A. above the material's plastic range and below the material's yield point
   B. above the material's yield point and below the material's elastic range
   C. above the material's yield point and below the material's fracture point

12. *True or False?* Product development includes designing the product and the production process.
13. An example of a(n) _____ in product development is the choice of materials or production processes versus the impact that these materials and processes may have on the environment.
14. _____ gauges can be placed at quality control checkpoints to determine whether a product meets the desired specification.
15. What is a manufacturing enterprise?

Critical Thinking

1. Choose a manufacturing industry and write a report that analyzes the impact that industry has on the environment.
2. Choose any type of product and describe the potential trade-offs the product designer had to consider when developing the product.
3. How is the engineering design process used in product development?

Chapter 15 Producing Products 293

STEM Applications

1. Select a simple product made from more than one part and do the following:
   A. List the parts the product is made of.
   B. Select one part. List the steps you think were used to manufacture it.
   C. Complete a form similar to the following for one of the steps.

Product:
Part name:
Production process:
Step needed to complete the process:

2. Investigate the concept of market research to explore the development of market surveys. Create a survey to determine consumer preferences for a specific product and collect responses to the survey from your school. Research how to analyze the results of the

Summary provides an additional review tool for you and reinforces key learning objectives.

Test Your Knowledge questions allow you to demonstrate knowledge, identification, and comprehension of chapter material.

Critical Thinking questions develop higher-order thinking, problem solving, personal, and workplace skills.

STEM Applications encourage application of concepts to real-life situations, developing skills related to chapter content.

Student Resources

Textbook

The *Foundations of Engineering & Technology* textbook provides an exciting, full-color, and highly illustrated learning resource. The textbook is available in print or online versions.

G-W Learning Companion Website

The G-W Learning companion website is a study reference that contains review questions, vocabulary exercises, and more! Accessible from any digital device, the G-W Learning companion website complements the textbook and is available to the student at no charge.

Lab Workbook

Chapter Review questions and activities are facilitated with the Lab Workbook. This comprehensive and flexible resource is a great value.

Online Learning Suite

Available as a classroom subscription, the Online Learning Suite provides the foundation of instruction and learning for digital and blended classrooms. An easy-to-manage shared classroom subscription makes it a hassle-free solution for both students and instructors. An online student text and workbook, along with rich supplemental content, brings digital learning to the classroom. All instructional materials are found on a convenient online bookshelf and are accessible at home, at school, or on the go.

Online Learning Suite/ Student Textbook Bundle

Looking for a blended solution? Goodheart-Willcox offers the Online Learning Suite bundled with the printed text in one easy-to-access package. Students have the flexibility to use the print version, the Online Learning Suite, or a combination of both components to meet their individual learning style. The convenient packaging makes managing and accessing content easy and efficient.

Instructor Resources

Instructor resources provide information and tools to support teaching, grading, and planning; class presentations; and assessment.

Instructor's Presentations for PowerPoint® CD

Help teach and visually reinforce key concepts with prepared lectures. These presentations are designed to allow for customization to meet daily teaching needs. They include objectives, outlines, and images from the textbook.

ExamView® Assessment Suite CD

Quickly and easily prepare, print, and administer tests with the ExamView® Assessment Suite. With hundreds of questions in the test bank corresponding to each chapter, you can choose which questions to include in each test, create multiple versions of a single test, and automatically generate answer keys. Existing questions may be modified and new questions may be added. You can prepare pretests, formative assessments, and summative assessments easily with the ExamView® Assessment Suite.

Instructor's Resource CD

One resource provides instructors with time-saving preparation tools, such as answer keys, lesson plans, correlation charts to *Standards for Technological Literacy*, Precision Exam standards, and other teaching aids.

Online Instructor Resources

Online Instructor Resources provide all the support needed to make preparation and classroom instruction easier than ever. Available in one accessible location, support materials include Answer Keys, Lesson Plans, Instructor Presentations for PowerPoint®, ExamView® Assessment Suite, and more! Online Instructor Resources are available as a subscription and can be accessed at school or at home.

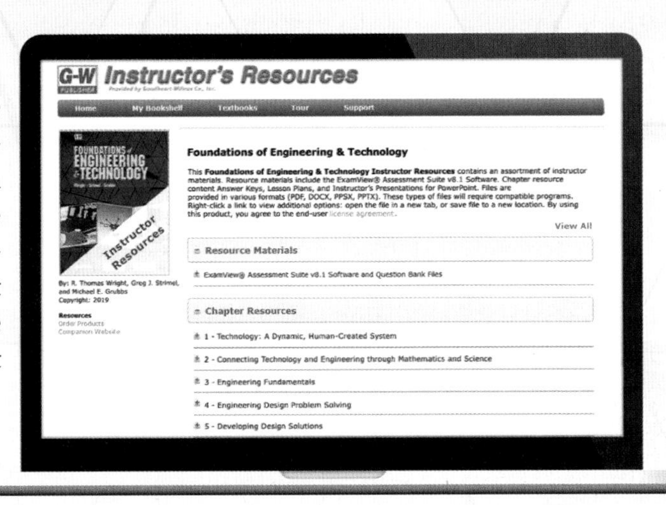

G-W Integrated Learning Solution

Together, We Build Careers

At Goodheart-Willcox, we take our mission seriously. Since 1921, G-W has been serving the career and technical education (CTE) community. Our employee-owners are driven to deliver exceptional learning solutions to CTE students to help prepare them for careers. Our authors and subject matter experts have years of experience in the classroom and industry. We combine their wisdom with our expertise to create content and tools to help students achieve success. Our products start with theory and applied content based on a strong foundation of accepted standards and curriculum. To that base, we add student-focused learning features and tools designed to help students make connections between knowledge and skills. G-W recognizes the crucial role instructors play in preparing students for careers. We support educators' efforts by providing time-saving tools that help them plan, present, assess, and engage students with traditional and digital activities and assets. We provide an entire program of learning in a variety of print, digital, and online formats, including economic bundles, allowing educators to select the right mix for their classroom.

Student-Focused Curated Content

Goodheart-Willcox believes that student-focused content should be built from standards and/or accepted curriculum coverage. Standards from *Standards for Technological Literacy* and standards from Precision Exams were used as a foundation in this text. ***Foundations of Engineering & Technology*** also uses a building block approach, with attention devoted to a logical teaching progression that helps students build on their learning. We call on industry experts and teachers from across the country to review and comment on our content, presentation, and pedagogy. Finally, in our refinement of curated content, our editors are immersed in content checking, securing and sometimes creating figures that convey key information, and revising language and pedagogy.

Precision Exams Certification

Goodheart-Willcox is pleased to partner with Precision Exams by correlating ***Foundations of Engineering & Technology*** to their Foundations of Technology standards. Precision Exams Standards and Career Skill Exams™ were created in concert with industry and subject matter experts to match real-world job skills and marketplace demands. Students who pass the exam and performance portion of the exam can earn a Career Skills Certification™. Precision Exams provides:

- Access to over 150 Career Skills Exams™ with pretest exams and posttest exams for all 16 Career Clusters.
- Instant reporting suite access to measure student academic growth.
- Easy-to-use, 100% online exam delivery system.

To see how ***Foundations of Engineering & Technology*** correlates to the Precision Exams Standards, please visit www.g-w.com/foundations-engineering-technology-2019 and click on the Correlations tab. For more information on Precision Exams, including a complete listing of their 150+ Career Skills Exams™ and Certificates, please visit www.precisionexams.com.

Billion Photos/Shutterstock.com

FOUNDATIONS *of* ENGINEERING & TECHNOLOGY

Seventh Edition

by

R. Thomas Wright
Professor Emeritus, Industry and Technology
Ball State University
Muncie, Indiana

Greg J. Strimel
Assistant Professor, Engineering and Technology Teacher Education
Purdue University
West Lafayette, Indiana

Michael E. Grubbs
Supervisor, Manufacturing, Technology, and Engineering Education
Baltimore County Public Schools
Baltimore, Maryland

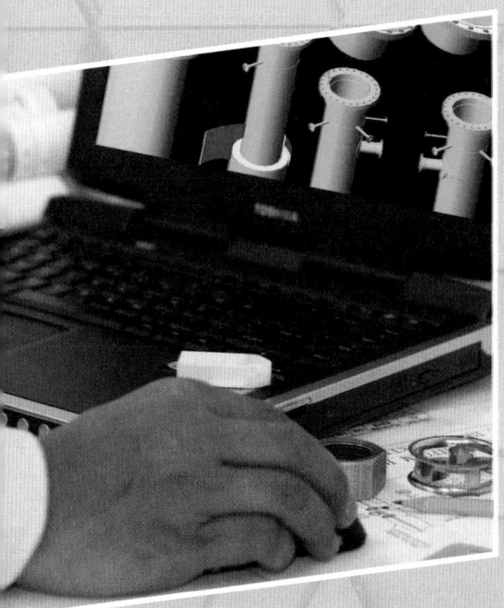

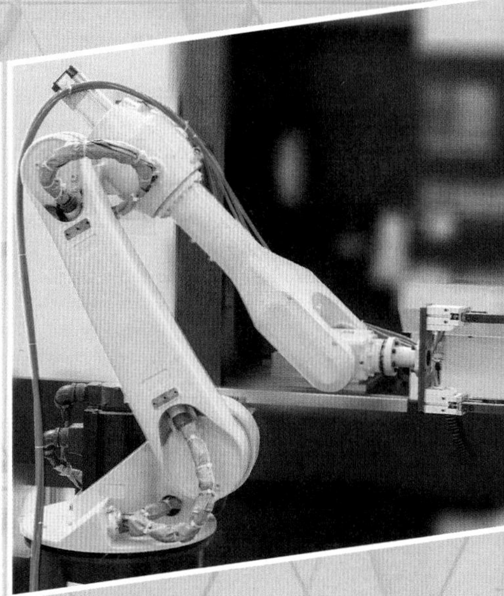

Publisher
The Goodheart-Willcox Company, Inc.
Tinley Park, IL
www.g-w.com

Previously published and copyrighted as *Technology & Engineering* by
The Goodheart-Willcox Company, Inc.

Manufactured in the United States of America.

ISBN 978-1-63126-886-1

1 2 3 4 5 6 7 8 9 – 19 – 22 21 20 19 18 17

Front cover images: RAGMA IMAGES/Shutterstock.com (left); LALS STOCK/Shutterstock.com (center); asharkyu/Shutterstock.com (right); Lena Serditova/Shutterstock.com (background)
Section 1 image: Photobank/Shutterstock.com; Section 2 image: Blazej Lyjak/Shutterstock.com; Section 3 image: asharkyu/Shutterstock.com; Section 4 image: cozyta/Shutterstock.com; Section 5 image: Scharfsinn/Shutterstock.com

INTRODUCTION

Foundations of Engineering & Technology will help you to understand the following:

- How people use technology to make our world work.
- Why technological systems work the way they do.
- In what ways technology affects both people and our planet.
- How engineering is needed to create technology.

Foundations of Engineering & Technology ties engineering into the seven areas of technological activity:

- Manufacturing.
- Construction.
- Energy.
- Information and communication.
- Transportation.
- Medicine and health technologies.
- Agriculture and biotechnology.

In this book, you will learn that technology is a reaction to problems and opportunities—a human adaptive system. You will learn about technological systems and how the engineering design process is used to create them. You will also learn about the problem-solving and engineering design process, including how to test, evaluate, and communicate design solutions.

Sections in the text explore the connection between technology and engineering, the steps in the engineering design process, the technological systems, the designed world, and the partnership between technology and society.

Foundations of Engineering & Technology contains many photographs, drawings, diagrams, and original artwork that help explain the concepts in the text. This material has been carefully selected to make technology and engineering understandable. Each chapter begins with objectives so you know what is covered. Technical Terms appear in bold italics in each chapter to increase your awareness of them. Test Your Knowledge questions, Critical Thinking activities, and STEM Applications improve your understanding of chapter materials. Engineering Design activities and TSA Modular Activities help you apply chapter content to real-world situations.

Impacts, both positive and negative, accompany the use of technology. The only way people in the modern world can choose and apply technology responsibly is to understand how technology develops, how the various technological systems work, and how engineering is applied to these systems.

ABOUT THE AUTHORS

Dr. R. Thomas Wright is one of the leading figures in technology-education curriculum development in the United States. Dr. Wright has served the profession through many professional offices. These offices include president of the International Technology and Engineering Educators Association (ITEEA) and president of the Council on Technology Teacher Education (CTTE; now known as the Council on Technology and Engineering Teacher Education [CTETE]).

His work has been recognized through the ITEEA Academy of Fellows award and Award of Distinction; the CTETE Technology Teacher Educator of the Year; the Epsilon Pi Tau Laureate citation and Distinguished Service citation; the Sagamore of the Wabash award from the governor of Indiana; the Bellringer award from the Indiana superintendent of public instruction; the Ball State University Faculty of the Year award and George and Frances Ball Distinguished Professorship; and the EEA-Ship citation. Dr. Wright's educational background includes a bachelor degree from Stout State University, a master of science degree from Ball State University, and a doctoral degree from the University of Maryland. His teaching experience includes three years as a junior high instructor in California and 37 years as a university instructor at Ball State University. In addition, he has also been a visiting professor at Colorado State University; Oregon State University; and Edith Cowan University in Perth, Australia.

Dr. Michael E. Grubbs is the supervisor for Manufacturing, Technology, and Engineering Education for Baltimore County Public Schools. Dr. Grubbs' educational degrees include a bachelor of science and a master of education in technology education from California University of Pennsylvania; an educational specialist in workforce education from the University of Georgia; and a doctorate of philosophy in curriculum and instruction with an emphasis on integrative STEM education from Virginia Tech. Dr. Grubbs has served as a middle school technology and engineering education teacher for Clayton County Public Schools, lead middle school teacher for Clayton County Public Schools, graduate research assistant for Virginia Tech, and adjunct instructor for New River Community College. His work has been recognized through the Donald Maley Spirit of Excellence: Outstanding Graduate Student citation; the CTETE International Travel grant and 21st Century Leader fellowship; Virginia Tech's Outstanding Student in Integrative STEM Education award; and the ITEEA Emerging Leader designation. Dr. Grubbs has served the profession in multiple roles, including secretary, marketing committee chair, and webmaster for CTETE. He has written numerous articles on technology and engineering education, and presented regionally, nationally, and internationally.

Dr. Greg J. Strimel is an assistant professor of Engineering and Technology Teacher Education at Purdue University in West Lafayette, Indiana. Dr. Strimel's educational degrees include a bachelor of science and a master of education in technology education from California University of Pennsylvania, and a doctorate of philosophy in occupational and technical studies, with an emphasis on engineering and technology education from Old Dominion University. His teaching experience includes serving as a high school engineering and technology teacher in Maryland, teaching assistant professor within the Benjamin M. Statler College of Engineering and Mineral Resources at West Virginia University, and adjunct instructor for engineering and technology teacher education at the University of Maryland Eastern Shore and California University of Pennsylvania. Dr. Strimel has also served as a high school career and technology education department chair, the director of K–12 initiatives for West Virginia University, and a professional development leader for multiple engineering and technology curriculum providers. His work has been recognized through the Technology and Engineering Education Association of Maryland's New Teacher of Excellence award; the governor of Maryland's citation for teaching excellence; Old Dominion University's Outstanding Graduate Student award; ITEEA's Emerging Leader designation and Donald Maley Outstanding Graduate Student citation; and the CTETE 21st Century Leader fellowship.

REVIEWERS

The author and publisher wish to thank the following industry and teaching professionals for their valuable input into the development of ***Foundations of Engineering & Technology***.

Luis Avila
Memorial High School
McAllen, Texas

Vinson Carter
University of Arkansas
Fayetteville, Arkansas

Adrienne Emerson
Greenville High School
Greenville, Texas

David T. Gluckner
Cleveland High School
Cleveland, Tennessee

Douglas Handy
Baltimore County Public Schools
Towson, Maryland

Kenneth R. Hanson
Marlborough High School
Marlborough, Massachusetts

Tyler S. Love
University of Maryland Eastern Shore
Princess Anne, Maryland

Robert Oehrli
Catonsville High School
Catonsville, Maryland

Mark Piotrowski
Lower Merion High School
Ardmore, Pennsylvania

David Quinton-Schein
Dulaney High School
Timonium, Maryland

Minda G. Reidy
Andover High School
Andover, Massachusetts

E. Clarke Stephens
Rancho Verde High School
Riverside, California

Kevin Sutton
North Carolina State University
Raleigh, North Carolina

Daniel Sweet
John Jay Science and Engineering Academy
San Antonio, Texas

ACKNOWLEDGMENTS

The author and publisher would like to thank the following companies, organizations, and individuals for their contribution of resource material, images, or other support in the development of ***Foundations of Engineering & Technology***:

Adolphe Ganot
Airbus Industries
Alaska Airlines
AMAX
American Electric Power
American Petroleum Institute
Arvin Industries
AT&T
Bell Helicopter
Bethlehem Steel Co.
Boeing
Boise Cascade
BP
Brush-Wellman
Buhler Group
Case IH
Caterpillar, Inc.
Cincinnati, Inc.
Cincinnati Milicron
Daimler
DaimlerChrysler
Dakota Growers Pasta Company
Deere & Co.
Delta
Erik Wannee
FMC Corp.
Ford
Freightliner Corp.
Gehl Co.
General Motors Corp.
GE Plastics
GM-Hughes
Goodyear Tire and Rubber Co.
Grumman Corp.
Gulf Oil Co.
Hasbro, Inc.
Inland Steel Co.
Japan Railways Group
Jeffboat Shipyard
Joseph O'Dea/NASA
J.R. Simplot
Library of Congress
L. S. Starrett Co.
MeadWestvaco
Mobil Oil Co.
NASA
NASA/JPL-Caltech
National Park Service
National Society of Professional Engineers (NSPE)
NOAA
North Dakota Mill
Northern Telephone Co.
Northwest Cherry Growers
Ohio Art Co.
PPG Industries
Renishaw, Inc.
Reynolds Metals Co.
Shell Oil Co.
Siemens
SpaceX
Tillamook County Creamery Association
Toyota
Transrapid International
United Parcel Service
United States Patent Office
US Department of Agriculture
US Department of Commerce
US Department of Energy
US Department of the Interior
Westinghouse Electric Corp.
Weyerhauser Co.
Wisconsin Milk Marketing Board

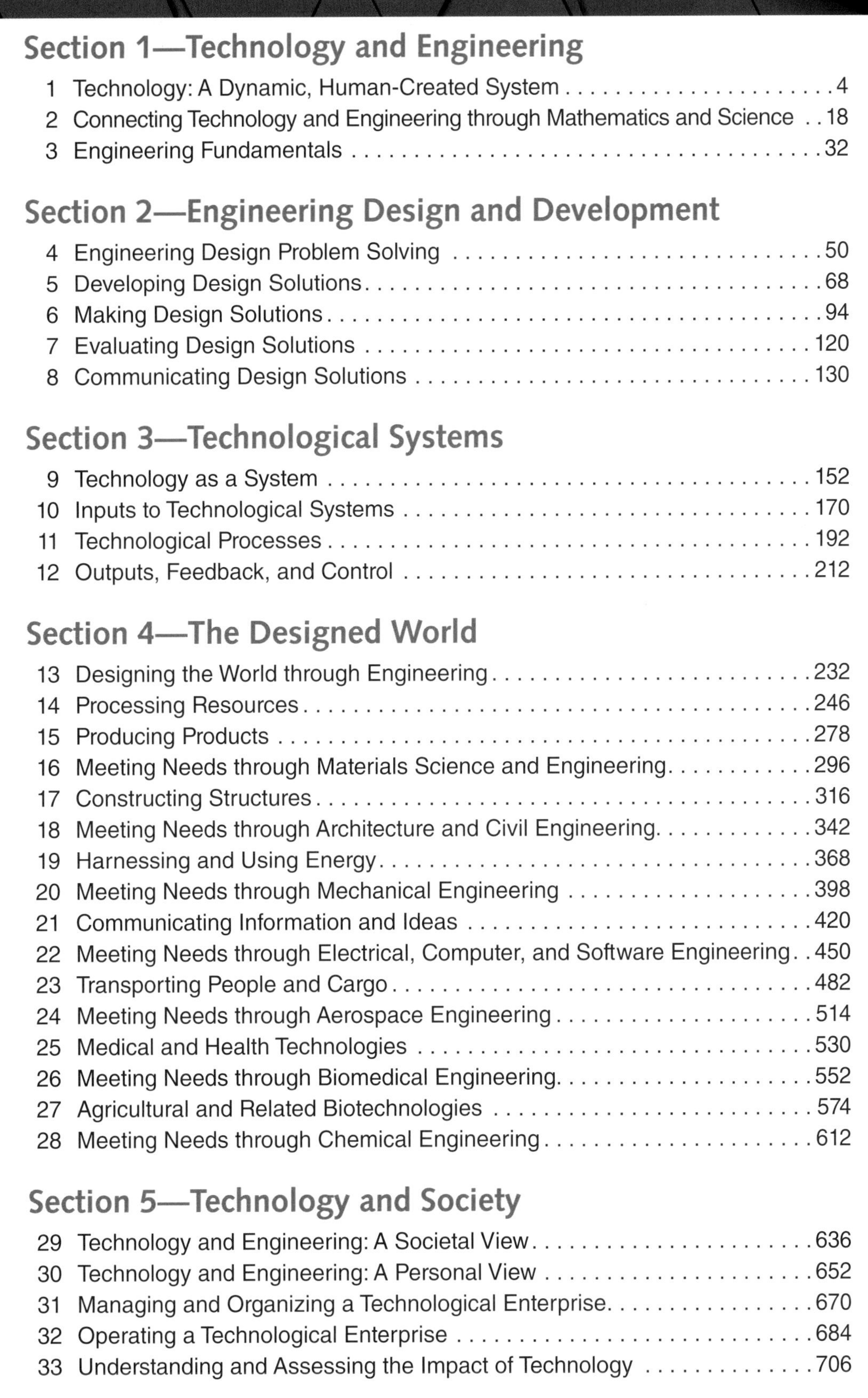

BRIEF CONTENTS

Section 1—Technology and Engineering

Section 2—Engineering Design and Development

Section 3—Technological Systems

Section 4—The Designed World

Section 5—Technology and Society

CONTENTS

Section 4
The Designed World

Section 5
Technology and Society

FEATURES

Tomorrow's Technology Today

Think Green

Career Connection

STEM Connection

Academic Connection

Technology Explained

Engineering Design Challenge

TSA Modular Activity

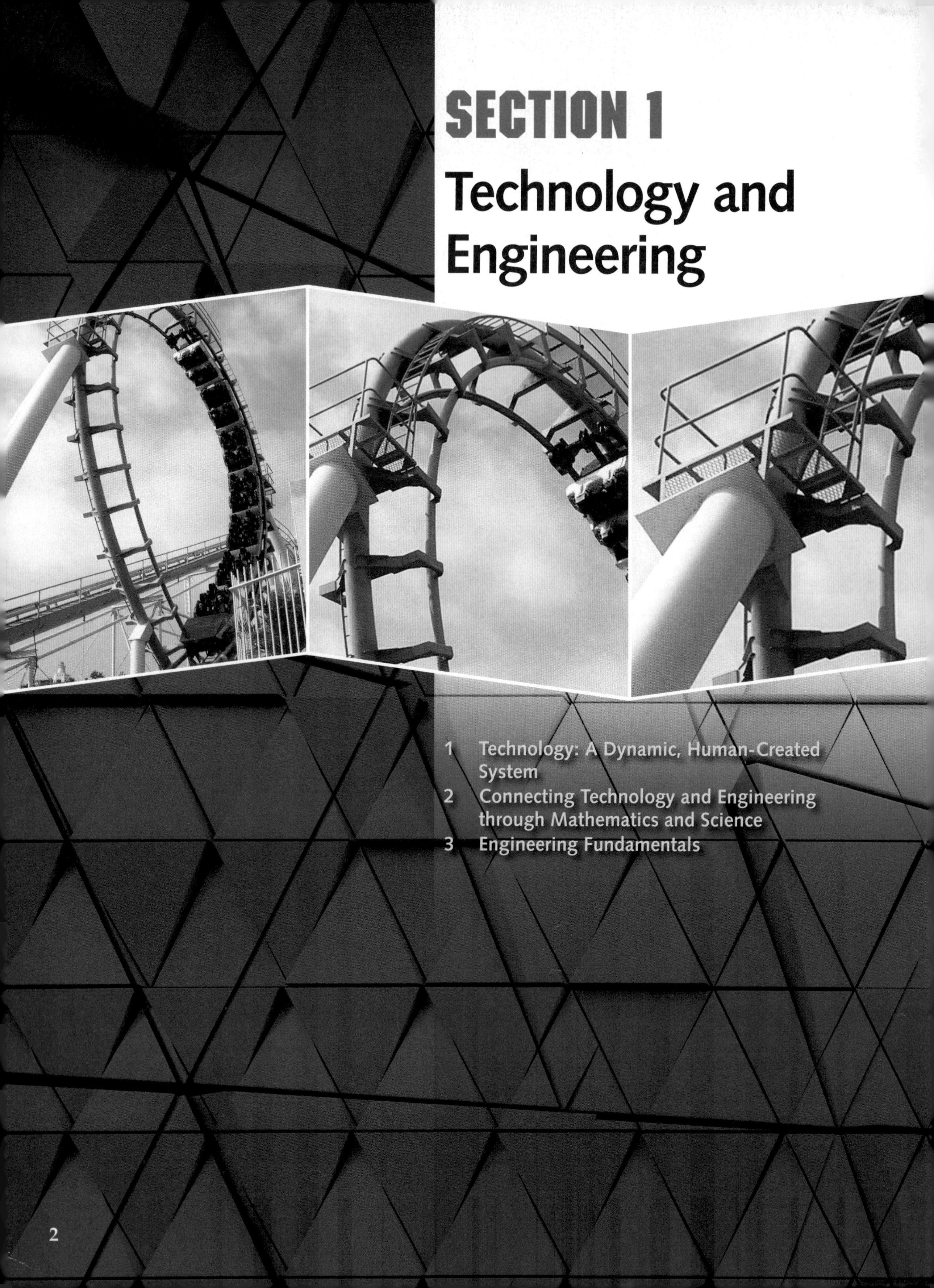

SECTION 1
Technology and Engineering

Tomorrow's Technology Today

Cloning

Identical twins often have stories of switching places with one another to avoid trouble—or to cause it. If you have ever wished for a twin to sit in your place on an exam day, your wish might not be altogether unrealistic. Modern science is now making it possible to create a human genetically identical to you, or a clone. A clone is, for all practical purposes, an identical twin of the original because every single gene is an exact copy of the original.

Recently, scientists have successfully cloned several types of plants and animals. These types include some endangered species. This is important because it opens up the possibility of repopulating species of plants and animals at risk of becoming extinct and maybe even species that already are! The main purpose of cloning, though, is to mass-produce plants and animals with specific genetic qualities, such as the best-tasting fruit or the ability to produce human insulin.

Cloning plants and animals has some undeniable benefits for humans and the quality of human life. Although there has been some debate about the ethical implications of plant and animal cloning, it is the idea of human cloning that is the most controversial among scientists, the government, and the public. Until recently, the idea of cloning humans was science fiction. Now, it is a real possibility.

Researchers are close to being able to successfully clone humans. Most cloning would not, however, result in entirely new humans. Scientists are studying the possibility of using cloning to produce repair and replacement kits for people with severe medical problems. It might be possible to clone healthy cells and to fix diseased cells. This type of cloning is currently being used in stem-cell research. A person's DNA is used to grow an embryonic clone. This clone is then used to grow stem cells. These cells can grow replacement organs and neurons to cure certain diseases.

Cloning, especially human cloning, creates many ethical dilemmas. The success rate of animal cloning is currently less than 3%. The process is very risky. Surviving clones often suffer from problematic genetic abnormalities. Some are born with defective lungs, hearts, blood vessels, or immune systems. It will be a more serious problem if human clones are created and born with these same problems. Cloning holds much promise. This solution, however, also holds many risks. As interest in human cloning continues to grow, the ethical debate certainly will as well.

While studying this chapter, look for the activity icon to:

- **Assess** your knowledge with self-check pretest and posttests.
- **Practice** technical terms with e-flash cards, matching activities, and vocabulary games.
- **Reinforce** what you learn by submitting end-of-chapter questions.

www.g-wlearning.com/technologyeducation/

CHAPTER 1

Technology: A Dynamic, Human-Created System

Check Your Engineering IQ

Before you read this chapter, assess your current understanding of the chapter content by taking the chapter pretest.
www.g-wlearning.com/technologyeducation/

Learning Objectives

After studying this chapter, you should be able to:

- ✔ Define *technology*.
- ✔ Identify four basic features of technology.
- ✔ Describe how technology is a dynamic, human-created process.
- ✔ Provide examples of positive and negative aspects of technology.
- ✔ Illustrate the major divisions in the evolution of technology.
- ✔ Classify technological developments in each period of technological history.
- ✔ Describe features of the information age.
- ✔ Identify the different major areas of technology.

Technical Terms

artifacts
Bronze Age
civilized conditions
development
dynamic process
Industrial Revolution
information age
Iron Age
Middle Ages
primitive conditions
Renaissance
Stone Age
technologically literate
technology

Close Reading

Before reading this chapter, write down what you think the term *technology* means. As you read the chapter, see how your perception of the term matches or differs from the description in the text. Also, consider the seven areas of technology described in the chapter and list examples from each that you encounter during daily life.

Do you know what technology is? Is it mainframe and desktop computers, industrial robots, laser scanners, fiber-optic communications, space shuttles, and satellites circling Earth? Does technology consist of people using tools and machines to make their work easier and better? Is it an organization devoted to operating agricultural and related biotechnology, communication and information, energy and power, medical, manufacturing, construction, or transportation systems? See **Figure 1-1**.

Chances are you are not sure if technology is one, two, or all three of these things. You are not alone in this lack of understanding. Throughout the world, people use technology every day. However, they understand very little about it. We go to school for years to learn how to read and write and to study mathematics, science, history, foreign languages, and other subjects. Few people, however, spend time learning about technology and the impact it has on everyday life.

This book has been developed to help you understand this vast area of human activity. After this study of technology, you will be able to find, select, and use knowledge about tools and materials to solve problems. This ability leads to an increased understanding of technology as it affects your life as a citizen, consumer, and worker. See **Figure 1-2**.

Eugenio Marongui/Shutterstock.com

Figure 1-1. We encounter technology in many areas of our lives.

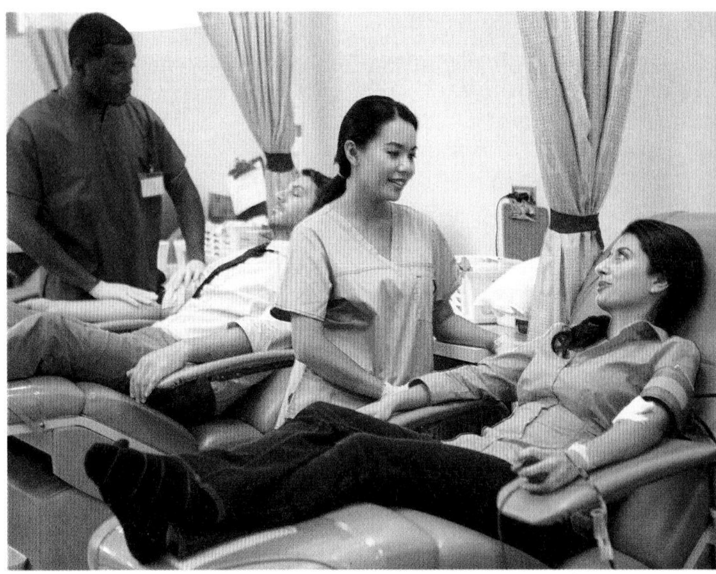

Monkey Business Images/Shutterstock.com

Figure 1-2. Technology affects our daily lives.

Technology Defined

Almost everyone uses the word *technology*. What does it really mean, however? To some people, it means complicated electronic devices and hard-to-understand equipment. To others, it means the source of the radical changes that are happening in all phases of life. Some people fear it. Others see it as the source of longer and more complete lives. Some people believe it to be a development of the twentieth century. Each of these views is partly correct.

Technology is a phenomenon that can be primitive and crude, or complex and sophisticated. This phenomenon has been here as long as humans have been on Earth. ***Technology*** is the application of knowledge, tools, and skills to solve problems and extend human capabilities. Technology is any modification of the natural world done to fulfill human needs and desires.

These actions have four basic features, **Figure 1-3**:

- Technology is human knowledge.
- Technology uses tools, materials, and systems.
- The application of technology results in ***artifacts*** (human-made things) and other outputs (pollution and scrap, for example).
- People develop technology to modify or control the environment.

These four characteristics suggest that the products and services available in society are the result of technology. The products and services

Technology...

Phovoir/Shutterstock.com

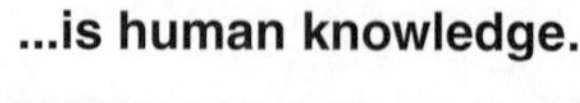

...is human knowledge.

B Brown/Shutterstock.com

...involves tools, materials, and systems.

Fotokostic/Shutterstock.com

...produces artifacts (outputs).

B Brown/Shutterstock.com

...modifies or controls the environment.

Figure 1-3. Technology has four characteristics.

are designed through ***development*** (technological innovation). They are produced by technological means (processes). These means are integrated with people, machines, and materials to meet an identified need (management). The products are distributed to customers using technology. They are maintained and serviced through technological actions. When the products fulfill their function or become obsolete, they are (or should be) recycled by technological means. As you can see, without technology, the world as we know it would not exist.

Technology as a Dynamic Process

All these points suggest that technology is a ***dynamic process***, meaning that it is constantly changing, often unforeseeable, and almost always seems to improve existing technology. See **Figure 1-4**. The typewriter is an improvement over handwriting. The electric typewriter is faster than the manual typewriter. The word processor and laser printer outperform the electric typewriter. Now, voice-recognition software enables humans to talk to computers, providing even faster communication. This is not the end. New devices will improve on this system.

Past technology improved productivity. Future technology will increase productivity even more. There is no turning back. We cannot feed the world's population using a plow pulled by mules. Commerce cannot be maintained using covered wagons. Housing for a growing population cannot be built using eighteenth-century construction methods. Technology is necessary for survival and is the hope for a better future.

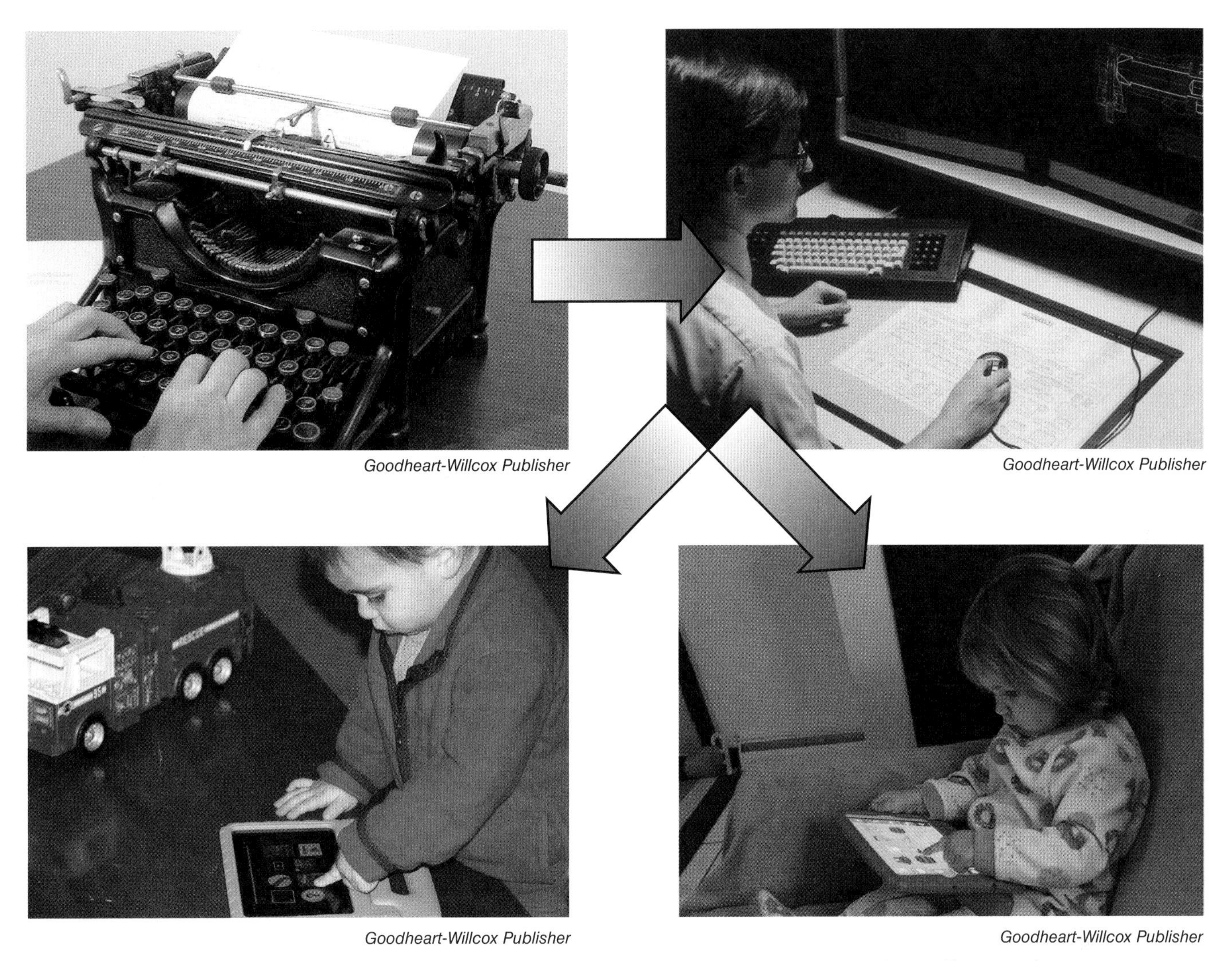

Goodheart-Willcox Publisher

Goodheart-Willcox Publisher

Goodheart-Willcox Publisher

Goodheart-Willcox Publisher

Figure 1-4. Technology is a dynamic process. People are always trying to improve existing artifacts and systems.

Career Connection Overview of Career Connection

As you may know, there is a growing need to fill jobs in the science, technology, engineering, and mathematics fields. There will be many opportunities in your future to obtain gainful employment in high-demand fields such as software engineering, database administration, mechanical engineering, accounting, construction management, and health care. These careers are accessible to those with a solid foundation of knowledge in science, technology, engineering, and mathematics. Throughout this text, we will look at careers related to the concepts covered in each chapter. The Career Connection features will highlight a specific career, provide a description of the job, describe what is required to obtain the job, explain the outlook for the specific career, and list professional organizations in that career area.

STEM Connection — Overview of STEM Connection

STEM is an acronym for science, technology, engineering, and mathematics. This acronym indicates how deeply intertwined these fields are in the world around us. Science, technology, engineering, and mathematics are all interrelated and knowledge from each field is required to operate as a scientist, engineer, technologist, technician, or mathematician. The STEM Connections features will present examples of science and mathematical concepts and practices being used in technology and engineering.

Positive and Negative Aspects of Technology

Technology increases human capabilities. Through technology, we can see more clearly and further by using microscopes and telescopes. Technology helps us to lift heavy loads by using hoists, pulleys, and cranes. This process allows us to communicate better and faster by radio, television, and telephone. Technology brings distant places closer when traveling by automobiles, trains, and aircraft. This process makes life more enjoyable with video games, motion pictures, and compact disc recordings. Technology brings us new products and materials, such as computers, optical fibers, and artificial human organs.

Technology, however, also has negative aspects. Poorly designed and used technology can cause air and water pollution and soil erosion, **Figure 1-5**. Technology can cause unemployment and radical changes in the ways people live. New technological devices might displace workers from traditional jobs. The consequences of technology are now more feared than natural events. Many people are more concerned about nuclear winter, acid rain, and air pollution than they are of earthquakes and tornadoes.

This good-news/bad-news view of technology requires a new type of citizen. Technology must be developed and used wisely and appropriately. This challenge requires individuals who understand and can direct new technology. People who have this understanding and ability are called ***technologically literate***.

Soil Erosion

US Department of Agriculture

Air Pollution

©iStockphoto.com/Anutik

Figure 1-5. Technology has positive and negative aspects. Poorly used technology can cause problems, such as soil erosion and air pollution.

Think Green Overview of Think Green

You have probably heard about ongoing efforts people are making to conserve resources or to be more environmentally friendly. There are many ways you can help by making your habits more "green." For example, you might recycle plastic, aluminum, and paper, or use reusable shopping bags rather than paper or plastic bags. Using compact fluorescent lights (CFLs) instead of incandescent lights saves energy. The Think Green feature will be looking at ways people can make personal and group choices to help the environment.

The Evolution of Technology

Many people see technology as new and dramatic. Technology is not new or dramatic. This knowledge is as old as humanity. The world is said to be about 5 billion years old. Humans have been on Earth for at least 2.5 million years. During this time, two major factors have distinguished humans from other species. These factors are humans' ability to make tools and humans' ability to use those tools. The level of this technology determines the type of life available.

Early humans were said to live in ***primitive conditions***. Nature determined these conditions. Primitive people tried to exist with nature. They did not attempt to control nature or improve the natural condition. For example, members of primitive cultures depended on harvesting natural vegetation and hunting game. Drought severely affected their food supply. They could not, however, do anything about it. If nature did not provide ample food, the people starved. Likewise, primitive cultures used only native materials (naturally found in that area) to build shelters and make clothing.

Civilized conditions are much different from primitive conditions. Civilized conditions allow people to exert their will on the natural scene. They make tools, grow crops, engineer materials, and develop transportation systems.

The evolution of civilization is directly related to the tools of the time. Many of the early periods of civilization are named for the materials used for tools. The products produced define others. See **Figure 1-6**. In each case, the technological developments during the period led to major changes in society and the creation of new needs and wants.

The Stone Age

The earliest period of known civilization was the ***Stone Age***. The Stone Age began about two million years ago. In this period, humans used stone as tools. These simple stone tools were used to cut and pound vegetables and cut meat from animal carcasses. From these modest beginnings came pointed stone hunting tools. These tools allowed people to obtain food more efficiently. Also during this period, humans learned how to harness fire for heating, cooking, and protection. The population became more productive. This meant that more people could live in a specific area.

The Bronze Age

As populations grew, new technology was needed to support the demand for food and shelter. This need was coupled with the discovery of copper. People learned that copper could be heated and then melted with other ores to produce bronze, a stronger metal than copper. These events allowed humans to enter the second major historical period, the ***Bronze Age***. The Bronze Age began around 3000 BCE. During this period, humans used copper and copper-based metals as the primary materials for tools because of the hardness and durability of these materials. Through such developments as large-scale irrigation systems, humans transformed agriculture to the extent that they did not have to depend on native vegetation and animals for survival. They also created better ways for storing food and developed writing, navigation, and other basic technologies.

Stone Age (1,000,000 BCE–3000 BCE)	Bronze Age (3000 BCE–1200 BCE)	Iron Age (1200 BCE–1300 CE)	Middle Ages (500 CE–1500 CE)	Renaissance (1300 CE–1600 CE)	Industrial Revolution (1750 CE–early 1900s)	Information Age (present time)
Stone tools, fire, cave paintings, pottery	Copper and bronze tools, smelting, frescoes, writing paper, ink	Iron tools, furnaces, aqueducts, body armor, ox-drawn plows, spinning wheels, windmills	Printing press, magnetic compass, paper money, waterwheel	Improved magnetic compass, telescope, hydraulic press, calculating machine, modern architecture	Steam engine, cotton gin, power looms, factories, electricity, automobile, airplane	Desktop computers, robots, solar energy, cell research, satellites

Iron Age

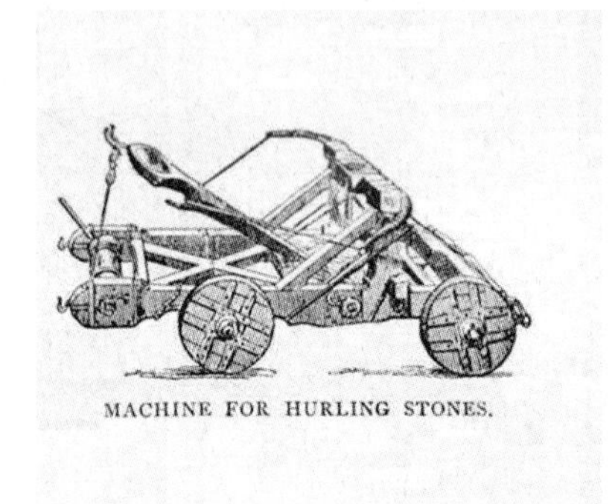

www.PDImages.com

Industrial Revolution

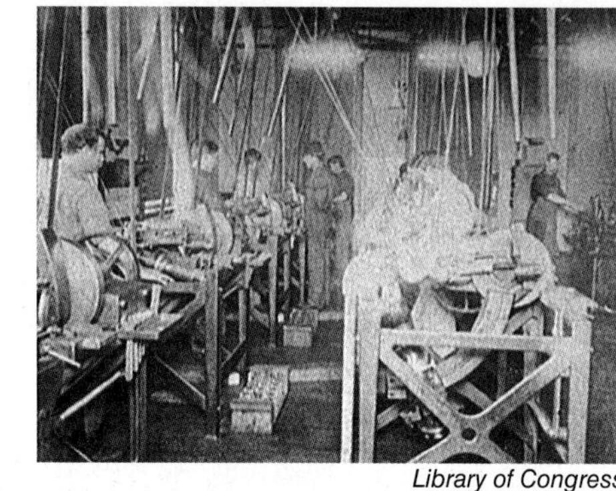

Library of Congress

Information Age

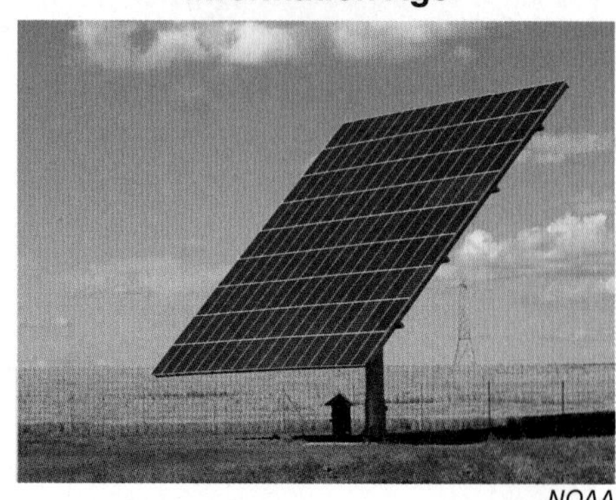

NOAA

Figure 1-6. The major divisions in the history of civilization are identified by the tools and products produced and used.

The Iron Age

The next historical period is called the ***Iron Age***. The Iron Age began around 1200 BCE. Iron and steel became the primary materials for tools during this period because they were more plentiful and less costly than copper and bronze. More people could afford these tools. This affordability created more opportunities for sustained technological advancement. During the Iron Age, the alphabet came into general use, as did coins. Trade, transportation, and communication all improved. Civilization expanded.

The Middle Ages

The ***Middle Ages*** began in Europe about 400 CE and continued until about 1500 CE. It began with the fall of the Roman Empire and ended with the Renaissance. The Middle Ages is known for its various upheavals. Tribes continually fought each other for territory. Technology, however, still

Academic Connection

Overview of Academic Connection

Working in the fields of science, technology, engineering, and mathematics requires a good understanding of social constructs, laws and regulations, and historical events. Additionally, it is necessary to know how to convey information in the best possible manner, both technically and creatively. The Academic Connection features will deliver opportunities for examining historical events, writing and analyzing technical reports, examining engineering case studies, and developing explanatory and argumentative essays to issues found in science, technology, engineering, and mathematics.

Technology Explained Overview of Technology Explained

Throughout this text, we will be looking at several different technological devices, such as 3-D printers, robots, and drones, to examine how they work. This information will be included in the Technology Explained sections within the various chapters. These sections will also provide a close look at the ways in which these technologies are created, how they operate, and what they are used for.

progressed. For example, one of the major inventions of this time was the printing press, developed by Johann Gutenberg around 1440. The printing press changed people's lives forever. Previously, to make books, words and illustrations were copied onto parchment by hand, a laborious process that made books very expensive. With the printing press, books could be produced more quickly and with less labor, thus making books less costly and available to more people.

Other items discovered and improved on during this time include the magnetic compass and the waterwheel. The waterwheel changed the power of water into mechanized energy. The adoption and use of paper money increased the amount of goods bought and sold.

The Renaissance

The ***Renaissance***, the period that began in the early 1300s in Italy and lasted until 1600, was a time of great cultural advancement. This period is known as a period of new ideas in art, literature, history, and political science. Technological developments also occurred. For example, because of improvements in ships, European voyagers traveled to America. They brought with them new crops and farming tools and learned new methods of farming from Native Americans, thus improving agriculture for both groups.

In Italy, Leonardo da Vinci, the famous artist, kept a notebook in which he drew plans for many inventions. These inventions included a flying machine and a movable bridge. The drawings greatly influenced future inventors. Other technologies developed during this period include the calculator and the telescope.

The Industrial Revolution

During the last 250 years, technology has dramatically changed the world. For example, humans applied technology to agriculture and went from cultivating food by hand to using horse and plow to adopting the mechanical reaper. Because of these developments, a smaller percentage of the population was needed to grow the food and clothing fiber needed to sustain the population. This allowed large numbers of people to leave farms and migrate to growing towns and cities. In these places, a new phenomenon was developing—the marvel of manufacturing.

The ***Industrial Revolution*** started in England in the late eighteenth century and is characterized by the introduction and improvements to power-driven machinery, as well as new developments in manufacturing processes. Tremendous changes in technology occurred during this period. For example, Edmund Cartwright changed weaving from a manual process to a mechanical one with his power loom. Joseph Jacquard further revolutionized weaving by creating a series of punched cards with recorded instructions. These cards controlled the intricate patterns the weaver designed, thus allowing the weaver to spend more time on management. (These cards later became the inspiration for computer programming.) James Watt improved the steam engine and made it available for the driving of machinery. This development revolutionized both the manufacturing process and methods of transportation. Eli Whitney, inventor of the cotton gin, developed the system of manufacturing standardized, interchangeable parts. These parts dramatically expanded manufacturing capabilities. As has been true throughout the history of technology, many of these developments were refinements of existing inventions. Here, however, scientific knowledge began to be applied to technological knowledge, greatly accelerating technical progress.

As manufacturing changed, so did the character of the workforce. Employees were divided into production workers and managers. Each division was given specific tasks. Efficiency of production became an area of serious study. For example, in the early twentieth century, Frederick Winslow Taylor developed his "four principles of scientific management" to reduce waste and increase productivity. Revolutionary for the time, Taylor's ideas included studying production workers' motions and rearranging equipment to decrease the time the workers spent not producing goods.

The various developments in industry led to continuous manufacture, **Figure 1-7**. Continuous manufacture is characterized by the following:

©iStockphoto.com/epixx

Figure 1-7. Some workers are involved with the continuous-manufacturing process.

- Improved machine life because of interchangeable parts.
- Division of jobs into tasks assigned to separate production workers.
- Creation of material-handling devices that bring the work to the workers.
- Classification of management as a professional group.

Later during the Industrial Revolution, enterprising people developed sophisticated transportation and communication systems to support the growing industrial activities. Dirt and gravel roads became paved highways. The diesel-electric locomotive replaced the steam locomotive. The motortruck and the airplane successfully challenged the dominance of transportation by railroads. The telegraph replaced pony express letter carriers.

Improved construction practices met the growing demand for shelter. Mass-produced dwellings replaced log cabins. Metal buildings became the factory of choice. Shopping centers and shopping malls made downtown shopping areas less important.

Efficient production was coupled with rising consumer demand. Extended free time was available for the first time, with the 40-hour workweek and annual vacations. Children could stay in school longer because they were not needed on the farm and in the factories. Universal literacy became a possibility, although it has yet to be reached.

The Information Age

The Industrial Revolution moved us from an agrarian era into the industrial era. We moved from a period when most people worked raising food and fiber to a period in which people worked in manufacturing. Technology is now moving many nations into a new period, the ***information age***, a time period characterized by information-based industries and the transmission of information through computer-based advances.

During the Industrial Revolution, the most successful companies processed material better than their competitors. The information age places more importance on information processing and cooperative working relations between production workers and managers. The information age has several characteristics, including:

- Wide use of automatic machines and information-processing equipment.
- High demand for trained technicians, technologists, and engineers.
- Blurring of the previously sharp line between production workers and managers.
- Constant need for job-related training and retraining of production workers.

These factors promise change. This is not, however, something new. Change takes place with every generation. For example, one generation traveled west in covered wagons during their youth, and their children saw an astronaut circle Earth in their later years. What will you see in your lifetime?

Types of Technology

There are many ways to examine technology. One approach to understanding technology is categorizing each area by the type of technology developed and used. This approach looks at technology as human actions and groups them accordingly. This grouping includes seven types of technology. See **Figure 1-8**. Humans have used all these types of technology throughout history to modify and control their environment:

- **Agricultural and related biotechnologies.** Used in growing food and producing natural fibers.
- **Communication and information technologies.** Used in processing data into information and in communicating ideas and information.
- **Construction technologies.** Used in building structures for housing, business, transportation, and energy transmission.
- **Energy and power technologies.** Used in converting and applying energy to power devices and systems.
- **Manufacturing technologies.** Used in converting materials into industrial and consumer products.
- **Medical technologies.** Used in maintaining health and curing illnesses.
- **Transportation technologies.** Used in moving people and cargo from one place to another.

In each technological system, inputs are processed and transformed into outputs. For example, in agricultural technology, materials (seeds and fertilizers, for instance) are changed into food and fibers. In medical technology, information and materials are changed into devices that aid in maintaining health. In transportation technology, energy is changed into a form of power that moves vehicles containing goods.

These systems can be looked at individually. Such a focus, however, gives an inaccurate view. The systems are all closely related and are part of a single effort. This effort is to help humans live better. These systems work together to support one another. In one case, manufacturing might be the focus, with the other systems in supporting roles. In another situation, manufacturing might be the supporting technological system, **Figure 1-9**.

Let us follow one material from its natural state to a finished product designed to make life better for us. We will follow iron ore on its journey to becoming a stainless steel cookie sheet. Our story starts with *manufacturing technology*. With regard to the cookie sheet, this technology system involves three major activities:

1. Locating and extracting the raw materials (iron ore, limestone, and coal) to make steel.
2. Producing strips of steel from the raw materials.
3. Making (stamping) the cookie sheet from the steel strips.

These manufacturing activities could not exist without the other technological systems. For example, *communication and information technologies* play a role at every point in the product's development. The need for the new product is communicated through sales orders. The specifications for the steel and the cookie sheet are communicated through engineering drawings and specification sheets. The availability of the product is communicated to potential customers through advertising. Sales reports communicate product success to the company's management.

Energy and power technologies help to power the melting furnaces and stamping presses, light the work areas, and heat the offices and control rooms. *Agricultural and related biotechnologies* facilitate the growth of trees. The trees are used to make the

Communication

IrinaK/Shutterstock.com

Construction

Alison Hancock/Shutterstock.com

Manufacturing

Matej Kastelic/Shutterstock.com

Transportation

Federico Rostagno/Shutterstock.com

Agriculture

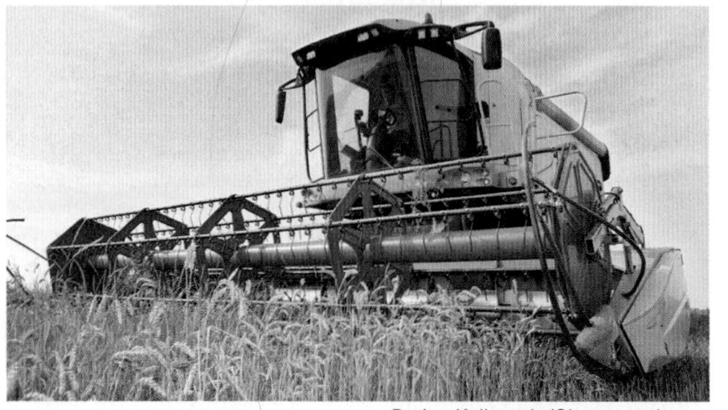

Dmitry Kalinovsky/Shutterstock.com

Medicine

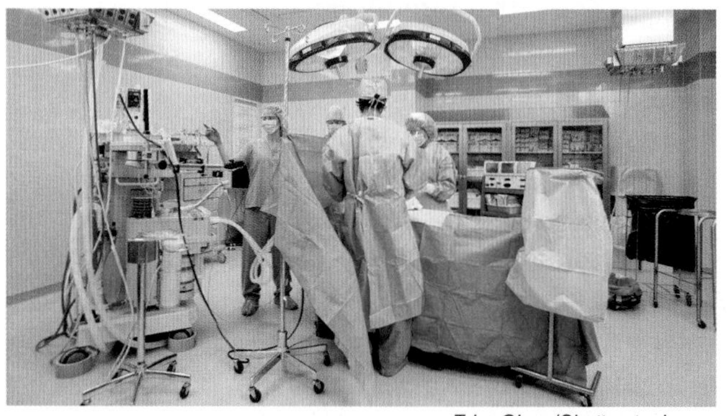

Tyler Olson/Shutterstock.com

Energy

pisaphotography/Shutterstock.comCommunication

Figure 1-8. Technology can be categorized by the type of product developed.

©iStockphoto.com/thegoodphoto

Figure 1-9. The production of each manufacturing product, such as these automotive wheels, involves a variety of steps. Each step depends on the seven major technological systems.

pallets containing the boxes of finished products. *Medical technologies* help to ensure that the workers remain in good health.

Likewise, *construction technologies* are essential. Constructed roads create access to iron-ore and coal mines and to the limestone quarries. Workers travel to the steel mill and the production factory on constructed roads and work in constructed buildings. Constructed power lines bring electricity to the various manufacturing sites. Constructed dams and pipelines make water available.

Transportation technologies also play a major role. They move the raw materials to the steel mill and the steel strips from the mill to the product manufacturer. These technologies deliver finished products to stores. Customers use private cars or public transportation systems to visit a store to purchase the product.

Throughout the rest of this book, we will focus on these seven major technological systems. We will examine the components for each of these systems. Also, we will explore the productive processes in greater depth.

Engineering Design Challenge

Overview of Engineering Design Challenges

Each Engineering Design Challenge included in the text will present a problem related to the material covered in the chapter and challenge you to use the engineering design process to solve the problem. The Challenge will also require you to pursue additional information to create a physical solution to the problem.

Summary

- ✔ Technology is humans using objects (tools, machines, systems, and materials) to change the natural and human-made (built) environments.
- ✔ Technology has four basic features: it is human knowledge; it uses tools, materials, and systems; the application of technology results in artifacts; people develop technology to modify or control their environment.
- ✔ Technology is a dynamic process. It changes constantly.
- ✔ Technology has both positive and negative aspects.
- ✔ Improperly used technology can cause serious damage to people, society, and the environment.
- ✔ Technologically literate people can properly develop, select, and responsibly use technology.
- ✔ Early humans lived in primitive conditions, existing with nature and its variations.
- ✔ Civilized conditions evolved from advances in civilization. Humans learn to exert their will on nature.
- ✔ Technological developments during the period led to major changes in society and the creation of new needs and wants.
- ✔ Civilization moved through a variety of eras that led to major societal changes: the Stone Age, Bronze Age, Iron Age, Middle Ages, Renaissance, Industrial Revolution, and information age.
- ✔ There are seven types of technology discussed in this text: agricultural and related biotechnologies; communication and information technologies; construction technologies; energy and power technologies; manufacturing technologies; medical technologies; transportation technologies.
- ✔ In each technological system, inputs are processed and transformed into outputs.

Check Your Engineering IQ

Now that you have finished this chapter, see what you learned by taking the chapter posttest.
www.g-wlearning.com/technologyeducation/

Test Your Knowledge

Answer the following end-of-chapter questions using the information provided in this chapter.

1. What is technology?
2. Which one of the following is *not* a feature of technology?
   A. Technology uses tools, materials, and systems.
   B. Technology results in artifacts and other outputs.
   C. Technology is found in nature.
   D. People develop technology to control their environment.
3. *True or False?* Technology almost always improves on existing technology.
4. Name one positive aspect and one negative aspect of technology.

Matching: Match each technological development with the correct historical period. Some letters will be used more than once.

5. Airplane
6. Waterwheel
7. Copper tools
8. Electricity
9. Fire
10. Ox-drawn plows
11. Paper money
12. Pottery
13. Printing press
14. Solar energy
15. Telescope

A. Stone Age
B. Bronze Age
C. Iron Age
D. Middle Ages
E. Renaissance
F. Industrial Revolution
G. Information age

16. James Watt's improvements to the _____ revolutionized both the manufacturing process and the transportation system.
17. *True or False?* The information age is characterized by the blurring of the line between production workers and managers.

Matching: Match each area of use with the correct technological system. Some letters will be used more than once.

18. Used to maintain health.
19. Used to build structures.
20. Used to convert materials into products.
21. Used to grow food.
22. Used to move people.
23. Used to process data.
24. Used to apply energy.
25. Used to produce natural fibers.
26. Used to communicate ideas.
27. Used to cure illness.
28. Used to transmit information.

A. Agricultural and related biotechnologies
B. Communication and information technologies
C. Construction technologies
D. Energy and power technologies
E. Manufacturing technologies
F. Medical technologies
G. Transportation technologies

Critical Thinking

1. Explain differences between the varying ages of human innovation.
2. Describe how one advancement has spurred innovation in other technological areas.
3. Defend whether or not technology advancement has reached its peak in terms of innovation.
4. What type of advancements do you foresee in the next age of human development?
5. Select one technology from each era and defend whether or not it was actually a positive influence on history.

STEM Applications

1. Develop a chart similar to the following one. For each type of technology, list five technological devices that you use daily.

Technology	Technological Device
Agricultural and related biotechnologies	
Communication and information technologies	
Construction technologies	
Energy and power technologies	
Manufacturing technologies	
Medical technologies	
Transportation technologies	

2. Create a new technological device you can use. Sketch what it would look like and how it would work.
3. Ask older people (grandparents, neighbors) to tell you about five technological devices they used in their lifetimes that are no longer used. List these devices, the devices that have replaced them, and the devices that you think will replace the current ones.

CHAPTER 2

Connecting Technology and Engineering through Mathematics and Science

Check Your Engineering IQ

Before you read this chapter, assess your current understanding of the chapter content by taking the chapter pretest.
www.g-wlearning.com/technologyeducation/

Learning Objectives

After studying this chapter, you should be able to:

- ✔ Define *engineering*.
- ✔ Describe the relationship between technology and engineering.
- ✔ Explore the relationship between technology, science, and mathematics.
- ✔ Compare and contrast the fields of science, technology, engineering, and mathematics (STEM).
- ✔ List the six basic steps for technological or engineering design processes.
- ✔ Describe the differences between inventions and innovations.
- ✔ Examine the impact of inventions and innovations.
- ✔ Evaluate the process of patenting to protect inventive or innovative ideas.

Technical Terms

engineering
engineers
experiments
hypothesis
innovations
inventions
mathematics
mechanical advantage
natural phenomena
optimization
patent
predictive analysis
quantify
quantities
science
scientific method
technological design
technology transfer
tests

Close Reading

As you read this chapter, determine what details illustrate the relationship between technology, engineering, science, and mathematics and prepare a description of the connections between these fields of study. Also, analyze the chapter's key ideas to develop an explanation of the way in which these fields together are responsible for the advancement of human civilization.

Technology often requires performing a task using an object that is not part of the human body. For example, suppose you want to crack the shell of a walnut. You could put it in your mouth and bite. That, however, is not technology, because the human body is doing the work. When you find that your jaw is not strong enough to do the job, you will need another way to open the walnut. Seeing some rocks, you place the shell on one rock and strike it with another rock. You have used a tool to improve your ability to open the walnut. This is considered technology. The rock extended your potential, or ability, to do a task. Technology is the development and application of knowledge, tools, and human skills to solve problems and extend human potential, brought about by needs and desires, **Figure 2-1**.

Another important feature of technology is the ability to develop new knowledge, tools, and skills. Many animals other than humans can use tools. For example, otters use rocks to open clams and monkeys use sticks to collect ants to eat, **Figure 2-2**. What makes technology a human process is that we use knowledge and resources to continuously refine or create new tools and practices, instead of restricting ourselves to natural artifacts. This is the intersection of technology and engineering. Broadly speaking, engineering involves the processes that we follow to create and improve technologies in order to better solve our problems. This chapter will further define and explain the connection between technology and engineering and examine the importance and relationship of mathematics and science.

Norma Cornes/Shutterstock.com

Figure 2-2. Monkeys use sticks as tools to help them collect ants.

Technology's Relationship to Engineering

Technology can be any type of tool or process that extends human capabilities and makes life easier by solving a problem. A hammer fastens materials by driving a nail into those materials. A rock can also drive a nail but a hammer can do the task more easily because of its structure. The hammer is a technological device because it was developed to make the task easier and more effective, and to extend our capabilities.

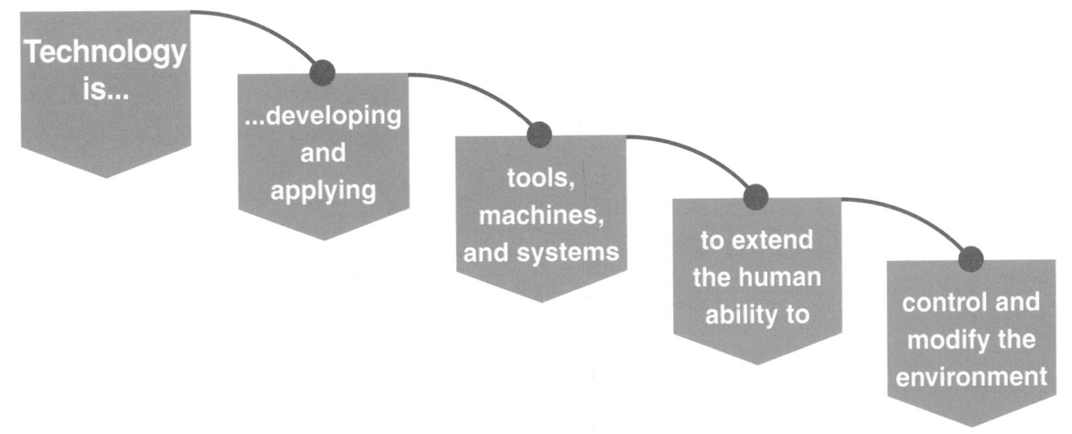

Goodheart-Willcox Publisher

Figure 2-1. Technology extends one's ability to do a specific task.

Hammers are also used to remove nails. This requires the exertion of heavy pressure. Using their knowledge of mathematics and science, early engineers looked to mechanical advantage to develop a way to remove nails. ***Mechanical advantage*** is a measure of the ability to amplify the amount of effort exerted using some type of mechanical device. Using mathematical practices to calculate mechanical advantage, engineers solved the problem by adding a claw to the back of the hammer, **Figure 2-3.** The claw is used as a lever, amplifying the ability to pry a nail out of the material without exerting a great deal of effort. This process of developing and improving technologies using knowledge of mathematics and science is engineering.

OlegSam/Shutterstock.com

Figure 2-3. An example of mechanical advantage is using the claw on a hammer to remove a nail.

STEM Connection: Science and Mathematics

Calculating Mechanical Advantage

Tools and machines are developed to extend human capabilities by achieving mechanical advantage. Recall that mechanical advantage is a measure of the ability to amplify the effort exerted using some type of mechanical device. When greater mechanical advantage is achieved, greater tasks can be accomplished. Determining the best ways to achieve mechanical advantage when using machines requires scientific knowledge of materials and forces and mathematical abilities to understand and calculate the level of mechanical advantage achieved.

Tools and machines are classified as one or more of six simple machines. Simple machines are the most basic mechanisms used to change a force being exerted to achieve some type of task. The six simple machines are the lever, wheel and axle, inclined plane, wedge, pulley, and screw. The claw found on a hammer or an adze is used to remove nails. It is considered a lever because it provides leverage to remove nails. A lever is comprised of a rigid bar that rests on a fulcrum, or a pivot point, to make moving objects easier.

Claws achieve mechanical advantage by converting a small input force into a larger output force. Mechanical advantage is calculated as a ratio of output force to the input force. Therefore, if a claw produces an output force 10 times greater than the force applied, it has a mechanical advantage of 10. To calculate the estimated mechanical advantage of a lever, use this formula:

$$\text{mechanical advantage (MA)} = \frac{\text{effort arm length}}{\text{load arm length}}$$

The effort arm length is the distance between the spot where effort is applied and the fulcrum. The load arm length is the distance between the load that will be moved and the fulcrum. See **Figure A.** The effort arm length is 5″ and the load arm length is 1/2″, providing a mechanical advantage of 10. This means that the claw multiplies the input force exerted 10 times.

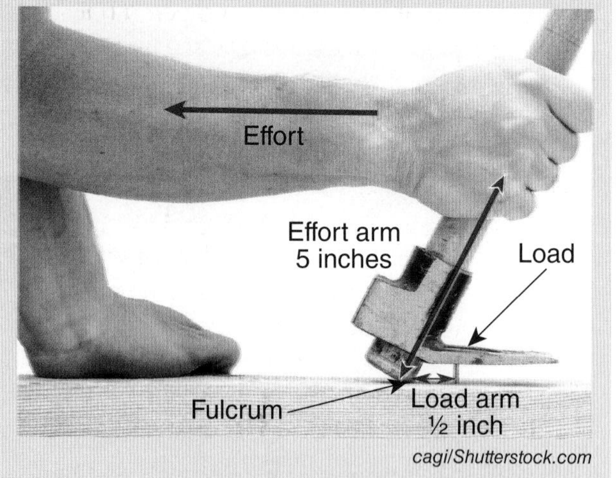

cagi/Shutterstock.com

Figure A. Calculating the mechanical advantage of a lever.

Engineering Defined

Engineering is the practical application of science, mathematics, and technological know-how to solve problems in the most efficient way possible. Engineering connects science and mathematics through a systematic and often iterative (repetitive) approach to designing, creating, evaluating, and improving technologies.

Engineers conduct research and apply scientific and technological knowledge to the design and development of products, structures, and systems, to solve real-world problems. They are problem solvers. They want to make things work better and faster. They want to modify the natural world to create things less expensively and with less environmental damage. Engineers are trained to design, create, evaluate, and improve the technologies that we use every day. These technologies include roads, vehicles, bridges, and navigation systems. Engineering is even part of the foods that we eat and the clothes that we wear. Engineering is part of everything we do on a daily basis. Engineering is a way to impact the world, help humanity, and advance civilization.

Connections between Technology, Engineering, and Other Fields of Study

There is a strong connection between technology, engineering, science, and mathematics, **Figure 2-4**. These fields of study are all related by

Career Connection Engineering Technician

Engineering technicians are trained in the skills and procedures required for equipment installation, repair, and maintenance in specific engineering fields. They often work in field service positions, traveling to different locations to solve technical problems with particular equipment on which they are trained. These individuals have a practical understanding of engineering concepts and assist engineers and engineering technologists.

Students interested in engineering technician jobs should explore high school career academy options, such as engineering technology or manufacturing. College students attend two-year technical institutions or community colleges to obtain an associate degree, such as an associate of applied science or associate of science in engineering and technology. These programs focus on learning the practical application of specific engineering tools. Individuals pursuing engineering technician careers should have a multidisciplinary skill set in problem solving, measurement, electronics, mathematics, and communication.

Engineering technicians are hired in a variety of technology fields ranging from manufacturing to medical industries. They install, maintain, troubleshoot, and repair equipment and systems. Engineering technicians might also work as lab technicians, test-engineering technicians, service technicians, engineering assistants, associate engineers, or field-engineering technicians.

Specific projects that engineering technicians may be involved in include:

- Assembling and setting up engineering equipment.
- Assisting engineers and engineering technologists with research and development projects.
- Producing technical drawings and technical manuals.

Engineering technicians are consistently in demand due to the growing importance of technological fields, such as petroleum and natural gas production and the revitalization of the manufacturing industry. The American Society of Certified Engineering Technicians is one professional organization that an engineering technician might join.

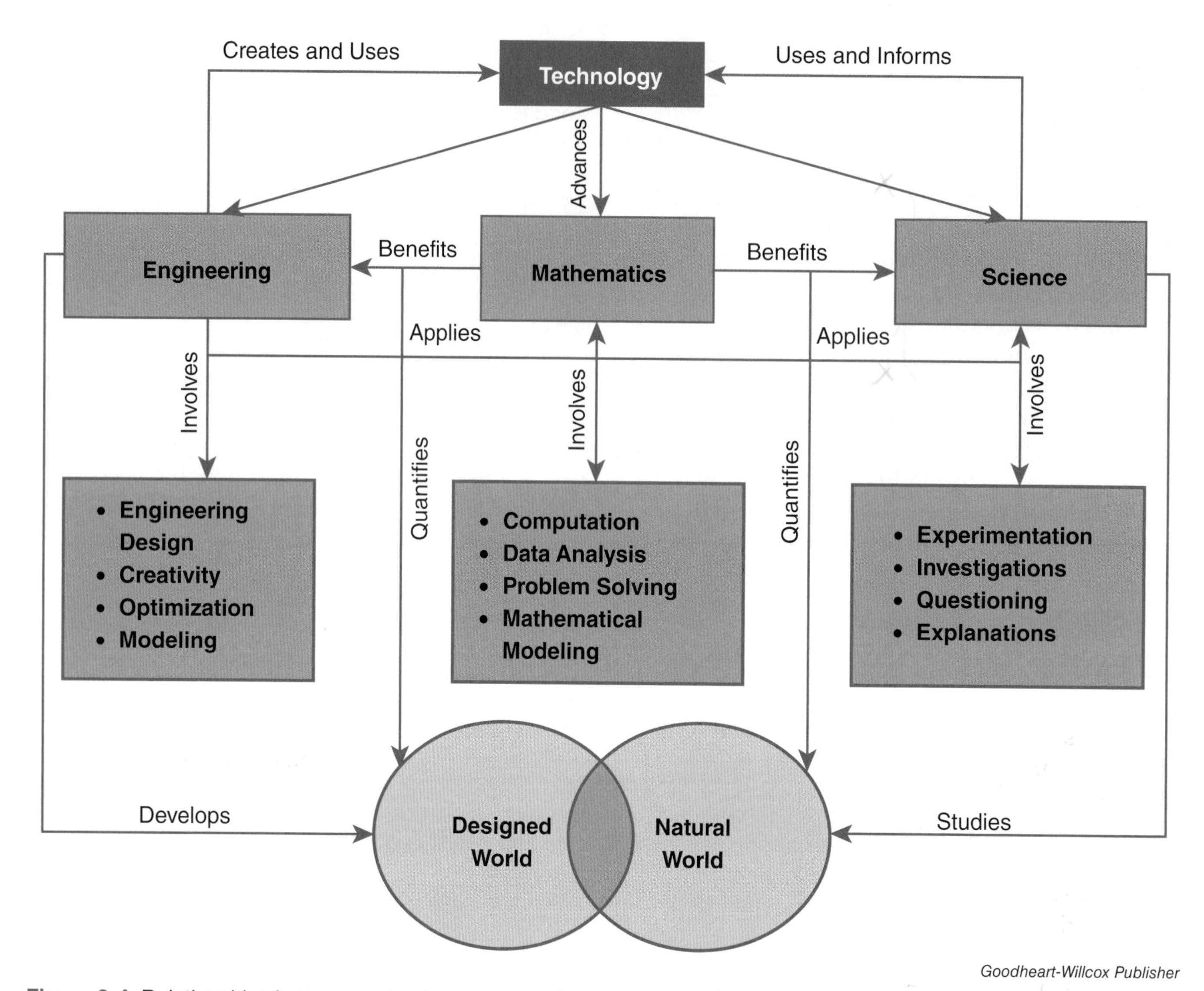

Figure 2-4. Relationships between technology, engineering, science, and mathematics.

the need and desire to enhance human capabilities through designing, creating, using, and evaluating technologies to solve problems. ***Science*** is the study of the natural world. Scientists obtain scientific knowledge by asking questions about natural phenomena, designing and conducting research experiments or scientific investigations to collect data about the phenomena, analyzing and interpreting the collected data, and then constructing explanations of the phenomena, **Figure 2-5**. ***Natural phenomena*** are any observable events that happen in nature and are not human made. Examples include wind, storms, tornadoes, and fog. The result of these practices is the development of new knowledge, scientific laws, and theories that engineers use as a foundation for designing and creating technology.

Scientific practices are also used to evaluate the effectiveness of a technological solution, meaning how well a technology solves a particular problem. These practices are also used to assess the impact that technologies have on the natural environment.

To evaluate technologies, engineers and scientists design and conduct tests and experiments. These will determine if the technology meets specified goals and what effects the technology may have on the environment in which it is used. In engineering, ***tests*** are procedures designed to examine a particular technological device to ensure it works as designed and that it meets the desired specifications. ***Experiments*** are established research procedures carried out in a controlled environment to prove or disprove a hypothesis. A ***hypothesis*** is a proposed explanation of a situation. For example, an

Monkey Business Images/Shutterstock.com

Figure 2-5. Scientists working in a research laboratory.

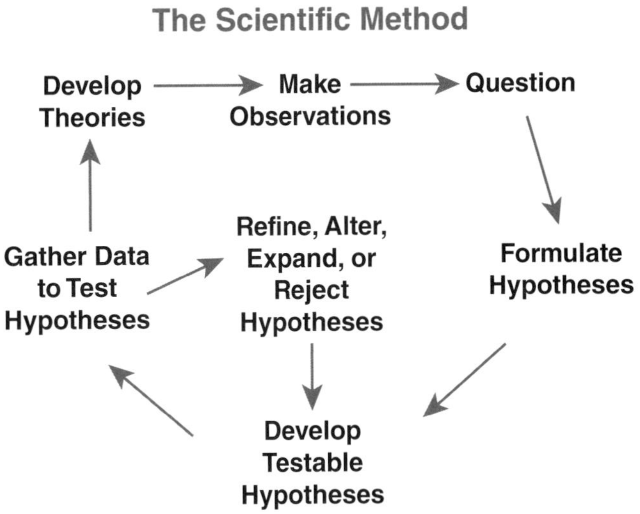

The Scientific Method

Develop Theories → Make Observations → Question → Formulate Hypotheses → Develop Testable Hypotheses → Gather Data to Test Hypotheses → Refine, Alter, Expand, or Reject Hypotheses → Develop Testable Hypotheses

Gather Data to Test Hypotheses → Develop Theories

Goodheart-Willcox Publisher

Figure 2-6. The scientific method.

experiment could be established to determine the effects of the development of a natural gas pipeline on wildlife in the surrounding area, such as white-tail deer. The hypothesis in this example may be that the construction of the natural gas pipeline will have no significant impact on the area's whitetail deer population. Scientists then develop and follow a procedure to collect and analyze the proper data to prove or disprove the hypothesis. These procedures are typically conducted using the ***scientific method***. This method structures scientific research in a way that ensures valid experimental results are obtained. Through the scientific method, scientists attempt to understand the world as it exists. The scientific method is described in **Figure 2-6**.

Scientists and engineers also use mathematical practices and procedures to collect, analyze, and interpret data about specific natural phenomena or technological devices. ***Mathematics*** is the study of measurements, patterns, and the relationships between quantities, using numbers and symbols. Mathematical practices and concepts commonly used by engineers include proper measurement techniques, estimation, reasoning with equations, scaling, proportion, and transformational geometry to design, model, develop, evaluate, and improve technologies. Mathematics is a tool that enables people to ***quantify*** the world by expressing situations as numbers, equations, and inequalities. This allows scientists and engineers to describe natural phenomena or other situations as ***quantities***, or numbers, and perform the necessary calculations to construct explanations of and make predictions about nature or technologies.

Think Green Carbon Footprint

A *carbon footprint* is a measurement of how much everyday behaviors of people, whether individuals or groups, impact the environment. It includes the average amount of carbon dioxide released into the air by energy and gas used at home and in travel. Engineering and technology contribute to carbon dioxide emissions. Electricity that goes into the production of technology and the by-products put into the air may create a larger carbon footprint.

Calculating either personal or group carbon footprints may be done with carbon calculators provided by various environmental organizations. By learning the details of their carbon footprints, people may be motivated to reduce them. This is important to keep in mind when developing new technologies.

Academic Connection: History

The Presidential Election of 1960

As vice president during the eight-year term of a popular president, most observers expected Richard Nixon to win the presidential election of 1960. Additionally improving his chances, he was running against a young and somewhat inexperienced senator, John Kennedy of Massachusetts.

Because of a relatively new invention called the television, however, the perceptions of the two candidates changed dramatically as the election grew closer. In a series of first-ever televised debates, Kennedy came across as confident and mature. Nixon was perceived as nervous and self-conscious. Interestingly enough, according to most people who listened to the debates on the radio, Nixon had won the contests. According to most of those who watched on television, however, Kennedy had won.

In the end, Kennedy won the election. The invention of the television had forever changed the political process. Can you name another technological invention or innovation that might have changed the course of history? In what way or ways did it do so? How do you think social media has impacted more current elections?

Technology and engineering also help advance the fields of science and mathematics. The development of new technologies can provide improved methods and enhanced abilities for observing nature, making accurate measurements, and conducting precise and extensive mathematical calculations. The James Webb Space Telescope, **Figure 2-7**, is a groundbreaking technology developed by teams of engineers. The telescope will orbit the sun, enabling scientists to observe the universe and study its history and evolution. Scientific knowledge and mathematical practices enlighten the processes of engineering, while technologies created through engineering drive new scientific breakthroughs. It is easy to see that science, technology, engineering, and mathematics are inseparable.

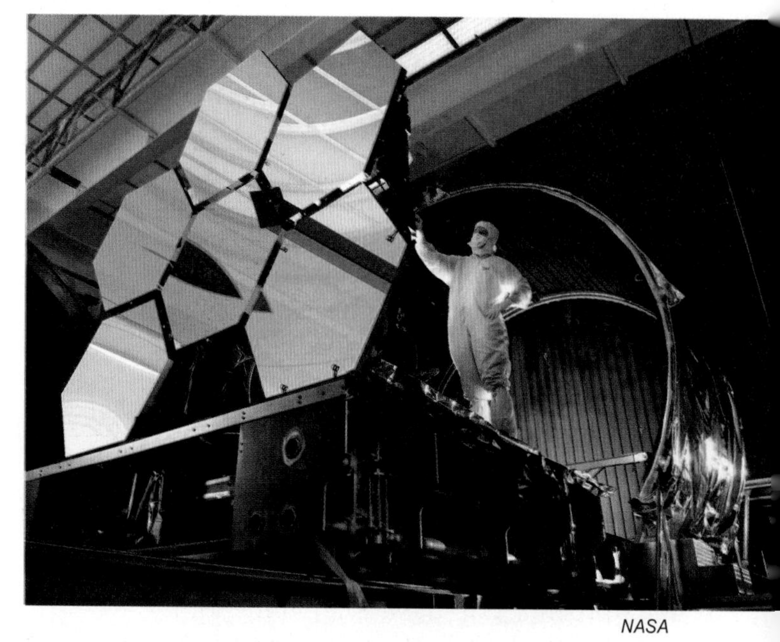

NASA

Figure 2-7. Construction of the James Webb Space Telescope requires the work of many engineers.

Technology versus Engineering

Technology is the application of knowledge, tools, and skills to solve problems and extend human capabilities. It is any modification of the natural world done to fulfill human needs and desires. Technology, as described earlier, also develops and uses tools and machines, systems, and materials to extend the ability to control and modify the environment. The key component in this definition is *develop*. Developing technology requires creative action. Early in human history, creative action to develop technologies was more primitive and was based mostly on technological know-how rather than on scientific knowledge. Technological know-how is the skills and knowledge that people have developed through a trial-and-error approach to solving problems.

The person cracking open the walnut used the trial-and-error approach based on technological know-how. The person used two rocks, knowing that the rock was harder than the walnut. This logic was not based on science; it was based on the person's experience with rocks. If the rock did not work, the person might have tried another object or materials to open the walnut. Once the object or material that worked was found, the person might try different types and shapes to find the ones that worked the best. This approach to solving problems combined with the creative actions of people to develop new technologies is considered a technological design process.

Design is simply the process of creating something. ***Technological design*** is the open-ended, trial-and-error process of creating a problem solution to meet needs and desires. Engineering differs from the technological design process, however, in that engineering is a more systematic approach to design solutions. Engineering is based on a solid foundation of scientific and mathematical knowledge. Engineering design is an explicit and intentional path to efficiently create a technological solution to a problem. This approach includes ***predictive analysis*** to determine how well a proposed solution solves a problem, before the solution is actually produced. The engineering design process also includes practices for optimizing a solution based on mathematical modeling and analysis. ***Optimization*** is the engineering practice of making something as fully effective or as perfect as possible using mathematical procedures.

Technological and engineering design processes include six basic steps to create technological solutions to problems, **Figure 2-8**.

1. **Identify and define a problem.** A person or group develops basic information about the problem and the design limitations.
2. **Gather information.** A person or group searches for all background information necessary to develop solution ideas for the problem.
3. **Develop a design solution.** A person or group develops and refines several possible solutions. They will then isolate, refine, and detail the best solution.
4. **Model and make a solution.** A person or group produces a physical, graphic, or mathematical model/prototype of the selected solution.

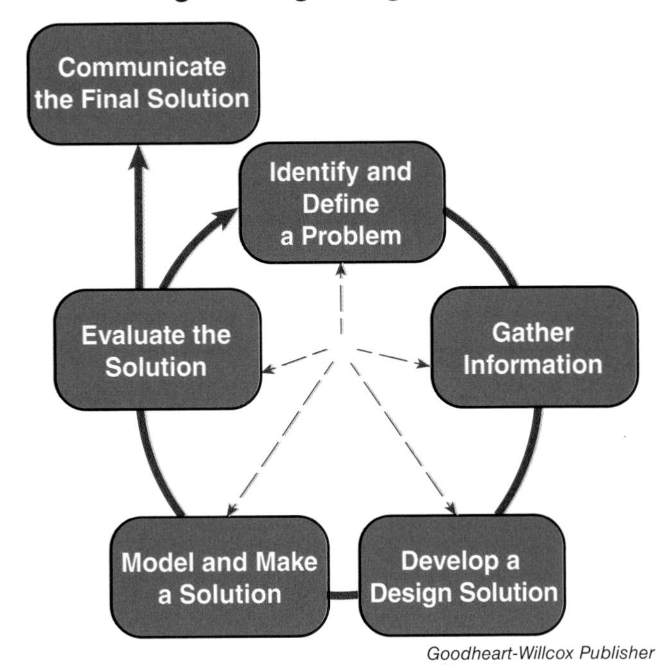

Figure 2-8. Six basic steps to designing and creating technological solutions.

5. **Evaluate the solution.** A person or group will test a physical, graphic, or mathematical model of the selected solution to the problem to determine how well it actually solves the problem. Based on this information, a person or group can optimize or redesign the solution to better solve the problem.
6. **Communicate the final solution.** A person or group selects a final solution and prepares documents needed to produce and use the device or system.

These steps in the design process are explained in depth in Chapter 4.

Achieving Goals through Technology and Engineering

Engineers have a goal in mind when developing technology. Each technological artifact, however, generally meets more than one goal. For example, engineers developed radar (*ra*dio *d*etection *a*nd *r*anging) to determine the position and speed of aircraft. Radar had a major impact on military strategies during World War II. Today, this technology contributes to safe, reliable air transportation. It is also used to predict the weather and to detect

Academic Connection: History

The Origin of Radar

As the threat of world war grew in the late 1930s, various countries realized that their radar devices were not reliable. Knowing that keeping track of an enemy's air and sea movements would be critical in a war, engineers were asked to improve radar technology. American and British researchers worked with each other to improve it. Their problem was increasing the reliability and accuracy of radar. With the problem defined, they developed and refined solutions using the engineering design process. For example, the British developed an improved vacuum tube that had enough microwave energy to be used in radar systems. Using that breakthrough, the Americans then developed units small enough for airplanes and boats. This improved radar became a key element in the Allies' eventual victory. In the past 75 years, researchers have further developed the capabilities of radar using the engineering design process. In what areas is radar used today?

Eugene Sergeev/Shutterstock.com

surface features on planets. This is an example of ***technology transfer***, or a technology that migrates from one field to another to solve a different type of problem, **Figure 2-9**.

Engineers must determine the best ways to achieve their goals while meeting the different objectives and concerns important to various groups. For example, suppose a company is developing a new natural gas-powered automobile. The primary goal in creating the automobile is to transport people. The automobile might also be an opportunity for the company to make money. Local government leaders might be concerned about the car's potential impact on economic growth in their city. The federal government might examine the automobile in terms of whether or not it will help reaffirm national technological leadership. The general public might consider whether or not this automobile is a good alternative to other fossil fuel-powered vehicles. Workers might wonder if development of the automobile will improve job security. Natural gas companies and oil companies might examine it with regard to whether it will increase or decrease their markets. Environmental groups might comment on how the car affects the environment. Note the variety and number of different goals and concerns highlighted here.

Gines Romero/Shutterstock.com

Figure 2-9. Radar technology can be used to track and predict weather.

Technology Explained Smart House

Smart house: A house that allows computers to control appliances and energy use.

Computer technology has been used to create an entirely new type of house called a *smart house*. The smart-house system makes home automation possible. This system allows the appliances and the electrical, telephone, gas, and communication systems in a house to interact. See **Figure A**.

The heart of a smart house is found in the system controllers. See **Figure B**. These units serve as the hubs for messages between products within the house. The system controllers constantly check the system for problems. Electrical power, telephone service, and cable television are brought into the house at a service center. A system controller is installed there.

The smart-house system uses microprocessors and a unique wiring system to allow the electrical appliances and products in a house to interact. Participating companies are developing products with built-in smart-house chips. These products can then interact with the system controllers and the rest of the smart-house system.

Smart houses allow homeowners to change how their houses work by programming the system controllers. Reprogramming a switch changes what the switch controls. The heating system can be programmed to vary the home's temperature during the day. There can be cool bedrooms at night and warm bathrooms in the morning. The water heater can be set to turn on at a particular time so hot water is available for morning showers. The electric range can be turned on to cook a meal in the oven. The dishwasher can be set to start at night, taking advantage of lower utility rates.

The wiring system in a smart house is unique. Smart houses use three types of cables. See **Figure C**. Hybrid branch cables carry 120-volt alternating current (ac) power and control signals. These cables provide power for standard appliances, such as coffeemakers, digital videodisc (DVD) players, and lamps. Communications cables carry audio, video, telephone, and computer data signals. Applications cables carry 12-volt direct current (dc) power and control signals. These cables provide power for low-voltage devices, such as smoke detectors, security systems, and switches. Large appliances, such as ovens and clothes dryers, use conventional high-amperage wiring. Radio frequency and infrared devices can be used to make a smart house wireless.

Goodheart-Willcox Publisher

Figure A. Smart-house systems control energy use, security, communications, lighting, and entertainment.

Goodheart-Willcox Publisher

Figure B. A series of smart-house controllers in a recently built home.

Goodheart-Willcox Publisher

Cables

Goodheart-Willcox Publisher

Convenience center

Figure C. The wiring used in a smart house is special. Three types of cables are used: applications, communications, and hybrid branch. Convenience centers are special outlets that can be changed to meet the needs of the homeowner.

Engineering and Technology Outcomes: Inventions and Innovations

Many achievements have been brought about by technology and engineering, along with the application of science and mathematics. These achievements, such as vaccines, space travel, and the Internet, are the result of technology and engineering and are made possible through an evolution of technological development.

The evolution of technological development is the outcome of engineers creating inventions and innovations. ***Inventions*** are new, useful products or processes that solve some type of problem and did not previously exist. ***Innovations*** are refinements or improvements made to preexisting products or processes that better solve a problem.

Inventions and innovations are a direct outcome of the work done by engineers. Engineers transform their understanding of mathematics and science into usable products and processes that improve our lives and increase the boundaries of human possibility. Despite the countless inventions and innovations, there are still many objects humans have yet to discover and create.

When engineers are designing inventions or innovations, they often keep detailed documentation of their process. This documentation is used to protect an engineer's work. It is important to protect inventions and innovations so that ideas are not stolen and inventions or innovations are not used without the permission.

Documentation of the process enables engineers to obtain patents to protect their work. A ***patent*** is a property right granted by the government that allows inventors or innovators exclusive use to create their designs and sell their ideas, **Figure 2-10**. Patents are granted for new, useful processes or products. The United States Patent and Trademark Office grants patents or trademarks. Applicants must prove that the idea for the invention or innovation was their creation and they must show documentation of the development process. Documentation should include dated sketches and explanations of how the invention operates and all related information, such as an engineering notebook or electronic portfolio.

The United States of America

The Commissioner of Patents and Trademarks

Has received an application for a patent for a new and useful invention. The title and description of the invention are enclosed. The requirements of law have been complied with, and it has been determined that a patent on the invention shall be granted under the law.

Therefore, this 5,860,492

United States Patent

Grants to the person(s) having title to this patent the right to exclude others from making, using, offering for sale, or selling the invention throughout the United States of America or importing the invention into the United States of America for the term set forth below, subject to the payment of maintenance fees as provided by law.

If this application was filed prior to June 8, 1995, the term of this patent is the longer of seventeen years from the date of grant of this patent or twenty years from the earliest effective U.S. filing date of the application, subject to any statutory extension.

If this application was filed on or after June 8, 1995, the term of this patent is twenty years from the U.S. filing date, subject to any statutory extension. If the application contains a specific reference to an earlier filed application or applications under 35 U.S.C. 120, 121 or 365(c), the term of the patent is twenty years from the date on which the earliest application was filed, subject to any statutory extension.

Acting Commissioner of Patents and Trademarks

Attest

United States Patent Office

Figure 2-10. A United States patent.

GaudiLab/Shutterstock.com

This engineer is verifying data from an online map while making measurements on a working drawing.

Summary

- ✔ Broadly speaking, engineering involves the processes that we follow to create and improve technologies in order to better solve our problems.
- ✔ Technology can be any type of tool or process that extends human capabilities and makes life easier by solving a problem.
- ✔ The process of developing and improving technologies using knowledge of mathematics and science is engineering.
- ✔ Engineering is the practical application of science, mathematics, and technological know-how to solve problems in the most efficient way possible.
- ✔ Engineers conduct research and apply scientific and technological knowledge to the design and development of products, structure, and systems.
- ✔ Technology, engineering, science, and mathematics are all related by the need and desire to enhance human capabilities through designing, creating, using, and evaluating technologies to solve problems.
- ✔ The result of scientific practices is the development of new knowledge, scientific laws, and theories that engineers use as a foundation for designing and creating technology.
- ✔ The scientific method structures scientific research in a way that ensures valid experimental results are obtained.
- ✔ Scientists and engineers use mathematical practices and procedures to collect, analyze, and interpret data about specific natural phenomena or technological devices.
- ✔ The development of new technologies can provide improved methods and enhanced abilities for observing nature, making accurate measurements, and conducting precise and extensive mathematical calculations.
- ✔ Technological and engineering design processes include six basic steps to create technological solutions to problems.
- ✔ Technological artifacts generally meet more than one goal.
- ✔ Technology transfer is technology that migrates from one field to another to solve a different type of problem.
- ✔ Engineers must determine the best ways to achieve their goals while meeting the different objectives and concerns important to various groups.

Check Your Engineering IQ

Now that you have finished this chapter, see what you learned by taking the chapter posttest.
www.g-wlearning.com/technologyeducation/

Test Your Knowledge

Answer the following end-of-chapter questions using the information provided in this chapter.

1. _____ can be any type of tool or process that extends human capabilities and makes life easier by solving a problem.
2. What is mechanical advantage?
3. _____ is the practical application of science, mathematics, and technological know-how to solve problems in the most efficient way possible.
4. The result of scientific practices is the development of new _____.
   A. knowledge
   B. scientific laws
   C. theories
   D. All of the above.
5. *True or False?* Tests are established research procedures carried out in a controlled environment to prove or disprove a hypothesis.
6. The study of measurements, patterns, and the relationships between quantities using numbers and symbols is known as _____.
7. In what way do technology and engineering help advance the fields of science or mathematics?
8. *True or False?* Technological know-how is the skills and knowledge that people have developed through a trial-and-error approach to solving problems.

9. _____ is an explicit and intentional path to efficiently create a technological solution to a problem.
   A. Predictive analysis
   B. Optimization
   C. Technological design
   D. Engineering design
10. Name the six basic steps for technological or engineering design processes.
11. Explain the difference between an invention and an innovation.
12. What is the purpose of a patent?

Critical Thinking

1. Create an illustration that provides a visual explanation of the relationship between technology, engineering, science, and mathematics.
2. Explain how the fields of mathematics and science are critical for technological development and engineering solutions to the world's problems.
3. How does technological design differ from engineering design when creating solutions to problems?
4. Choose a technology and develop and describe three creative ideas for how it could be used for a different purpose.

STEM Applications

1. Build a simple machine (lever, screw, wheel and axle, inclined plane, pulley, or wedge) and use it to achieve a simple task, such as lifting a chair. Then research the mathematical formulas for calculating the mechanical advantage that is achieved by the simple machine that you built. Using the formula, calculate the mechanical advantage that is achieved. Then determine ways in which you can alter the simple machine to increase the mechanical advantage.
2. Select an early technological advancement (invention). Prepare a short report including the inventor's name, the nation from which the invention came, events that led up to the invention, the invention's impact on life at the time, and the invention's later refinements. Include a sketch of the item.

CHAPTER 3

Engineering Fundamentals

Check Your Engineering IQ 

Before you read this chapter, assess your current understanding of the chapter content by taking the chapter pretest.
www.g-wlearning.com/technologyeducation/

Learning Objectives

After studying this chapter, you should be able to:

- ✔ Document historical engineering feats.
- ✔ Describe the role of the engineering profession.
- ✔ Distinguish between engineering disciplines.
- ✔ List several qualities and practices of engineers.
- ✔ List specific engineering characteristics.
- ✔ Explain the purpose of ethics in engineering.
- ✔ Describe common engineering skills used in the profession.

Technical Terms 

chemical engineering
civil engineering
creativity
creed
dimension
electrical engineering
engineering design
engineering notation
estimation
ethics
Gantt chart
iterative
mechanical engineering
optimism
optimize
scientific notation
significant figures
systems
technical writing
unit

Close Reading

As you read this chapter, think about the following questions and what you believe to be the defining features of an engineer. What is the purpose of engineering? What are the fundamental characteristics of an engineer? What do they do day to day? What are their responsibilities? How has engineering changed over time?

As engineering continues to help create new technologies, it is necessary to recognize the historical impacts of engineering over time and how the profession has changed. After reading this chapter, you will be able to describe the role of the engineering profession and understand the many responsibilities, traits, and skills needed by engineers.

Engineering in History

Engineering as a profession can be traced back as early as 6000 BCE, **Figure 3-1.** This Egyptian painting shows a basic boat design. Since that time, engineering advancements have increased exponentially. This can be attributed to contributions from mathematics and science and the engineering of new technologies, which resulted in more innovative methods and designs of more efficient technologies. For example, stronger concrete engineered during the Roman Empire, around 30 BCE–4756 CE, aided in the construction of stronger structures once made of stone. Reinforced concrete, developed in the mid-1800s by Joseph Monier, resulted in stronger bridges, structures, and dams. Humans have engaged in engineering for thousands of years and each advancement results in newer technologies being created. Consider the world around you and how engineering has changed since you have been alive.

BasPhoto/Shutterstock.com

Figure 3-1. This rock painting of an ancient Egyptian sailing vessel depicts a device created through early engineering processes.

The Engineering Profession

Although engineering has contributed to the creation of technology throughout human history, it has taken thousands of years to become the organized, structured, and professional industry it is today. An early engineer was a combination of artist, craftsperson, and tinker who solved problems by designing and creating physical solutions. Many of these individuals were not formally trained.

One notable engineer was Leonardo da Vinci, who lived during the Renaissance, **Figure 3-2.** Despite having no formal education in engineering, he used his artistic skills and imagination to conceptualize many engineering ideas. For example, his notebook contains drawings and illustrations of flying technologies, boat designs, and solar power contraptions, **Figure 3-3.**

Georgios Kollidas/Shutterstock.com

Figure 3-2. Leonardo da Vinci created several inventions using early engineering practices.

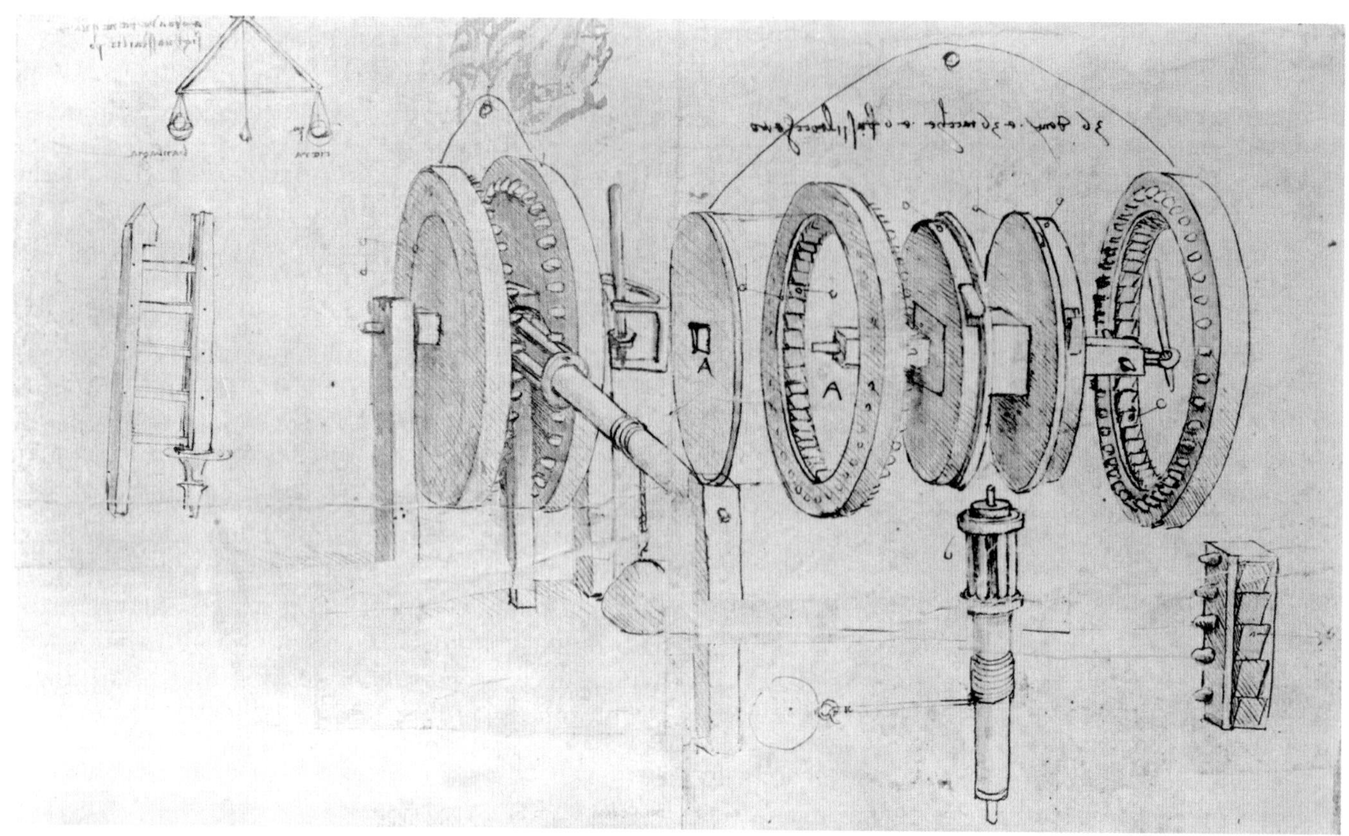

Everett Historical/Shutterstock.com

Figure 3-3. This page from Leonardo da Vinci's notebook shows drawings of gears.

Engineers of the twentieth century, including Dr. Robert Goddard, Edith Clarke, Nikola Tesla, and the Wright brothers, solved engineering problems with a much more complete understanding of mathematics, science, and technological knowledge than engineers of previous centuries. Yet even 100 years ago, many of these engineers solved problems using creativity and knowledge that spanned minimally across fields. More recently, twenty-first century engineers, such as Elon Musk, have created new technologies that require highly specialized knowledge in each field. His company, SpaceX, designs and produces spaceships that require the intentional application of mathematical and scientific knowledge, **Figure 3-4**.

The well-established engineering profession now takes a more organized and disciplined approach to solving problems, which leads to more effective and efficient solutions. This differs from the trial-and-error method typically used by nonengineers. In a trial-and-error process, solutions are implemented without considering all design

Think Green Environmental Engineering

Environmental engineering is an engineering specialization that involves the management of hazardous materials and the control of water and air pollutions, waste disposal, and various public health concerns. Environmental engineers use the principles of engineering and scientific knowledge to develop solutions to environmental problems. These engineers are responsible for minimizing the impact that all human activities have on the environment. Additionally, they work to develop laws for technological development to help ensure all other engineering activities are done in a sustainable manner.

SpaceX

Figure 3-4. The SpaceX Crew Dragon is designed to carry astronauts to earth's orbit and beyond. It will be capable of refueling and flying again for rapid reusability.

aspects, and then the design is altered as solutions fail. This is a costly method of problem solving that engineers avoid by using their knowledge of mathematics and science.

Engineers take on many roles, **Figure 3-5**. Some engineers work on structures, such as bridges, dams, and skyscrapers. Other engineers solve mechanical problems related to automobiles, engines, and airplanes. Engineering has become a highly specialized field. This is the result of professionals who are highly knowledgeable in solving specific problems.

The engineering profession can be grouped into disciplines. Within each discipline, professional engineers have different responsibilities. Some engineers are designers, which require brainstorming abilities and creativity to design new solutions to problems. Other engineers may focus specifically on testing and evaluating. And others may cover the entire process from start to finish.

Engineering Disciplines

There are four overarching engineering disciplines. ***Civil engineering*** focuses on the design and improvement of structures such as bridges, skyscrapers, roadways, and dams. These engineers apply mathematical and scientific knowledge to ensure that structures are designed safely and economically, and are environmentally sound. ***Electrical engineering*** focuses on electricity and electromagnetism. Electrical engineers understand electrical charge, forces of electricity, and how to safely design electrical technologies. They work on both small- and large-scale electrical systems. This may include the design and maintenance of a circuit board or determining how to transfer energy from a dam to a local home. ***Mechanical engineering*** involves designing and improving mechanical systems and components. A mechanical engineer works with the assembly of parts to achieve a functioning mechanism.

Chemical engineering involves the conversion of raw materials into useful products, **Figure 3-6**. Chemical engineers need a background in chemistry to design and improve existing materials. This field often collaborates and is closely aligned with the scientific community.

These four overarching disciplines of engineering can be broken down into more specialized subdisciplines. For example, within civil engineering, there exist railway, environmental, mining, and even hydraulic engineering. There also exist integrated disciplines that do not fit neatly within any one discipline, such as aerospace engineering, industrial engineering, systems engineering, and biomedical

ndoeljindoel/Shutterstock.com

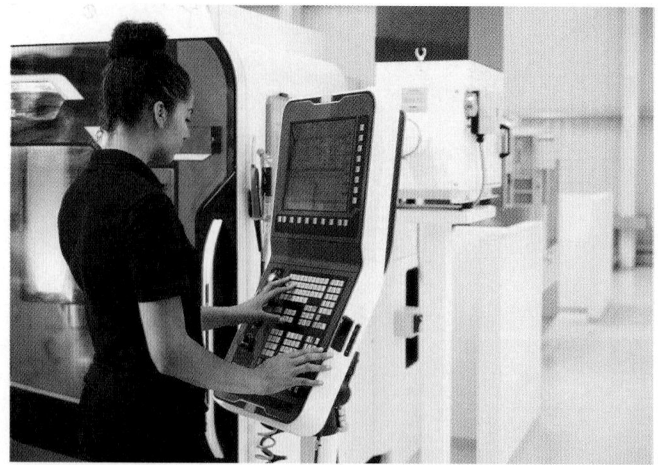

Monkey Business Images/Shutterstock.com

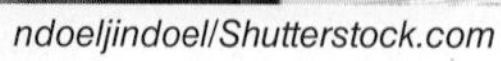

Monkey Business Images/Shutterstock.com

Figure 3-5. Engineers work in a variety of environments.

TTstudio/Shutterstock.com

Figure 3-6. Many chemical engineers work at petrochemical plants.

Career Connection Engineering Professionals

An engineer is a person trained in the skills and procedures for applying scientific, mathematical, and technical knowledge to design and develop systems, products, structures, tools, and equipment to solve societal or commercial problems. Engineers specialize in specific engineering disciplines. Within each specialization, engineers work to develop new technological solutions to problems using the engineering design process and analyze technological problems through testing procedures, all while maintaining the highest standards of ethics and integrity.

High school students interested in engineering should take a variety of technology and engineering courses, along with the highest-level mathematics and science courses available. Students should also complete college-level calculus and physics while still in high school. Career academies designed for students interested in pursuing engineering can help prepare students for postsecondary engineering programs.

In order to be professional engineers, students must complete a four-year college degree program to earn a bachelor of science degree in engineering. There may be some alternative routes to become an engineer that involve completing an engineering technology degree and completing an engineering certification program. Postsecondary engineering programs are geared toward development of conceptual skills, and consist of a sequence of engineering fundamentals and design courses built on a foundation of complex mathematics and science courses.

Individuals pursuing a career as an engineer must have a strong foundation of mathematical and scientific knowledge. These individuals should also be able to collaborate with others, communicate effectively, think creatively, maintain an optimistic outlook, and make ethical decisions.

Engineers can choose from a variety of specializations to gain them employment in almost any industry. Chemical engineers might study and develop new materials. Civil engineers can plan and design new buildings or bridges. Engineers can also work toward management positions in a variety of industries.

The job outlook for engineers is generally good. Engineers are always in high demand because there is a lack of qualified engineers around the world to work in increasingly expanding technological industries. There are a variety of professional organizations for each engineering specialization.

engineering. These can be considered *interdisciplinary* branches of engineering.

Engineering Qualities

Although there are many disciplines and subdisciplines of engineering, engineers tend to have certain qualities. Engineers have extensive background knowledge in science, mathematics, and technical disciplines. Additionally, they are problem solvers and logical thinkers, are attentive to detail, and are able to communicate their knowledge and solutions to a wide range of people. Engineers are resourceful, creative, and able to visualize ideas or objects apart from their actual physical presence. Engineers are very motivated individuals. They often take continuing education courses to keep up with advances in the field, and many engineers need licenses to work as engineers. Engineers must be able to work with others, including people from other fields.

Engineering Practices

Engineers work to solve problems using a specific set of practices. An engineer practices defining problems, developing and using models, planning and conducting investigations, analyzing and interpreting data, using computational thinking, designing solutions, arguing based on evidence, and collecting, assessing, and communicating information, **Figure 3-7**. This problem-solving process is called ***engineering design***.

wavebreakmedia/Shutterstock.com

Figure 3-7. Problem solving is a common task for many engineers.

Engineering Design

Engineers use a specific process for solving problems called engineering design. The engineering design process involves creating solutions to problems following specified criteria (guidelines) and under specific constraints (limitations) in an ***iterative*** (repetitive) manner. The iterations (repetitions) of this process are done to continually ***optimize*** (improve) the design solution. The engineering design process is not the same as a trial-and-error process. The engineering design process is more purposeful in developing an optimal solution design before building and testing it. The development of optimal solution designs requires a combination of engineering practices and engineering concepts, and the application of mathematical, technical, and scientific knowledge. Additionally, as engineers work through the engineering design process, they must think critically and innovatively, consider solutions as a part of the larger man-made and natural environments, and use failures as learning experiences. The engineering design process will be discussed fully in Chapter 4.

Engineering Characteristics

Engineers generally have specific ways of thinking when they are creating optimal solutions to problems. These ways of thinking can be considered the characteristics of quality engineers. These characteristics influence the ways engineers approach problems where resolutions are not immediately apparent. These characteristics include but are not limited to creativity, collaboration, communication, optimism, ethical behavior, and systematic thinking, **Figure 3-8**.

Creativity

Creativity is a characteristic that is evident during the engineering problem-solving process. Creativity is not about receiving the best score on an exam or solving a given problem. Most people who study hard can get a high score on an exam, and most people can solve a problem given enough resources. ***Creativity*** is thinking in a way that is different from the norm. Creative people analyze problems and attempt to solve them in a way that is not the most obvious. Creativity is important in engineering because it is the type of thinking that leads people to new and better ideas, innovations, and inventions.

Sofi photo/Shutterstock.com

Figure 3-8. Do you possess any of these engineering characteristics?

When everyone thinks exactly the same way, there is a lack of innovation. Great engineers, designers, or problem solvers are able to think creatively.

Collaboration and Communication

To successfully solve problems, engineers must work collaboratively with others, **Figure 3-9**. Collaboration is important because most engineering design projects are undertaken as a team. This team-oriented approach also requires the ability to effectively communicate ideas and outcomes through various means, such as technical reports; conceptual, physical, and mathematical models; and graphical representations of data. Collaboration and communication are essential for working with others, directing people in the engineering design process, understanding the needs of customers, effectively communicating ideas to others, and explaining and justifying design solutions. Effective collaborative and communication skills are critical for every engineer to develop because their work is only as good as their ability to communicate.

Optimism

Optimism enables a person to view challenging situations as opportunities to learn and improve or as chances to develop new ideas. ***Optimism*** is the tendency to look at the more favorable side of events or to expect the best outcomes in various situations. Optimism enables an engineer to be persistent in looking for the optimal solutions to problems. When solving technological problems, engineers often experience repeated failures or unfavorable situations. An optimistic outlook can provide the motivation to succeed and the dedication to focus on solving the problem at hand.

Ethics

Through their work, engineers help people and improve society, **Figure 3-10**. Because of their work, engineers are often highly regarded members of society who are trusted to create solutions to the world's problems. Members of the engineering profession are expected to maintain the highest standards of integrity and honesty when making decisions about design solutions. Their decisions and designs must be ethical to protect the public's safety, health, and welfare. ***Ethics*** are the principles of conduct that govern the actions of an individual or group. These principles can include personal morals, personal preferences, professional codes of ethics, and established laws.

Engineering ethics mean that engineers make the best choices and do the "right" thing. However, knowing what is the right thing can sometimes be difficult, and it often involves making choices between conflicting alternatives. For example, when

Dragon Images/Shutterstock.com

Figure 3-9. These engineers are discussing plans for a new construction project.

Riccardo Mayer/Shutterstock.com

Figure 3-10. Engineers often help communities through efforts such as designing and creating better ways to access clean drinking water.

choosing the material for a project, factors concerning environmental impacts and affordability must be considered. The right choice is not always clear. Rather than choosing what is right or wrong, the choice often involves finding some type of balance or middle ground. To assist in making ethical decisions, professional engineering organizations have established a code of ethics and the Engineers' Creed.

Code of Ethics

The National Society of Professional Engineers (NSPE) publishes a code of ethics for practicing engineers to follow. The code contains a preamble, the fundamental canons, rules of practice, and professional obligations, **Figure 3-11**.

The fundamental canons describe what engineers must do as they fulfill their obligations. This includes holding public health and safety in the highest regard; doing work only within their areas of expertise; sharing information in an objective truthful manner; acting as a faithful agent to their employer or client; avoiding any deceptive acts; and always acting in a professional manner. The rules of practice describe general situations that engineers face and how they should act in each situation. The last section of the code of ethics describes the professional obligations of every engineer. This includes holding themselves to the highest standards of integrity and honor, always keeping the public's best interest in mind, avoiding all deceptive behavior, not disclosing confidential information, not being influenced by conflicting interests, accepting personal responsibility for their activities, and giving credit for work where credit is due.

The Engineers' Creed

To assist in engineering ethics, the National Society of Professional Engineers has also established the Engineers' Creed. A ***creed*** is a set of beliefs that is established to guide someone's actions. Therefore, the Engineers' Creed is a statement that describes what an engineer must always do. Professional engineers must dedicate themselves to bettering the human race and advancing society. To do so, they must work in accordance with all applicable laws and professional expectations and pledge to give their utmost and honest performance. These are the ideals that must be followed if they wish to practice as a professional engineer.

NSPE Code of Ethics

Preamble

Engineering is an important and learned profession. As members of this profession, engineers are expected to exhibit the highest standards of honesty and integrity. Engineering has a direct and vital impact on the quality of life for all people. Accordingly, the services provided by engineers require honesty, impartiality, fairness, and equity, and must be dedicated to the protection of the public health, safety, and welfare. Engineers must perform under a standard of professional behavior that requires adherence to the highest principles of ethical conduct.

Reprinted by Permission of the National Society of Professional Engineers (NSPE) www.nspe.org;Stephen Coburn/Shutterstock.com

Figure 3-11. The preamble for the NSPE Code of Ethics summarizes the important points of their code.

STEM Connection: Mathematics

Designing the Pyramids

The application of mathematics to solve engineering problems has been around for thousands of years. The mathematical knowledge that you have learned in your various classes can be applied to solve major engineering challenges. One example is using the Pythagorean theorem to design a pyramid. This theorem can be used if we have right triangles, and know the lengths of at least two sides, but need to find the length of the third side. The equation used to find the missing side is:

$$a^2 + b^2 = c^2$$

where *c* is the hypotenuse and *a* and *b* are the remaining two sides of the right triangle.

To apply this principle, see **Figure A**. Presume the pyramid is 100 feet tall (represented as *h* in the figure) and with a width of 50 feet (represented as *r* in the figure). What would the hypotenuse be? To solve, we would use the following equation:

$$50^2 + 100^2 = c^2$$
$$2500 + 10000 = c^2$$
$$12{,}500 = c^2$$
$$c = \sqrt{12{,}500}$$
$$c = 112 \text{ feet}$$

Goodheart-Willcox Publisher

Figure A. Finding the hypotenuse.

Think of a situation in your everyday life in which you can apply the Pythagorean theorem to help you design a new product or solve an engineering challenge.

Systematic Thinking

When engineers work to solve problems, they must think in terms of systems. ***Systems*** are sets of interacting elements that work interdependently to form a complete entity that has an overall function or purpose. These systems can be either man-made or natural. An example of a man-made system is the interstate highway system, **Figure 3-12**. Interstates are a network of controlled-access, interconnected primary highways designed for safe and speedy travel between cities and states. One example of a natural system is the water cycle, **Figure 3-13**. The water cycle involves many components that enable the continuous movement of water on, below, and above the earth's surface.

Thinking in terms of systems means understanding how each component of a solution design or idea fits with other components while forming a complete design or idea. Additionally, systems thinking involves considering how a solution idea or design interacts as a part of the larger man-made and natural systems. For example, an engineer may need to consider how a design solution can impact the country's transportation systems or how it affects the local ecosystem. It is important to recognize that

trekandshoot/Shutterstock.com

Figure 3-12. The interstate highway system is a network of controlled-access primary highways designed to allow safe, speedy travel within the United States.

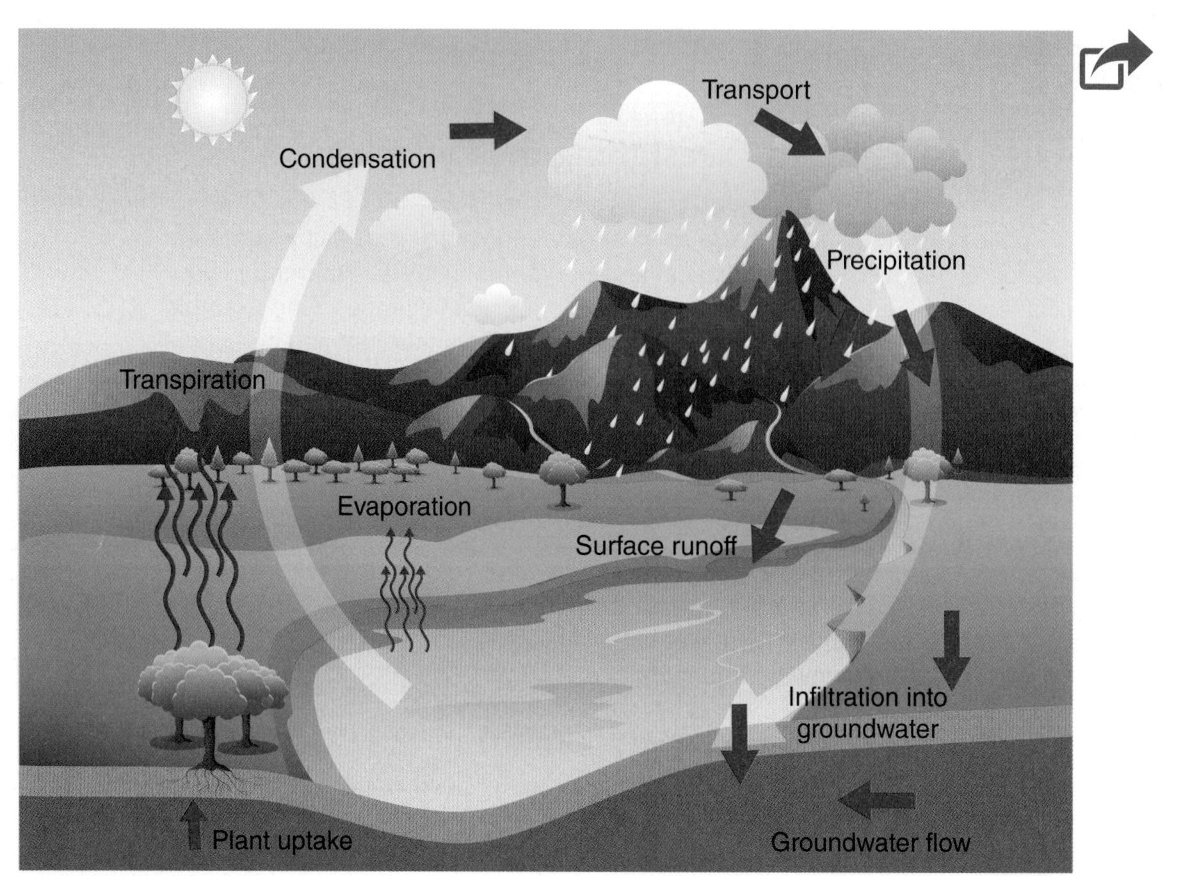

stockshoppe/Shutterstock.com

Figure 3-13. The water cycle involves the continuous movement of water on, below, and above the earth's surface.

there are all types of systems and they are all interrelated in some manner. Therefore, it is important to recognize that all solutions are systems of interacting elements that are all embedded within larger man-made and/or natural systems.

Engineering Skills

The following list describes some of the basic skills that effective engineers use throughout their careers.

Make Estimations

Engineers often need to solve open-ended problems, where insufficient information is provided to design a solution. Therefore, engineers must be able to make accurate ***estimations*** or rough calculations. Using logical thinking and mathematical computations, engineers determine the critical information that is missing and decide whether a particular solution is reasonable. For example, an engineer might need to determine how much energy heavy machinery will consume when clearing a foundation for a new structure.

Determine Significant Figures

Engineers need to be able to perform mathematical calculations using significant figures. ***Significant figures*** are the digits of a number that are considered to be reliable as a result of a calculation or measurement. They are not simply the number or decimal places in a number. Determining significant figures is important because it is associated with accuracy in measurements and calculations. This accuracy level can significantly impact the ability of a design to be produced.

Use Scientific and Engineering Notation

Engineers often work with very large or very small numbers. These numbers can be expressed much more clearly as scientific notation. ***Scientific notation*** includes just the digits of a number with the decimal point placed after the first digits, followed by a multiplication of a power of ten that will put the decimal back in the correct place. For example, 4,750,000,000.00 in scientific notation is 4.75×10^9. Additionally, the number 0.00000567 would be 5.67×10^{-6}. ***Engineering notation*** is a modification of scientific notation in which all of the powers of 10 are multiples of 3. This allows numbers to more easily conform to International System of Units prefixes. We can have numbers like $56{,}700{,}000 = 56.7 \times 10^6$. With 10 to the power of 6, you can easily recognize that the number is in the millions. So, quickly you can say that 56.7×10^6 is 56.7 million.

Manage Projects

Engineers are often chosen as project leaders or managers. They need to develop skills that will ensure the project is completed on time. One managerial tool that engineers use is a Gantt chart, **Figure 3-14**. A ***Gantt chart*** is a tool used to illustrate a project schedule using horizontal bars that represent each phase and activity that makes up the project. Developed by Henry Gantt in 1917, these charts are also ideal for allocating the proper resources throughout a project.

Write Technical Reports and Papers

Engineers must communicate information about their projects and share the results of their work with others. They do this by writing technical reports, which require strong writing skills. ***Technical writing*** is a form of expository, informative, or explanatory writing used to communicate complex information to those who need it for a specific reason. The purpose of technical writing is to provide facts, data, and results in a concise and understandable manner.

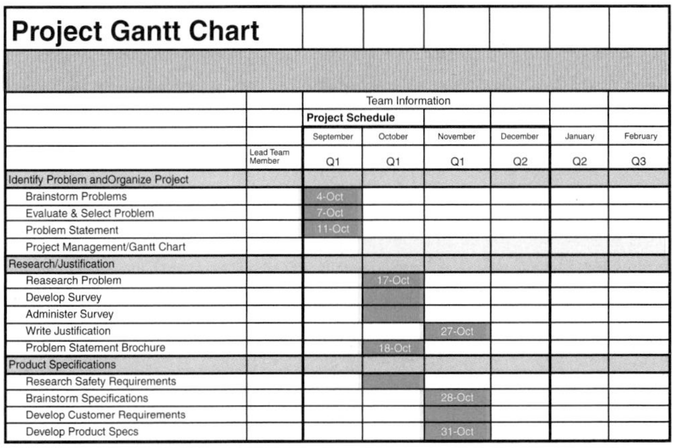

Project Gantt Chart							
		Team Information					
		Project Schedule					
		September	October	November	December	January	February
	Lead Team Member	Q1	Q1	Q1	Q2	Q2	Q3
Identify Problem andOrganize Project							
Brainstorm Problems		4-Oct					
Evaluate & Select Problem		7-Oct					
Problem Statement		11-Oct					
Project Management/Gantt Chart							
Research/Justification							
Reasearch Problem			17-Oct				
Develop Survey							
Administer Survey							
Write Justification				27-Oct			
Problem Statement Brochure			18-Oct				
Product Specifications							
Research Safety Requirements							
Brainstorm Specifications				28-Oct			
Develop Customer Requirements							
Develop Product Specs				31-Oct			

Goodheart-Willcox Publisher

Figure 3-14. A Gantt chart is a project management tool that engineers often use.

Make Presentations

Engineers routinely make presentations to their clients. They must be able to share information in a manner that is understandable to the audience. Presentations must be balanced with visual elements.

Visualize Scientific Laws

Engineers understand various scientific laws and how they affect their design solutions. For example, engineers must understand how Newton's laws of motion or the laws of thermodynamics factor into their proposed designs or design modifications.

Create Engineering Graphics and Drawings

Engineering graphics and drawings follow a specific set of rules and guidelines to communicate a design or idea. These graphics and drawings are sometimes created by hand but almost always are transferred to digital graphics or models using three-dimensional design software. It is important that engineers develop the skills to accurately convey information about their designs to others, **Figure 3-15**.

Dimension and Measure

Engineers need to accurately measure and then dimension their designs. A ***dimension*** is a measurable feature of an object such as length, width, or height. Establishing accurate dimensions of a design and making precise measurements are essential for an engineer to develop a successful solution to a problem.

Convert Units

A ***unit*** is a physical quantity that is consistent to a standard form of measurement. Engineers are often faced with situations where quantities or measurements are provided in many different units.

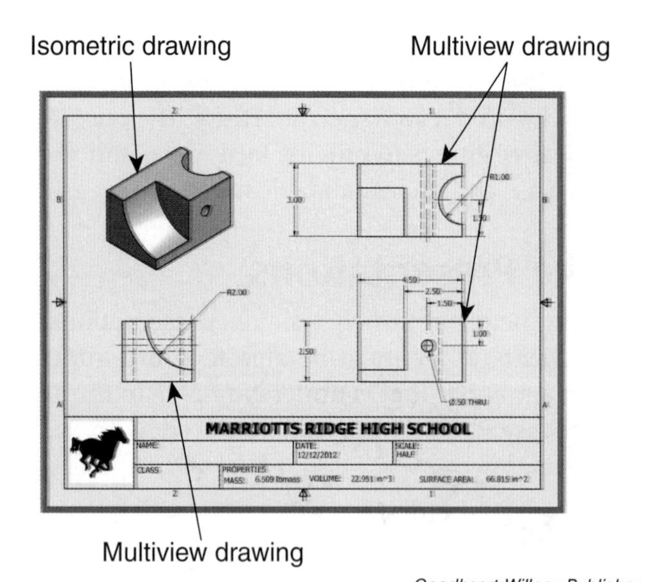

Goodheart-Willcox Publisher

Figure 3-15. This student engineering drawing contains both an isometric drawing and multiview drawings of the object.

Engineers need the mathematical skills to convert all measurements into similar units. Errors in unit analysis, conversions, or the lack of units altogether in problem solving can lead to inaccuracy and error in generating solutions to problems.

Analyze Data Using Software

Engineers record, organize, and analyze data when conducting experiments and tests. To do this, they use data analysis software. These software packages enable users to make the statistical calculations to determine how well a solution performed or why a solution performed in the way that it did.

Academic Connection: History

Engineering the Pyramids versus the Burj Khalifa

The design, construction, and engineering of structures have changed dramatically over the course of human existence. From the earliest bridges and structures to the most advanced structures of today, the engineering process and methods used have evolved over time. For example, the Great Pyramid of Giza, the oldest and largest of the three pyramids in Giza, Egypt, took between ten and twenty years to construct, **Figure A**. The pyramid consists of millions of stone blocks. Theories on how the stones were put into place vary. It is thought that it took 200,000 men to construct.

Dorling Kindersley/Thinkstock.com

Figure A. Ancient Egyptians using machinery to aid in the construction of the pyramids.

Compare this process to the construction of the tallest structure in the world, the Burj Khalifa, **Figure B**. Construction lasted about five years and took considerably fewer than 200,000 workers to build. The materials used in this structure were state of the art.

Engineering continues to push the limits of structures. Consider how the design, construction, and engineering process of construction have changed over time. You may compare the Great Pyramid of Giza to the construction of the Burj Khalifa, or compare and contrast two different structures (one ancient and one modern) that have been built. Support your comparisons with evidence from reputable websites and identify illustrations that capture how the processes have changed.

Ilona Ignatova/Shutterstock.com

Figure B. The Burj Khalifa in Dubai, the United Arab Emirates, is the tallest man-made structure in the world.

Summary

- ✔ Thousands of years of engineering advancement have resulted in newer technologies being created.
- ✔ Early engineers solved problems by designing and creating physical solutions.
- ✔ Engineers of the twentieth century solved engineering problems with a more complete understanding of mathematics and science and more technological knowledge than earlier engineers.
- ✔ The engineering profession can be grouped into four overarching disciplines: civil, electrical, mechanical, and chemical engineering.
- ✔ Engineering subdisciplines include environmental, aerospace, and biomedical engineering.
- ✔ Some qualities of engineers include knowledge of mathematics, science, and other technical disciplines; ability to solve problems, think logically, attend to details, communicate clearly, and visualize ideas without physical examples.
- ✔ Engineers use the engineering design process to solve problems.
- ✔ Engineering characteristics include creativity, collaboration, communication, optimism, ethical behavior, and systematic thinking.
- ✔ Engineering is a dynamic field that requires professionals who can adapt to new challenges, techniques, and technologies by developing a strong foundation of technical and multidisciplinary skills.
- ✔ Skills used by effective engineers include making estimations, determining significant figures, using scientific and engineering notation, managing projects, writing technical reports and papers, making presentations, visualizing scientific laws, creating engineering graphics and drawings, dimensioning and making measurements, converting units, and analyzing data.

Check Your Engineering IQ

Now that you have finished this chapter, see what you learned by taking the chapter posttest.
www.g-wlearning.com/technologyeducation/

Test Your Knowledge

Answer the following end-of-chapter questions using the information provided in this chapter.

1. Engineering advancements can be attributed to contributions from _____ and _____ and the engineering of new technologies which resulted in more innovative methods and designs of more efficient technologies.
2. *True or False?* Early engineers used a trial-and-error approach, rather than mathematical and scientific principles when solving problems.

Matching: Match each of the following engineering disciplines with the correct definition.

3. Civil
4. Electrical
5. Mechanical
6. Chemical

A. The design and improvement of mechanical systems and components.
B. Focuses on electricity and electromagnetism.
C. Focuses on the design and improvement of structures.
D. Involves the conversion of raw materials into useful products.

7. What is engineering design?
8. _____ is thinking in a way that is different from the norm.
9. *True or False?* Most engineering design projects are done individually.
10. What organization publishes a set code of ethics?
11. A(n) _____ is a set of interacting elements that work interdependently to form a complete entity that has an overall function or purpose.
12. In _____, all powers of 10 are multiples of 3.
    A. scientific notation
    B. estimations
    C. engineering notation
    D. None of the above.
13. What is a Gantt chart?

Critical Thinking

1. Explain the differences you might find between an engineer of the 1800s and a modern-day engineer.
2. List the benefits of using mathematical and scientific principles rather than a trial-and-error approach when solving engineering challenges.
3. Identify several ethical considerations that were not identified in the book. What might engineers consider an ethical decision that someone else, such as you, might not?
4. How have you used mathematics and science principles to solve an authentic problem, either in school or at home?
5. What engineering qualities did you not expect to read about? What are some qualities that were not included that you think should be included?

STEM Application

1. Engineers often need to solve open-ended problems, where insufficient information is provided to design a solution. Therefore, engineers must be able to make accurate estimations, or rough calculations. Using logical thinking and mathematical computations, engineers determine the critical information that is missing and decide whether a particular solution is reasonable.

   On the average day, humans consume a significant amount of water for various applications, such as washing clothes or dishes. For this activity, gather information on how much water is consumed per week through various sources, and estimate how much water you use in a week and how much this costs. Determine ways you could reduce water consumption to save money.

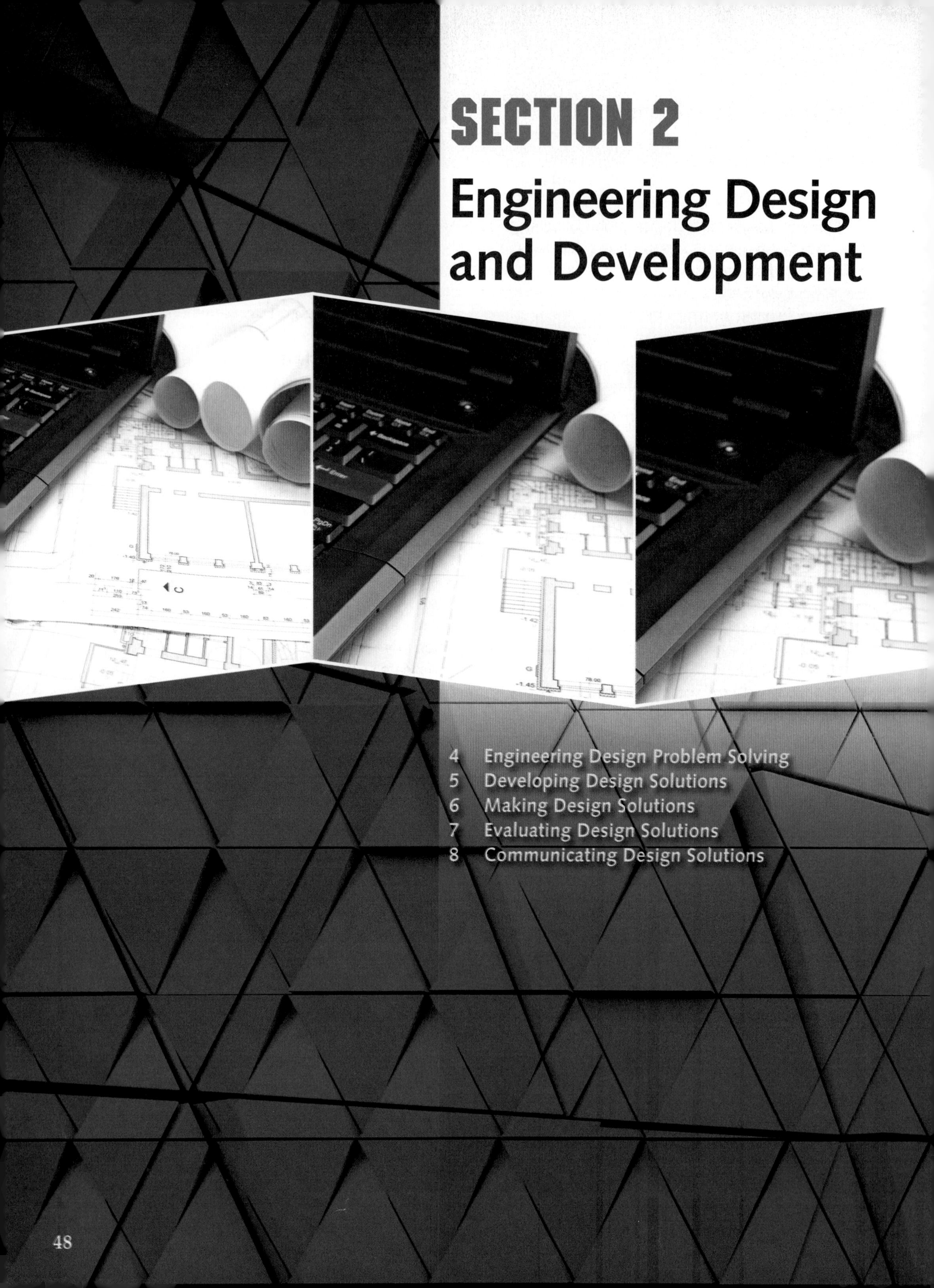

SECTION 2
Engineering Design and Development

Tomorrow's Technology Today

Nanotechnology

Many of today's technological products seem to be getting smaller and smaller—from machines to computer chips. It is estimated that, within the next 50 years, these products will continue to decrease in size so much that thousands of machines will be able to fit on the point of a pencil! In the coming decades, nanotechnology, or "the science of the small," will use these tiny machines for many amazing applications. One nanometer equals one-millionth of a millimeter. We are talking about extremely tiny machines.

All matter in the universe is made up of atoms. Cells are nature's tiny machines. We still have a lot to learn about constructing materials on such a small scale. With new developments in nanotechnology, however, we are getting closer. Many scientists believe we will be able to manipulate each individual atom of an object in the near future.

Nanotechnology combines engineering with chemistry. The goal of nanotechnology is to manipulate atoms and place them in a specific pattern. These patterns will produce a certain kind of structure. Depending on the atoms used and the patterns created, nanotechnology can be applied in several different areas. The most promising uses for this technology are in medicine, construction, computers, and the environment.

This branch of technology is expected to greatly impact the medical industry. Nanotechnology devices will be programmed to attack cancer cells and viruses and rearrange their molecular structures to make them harmless. The tiny robots might also be able to slow the aging process. This might increase life expectancy. Since these machines will be able to perform work a thousand times more precisely than current methods, they can also be programmed to perform delicate surgeries.

The construction industry will benefit from nanotechnology as well. The tiny machines will be used to make much stronger fibers. There is hope that, one day, we will be able to replicate anything, from diamonds to food.

Computer parts continue to get smaller all the time. They will, however, soon reach their limit. Nanotechnology will be used to make much smaller storage devices with the capability to hold much more information.

Finally, nanotechnology can be used to better the environment. We are currently running out of many natural resources. With nanotechnology, we will be able to construct more of them. Manufacturing with nanotechnology produces much less pollution than traditional manufacturing. In addition, these devices can potentially be programmed to rebuild the ozone layer and clean up oil spills.

Nanotechnology has many exciting potential uses. This technology, however, also has its challenges. Similar to all new technology, it raises questions about its impacts on people, society, and the environment.

While studying this chapter, look for the activity icon to:

- **Assess** your knowledge with self-check pretest and posttests.
- **Practice** technical terms with e-flash cards, matching activities, and vocabulary games.
- **Reinforce** what you learn by submitting end-of-chapter questions.
  www.g-wlearning.com/technologyeducation/

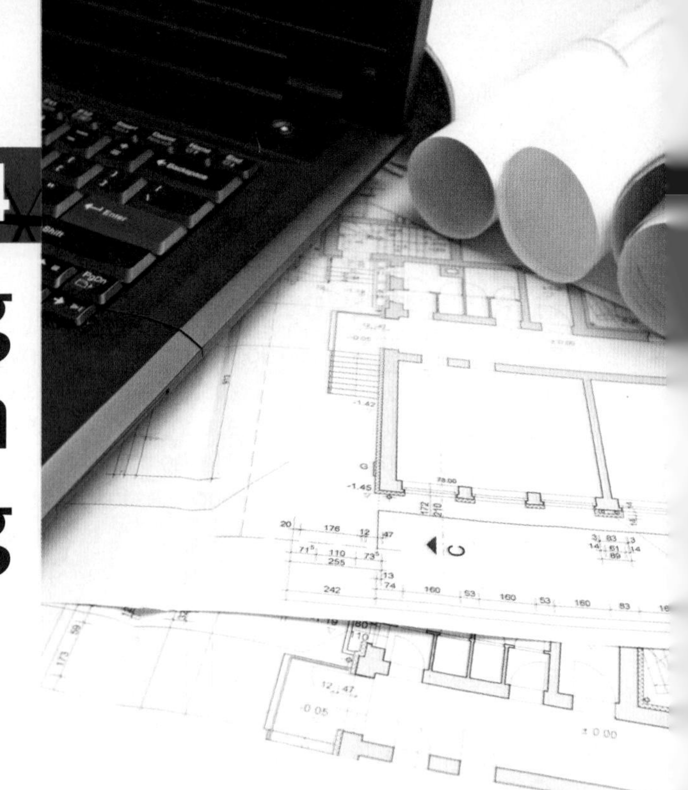

CHAPTER 4

Engineering Design Problem Solving

Check Your Engineering IQ

Before you read this chapter, assess your current understanding of the chapter content by taking the chapter pretest.
www.g-wlearning.com/technologyeducation/

Learning Objectives

After studying this chapter, you should be able to:

- ✔ Compare and contrast problem solving and design.
- ✔ Explain the differences between general problem solving and the engineering design process.
- ✔ Recall the three ways that problems vary.
- ✔ Describe an engineering design problem.
- ✔ Define *engineering design*.
- ✔ Explain the difference between system design and product design.
- ✔ Explain how the engineering design process is a systematic and iterative approach to solving design problems.
- ✔ Identify the major steps involved in solving engineering design problems.
- ✔ Recall strategies for using the engineering design process.

Technical Terms

concept map
critical thinking
decision matrix
design
engineering design process
mind map
problem-solving process
product design
system design

Close Reading

As you read this chapter, make an outline of the details of the topics discussed, such as the different steps involved in the engineering design process.

All technology is created for a purpose. People develop technological products or systems as a means to meet needs and wants by solving some type of problem. For example, some people have developed technologies that protect us from the dangers of the physical environment. Others have developed technological artifacts that have helped us become more informed citizens. Still others have created technological systems that allow us to travel faster, cheaper, and safer.

Most modern technological products and systems are not developed or discovered by accident. As noted in earlier chapters, people use problem-solving processes and engineering design to create, produce, and improve technologies. Section 2 of this textbook focuses on the activities that take place in engineering design problem solving. In this chapter, you will learn about problem solving in general and the engineering design process steps followed to effectively and efficiently solve problems.

Problem Solving and Design

The terms *problem solving* and *design* should not be used interchangeably. Problem solving is a process that can be applied to a broad range of situations. There are many types of problems. For example, you might have trouble interacting with some of your friends. This is a social problem. You might have problems making your allowance stretch through the week. This is a financial problem. The ***problem-solving process*** requires people to employ thinking and reasoning skills to consider and find solutions for a situation that has an identified goal state, readily available solutions, and one or more reasonable pathways for solving. ***Design*** is also a thinking and reasoning process, but it differs from problem solving because it involves devising a new solution, one that has not previously been found or defined.

Problems are solved through design every day. People design the layout of their bedroom, their hairstyle for the day, or a new way to harness energy. See **Figure 4-1.** Design can range from creative endeavors, such as composing new pieces of artwork, to analytical engineering design for solving global issues.

Both problem solving and design require ***critical thinking***. This type of thinking uses mental abilities to identify, analyze, and evaluate a situation or problem. Using these processes, a person considers possible solutions to form a judgment about a course of action.

General Problem-Solving Process

People usually realize a problem exists when they encounter a difficult situation and are not sure how to resolve it. A problem can be as simple as opening a walnut when tools are unavailable. The problem can be as complex as moving 500 people from Los Angeles to Tokyo in two hours. See **Figure 4-2**. These problems have two things in common. In both cases, an identifiable goal exists (cracking a walnut or moving people). In addition, no clear path is apparent from the present state (an intact walnut or people who need to be moved) to the goal state.

Problem solving is a common human activity. The following sequence of steps is typically used to solve problems:

1. Develop an understanding of the problem through observation and investigation.
   - Example: You recognize the chain on your pedal bike has come loose. As a result, you observe where the chain became detached from the sprocket, and you begin to move the pedals to see how it was spinning and where it may possibly have broken free.
2. Devise a plan for solving the problem.
   - Example: You identify if a part is broken, such as the chain tensioner, then identify steps for placing the chain back on the sprocket.
3. Implement the plan.
   - Example: Obtain the correct tools, and begin placing the chain back on the bike.
4. Evaluate how the plan was implemented.
   - Example: Examine how next time you may have the correct tools, or an alternative step for ensuring the chain works effectively.

Problem Variations

Problems vary in complexity, structure, and context. Complexity refers to how complicated and difficult a problem is to solve. The structure of the problem refers to how much information is

Photographee.eu/Shutterstock.com

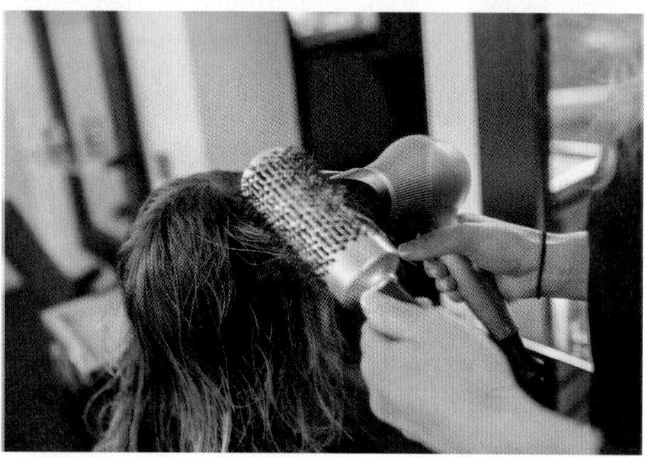
Tyler Olson/Shutterstock.com

bikeriderlondon/Shutterstock.com

Figure 4-1. Designing is a part of everyday life. For example, people design the layout of their bedroom, their hairstyle, or even a new way of harnessing energy.

kavram/Shutterstock.com

NASA/JPL-Caltech

Figure 4-2. Technology is born from a problem, which can be as simple as how to clear snow from a road or as complex as how to build and test a Mars exploration vehicle.

provided that implies what a successful solution will look like. A well-structured problem provides the problem solver with a well-defined list of guidelines and limitations for solving the problem. On the other hand, a problem that is not well structured provides limited directions for solving and often has multiple and conflicting goals for a solution. Context variations involve the set of circumstances or facts that surround a specific problem.

Engineering design problems are considered to be the most complex and least structured types of problems. See **Figure 4-3**. Engineering design involves the need to design either a technological product or system as a solution to a problem. See **Figure 4-4**.

Problem Variations

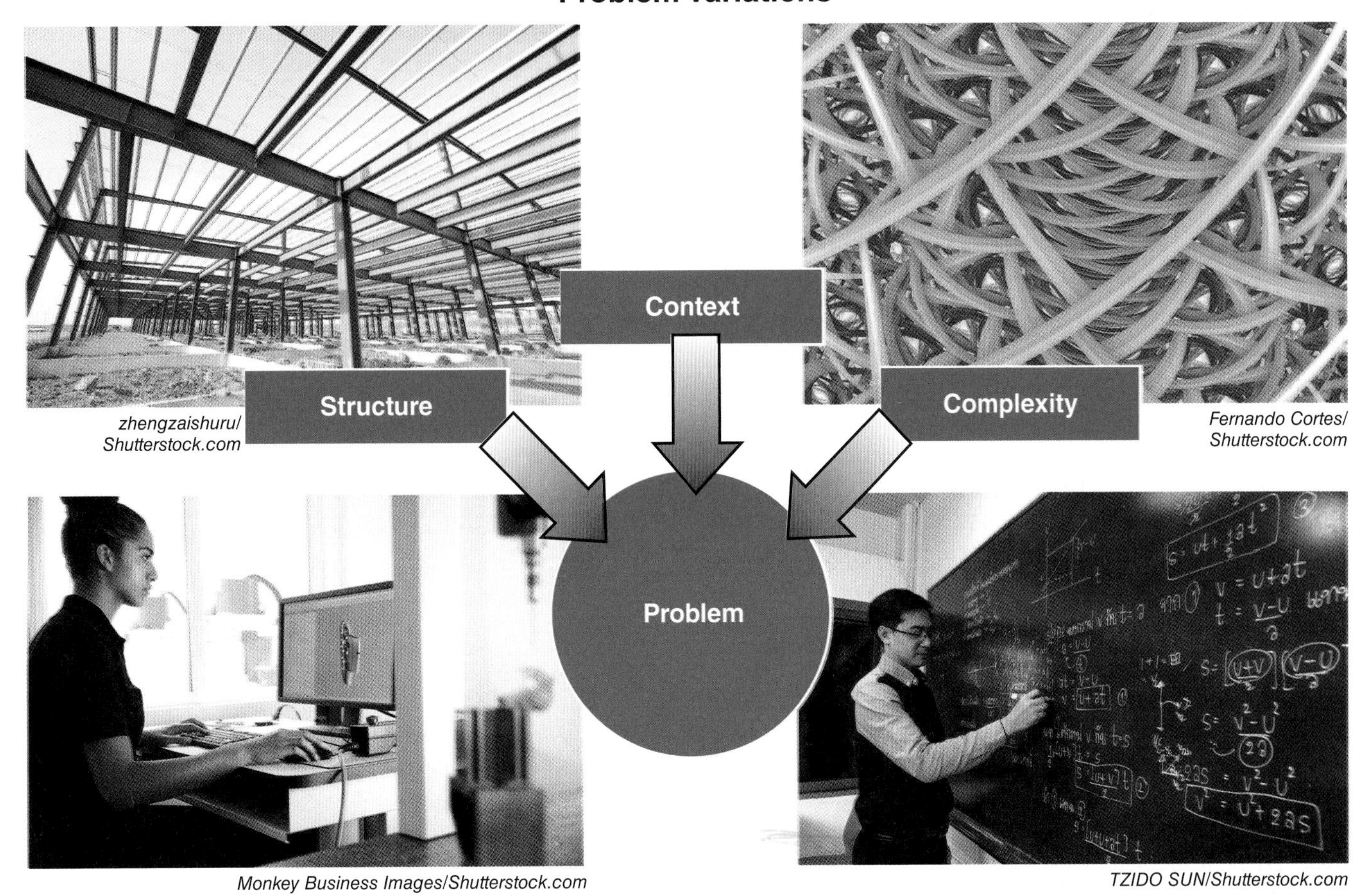

Figure 4-3. Problems can vary in three ways: complexity, structure, and context.

Types of Design

System Design

Product Design

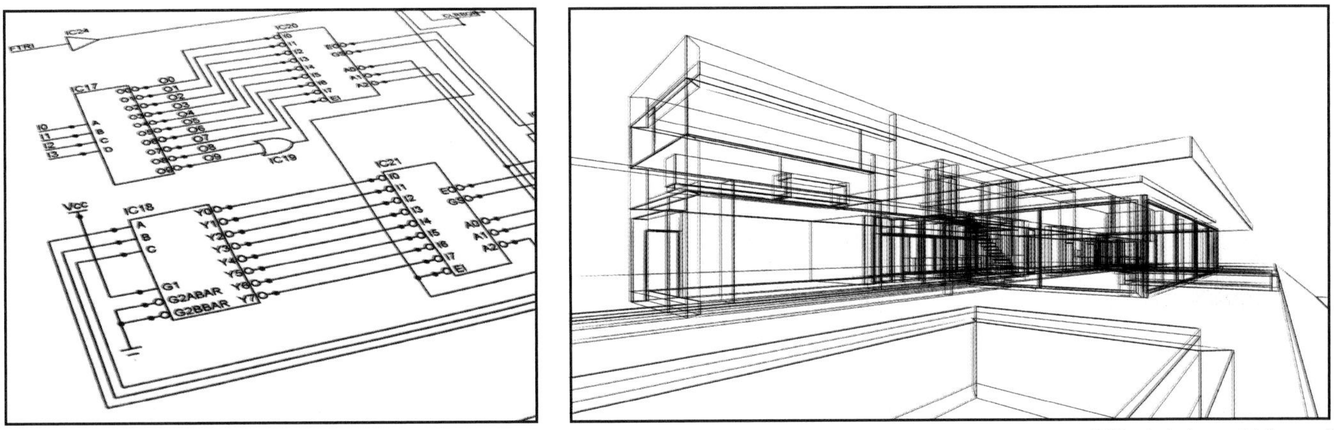

©iStockphoto.com/calvinng

©iStockphoto.com/deliormanli

Figure 4-4. Engineers design both systems and products.

System and Product Design

System design involves the arrangement of components to produce a desired result. For example, automotive braking systems are a result of system-design efforts. The design of a drum brake system, **Figure 4-5**, brings together mechanical and hydraulic components into a speed-reduction system. The brake pedal unit is a mechanical linkage. When the pedal is depressed, a plunger in the master cylinder is moved. This motion causes the fluid to move in the hydraulic system connecting the master cylinder to the wheel cylinders. The fluid movement pushes the pistons outward in the wheel cylinders. These pistons are attached to the brake shoes. The piston movement causes the shoe to be forced against the brake drum. The mechanical action creates friction between the shoe and the drum, which slows the automobile.

System design can be used in all technological areas. For example, it is an important part of construction technology. See **Figure 4-6**. Electrical, heating and cooling, plumbing, and communication systems are designed for buildings. In manufacturing, the methods of production, warehousing, and material handling must be designed. Communication systems are designed to transmit messages using fiber optics, cable lines, microwaves, etc. Transportation systems combine manufactured vehicles and other components to move goods and passengers from place to place. Irrigation systems are used to water crops. Pipelines are part of natural gas distribution systems. Doctors and hospitals provide patient care in health care systems.

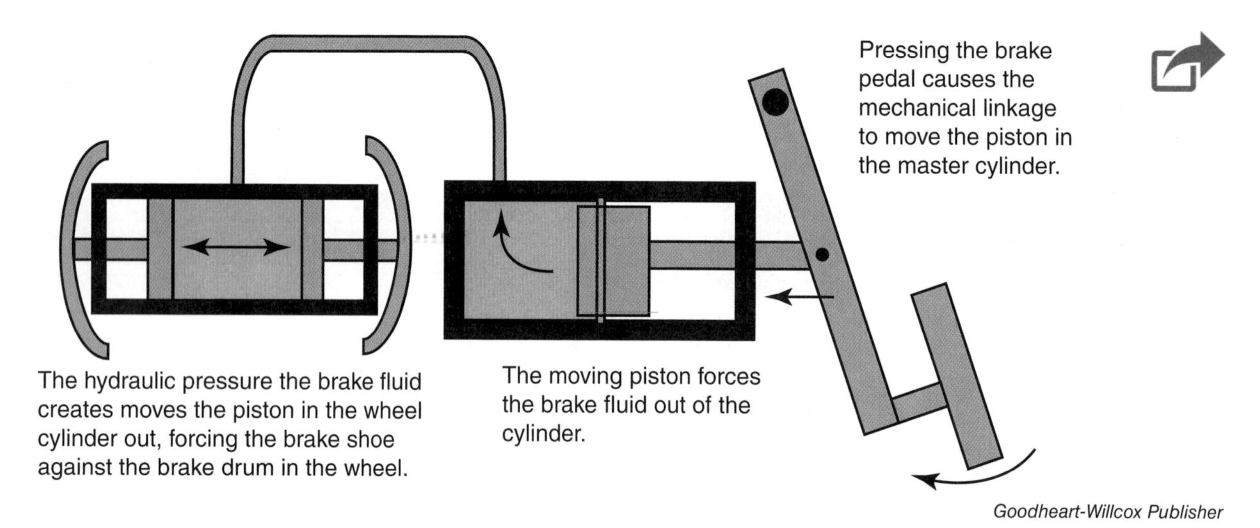

Goodheart-Willcox Publisher

Figure 4-5. This brake system was the result of system design efforts.

Marcin Balcerzak/Shutterstock.com

Scanrail1/Shutterstock.com

Figure 4-6. System engineers can devise the security, networking, and power systems for large commercial construction projects.

Think Green Sustainable Design

Sustainable design is the idea of creating technologies while being conscious of the impact they will have on the environment. It is important for engineers to understand practices of sustainability to ensure that our planet will continue to support life.

The word *sustainability* is used to describe the idea of the earth's ongoing ability to produce natural resources. The resources in question may be anything from trees to petroleum. Several natural resources are known as *exhaustible resources*. When these resources are used up, they will not grow back and will be gone forever. In some instances, these resources are not being depleted due to consumption, but rather by the way humans interact with the environment. For example, the use of pesticides may lead to groundwater contamination.

Individuals, companies, and nonprofit and government organizations are working to encourage sustainability and to prevent the loss of our natural resources. They are looking for ways to reduce the use of exhaustible resources while finding alternative renewable resources. With a common goal, people are making an effort to help preserve our natural resources.

Product design involves the development of a product or structure to meet the needs of customers. In manufacturing, products are designed by designers. In construction, structures are designed by architects. See **Figure 4-7**. The product or structure must function well, operate safely and efficiently, be easily maintained and repaired, have a pleasant appearance, and deliver good value.

In addition, products and structures must be designed so they can be produced economically and efficiently. They must also sell in a competitive environment. In short, the product or structure must be designed for the following:

- **Function.** The product or structure must be easy and efficient to operate and maintain.
- **Production.** The product or structure must be easy to manufacture or construct.
- **Marketing.** The product or structure must be appealing to the end user.

Types of Product Design

Manufactured Products

Stason4ik/Shutterstock.com

Constructed Structures

B Brown/Shutterstock.com

Figure 4-7. Manufactured parts and structures are a result of product design.

Problem Solving and Design in Engineering

As you have learned, people face an engineering design problem when they need to develop technological products or systems to fulfill a need or desire. An agricultural example illustrates this point. Until the 1940s, most farmers raised their own hay and used it as feed for the animals on the farm. However, in recent decades, many self-sufficient family farms have been replaced by larger farms specializing in a specific crop. Some farmers operate large dairy farms that do not raise hay. Other farmers specialize in raising and selling hay to these dairy farmers and beef cattle feedlots. Hay must be shipped to other farms and to the feedlots. Having people gather and ship loose hay, however, is not very productive. In addition, farm labor became increasingly expensive. To solve this problem, farm implements were developed that allowed one person to do more work with the assistance of machines. See **Figure 4-8**. People learned how to collect hay into bales for easier transportation. This process included the following steps:

1. The field of hay was cut with a mowing machine and allowed to dry on the ground.
2. A machine called a *side-delivery rake* was used to collect the hay into a narrow pile called a *windrow*.
3. A tractor pulled the baler along the windrows. Here, it collected, compacted, and tied the hay into bales.
4. The bales were then pushed out of the machine onto the ground. Farmworkers picked them up from the ground and loaded the bales onto a truck.

Stanislav Smoliakov/Shutterstock.com

Figure 4-8. Technology changed farming from a labor-intensive activity to an equipment-intensive activity. A single farmer can now do the work of many people.

More recent engineering developments have improved the machines used in harvesting hay. Harvesting hay was once labor-intensive. A swather has replaced mowing machines and side-delivery rakes. A swather cuts, conditions, and windrows hay in one operation. The hand loading and hauling of baled hay is being replaced with bale wagons that automatically pick up and stack bales into cubes. Hay is also gathered into large rolls. These cubes or rolls are loaded with forklifts onto trucks that haul the hay to customers. Thus, harvesting hay has become equipment intensive through the use of engineering design problem solving.

An examination of other technological areas reveals the same pattern. For example, in manufacturing operations, computer-aided design systems have increased the productivity of drafters. Robots perform many routine manufacturing tasks without human interference or monitoring. See **Figure 4-9**. In transportation, computer-controlled, driverless people movers speed people between terminals at airports. Hybrid-bus systems promise to increase the efficiency of urban transport, **Figure 4-10**. Auto-

Nataliya Hora/Shutterstock.com

Figure 4-9. Machines, such as these robots, can perform repetitive tasks with high accuracy.

Daimler

Figure 4-10. Modern transportation systems, such as this hybrid bus, reduce energy use and emissions.

mated equipment on new commercial aircraft has reduced the number of flight deck officers from three to two.

Problem solving in engineering involves product and system design, development, integration, and implementation. To solve an array of problems involving products and systems, engineers and other people in technological areas take a more disciplined, informed, and organized approach than is used in the general problem-solving process. This approach, called *engineering design*, enables the development of more efficient and effective solutions.

Engineering design is described as designing solutions to problems under constraints (limitations or restrictions). Engineering design problem constraints can include restrictions related to the following areas:

- Deadlines.
- Safety.
- Material cost and availability.
- Machine and tool use.
- Energy consumption.
- Manufacturability.
- Funding.
- Environmental regulations.

Engineering design also involves consideration of the full impacts of a design solution. A variety of tools and predictive analysis through mathematical and scientific modeling are used to analyze design and optimize solutions before producing them. Additionally, when solving engineering problems, engineers and other professionals apply experience and knowledge gained from solving similar problems.

As a result, engineers and other professionals in technological areas have created a specific approach to solve problems in the most efficient and effective way possible. This approach is called the *engineering design process*. The ***engineering design process*** is a method used to solve problems by designing a product or system that meets a desired goal, while adhering to established constraints and taking into consideration many factors, such as potential impacts, risks, and benefits.

In the engineering design process, the term *design* is used in two ways. First, it describes the action, or process, used to create the appearance or operation of a technological product or system. Second, it describes the product of the design process. In this way, the word *design* is used to describe the plan, or drawing, that shows the appearance or operation of a technological device or system.

Steps in the Engineering Design Process

The engineering design process is not used only by engineers, designers, or other technical professionals. Anyone can use this process to solve problems.

The engineering design process does not proceed in a direct path from start to finish. Instead, it often involves many twists and turns along the path to developing a solution. The steps are taken in multiple iterations or loops. It is often necessary to move back and forth between steps, or even completely restart the process, to develop the most optimal solution. Additionally, the steps can overlap. However, this process does require some specific steps or actions. Generally, a cycle of six steps is involved. See **Figure 4-11**. These six steps are as follows:

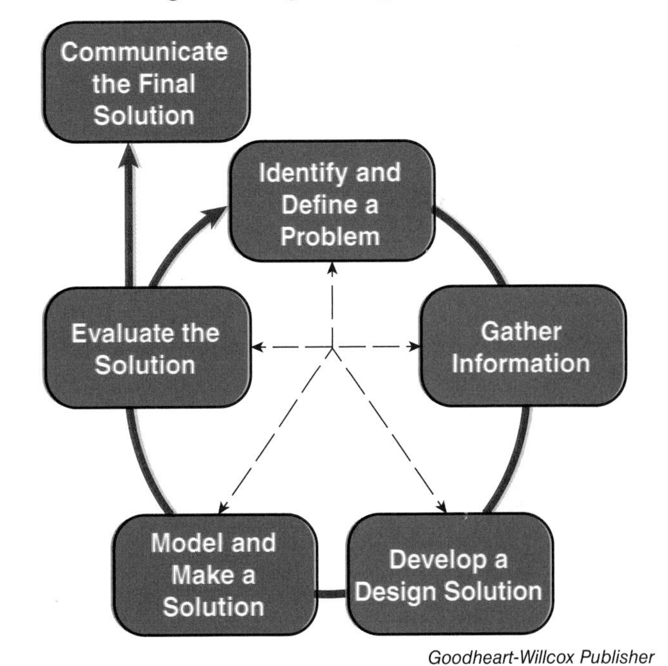

Figure 4-11. The engineering design process is a cyclical and iterative series of steps and actions used for solving engineering design/technological problems.

1. Identify and define a problem.
2. Gather information.
3. Develop a design solution.
4. Model and make a solution.
5. Evaluate the solution.
6. Communicate the final solution.

Identify and Define a Problem

The *identify and define a problem* step of the engineering design process involves a person or group acquiring the basic information about the problem and establishing the design limitations.

Tasks in This Step

Identifying and defining a problem can include the following tasks:

- **Identifying and validating a problem.** Recognizing a problem and determining if the problem is reasonable and worthy of solving.
- **Formulating a problem statement.** Developing a problem statement, which is a clear and concise description of the problem. The problem statement should include who the problem affects, what the problem is, when the problem occurs, where the problem happens, and why it is important to solve. See **Figure 4-12**.
- **Understanding and establishing criteria and constraints.** Determining the guidelines and limitations for developing a successful solution to the problem.
- **Planning and managing the process.** Developing a detailed work plan to solve the problem based on the available resources before progressing in the engineering design process.

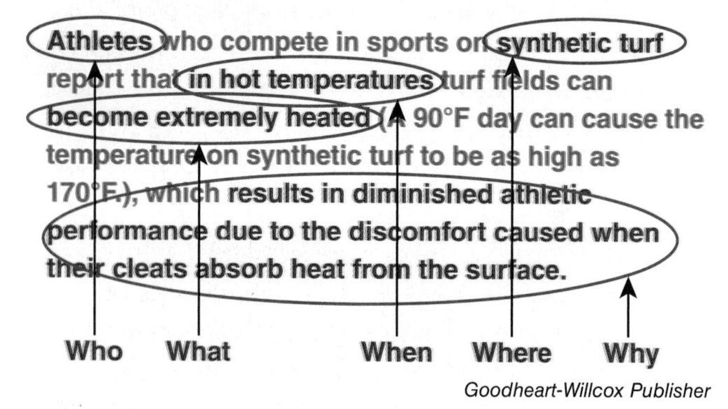

Goodheart-Willcox Publisher

Figure 4-12. A good problem statement includes (a) who the problem affects, (b) what the problem is, (c) when does the problem occur, (d) where does the problem happen, and (e) why is it important to solve.

Strategies for This Step

The following strategies are beneficial for defining the problem:

- Separate the engineering design project into smaller, more manageable tasks.
- Establish a set of benchmarks for project completion.
- Develop a time line for solution development.
- Evaluate the resources available to develop a solution to the problem.
- Use project management tools, such as a Gantt chart or project management software, to aid in the problem-solving process. See **Figure 4-13**.

Gather Information

The *gather information* step of the engineering design process involves a person or group obtaining the knowledge necessary to develop a design solution by studying the available information and conducting research.

Tasks in This Step

Gathering information can include the following tasks:

- **Researching current solutions.** Locating and analyzing what others are doing or have done

Project Schedule

2011-2012
MST-ED
Tiffany, Danielle, Mark, Morgan, Dillon J, Chris, Dylan V, Kirsten

Lead Team Members
Tiffany-L Dillon-MI
Kirsten-W Morgan-E
Mark-P
Danielle-R

green means finished
red means haven't started
yellow means in progress

Project Schedule

	Lead Team Member	November 4th-10th	November 11th-20th	November 21st-30th	December 1st-7th	December 8th-15th	December 16th-22th	December 23th-31st	January 1st-18th	January 19-Jan
Identify Problem and Organize Project										
Brainstorm Ideas	L									
Evaluate & Select Problem	L									
Website	W									
Project Management / Gantt Chart	P									
Research										
Research Rules	R									
Research Problem	L									
Product Specifications										
Research Safety Requirements	R									
Brainstorm Specifications	L									
Design Selection										
Brainstorm Possible Solutions	L									
Create Sketches of Solutions	R									
Develop Design Evaluation Method	MI									
Perform Design Evaluation	MI									
Decision Matrix	P									
Select final solution from Matrix	L									
Design and Optimize										
Prototype Development	MI									
Create Working Drawings	R									
Build Prototype										
Wire	E									
Program	P									
Fabricate and Assemble	MI									
Test Prototype										
Develop Test Procedures	L									
Test	L									
Analyze Test Results	L									
Final Deliverables										
Final Display	L									
Finished Website	R									
Electronic Copy of Whole Project	L									

Goodheart-Willcox Publisher

Figure 4-13. A Gantt chart is often used as a project management tool for solving engineering problems.

to solve similar problems in order to generate ideas for developing a solution.

- **Exploring concepts.** Establishing the necessary knowledge base for designing a solution to the problem and investigating possible concepts that can be applied to the problem in a novel or innovative way.
- **Documenting details.** Recording research as well as all of the important aspects of the problem-solving process. This can be done using an engineering notebook or electronic portfolio. See **Figure 4-14**. Details should be documented throughout the entire engineering design process.

Strategies for This Step

The following strategies are beneficial for gathering information:

- Categorize the underlying principles necessary for solving the problem and ignore superficial details.
- Recognize relevant information and organize it into "chunks" or patterns.

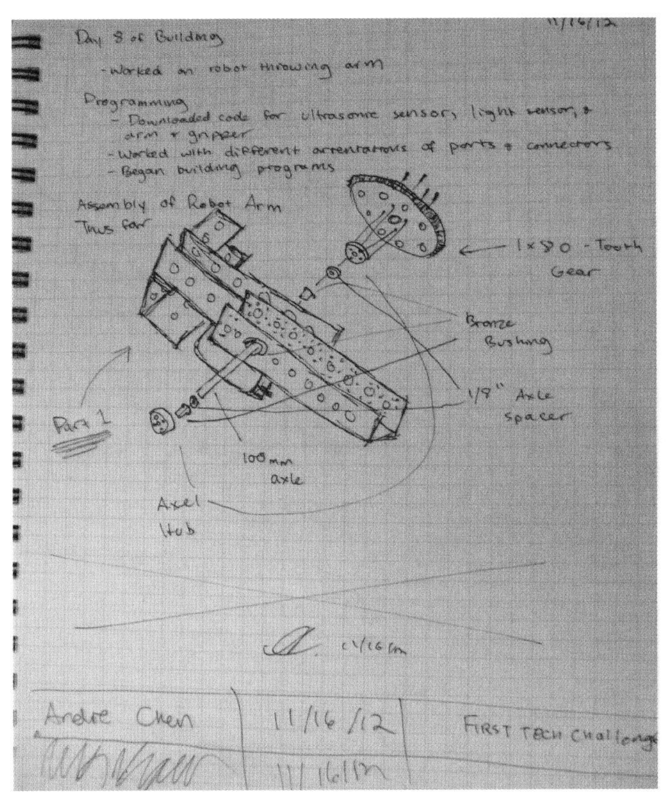

Goodheart-Willcox Publisher

Figure 4-14. An engineering notebook or journal is a tool used to document the engineering design process.

- Draw inferences to other potential ideas or concepts using past and current solutions for similar problems.
- Monitor solution progress based on the established benchmarks for project completion.

Develop a Design Solution

The *develop a design solution* step of the engineering design process involves a person or group creating and refining several possible solutions. This step also includes selecting and refining the most promising solution ideas and then detailing (describing) the best design solution.

Tasks in This Step

Developing a design solution can include the following tasks:

- **Searching for solutions.** Combining ideas and concepts to generate unique solution ideas.
- **Developing possible solutions.** Originating a number of different solutions that can solve the problem.
- **Isolating, refining, and detailing the best solution.** Selecting the most promising solution; integrating, modifying, and improving the solution; and creating detailed sketches of the best solution.
- **Finalizing design specifications.** Clearly delineating the features and performance expectations necessary for a design to successfully meet the established criteria and constraints.
- **Creating a detailed solution representation.** Communicating the features and functions of a design solution idea through detailed and well-annotated visual representations. An example of a detailed solution representation is a technical drawing created with 3-D design software, **Figure 4-15**.
- **Developing a production strategy.** Outlining a procedure for creating a model or prototype of the design solution.
- **Conducting a predictive analysis.** Thoroughly analyzing detailed technical drawings of designs to ensure that the design meets the desired specifications. This also includes conducting a thorough mathematical analysis of designs to predict how they will perform. Using three-dimensional (3-D) modeling or computer-aided design software can

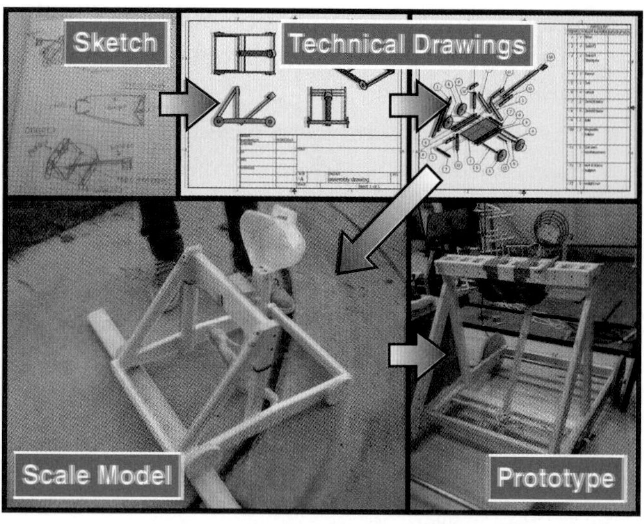

Goodheart-Willcox Publisher

Figure 4-15. With three-dimensional modeling software, you can convert a hand sketch of a design into a technical drawing used for making design models and prototypes.

support this activity. When conducting a predictive analysis, it is important to analyze a design solution with the recognition that all technologies are systems of interacting parts that are, in turn, embedded in larger systems.

- **Optimizing.** The ongoing process of evaluating whether the solution model or prototype meets the design specifications with the purpose of enhancing the effectiveness of the final solution.
- **Selecting the proper tools, equipment, and materials.** Selecting, acquiring, and planning the use of industry quality tools and equipment for manipulating materials to produce a model or prototype of the solution.

Strategies for This Step

The following strategies are beneficial for developing a design solution:

- Use creativity in devising solutions to the problem. Creating a mind map can aid in this process. A *mind map* is a diagram that helps people visually organize information or ideas around a central topic or theme. Engineers and designers can use mind maps as a tool for brainstorming and generating ideas for developing a solution to a problem. See **Figure 4-16**.
- Explore multiple solution possibilities—the first idea is not always the best.
- Select the best solution approach using a decision matrix. A ***decision matrix*** is a chart used to record a rating for how well a design meets a desired criterion. See **Figure 4-17**.
- Employ computer-aided design software to help in design creation and concept analysis.

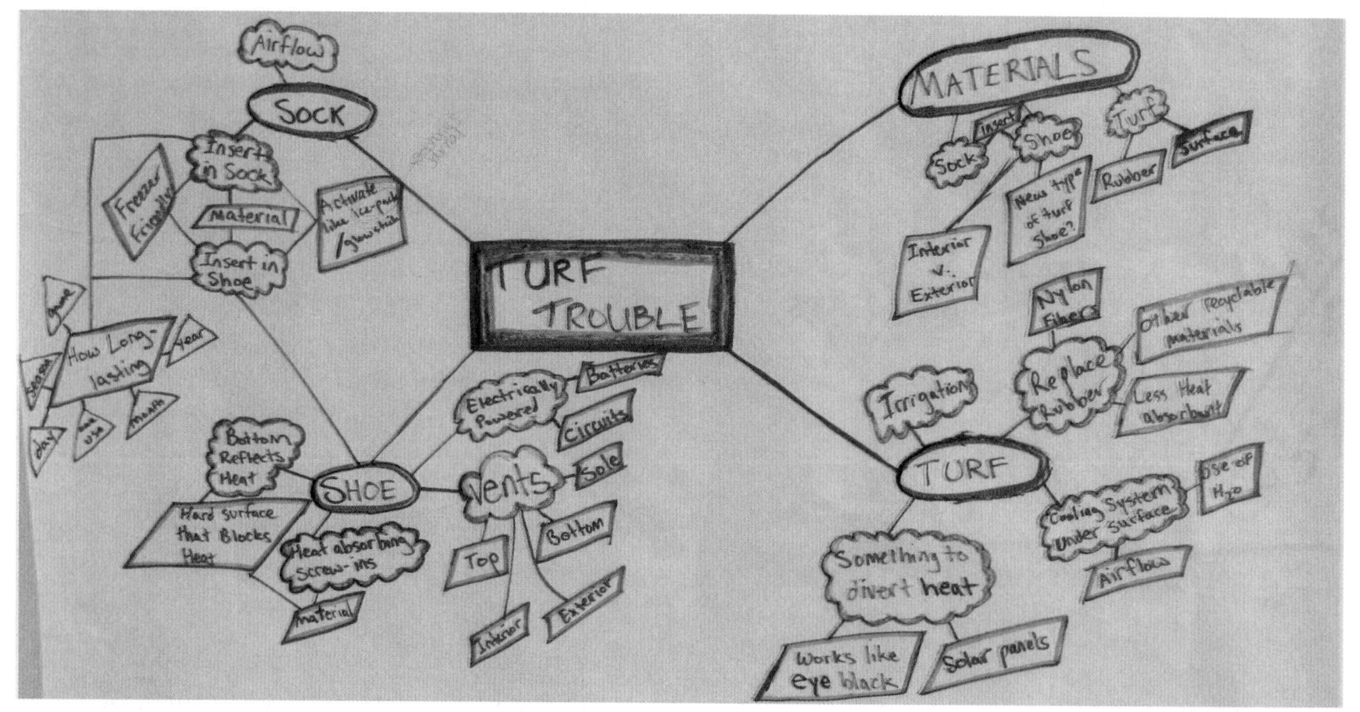

Goodheart-Willcox Publisher

Figure 4-16. A mind map that engineering students used for generating solution ideas for a problem dealing with synthetic turf becoming too hot for athletes in warm weather.

Options for What Type of Swing Check Valve to Use	Time?	Cost?	Effective?	Labor?	Total?
Continue working on a 1/2″ valve based system: the valve had already been purchased and was attached to a PVC component, ready to be attached.	5	4	3	5	17
Source a 1″ valve: this had so far proved unsuccessful, although there were some options that had not yet been explored. Would allow the team to move to a 1″ system.	3	4	5	4	16
Fabricate a 1″ valve: rather than continuing to try to find an appropriate premade valve, the team would construct one from scratch.	2	5	3	1	11

Explanation of Criteria:

Time—How long would it take to acquire or fabricate this option. The project cycle is only four weeks long, making time a major factor.

Cost—The cost associated with pursuing this option. One of the restrictions placed on the project is that no part may cost more than $20.

Effective—How effective would this option be at filling its role in the system? The team had anecdotal evidence that some of these options worked in similar systems, though they are all untested in this system.

Labor—Although this is partially tied to the criteria of time, this criteria was used to judge the amount of time that would be spent directly working on this component. Time spent working on this component could otherwise be spent on the system as a whole, or on testing upon completion.

Goodheart-Willcox Publisher

Figure 4-17. A decision matrix.

- Evaluate multiple industry quality tools when developing a plan to make a solution.
- Create a bill of materials for the proposed solution, **Figure 4-18**. A bill of materials is a list of all the components needed to produce a product or system. This list should provide a description of the component, the manufacturer or vendor, the item or part number, the quantity needed, and the cost.
- Create a list of the necessary industry quality tools and equipment needed to make the solution.
- Monitor solution progress based on the established benchmarks for project completion.

Bill of Material

Customer name: ABC Company
Customer address: 10 Jane Avenue, New York, New York
Date: 1/1/2020

Item #	Description	Vendor	Part Number	Quantity
1	FGHT sun model	MNF	GHY253	7
2	120 w photovoltaic module	Food FG	2543-I	14
3	Hardware	Mottle	1452-UKJ	2
4	4-module for GHN	IFS	256BN	1
5	822 air G	Mood plc	947673 IOL	4
6	SMA Model Sunny Boy Inverter	MNF	8 GH	2
7	9B, 700 GHT	IFS	f58	1
8	Fused Disconnect Switch	Food FG	56e	1

Goodheart-Willcox Publisher

Figure 4-18. A bill of materials.

Model and Make a Solution

The *model and make a solution* step of the engineering design process involves a person or group producing physical, graphic, computer, or mathematical models of the selected solution.

Tasks in This Step

Modeling and making a solution can include the following tasks:

- **Acquiring the appropriate resources.** Obtaining the materials that will enable the production of a quality model or prototype.
- **Modeling/prototyping the solution.** Using a developed design strategy/procedure to produce a quality model or prototype of the design solution.
- **Modifying the design solution.** Continuously making improvements to the model or prototype to optimize the solution's effectiveness.
- **Attending to quality.** Focusing on the production of a well-built solution model or prototype by conducting accurate measurements and using the proper tools, equipment, and materials. Additionally, this includes continuously referring to the design specifications to ensure that the solution meets all criteria and constraints. Attending to quality also helps ensure that the testing of the solution and analysis of the testing data is realistic and valuable.

Academic Connection: Literacy

Reading in Technical and Scientific Subjects

Those who work in a technical or scientific profession must be able to think deeply about their specific discipline. They develop competency in their particular area of focus through literacy—the ability to read and write effectively.

When reading technical or scientific texts, you may find the following strategies helpful:

- Determine the central ideas or most important points of the text.
- Explain or depict a complex process.
- Write an accurate summary of the text.

In this chapter, you are reading about the complex process of engineering design. To aid in your understanding of this process, complete the following tasks:

- Create a list of the most important points about the engineering design process. Start by looking at the headings and subheadings in the chapter text and write down the main point of each section.
- Make a sketch, graphic, or visual of how you would enact the engineering design process to solve a particular problem.
- Develop a concept map to summarize all that you have learned about engineering design after reading this chapter. A ***concept map*** is a tool that you can use to graphically organize the information that you have obtained. See **Figure A.** To create a concept map, write *engineering design* on the center of a piece of paper. Connect this term to the most important points of the chapter, and then connect those main points to their supporting concepts.

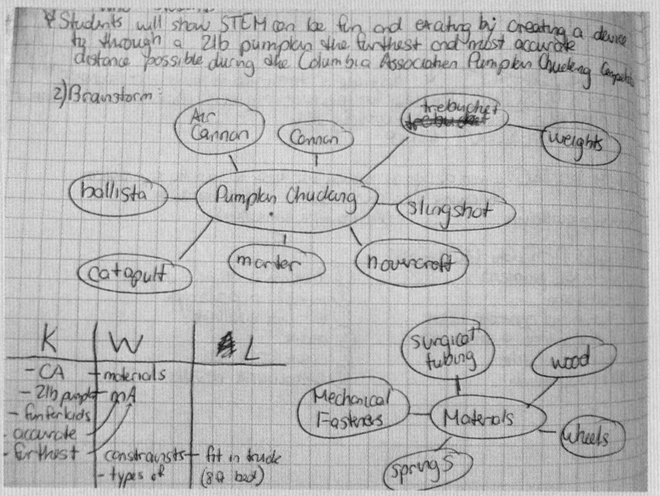

Goodheart-Willcox Publisher

Figure A. A concept map.

Completing these tasks will help you to precisely follow a multiple-step procedure. These strategies are useful when you are developing solutions to complex problems through the engineering design process.

A strategy to help you use the engineering design process is an engineering notebook or journal. In this notebook or journal, write your proposed set of procedures to solve the problem. Base the procedures on the information in this chapter. (Remember that the engineering design process is not a linear process—your procedure can continuously change.) As you complete the engineering design process, be sure to record all of the results in your engineering notebook or journal.

- **Conducting material experimentation.** Continuously evaluating a variety of materials used for the production of a model/prototype through scientific experiments.
- **Optimizing.** The ongoing process of evaluating whether the solution model or prototype meets the design specifications with the purpose of enhancing the effectiveness of the final solution.

Strategies for This Step

The following strategies are beneficial for making a solution:

- Follow the developed design strategy to ensure that the solution meets the design specification.
- Draw on the perspectives, knowledge, and capabilities of team members to address design challenges.
- Maintain an optimistic outlook to persist in creating the solution.
- Continue to look for improvements to the solution.
- Document any changes made to the design solution while producing the model or prototype.
- Monitor solution progress based on the established benchmarks for project completion.

Evaluate the Solution

The *evaluate the solution* step of the engineering design process involves a person or group testing the physical, graphic, computer, or mathematical models of a selected solution.

Tasks in This Step

Evaluating the solution can include the following tasks:

- **Determining test criteria.** Clearly defining what types of data need to be collected to evaluate the solution based on the design specifications.
- **Establishing a testing procedure/experiment.** Developing a procedure for each test or experiment to be performed to collect the desired data.
- **Conducting scientific investigations.** Testing and evaluating design solutions in a scientific manner to collect the proper data to inform solution redesign. Also, investigating how a solution impacts people, systems, and the environment. Designs or products that have unexpected and undesirable impacts on people, systems, or the environment must be corrected.
- **Conducting mathematical examinations.** Using statistical procedures to analyze the testing data in order to evaluate and improve their designs.
- **Collecting data.** Conducting the design tests/experiments to evaluate how well the solution solves the problem.
- **Analyzing data.** Drawing conclusions based on the test/experiment results.
- **Refining the solution.** Improving the solution based on the conclusions drawn from the testing results.

Strategies for This Step

The following strategies are beneficial for evaluating the solution:

- Develop a detailed description of the testing procedure to ensure that data is collected in a controlled environment.
- Attend to ethics, such as privacy of information or coercion, when collecting, analyzing, and sharing data related to the effectiveness and impact of the design.
- Use software, such as SPSS Statistics or statistical software for Microsoft® Excel, to help analyze testing data.
- Document any modifications made to the design solution.
- Conduct a critical design review with external stakeholders, such as consumers who may be impacted by this product, to evaluate the solution.
- Monitor solution progress based on the established benchmarks for project completion.

Communicate the Final Solution

A person or group selects a final solution and shares the results of the engineering design process by preparing the documents and presentations needed to share the evaluation outcomes, to produce the final solution, and to use the final solution.

Tasks in This Step

Steps include the following tasks:

- **Studying the solution results.** Drawing conclusions about the solution to a problem from an analysis of the entire engineering design process.
- **Sharing the results and conclusions.** Presenting the results and conclusions, as well as the final design solution, to the proper audience.
- **Interpreting the solution.** Communicating the final solution through detailed documents and reports, such as engineering drawings, bill of materials, and specification sheets.
- **Presenting the solution for approval.** Presenting written and oral reports to obtain the appropriate approval (from management or government, for example) for implementing the solution.

Strategies for This Step

The following strategies are beneficial for communicating results:

- Examine all documentation of the design process to ensure that all key elements of the problem-solving process are shared.
- Convey information using multiple forms of media.
- Use presentation software to assist in communicating results.
- Attend to ethics, such as being honest and truthful, when collecting, analyzing, and sharing data related to the effectiveness and impact of the design.

Enacting the Engineering Design Process

The steps and strategies used in solving engineering design problems are used in different phases of the engineering design process. For example, after an engineer identifies a problem, writes a problem statement, and begins to gather information and develop solutions, he or she incorporates testing. This testing is preliminary and is often used as a predictive measure. Yet, testing is also relevant and necessary during the stage of the engineering design process in which the solution is evaluated. In this stage, a model or prototype would be tested.

As you read the remaining chapters of this section, remember that many design actions are iterative and used throughout all steps of the design process. The same tasks and strategies may vary in purpose and intent in the different steps.

NakoPhotography/Shutterstock.com

Once a final solution is selected, it must be communicated to the proper audience. However, this is not the end of the design process.

Summary

- ✔ Both problem solving and design require thinking and reasoning skills.
- ✔ Problem solving involves finding solutions for a situation that has an identified goal state, readily available solutions, and one or more reasonable pathways for solving. Design differs from problem solving because it involves devising a new solution, one that has not previously been found or defined.
- ✔ The sequence of steps used to solve problems is to (1) develop an understanding of the problem through observation and investigation; (2) devise a plan for solving the problem; (3) implement the plan; (4) evaluate the plan.
- ✔ Problems vary in complexity, structure, and context.
- ✔ Engineering design problems involve the need to design either a technological product or system as a solution.
- ✔ System design involves the arrangement of components to achieve a desired result. Product design involves the development of a product or structure to the needs of customers.
- ✔ *Engineering design* is the designing of solutions to problems under constraints (limitations or restrictions).
- ✔ The steps involved in solving engineering design problems are: (1) identify and define a problem; (2) gather information; (3) develop a design solution; (4) model and make a solution; (5) evaluate the solution; (6) communicate the final solution.

Check Your Engineering IQ

Now that you have finished this chapter, see what you learned by taking the chapter posttest.
www.g-wlearning.com/technologyeducation/

Test Your Knowledge

Answer the following end-of-chapter questions using the information provided in this chapter.

1. *True or False?* Problem solving involves devising a new solution that has not previously been found or defined.
2. Critical thinking is a type of thinking that uses mental abilities to identify, _____, and evaluate a situation or problem.
3. What is the first step in the general problem-solving process?
4. What are the three ways in which problems vary?
5. _____ design problems are the most complex and least structured types of problems.
6. Designing the arrangement of components to produce a desired result is called _____.
7. Deadlines, energy consumption, and funding are examples of engineering design _____.
8. *True or False?* It is often necessary to move back and forth between the steps of the engineering design process.
9. Engineering design problems involve the need to create either a technological _____ or _____ as a solution.
10. List, in order, the six major steps used to solve engineering design problems.
11. What five things should a problem statement contain?
12. What is a *mind map* and how is it used in the engineering design process?

Critical Thinking

1. Explain the difference between general problem solving and the engineering design process.
2. Why is the engineering design process important to all people, not just engineers?
3. Present a realistic example of an engineering design problem.
4. Why is project management important to the engineering design process?

STEM Applications

1. Create a poster of the engineering design process that uses images to depict each step of the process.
2. Identify a potential problem that is in need of a technological solution. Then create a tentative procedure to follow to solve this problem. Use the steps of the engineering design process.

Engineering Design Challenge

Temporary Shelter

Background

WeStudio/Shutterstock.com

Figure A. A temporary shelter.

In 2010, a catastrophic earthquake occurred approximately 16 miles west of Port-au-Prince, Haiti's capital. The earthquake had a magnitude of 7.0 on the Richter scale. Within two weeks after the major quake, 52 aftershocks measuring 4.5 or greater on the Richter scale had been recorded. An estimated three million people were affected by the quake. An estimated 250,000 homes and 30,000 commercial buildings had collapsed or were severely damaged, leaving about 1,000,000 people homeless. In addition to loss of shelter, widespread disease and malnourishment resulted.

Situation

In emergencies like this, inexpensive, temporary shelters are urgently needed to house the many people who have lost their homes in the disaster. The shelters are used temporarily while the homes are rebuilt.

The Air House Inflatable Structure Company wants to design a temporary inflatable structure that could be deployed in an emergency, such as the Haiti earthquake. You have been tasked by the Air House Company to design an inflatable shelter. This shelter must be able to house a minimum of four people for an extended time period after a natural disaster. The shelter should be an air-inflated or air-supported structure that is easy to set up and take down. Additionally, the structure must be portable and able to be carried by one or two people. Use the engineering design process to develop a solution.

Desired Outcomes

- Detailed design of an air-inflated or air-supported structure that meets the following criteria:
  - Accommodates at least four people.
  - Transportable by one or two people.
  - Quickly deployable.
  - Portable.
  - Easy to take down.
  - Inexpensive.
- Explanation of how the engineering design process was used to develop the design.

Materials and Equipment

- Computer.
- Internet access.
- Sketch paper and pencils.
- Design software, if available.

Procedure

- Use the engineering design process to define the problem and create a viable solution to it.
- Go back and forth between the different steps as needed to solve the problem.
- Define the problem based on the background and situation provided.
- Gather information related to the problem that has been defined.
- Apply the knowledge you obtain from the information you gathered. Employ the engineering design process in order to propose a new or improved solution to the problem. There will be a natural progression from brainstorming, to generating ideas, to developing a design, to evaluating the design's potential success.
- Document all of your work.

CHAPTER 5

Developing Design Solutions

Check Your Engineering IQ 

Before you read this chapter, assess your current understanding of the chapter content by taking the chapter pretest.
www.g-wlearning.com/technologyeducation/

Learning Objectives

After studying this chapter, you should be able to:

- ✔ Explain the major steps followed in developing design solutions.
- ✔ Describe some advantages of using a design team over an independent designer.
- ✔ List the ways to generate ideas when developing design solutions.
- ✔ Distinguish between the different kinds of design sketches.
- ✔ Differentiate the three major types of information required for detailed sketches when building models.
- ✔ Illustrate the three types of pictorial sketches used in product design.
- ✔ Explain the advantages of using predictive analysis through scientific and mathematical models.

Technical Terms 

assembly drawings
auxiliary view
brainstorming
cabinet oblique drawing
cavalier oblique drawing
charts
classification
computer models
conceptual models
constraints
convergent thinking
criteria
descriptive methods
detailed sketch
diagrams
divergent thinking
experimental methods
graphic/computer models
graphs
historical methods
isometric sketches
mathematical models
modeling
multiview drawing
oblique sketches
orthographic assembly drawings
orthographic projection
perspective sketches
pictorial assembly drawings
predictive analysis
refined sketch
rough sketches
scientific models
solid models
surface models
synergism
systems drawings
trade-offs
what-if scenarios
wire-frame models

Close Reading

As you read this chapter, make a list of the terms that are used in developing design solutions. Can these terms be used in other contexts, or are they concepts specific to design?

Addressing engineering design problems is a process that is done in steps, **Figure 5-1**. Developing design solutions is the first step and guides the remainder of the design process. Engineers must spend all the time necessary to specifically understand what the problem is and to gather information to solve the problem. Engineers often spend more time examining and defining the problem and investigating possible ideas than actually constructing a final solution. The focus of this chapter will be these steps in the engineering design process:

1. Identify and define a problem.
2. Gather information.
3. Develop solutions.

The Design Team

Using a design team allows various people to contribute their talents to a project. For example, no one person could design, plan, and construct a building alone. The task would be too large and require too much knowledge. For the team, however, the job is manageable, **Figure 5-2**.

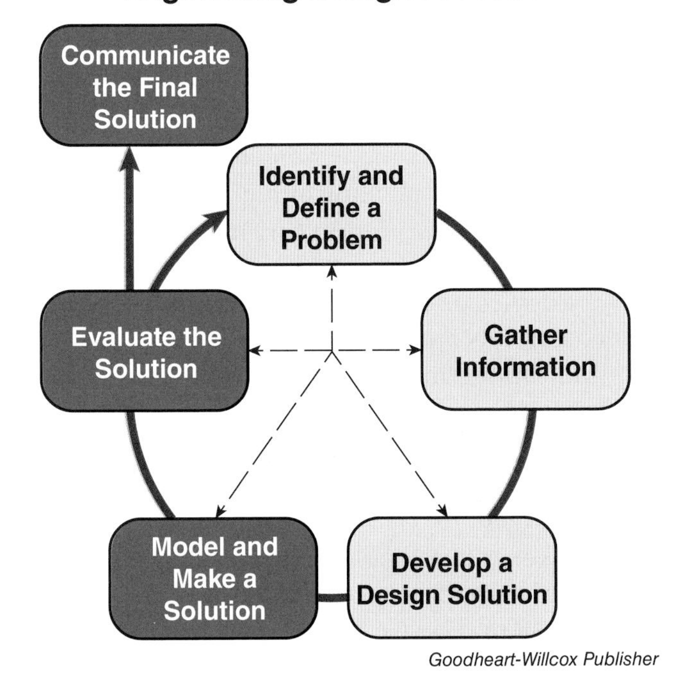

Goodheart-Willcox Publisher

Figure 5-1. Steps in the engineering design process.

©iStockphoto.com/MarcusPhoto

Figure 5-2. Members of an architectural design team look over a model of the building they are designing.

Identify the Problem

Technological products or systems are devised to fulfill our needs and wants that make life easier, enhance happiness or well being, or extend human capabilities. To achieve this goal, the problem must first be identified. This is not always simple. For example, holes have developed in the ozone layer over the Antarctic and Arctic regions. The holes are not the real problem, however. Instead, their presence indicates that a problem might exist. After gathering information, scientists learned that the holes were created by the by-products of fossil fuels and aerosol products. The problem, then, is to create items, such as vehicles with cleaner fuels and products without the aerosol component.

There are times when engineers attempt to bring innovation to a market or product and have to identify a problem that exists across multiple technologies. For example, they may ask, "What is missing from existing technologies?" This problem can then be addressed through the engineering design process by first defining what the problem is and then gathering the information needed to begin developing solutions. In each of these steps, designers follow certain procedures.

Define the Problem

When defining the problem, engineers must:

1. Describe the situation that needs a technological solution through a depth-first approach. Consider multiple views and representations of the problem.
2. Write a clear, concise problem statement.
3. Establish the criteria and constraints under which the artifact or system must operate.

Describe the Situation: A Depth-First Approach

Defining a problem begins by describing it in general terms. The description can be refined after the situation has been studied thoroughly. Engineers take time to consider what the problem actually is. They must consider multiple questions and may begin with a mind map to consider options, **Figure 5-3**. Engineers may ask *what* is the problem, *who* has the problem, *where* is the problem located, *why* is it a problem. This process is often collaborative and iterative.

Write a Clear, Concise Problem Statement

As engineers explore the situation, they should be careful not to confuse the problem with the solution. For example, a bookend is a solution, not a problem. The problem is to hold books in an organized manner on a flat surface. Descriptions can restrict creativity or open doors to a variety of solutions. Look back at the ozone layer problem. Closing the hole is the solution. The problem is to develop technological devices that emit fewer pollutants into the atmosphere.

A problem statement should be only a few sentences long, identify a specific need, be clear and concise, and not provide criteria or constraints at this point. The problem statement should also reduce any restrictions to possible solutions. Here is an example of a strong problem statement:

A new medicinal product contains a contaminant as a result of the manufacturing production process and must be removed to ensure patients are not infected.

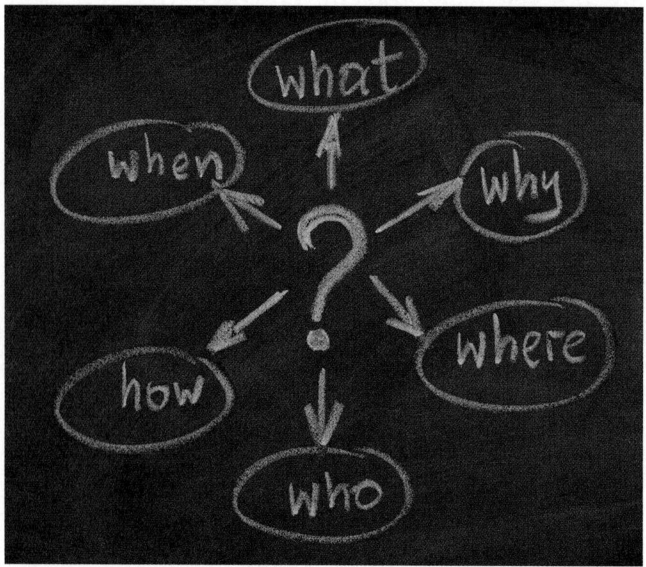

marekuliasz/Shutterstock.com

Figure 5-3. Mind maps help engineers to formulate problem statements.

Establish Criteria and Constraints

The problem statement leads directly to the next step, establishing a set of criteria and constraints. Designers must know how the effectiveness of the new technology will be evaluated. To do this, they establish criteria and constraints under which the product or system must operate. ***Criteria*** include a listing of the features of the artifact or system. ***Constraints*** deal with the limits on the design, **Figure 5-4**. These criteria and constraints can be grouped into general categories:

- Technical or engineering criteria and constraints describe the operational and safety characteristics the device or system must meet. These criteria and constraints are based on how, where, and by whom the product will be used. An example of an engineering criterion for a new windshield wiper might be, "Must effectively clear the windshield of water when the vehicle is traveling at highway speeds."
- Production criteria and constraints describe the resources available for producing the device or system. These criteria and constraints are based on the natural, human, and capital (machine) resources available for the production of the device or system. A production constraint for a product might be, "Must be manufactured using existing equipment in the factory."
- Market criteria and constraints identify the function, appearance, and value of the device or system. These criteria and constraints are derived from studying what the user expects from the device or system. An example of a market constraint might be, "Must be compatible with mid-century modern decor."

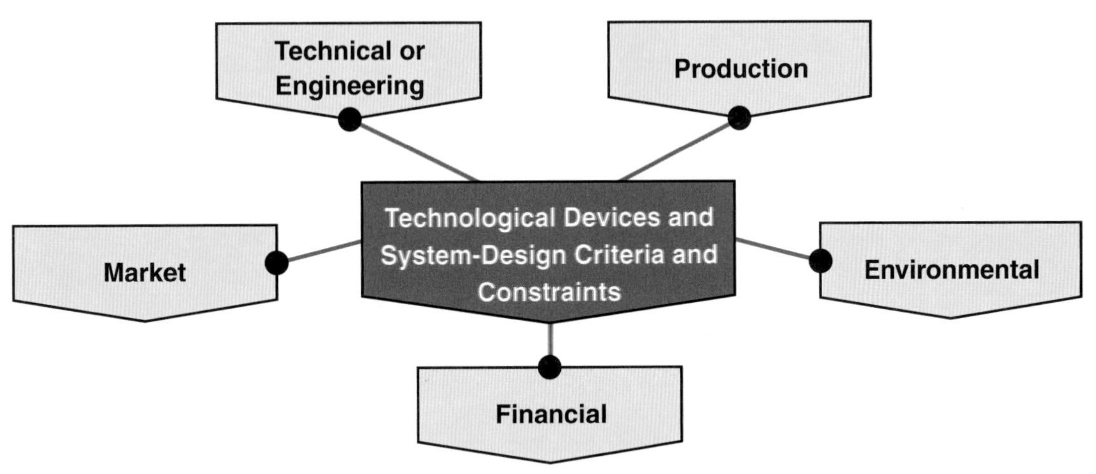

Goodheart-Willcox Publisher

Figure 5-4. Technological devices and systems must meet both individual and societal expectations. These expectations can be communicated through five types of criteria and constraints.

- Financial criteria and constraints establish the cost-benefit ratio for the device or system. These criteria and constraints address the money required to develop, produce, and use the technological device or system and the ultimate benefits from its use. An example of a financial constraint might be, "Must be priced at 5% less than the major competitor's product."
- Environmental criteria and constraints indicate the intended relationship between the device or system and the natural and social environments. These criteria and constraints deal with the impacts of the technological device or system on people, societal institutions, and the environment. An environmental criterion might be, "Must remove 95% of all sulfur dioxide emitted from the electric generating plant smokestacks."

The problem statement and set of criteria and constraints allow designers to move to the next step in the identification process, gathering information.

Gather Information

Designing technological products and systems requires many steps, beginning with researching current solutions. This step allows engineers to locate and analyze the knowledge needed to solve a problem. This knowledge is derived from obtaining and studying information. A wide variety of information might be needed, **Figure 5-5**. Typically, this information includes:

- Historical information about devices and systems developed to solve similar problems.
- Scientific information about natural laws and principles that must be considered in developing the solution.
- Technological information about materials and energy-processing techniques that can be used to develop, produce, and operate the device or system.
- Human information affecting the acceptance and use of the device or system. This information might include such factors as ergonomics, body size, consumer preferences, and appearance.
- Legal information about the laws and regulations controlling the installation and operation of the device or system.
- Ethical information describing the values people have toward similar devices and systems.

Designers can obtain information through historical-, descriptive-, or experimental-research activities. They use ***historical methods*** to gather information from existing records. For example, they might consult books, magazines, and journals, **Figure 5-6**. They can review sales records, customer complaints, and other company files. Designers might also check judicial codes describing laws that could affect the development project.

Designers use ***descriptive methods*** to record observations of present conditions. For example, they might survey people to determine product preferences, opinions, or goals. They might observe and describe the operation of similar devices. Designers

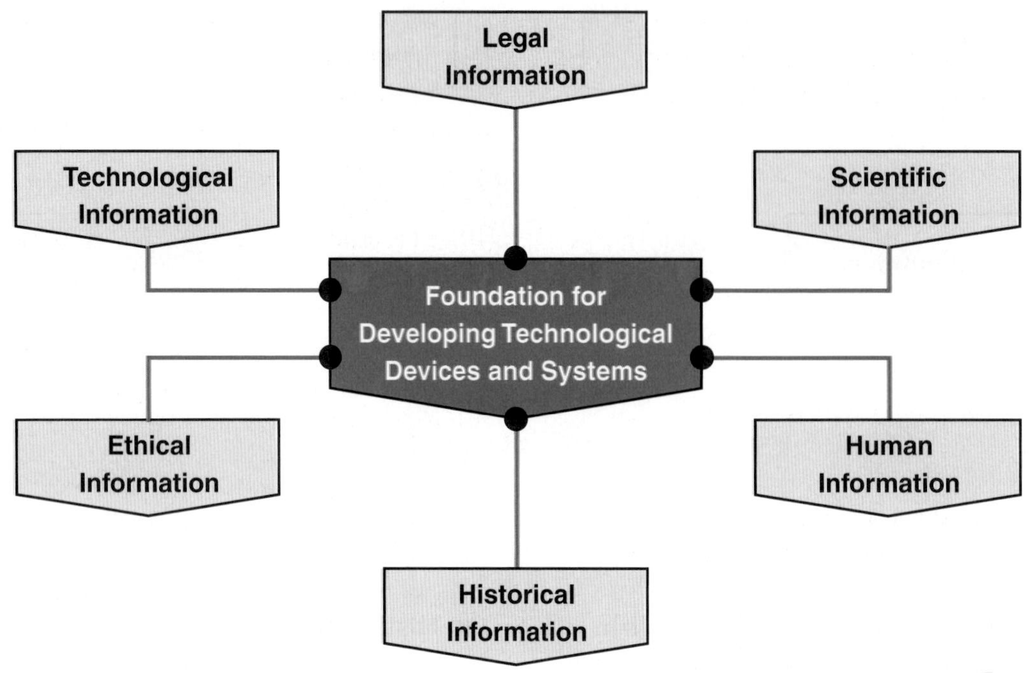

Goodheart-Willcox Publisher

Figure 5-5. Many types of information are used as a foundation for technological-development activities.

might measure and record physical qualities, such as size, weather conditions, and weight.

Finally, designers use ***experimental methods*** to compare different conditions. One condition is held constant. The other is varied. Designers use this method to determine scientific principles, assess the usefulness of technological processes, or gauge human reactions to situations. See **Figure 5-7**.

All of these methods are used to gather information. Once engineers have gathered the informa-

©iStockphoto.com/dlewis33

Figure 5-6. Books, magazines, and journals are a valuable source of information for solving technological problems.

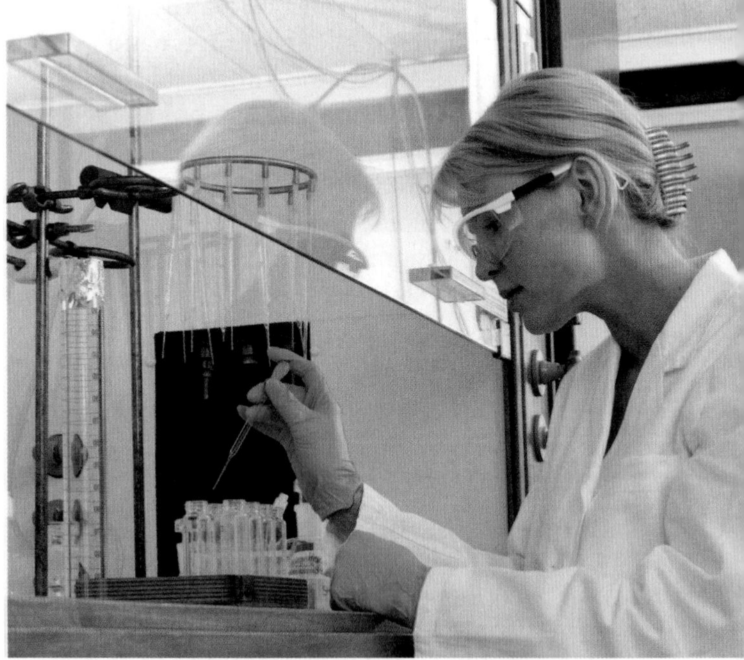

©iStockphoto.com/pidjoe

Figure 5-7. Laboratory experiments provide valuable information that can be used to solve technological problems.

tion and knowledge required to develop a solution, they can explore concepts and document details that may be used to develop the solution.

Develop Solutions

After a clear definition of the situation is established, based on the procedures for defining the problem or situation, solutions often evolve through multiple phases including:

1. Develop preliminary solutions.
2. Isolate and refine the best solution.
3. Detail the best solution.
4. Model the solution using a computer.
5. Apply predictive analysis: mathematical and scientific principles.
6. Create the best solution based on the specifications and constraints of the design. This is called *optimization*.
7. Develop a production procedure.

As these phases are completed, the problem is often redefined or refined. The phases can then be repeated until a final design is developed. This circular system is shown in **Figure 5-8**.

Divergent thinking seeks to imagine as many different (divergent) solutions as possible. The most promising solutions are then refined and reduced until one "best" answer is found. The refinement of ideas requires ***convergent thinking***. The goal of this thinking is to narrow and focus (converge) the ideas until the most feasible solution is found.

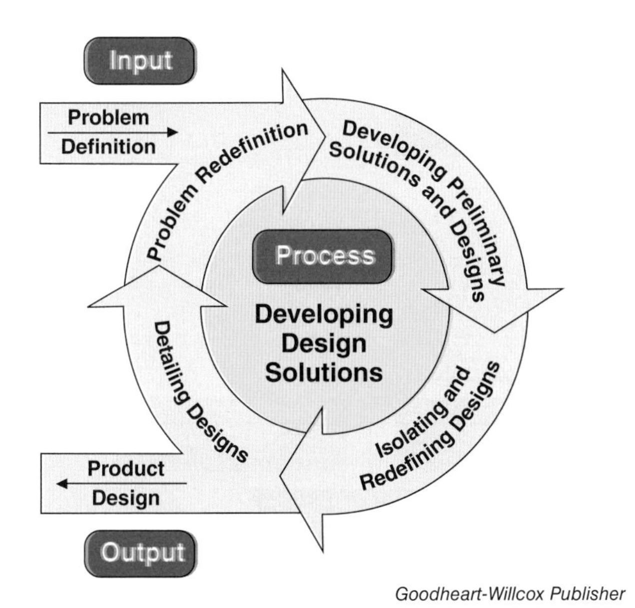

Goodheart-Willcox Publisher

Figure 5-8. The process of developing design solutions has seven major phases, plus a problem-redefinition phase.

The best solution is not always the one that works best or is the least expensive. As noted earlier, criteria and constraints can compete with one another. ***Trade-offs***, which are decisions that arise in which a choice must be made between two competing items, often occur among appearance, function, and cost. On an individual level, many people cannot afford the most expensive product. Budget and the length of time a person expects to keep the product can influence product choices. For example, buying a $900 racing bicycle might be unwise for a person who will only ride the bicycle occasionally to the park. Someone who regularly rides a bicycle to work or goes on cycling vacations might be able to justify the expensive model. Product-design activities, or the consideration of the form and function of a product, result in the creation of a wide range of products. For example, time is spent during the design process considering what takes priority for the final product. This may include cost, aesthetics, ergonomics, speed, marketing costs, and other design considerations. Often, the type of product produced will even vary across markets. This variety allows consumers to select one that meets their performance needs and financial resources.

Develop Preliminary Solutions

Ideas can be generated in various ways. Three popular techniques are brainstorming, classification, and what-if scenarios.

Brainstorming

Brainstorming is a process in which two or more people work together to generate ideas in order to solve an identified problem. Members of the group offer solutions, which in turn can cause other group members to think of additional ideas, **Figure 5-9**. This concept is called synergism. ***Synergism*** is the working together of two or more individuals that produces an effect greater than the sum of individuals. The number of ideas generated is greater than the number generated if each person worked alone.

Brainstorming activities work best when the group follows some basic rules:

- **Encourage unconventional, unusual ideas.** There are no bad ideas. Unconventional, but promising, ideas can always be engineered back to real-life possibilities.

©iStockphoto.com/track5

Figure 5-9. Members of a graphic design team brainstorm ways to improve a design for a national advertisement.

- **Record the ideas without reacting to them.** Many people will stop offering ideas if those ideas are criticized. Record ideas without judgment.
- **Seek quantity, not quality.** The chances of good ideas emerging are increased as the number of ideas increases.
- **Maintain a rapidly paced session.** This keeps minds alert and reduces the temptation to judge ideas.

Classification

Classification involves dividing the problem into major segments. Each segment is then reduced into smaller parts. For example, buildings can be classified as business and commercial, homes, and industrial. Homes can be further classified as houses, apartments, and condominiums, for example. A house can be classified by its major features: foundations, floors, walls, ceilings, roof, doors, windows, and so on. Foundations can then be classified as poured concrete, concrete block, wood posts, timber, and so on. This process might result in a classification chart. One type of classification chart is a tree chart, with each level having a number of branches below it, **Figure 5-10**.

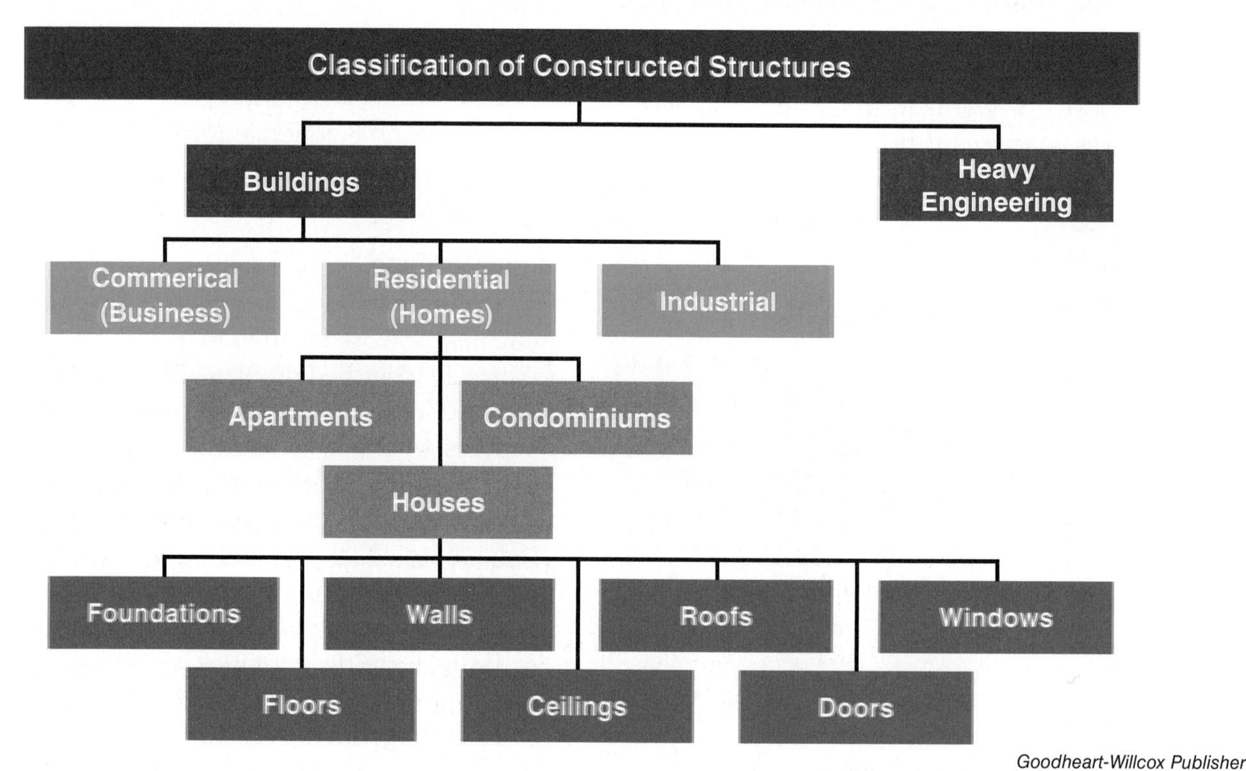

Goodheart-Willcox Publisher

Figure 5-10. This tree chart divides the design problem into segments to develop solutions more easily.

What-If Scenarios

What-if scenarios start with a wild proposal. The proposal's good and bad points are then investigated. The good points are then used to develop solutions. For example, peeling paint is a problem for housepainters. They must remove the old paint from a house before repainting. A wild solution suggests mixing an explosive material with the paint before it is applied. Whenever the building is ready to be repainted, the old paint can be blown off the building. Obviously, exploding house paint is ridiculous and unsafe. The proposal, however, can lead to a solution. Paint sticks to a house through the adhesion between the paint and the siding. Maybe a material can be mixed with the paint that causes the paint to lose adhesion when a special chemical is applied. At repainting time, the chemical can be sprayed on to loosen the paint. The paint can then be easily removed from the siding.

Rough Sketches

As designers have generated a number of ideas, they are also recording them throughout the process. The most common recording method is to develop rough sketches of the products, structures, or system components, **Figure 5-11**. ***Rough sketches*** are incomplete, unrefined sketches used to communicate design solutions. The term *rough* is not used to describe the quality of the drawing. Rough sketches are not necessarily crude. They often represent good sketching techniques.

Rough sketches are as much a part of the thinking process as they are a communication medium. Designers are forced to think through concepts such as size, shape, balance, and appearance. The sketches then become a library of ideas for later design efforts.

©iStockphoto.com/nicolas_

Figure 5-11. Designers use rough sketches to record their ideas.

Isolate and Refine Design Solutions

Rough sketches allow designers to capture a wide variety of solutions for the design problem. These sketches are similar to books in a library. They contain a number of different thoughts, views, and ideas. These sketches can be selected, refined, grouped together, or broken apart.

Isolating and refining original designs in this library of ideas is the second step in developing a design solution. Promising ideas are chosen and then studied and improved. This process might involve working with one or more good rough sketches. The size and shape of the product or structure might be changed and improved. Details might be added, and the shape might be reworked. In short, the design is being refined as problems are worked out and the proportions become more balanced.

Refined design ideas might also be developed by merging ideas from two or more rough sketches into a refined sketch, **Figure 5-12**. A ***refined sketch*** typically merges ideas from two or more rough sketches. The overall shape might come from one sketch, and specific details might come from others. This approach is one of integration, which blends the different ideas into a unified whole. The new idea might not look at all similar to the original rough sketches.

Detail Design Solutions

Rough and refined sketches do not tell the whole story. Refer again to **Figure 5-12**. What size is

©iStockphoto.com/nicolas_

Figure 5-12. Refined sketches are used to develop ideas captured with rough sketches.

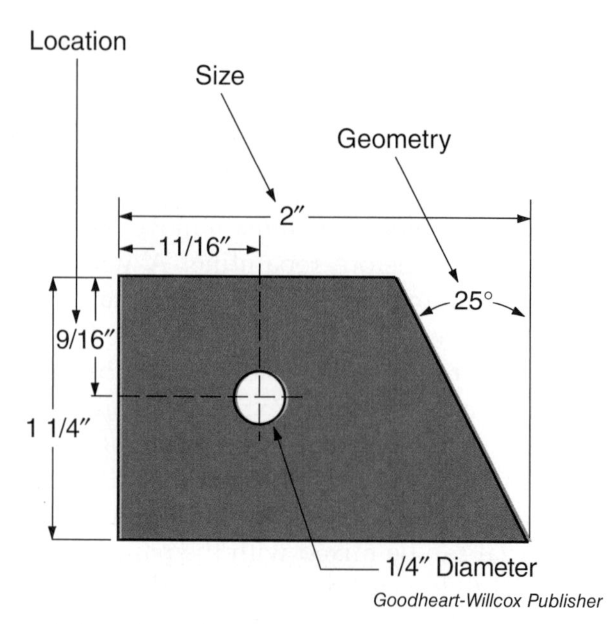

Goodheart-Willcox Publisher

Figure 5-13. Information provided on a detailed sketch includes size, location, and geometry.

the product in the sketch? You cannot tell. Refined sketches communicate shape and proportion, not size. For this task, more detail must be added in a third type of sketch called a detailed sketch. ***Detailed sketches*** communicate the information needed to build a model of the product or structure. They can also be used as a guide to prepare engineering drawings for manufactured products and architectural drawings for constructed structures.

Detailed sketches are helpful when models of products or structures are made. Building models requires three major types of information, **Figure 5-13**:

- **Size.** Size information explains the overall dimensions of the object or the sizes of features on an object. Size information might include the thickness, width, and length of a part, the diameter and depth of a hole, or the width and depth of a groove.
- **Location.** Location information gives the positions of features within the object. Location information might establish the location of the center of a hole, the edge of a groove, or the position of a taper.
- **Geometry.** Geometric information describes the geometric shapes or relationships of features on the object. Geometric information can communicate the relationship of intersecting surfaces (square or 45° angle, for example), the shapes of holes (rectangular or round, for example), or the shapes of other features.

Designers often use pictorial-sketching techniques to capture and further refine product-design ideas. These techniques show the artifact similarly to how the human eye will see it. A single view is used to show how the front, sides, and top will appear. Designers produce three different kinds of pictorial sketches when refining ideas:

- Oblique sketches.
- Isometric sketches.
- Perspective sketches.

Oblique Sketch

An ***oblique sketch*** shows the front view of an object as if a person is looking directly at it. The sides and top extend back from the front view. They are shown with parallel lines, called depth lines. They are generally drawn at a 45° angle to the front view, but may also be drawn at a different angle, such as 30° or 60°, depending on the intended result.

To produce an oblique sketch, the designer completes steps similar to those shown in **Figure 5-14**.

1. Lightly draw a rectangle that is the overall width and height of the object.
2. Lightly extend parallel lines from each corner of the box. Draw the depth lines at 30°, 45°, or 60°.

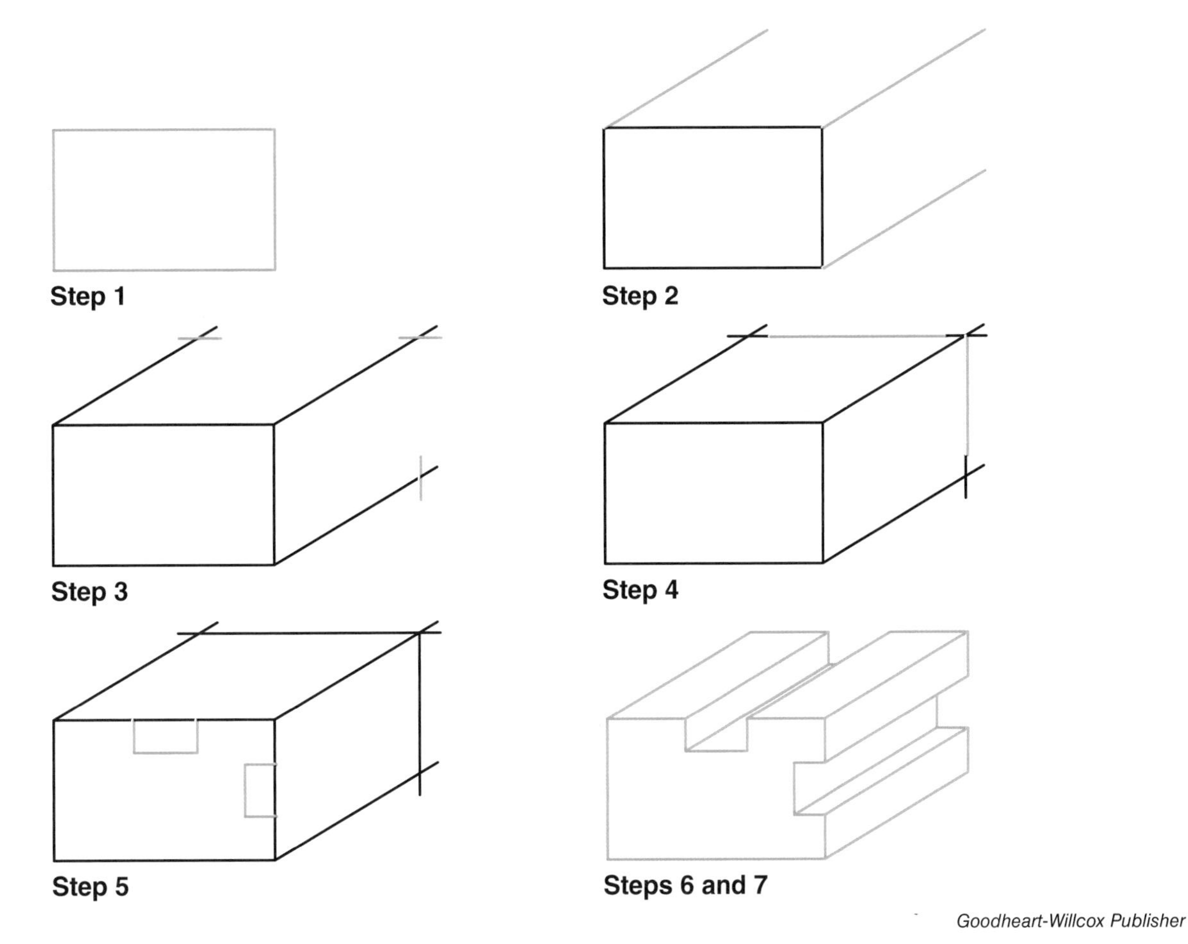

Goodheart-Willcox Publisher

Figure 5-14. Follow these steps to create an oblique sketch.

3. Lightly mark extension lines (which are lightly drawn lines that indicate the points between which the dimension figures will be presented) at a point equal to the depth of the object.
4. Lightly connect the depth lines to form a box.
5. Add any details, such as holes, notches, or grooves, onto the front view.
6. Extend the details the depth of the object.
7. Complete the sketch by darkening in the object and detail outlines.

The procedure listed above produces a ***cavalier oblique drawing***. This type of drawing makes the sides and top look deeper than they are. To compensate for this appearance, designers often use ***cabinet oblique drawings***. See **Figure 5-15**. This type of

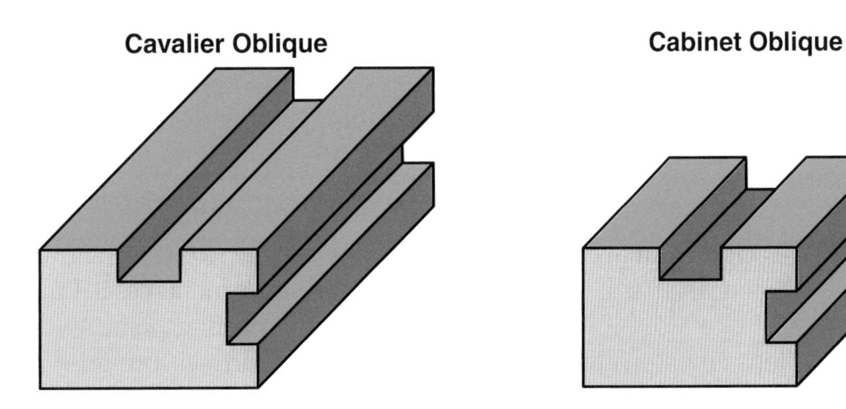

Goodheart-Willcox Publisher

Figure 5-15. Note the differences between these cavalier and cabinet oblique drawings.

Career Connection Drafting and Design Engineering

Drafting and design engineers prepare engineering and architectural drawings and plans used by manufacturing and construction workers. Drafting and design engineers communicate technical details using rough sketches, drawings, and specifications. Most use computer-aided design (CAD) systems to prepare their drawings. Types of drafting and design engineers include aeronautical, architectural, electrical and electronic, and mechanical. Drafting and design engineers work across many engineering disciplines and work on a variety of projects, both large and small.

Monkey Business Images/Shutterstock.com

Most drafting and design positions require postsecondary-school training in drafting, which can be done at technical institutes, community colleges, and four-year colleges and universities. Entry-level drafting and design engineers usually do routine work under close supervision. More experienced professionals work with less supervision.

High school students interested in drafting and design engineering should take technology and engineering courses offered at their schools. These courses provide introductory computer-aided drafting skills, including 2- and 3-D modeling. Students also benefit from taking art and geometry classes to gain knowledge in drawing basics and how to lay out and dimension shapes.

College students pursuing a career in drafting and design engineering should have strong mathematical and scientific skills, excellent communication and problem-solving skills, and be visually creative.

Drafting and design engineers can choose a variety of specializations in almost any industry. They can also work toward management positions in a variety of industries.

Drafting and design engineers are consistently in high demand due to a lack of highly qualified engineers around the world and increasingly expanding technology industries. There are a variety of professional organizations for each engineering specialization.

drawing shortens the lines projecting back from the front to one-half their original lengths.

Isometric Sketches

Isometric sketches are another type of pictorial drawing used to produce refined sketches. The word *isometric* means "equal measure." ***Isometric sketches*** get their name from the fact that the angles formed by the lines in the upper-right corner are equal. Each angle is 120°. Designers use isometric sketching when the top, sides, and front are equally important. The object is shown as if it is viewed from one corner.

Designers follow four steps when producing an isometric drawing, **Figure 5-16**. These steps include:

1. Lightly draw the upper-right corner of an isometric box that will hold the object.
2. Complete the box by lightly drawing lines parallel to the three original lines.
3. Locate the major features, such as notches, tapers, and holes.
4. Complete the drawing by darkening the features and darkening the object outline.

Perspective Sketches

Perspective sketches show the object as the human eye or a camera sees it. Lines are drawn to meet at a distant vanishing point on the horizon. Looking down a railroad track provides a similar effect. The rails remain the same distance apart, but they appear to converge (come together) in the distance.

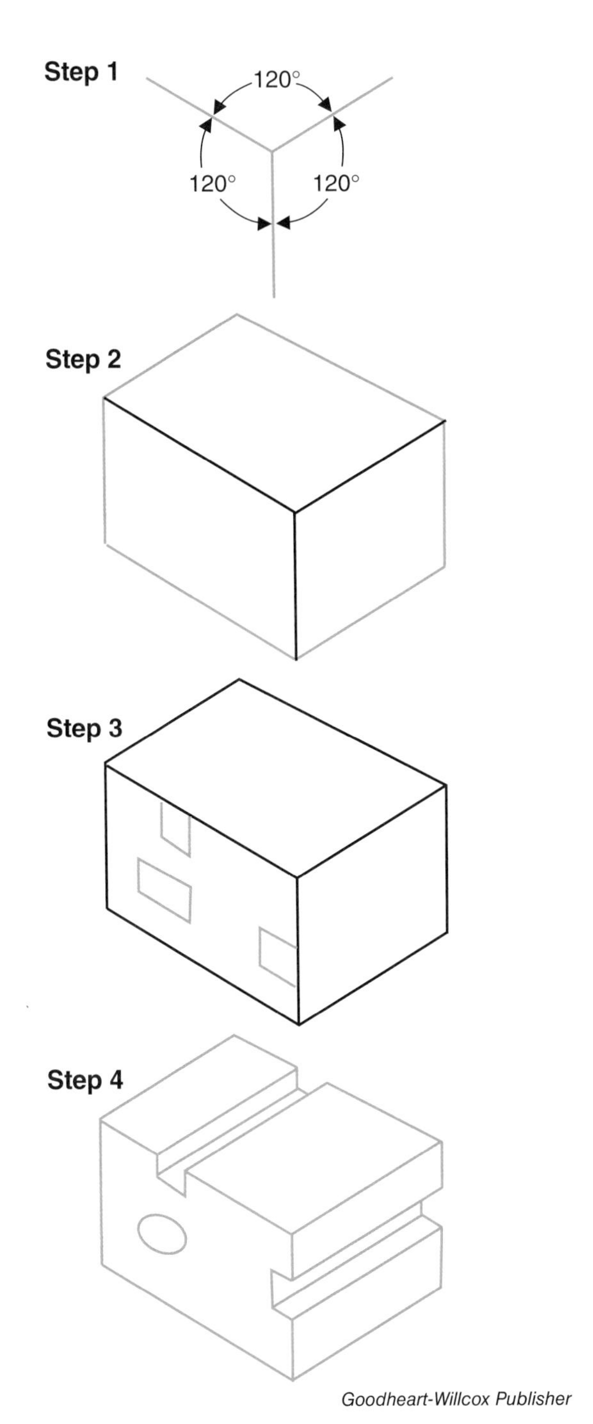

Figure 5-16. Follow these four steps to create an isometric sketch.

There are three major types of perspective views: one point, two point, and three point. The number of vanishing points used determines the type of view, **Figure 5-17**.

A one-point perspective shows an object as if standing directly in front of the object. All the lines extending away from the viewing plane converge at one point. The one-point perspective is similar to an oblique drawing with tapered sides and a tapered top.

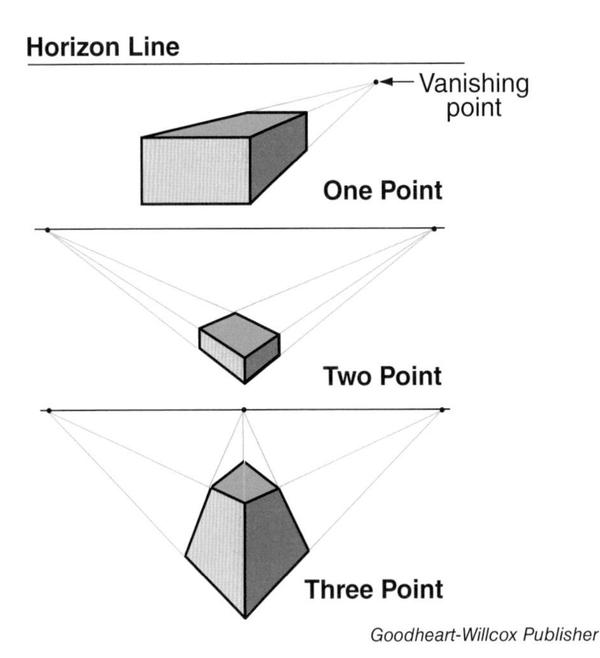

Figure 5-17. Three types of perspective drawings.

A two-point perspective shows how an object appears when standing at one corner. This perspective is constructed similarly to an isometric drawing. Again, the sides are tapered as the lines extend toward the vanishing points. This means the sides extend gradually from a larger size to a smaller size.

A three-point perspective shows how the eye sees the length, width, and height of an object. All lines in this drawing extend toward a vanishing point. The appearance of a perspective drawing changes as the horizon changes, **Figure 5-18**. Changing the position of the horizon line can cause the object to be seen as if the observer is looking down on the object (aerial view), directly at it (general view), or up at it (ground view). The designer must decide which of these views best suits the object and the audience who will see the sketch.

When developing the basic structure for one-, two-, or three-point perspective sketches, designers follow the same basic steps, **Figure 5-19**, including:

1. Establish the horizon line, vanishing point(s), and front of the object. Connect the front line(s) to the vanishing point(s).
2. Establish the depth of the objects along the lines extending to the vanishing point(s).
3. Connect the depth lines to the vanishing point(s). Darken all lines.

Designers then add details to complete the sketch. Perspective sketches are often shaded to add to their communication value. Developing the

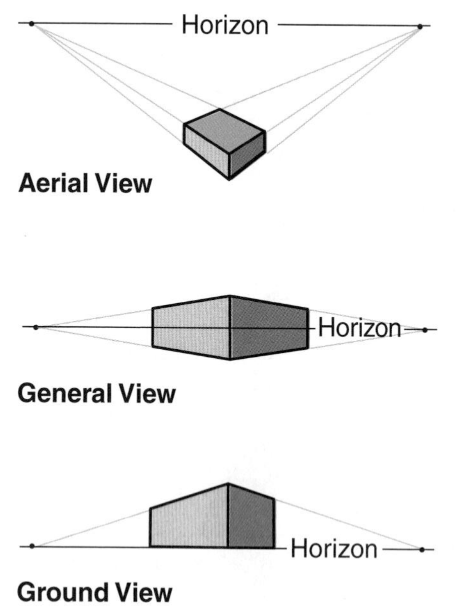

Figure 5-18. Changing the horizon location changes the appearance of a perspective drawing.

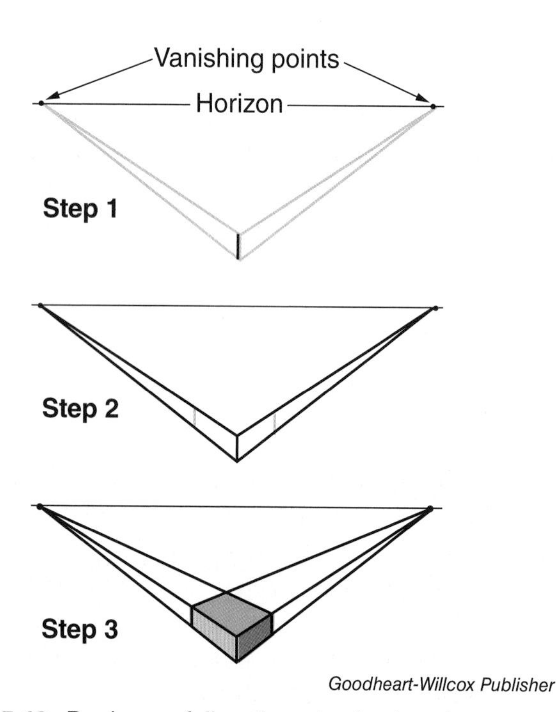

Figure 5-19. Designers follow three basic steps in developing perspective sketches.

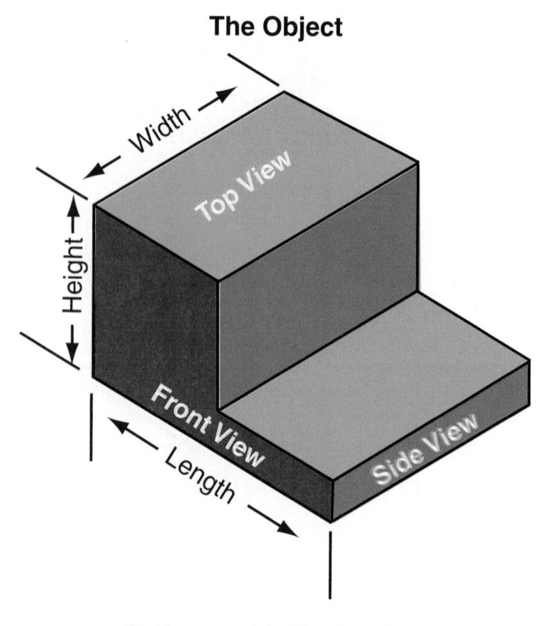

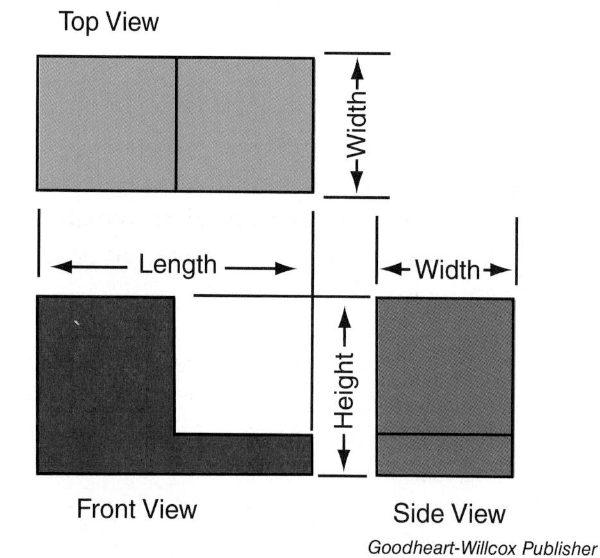

Figure 5-20. The orthographic projection shown on the right represents the object shown on the left.

perspective view is more difficult than developing the oblique or isometric views. Perspectives are, however, the most realistic of the three pictorial sketches.

Multiview Drawings

Most detail drawings are prepared using the multiview (multiple views) method. ***Multiview drawings*** include one or more views of the object in one drawing in order to fully communicate the shape and size of that object, **Figure 5-20**. The number of views depends on the complexity of the object. The most common multiview drawings are:

- **One-view drawings.** These drawings show the layout of flat pieces, which usually have standard or predetermined thicknesses, **Figure 5-21A**.
- **Two-view drawings.** These drawings show the size and shape of cylindrical parts or objects with round shapes. The front and top views are generally identical. In those

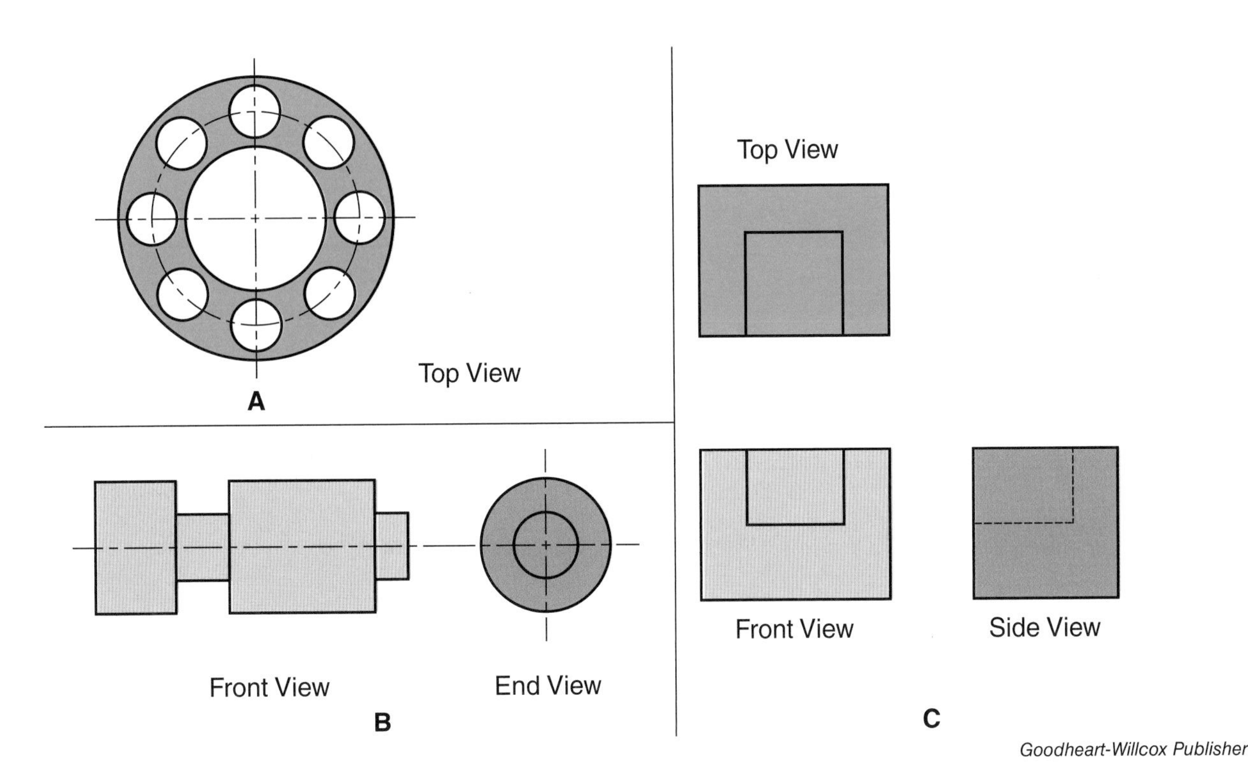

Figure 5-21. Commonly used multiview drawings. A—One-view drawing. B—Two-view drawing. C—Three-view drawing.

instances, only one view is needed. The two views shown are the front view, which shows the features along the length of the part, and the end view, **Figure 5-21B**.

- **Three-view drawings.** These drawings show the size and shape of rectangular and complex parts. Generally, a top, right-side, and end view are shown, **Figure 5-21C**. This arrangement is called *third-angle projection*. Complex parts might need to add more views, called ***auxiliary views***.

In all cases, the least number of views is used because creating unnecessary views costs time and money. Each drawing must fully communicate all the information needed to make the object.

Multiview drawings use orthographic projection to present the information in two or more views of the object. ***Orthographic projection*** is projection of a single view of an object onto a drawing surface in which the lines of projection are perpendicular to the drawing surface. Refer again to **Figure 5-20**. Before beginning a multiview drawing, designers select the surface of the object to be shown in the front view. The surface that has the most detail is chosen as the front view. The goal is to have as few hidden details as possible. Refer to **Figure 5-22** as you read about preparing a three-view drawing.

1. Draw the front view in the lower-left quadrant of the paper. An accepted practice is to construct a box enclosing the object using light construction lines. Locate and lightly draw details on the object.
2. At the same distance above and to the right of the front view, draw light lines as the place to start the right-side and top views.
3. Lightly draw projection lines above and to the right of the front view. These lines are used to project the size of the object to the other views.
4. Lightly draw the outline and details of the top view.
5. Draw a 45° line in the upper-right quadrant. The outline and features of the top view are projected to the side view. This is done by projecting sizes and features to the 45° line

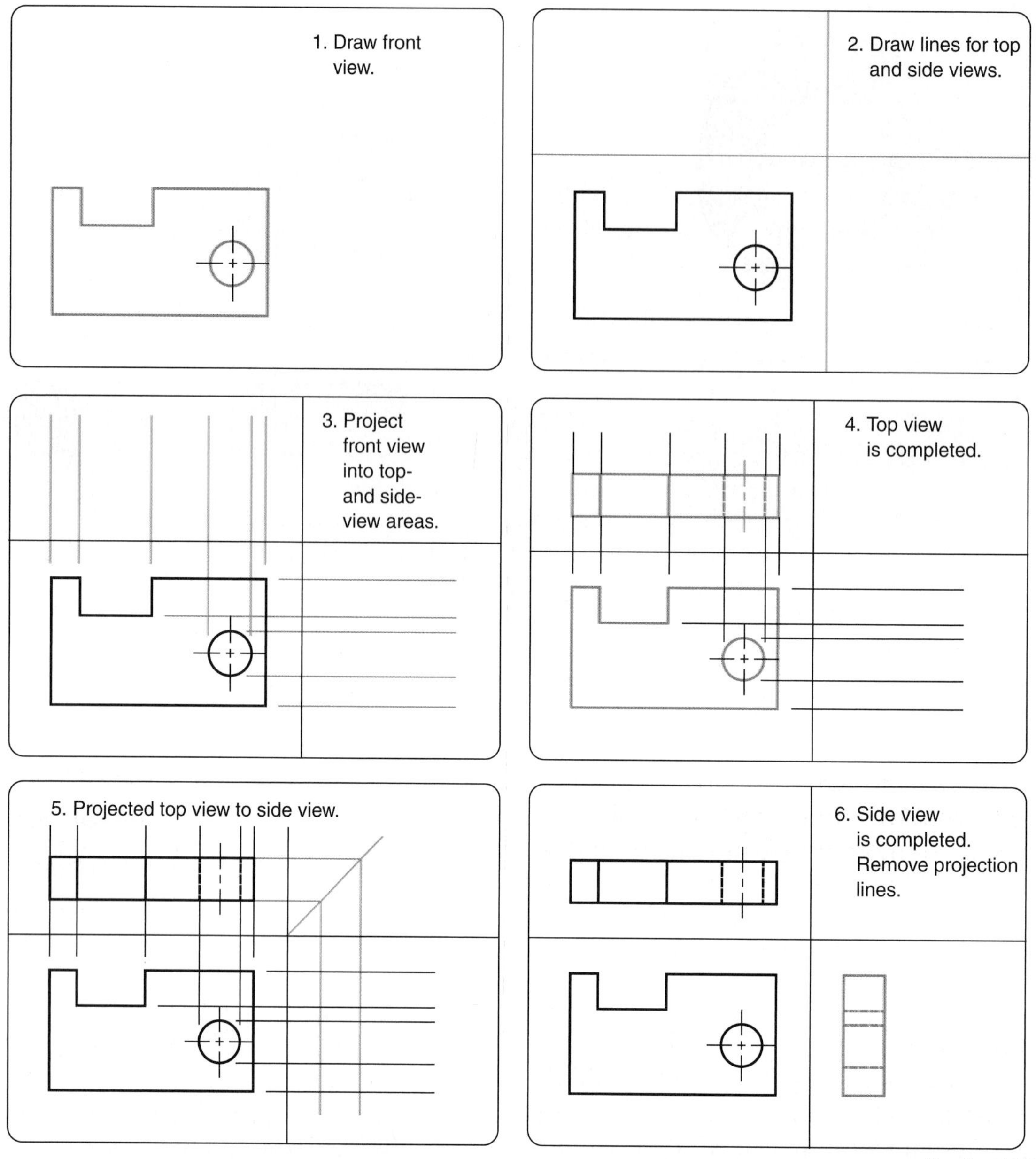

Goodheart-Willcox Publisher

Figure 5-22. Creating a three-view drawing is a multistep process.

and then down to cross the projection lines drawn earlier from the front view. The side view appears where these projection lines cross. As in Step 2, draw these lines lightly so they can be erased.

6. Complete the side view by constructing the overall shape from the front and top projections. Locate details and darken object lines in all views.

After drawing the three views of the object, dimension and extension lines are added. All these dimensions must be included on the drawings. To ensure all dimensions are included, follow this technique:

1. Dimension the size of the object first, followed by the sizes of all major features.
2. Dimension the locations of all features.
3. Indicate any necessary geometric dimensions. (Angles not indicated are assumed to be 90°.)

Dimensioning is done using two types of lines. Extension lines indicate the points at which the measurements are taken. Between the extension lines are dimension lines. Dimension lines have arrows pointing to the extension lines indicating the range of the dimension. The actual size of the dimension is shown near the center of the dimension line. These dimensions can be fractions or

STEM Connection: Mathematics

Solid Geometry

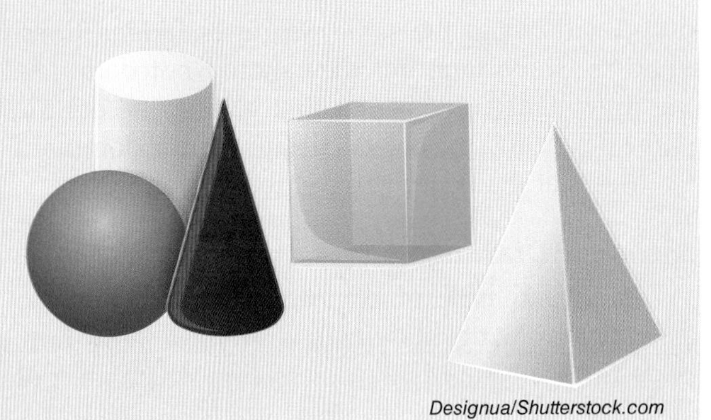
DesignuaI/Shutterstock.com

Designers should be familiar with some basic geometric concepts in order to create effective pictorial sketches. They use the concepts of solid geometry, for example, when drawing such three-dimensional images as pyramids, cones, cylinders, and cubes. A pyramid, for example, is a solid figure with a polygon as a base. A polygon is a closed plane figure bounded by three or more straight lines. (The word *polygon* comes from the prefix *poly-*, meaning "many," and the suffix *-gon*, meaning "angle." Thus, a polygon is a many-angled figure.) The faces (surfaces) of the pyramid are triangles with a common vertex, or point where they intersect. In a regular pyramid, the base is a regular polygon. The faces are *congruent* triangles, meaning they are equal in size and shape.

Other three-dimensional images share some characteristics, but differ in others. For example, in what way is a cone similar to a pyramid? In what way is it different?

decimals of an inch. In addition, a tolerance might be included with the dimension. The tolerance indicates the amount of allowed deviation in the dimension. For example, +/– (plus or minus) 1/64″ (0.4 mm) after the dimension indicates that the size can be 1/64″ larger or 1/64″ smaller than the dimension and still be acceptable.

Drafting Standards

Drafting standards, such as ANSI/ASME, have been developed so all drawings are easily understood. The standards allow each user to interpret the drawing in a similar manner. One set of essential drafting standards deals with lines and line weights, **Figure 5-23**. The shape of the object is of primary importance, so the lines outlining the object and its major details must stand out. These solid lines are called object lines and are the darkest on all drawings.

Some details are hidden in one or more of the views. Their shapes and locations, however, are important to understanding the drawing. Therefore, they are shown, but with lighter, dotted lines called hidden lines.

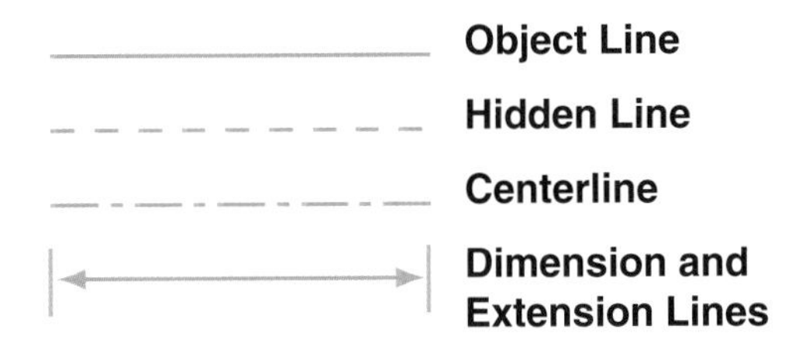

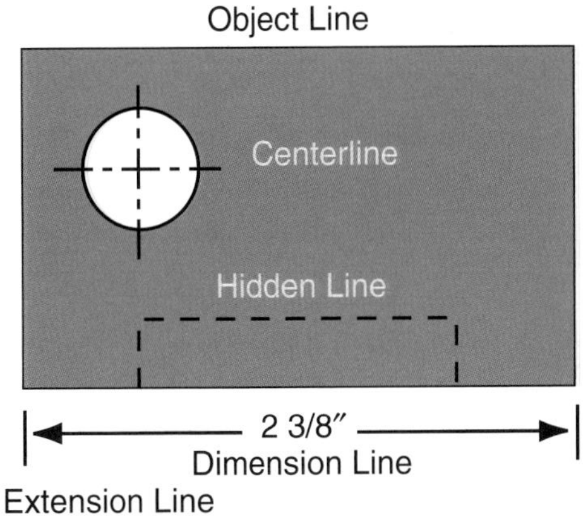

Goodheart-Willcox Publisher

Figure 5-23. The alphabet of lines shows the types and weight of lines used in drawings.

A third type of line indicates holes in the part. These lines pass through the center of the hole and are called centerlines. Centerlines are constructed of a series of light long and short dashes.

Dimension and extension lines, which were introduced earlier, are the same weight as hidden lines. They are important, but they should not dominate the drawing. Remember, at first glance, the outline of the object and the object's details are the dominant features that should jump out at the viewer.

Assembly Drawings

A second type of engineering drawing is the assembly drawing. ***Assembly drawings*** show how parts fit together to make assemblies, which are put together to make products. Two types of assembly drawings are orthographic assembly drawings and pictorial assembly drawings, **Figure 5-24.** ***Orthographic assembly drawings*** use standard orthographic views to show parts in their assembled positions. ***Pictorial assembly drawings*** show the assembly using oblique, isometric, or perspective views. Assembly drawings do not typically contain dimensions.

Each type of assembly drawing can be a standard view or an exploded view, **Figure 5-25.** Standard views are constructed using the normal techniques for constructing orthographic or pictorial drawings. These views show the product in one piece, as it will be after it is assembled. Exploded views show the parts of the object as if it was taken

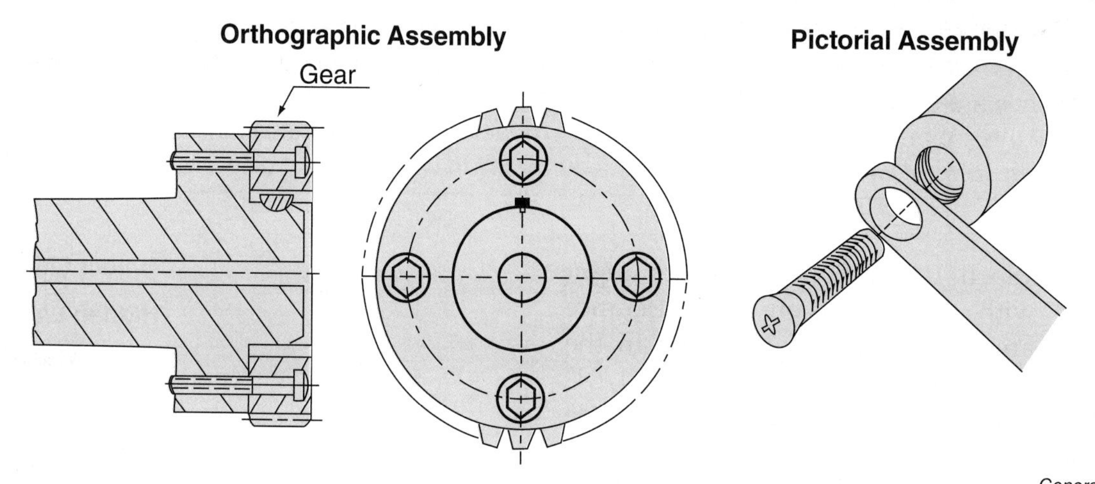

Figure 5-24. Assembly drawings can be orthographic or pictorial.

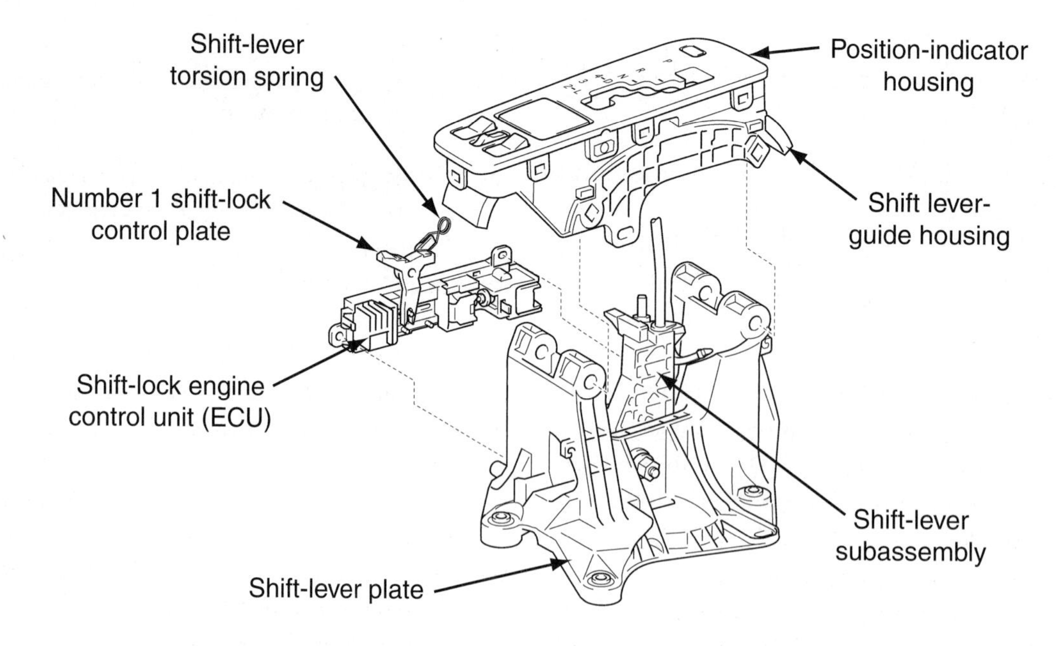

Figure 5-25. An exploded-view assembly drawing shows the parts of a product as if the product was disassembled.

apart. The parts are arranged in the proper relationships to each other on the drawing. This type of drawing is often used in repair manuals.

Systems Drawings

Systems drawings are used to show the way parts in a system relate to each other and work together. They are used for electrical, hydraulic (fluid), and pneumatic (gas) systems. They do not show the actual positions of the parts in a product. Assembly drawings do this. Systems drawings are designed to show the connections for wires, pipes, and tubes.

These drawings use symbols to represent the components. Standard symbols have been developed for electrical, pneumatic, and hydraulic parts. **Figure 5-26** includes some common symbols for electrical and electronic components.

Systems drawings are developed by first arranging the major components on the sheet. Connecting wires, pipes, and tubes are then indicated. Special drawing techniques are used to indicate whether the lines connect or simply cross each other.

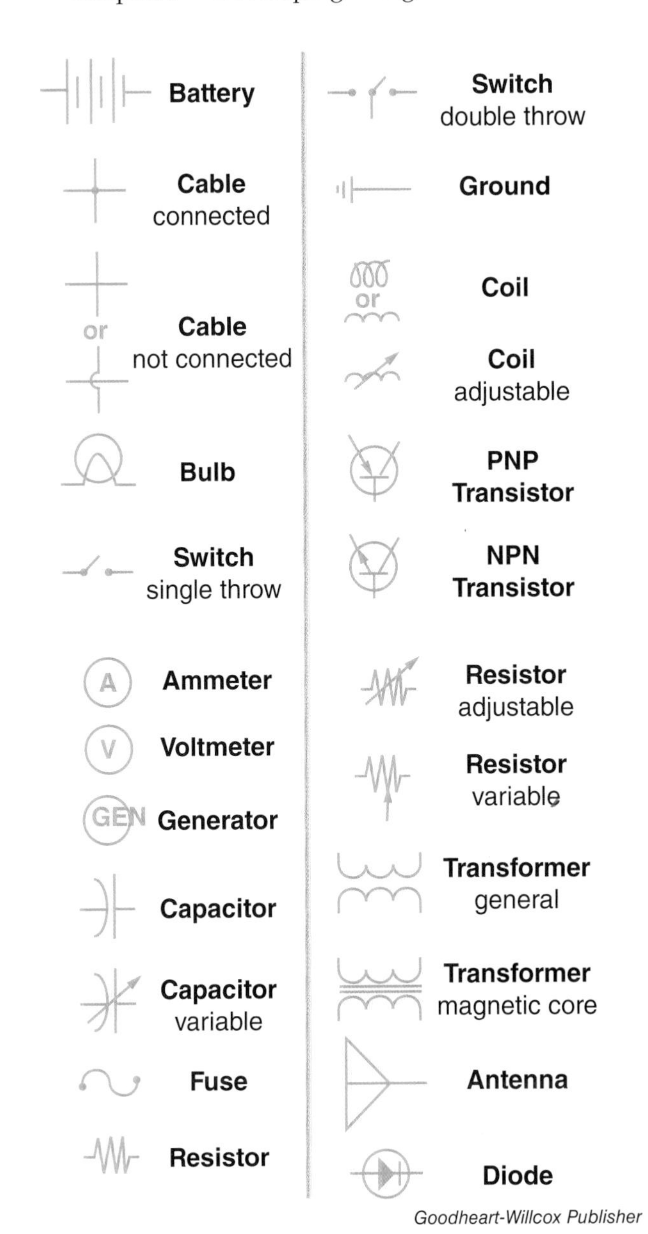

Figure 5-26. Commonly used symbols on electrical and electronic systems drawings.

Model Design Solutions

Design solutions are also modeled during this phase of the design process to determine how well the solution will perform. ***Modeling*** is a simulation of actual events, structures, or conditions. This helps engineers or designers determine the effectiveness of their design solution prior to making it, saving valuable resources. There are three types of models, **Figure 5-27**:

- Graphic/computer models.
- Mathematical models.
- Physical models.

Graphic/computer and mathematical models are generally created prior to making physical models of the solution. Physical models are the focus of Chapter 6.

Graphic/Computer Models

Designers cannot make physical models early in the product- and structure-development process. They do not have enough information about the design to construct a physical model. The designers must explore ideas for components and systems. One way to do this is by creating graphic models. Typical ***graphic/computer models*** are conceptual models, graphs, charts, and diagrams. Each of these graphic models serves a specific purpose.

Conceptual models capture the designer's ideas for specific structures and products. They show a general view of the components and their relationships, **Figure 5-28**. Conceptual models are often the first step in evaluating a design solution. Relationships and working parameters of systems and components can be studied, modified, and improved using conceptual models. For example, conceptual models for a toy train might explore ways to connect cars together, fabricate wheel-and-axle assemblies, and attach car bodies to chassis assemblies. Refined and detailed sketches can serve as conceptual models.

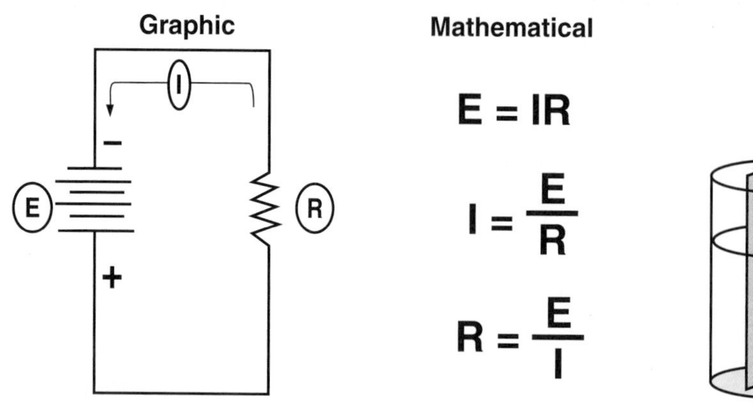

Goodheart-Willcox Publisher

Figure 5-27. Models can be graphic, mathematical, or physical.

Daimler

Figure 5-28. This conceptual model shows a designer's ideas for an automotive system.

Graphs allow designers to organize and plot data. They display numerical information that can be used to design products and assess testing results. For example, a graph can be developed showing vehicle speed and braking distance for different types of brakes. The data can be charted on a line graph. This information helps designers select the type of braking system to be used in a specific vehicle. Likewise, plotting data on the colors of shirts purchased during a specific period can help designers select colors for next year's products. This type of data can be shown on a line graph, a bar graph, or a pie graph. See **Figure 5-29**.

Charts show the relationship among people, actions, or operations. They are useful for selecting and sequencing tasks needed to complete a job. Various charts are used for specific tasks. Flowcharts help computer programmers write logical programs and help manufacturing engineers develop efficient

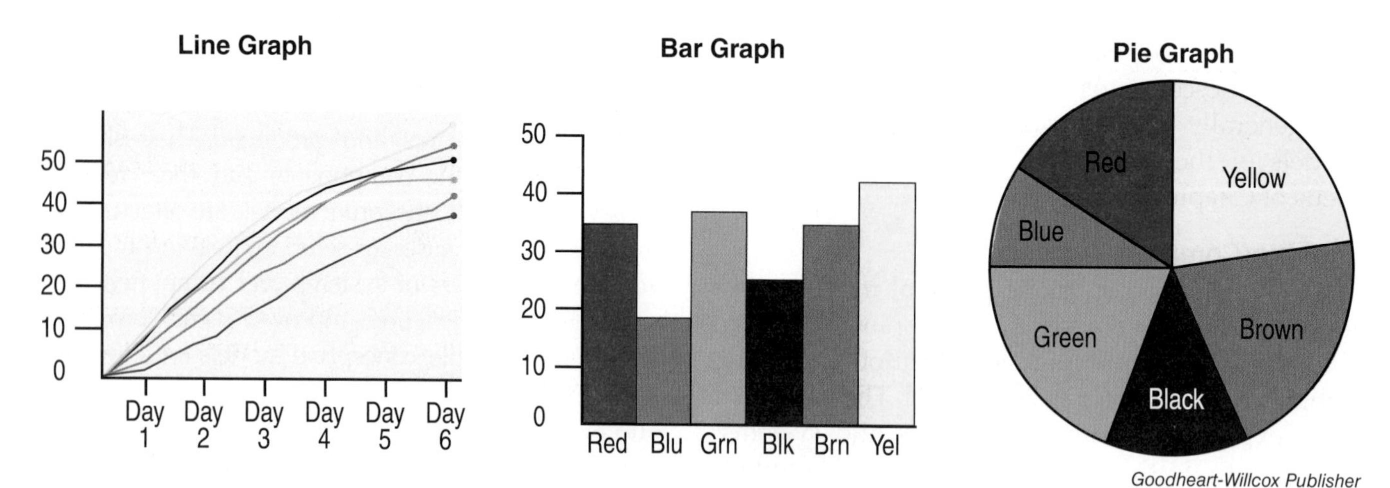

Goodheart-Willcox Publisher

Figure 5-29. Graphs show the relationship among numerical data gathered about specific factors.

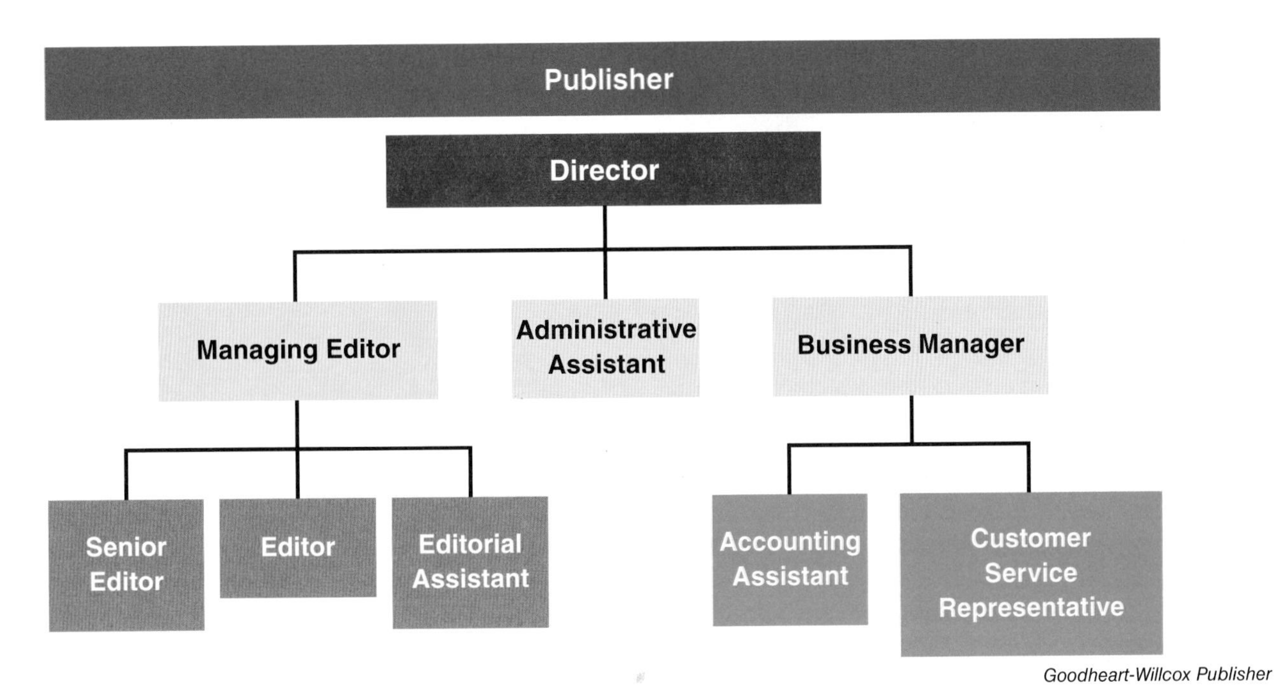

Figure 5-30. This chart shows the structure of an organization.

manufacturing plants. Organization charts show the flow of authority and responsibility within a company, school, or business, **Figure 5-30**.

Diagrams show the relationship among components in a system. A schematic diagram can be used to indicate the components in electrical, mechanical, or fluid (hydraulic or pneumatic) systems, **Figure 5-31**. Schematics do not, however, show the specific locations of parts. The relationship of the parts in the system is the information that needs to be communicated. Flow diagrams show how parts move through a manufacturing facility. Lines and arrows indicate the path the material takes as it moves from operation to operation.

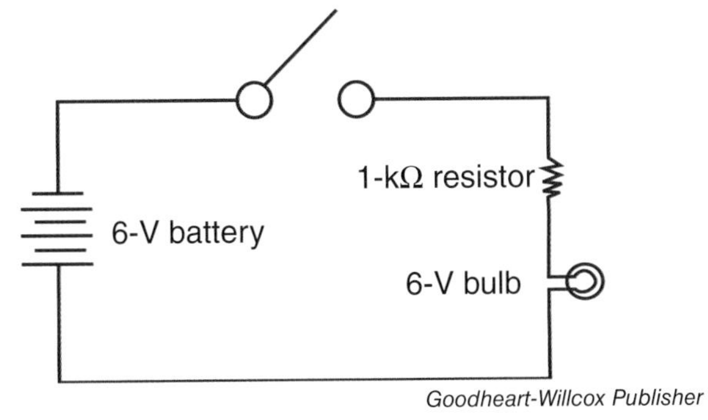

Figure 5-31. This schematic diagram shows the relationship among components in an electrical system.

Another type of diagram that many people are familiar with is a play diagram. Football coaches use play diagrams to show the players (components) how to interact during a play (system). The play diagram summarizes the purpose of all diagrams. The purpose is to show how something is designed to happen or how an event took place.

Computer Models and Simulations

Computer models are models produced by computers. They allow designers to test the strength of materials and structures, and to observe products during normal operation. Computer modeling techniques can replace one or more traditional models. In many cases, designers no longer need physical models. They can develop and analyze the structure or product with computer modeling and simulation.

Computers and computer models have affected the way we develop solutions to problems. Computers can be used to develop three types of three-dimensional models: wire frame, surface, and solid, **Figure 5-32**.

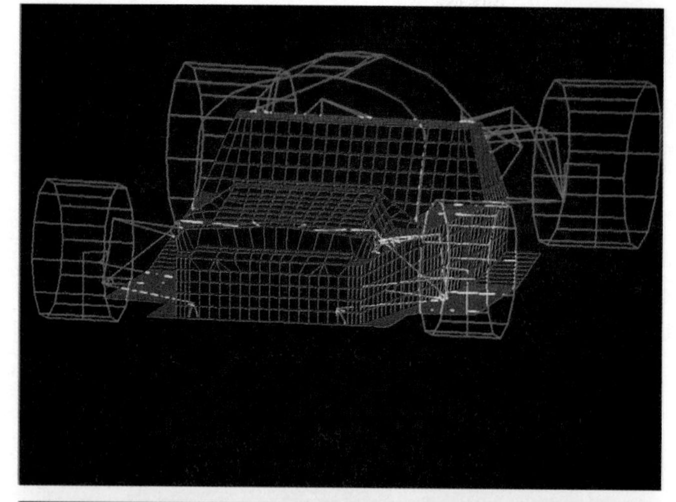

DaimlerChrysler

Figure 5-32. Computer models. Top—Wire-frame model. Bottom—Surface model.

A *wire-frame model* is developed by connecting all the edges of the object. The process produces a structure made up of straight and curved lines.

A surface model can be thought of as a wire frame with a sheet of plastic drawn over it. ***Surface models*** show how the product will appear to an observer. Surface models are used to develop sheet metal products. The surface model is first developed. The computer then unfolds the model to produce a cutting pattern for the metal.

Solid models are complex computer models that take into account both the surfaces and interior substance of object. Solid models look similar to surface models, except the computer thinks of them as solids. This allows the designer to direct the computer to cut away parts, insert bolts and valves, and rotate moving parts, for example. Solid models can also be used to establish fits for mating parts and to set up the procedures needed to assemble parts into products.

Other computer simulations allow designers to test strengths of materials and structures, **Figure 5-33**. The designers create the structure on the computer and then apply stress. They might simulate forces created by weight, operating conditions, or outside conditions, such as wind or earthquakes.

Finally, designers can use computer models to observe a product during normal operation. For example, computer programs test air circulation in buildings and vehicles, **Figure 5-34**. This information is essential for the designers of heating and cooling systems.

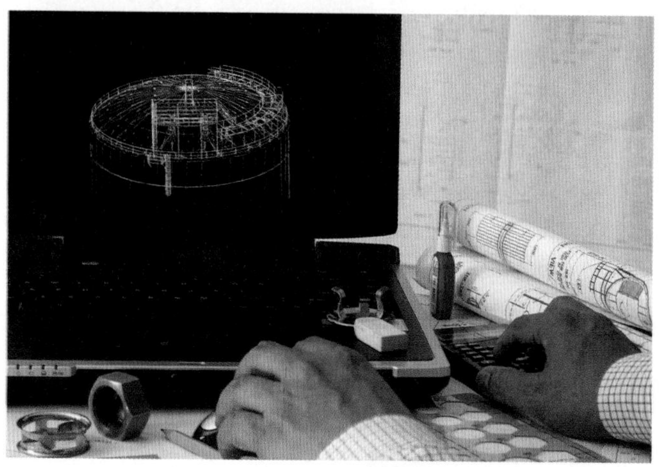

RAGMA IMAGES/Shutterstock.com

Figure 5-33. This computer simulation allows the designer to test a material's strength.

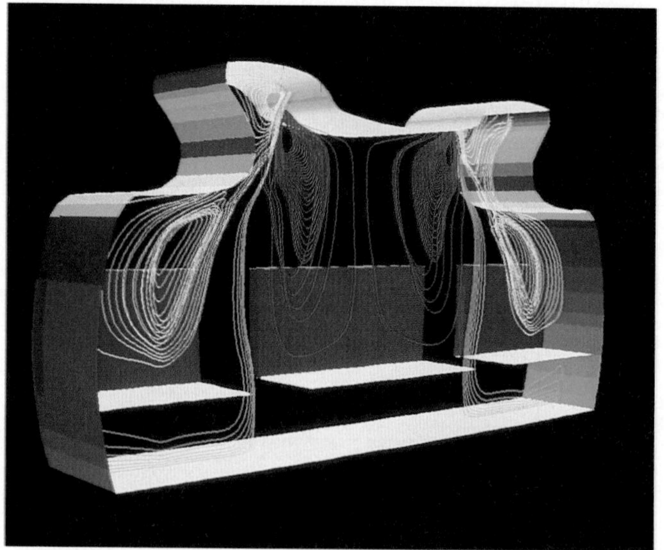

Boeing

Figure 5-34. This computer program shows a simulation of air circulation in an aircraft interior. Note the seat (purple) and the shape of the overhead bins.

Predictive Analysis

Predictive analysis is a method used to solve problems and test initial ideas through the application of mathematical or scientific concepts, often through the use of computers, before actually building models, prototypes, or final designs. This is an essential part of the engineering design process and necessary for developing solutions. It allows engineers to increase the effectiveness of the design, reduce errors, and increase safety. Engineers apply knowledge of scientific and mathematical principles prior to developing a final design solution.

Mathematical Models

Mathematical models show relationships through formulas. For example, the relationship between the force needed to move an object and the distance the object is lifted is shown in the formula for work. This formula is work (w) = force (f) × distance (d). The formulas used to explain chemical reactions are also mathematical models. The following formula shows the chemical reactions that take place as an automotive lead-acid storage battery is charged and discharged:

$$PbO_2 + Pb + 2H_2SO_4 \underset{\text{charge}}{\overset{\text{discharge}}{\rightleftarrows}} 2PbSO_4 + 2H_2O$$

Many more complex models with thousands of formulas are used to predict the results of complex relationships. These formulas might predict economic growth for a period of time, track storms and spaceflights, predict ocean currents and land erosion, or help scientists conduct complex experiments.

Scientific Models

Scientific models are simulations based on existing or accepted findings related to the natural world. They are made before creating a physical model or prototype. Models can be created when it is difficult or impossible to create true conditions that an engineer can measure. While getting measurements from an actual experiment is often the best method, it is not always the most practical method. Simulations can provide data on material reaction to the natural environment over time, without losing money and time waiting for the actual results. This data can then be used to inform designers as they develop and evaluate solutions.

Optimization

Predictive analysis enables the next task: developing a design solution, or optimization. Though optimization can be present in other stages of designing, it is a key segue between developing and making design solutions. Engineers often use decision matrices to choose and justify a solution, as discussed in Chapter 4. Decision matrices allow engineers to list multiple options in a table and list criteria and factors in the columns. A score is then given for each design option and total scores can be calculated. Factors often used include strength, efficiency, cost of production, and ergonomics.

Develop a Production Procedure

Developing a production procedure involves outlining the necessary steps for making or producing a physical model or prototype of a design. A physical model or prototype is needed to test functionality or to see the aesthetic properties of a design. Recording the necessary steps also allows designers to identify and obtain the proper tools, equipment, and materials needed for production. The list of needed materials is called a *bill of materials*.

Summary

- ✔ Design solution development includes identifying and defining a problem, gathering information, and developing solutions.
- ✔ As technology advanced, the single-person approach to design was replaced with design teams.
- ✔ Design teams allow a variety of people to contribute to a project, creating a stronger project.
- ✔ Defining a design problem begins by describing it in general terms; writing a clear, concise problem statement; and establishing criteria and constraints.
- ✔ Gathering information requires researching current solutions, which allows engineers to locate and analyze the knowledge needed to solve a problem.
- ✔ Designers can obtain information through historical-, descriptive-, or experimental-research activities.
- ✔ Steps for developing design solutions include developing preliminary solutions, isolating and refining the best solution, detailing the best solution, computer modeling of the best solution, applying predictive analysis, optimization, and developing a production procedure.
- ✔ As steps are completed, the problem may be redefined or refined and the steps repeated until a final design is developed.
- ✔ Preliminary solutions can be generated through brainstorming, classification, and what-if scenarios.
- ✔ A common way to record ideas generated is by developing rough sketches of them.
- ✔ Isolating and refining design ideas is the second step in developing a design solution.
- ✔ A refined sketch merges ideas from two or more rough sketches.
- ✔ Detailed sketches communicate the information needed to build a model of the product or structure.
- ✔ Building models requires information about the overall dimensions of the object, the location of features within the object, and geometric information about features of the object.
- ✔ Designers use oblique, isometric, and perspective sketches when refining ideas.
- ✔ Most detail drawings are presented as multiview drawings. The number of views depends on the complexity of the object.
- ✔ Modeling is a simulation of actual events, structures, or conditions.
- ✔ Types of models include graphic/computer models, mathematical models, and physical models.
- ✔ Predictive analysis is used to solve problems and test initial ideas by applying mathematical or scientific concepts.
- ✔ Optimization is a process in which the ideal solution is created based on the specifications and constraints of the design.
- ✔ Developing a production procedure involves outlining the necessary steps for making or producing a physical model or prototype of a design.

Check Your Engineering IQ 

Now that you have finished this chapter, see what you learned by taking the chapter posttest.
www.g-wlearning.com/technologyeducation/

Test Your Knowledge

Answer the following end-of-chapter questions using the information provided in this chapter.

1. What is the advantage of using a design team versus a single-person approach to design?
2. List the steps followed in developing design solutions.
3. _____ thinking seeks to imagine as many different solutions as possible. _____ thinking narrows and focuses ideas until the most feasible solution is found.
4. *True or False?* The best solution is always the least expensive one.
5. Name one of the rules used for effective brainstorming.
6. Define the term *classification* as it is used in developing design solutions.
7. What is a *what-if scenario*?
8. _____ sketches are incomplete, unrefined sketches used to communicate design solutions.
   A. Detailed
   B. Unified
   C. Rough
   D. None of the above.

9. Which of the following statements regarding refined sketches is *not* true?
   A. They merge ideas from two or more rough sketches.
   B. The shape may come from one sketch and details from another sketch.
   C. Different ideas are blended into a unified whole.
   D. They are exactly like the original rough sketches.
10. _____ sketches communicate the information needed to build a model of the product or structure.
11. List the three types of information required in detailed sketches when building models.
12. What are the primary differences among one-point, two-point, and three-point perspective sketches?
13. _____ drawings include one or more views of the object in one drawing in order to fully communicate the shape and size of that object.
14. _____ lines indicate the points at which the measurements are taken.
   A. Extension
   B. Dimension
   C. Tolerance
   D. None of the above.

Matching. Match each description with the type of pictorial sketch being defined. Letters will be used more than once.

A. Oblique
B. Isometric
C. Perspective

15. Used when the top, sides, and front are equally important.
16. Shows the object as the human eye sees it.
17. The angles formed by the lines in the upper-right corner are equal.
18. The front view appears as if a person is looking directly at the object.
19. Parallel lines meet at distant vantage point.
20. Always has one corner made up of three 120° angles.
21. The two types are cavalier and cabinet.
22. Work (*w*) = force (*f*) × distance (*d*) is an example of a(n) _____ model.
23. Scientific modeling _____.
   A. is an activity in which engineers create simulations based on existing or accepted findings related to the natural world
   B. can be used when it is difficult or impossible to create true conditions that can be measured
   C. can provide data on material reaction to the natural environment over time, which can then be used to develop and evaluate solutions
   D. All of the above.
24. _____ is a process in which the ideal solution is created based on the specifications and constraints of the design.

Critical Thinking

1. Explain the purpose of predictive analysis and why it should be used instead of trial-and-error design approaches.
2. Consider ways to use predictive analysis to solve a specific design problem.
3. Explain the importance of working through multiple phases of drawings when creating a design.
4. Why is optimization necessary? Why not pursue every possible idea when solving a design problem?

STEM Applications

1. Develop a set of rough sketches for the following design problem.
   The director of the school cafeteria would like you to design a lunchroom-table organizer. It should be able to hold one saltshaker, one pepper shaker, 20 rectangular (1″ × 1 1/2″) sugar packets, and one bottle of ketchup. The holder should be easy to remove from the table.
   A. Refine the best sketch you produce for the lunchroom-table organizer.
   B. Develop a detailed sketch for your lunchroom-table organizer.
2. Select a device in the technology laboratory and develop an oblique, isometric, and perspective sketch of it.

Engineering Design Challenge

Design Problem

Background

All technological products are the results of design efforts by people. People define problems, think up many solutions, and model the solutions selected. Finally, designers communicate the designs to production personnel.

Situation

Road Games Inc. is a company specializing in designing small portable games. You have been recently hired in the creative-concepts department of the company.

Challenge

Design a portable game using pegs and a 3/4″ × 3 1/2″ × 3 1/2″ wood board. Dice can be used in the game, but they are not required. Your boss expects you to produce rough sketches for five different ideas, a refined sketch for the best idea, a prototype, and a detailed drawing of the game board.

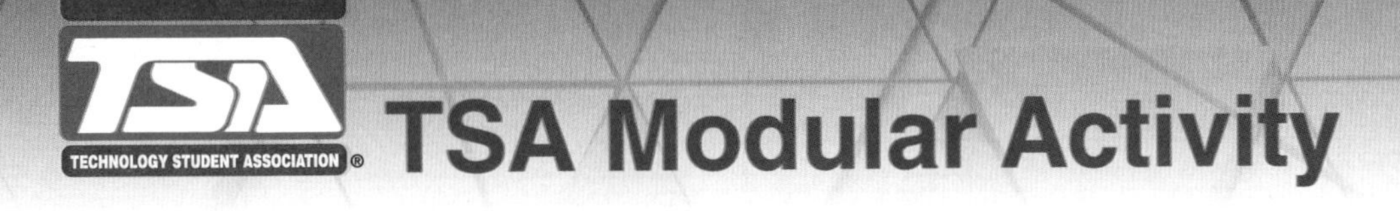

This activity develops the skills used in TSA's Computer-Aided Design (CAD) 3-D, Engineering event.

Computer-Aided Design (CAD), Engineering with Animation

Activity Overview

In this activity, you will create an animation illustrating the assembly sequence for a product.

Materials

- Paper.
- Pencil.
- Computer with CAD and animation software.

Background Information

- **Product selection.** Use brainstorming techniques to develop a list of possible products to model in your animation. Part modeling is easier if you have a sample of the product to use for measurements or if you have actual part drawings. Some possible items to animate include:
    - A piece of self-assembled furniture, such as a computer desk.
    - A ballpoint pen or mechanical pencil.
    - A small construction project, such as a doghouse or shed.
    - Sports equipment, such as a swing set, tennis racket, or weight bench.
    - A mechanical device, such as a wheel-and-axle assembly.
- **Part modeling.** After selecting the object, create models for each part.
- **Animation.** Your animation is intended to illustrate the assembly sequence. You can begin the illustration with all parts shown or you can have the parts appear as they are assembled. To help the viewer anticipate the action, highlight a part before it is moved into place in the assembly. You can highlight the part in several ways, such as changing its color, outlining it, momentarily enlarging the part, or momentarily stretching the part.
- **Output.** Output the animation into a video file. Using a medium color depth (such as 16-bit) and low resolution (approximately 320 × 200 pixels) speeds the processing of the animation.

Guidelines

- Your assembly must include at least five components. There must be at least five steps in the assembly procedure.
- Your animation should clearly show the assembly procedure.
- Your animation must be at least 10 seconds long.

Evaluation Criteria

Your project will be evaluated using the following criteria:

- Originality and creativity in design.
- Effectiveness of animation in illustrating the assembly sequence.
- Use of animation features.
- Technical animation skills.
- Smoothness and timing of animation.

CHAPTER 6

Making Design Solutions

Check Your Engineering IQ 

Before you read this chapter, assess your current understanding of the chapter content by taking the chapter pretest.
www.g-wlearning.com/technologyeducation/

Learning Objectives

After studying this chapter, you should be able to:

- ✔ Explain how physical models are used in the design of technological products and systems.
- ✔ Recall the categories of production tools.
- ✔ Describe the characteristics and types of machine tools.
- ✔ Contrast the US customary and the metric measurement systems.
- ✔ Differentiate between standard and precision measurements.
- ✔ Describe how common measuring tools are used to measure linear distances, diameters, and angles.
- ✔ Explain the relationship between measurement and quality control.

Technical Terms 

arbors
band saws
broach
chucks
circular saws
computer
cutting motion
cylindrical grinder
direct-reading measurement tools
drilling machines
feed motion
Forstner bits
grinding machines
indirect-reading measurement tools
lathes
linear motion
machine tools
material processing
measurement
metric system
micrometer
mock-up
planing machines
precision measurement
prototype
quality control
reciprocating motion
rotary motion
rule
sawing machines
scale model
scroll saws
shaping machines
spade bits
square
standard measurement
surface grinder
turning machines
twist drills
US customary system

Close Reading

As you read this chapter, make an outline of the various types of production and measurement tools and note how each can play a role in producing a physical model of a design solution.

This chapter describes the actions that take place during the *Model and Make a Solution* step of the engineering design process. See **Figure 6-1**.

We use the technological products of engineering every day. Each product started with a problem that could be defined. Engineers or designers were then challenged to solve this problem by developing a design solution. They explored various solutions that began with ideas that were recorded on paper and/or digitally. The engineers or designers developed rough, refined, and detailed technical drawings of their design ideas. At some point, they changed their detailed drawings into three-dimensional (3-D) computer models for the purpose of predictive analysis using modeling and mathematical software. See **Figure 6-2**.

Following the predictive design analysis, the optimal design often needs to be converted from a technical drawing or 3-D computer model to a physical model for further testing and evaluation. This conversion involves the following:

- Physical modeling of design solutions.
- Use of tools and machines to produce quality physical models.

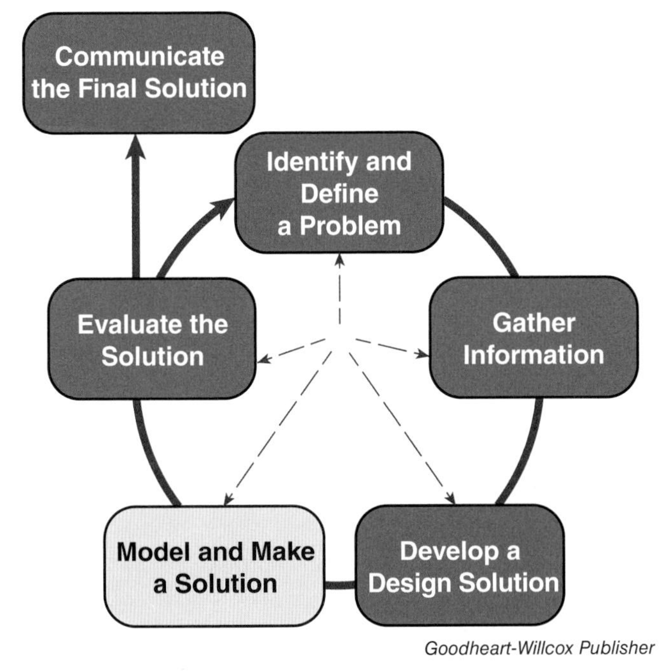

Figure 6-1. Designs are modeled in the *Model and Make a Solution* step of the engineering design process.

Sketches

Ford

Models

Ford

Figure 6-2. The design process changes product sketches into models that can be evaluated.

- Accurate measurement and use of tools.
- Experimenting with materials and tools for optimization.

Physically Modeling Design Solutions

Everyone is familiar with models. Children play with dolls, model cars, and toy trains. They build towns out of wooden blocks, convert cardboard boxes into frontier forts, and have grand prix races with their tricycles. These children pretend they are dealing with real-life situations. Models are used in classrooms to help students learn about a variety of structures and systems. See **Figure 6-3**.

This activity of imitating reality is widely used in product and system design. Simply stated, modeling is the simulation of actual events, structures, or conditions. Modeling allows people to simulate expected conditions in order to test their design ideas. For example, architects build models of buildings to show clients how they will look. They might also use structural models to test a building's ability to withstand an earthquake or wind forces. Economists devise models to predict how the economy will react to certain conditions. A model reduces complex mechanisms and events into a more easily understandable form.

Models allow designers and engineers to focus on important parts of the total problem and build understanding one part at a time. For example, an automobile is a complex artifact that is almost impossible to study and understand as a whole. If you look at and understand the systems that make up the automobile one at a time, however, the whole becomes clear. For example, you can model and study its power train, cooling system, electrical system, lubricating system, or suspension system. See **Figure 6-4**.

Rawpixel.com/Shutterstock.com

Figure 6-3. Students study models to learn about structures and systems.

DaimlerChrysler

Figure 6-4. This graphic model shows an automotive torsion bar rear suspension system.

Physical Models

Although computer modeling techniques can replace physical models, there is often a need to physically model design solutions. The three traditional types of physical models are mock-ups, scale models, and prototypes.

Mock-Ups

A physical model that represents only how a product or structure will look is called an *appearance model*, or ***mock-up***. This type of model is used to evaluate the styling, balance, color, and other aesthetic features of a technological artifact. See **Figure 6-5**. Mock-ups are not used to analyze the behaviors or properties of a design because they are not made of materials that enable the model to be fully functional.

Mock-ups are generally constructed of materials that are easy to work with. These materials include wood, clay, Styrofoam™ product, paper, and paperboard (such as cardboard and poster board). See **Figure 6-6**. A mock-up can be built either to scale or full-size.

Scale Models

Sometimes creating full-size physical models is impractical, such as building a full-size model of a new skyscraper. In engineering and technology, a scale model is built when the product or structure is too large or expensive to construct in full size for testing purposes. A ***scale model*** is a three-dimensional miniature physical representation of a design solution that maintains accurate relationships between all aspects and features of the design. Scale models are also built with materials that provide form and shape, which preserves the physical properties of the full-size design.

Goodheart-Willcox Publisher

Figure 6-5. This mock-up shows the relationship among the different buildings.

Hasbro, Inc.

Figure 6-6. These designers are developing mock-ups of new toys.

A scale model is proportional to actual size. This means the model's size is related to actual size by a ratio. A ratio of 1 to 4, written as *1:4*, means that one unit on the model equals four units in actual size. In other words, the model is a quarter of the size of the actual product. A ratio of 4 to 1 (4:1) would indicate that the model is four times the size of the actual product.

A scale model of a new building shows clients how the structure will look, how it will fit on the site, and how it will be landscaped. The engineer and other interested people can observe, handle, analyze, and sometimes test the functionality of these models. Scale models enable the demonstration of behaviors or properties of a design.

Prototypes

A ***prototype*** is a full-size and fully functional model of a system, assembly, or product. Prototypes are built to test the operation, maintenance, or safety of the design. See **Figure 6-7**. They are generally built of the same material as the final product. In some cases, however, substitute materials are used. For example, some automobile manufacturers have found that a specific plastic reacts to external forces in the same manner as steel does. Plastic is used in place of steel in prototypes because it is easier to fabricate.

Production of Physical Models

Producing accurate and high-quality physical models of a design solution requires the knowledgeable use of tools and machines. A focus on quality is crucial when making physical models. Physical models should be created as accurately and precisely as possible using materials that maintain the desired properties of the design solution. Otherwise, testing a physical model will not provide reliable results.

In order to make quality physical models, an understanding of industry-quality tools and

Career Connection Machine Technician

A machine technician sets up, operates, maintains, and repairs machinery. These technicians also monitor the operation of machines and equipment, as well as inform others how to properly use and maintain them. A machine technician can choose a specialization area based on interests and skills.

High school students interested in a machine technician career should take a variety of technology and engineering courses. Additionally, they should seek out career or technical academies or apprentice programs focused on manufacturing or engineering technology.

Some machine technician positions are available through apprenticeship programs. However, attending a community or technical college can result in a wide range of opportunities to become a higher paid technician. Technicians with only a high school diploma typically earn a lower pay rate.

People interested in a machine technician career should have mechanical aptitude and be comfortable with using machines, tools, and equipment. They should also know how to read and interpret all types of engineering drawings. Additionally, they should be good troubleshooters and problem solvers.

As machines become more sophisticated and complex, it is expected that additional machine technician jobs will be created. Machine technicians can advance by obtaining additional education or certifications.

A machine technician may work on the following types of projects:

- Designing and setting up new machinery.
- Diagnosing machine malfunctions.
- Recommending machines for different purposes and operations.
- Testing machinery and equipment.
- Establishing and enforcing safety guidelines.

DaimlerChrysler

Figure 6-7. This prototype is being used to test the aerodynamics of an aircraft.

machines is required. A discussion of the thousands of different tools and machines used today is beyond the scope of this textbook. Three major types of processing tools are material-, energy-, and information-processing tools. This chapter will focus specifically on material-processing tools. Energy- and information-processing tools will be discussed in subsequent chapters.

Understanding tools and machines involves knowing not only their particular features but also how they should be handled. Before working with any tool or machine, you must complete specific safety training and follow formal guidelines such as:

- Use only sharp tools and well-maintained machines.
- Return all tools and machine accessories to their proper places.
- Use the right tool or machine for the right job.
- Do not use any tool or machine without permission or proper instruction.

Material-Processing Tools and Machines

The world around us is full of artifacts. Each was made using material-processing tools and machines. The term ***material processing*** refers to changing the form of materials. Tools and machines are used to cast, form, and machine materials into specific shapes. They also help assemble products and apply protective coatings.

Fundamental to all material processing is a group of tools called ***machine tools***, which are the machines used to cut or shape rigid materials such as metal or wood. The following sections explain machine tool characteristics and types.

Characteristics of Machine Tools

As shown in **Figure 6-8**, machine tools have the following common characteristics:

- A method of cutting materials to produce the desired size and shape. The new size and shape are achieved with a cutting tool.
- A series of motions between the material and the tool. This movement causes the tool to cut the material.
- Support of the tool and the workpiece (the material being machined).

Method of Cutting

Most cutting actions require a cutting tool. Cutting tools come in many sizes and shapes, but they must meet certain requirements. First, they must be harder than the material they are cutting. For example, a diamond cuts steel, steel cuts wood, and wood cuts butter. Second, the cutting tool must have the proper shape. That is, it needs a sharpened edge and relief angles. See **Figure 6-9**. The sharpened edge allows the tool to cut into the material. The relief angles keep the sides of the tool from rubbing against the material as the material is cut. The rake angle also helps to create a chip as the material is cut. This action allows waste material to be carried away efficiently.

Series of Motions

Movement between the tool and the workpiece must occur before cutting can take place. All cutting operations have two basic motions. These motions are cutting motion and feed motion. ***Cutting motion*** is the action causing material to be removed from the work. This motion causes the excess material to be cut away. ***Feed motion*** is the action bringing new material into the cutter. This motion allows the cutting action to be continuous. See **Figure 6-10**.

To understand these movements, consider a band saw. Imagine a piece of wood placed against

Elements of Machine Tools

Cutting Tool
Single Point
Multiple Point

Motion
Rotary
Reciprocating
Linear

Support
Tool
Work
Rotary
Work
Tool
Linear and Reciprocating

Nagy-Bagoly Arpad/Shutterstock.com

Figure 6-8. All machine tools have three basic elements—a cutting tool, a method of moving the tool or workpiece, and a method of supporting the tool and the workpiece.

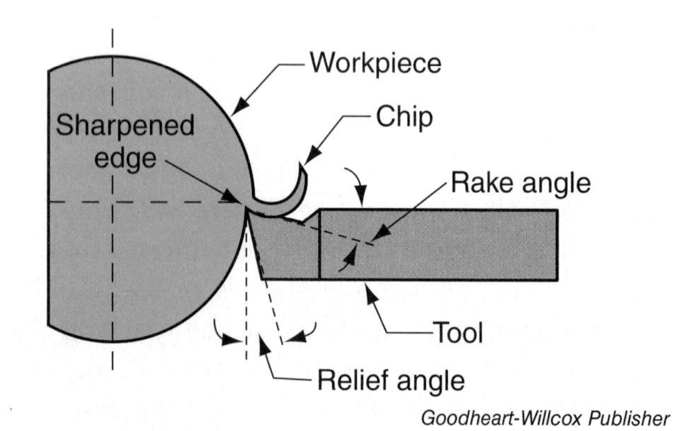

Goodheart-Willcox Publisher

Figure 6-9. Cutting tools have a sharpened edge, a relief angle, and a rake angle.

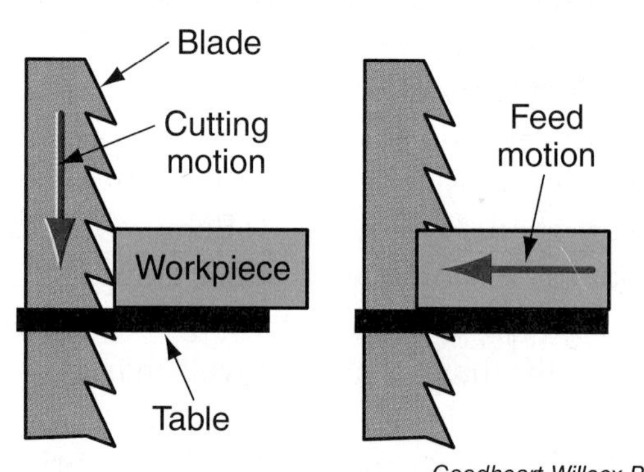

Goodheart-Willcox Publisher

Figure 6-10. All cutting operations include cutting motion and feed motion.

the blade. When the machine is turned on, the first tooth cuts the material. This is the cutting motion. The next tooth follows the path of the first tooth, but does not produce a chip. The first tooth has already removed the material in the kerf (the slot a blade cut). The cutting action continues only if the wood is pushed into the blade. This makes new material available for the next tooth to cut. Moving the wood is the feed motion.

Cutting and feed motions can be rotary, linear, or reciprocating. ***Rotary motion*** uses round cutters or spins the work around an axis. ***Linear motion*** moves a cutter or work in one direction along a straight line. ***Reciprocating motion*** moves the tool or the work back and forth. Refer to **Figure 6-8** for a diagram of these three types of motion.

Support of the Tool

The types of cutting and feed motions determine the support system needed for the tool and the workpiece. Rotary motion requires a holder revolving around an axis. ***Chucks*** are attachments used to hold and rotate drills and router bits. ***Arbors*** are spindles, or shafts, used to hold table-saw blades and milling cutters. Parts are placed between centers on lathes and cylindrical grinders.

Linear motion is produced in several ways. A tool can be clamped or held on a rest and then be moved in a straight line. This practice is common for wood and metal lathes and hand wood planes. Band saw blades travel around two or three wheels to produce a linear cutting motion. Material can be pushed through a saw blade while the machine table supports it.

Reciprocating motion is common with scroll saws and hacksaws. The blade is clamped at both ends into the machine and then moved back and forth. Likewise, a workpiece can be clamped to a table reciprocating under or across a cutter. This action is used in a bench grinder. The operator moves a workpiece back and forth across the face of a grinding wheel. This action is used to sharpen hand tools, lawn mower blades, and kitchen knives.

Types of Machine Tools

Hundreds of different machine tools exist. All machine tools can be grouped into six categories. See **Figure 6-11**. These categories are as follows:

- Turning machines.
- Sawing machines.
- Drilling machines.
- Shaping machines.
- Planing machines.
- Grinding machines.

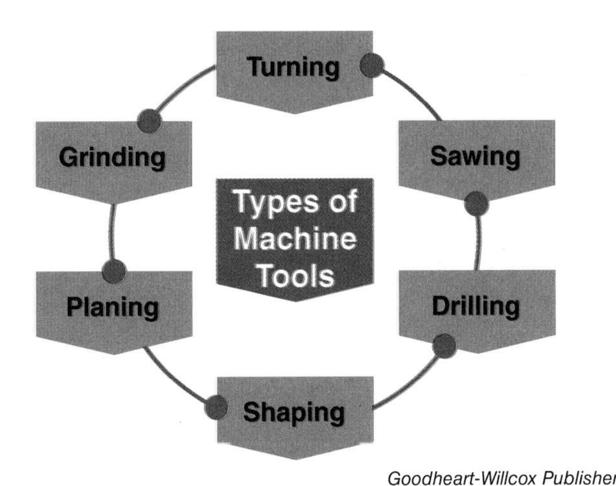

Goodheart-Willcox Publisher

Figure 6-11. The six categories of machine tools.

Turning Machines

Turning machines use a process in which a workpiece is held and rotated on an axis. This action is produced on machines called ***lathes***, **Figure 6-12**. All lathes produce cutting motion by rotating the workpiece. Linear movement of the tool generates the feed motion.

Lathes are primarily used to machine wood and metal; however, plastics can be machined on lathes as well. Wood and metal lathes contain four main parts. A headstock, which contains the machine's power unit, is the heart of the lathe. The power unit has systems to rotate the workpiece and adjust the speed of rotation. At the opposite end of the machine is a tailstock, which supports the opposite end of the part that is gripped at the headstock. The headstock and tailstock are attached to the bed of the lathe. Finally, a tool rest or toolholder is provided to support the tool. Tool rests on metal lathes clamp the tool in position and feed it into or along the work. Wood lathes commonly have a flat tool rest, along which the operator moves the tool by hand.

Turning Machines

Vereshchagin Dmitry/Shutterstock.com

Metal Lathe

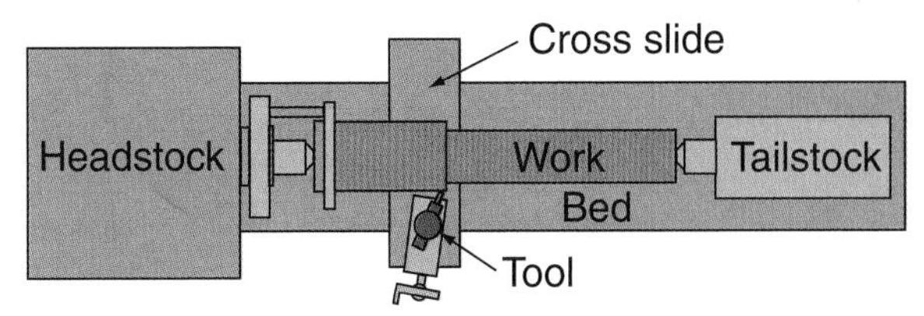

Cutting Motion: Rotating Work
Feed Motion: Linear Tool Movement

Goodheart-Willcox Publisher

Figure 6-12. Lathes are used for turning operations.

The work can be held in a lathe in two basic ways. Work can be placed between centers. One center is in the headstock, and the other is in the tailstock. These centers support the workpiece and can be of two types. Live centers rotate with the workpiece, whereas dead centers are fixed as the work rotates around them.

Both wood lathes and metal lathes have centers, but they differ in the way the turning force is applied to the workpiece. Wood lathes apply the force through the headstock using a spur center (a center with cross-shaped extensions that seat into the wood). Metal lathes use a device called a *dog* to rotate the workpiece. The dog clamps onto the work and has a finger that engages the faceplate attached to the drive spindle.

A small workpiece can be rotated using only the headstock in the device. The headstock is called a *chuck*. See **Figure 6-13**. Three-jaw, four-jaw, or collet chucks are commonly used to grip and rotate the part. Chucks are found on metal lathes. These chucks have jaws that squeeze and hold the part.

Goodheart-Willcox Publisher

Figure 6-13. Chucks can grip and rotate parts.

Lathes are used to perform many operations, **Figure 6-14**. In addition, drills and reamers can be placed in the tailstock to produce and finish holes in the workpiece. The most common lathe operations are as follows:

- **Turning.** Cutting along the length of a workpiece. Turning produces a cylinder of uniform diameter.
- **Tapering.** Cutting along the length of a cylinder at a slight angle to produce a cylindrical shape with a uniformly decreasing diameter.
- **Facing.** Cutting across the end of a rotating workpiece. Facing produces a true (or square) end on the workpiece.
- **Grooving (shouldering).** Cutting into a workpiece to produce a channel with a diameter less than the main diameter of the workpiece.
- **Chamfering.** Cutting an angled surface between two diameters on the workpiece.
- **Parting.** Cutting off a part from the main workpiece.
- **Threading.** Cutting threads along the outside diameter or inside a hole in the workpiece.
- **Knurling.** Producing a diamond pattern of grooves on the outside diameter of a portion of the workpiece. Knurling produces a surface that is easier to grip and turn.

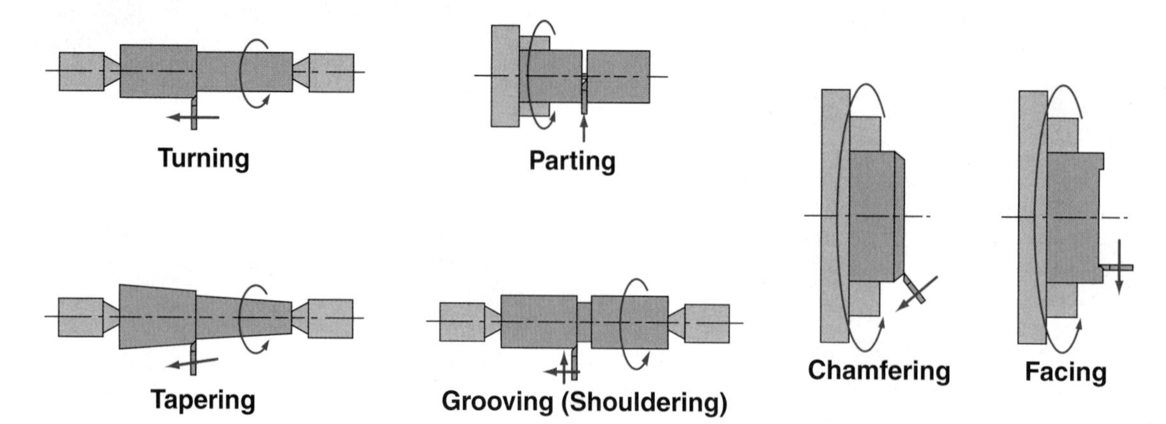

Goodheart-Willcox Publisher

Figure 6-14. Common lathe operations.

Think Green Recycling

Recycling is the reprocessing of used materials to help create new objects. Keep recycling in mind when making physical models of design solutions. Any wasted materials should be recycled, if possible. Many materials can be recycled, including the following:

- Most types of glass.
- Aluminum.
- Most plastics.
- Paper/cardboard.

When possible, use recycled or recyclable materials to make models. Also, consider materials that cannot be recycled, or are difficult to recycle, such as cell phone batteries, Styrofoam™, or certain lightbulbs.

Recycling greatly benefits the environment by reducing the use of raw materials. This helps reduce the consumption of exhaustible resources, which are materials that cannot be replaced once they have been consumed. These materials include coal, natural gas, copper, and nickel. Recycling saves energy, because raw materials take more energy to process than recycled materials. Recycling also allows these materials to be reused and not consumed as much.

Recycling also reduces the amount of waste collecting in landfills. Some of the materials that can be recycled (such as glass bottles, aluminum cans, and plastic packaging) do not decompose well, causing landfills to grow.

Many communities participate in recycling programs. Learn what materials are accepted for recycling in your area. When you go to the store, look at the packaging of the products you intend to buy. Many types of packaging contain a percentage of recycled material.

Sawing Machines

Sawing machines use teeth on a blade to cut material to a desired size and shape. These machines are designed to perform a number of different cutting actions, **Figure 6-15**. The actions include the following:

- **Crosscutting (or cutoff).** Reducing the length of a material.
- **Ripping (or edging).** Reducing the width of a material.
- **Resawing.** Reducing the thickness of a material.
- **Grooving, dadoing, and notching.** Cutting rectangular slots in or across a part.
- **Chamfering and beveling.** Cutting an angled surface between two primary surfaces of a material.

Sawing machines can be grouped according to the type of blade and the methods used to produce the cutting action. See **Figure 6-16**. This grouping identifies three basic types of saws:

- *Circular saws* use a blade in the shape of a disk with teeth arranged around the edge. These teeth vary in shape and arrangement, depending on the operation to be performed.

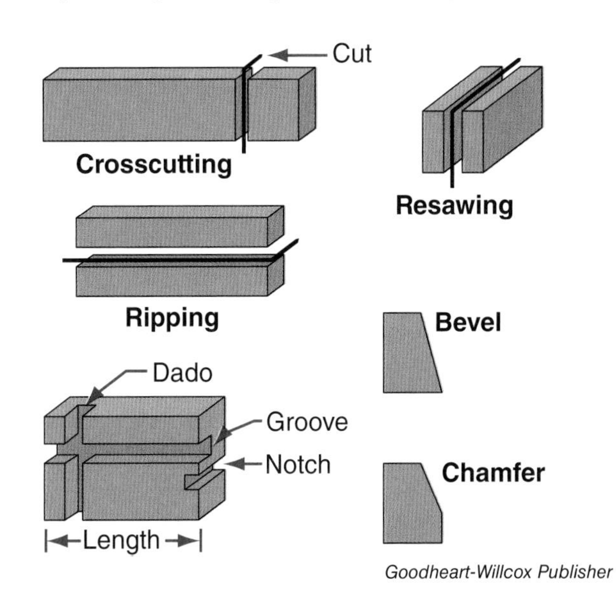

Figure 6-15. Typical sawing operations.

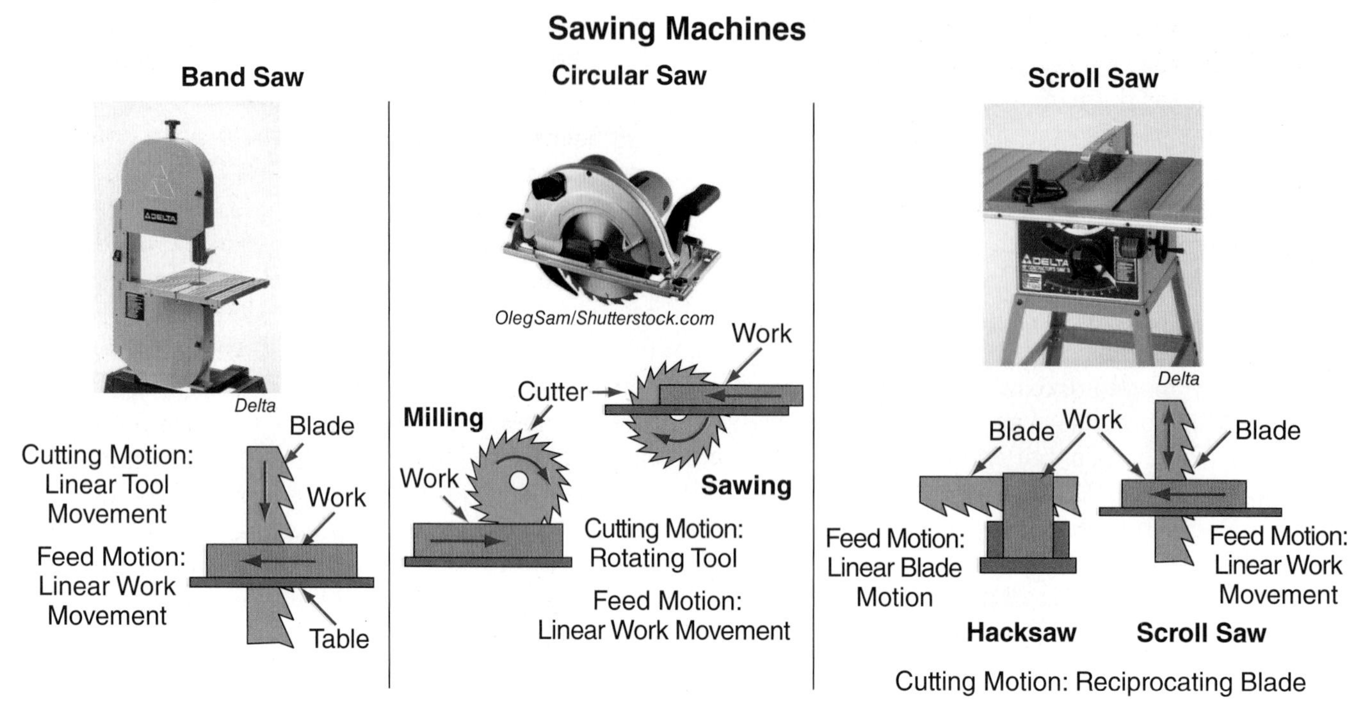

Figure 6-16. The three basic types of saws are circular saws, band saws, and scroll saws.

Common blades are available for crosscutting, ripping, and combination cutting (crosscutting and ripping).

- ***Band saws*** use a blade made of a continuous strip, or band, of metal. Most of these bands have teeth on one edge. The band generally travels around two wheels, which gives a continuous linear cutting action. The band can be vertical or horizontal. Horizontal band saws are usually used as cutoff saws for metal rods and bars. Vertical band saws are widely used to cut irregular shapes from wood or metal sheets.
- ***Scroll saws*** use a straight blade that is a strip of metal with teeth on one edge. The blade is clamped into the machine. The machine then moves the blade up and down to produce a reciprocating cutting motion. The material is placed on the table and manually fed into the blade.

Drilling Machines

Holes in parts and products are common and can be produced in various ways. They can be punched into sheet metal using punches and dies. Holes can be burned into plate steel using an oxyacetylene cutting torch. They can be produced in the part directly as it is cast from molten material. Holes can be produced with powerful beams of light (laser machining) or electrical sparks (electro-discharge machining).

Many cylindrical holes, however, are produced through drilling operations. ***Drilling machines*** produce or enlarge holes using a rotating cutter. Generally, the cutting motion is produced as the drill, or bit, rotates. The drill is moved into the work to produce the feed motion. The drill press is the most common machine that uses these cutting and feed motions. See **Figure 6-17**.

Drilling can also be done by rotating the work to produce the cutting motion. The stationary drill is then fed into the work. This practice is common with drilling on a metal lathe or a computer-controlled machining center.

Many operations can be completed on drilling machines, **Figure 6-18**. Common drilling machine operations are as follows:

- **Drilling.** Produces straight, cylindrical holes in a material. These holes can be used to accommodate bolts, screws, shafts, and pins for assembly. They also are a functional feature of products, such as furnace burners and compact disc recordings.

- **Counterboring.** Produces two holes of different diameters and depths around the same center point. Counterbores are used to position shafts and recess heads of fasteners, among other purposes.
- **Countersinking.** Produces a beveled outer portion of a hole. Most often, countersinking is used with flathead wood and metal screws. Countersinking holes allows screw heads to be flush with the surface of the part.
- **Reaming.** Enlarges the diameter of a hole. The action is generally performed to produce an accurate diameter for a bolt hole.

Drilling Machines

Feed
Chuck
Drill
Motor
Table
Column
Base
Cutting Motion: Rotating Tool
Feed Motion: Linear Tool Movement

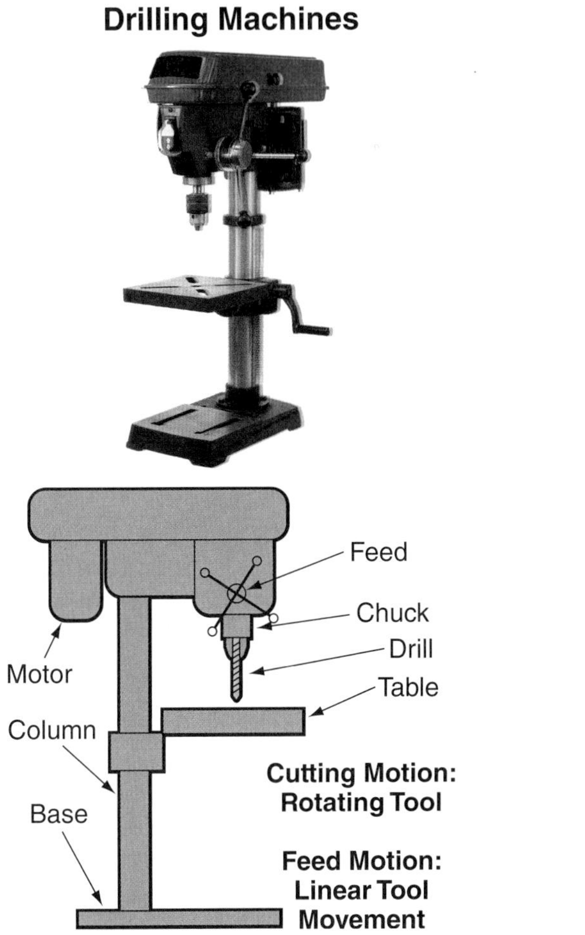

Photoexpert/Shutterstock.com; Goodheart-Willcox Publisher

Figure 6-17. A drill press is the most common drilling machine.

Various drilling tools are used for drilling operations, including twist drills, spade bits, and Forstner bits. See **Figure 6-19**. ***Twist drills*** are shafts of steel with points on the end to produce a chip. These chips are carried from the work on helical flutes circling the shaft. ***Spade bits*** are flat cutters on the end of a shaft. The bottoms of the cutters are shaped to produce the cut. ***Forstner bits*** are two-lipped woodcutters that produce a flat-bottomed hole. Hole saws, which use sawing machine action, and fly cutters, which use lathe-type tools, can also be used to produce holes.

Shaping and Planing Machines

Shaping machines and ***planing machines*** are two metalworking machine tools that produce flat surfaces. Do not confuse these machines with the woodworking shaping and planing machines, which operate on the same principles as sawing machines. Wood shapers and planers use a rotating cutter, into which the wood is fed.

Both the metal shaper and the metal planer use single-point tools and reciprocating motion to produce the cut. See **Figure 6-20**. Their difference lies in the movements of the tool and the workpiece. The metal shaper moves the tool back and forth over the workpiece to produce the cutting motion. The work is moved over after each forward-cutting stroke to produce the feed motion. The metal planer reciprocates the workpiece under the tool to

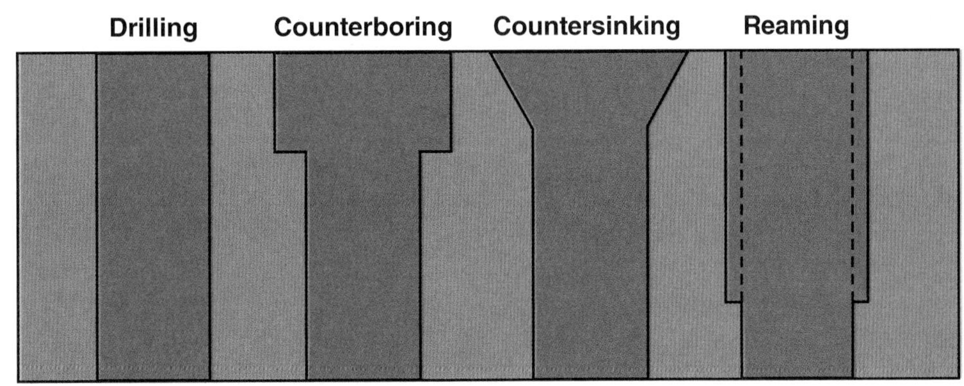

Goodheart-Willcox Publisher

Figure 6-18. Operations commonly performed on drilling machines.

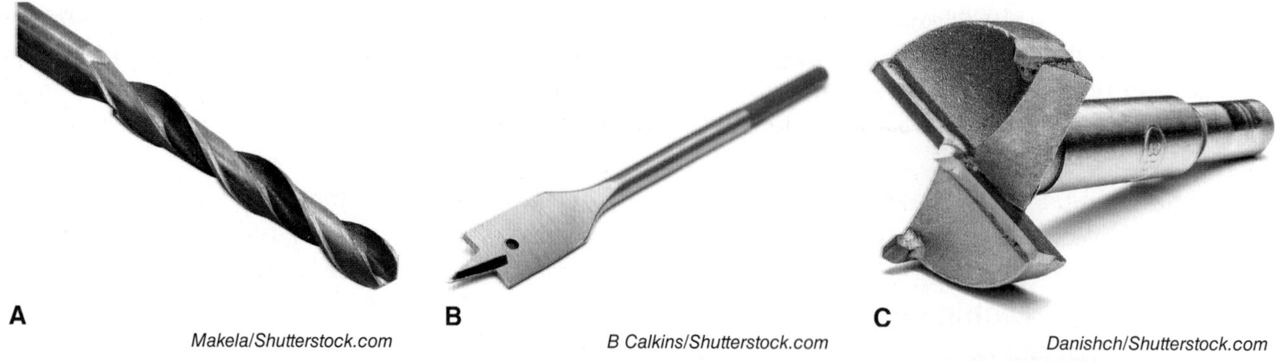

Figure 6-19. Common drilling tools. A—Twist drill. B—Spade bit. C—Forstner bit.

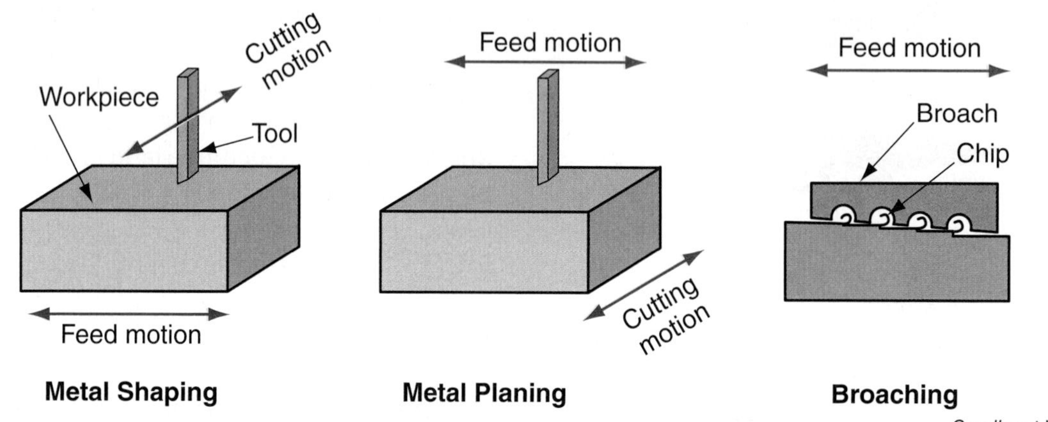

Figure 6-20. Shaping and planing machines use different feed and cutting motions.

generate the cutting motion. The tool is moved one step across the work for each cutting stroke.

Shapers and planers both can cut on the face or side of the part. They can also be used to machine grooves into the surface. Both machines have limited use in material processing.

A ***broach*** is closely related to shapers and planers. This machine uses a tool with many teeth. Each tooth sticks out slightly more than the previous tooth. As the broach tool is passed over a surface, each tooth cuts a small chip. When all the teeth pass over the work, however, a fairly deep cut is possible. Broaches are often used to machine a keyway (rectangular notch) in a hole. Keyways are widely used to assemble wheels and pulleys with axles.

Grinding Machines

Grinding machines use bonded abrasives (grinding wheels) to cut the material. These wheels have cutting surfaces that remove the material in the form of tiny chips. The two most common types are cylindrical grinders and surface grinders. See **Figure 6-21**. A ***cylindrical grinder*** uses the lathe principle to machine the material. The workpiece is held in a chuck or between centers and rotated. A grinding wheel is rotated in the opposite direction. The opposing rotating forces produce the cutting motion. The grinding wheel is fed into the work to produce the feed motion.

A ***surface grinder*** works on the metal-planer principle. A rotating grinding wheel is suspended above the workpiece. The work is moved back and forth under the wheel to produce the cutting motion and moved slightly, or indexed, after each grinding pass to produce the feed motion.

A third type of grinder is the pedestal grinder. The rotating grinding wheel produces the cutting motion. The operator manually moves the workpiece across the face of the wheel to produce the feed motion.

Measurement Systems

Making quality physical models requires an understanding of measurement systems and

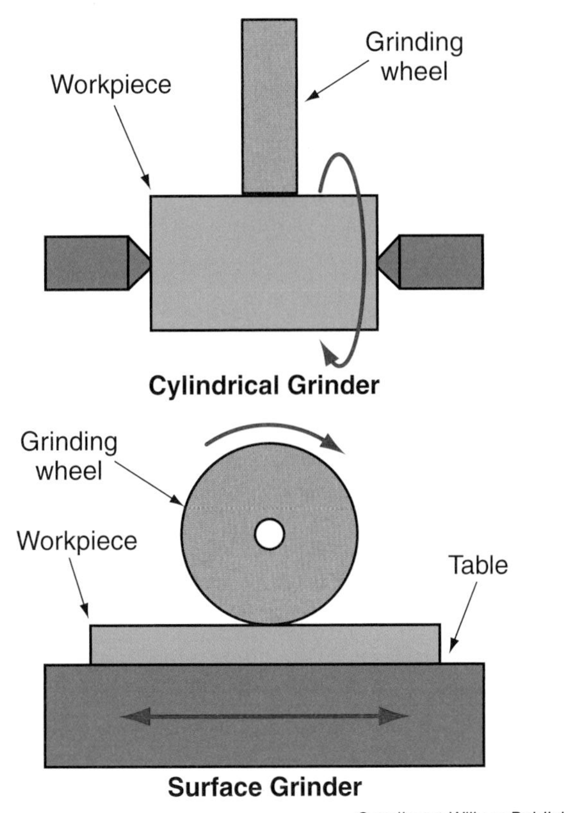

Goodheart-Willcox Publisher

Figure 6-21. Two common grinding machines are the cylindrical grinder and the surface grinder.

industry-quality measurement tools. A common reference or standard for producing a physical model of a design is needed. We use measurement to describe objectively the physical qualities of an object. ***Measurement*** is the practice of comparing the qualities of an object to a standard. See **Figure 6-22**.

All measurements compare the quality being described against a standard. Each major physical quality is compared to a standard measurement that governments and international agreements have set.

There are two major measurement systems used today. One is the US customary system. The other is the more widely used International System of Units. This system, abbreviated *SI* from the French name *Système international d'unités,* is more commonly known as the *metric system*. See **Figure 6-23**. The United States is the only industrialized country that has not widely converted to the metric system for everyday use.

The ***US customary system*** is based on the measurement system that developed in England from approximately the 1100s to the 1500s. As in earlier times, some of this system was based on the sizes of human body parts. For example, the term *inch,* a word meaning *thumb* in some languages, was used to represent the width of a human thumb. The *foot* was originally based on the length of a human foot. Other

Goodheart-Willcox Publisher

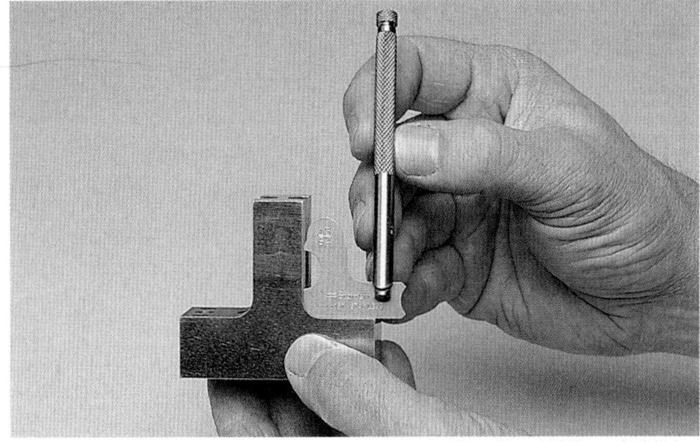

Goodheart-Willcox Publisher

Goodheart-Willcox Publisher

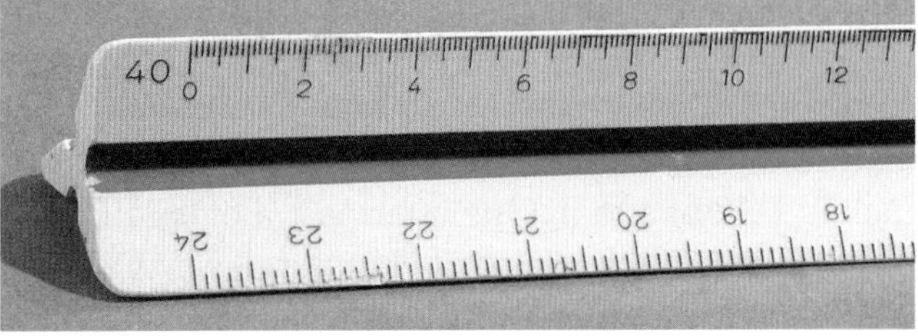

Goodheart-Willcox Publisher

Figure 6-22. Four different standards of measurement.

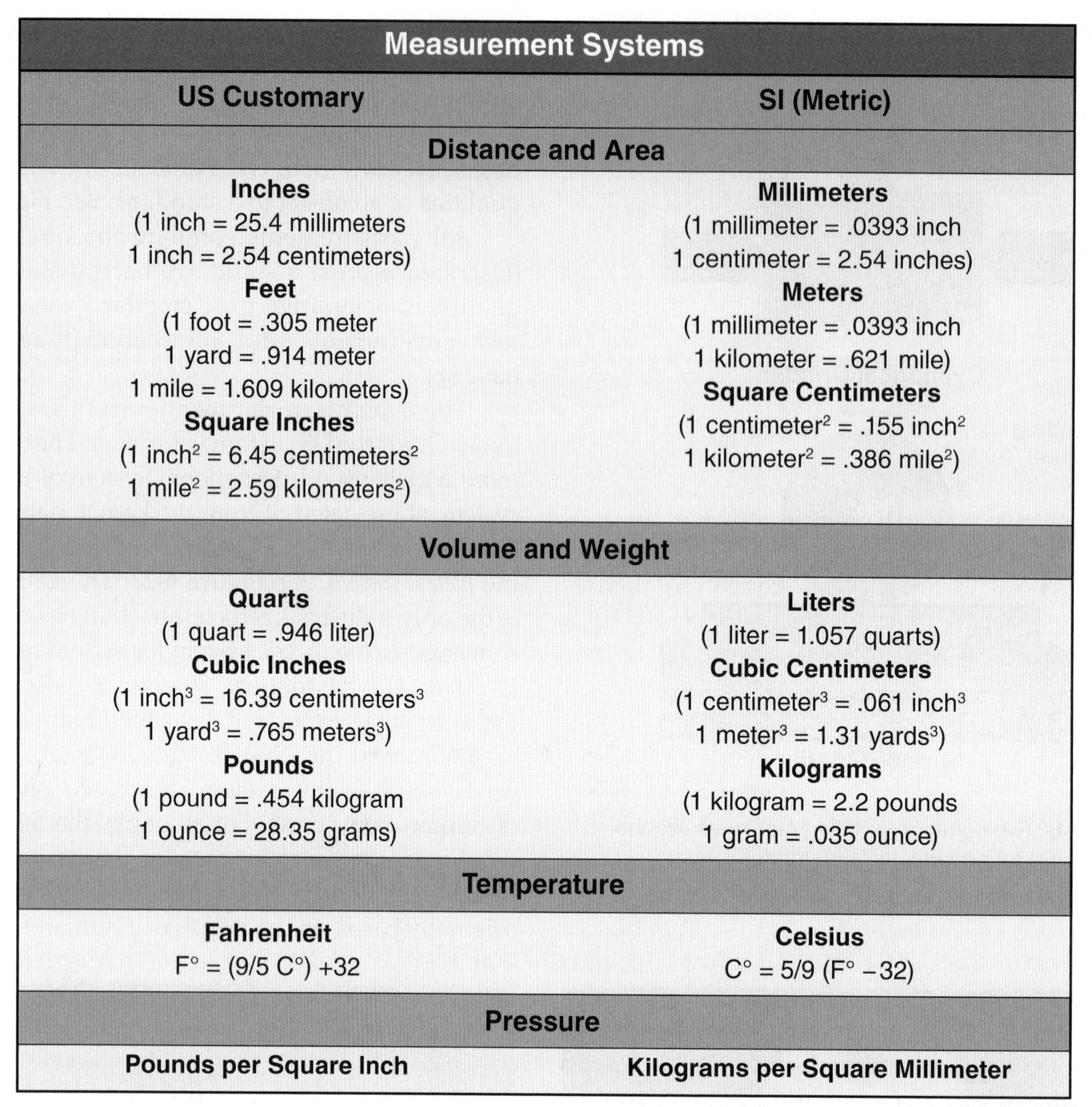

Measurement Systems	
US Customary	**SI (Metric)**
Distance and Area	
Inches (1 inch = 25.4 millimeters 1 inch = 2.54 centimeters) **Feet** (1 foot = .305 meter 1 yard = .914 meter 1 mile = 1.609 kilometers) **Square Inches** (1 inch^2 = 6.45 centimeters^2 1 mile^2 = 2.59 kilometers^2)	**Millimeters** (1 millimeter = .0393 inch 1 centimeter = 2.54 inches) **Meters** (1 millimeter = .0393 inch 1 kilometer = .621 mile) **Square Centimeters** (1 centimeter^2 = .155 inch^2 1 kilometer^2 = .386 mile^2)
Volume and Weight	
Quarts (1 quart = .946 liter) **Cubic Inches** (1 inch^3 = 16.39 centimeters^3 1 yard^3 = .765 meters^3) **Pounds** (1 pound = .454 kilogram 1 ounce = 28.35 grams)	**Liters** (1 liter = 1.057 quarts) **Cubic Centimeters** (1 centimeter^3 = .061 inch^3 1 meter^3 = 1.31 yards^3) **Kilograms** (1 kilogram = 2.2 pounds 1 gram = .035 ounce)
Temperature	
Fahrenheit F° = (9/5 C°) +32	**Celsius** C° = 5/9 (F° −32)
Pressure	
Pounds per Square Inch	**Kilograms per Square Millimeter**

Goodheart-Willcox Publisher

Figure 6-23. Common measurements in the US customary system and metric system.

terms in the system were developed from Roman measurements. For example, the word *mile* originally meant the length of 1000 paces of a Roman legion.

The US customary system can create confusion because numerous terms are used to describe the same kind of measure. Twelve inches equal one foot, but thirty-six inches equal either three feet or one yard. The system can also be challenging because of inconsistencies in computing fractions and multiples of different base measurements. For example, one inch is 1/12 of a foot. One quart is 1/4 of a gallon.

In the 1790s, a group of French scientists assembled to create a more logical and exact measurement system. The SI system, or ***metric system***, is based on decimals (multiples of the number 10). There is a logical progression from smaller units to larger ones throughout the system because the base unit always decreases or increases by values of 10. Smaller units are decimal fractions of the base unit (1/10, 1/100, 1/1000, and so on). Larger units are multiples of 10 (10, 100, 1000, and so on) times the base unit.

In the metric system, a prefix is used to describe how the base unit is being changed. See **Figure 6-24**. For example, the base unit for distance is the meter. For large distances, we use the word *kilometer.* The prefix *kilo-* means *1000,* so seven kilometers are equal to 7000 meters. For small distances, we use the word *millimeter.* The prefix *milli-* means *1/1000th.* Twelve millimeters are 12/1000 of a meter. The metric system is simple to use in mathematical formulas.

SI-Measurement Prefixes			
Factor	**Magnitude**	**Prefix**	**Symbol**
10^9	1,000,000,000	giga-	G
10^6	1,000,000	mega-	M
10^3	1000	kilo-	k
10^2	100	hecto-	h
10^1	10	deka-	da
10^{-1}	1/10	deci-	d
10^{-2}	1/100	centi-	c
10^{-3}	1/1000	milli-	m
10^{-6}	1/1,000,000	micro-	μ
10^{-9}	1/1,000,000,000	nano-	n
10^{-12}	1/1,000,000,000,000	pico-	p

Goodheart-Willcox Publisher

Figure 6-24. Metric prefixes are useful in describing quantities of the SI system.

Engineering notation, described in Chapter 3, is used by engineers because it is a quick way to identify the prefix of a number. For example, 21.2×10^6 can be quickly recognized as 21.2 million because 10 to the power of six is equal to 1 million.

The seven standards for physical qualities are size and space, mass, time, temperature, number of particles, electric current, and light intensity. These qualities are measured as technology is developed and used.

Types of Measurement

The measurement of part and product size is important in technological design and production activities. This type of measurement can be divided into two levels of accuracy—standard and precision. See **Figure 6-25**.

Many production settings do not require close measurements. The length of a house, the width

Types of Measurement

Standard

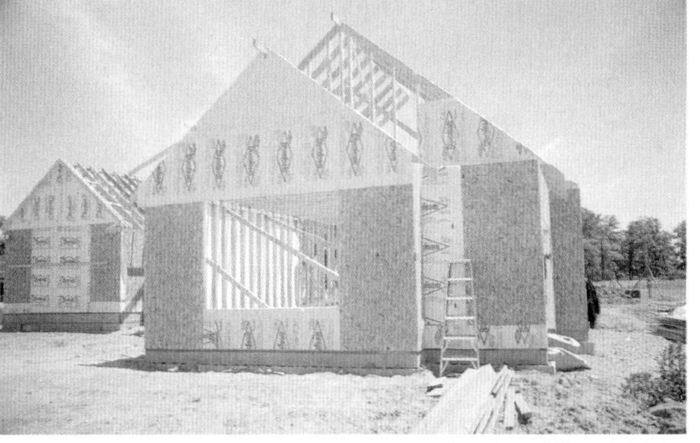

Goodheart-Willcox Publisher

Precision

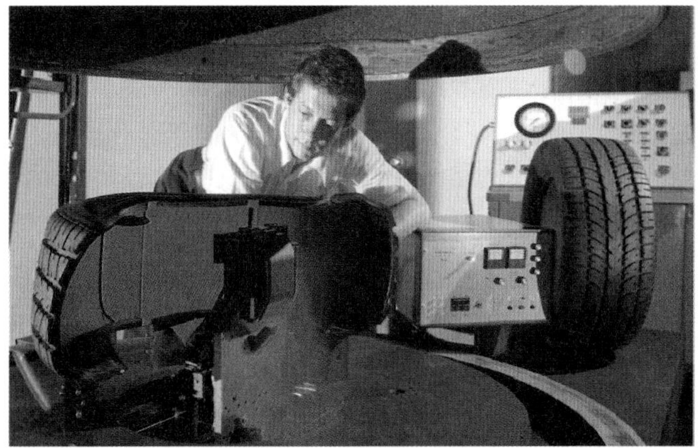

Goodyear Tire and Rubber Co.

Figure 6-25. Standard measurement is widely used in the construction industry. Precision measurement is found in many manufacturing applications.

STEM Connection: Mathematics

Measuring Area

Why do we need to use measurement systems in technology? One answer is that measurement systems provide a common terminology when we wish to change or improve our environment. Suppose, for example, you want to paint your bedroom a different color. How much paint do you need to buy to cover all four walls?

To calculate the amount of paint needed, you must determine the area to be covered. Area equals length times width, and the resulting amount is expressed in square feet. See **Figure A**.

Two walls of your bedroom are 9′ long and 8′ high and the other two walls are 11′ long and 8′ high. There is one window, which measures 36″ by 48″, and two doors measuring 36″ by 84″ each. Calculations needed to determine the area to be covered are as follows:

One wall at 9′ long × 8′ high = 72 ft^2
72 square feet × 2 walls = 144 ft^2
One wall at 11′ long × 8′ high = 88 ft^2
88 ft^2 × 2 walls = 176 ft^2
Total room area = 320 ft^2
Subtract:
One window opening at 36″ × 48″ (3′ × 4′) = 12 ft^2
Two door openings at 36″ × 84″ (3′ × 7′) × 2 doors = 42 ft^2
Total opening area = 54 ft^2
Total area to be painted = 320 ft^2 – 54 ft^2 = 266 ft^2

Pressmaster/Shutterstock.com

Figure A. Area measurements help painters estimate the cost and time for a painting job.

At the paint store, you find that one gallon of the paint you chose covers 250 ft^2. Therefore, you have to buy one gallon, plus another smaller can to cover the 266 square feet.

Assuming one gallon of paint covers 250 ft^2, calculate how much paint you should purchase if the dimensions of your bedroom are as follows:

Two walls are 10′ long and 8′ high
Two walls are 13′ long and 8′ high
Two windows each measure 32″ by 48″
Two door openings measure 30″ by 84″

of a playing field, and the angle of the leg on a playground swing set need not be accurate to the utmost degree. If the product is within a fraction of an inch or a degree of angle, it works fine. This type of measurement, called ***standard measurement***, is often made to the foot, inch, or fraction of an inch in the customary system or to the nearest whole millimeter in the metric system.

In addition to type of production, the material being measured also determines the measurement type used. For example, wood expands or shrinks with changes in its moisture content and the atmospheric humidity. Measurements closer than 1/32″ or 1 mm are not useful. Wood can change more than that amount in one day. Standard measurements are common in cabinet- and furniture-manufacturing plants, construction industries, and printing companies. The printing industry uses its own system of standard measurement based on the pica (1/6″) and the point (1/72″).

Standard measurement is not accurate enough for many production applications. Watch parts and engine pistons are useless if they vary by as much as 1/32″ (0.8 mm). These parts must be manufactured to a more precise size. For this type of production, ***precision measurement*** is required.

Precision measurement measures to 1/1000″ (one one-thousandth of an inch) to 1/10,000″ (one ten-thousandth of an inch) in the US customary system. Metric precision measurement measures to 0.01 mm (one one-hundredth of a millimeter). Workers in manufacturing activities involving metals, ceramics, plastics, and composites use precision measurements. Precision measurements are also used in laboratory testing, material research, and scientific investigation.

Measurement Tools

The measurement tools an operator manipulates and then reads are called ***direct-reading measurement tools***. Newer measurement tools and machines, called ***indirect-reading measurement tools***, bring together sensors and computers to automate measurement. Sensors gather the measurement data, which is processed by computers or other automatic devices. The final measurement can be displayed on an output device, such as a digital readout, computer screen, or printout. These systems include laser measuring devices, optical comparators, and direct-reading thermometers. See **Figure 6-26**. If you have weighed yourself on a digital scale, you have used an indirect-reading measuring device.

Measurement is commonly used to find linear dimensions, diameters, and angles. Standard or precision devices can be used for these measurements. See **Figure 6-27**.

Linear Measurement Devices

The most common linear measurement device is the rule, **Figure 6-28**. A ***rule*** is a rigid or flexible strip of metal, wood, or plastic with measuring marks on its face. The two types of rigid rules used in linear measuring are the machinist's rule and the woodworker's or bench rule. A bench rule is generally divided into fractions of an inch. The most common divisions are sixteenths (1/16″). Metric bench rules are divided into whole millimeters. Machinist's rules are designed for finer measurements. Customary machinist's rules are divided into sixty-fourths (1/64″) or into tenths (1/10″) and hundredths (1/100″). Metric machinist's rules are divided into 0.5 mm increments.

A part is measured with a rule by first aligning one end of the part with the zero mark usually found on the left side of the rule. The linear measurement is then taken by reading the rule division at the other end of the part.

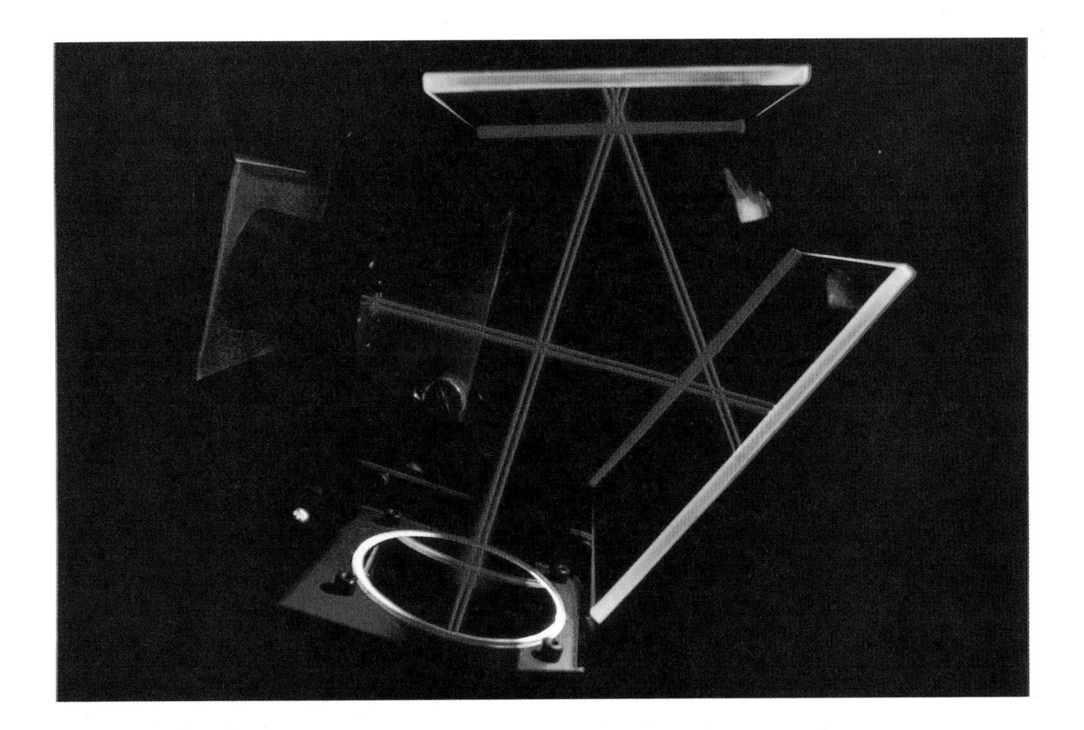

GM-Hughes

Figure 6-26. Machine operators use manual measurements for routine work. Indirect-measurement systems, such as this laser measurement system, are often built into continuous-processing and assembling operations.

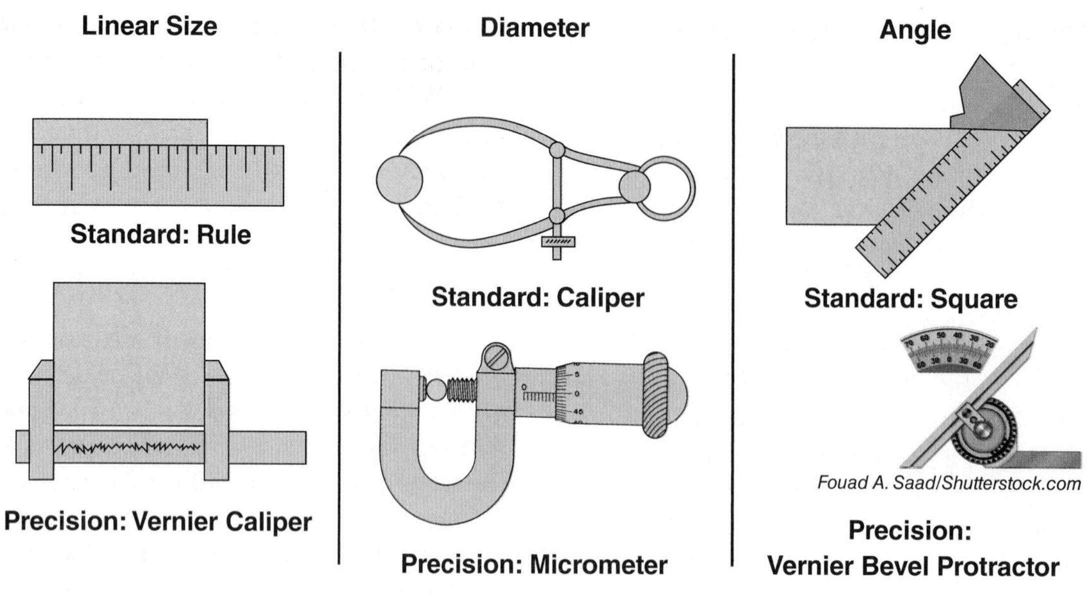

Goodheart-Willcox Publisher

Figure 6-27. Both standard- and precision-measurement tools are used to determine linear sizes, diameters, and angles.

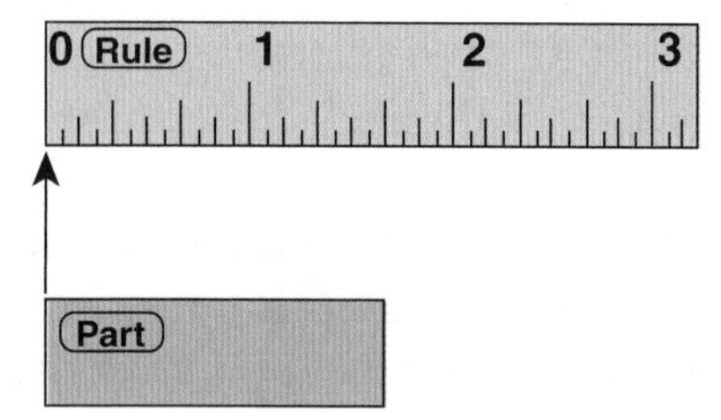

Goodheart-Willcox Publisher

Figure 6-28. Most common linear measurements are made with a rule.

Flexible rules, often called *tape rules*, are used in woodworking and carpentry applications. The catch at one end of the rule is hooked to one end of the board or structure. The tape is pulled out until it reaches the other end of the board or structure. A measurement is then taken.

Metric tape rules are divided into 1 mm increments. Commonly, the smallest division on a customary tape rule is 1/16″. Some tape rules highlight every 16″, which is the common spacing of studs in home construction. Tape rules are generally available in lengths from 8′ to 100′ or from 2 m to 30 m.

Various machinist's tools can be used to make precision linear measurements. The most common are inside and outside micrometers, depth gauges, and Vernier calipers. See **Figure 6-29**. These tools measure physical qualities such as length and thickness.

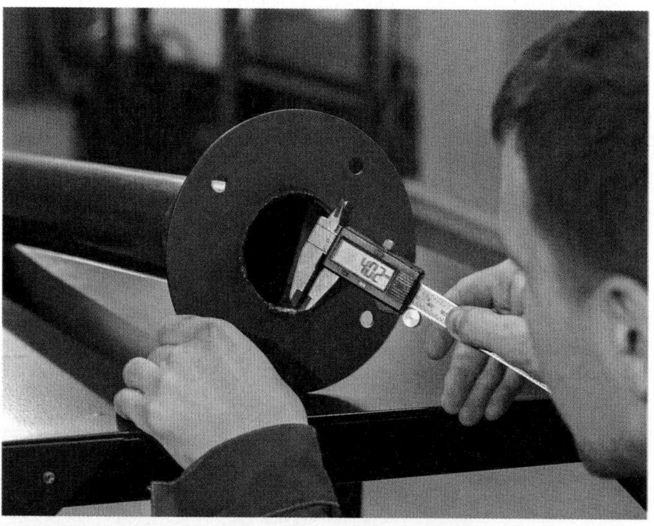

sandrexim/Shutterstock.com

Figure 6-29. A mechanical engineer is using a Vernier caliper to measure a metal part.

Diameter-Measuring Devices

Determining the diameter of a round material or part is a common measurement task. A simple, rough measurement can be made with hole gauges or circle templates. These devices have a series of holes into which the stock can be inserted. The smallest hole into which the material fits establishes the approximate diameter of the item.

More precise diameters can be established using a ***micrometer*** that is either mechanical or digital. See **Figure 6-30**. The part to be measured is placed between the anvil (fixed part) and the spindle (movable rod). Turning the barrel moves the spindle forward. When the spindle and the anvil touch the part, a reading is taken on the barrel. Most customary micrometers measure to within 1/1000 of an inch. Metric micrometers measure to 0.01 mm.

Angle-Measuring Devices

The angle between two adjacent surfaces or intersecting parts is important in many situations. The legs of a desk are generally square (at a 90° angle) with the top. The ends of picture-frame parts must be cut at a 45° angle to make a square frame.

Measurements at 90° angles are commonly done with a measuring tool called a ***square***. See **Figure 6-31**. These tools have a blade that is at a right angle to the head. The head is placed against one surface of the material. The blade is allowed to rest on an adjacent surface. If the blade touches the surface over its entire length, the part is square. Parts that are not square allow light to pass under the blade.

Some squares have a shoulder on the head that allows the square to be used to measure 45° angles. This angle is important in producing mitered corners on furniture, boxes, and frames. Common types of squares are the rafter or carpenter's square, the machinist's square, the try square, and the combination (90° and 45°) square.

Measuring with a Micrometer

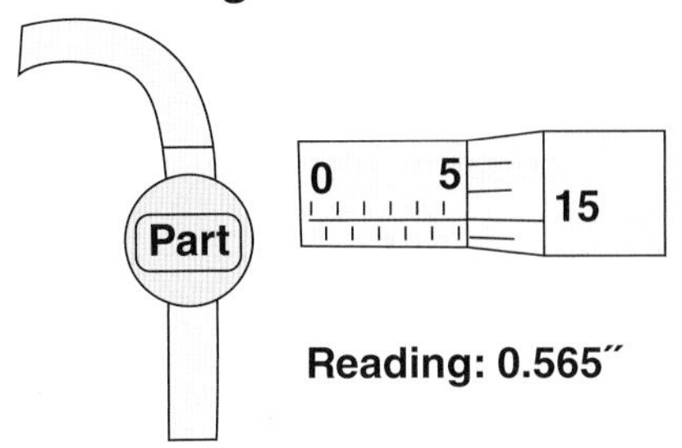

1. Align the part between the spindle and the anvil of the micrometer.
2. Gently move the spindle against the part.
3. Read the measurement on the barrel.
   a. Read to the nearest .050″ (each mark on the barrel).
   b. Add the fraction from the thimble –each mark is .001″.

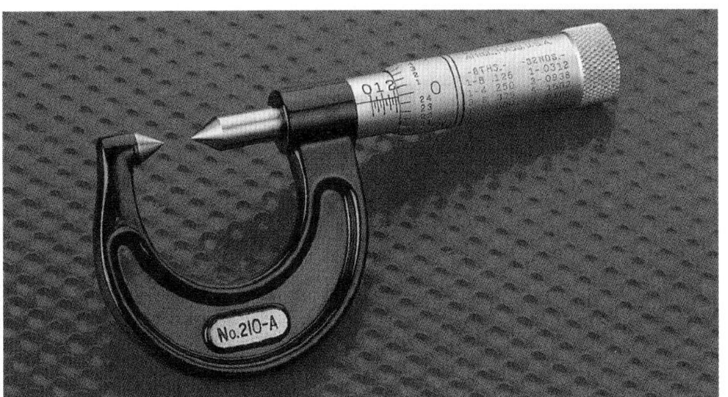

L.S. Starrett Co.

Figure 6-30. Micrometers are used to obtain precise measurements of diameters.

Checking Angles with a Square

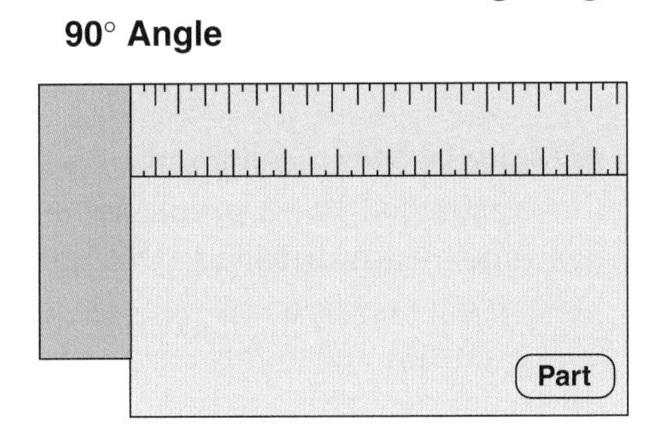

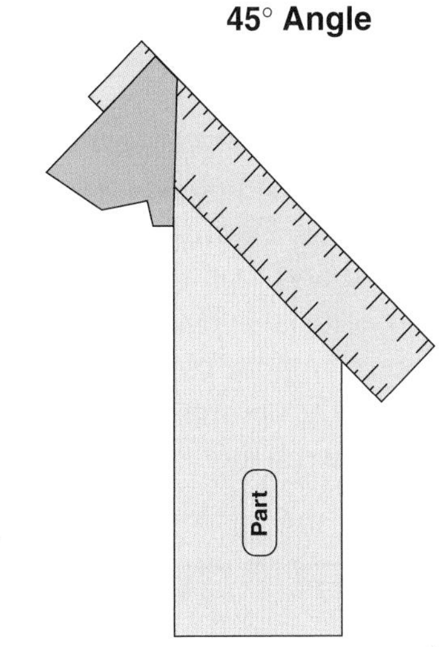

1. Place the head of the square against one surface of the part.
2. Hold the part and square up to the light.
3. Look for gaps along the blade of the square.

Goodheart-Willcox Publisher

Figure 6-31. Angles that are 90° can be measured with a square. When an angle is not 90°, the blade of the square does not touch the surface of the work.

Technology Explained Computer Numerical Control

Computer numerical control: Using a computer and sequenced instructions to control a machine tool.

The Industrial Revolution was based on producing large numbers of uniform products using a continuous manufacturing system. Any change in the design of a product necessitated expensive changes in the machines that made the product.

The development of numerical control (NC) in the 1950s changed how objects were made. Programmers punched holes in a paper tape to store the information needed for the commands used to control the machine tools.

Numerical control evolved into computer numerical control (CNC) with the development of small computers that could be used with machine tools. This greatly changed how manufacturing systems work. It became possible for information to be entered into the machine on a computer. In CNC, the machine is automated and monitored by the computer. See **Figure A**. The design for a product is expressed in computer language that can be used to control the machine. Computers control how products move through a sequence of operations by feeding codes into the machine tools, which then cut the objects. See **Figure B**. Computers are also used to design, test, and refine products, as well as inspect product quality.

CNC systems use a number of technologies, including computer-aided design (CAD), to develop and specify a product's size and shape. Robots load and unload machines and move products from machine to machine. See **Figure C**. CNC is used to guide machine operation. Electronic measurement and computer data processing help to improve inspection and quality control.

Timofeev Vladimir/Shutterstock.com

Figure A. CNC manufacturing is based on computer control.

Aumm graphixphoto/Shutterstock.com

Figure B. CNC machining of a precision part.

Maksim Dubinsky/Shutterstock.com

Figure C. Robots are used with CNC systems for moving products.

Protractors and sliding T-bevels can be used to measure angles that a square cannot. A protractor allows for direct reading of angles. Place the protractor over the angle to be measured and then read the angle. The sliding T-bevel has an adjustable blade. Place the head on one surface and then clamp the blade along the angle of the second surface. With a protractor, or by using a mathematical formula, you can then measure this angle.

Measurement and Control

All technological processes produce products or services, such as goods, buildings, or communication media. Measurement is necessary in designing these artifacts. Their sizes, shapes, or other properties are communicated through measurements. Processing equipment is set up and operated using design measurements. Materials needed to construct the items are ordered using measurement systems. All personal or industrial production is based on measurement systems.

In industrial settings, measurement is the key to quality control in the production process. ***Quality control*** is the process of setting standards, measuring features, comparing them to the standards, and making corrective actions. See **Figure 6-32**. This

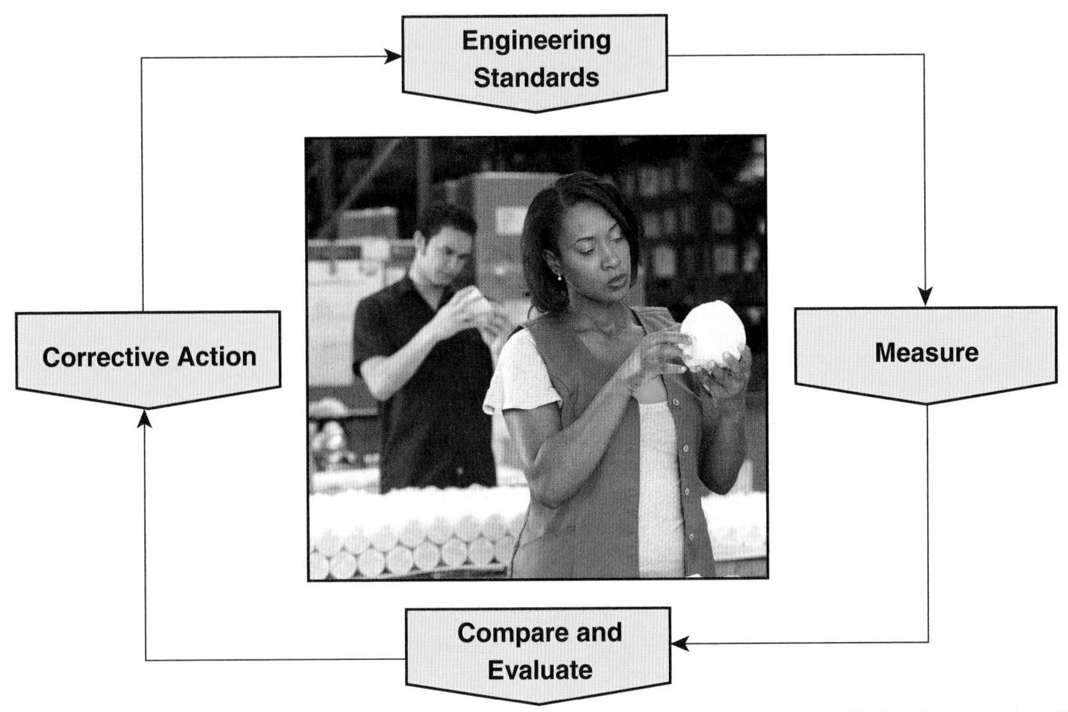

Monkey Business Images/Shutterstock.com

Figure 6-32. The foundation of a quality control system is measurement and analysis.

process is designed to ensure that products, structures, and services meet the needs of the users.

Quality control involves measuring and analyzing materials entering the system, work in process, and the outputs of the system. This control is an ongoing process designed to ensure that resources are efficiently used and that customers receive functional products.

Measurement can also be used to compare the present condition with a desired condition. For example, imagine that you are building a bookcase and need a shelf that is 24″ (610 mm) long. To cut the shelf from a longer board, you first measure and mark the location for the cut. You then saw along the line. If the board is too long, it will not fit into the case. A short board will fail to rest on the shelf supports. Measurement can confirm that you have produced a 24″ long shelf. A board of the correct length fulfills the intended purpose, and the shelf meets your quality standards.

Experimenting with Materials and Tools for Optimization

Optimization means making something, such as a technological product or system, as effective and as close to perfect as possible. Optimization is critical to the engineering of solutions. Designers and engineers must continuously identify ways to enhance or improve the solution as they progress through the engineering design process. Therefore, optimization should take place during the *Model and Make a Solution* step. This involves the ongoing practice of evaluating whether the physical model of the solution meets the desired design specifications and determining ways in which the design could be altered to better meet these specifications.

A series of scientific experiments can be conducted during the creation of a physical model of a solution. These experiments can help determine which materials or tools can most effectively produce the different components of a design. For example, an engineer creating a new space telescope may set up a scientific experiment to test which materials are the most heat resistant. Another example would be experiments to determine which material and tools can be used to achieve the desired shape of a product design. Experimenting with both minor and major design modifications while making a solution can lead to new procedures for solving the problem. These results can then be used to refine or optimize the design solution. The results may also be useful in future problem-solving ventures.

Summary

- ✔ A model simulates actual events, structures, or conditions. Models reduce complex mechanisms and events into more easily understandable form.
- ✔ Three types of physical models are mock-ups, scale models, and prototypes.
- ✔ Making quality physical models requires an understanding of measurement systems and industry-quality measurement tools.
- ✔ Material-processing tools and machines cast, form, and machine materials into specific shapes, help assemble products, and apply protective coatings.
- ✔ Machine tools include turning, sawing, drilling, shaping, planing, and grinding machines.
- ✔ Measurement is used to objectively describe the physical qualities of an object and provides a common reference or standard for producing a physical model of a design.
- ✔ The two measurement systems in use today are the US customary system and the SI (metric) system.
- ✔ Measurement of part and product size can be divided into two levels of accuracy—standard and precision. Standard and precision tools are used to find linear dimensions, diameter, and angles.
- ✔ In industrial settings, measurement is the key to quality control in the production process.
- ✔ Designers and engineers must continuously identify ways to enhance or improve (optimize) the solution as they progress through the engineering design process. Experiments can be conducted during the modeling phase to help determine what materials or tools work best to produce the different components of a design.

Check Your Engineering IQ

Now that you have finished this chapter, see what you learned by taking the chapter posttest.
www.g-wlearning.com/technologyeducation/

Test Your Knowledge

Answer the following end-of-chapter questions using the information provided in this chapter.

1. Explain why mock-ups are *not* used to analyze the behaviors or properties of a design.
2. Tools and machines are used in what three major types of processing?
3. What three characteristics are common to machine tools?
4. What is the difference between feed motion and cutting motion?
5. Attachments used to hold and rotate drills and router bits are called _____.
6. *True or False?* Plastics can be machined on lathes.
7. The three basic types of saws are the circular saw, the band saw, and the _____ saw.
8. Define the term *counterboring*.
9. Describe the difference between shaping machines and planing machines.
10. What is measurement?
11. *True or False?* Precision measurement is often made to the inch or fraction of an inch.
12. *True or False?* In the metric system, the same prefixes are used for all base units.
13. Diameters can be precisely measured with a device called a(n) _____.
14. To measure 90° angles, a tool called a(n) _____ is commonly used.

Critical Thinking

1. Explain how physical models are used in the engineering design process.
2. Why is it important to understand the different types of production tools and machines and measurement devices when enacting the engineering design process?
3. What is the relationship between measurement and quality control?

STEM Applications

1. A. A scale model built to a 1:12 ratio is 6″ long and 4″ wide. What are the dimensions, in feet, of the object represented by the scale model?
   B. A structure is 180′ long and 60′ wide. Your scale model of this structure is 3′ long and 1′ wide. What is the ratio of your model to the original structure?
2. Select a major development in material processing, energy processing, or information processing. Write a two-page report on the development and its applications.
3. List the metric unit and the US customary unit of measurement that would be used to measure each of the following:
   A. Distance from Los Angeles to New York
   B. Outdoor temperature
   C. Weight of a loaf of bread
   D. Length of a pencil
   E. Volume of a container

Engineering Design Challenge

Making a Board Game

Background

We are surrounded by products that have been developed and produced using material-processing technology. In this activity, you will change the form of materials to make a product that people can use.

Challenge

Work with a partner to make a game that can be given to a local charity. See the layout for the game in **Figure A**.

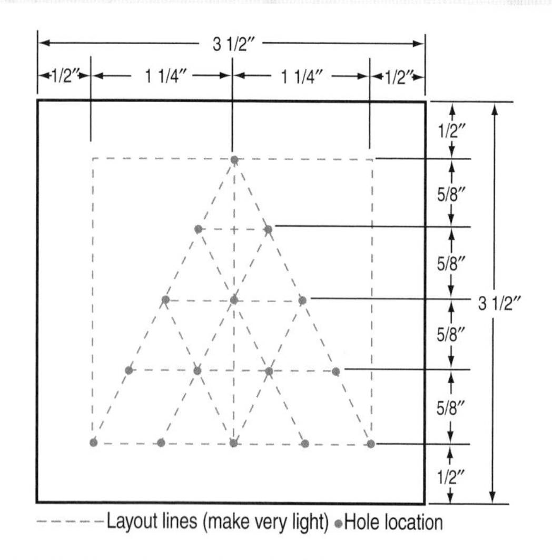

Figure A. The game-board layout.

Materials and Equipment

- One piece of 3/4″ × 3 1/2″ × 3 1/2″ clear pine, redwood, or western red cedar.
- 12 wood pegs or golf tees (1/8″ diameter × 1″).

Procedure

1. Carefully watch your teacher's demonstration of the proper use of tools and machines needed to make the product.
2. Follow the procedure given on the operation-process chart to construct the game. See **Figure B.** Follow all safety rules your teacher discussed during the demonstration.
3. Photocopy the directions for playing the game. The directions will be packaged with the game.

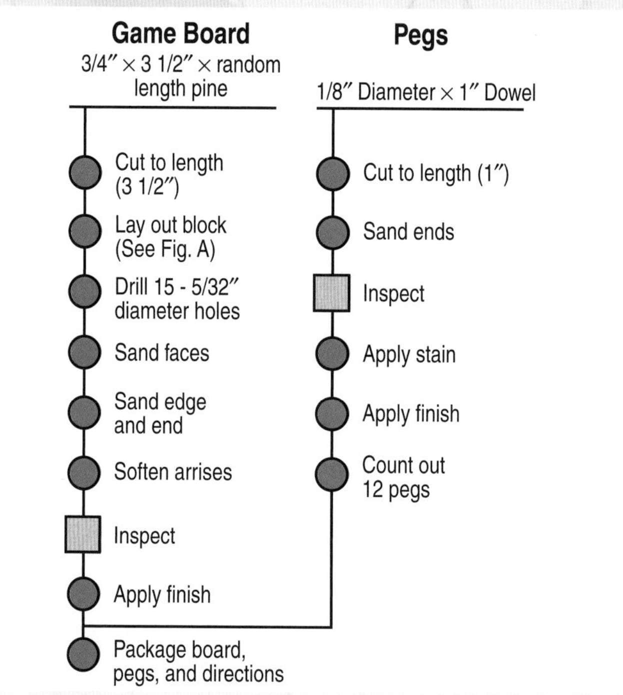

Figure B. The operation-process chart.

Analysis

Meet as a class to analyze the material-processing activities you completed. Answer the following questions:

- Which material-processing tools did you use?
- Which measuring tools did you use?
- How could you have increased the speed of the manufacturing process?
- What changes in the material-processing actions would you make to improve the quality of the product?

Mind Challenge

Directions

1. Put a peg in each hole, except one.
2. Select one peg. Jump an adjacent peg, ending in an empty hole.
3. Remove the jumped peg.
4. Continue jumping pegs until only one remains or no more jumps are possible.

Scoring

One peg remains—Terrific—30 points.
Two pegs remain—Good—20 points.
Three pegs remain—Fair—10 points.
Four pegs remain—Poor—0 points.

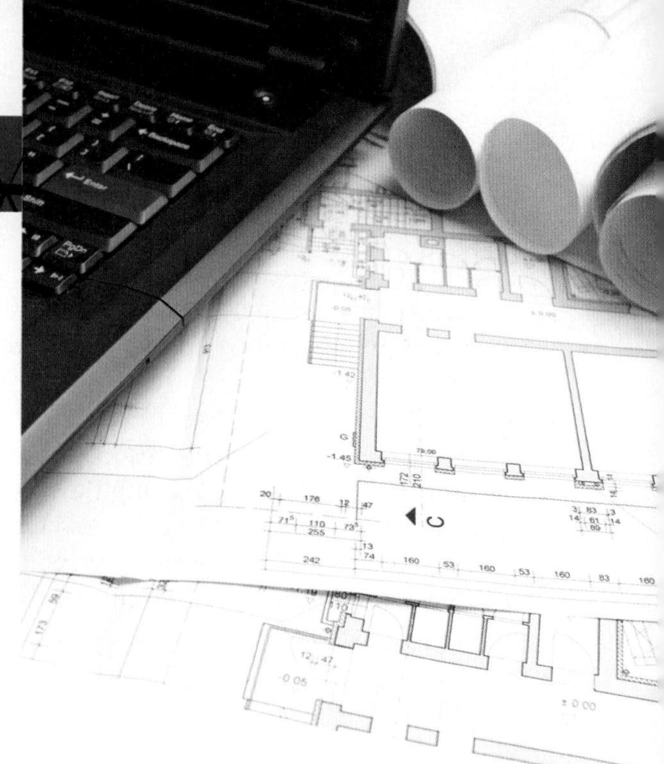

CHAPTER 7

Evaluating Design Solutions

Check Your Engineering IQ

Before you read this chapter, assess your current understanding of the chapter content by taking the chapter pretest. www.g-wlearning.com/technologyeducation/

Learning Objectives

After studying this chapter, you should be able to:

- ✔ Describe the types of analysis used to identify test criteria.
- ✔ Describe three types of design analysis.
- ✔ Explain how to establish a testing procedure/experiment.
- ✔ Identify the steps involved in conducting a scientific investigation.
- ✔ Describe the role of refining solutions.

Technical Terms

design analysis
ergonomics
functional analysis
performance tests
physical tests
qualitative
quantitative
specifications
technical standards
tolerance

Close Reading

Before you read this chapter, consider technological products that you use every day. Which products meet your needs and which do not? Are the products high-quality and dependable? Do they serve their intended purpose? As you read this chapter, reflect on how engineers and designers must evaluate design solutions to high standards to ensure that consumers are able to use them effectively.

This chapter describes the actions that take place during the *Evaluate the Solution* step of the engineering design process. See **Figure 7-1**.

During all stages of the engineering design process, a technological product is evaluated for how successfully it meets the goals of the engineer(s). The technological products we use each day began as a problem. Engineers first evaluate whether the problem exists or not. Engineers and designers are then challenged to solve this problem. After exploring various solutions, the engineers evaluate and choose a solution or select a few solutions that they will dedicate their time and energy to. They move from creating drawings to models to prototype design solutions. The design solutions are then evaluated.

Evaluating Design Solutions

Evaluation of design solutions is crucial to a product's success. Without this stage, costly and irreversible errors can be made. The actions in the evaluation phase require mathematical and predictive analysis. The following steps are involved:

- Determine test criteria.
- Establish a procedure for testing/experimentation.
- Conduct scientific investigations.
- Conduct mathematical examinations.
- Collect data.
- Analyze data.
- Refine the solution.

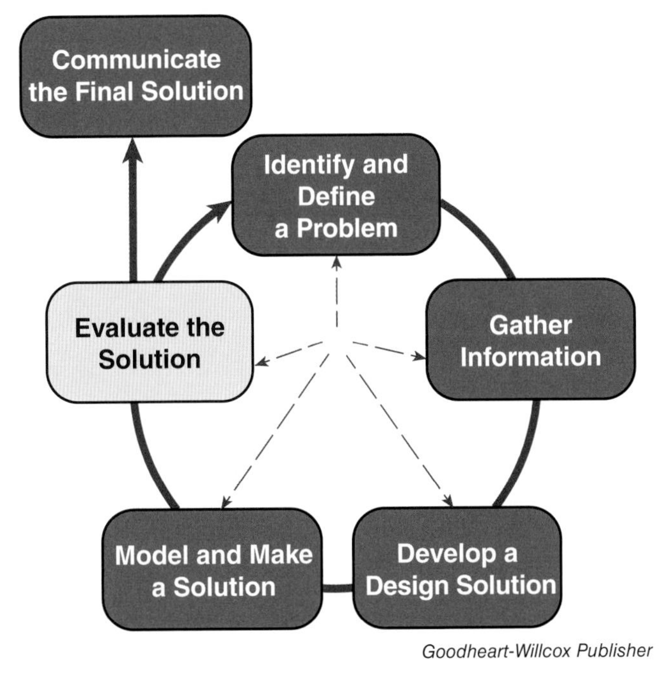

Figure 7-1. Tests, experiments, and analysis are part of the *Evaluate the Solution* step of the engineering design process.

Determine Test Criteria

One of the first steps in evaluating design solutions is to clearly define the criteria required for testing the product or structure. Engineers identify test criteria through specification analysis, functional analysis, and design analysis.

Specification Analysis

Every product must meet certain specifications. ***Specifications*** are criteria or standards, such as quality of craftsmanship, materials, or precision, required to complete a goal. Common examples of specifications are size, weight, speed, accuracy, and strength.

The specifications for all new products and structures must be analyzed to ensure that they are adequate for the function of the product. Excessive specifications, however, add to the cost of the product.

Tolerance is the amount a dimension can vary and still be acceptable. Tolerances vary among products. For example, holding bicycle handlebar diameters to 1/1000″ (0.025 mm) is foolish. That level of precision is not needed for the part to function. However, tolerances must be close for spark plug threads, **Figure 7-2**. The goal is to produce an economical, efficient, and durable product that operates properly and safely.

Specifications relate to the material and manufacturing processes to be used. Holding wood parts to tolerances smaller than 1/64″ (0.397 mm) is impossible because normal expansion and contraction due to changes in humidity can cause this much change. Specifying aluminum for the internal parts of a jet engine is unwise because temperatures inside the engine would melt aluminum parts. Specifying

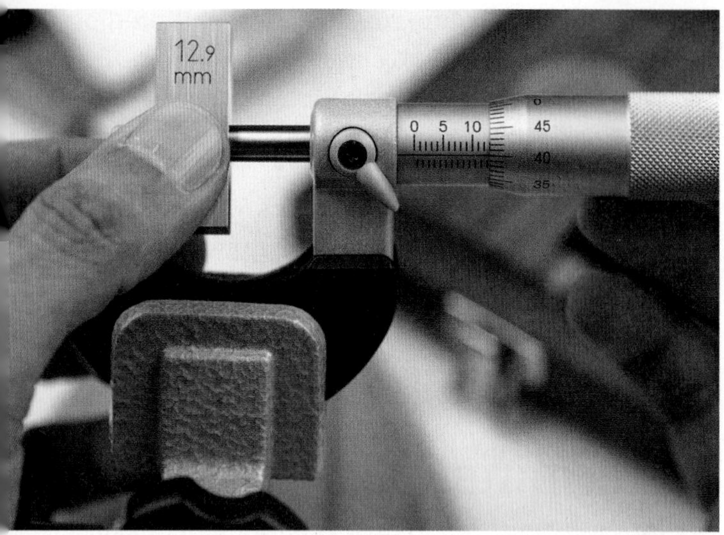

Kimtaro/Shutterstock.com

Figure 7-2. Engineer testing the tolerance of a design solution.

green sand casting as a process to produce precision parts is a mistake. Although green sand casting produces low-cost parts, the parts cannot be held to close tolerances.

Functional Analysis

Functional analysis evaluates the degree to which a product, structure, or system operates effectively under the conditions for which it was designed. Examples of these conditions include the outside environment (such as weather, terrain, or water), operating conditions (such as stress, heat, or gases produced during use), normal wear (such as material fatigue or part distortion), and human use and abuse. See **Figure 7-3**.

jahawk1/Shutterstock.com

Figure 7-3. Start of a long-term environmental effect on a copper penny. Engineers can test materials to determine methods of reducing environmental effects.

Design Analysis

Design analysis also helps engineers determine test criteria and evaluate a design solution. ***Design analysis*** helps engineers, designers, and decision makers assess the design's purpose and likely acceptance in the marketplace in order to choose the best design. This type of analysis does not require as much engineering, scientific, and mathematical knowledge as functional and specification analysis. See **Figure 7-4**. Three common types of design analysis are as follows:

- Human-factors analysis.
- Market analysis.
- Economic analysis.

Human-Factors Analysis

To meet human needs, we design products, systems, and structures for the people who will use them, travel in them, or live and work in them. Designing technological artifacts around the people who use them is the focus of human-factors analysis, more commonly known as ***ergonomics***. This science considers the size and movement of the human body; mental attitudes and abilities; and senses, such as hearing, sight, taste, and touch.

Ergonomics also considers the type of surroundings that are the most pleasing and help people to become more productive. A good example of matching the environment to humans is an aircraft flight deck, **Figure 7-5**. All the controls are within easy reach. Dials and indicator lights are within the pilot's field of vision. Windows are located so the pilots have a clear view of the sky ahead and above them.

Market Analysis

Most products of technology are sold to customers, such as the general public, government agencies, or businesses. During design activities, the market for the product is studied. The designs are analyzed in terms of that market. Market analysis includes determining customer expectations for the product's appearance, function, and cost. This analysis also includes studying present and anticipated competition.

Ford

Figure 7-4. Product designs must be analyzed in five different areas.

Airbus

Figure 7-5. This aircraft flight deck was designed with human movement and senses in mind.

Market analysis is often conducted by taking surveys of potential customers and analyzing competing products available on the market. This data is then analyzed by market researchers. See **Figure 7-6**.

Economic Analysis

Private companies risk money to develop, produce, and market products. In turn, they hope to make a profit as a reward for their risk taking. To increase the chances

Andrey_Popov/Shutterstock.com

Figure 7-6. Market researchers plan and interpret consumer data.

of success, an economic analysis for a new product is conducted. The product is studied in terms of the costs of development, production, and marketing. This data is compared with expected sales income to determine the financial wisdom of producing the product.

Often, a product is judged on what is called its *return on investment (ROI).* ROI indicates the percentage of return based on the money invested in developing, producing, and selling the product. The higher the ROI, the better the anticipated financial returns to the company will be. An example is that it takes $1,000 to research, design, and produce a part, while the company will receive $2,000 from selling that part. This would be a high ROI.

Establish a Procedure for Testing/Experimentation

Once test criteria are determined, an engineer or designer establishes a procedure for each test or experiment to be performed. Specific testing procedures may be predetermined for an industry, company, or designer. An engineer may be required to follow ***technical standards***, which are previously established norms and guidelines. Sometimes custom plans, which are unique and not standard, are put in place.

Types of tests include physical tests and performance tests. ***Physical tests*** assess whether features of a new technological product can withstand physical demands such as:

- **Tensile stress.** Pulling motion on an object in opposite directions.
- **Compression.** Pushing motion on an object from two sides toward each other.
- **Torsion.** Twisting motion on an object.

Performance tests allow an engineer to examine how a product functions in an authentic scenario. For example, performance tests can be used to identify the final characteristics of a chair or the average life of an escalator over extended periods of use. Chemical tests assess the effect of chemicals on or within a design solution. For example, an engineer designing a rubber gasket may test the design with varying amounts of chemicals to determine the material's resistance to acid rain.

Documented procedures are important not only to establish guidelines for a test or experiment but also as a form of review for analysis and replication. Therefore, it is vital to fully document testing, experiments, and measurements.

Engineers must document all initial test plans to testing procedures. Notes on what the engineer seeks to document as the test is conducted, calculations that are expected to be used, data that needs to be collected, and forms of analysis, such as graphs, should be prepared in advance. The reason a particular test method was chosen and how the results will be used should be noted. The engineer should address any concerns regarding the testing environment or sampling procedures prior to the start of testing.

Documentation includes titles (of the projects, tests, or other identifiable classification), dates, and individuals or offices responsible for overseeing the testing. All apparatus, measuring instruments, and tools used in testing, as well as safety precautions taken, should be documented ahead of time and concurrently. See **Figure 7-7**. All terminology should be clearly defined. All subsequent revisions to planned testing procedures should be documented.

Testing is done throughout each stage of the engineering design process. For example, during the *Develop Design Solutions* step, which includes predictive analysis and computer modeling, rigorous testing is a proactive measure to reduce time spent on solutions. However, during the *Evaluate the Solution* step, testing is more controlled.

Chutima Chaochaiya/Shutterstock.com

Figure 7-7. An engineer using testing apparatus.

wavebreakmedia/Shutterstock.com

Figure 7-8. This scientist is examining how a solution impacts the natural world.

Conduct Scientific Investigations

Scientific investigations are an essential component of the testing and evaluation of design solutions. This step includes investigating how a solution is impacted by or impacts the natural world. See **Figure 7-8**. Any designs or products that have unexpected and undesirable impacts on people, systems, or the environment need to be corrected.

The knowledge gained from a scientific investigation can be used to refine the design solution, eliminate a chosen design solution, and provide information for future engineering practices. The scientific method is used to conduct a scientific investigation. The steps are as follows:

1. Ask a question.
2. Complete background research.
3. Construct a hypothesis.
4. Test the hypothesis with an experiment.

STEM Connection: Science

Conducting an Experiment

When conducting an experiment, it is important to understand the terminology used in experiments. Specifically, you must understand the concept of variables and how they are used to investigate an issue, as well as control them throughout the experiment. First, a variable is a category used to describe a particular idea, such as an event, an individual's feelings, time, or an object.

Additionally, there are primarily two types of variables—independent and dependent. An independent variable is one that is not changed by the other variables being measured in the experiment. For example, an individual's age could be an independent variable. Conversely, a dependent variable is one that depends on other factors, such as test scores, which may depend on the amount of study time or sleep involved. As you are conducting an experiment, you typically are looking at how the dependent variable changes.

An example of an experiment that illustrates these variables is to determine the effect of acid rain on a copper penny. The independent variable would be the amount of acid rain that impacts the copper penny, while the dependent variable would be the discoloration of the penny.

5. Analyze the data.
6. Draw a conclusion.

Complementing scientific investigations is the use of mathematical examinations during the evaluation of design solutions. Mathematical and statistical procedures are used to analyze test data in order to evaluate and improve a design.

Collect and Analyze Data

Tests, scientific investigations, and mathematical examinations performed to evaluate a design solution all produce data. See **Figure 7-9**. Collection of relevant data allows engineers to answer research questions, test hypotheses, and determine if initial design solutions are viable or need refinement.

There are many different types of data collection methods. These methods can largely be divided into ***quantitative*** methods, which involve numbers and objective data, and ***qualitative*** methods, which involve words or subjective data. Quantitative methods are also typically confirmatory, meaning they are evaluating how a product worked, did not work, or performed overall. Qualitative methods are used to explore and describe products.

Once data is collected, the next step is to analyze it so conclusions can be drawn based on the test and experiment results. To do so, engineers prepare graphs and charts (**Figure 7-10**) to show the relationship among numerical data gathered about specific factors:

- **Comparison/relationships.** Used when data is available for two or more variables, and when comparisons/relationships are to be shown. Examples include line charts, bar charts, and Venn diagrams.
- **Distribution charts.** Used to visualize how a group of data is distributed throughout the group. Examples include column and line histograms and scatter plot charts.
- **Trend charts.** Used to show how data changes over time. Examples include a line chart or graph.
- **Composition charts.** Used to show how data is composed. Examples include pie charts.

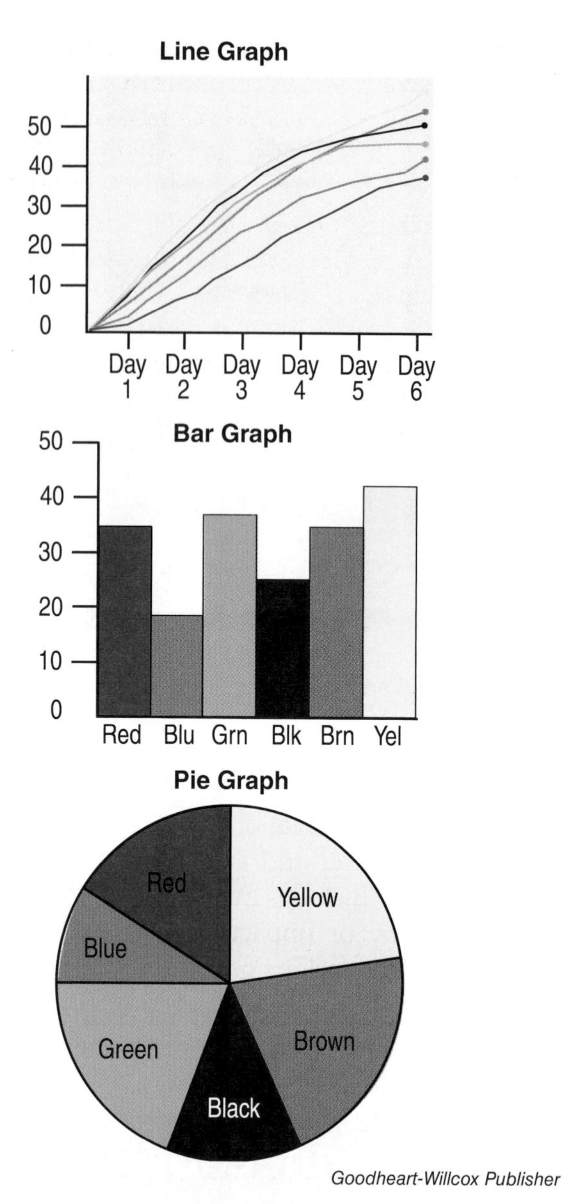

Goodheart-Willcox Publisher

Figure 7-10. Graphs show the relationship between numerical data gathered about specific factors.

RAGMA IMAGES/Shutterstock.com

Figure 7-9. This computer program allows the designer to test a material's strength.

Refine the Solution

A final and often continuous stage of evaluating design solutions is the improving of a solution based on the conclusions drawn from the testing results. The goal of all technology development activities is to design an artifact that will help people control or modify the environment. This goal requires the problem-solving and engineering design process to be continual. Needs are identified and defined. Solutions are designed and modeled. The designs are analyzed. Flaws are identified. Redesign is then often required. Problems in the original design must be solved. New designs might be developed. The original designs might be altered. New models are built and evaluated until an acceptable product or system emerges. See **Figure 7-11**.

Products have a life expectancy. New technologies and changes in people's needs and attitudes make some products obsolete. Old products need redesigning. New product definitions appear. Product and system design is a never-ending process.

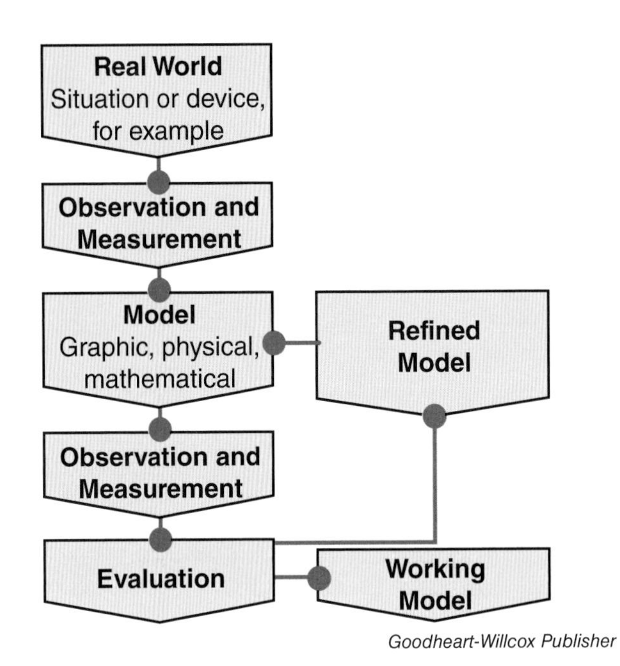

Goodheart-Willcox Publisher

Figure 7-11. The role of a model in product design and redesign.

Career Connection Market Research Analyst

Market research analysts gather and analyze information about the potential sales of products and services. They research past sales data, information about competitors, and market trends to predict the market success of products. They then evaluate the information they gather and make recommendations based on their findings.

High school students interested in a market research analyst career should take courses offered by their school's Business Education department. They can also join student organizations such as FBLA (Future Business Leaders of America) or DECA and take statistics courses to better prepare them with the necessary skills.

To become a market research analyst, college students should first complete a bachelor's degree that requires courses such as marketing, business, psychology, economics, sociology, mathematics, statistics, and computer science. People who hold bachelor's degrees in marketing and related fields, such as business administration, statistics, or communications, may qualify for entry-level positions. However, a master's degree is required for many private-sector market research positions. Market research analysts with advanced degrees and strong quantitative skills are more likely to obtain the best positions. Experienced market research analysts can obtain Professional Researcher Certification through the Marketing Research Association (MRA).

Career titles include associate research analyst, marketing analyst, research analyst, senior marketing analyst, business intelligence manager, and senior manager of survey research. Market research analysts face a very competitive marketplace. Some may have to relocate to find a position or work an industry that is not specifically related to their interests to begin their career.

A market research analyst may work on the following types of projects:

- Collecting and analyzing data related to customers, including demographics or buying preferences.
- Assessing customer satisfaction.
- Assessing the effectiveness of a specific marketing strategy.
- Preparing reports related to findings from data analysis.

Lisa photographer/Shutterstock.com

Summary

- ✔ Engineers identify test criteria through specification analysis, functional analysis, and design analysis.
- ✔ Three types of design analysis are human factors analysis (ergonomics), economic analysis, and market analysis.
- ✔ After test criteria are determined, an engineer or designer establishes a procedure for each test or experiment to be performed.
- ✔ Engineers perform physical tests (tests that assess whether features of a technical product can withstand physical demands) and performance tests (tests that examine how a product functions in an authentic scenario).
- ✔ Testing is done throughout each stage of the engineering design process.
- ✔ Scientific investigations are performed in the testing and evaluation of design solutions in order to determine how a solution is impacted by or impacts the natural world.
- ✔ Collection of relevant data allows engineers to answer research questions, test hypotheses, and determine if initial design solutions are viable or need refinement.
- ✔ Engineers prepare graphs and charts to show the relationship among numerical data gathered about specific factors.
- ✔ A final and often continuous stage of evaluating design solutions is the improving of a solution based on the conclusions drawn from the testing results.

Check Your Engineering IQ 

Now that you have finished this chapter, see what you learned by taking the chapter posttest.
www.g-wlearning.com/technologyeducation/

Test Your Knowledge 

Answer the following end-of-chapter questions using the information provided in this chapter.

1. What is *tolerance*?
2. Evaluating whether or not a product or system will operate effectively under the conditions for which it was designed is the goal of _____ analysis.
3. The analysis of how a new technological artifact or structure will affect humans is commonly called _____.
4. Market analysis involves learning customer expectations for a product's function, _____, and cost.
5. *True or False?* In market analysis, a product is often judged by its ROI.
6. When establishing testing procedures, an engineer may be required to follow previously established norms and guidelines called _____.
7. *True or False?* A compression test is an example of a performance test.
8. List the six steps of the scientific method.
9. _____ data collection methods involve numbers and objective data.
10. *True or False?* Product redesign is a common activity.

Critical Thinking

1. The engineering design challenge in Chapter 4 asked you to develop an inflatable shelter. Review the criteria required to be met in the design solution for that engineering design challenge and complete the following tasks:
   A. Explain how engineers would determine the test criteria for the inflatable structure design.
   B. Create a testing procedure or experiment that could be used to evaluate your inflatable structure design.
   C. List the data that would be collected and describe how it would be analyzed to evaluate the design solution.
2. Explain why collecting and analyzing data is important to refining a design solution.

STEM Applications

1. Select a simple game on the market. Examine the game and develop a set of specifications for its manufacture. Play the game with classmates to analyze the game's function and market acceptance. Write a brief report summarizing your analysis.
2. Find a building material or consumer product. Conduct a scientific investigation to determine the effect of acid rain or other environmental issues. Write a documented report on what you discovered.

Rawpixel/Shutterstock.com

This class is working together to create design solutions.

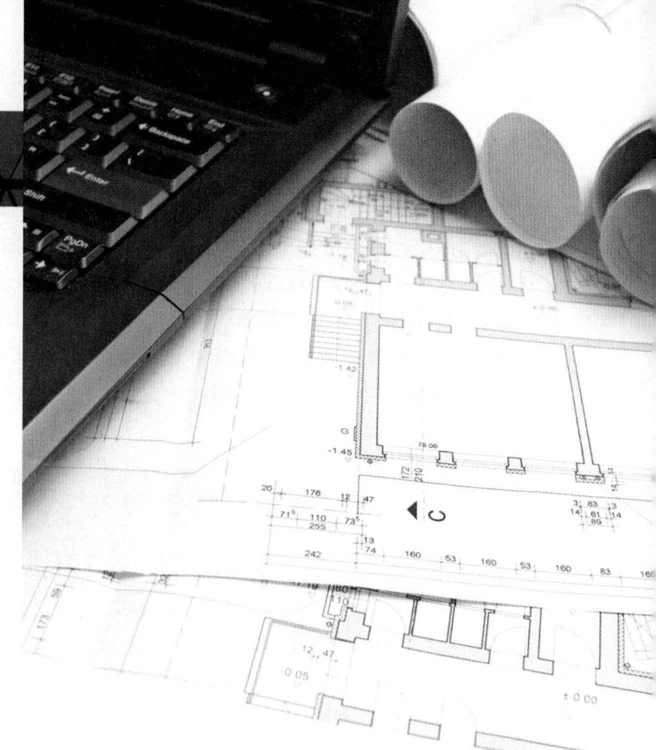

CHAPTER 8
Communicating Design Solutions

Check Your Engineering IQ 

Before you read this chapter, assess your current understanding of the chapter content by taking the chapter pretest.
www.g-wlearning.com/technologyeducation/

Learning Objectives

After studying this chapter, you should be able to:

- ✔ Explain why communication is vital to the engineering design process.
- ✔ Properly document work in an engineering notebook.
- ✔ Describe the techniques for writing technical documents.
- ✔ Explain the importance of documenting all engineering endeavors.
- ✔ Identify the different types of technical documents.
- ✔ Write a technical report.
- ✔ Recall the types of engineering design presentations.
- ✔ Explain effective ways to develop and deliver oral presentations.

Technical Terms 

active voice
empirical research
engineering notebooks
executive summary
figure
grant
intellectual property
literature review
passive voice
patent
stakeholder
table
technical writing

Close Reading

As you read this chapter, make a list of the different types of written and oral communication and write specific strategies to effectively deliver them.

Communicating design solutions is a key element in the engineering design process. See **Figure 8-1**. Engineers and other professionals must be able to study their design solution, interpret the solution evaluation data, and then share the results and conclusions in a way that can be understood by the intended audience. The work of engineers and designers must be effectively communicated in order to be useful. In addition, properly communicating the solution enables technology and society to continue to advance by building on the work of others.

Communication does not occur only at the end of the engineering design process—it takes place throughout the entire process. Engineers and other professionals work as a team to solve problems by communicating ideas and information with one another. Engineers must develop the skills needed to effectively communicate. Any miscommunication can lead to mistakes.

Engineers may be asked to share progress updates with various stakeholders. A ***stakeholder*** is any person or group that has a vested interest in the project, such as managers, government agencies, creditors, owners, or funders. Communicating the progress of the project is crucial in order to receive continued support from stakeholders for engineering design efforts. If progress is not effectively communicated, the project may even be shut down.

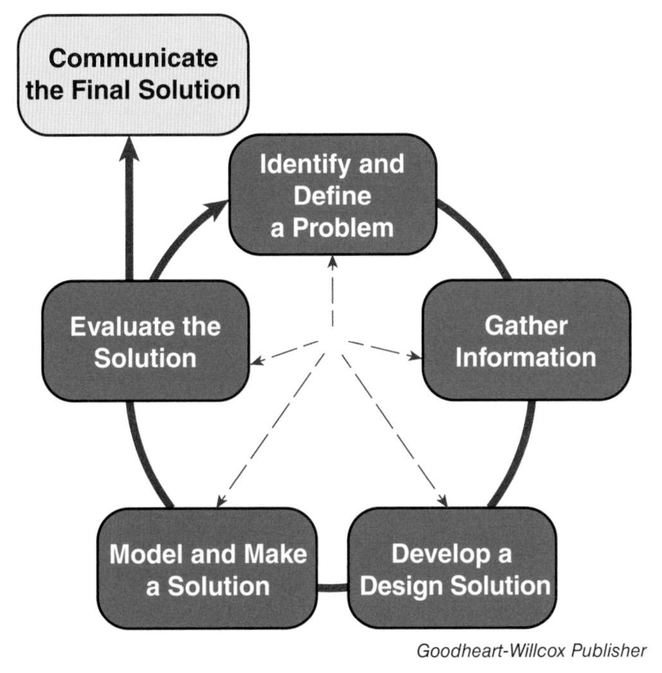

Figure 8-1. Communication takes place throughout the entire engineering design process.

Communicating design solutions involves two areas of communication. Written communication is the creation of technical documents and reports specifying all the details of the solution. Oral communication involves collaborating with others and making presentations to communicate necessary information.

Written Communication

All engineering endeavors must be written and reported in order to share meaningful information. If this information is not shared, there is no reason for collecting it. Both informal and formal, or technical, written communication are used in engineering. In the *Develop the Design Solution* step, processes, procedures, and information are often informally documented in engineering notebooks. These notebooks are then used to develop more formal or technical types of written communication for sharing design solutions.

Engineering Notebooks

Professionals use ***engineering notebooks***, sometimes called *engineering journals*, as a means to document their work toward solving a problem. All notes, research, brainstorming work, sketches, test procedures, test data, and any other thoughts or ideas related to the project are recorded. The entries are made in chronological order. The information is generally handwritten in ink and never erased. However, additional items can be glued into the notebook. See **Figure 8-2**. Furthermore, it is becoming common to keep electronic engineering journals or portfolios as well. Using electronic means enables engineers to capture videos, 3-D renderings, and other items in addition to notes and sketches.

The following practices should be followed when keeping an engineering notebook:

- Use a notebook that contains quadrille-lined or grid paper.
- Use a bound notebook. Pages should not be added or removed.

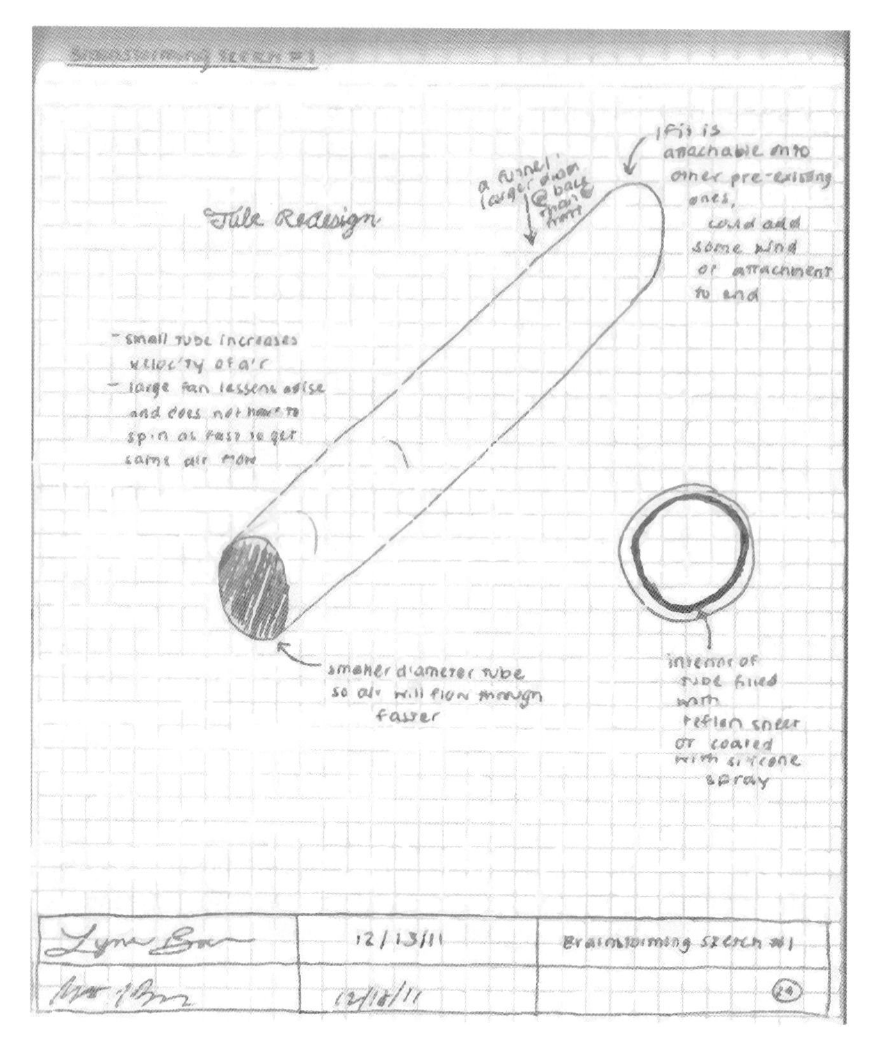

Goodheart-Willcox Publisher

Figure 8-2. An engineering student's engineering journal.

- Always use an ink pen. Do not use a pencil or marker.
- Place a sequential page number on the top outside edge of each page. See **Figure 8-3**.
- Place your signature and the date at the bottom of every page.
- Provide a location on each page for a collaborator to sign off on work.
- Store the notebook in a secure location.
- Cross out any remaining space on a page so work cannot be inserted after the date the page was created.
- If you make a mistake, cross it off and add the corrections along with your initials. Do not erase the mistake or remove the page.
- Annotate all drawings and figures.
- Be consistent in recording all work.
- If you fill a notebook, start a new notebook where the other left off.

Documenting all work in an engineering notebook serves three main purposes. First, it authenticates an engineer's ***intellectual property***. Intellectual property is any creation of the mind that can be legally protected. An engineering notebook can be used as a legal document to prove ownership of an original solution idea. This legal document is needed to obtain a ***patent*** for a design solution. A patent is a government granted right to someone who has devised a solution. The patent excludes all others from making or selling the originator's invention or innovation.

Second, engineering notebooks are essential for revisiting work that has been completed. As engineers progress toward a solution, they may need to refer to completed work in order to move forward. Some problem-solving projects take a long time to complete, with many pauses in the process. A notebook enables engineers to pick up where they left

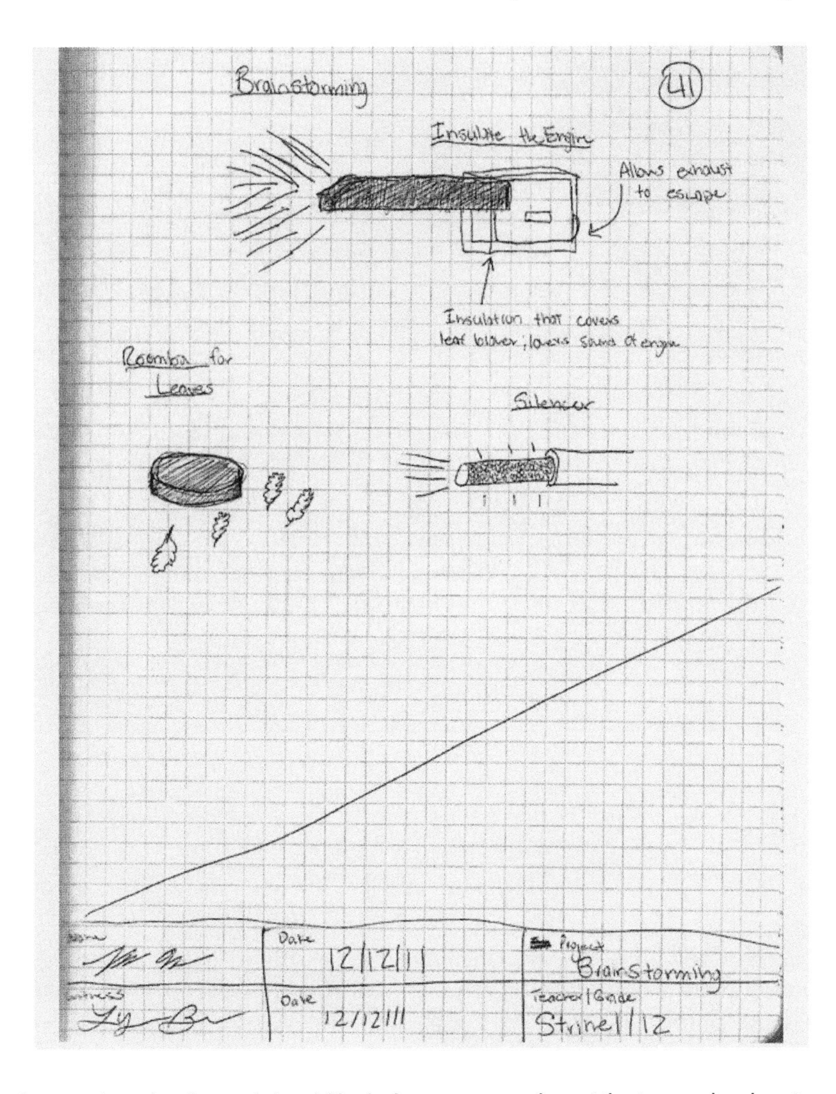

Goodheart-Willcox Publisher

Figure 8-3. Each page of an engineering journal should include a page number at the top and a signature and date at the bottom.

off. If a new person takes over a project, he or she can refer to the notebook to continue the work.

Third, an engineering notebook is used to create the technical documents and reports that are needed to share information with stakeholders. The notebook contains essential information and data used to develop other forms of written communication that use a specific formal style of writing called *technical writing.*

Technical Writing

Technical writing is a type of *expository writing,* which is a form of written communication used to explain, describe, or inform. In engineering and other technical subjects, technical writing is used to convey complex information about a project in the most clear and concise manner possible. Technical writing is different from other forms of writing, such as creative writing. The purpose of technical writing is not to entertain but to share factual and detailed information. Technical writing must be focused, accurate, and easy to understand.

Technical writing is almost always done in the ***passive voice***. The passive voice writing style focuses on the subject that experiences an action rather than the subject that performs the action. In a sentence using passive voice, the subject is acted on.

An example of passive voice is *The design solution was tested for durability by the technician.* The subject of the sentence (the design solution) is being acted on (tested) by another person (technician). This is in contrast to ***active voice***, in which the subject of the sentence performs the action. An example of active voice is *The technician tested the design solution for durability.* In this sentence, the technician is the subject of the sentence instead of the design solution. Another example of active voice is

Academic Connection: History

Thomas Alva Edison

Thomas Alva Edison, born in 1847, is regarded as one of the greatest inventors in American history. He is credited for creating devices that influenced our modern technological world, such as the phonograph, the fluoroscope, the electric lightbulb, and the motion picture camera. Some of his inventions led to the establishment of new industries, such as the movie industry and power utilities.

Edison holds over a thousand US patents. However, some people question if he created all of these inventions himself. For example, the fluoroscope began through the work of Wilhelm Rontgen. Rontgen's work then led Edison to design the first commercial fluoroscope. Which of these men should be credited with the invention? Edison helped to prove his inventions by keeping an inventor's notebook similar to an engineering notebook. He relentlessly recorded and illustrated all of his work toward inventing new technologies. Over 3,000 of Edison's journals have been discovered since his death in 1931. Therefore, his written communication efforts allowed him to continue to hold the claim for his many inventions to this day.

The team designed a mechanism to solve the problem. In a technical document, this sentence would be written as *A mechanism was designed by the team to solve the problem.*

Additionally, most technical writing is done in past tense because technical documents typically report what has been done. The purpose of these documents is to present information that is factual. Therefore, a technical paper would contain a sentence such as *The design solution was improved by changing the types of materials* instead of *The design solution will be improved by changing the materials.*

Technical writing also includes figures and tables. A ***figure*** is any type of graph, chart, drawing, photograph or graphic material that depicts some type of information necessary to the project. A ***table*** generally displays numerical data or textual information in column format. All figures and tables must be labeled following the proper format for academic papers. The most common formats are APA (American Psychological Association) and MLA (Modern Language Association). These formats are used to style the layout of most technical documents.

Technical Documents

Technical writing is used to produce technical documents, such as technical reports, progress reports, specification sheets, and proposals. Technical documents are organized around one central topic, such as an engineering project. These documents provide specific information to stakeholders and technical professionals. For example, an engineer may need to communicate the technical design specifications to the technicians who will fabricate the design. These technicians must have knowledge of the size, shape, and all other aspects of the design in order to set up and operate machines and perform assembly operations. The engineer would provide this information in a technical document.

The information about a design solution is delivered through different types of documents for a variety of purposes. The following sections discuss how these documents are developed and the information each one communicates.

Technical Reports

Technical reports are important documents in engineering and technology. Engineers, technicians, technologists, and designers dedicate a significant amount of time to writing technical reports. These reports provide project information to stakeholders, many of whom have an engineering or technical background.

Several types of formats are used for technical reports. However, they generally consist of three main parts—the front matter, the body, and the back matter.

The front matter of a technical report typically contains the following sections:

- **Cover page.** The cover page typically provides a descriptive title of the project, the names of the authors, the name of the authors' organizations of employment, and the date the report was published. This page should not contain any pictures or figures. See **Figure 8-4**.
- **Abstract.** The abstract is the first section of the report following the cover page. This short summary of the whole project briefly states what was done, how it was done, and what the results were. The abstract is typically about six to ten sentences and is written after the rest of the technical report is completed. The abstract must be written so it can stand alone, meaning that readers do not have to read the entire report to understand what is written in the abstract. See **Figure 8-5**.
- **Table of contents**. The table of contents is a list of all the major sections of the report, along with the beginning page number of each. See **Figure 8-6**.
- **Lists of tables and figures.** Following the table of contents section is a list of tables and a list of figures. These lists provide the page number where each figure and table in the report is located. See **Figure 8-7**.

The body of the technical report typically includes the following sections:

- **Introduction.** The introduction provides a broad overview of the project, focusing on the purpose of the project. This section should not include an explanation of how the project was completed or any final results and conclusions.
- **Problem statement.** The problem statement focuses on describing the problem the project intended to resolve. This section is typically only one to three sentences in length. The problem statement describes who the problem affects, what the problem is, when the problem occurs, where the problem happens, and why the problem is important to solve.
- **Objective.** This section defines the specific objectives or the results that a project aims to achieve in solving the problem. The objective section focuses on defining the scope and boundaries of the project. It does not include the work that has been done or a detailed work plan.
- **Background.** The background section provides a detailed explanation of what is already known in relation to the stated problem. It provides the information needed to begin thinking about how the problem could be solved, but does not describe how the

CHEMICAL HANDLING SYSTEMS ANALYSIS AND DESIGN

by

John H. Smith P.E.

Jane A. Doe, Ph.D

Department of Chemical Engineering
NORTH WEST CHEMCO

16 October

Goodheart-Willcox Publisher

Figure 8-4. An example of a technical report cover page.

ii

ABSTRACT

The improper handling of flowback water used in hydraulic fracturing can impede the efficient extraction of natural gas. The review of literature showed that it is unclear as to the best way to collect, transport, and dispose of flowback water. Therefore, the first objective of this project was to identify the issues that take place while collecting, transporting, and disposing of flowback water. The second objective was to design a technological system for effectively and efficiently collecting, transporting, and disposing of flowback water. The methods and procedures included multiple forms of data collection during the hydraulic fracturing process. The analysis of these data was then used to identify key areas for improvement in this process. The results enabled the development of an innovative system for handling the flowback water for the enhanced efficiency in the extraction of natural gas.

Goodheart-Willcox Publisher

Figure 8-5. An example of an abstract for a technical report.

iv

TABLE OF CONTENTS

Page

ABSTRACT.. ii

LIST OF TABLES.. vi

LIST OF FIGURES.. vii

CHAPTERS

I. INTRODUCTION.. 1

STATEMENT OF THE PROBLEM.. 2

OBJECTIVES.. 3

BACKGROUND AND SIGNIFICANCE.. 3

II. REVIEW OF LITERATURE.. 10

III. METHODS AND PROCEDURES.. 12

IV. RESULTS.. 12

V. DISCUSSION, CONCLUSIONS, AND RECOMMENDATIONS........ 15

REFERENCES.. 18

APPENDICES.. 20

A. TECHNICAL DRAWING 1.. 21

B. TECHNICAL DRAWING 2.. 22

Goodheart-Willcox Publisher

Figure 8-6. An example of a table of contents for a technical report.

iv

LIST OF TABLES

LIST OF FIGURES

Goodheart-Willcox Publisher

Figure 8-7. An example of a list of tables and figures for a technical report.

problem will be solved. This section contains many references and citations.

- **Methods and procedures.** This section describes the procedure used to develop a solution to the problem. It focuses on the ways in which the solution was evaluated. The methods and procedures are written in a way that allows a reader to duplicate the procedures of creating and testing the solution, if necessary, to check the results of the project. This section, however, does not include the results.
- **Results.** The results section provides only the results of testing the solution. It does not provide any interpretation of the results.
- **Discussion, conclusions, and recommendations.** This section first discusses an interpretation of the results and then describes the final conclusions of the project based on the analysis and interpretation of the results. This section concludes with recommendations for conducting future work related to the project results.

The back matter of a technical report typically contains the following sections:

- **References.** This section contains all of the citations referenced throughout the report. It is illegal to use someone else's words as your own. Therefore, any information from another source must be cited in the text and referenced in this section. There are several methods for citing work in a report. The most popular guidelines are the APA (American Psychological Association) and MLA (Modern Language Association) procedures. See **Figure 8-8**.
- **Appendices.** The appendices include any additional information needed to understand what is written in the body of the report. This can include engineering drawings, mathematical analyses, illustrations, or photographs.
- **List of symbols and abbreviations.** This section lists and describes symbols and abbreviations used throughout the report.

Executive Summaries

Engineers write an executive summary to briefly inform business administrators, executives, funders, and other interested parties of a project that is being proposed or is underway. An ***executive summary*** provides only the information needed to broadly describe a project and does not include too much detail. An executive summary is carefully crafted to summarize a project in only a couple of paragraphs.

76

REFERENCES

Frank, S., Love, K. B., & White, L. (2014). Understanding the differences between material properties. *Journal of Material Research, 4*(1) 8-19.

Anderson, R. (1998). *Materials and Their Properties.* New York, NY: Sample Press.

Smith, J. R. (2015). *Implications for Material Science*. San Francisco, CA: The Best Publishers.

Cooper, C. F., Anderson, J. R., Smith, K. M., & Doe, H. L. (2013). A comparison of material testing procedures. *Journal of Material Testing, 21*(2), 123-154.

Goodheart-Willcox Publisher

Figure 8-8. An example of a reference section using APA guidelines.

Status and Progress Reports

Engineers and other technical professionals regularly report the progression of their projects by creating a status report, or progress report. These documents keep stakeholders informed of the status of the work being completed in order to assist in tasks such as planning, scheduling, and budgeting. A status or progress report can take on many forms based on the requirements of the company or industry involved. They can be short summaries or longer detailed descriptions. However, these reports typically include the project objectives that have been met since the previous report, any issues that were encountered, and the next steps toward completing the project.

Design Proposals

Before a product or structure can be built, someone in charge must approve it. For example, approval may be required by company management, government agencies, or the customer. The approval process generally requires designers to prepare a design proposal. These proposals can include need statements, proposed design solutions, cost estimates, marketing strategies, economic forecasts, and environmental-impact statements.

Emails, Social Media, and Letters

People in engineering and technology professions are expected to communicate in a professional manner. This applies to communicating through emails and social media, as well as through traditional letter writing. In written communication with others, it is important to use all formal writing conventions and professional language.

Journal Articles

Engineering and technology professionals often prepare and submit articles for publication in academic journals. The majority of these journals share empirical research results. ***Empirical research*** is based on the scientific method, which is the process of conducting experiments and observations of events to determine desired information. This type of research generates new knowledge about particular phenomena. Writing journal articles is how this new knowledge is shared with professionals all over the world. A research journal article typically follows the same format as a technical report—an abstract, introduction, problem statement, research objectives, background, research methodology, results, discussions, conclusions, and recommendations.

Literature Reviews

A ***literature review*** is a scholarly paper that presents an extensive explanation of the current knowledge related to a topic based on a thorough examination of related literature, or written works. A literature review provides only information based on current sources and does not report new or original research. This type of document may be necessary for developing a design proposal, creating a final technical report, and working in an academic setting.

Grant Proposals

Engineering projects often need some type of external funding (money to support the project from someone outside of the company or organization). Professionals tasked to seek this funding in order to begin or continue a project may develop and submit a grant proposal, or a formal request for receiving a grant. A ***grant*** is a sum of money that is given by the government, a private foundation, or a public corporation for a specific purpose. Grant proposals are written in a variety of ways. Some grantors use a standard application form, while others require a full written report that completely describes the project and details all of the costs involved.

Bills of Materials

The name of this document causes some confusion. We pay bills for things we buy and use. A bill of materials, however, is a list of the materials needed to make one complete product. See **Figure 8-9**. A bill of materials generally contains the following information for each part of the product:

- A part number that can be used on assembly drawings and for ordering repair parts.
- A descriptive name for the part.
- The number, or quantity (abbreviated *qty*.), of parts needed to manufacture one product.
- The size of the part, indicating the part's thickness, width, and length (for rectangular parts) or the part's diameter and length (for round parts). Sizes are given in the order shown:

  T × W × L or Dia. × L
- The material out of which the part is to be made.

The items on a bill of materials are listed in a priority order. Manufactured parts are listed first. Parts purchased ready to use and fasteners are listed after the manufactured parts.

Specification Sheets

Specification sheets communicate the important properties a material must possess for a specific application. See **Figure 8-10**. Not all materials can be shown on a drawing. A drawing of engine oil, an adhesive, or sandpaper would be of little value. These and thousands of other items are not chosen for their size and shape. Other properties are important in their selection.

For example, some important factors in adhesives are the working time (time between application and clamping), clamping time (time the work is held together for the glue to set), and shear strength. Window glass must be transparent. Insulating materials must stop heat from passing through them.

The properties provided on a specification sheet might include the following:

- Physical properties, such as moisture content, porosity, and surface condition.
- Mechanical properties, such as strength, hardness, and elasticity.
- Chemical properties, such as corrosion resistance.
- Thermal properties, such as resistance to thermal shock, thermal conductivity, and heat resistance.
- Electrical properties, such as resistance and conductivity.
- Magnetic properties, such as permeability.
- Acoustical properties, such as sound absorption and sound conductivity.

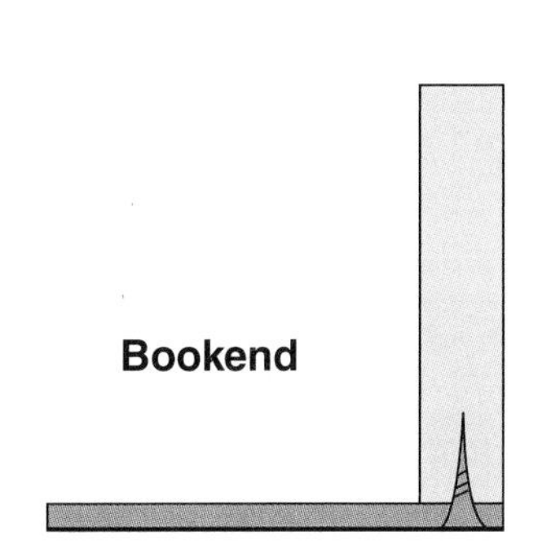

Bill of Materials						
			Size			
Part #	Part Name	Qty	T	W	L	Material
BE-1	End	1	3/4	6	8	Oak
BE-2	Bottom	1	28ga	6	6	Steel
BE-3	Protection	1	1/32	6	6	Felt
	FH Screws	3	#6		1	Steel

Goodheart-Willcox Publisher

Figure 8-9. A bill of materials for a simple bookend.

GASOLINE
Unleaded

GM 6117-M
(For Factory Fill)

GM 6118-M
(For Immediate Use)

1. Scope.
These specifications cover two types of unleaded gasoline: one for factory-fill and one for immediate or normal use. The factory-fill gasoline is intended for use in vehicles which are stored for extended periods of time, or in tanks where the rate of turnover is low.

1.1 These specifications apply to samples taken directly from the tank car or tank wagon.

2. Chemical and Physical Properties

2.1 **LEAD** Lead shall not exceed 13.209 mg per liter (.05 gram per gallon).

2.2 **PHOSPHORUS** Phosphorus shall...

...cupric acetate in benzene containing 0.20 milligram of copper, using the apparatus and procedure described in the standard method of test for oxidation stability of gasoline (induction period method).

Note 2: The oxidation stability test is designed to ensure that gasoline purchased under these specifications is sufficiently stable in the presence of metal. If metal deactivator is added to the gasoline in order to meet the specification, extreme care is necessary to ensure adequate blending of the metal deactivator with the gasoline.

In sampling gasoline for this test, it is desirable to collect the...

Goodheart-Willcox Publisher

Figure 8-10. The characteristics of materials are described on specification sheets.

- Optical properties, such as color, transparency, and optical reflectivity.

These properties are further discussed in Chapter 10 and Chapter 19.

Technical Data Sheets

The specifications are included on two types of sheets. In the first type, manufacturers prepare technical data sheets to communicate the specifications for products they have on the market. These kinds of products are called *standard materials and components* or *off-the-shelf materials and components.* They are generally kept in stock by the manufacturer and are often listed in a supplier's catalog.

For example, you can write to a manufacturer regarding your need for an adhesive. You would probably receive technical data sheets on several adhesives the manufacturer makes that might meet your needs. You can also research online to view and possibly download the technical data sheets. For each product, you would study the specifications and choose the one that meets your needs.

In the second type, large organizations prepare their own specifications for materials and products they need. They send the specifications to suppliers, who then compete to supply a specific item. One example is the military specification (Mil-Spec) system, in which the government lists its specifications. Large manufacturing companies also have specification systems.

Engineering Drawings

As you learned in Chapter 5, engineering drawings are used to communicate a design solution idea or the information needed to produce the design solution. Engineering drawings are an essential component of technical documents and technical presentations. Engineering drawings are primarily produced through computer-aided design (CAD) software. The drawings graphically communicate the details of a design that would be difficult to describe in writing or verbally. See **Figure 8-11**.

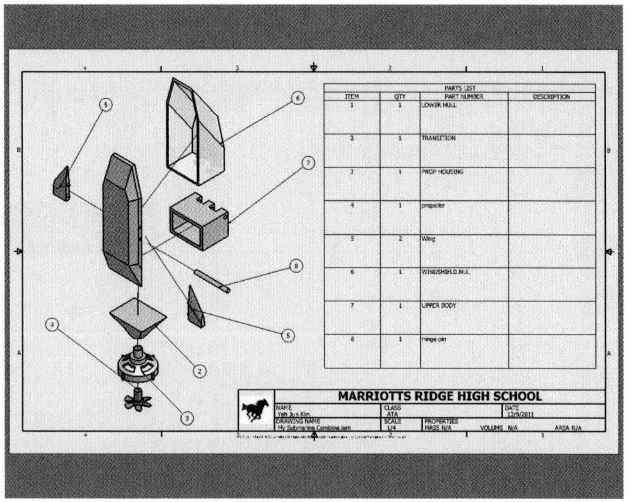

Goodheart-Willcox Publisher

Figure 8-11. An engineering drawing used in a technical report.

Oral Communication

Oral communication in the engineering design process can be formal or informal. Informal communication takes place during phone calls or at meetings, while formal communication is done through professional speeches or presentations. These types of communication are necessary to convey the status of work to colleagues and stakeholders who may control certain aspects of the project. Therefore, it is important to be able to effectively share your work verbally to a variety of audiences with confidence and professionalism.

Engineering Design Presentations

Engineers and other technical professionals deliver different types of oral presentations to diverse audiences, **Figure 8-12**. However, engineers typically deliver four types of presentations:

- **Problem proposals.** The purpose of a problem proposal is to present a valid problem in need of a technological solution to potential stakeholders as a means to justify a potential project. This presentation should provide an explanation of a problem statement, a justification as to why the problem needs a solution, and any market research related to the problem. Potential stakeholders can then provide feedback as to whether the problem proposed is worth working on to solve.
- **Preliminary design reviews.** The purpose of a preliminary design review is to present a developed solution pathway for solving a proposed problem. The presentation is made to a panel of team members and experts as a means to justify and evaluate the decisions made for the solution pathway. This presentation should include a detailed explanation of a design solution concept and the manner in which the concept was developed. The panel then provides feedback on the design concept to be implemented.
- **Critical design reviews.** The purpose of a critical design review is to present a final design solution or prototype to a panel of experts and stakeholders as a means to formally assess

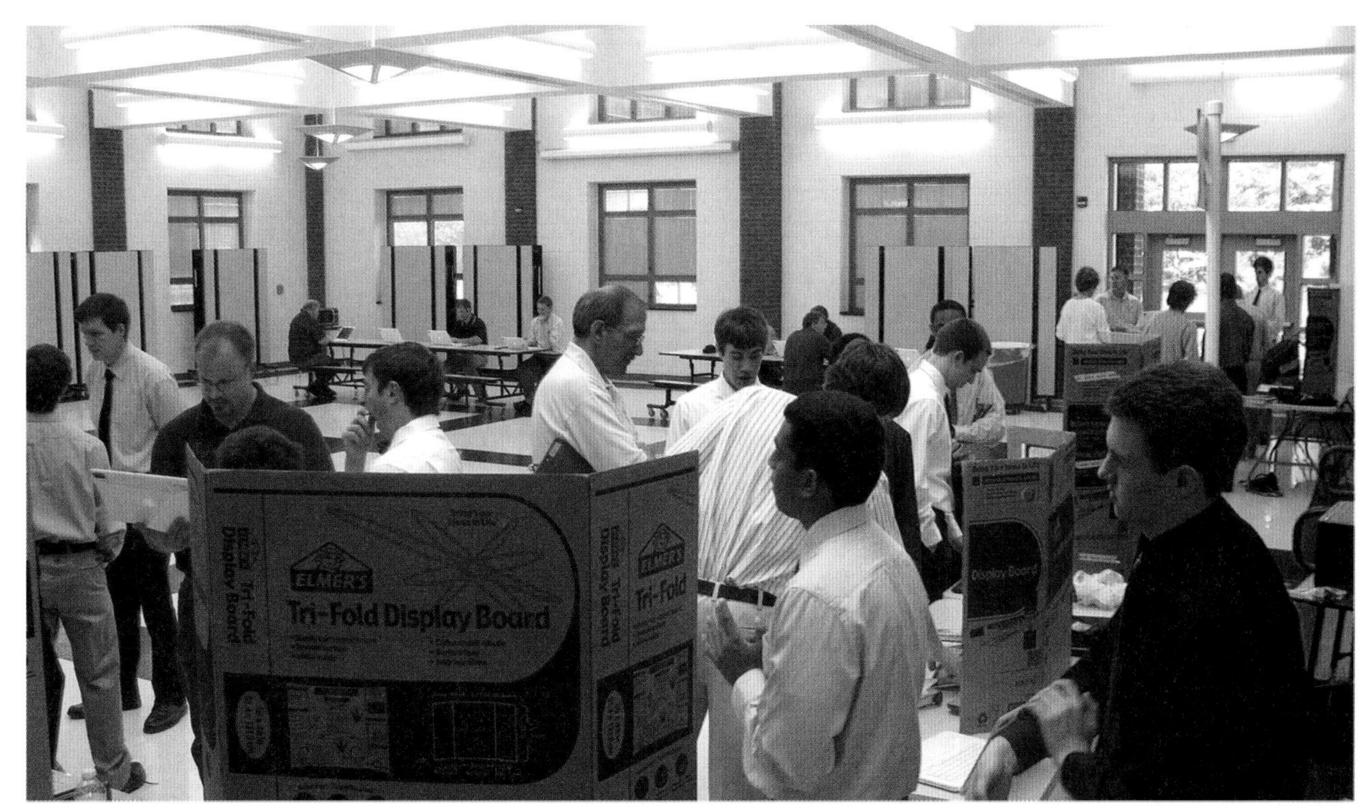

Goodheart-Willcox Publisher

Figure 8-12. Engineering students must be able to present to a panel of experts.

Career Connection Technical Writer

Undrey/Shutterstock.com

Technical writers produce technical documents in which complex information is communicated in the most comprehensible way possible. These individuals create, gather, and share technical information between engineers, designer, customers, manufacturers, and business executives.

High school students interested in becoming a technical writer should focus on their coursework in English. Additionally, many high schools offer courses in journalism and web design, which can be very beneficial for a future technical writer. Students may also want to explore engineering and technology coursework to establish foundational knowledge in technical subjects.

Most technical writer positions require a bachelor's degree in English, communication, or journalism. Certain positions may also require a degree or certification in a technical field, such as engineering or computer science. Education in web design and multimedia is also important.

Technical writers must have excellent writing and communication skills in order to clearly explain complex information. They need to be detail oriented to ensure that all written communication is correct and easy to read. Additionally, technical writers need to be able to work well with others because their job involves communicating with technical professionals and other writing team members.

Technical writers work in a variety of companies, including information technology, scientific, and technical research companies. As they learn more about the company's products and become more specialized in their activities, technical writers can advance to more senior level positions. In higher level positions, they work on complex projects and may take on the responsibility of training and leading junior staff. Some technical writers use their skills to become freelance writers who may be self-employed or work for a technical consulting firm.

Employment opportunities for technical writers are expected to continuously grow as more industries become increasingly scientific and create more highly technical products. For example, the electronics industry will need more people who can explain highly sophisticated products in simpler terms. Additionally, the information age requires all companies and industries to a have a web presence that requires additional trained technical writers.

A technical writer may work on the following types of projects:

- Collaborating with engineering and technology professionals to organize and write supporting technical documents for products and systems.
- Studying the needs of product and system end users in order to create technical documentation that better addresses their needs.
- Writing instruction manuals or journal articles.
- Revising technical documents.
- Generating graphics to support technical documentation.
- Assessing customer satisfaction.

Students and technical communications professionals can join the Society for Technical Communication. This organization provides educational opportunities and helps its members keep up with changes in the field.

the work. This formal assessment enables engineers and technical professionals to evaluate their efforts, determine any necessary revisions, and establish a plan for completing the project. These presentations should include an explanation of design specifications, the procedure used to develop the solution, tests performed to evaluate the solution, the initial testing results, an explanation of any issues or challenges faced, and any changes made to the solution. The panel then provides input regarding revisions to be made before the solution is finalized.

- **Final design presentation.** A final design presentation can be delivered to stakeholders as a means to provide the results of the entire engineering design process. These presentations broadly describe the problem, the design specifications, the procedure used to develop the final solution, and the revisions made throughout. Additionally, a final presentation should include an in-depth description of how well the solution meets the desired goals, as well as the project timeline and expenditures.

Understanding Your Audience

Regardless of the type of presentation, it is important to understand the audience. Engineers present information to other engineers or technicians as well as business executives, potential funders, or customers who may or may not have a technical background. Therefore, it is important to tailor all presentations to your audience by adjusting the terminology and the amount of detailed information provided.

For example, if you are presenting to engineers, include a detailed and thorough theoretical explanation of the design solution using a full range of technical language. If you are presenting to technicians, focus less on theory and more on technical specifications, materials, and engineering drawings. Technicians are talented in their specialties and already have solid background knowledge in their areas of expertise. If you are presenting to business executives or funders, supply succinct information and focus more on budgets and timelines. When presenting to customers, truthfully discuss your commitments, timelines, and the capabilities of your products or systems.

Developing a Presentation

Presentation software, such as Microsoft® Office PowerPoint®, is commonly used to present information and enhance a presentation. In addition to text, PowerPoint® slides can contain visual aids such as drawings, videos, images, photographs, diagrams, graphs, and charts. However, presentation software must be used wisely. A poorly developed slideshow can distract the audience and hinder their comprehension. Follow these rules when using presentation software:

- Use bullet points instead of sentences or paragraphs. Only key information should be provided on a slide. Keep bullet points to a maximum of six per slide. Ensure there is proper spacing between bullets to allow the audience to distinguish between bullet points easily.
- All text needs to be large enough to be visible by the audience members in the back of the room. Generally, text should not be smaller than 24 point font.
- Use visual aids whenever possible to minimize the use of text. One of the biggest complaints about presentations is that there is too much text, which disengages the audience from the presentation. Visual aids, such as engineering drawings, videos, graphs, and photographs, engage the audience and enhance the comprehension of the concepts presented.
- All visual aids must be large enough to be seen by all members of the audience. Visual aids and graphics that contain text should be large enough so that the audience can read it.
- Design presentation slides in a way that is stimulating and professional. Ensure that the colors used are not overwhelming and distracting. The slide design should provide a lot of contrast between the slide background and the text for readability. See **Figure 8-13**.
- Be consistent throughout the presentation. The presentation will look more professional if the same slide design, font, and colors are used for the entire presentation.
- Use animations and sounds only if they help the audience understand concepts. Unnecessary sounds and animations can distract the audience and draw their attention away from the presenter.

PRESENTATION OVERVIEW
Background Information
Study Purpose
Methodology
Preliminary Findings
Discussion
DGLimages/Shutterstock.com

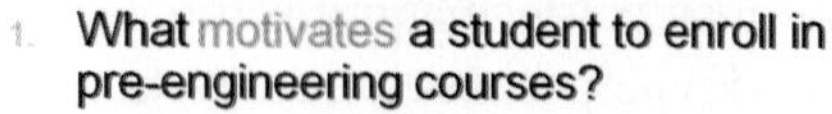

STUDY PURPOSE
Identify and differentiate the motivational factors of male and female ninth grade students for enrolling in a four-year pre-engineering academy in order to increase enrollment in engineering
1. What motivates a student to enroll in pre-engineering courses?
2. What could be done to encourage other male and female students to enroll in pre-engineering courses?
3. Determine the current pre-engineering course experience of these students.
This population was targeted because...
Enrollment tripled overall
Increase from one female student to eighteen
dotshock/Shutterstock.com; Monkey Business Images/Shutterstock.com; DGLimages/Shutterstock.com

Figure 8-13. A comparison of a well-designed presentation slide (left) and a poorly designed presentation slide (right).

Delivering the Presentation

Delivering a presentation can be a difficult task for engineering and technology professionals. Many people are not practiced in public speaking and are frightened of speaking in front of audiences. The following practices can aid in delivering a successful presentation.

- Tailor your presentation to the audience. Make sure that you deliver information in a manner that is comprehensible to your audience.
- Practice delivering a presentation in the mirror as well as in front of friends and family prior to delivering the actual presentation. With practice, the real presentation will likely be easier to deliver.
- Speak naturally. Do not read from your presentation slides or memorize a script. The audience does not want to listen to you read something that they could read on their own.
- Speak loud enough so your entire audience can hear you clearly. Keep an eye on your audience to see if they are straining to hear you. If so, then speak up. If you feel that you cannot speak loudly enough for the size of the room, use a microphone.
- Wear professional attire when delivering a presentation. This will help the audience take you seriously. You will also feel more confident while presenting if you are well dressed. See **Figure 8-14**.
- Stay focused on the topic. Do not ramble on about information that is not important to your audience. Pay attention to the body language of your audience to be sure that you are not losing their interest with information that is not important to them.
- Begin and end your presentation on time. If the presentation is supposed to last 15 minutes, then it should end at 15 minutes. If you go past the planned time, your audience may disengage from your presentation. Practice your timing prior to the presentation.
- Check the technology you are using prior to your presentation to ensure that it is functional. Ensure that the projector, computer, and any software, is compatible with your presentation materials. Be prepared by having a backup idea for the presentation if the technology fails. For example, if you are using a video during your presentation, make sure you have it in a variety of standard formats so it is likely to work with any software.
- Answer questions honestly and professionally. When you are asked a question, repeat it so you can affirm that you heard the question correctly and so the entire audience can hear the question. Then answer the question to the best of your ability. If you do not know an answer, say that you do not have an answer at this time but you will look into it and answer the question when you have the information.

Monkey Business Images/Shutterstock.com

moodboard/Thinkstock.com

Figure 8-14. When delivering a presentation, wear professional attire.

Summary

- ✔ Engineers and other professionals must be able to share design information in a way that can be understood by the intended audience.
- ✔ Engineers share progress updates with stakeholders who have a vested interest in the project in order to obtain their continued support.
- ✔ Notes, research, brainstorming work, sketches, test procedures, test data, and any other thoughts or ideas related to a design project are recorded in an engineering notebook.
- ✔ Technical writing is used to explain, describe, or inform. Technical writing is focused, accurate, easy to understand, and written in the passive voice.
- ✔ Technical writing is used to produce technical documents, such as technical reports, progress reports, specification sheets, and proposals.
- ✔ Technical reports consist of three main parts—the front matter, the body, and the back matter, each containing multiple sections.
- ✔ In addition to technical reports, engineers and technical professionals write documents such as executive summaries, status and progress reports, design proposals, and journal articles.
- ✔ Grant proposals are written in an effort to obtain funding for a project. A grant is money given by the government, a private foundation, or a public corporation for a specific purpose.
- ✔ Specifications communicate the important properties a material must possess for a specific application. Specifications are included on technical data sheets.
- ✔ Formal communication, through speeches or presentations, is used to convey the status of work to colleagues and stakeholders.
- ✔ Presentations should be tailored to the audience. Visual aids, such as engineering drawings, videos, graphs, and photographs, engage the audience and enhance the comprehension of the concepts presented.

Check Your Engineering IQ

Now that you have finished this chapter, see what you learned by taking the chapter posttest.
www.g-wlearning.com/technologyeducation/

Test Your Knowledge

Answer the following end-of-chapter questions using the information provided in this chapter.

1. Engineers document their work in a(n) _____ that can be used to prove ownership of an original solution idea.
2. The purpose of _____ writing is to convey complex information about a project in a clear and concise manner.
3. Technical documents are written in the _____ voice and in _____ tense.
4. The section of a technical report that contains a short summary of the whole project is the _____.
   A. objective
   B. introduction
   C. background
   D. abstract
5. *True or False?* The background section of a technical report describes how a problem will be solved.

Matching questions: Match each technical document to its description.

A. Technical report
B. Specification sheet
C. Grant proposal
D. Literature review
E. Executive summary
F. Bill of materials

6. Communicates the important properties a material must possess for a specific application.
7. A list of the materials needed to make one complete product.
8. A formal request for funding.
9. A scholarly paper that presents an extensive explanation of the current knowledge related to a topic.
10. Explains in depth information of a project to a variety of stakeholders who mostly have an engineering or technical background.
11. Briefly informs business administrators, executives, funders, and other interested parties of a project that is being proposed or a project that is underway.

12. What is the purpose of a critical design review?
13. *True or False?* When preparing a presentation using presentation software, you should minimize the use of text.

Critical Thinking

1. How is communication vital to the engineering design process?
2. What are the important reasons for documenting all engineering endeavors?
3. Explain how you would prepare a presentation on a complex technical topic for an audience with little technical background. What steps would you take to aid their comprehension of the topic?

STEM Applications

1. In Chapter 4, you completed an engineering design challenge in which you designed an inflatable shelter. In Chapter 7, you established a procedure for evaluating your solution. Using the information from these activities, write a technical report. Review the information in this chapter regarding the sections of a technical report and ensure that your report contains all of the sections.
2. Develop a preliminary design review presentation and deliver it to your classmates. This presentation should include a detailed explanation of a design solution concept as well as the manner in which the concept was developed.
3. Create an online version of an engineering notebook using a free website builder.

Engineering Design Challenge

Bookend Design

Background

Product designers often work to improve an existing product. They also develop new and improved products to meet the changing demands and requirements of customers.

Challenge

You are a product designer for the Acme Bookend Company, which makes bookends for different markets. Each market gets unique graphics and special shapes for the bookends. Your boss has asked you to modify an existing product to meet the following criteria:

- A new shape.
- New graphics appealing to high school students.

Material and Equipment

- Sketch paper.
- Drawing paper.
- Pencils.
- Felt-tip pens.
- A drafting ruler.
- A T-square.
- Triangles.

Procedure

Redesign the bookend to meet the new criteria by designing the new product shape and developing a new decoration for the product. Prepare a set of drawings to communicate your new design. See **Figure A**.

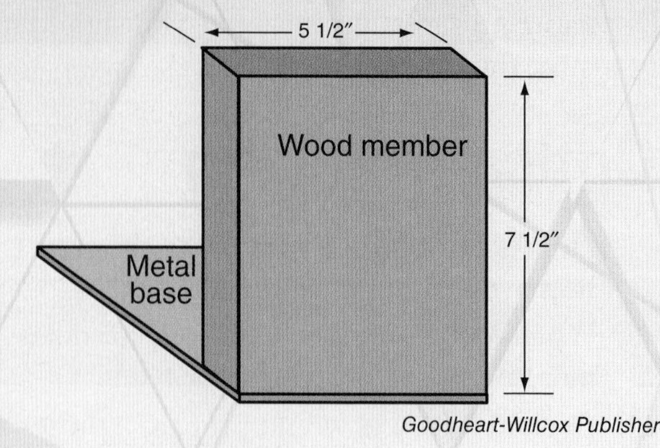

Figure A. A basic bookend.

Designing the Shape

1. Photocopy the bookend layout sheet. See **Figure B**.
2. Sketch four new shapes for the bookend.
3. Select the best shape. Circle it with a felt-tip pen.

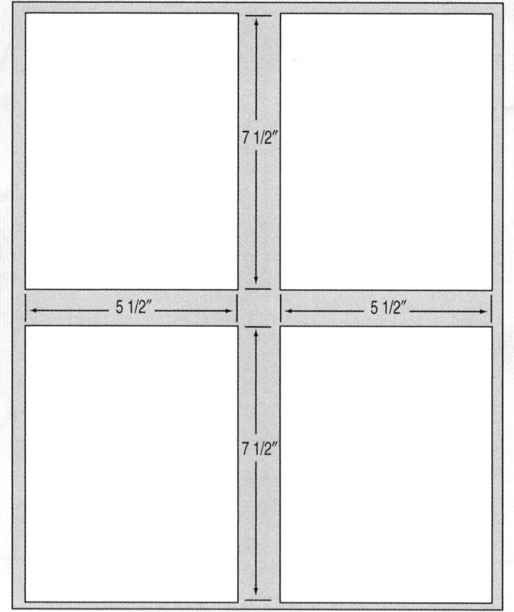

Figure B. A bookend-design layout sheet.

Decorating the Product

1. Make four layouts using the shape you chose.
2. Sketch four new graphic designs. Make sure the design appeals to high school students by conducting a survey or taking a poll of other students.
3. Circle the best design with a felt-tip pen.

Communicating the Design

1. On a piece of drawing paper, draw a border 1/2″ in from all edges of the paper.
2. Draw a title box. See **Figure C**.
3. Produce a dimensioned two-view orthographic drawing of your design for the new bookend.

School:	Part Name:	Drawn By: Date:
Class:	Part Number:	Checked By: Date:

Goodheart-Willcox Publisher

Figure C. A title block.

SECTION 3
Technological Systems

Tomorrow's Technology Today

Home Fuel Cells

Have you ever thought about where electric power comes from? It is transmitted into homes, schools, and many other buildings. Our society would be very different without it. It provides light in the dark, and it powers everything from refrigerators to computers. Nearly half of the energy consumed in the United States is used for electricity.

Steam-powered electrical plants are commonly used to generate electricity. Fossil fuels are frequently used. The fuels used in this type of energy conversion are coal, natural gas, or fuel oils. These resources are burned to produce thermal energy. The thermal energy must then be converted into mechanical energy in order to provide power to a generator.

This method of electricity generation consumes large amounts of fossil fuels while emitting harmful carbon dioxide into the atmosphere. One alternative to this conventional method of electrical energy production is the use of home fuel cells. A fuel cell uses a chemical reaction to produce electricity. Although home fuel cells have been in the works for some time, they are now becoming a reality.

There are several companies attempting to create home fuel cells to generate electricity. The fuel cells use the reaction between oxygen and natural gas to generate power. Biogas can be used instead of natural gas, completely eliminating any carbon dioxide emissions. Solid oxide fuel cells cost less and are more efficient than proton exchange membrane (PEM) fuel cells. Solid oxide fuel cells use a ceramic electrolyte. Companies working to produce home fuel cells have begun working on more efficient alternatives to this electrolyte, however. They have been looking for lower-cost electrolytes to make home fuel cells more affordable for the general public.

Attempts have been made in the past to manufacture inexpensive home fuel cells, but the cost remains high. As companies work to find less expensive ways to create home fuel cells, however, we may see them in residential use soon.

While studying this chapter, look for the activity icon to:

- **Assess** your knowledge with self-check pretest and posttests.
- **Practice** technical terms with e-flash cards, matching activities, and vocabulary games.
- **Reinforce** what you learn by submitting end-of-chapter questions.
  www.g-wlearning.com/technologyeducation/

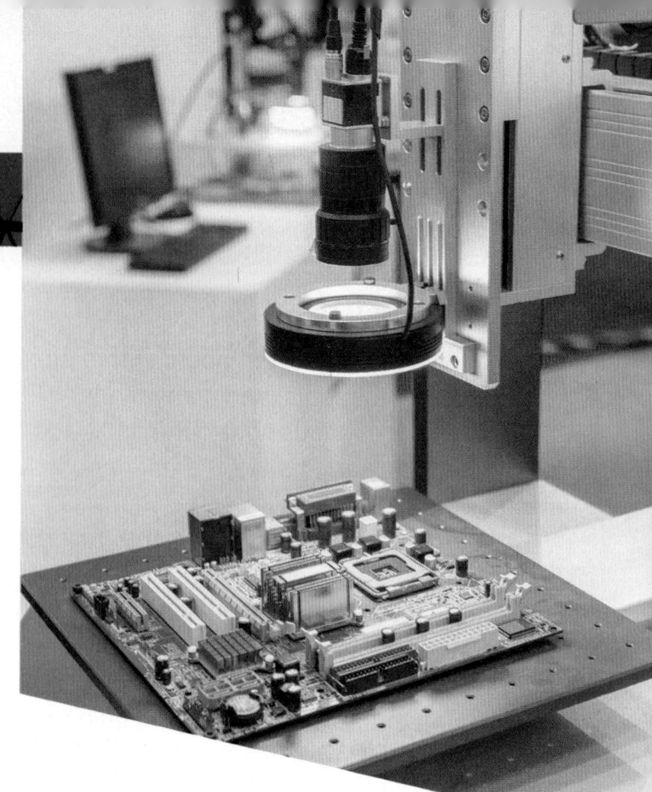

CHAPTER 9
Technology as a System

Check Your Engineering IQ 

Before you read this chapter, assess your current understanding of the chapter content by taking the chapter pretest.
www.g-wlearning.com/technologyeducation/

Learning Objectives

After studying this chapter, you should be able to:

- ✔ Explain how technology is a system.
- ✔ Discuss how "systems thinking" influences technological development.
- ✔ Generalize the major components of a technological system.
- ✔ Analyze the inputs to a technological system.
- ✔ Summarize the steps in the engineering design problem-solving process.
- ✔ Recognize the major activities involved in management processes.
- ✔ Evaluate examples of positive and negative technological outputs.
- ✔ Explain feedback and control.

Technical Terms 

control
energy
feedback
finances
information
inputs
machines
management processes
materials
natural systems
outputs
processes
production processes
profit
social systems
system
systems thinking
technological systems
time
tools

Close Reading

Engineers often create graphics and diagrams while reading to visualize their thoughts and conceptual understanding of ideas. As you read this chapter, make a diagram of how everyday technologies relate to the universal systems model.

Engineering and technology involve the development and application of knowledge, tools, and human skills to solve problems and extend human potential, **Figure 9-1**. Technology starts from and grows out of human wants and needs. The development of technologies is designed and evaluated by people and, in time, is improved or abandoned. Every technology, whatever the force behind it, is intertwined with a variety of natural, social, and technological systems; is a system itself; and is developed through a system. A ***system*** can be described as any entity or object that consists of parts, each of which has a relationship with all other parts and to the entity as a whole. These parts work together in a predictable or planned way to achieve a specific goal.

Technological systems are a group of components designed by people to function together to complete a desired task, **Figure 9-2**. These systems solve a variety of problems by employing many engineering practices. They are the result of the engineering design process and are often created using other technological systems, such as production tools and machines. Technological systems also interact with other technological, natural, and social systems.

Technological systems vary in size and complexity. Some technological systems can be the minor systems or subsystems that operate in or are created through larger technological systems. For example, the automobile is a technological system that is comprised of other subsystems, such as guidance and control systems. These subsystems all interact together to form the larger automobile system. These subsystems and systems can also function together with other related systems or subsystems to compose major technological systems, such as transportation, manufacturing, communication, and construction systems.

Technological systems also interact with systems that are found in nature. ***Natural systems*** are comprised of elements that work together in nature in a predictable, but at times unexpected, manner to perform some type of function not under the direct control of humans, **Figure 9-3**. Natural systems can include things like ecosystems, the water cycle, the nitrogen cycle, and biological systems, such as the nervous, respiratory, and reproductive systems.

mandritoiu/Shutterstock.com

Figure 9-2. Technological systems are often designed to complete complex tasks, such as producing energy through nuclear fission.

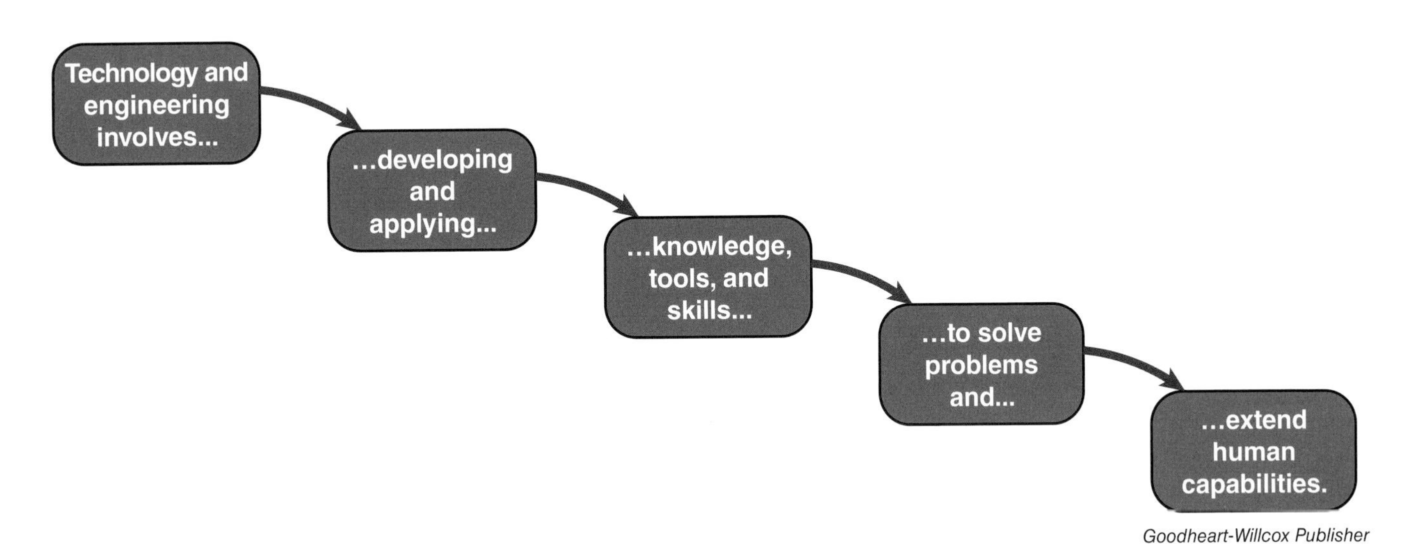

Goodheart-Willcox Publisher

Figure 9-1. Technology and engineering extend human capabilities.

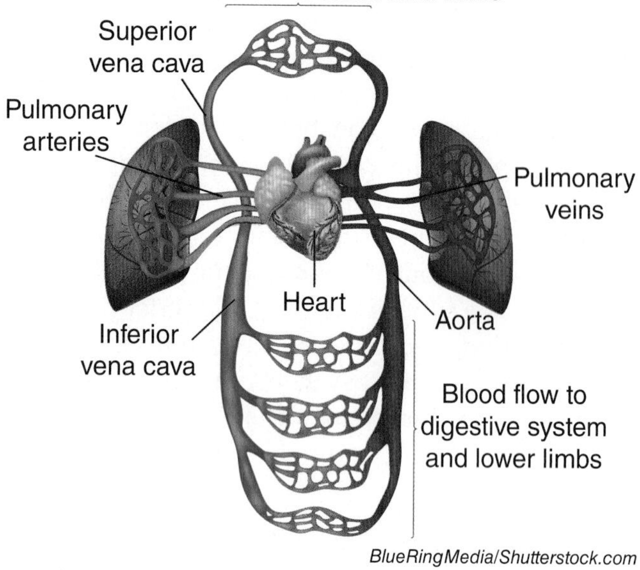

BlueRingMedia/Shutterstock.com

Figure 9-3. Nature operates in systems, such as the circulatory system.

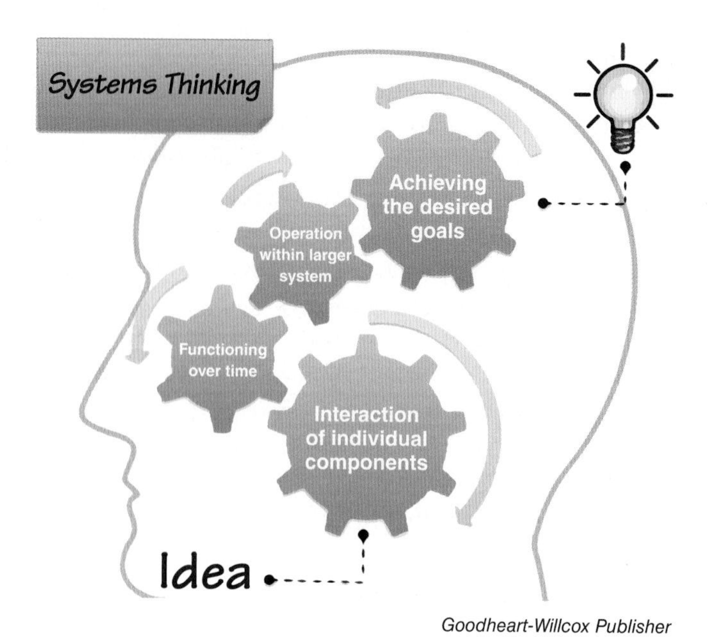

Goodheart-Willcox Publisher

Figure 9-4. Systems thinking is a holistic way to determine how a system's components interact with each other, how a system will function over time, and the way in which the system operates within the context of larger systems.

Technological systems operate in, affect, and are influenced by social systems. ***Social systems*** are organized groups of individuals that have a variety of goals, functions, or characteristics. These systems may involve the ideas and values of different cultures and the political operations of different governments that drive the use and development of technology.

Understanding how all types of systems work and interact with each other is important for engineers. Understanding systems enables them to develop and create new and successful designs to solve engineering problems. The ability to think about the design, function, and interaction of systems is called systems thinking. The ability to think in terms of systems requires an understanding of the general components of the universal systems model for technological systems.

Systems Thinking

Systems thinking is the process of understanding and mentally exploring systems in multiple ways, **Figure 9-4**. It is a method that scrutinizes the way in which a system's components interact with each other, how a system functions over time, and the way in which the system operates within the context of larger systems.

Systems thinking begins by examining the way the parts of a system interact with each other as one entity to achieve a desired goal. When designing a technological system, engineers need to visualize how each design idea or modification, no matter how small, affects the overall function of the system design itself. This requires in-depth knowledge of the concepts involved in areas such as material properties and the interaction of forces and motion, as well as the skills to conduct predictive analyses of designs.

Systems thinking also involves understanding and examining the way a technological system influences and interacts with other technological systems or subsystems. For example, designing a new battery system to increase the distance an electric vehicle can travel might also impact the structural system if the new battery system weighs more than the old system. The structural system may need to be changed or modified to support the new, heavier battery system. A new battery system can also influence the transportation system. New charging stations may need to be constructed, and may also change the need for and production and transportation of gasoline. These complexities must be considered when developing and producing all technologies.

Systems thinking also involves studying scientific research related to natural systems and understanding the culture in which the system operates.

STEM Connection: Science

Newton's Three Laws of Motion

Georgios Kollidas/Shutterstock.com

Sir Isaac Newton (1642–1726) was one of the most influential scientists of all time. His contributions to understanding the natural world laid the foundation for engineering and technology. Newton was an English physicist and mathematician who discovered the laws of gravity, is often credited with developing calculus, and formulated the laws of motion.

Newton's three laws of motion are extremely important when thinking in terms of technological systems. These laws describe the relationships between an object and the forces that act on it that result in motion. Understanding these laws can aid in the design of technological systems to achieve a desired effect. For example, applying this scientific knowledge can help develop a safety system that could minimize the force on a person during a collision in an automobile.

Newton's first law of motion states that an object at rest will remain at rest and an object in motion will remain in motion at a constant velocity, unless acted on by an external force. This law is also known as the law of inertia. Inertia is the tendency of an object to resist change in motion. So what does this mean? Think about a golf ball sitting on a tee. The golf ball will remain there forever, unless some type of force is applied to it, such as a golf club or the wind. Once the ball is hit with a golf club, it will remain moving forward at the same velocity of the golfer's swing until it is acted on by some type of force—friction from the air, the pull of gravity, a tree or building in its path. If the ball was hit in space where air or gravity is not present, then the ball would continue on at the same velocity forever. When designing different technological systems, one must keep this law in mind. Think about the design of an automobile. If a vehicle traveling 60 miles per hour hits a brick wall, any vehicle occupants will continue to travel at 60 miles per hour until they are acted on by some type of force. This is why safety restraints have been designed. They are the force that resists the motion of a person in an automobile collision.

Newton's second law of motion mathematically describes the relationship between the force applied to an object, the object's mass, and its acceleration. This relationship is described as *the force of an object is equal to its mass multiplied by its acceleration*, or force = mass × acceleration. What this really means is the movement of an object depends on the size and direction of the force applied to it as well as the mass of the object itself. To understand this law, think about two objects of different masses, such as a golf cart and a car. If you push the golf cart and the car in the same direction and with the same amount of force, they will both move in the same direction they are being pushed. However, based on this law, the acceleration of each object is equal to the force that you pushed them with divided by the mass of the object. Therefore, the object with more mass (car) will move more slowly than the object with less mass (golf cart).

Newton's third law of motion states that for every action by a force, there is an equal and opposite reaction by another force. This is an extremely important concept to understand when designing or operating a variety of technological systems. For example, we can see this law at work with a jet engine. The engine is designed to shoot out a mass of hot gas. The force of this gas in one direction creates an equally strong force in the opposite direction. Thus, when the hot gas shoots backward, the jet plane shoots forward. We can also see this law at work when a book is lying on a table. It is actually exerting a force on the table because of the pull of gravity. In turn, the table exerts an equal and opposite reaction on the book. What might be another example of this law at work?

What other ways can these scientific principles and engineering practices be applied to design or improve technological systems?

Engineering and technology can certainly affect the natural world and the ways in which people live. All engineering endeavors have short-term and long-term effects for people and the natural environment. These effects can be positive or negative. Because engineering and technological activity can draw on natural resources, it can impact the earth's natural systems. It is important to understand how technological and natural systems interact when developing solutions to problems. This understanding will aid in the design, production, and use of technological systems so any negative impact on the natural environment is minimized.

Constant advancement of engineering, technology, and science not only changes the way people interact with their natural environment but also the way in which people live and interact with one another. These advancements, especially within the agriculture, information and communication, energy, and medical systems, have drastically changed society and the systems within it. Because of these advancements, people now live longer and can make more conscientious and informed decisions about the ways in which technological systems are designed and used. Technological systems impact and are influenced by both natural and social systems. As a result, it is important for all humans to think in terms of systems and understand the interworking of their individual components when developing and using all technological systems.

Universal Systems Model

The general components of a technological system can be visualized using the universal systems model, **Figure 9-5**. Technological systems are designed to meet a goal and do so by inputting resources and using processes to produce a desired output. Technological systems also collect feedback as a way to control the system to achieve the desired output. To be able to think in terms of systems, these general components must be understood in depth.

- Goals.
- Inputs.
- Processes.
- Outputs.
- Feedback and control.

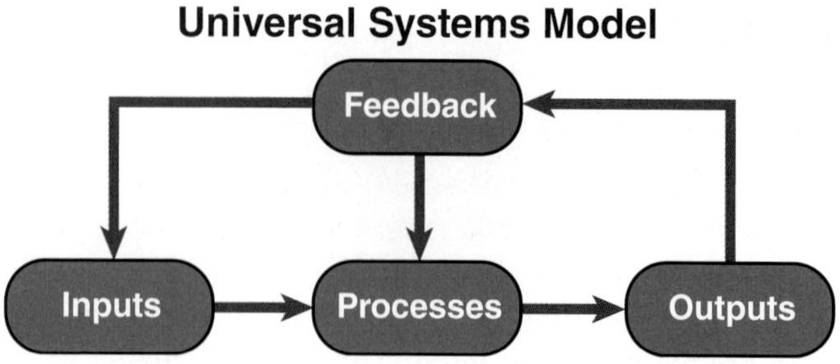

Universal Systems Model

Feedback

Inputs → Processes → Outputs

Goodheart-Willcox Publisher

Figure 9-5. Technological systems consist of general components, which are represented in the universal systems model.

Goals

Humans develop technological systems to achieve desired tasks. This means humans have a goal in mind when developing a technological system. Each system, however, generally meets more than one goal. For example, humans developed global positioning systems (GPS), **Figure 9-6**, which aids in navigation by determining the location of vehicles moving over land, through air, and across oceans. This technological system was

©iStock/Thinkstock.com/ChamilleWhite

Figure 9-6. This GPS navigation unit is comprised of the general components of the universal systems model and interacts with major information and communication systems and transportation systems.

Career Connection Systems Engineer

Systems engineering is the science of designing functional systems to meet specific goals while operating within established constraints. A systems engineer designs, manages, improves, and maintains complex technological systems. This work includes leading large and complex engineering projects as well as conducting optimization operations to improve system functionality.

Students interested in becoming systems engineers should prepare for college programs by taking advanced coursework in physics, calculus, and statistics. They should also take any courses or programs of study available in engineering.

To become a professional systems engineer, a four-year degree in engineering is required. An engineering program consists of multiple courses in system engineering, engineering management, risk analysis, modeling and simulation, design, and testing and evaluation, in addition to extensive mathematics and science courses.

Individuals pursuing careers as systems engineers should have strong mathematical and scientific abilities, and excellent communication and problem-solving skills, and be visually creative.

Systems engineers can choose a variety of specializations in almost any industry. For example, they can be hired as managers in manufacturing organizations, electrical systems engineers, or database managers in software engineering companies. Other potential career paths include:

- Applying mathematical, scientific, and engineering knowledge to the development and evaluation of technological systems.
- Managing the cost effectiveness of system operations.
- Utilizing engineering tools and techniques to create and maintain technological systems.
- Developing written documents to support the development and operation of technological systems.

Systems engineers are consistently in high demand due to the lack of qualified engineers around the world and increasingly expanding technological industries. Systems engineers typically join professional organizations such as the International Council on Systems Engineering, the Society for Engineering and Management Systems, or the Society for Industrial and Systems Engineering.

developed specifically for military use and was designed to meet additional goals. GPS satellites house sensors to detect the launch of nuclear missiles and carry atomic clocks to precisely measure time. GPS were later expanded to support the navigation of all civilian and commercial vehicles. This technological innovation now contributes to safe, reliable air transportation. We also use it to help predict weather and to detect surface features on planets.

As another example, suppose a company is developing a new hydrogen fuel cell-powered vehicle. The primary goal is to transport people. But developing the vehicle might also meet other goals and concerns. It might be an opportunity for the innovators to make money. Local government leaders might look at its potential impact on economic growth in their city. The federal government might examine the automobile in terms of whether or not it will help reaffirm national technological leadership. The general public might consider whether or not this vehicle is a good alternative to fossil fuel-powered vehicles. Workers might review whether or not developing the vehicle will improve job security. Oil companies might examine it with regard to whether it will decrease their markets. Environmental groups might comment on how the vehicle affects the environment. Note the number of different goals and concerns mentioned. As you can see, a technological development can meet a number of different goals and concerns that are important to different groups.

Inputs

All natural and human-made systems have inputs. ***Inputs*** are the resources that go into the system and are used by the system. Technological systems have at least seven basic inputs, **Figure 9-7**:

- People.
- Materials.
- Tools and machines.
- Energy.
- Information.
- Finances.
- Time.

People

People are an essential input to technological systems, **Figure 9-8**. Human needs and wants give rise to the systems. Human will and purposes decide the types of systems that will be developed. People bring specific knowledge, attitudes, and skills to the systems. They provide the management and technical know-how to design and direct the systems. Their labors make the systems function. Human ethics and values control and direct the systems. People make policies that promote or hinder technological systems. Finally, people are the consumers of technological outputs. They use the products and services that the systems provide.

Materials

All technology involves physical artifacts. Artifacts are objects humans make. Tools, buildings, and vehicles are some examples of artifacts. ***Materials*** are the substances from which artifacts are made. They can be categorized as natural (found in nature, such as stone and wood), synthetic (made by humans, such as plastics and glass), or composite (modified natural products, such as paper and leather). See **Figure 9-9**. Some of these materials provide the mass and structure for technological devices. Others support the productive actions of the system. For example, some materials lubricate machines. Others contain data. Still others package and protect products.

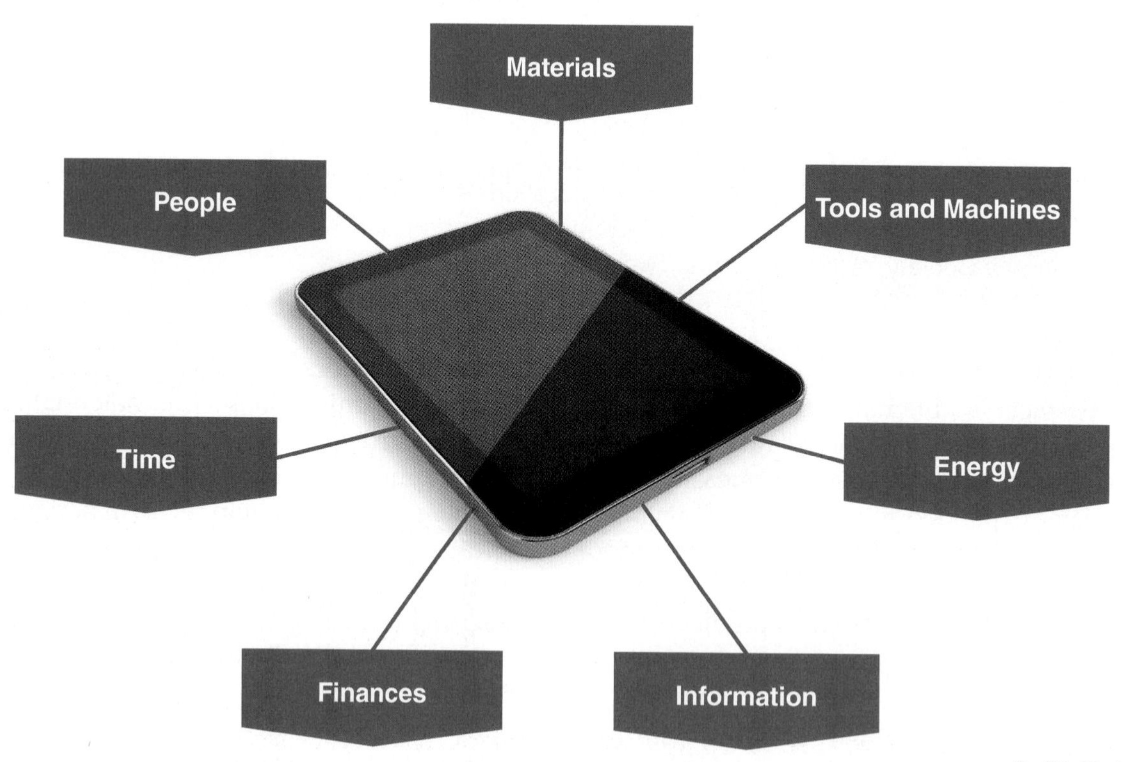

Sashkin/Shutterstock.com

Figure 9-7. Technological systems have at least seven inputs.

wellphoto/Shutterstock.com

science photo/Shutterstock.com

Rawpixel/Shutterstock.com

Figure 9-8. People bring knowledge and skills to each technological system. For example, they operate machines, design devices, and provide management.

Tools and Machines

Tools are the artifacts we use to expand what we are able to do. They include both hand tools and machines. Hand tools expand muscle power and include such artifacts as hammers and screwdrivers.

yoeml/Shutterstock.com

Figure 9-9. Materials, such as stone, can be found in nature. It can be manufactured into countertops, flooring, and other products.

Machines amplify the speed, amount, or direction of a force. Early humans relied on power from animals and water to affect force. Now, however, we use energy from such sources as electricity to power machines. We use tools and machines to do many things, including building structures, communicating information and ideas, converting energy, improving our health, growing crops, diagnosing illnesses, and transporting people and goods.

Energy

The development and application of knowledge, tools, and skills involve doing something. All technological activities require energy. ***Energy*** is the ability to do work. The form the energy takes varies. This energy form can range from human muscle power to nuclear power, and from heat energy to sound energy.

Technological systems require energy to be converted, transmitted, and applied. For example, a hydroelectric generator might convert the energy of falling water into electrical energy, **Figure 9-10.** The electricity might then be carried to a factory over transmission lines. At the factory, motors might convert the electrical energy, into mechanical motion. Halogen and fluorescent bulbs might convert it into light energy, or radiant heaters might convert it into heat energy.

©iStock/Thinkstock.com/mtn05

Figure 9-10. The energy of falling water can be converted into electrical energy.

Information

Everywhere we turn, we encounter facts and figures called *data*. When we organize this data and group it according to its type, we create ***information***. Information is essential for operating all technological systems. There is a difference between data and information. Imagine you measure the size and weight of everything you can find in a single room. This is data because it is random and assorted. If you sort the data so the size and weight of like objects, such as books, are grouped together, you have information. This information is used to establish relationships and draw conclusions. For example, adults are generally taller than children. With this final step, you have developed knowledge. Knowledge is using information to understand, interpret, or describe a specific situation or series of events.

Finances

Technology system inputs include people, materials, tools and machines, energy, and information. These resources have value and so must be purchased. The outputs of technology systems also have value and can be sold. The money and credit necessary for the economic system to operate are known as ***finances***, and are another type of technology input.

Imagine that a construction company decides to build a new house to sell. The company must purchase the land on which to build the house and materials to build it. The company must also purchase plans for the house. The construction company must rent or buy the equipment needed for construction. Employees of the company build the house. All these actions require money. Ideally the company then sells the finished house at a price that covers all these expenses. See **Figure 9-11.** Money remaining after all expenses are paid is the ***profit***. Profit is the goal of most economic activities.

Goodheart-Willcox Publisher

Figure 9-11. The construction company selling this newly built home wants to cover the costs of building the home and to earn a profit.

Technology Explained Global Positioning System (GPS)

Global positioning system (GPS): A system that relies on satellites to provide accurate time and location information to receivers on Earth.

The global positioning system (GPS) is a satellite system developed in the early 1970s by the US Department of Defense. The satellite system is now operated by the US Air Force. GPS was originally developed for military use but has now become a civilian tool. Nearly every cell phone or smartphone contains GPS receivers that function in a variety of applications.

Satellites send signals down to the receivers on Earth, **Figure A**. There are typically 24 satellites in operation. These satellites are able to provide accurate time and location information. At least four of these satellites are used to determine three-dimensional positions. The satellites then send a one-way signal back to Earth with location and time information.

The satellites work by knowing their own location information. Their locations, relative to one another, determine the location of the GPS receiver. The satellites also must be able to determine the distance between themselves and the receiver to give exact location information. The satellites beam down their information using radio signals.

One common use of GPS receivers is navigation, **Figure B**. People use these receivers to get directions from one location to another. Most receivers can also provide audio narration of the directions. And some receivers inform the user of upcoming traffic congestion, accidents, and construction.

Location radar uses a narrow, flashlight-type beam. The beam is focused on an object so accurate elevation, distance, and speed data can be obtained. This type of radar has many applications. Most people are familiar with its use by police officers to enforce speed limits.

Applications for GPS are always changing and improving. GPS receivers are improving the efficiency with which traffic incidents are reported. Other uses for GPS receivers include collecting environmental and agricultural data, and air and water navigation.

©iStock/Thinkstock.com/koto_feja

Figure A. Satellites orbit the Earth and send signals to receivers on the planet.

Beyla Balla/Shutterstock.com

Figure B. GPS receivers are commonly used in navigation applications.

Time

Time, a measurement of how long an event lasts, is a key resource in developing and operating technological systems. All jobs and activities take time. Time is allotted each day to work, eat, sleep, and relax. When working, time must be allotted for all technological endeavors done during a workday. The most important tasks are most likely completed. If time is not available, less critical tasks are left undone or postponed to a later date. Therefore, not all needed technology can be developed immediately. Some activities have to wait until time and other resources are available.

Processes

All technological systems require that a series of tasks be completed. The steps needed to complete these tasks are called ***processes***. Technological systems use three major types of processes: problem-solving and engineering design processes, production processes, and management processes, **Figure 9-12.**

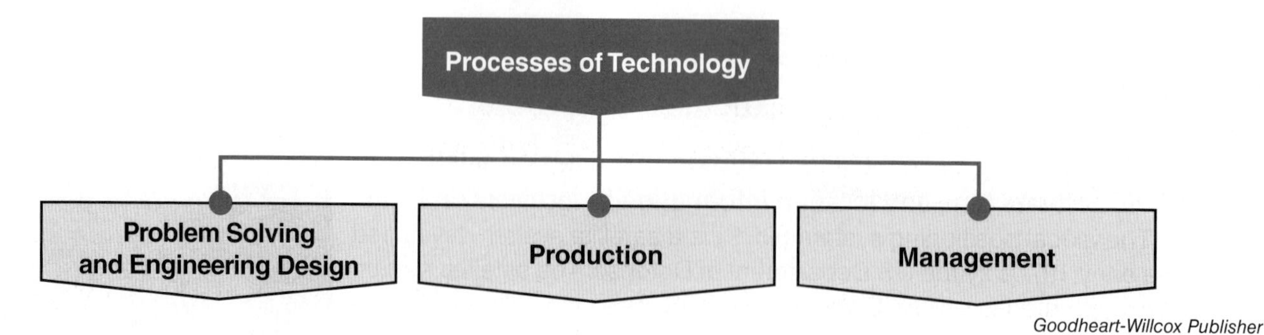

Figure 9-12. Engineering uses problem solving and engineering design, production, and management processes to complete tasks.

The Problem-Solving and Engineering Design Process

In Chapter 2, you learned that science involves activities that study and interpret the natural world. Scientists carry out their work through a set of procedures called the scientific method. This method structures research in a manner to help ensure valid study results are obtained. These results then help scientists attempt to understand the world that now exists.

Unlike science, engineering develops solutions to problems by creating and using technological products and systems to extend the human ability to control and modify the environment. Developing solutions requires creative action. The procedure used to develop technological products and systems is called the engineering design process, **Figure 9-13**. Recall these steps:

1. **Identify and define a problem.** A person or group develops basic information about the problem and the design limitations.
2. **Gather information.** A person or group obtains the knowledge necessary to develop a solution design by studying information and conducting research.
3. **Develop a design solution.** A person or group develops and refines several possible solutions. A person or group then selects and refines the most promising ideas and then details (describes) the best solution.
4. **Model and make a solution.** A person or group produces physical, graphic, computer, and/or mathematical models of the selected solution.
5. **Evaluate the solution.** A person or group tests the physical, graphic, computer, or mathematical models of the selected solution.

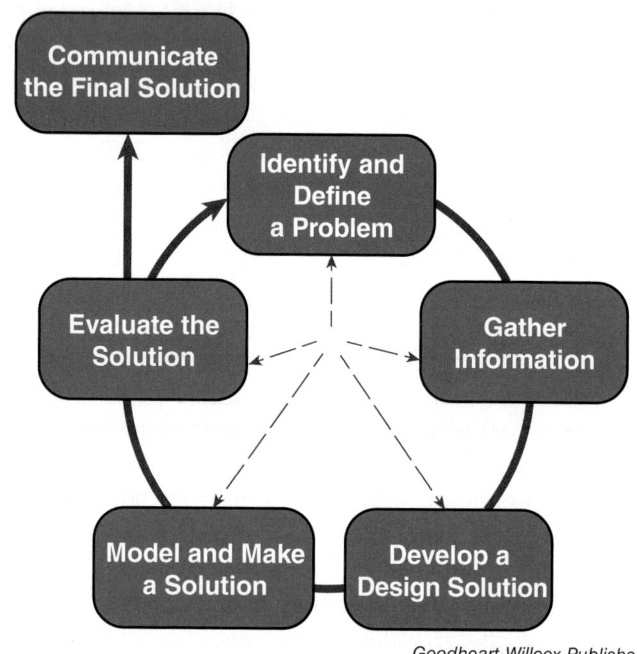

Figure 9-13. The engineering design process involves six major steps to reach a solution.

6. **Communicate the final solution.** A person or group selects a final solution and shares the results of the engineering design process by preparing the documents and presentations needed to share the evaluation outcomes, to produce the final solution, and to use the final solution.

Production Processes

Production processes are actions completed to perform the function of the technological system. For example, a company might use a series of production processes to produce an informational

STEM Connection: Mathematics

Calculating Acceleration Using Newton's Second Law

Mathematics and computational thinking are essential to the engineer and design of technological systems. Mathematics enables the numerical representation of the relationships between physical entities, such as the components of a system based on scientific theories and physical laws. These numerical representations allow engineers to make the calculations for all sorts of quantities, such as force, work, and power, to help predict what the outcomes of a system will be.

Calculating acceleration is necessary for a variety of engineering applications, such as testing system designs and the materials used in a system design. Newton's second law of motion can be used to calculate the acceleration of a component of a system. The law mathematically describes the relationship between the force applied to an object, the object's mass, and its acceleration. This relationship is described mathematically as force = mass × acceleration or $F = ma$. Force can be described as a push or pull that is applied to an object to make it move. Mass is the measure of the amount of matter in an object. It is not the same as weight. Weight changes depending on location, but matter does not. For example, your weight on Earth and your weight on the moon are different because the force of gravity is different. However, your mass is the same in both locations because you are still made up of the same amount of matter. Mass is usually measured using a balance scale, which compares amount of matter between objects. It is typically measured in grams. Acceleration is the rate at which an object changes its velocity. Velocity is the speed and direction of motion.

Knowing Newton's second law means we can begin to calculate the acceleration of components of a system. In this example, a string is placed over a massless and frictionless pulley. A mass of 8 kg is suspended at one end while a mass of 5 kg is suspended from the other end. What is the acceleration of the system?

m_1 = Mass 1
m_2 = Mass 2
g = Gravity = 9.8 m/s^2
F_{net} = Total net force
m_T = Total mass
F_{net} = Total net force
5 kg
8 kg

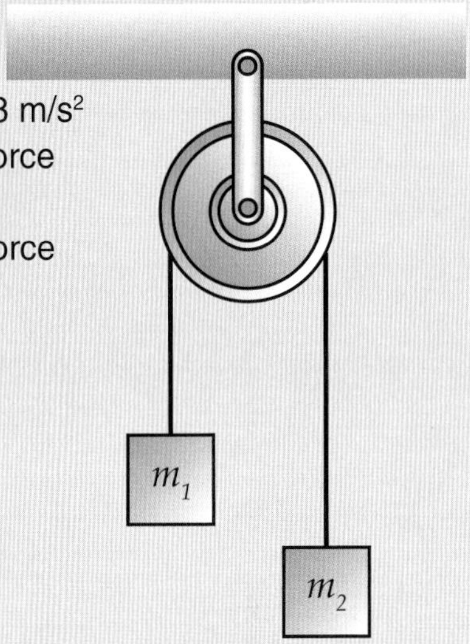

m_1
m_2

Goodheart-Willcox Publisher

booklet, **Figure 9-14**. The company writes and edits the message or copy to be communicated. Members of the company produce photographs to illustrate the document. They put the photographs and copy together into a page layout. The members of the company convert the layout into printing plates. Copies of the booklet are printed from these plates.

Each technological system has its own production processes specific to each of its associated tasks. For example, production processes are used to grow and harvest crops in agriculture, change natural resources into industrial materials in manufacturing, prevent and treat illnesses in medicine, convert materials into structures in construction, transform information into media messages in communications, convert forms of energy in energy and power systems, and move people or goods in transportation.

Management Processes

Management processes are all the actions people use to ensure that the production processes operate efficiently and appropriately. People use

wavebreakmedia/Shutterstock.com

Kim Steele/DigitalVision/Thinkstock.com

Figure 9-14. This production process converts page layouts into booklets.

these processes to direct the design, development, production, and marketing of the technological device, service, or system. Management activities involve four functions, **Figure 9-15**.

- **Planning.** Setting goals and developing courses of action to reach the goals.
- **Organizing.** Dividing the tasks into major segments so the goals can be met and resources can be assigned to complete each task.
- **Actuating.** Putting the system into operation by assigning and supervising work.
- **Controlling.** Comparing system output to the goal.

Individuals and groups use management processes to organize and direct their activities. To illustrate this process, imagine you need to write a term paper. Planning the activity involves selecting a topic, establishing major steps to be completed, and setting deadlines for each task. Organizing involves finding resources and securing reference materials, acquiring writing or word processing equipment, and allocating time. Actuating involves reading and viewing reference material, taking notes, preparing a draft of the paper, and editing the draft into final form. Your instructor completes the control step by comparing your paper to established standards and giving you the results. Control includes evaluation, feedback, and corrective action.

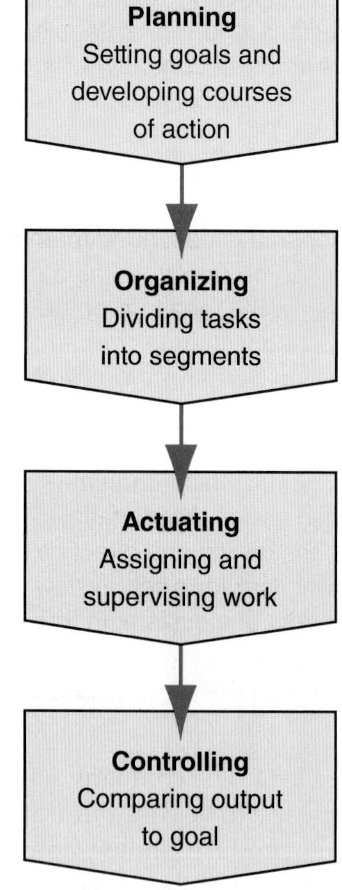

Goodheart-Willcox Publisher

Figure 9-15. Management activities direct the operation of the production process.

Outputs

All technological systems are designed to produce specific outputs. ***Outputs*** are the results, good or bad, of the operation of any type of system. They include manufactured products, constructed structures, communicated messages, or transported people or goods. These primary outputs are not, however, the only ones that come from technology.

In operating technological systems, there can be less direct and unwanted outputs. For example, manufacturing and construction activities generate scrap and waste. Manufacturing operations also create chemical by-products, fumes, noise, and other types of pollution. Cars, airplanes, trains, and trucks often produce noise, air pollution, and congestion. Poorly designed housing developments, industrial parks, and shopping centers can contribute to soil erosion, as can careless farming practices.

Technological systems also have social and personal impacts. The products of technology shape and are shaped by society. For example, the automobile was a novelty in the early 1900s. The wide acceptance of the automobile, however, has greatly changed where and how people live. We live farther from our jobs. Sprawling cities have replaced small, compact towns. Shopping is no longer close to the neighborhood. But, we have also shaped the automobile. Rising gasoline prices and government fuel-economy standards have resulted in improved fuel efficiency with fewer exhaust products. Personal buying habits have also dictated the types of cars that are built.

Feedback and Control

Feedback is the process of using information about the outputs of a system to regulate the inputs. ***Control*** is the feedback loop that causes management and production activities to change through evaluation, feedback, and corrective action. All systems are characterized by feedback and control. For example, many homes have heating systems that are controlled automatically with a thermostat. See **Figure 9-16**. The thermostat is set at the desired temperature. When the temperature is too low, the thermostat closes a switch that turns on the furnace. As the room is heated, the thermostat monitors the temperature. The thermostat is using the output of the furnace to determine the necessary adjustments for the system. When this output (heat) warms the room to the proper temperature, the thermostat changes the system's operation. The thermostat opens the switch and stops the furnace. The cycle is repeated when the room cools again to a specific temperature. Thus, through feedback and control (in this case, automatic control), the thermostat regulates the heating system.

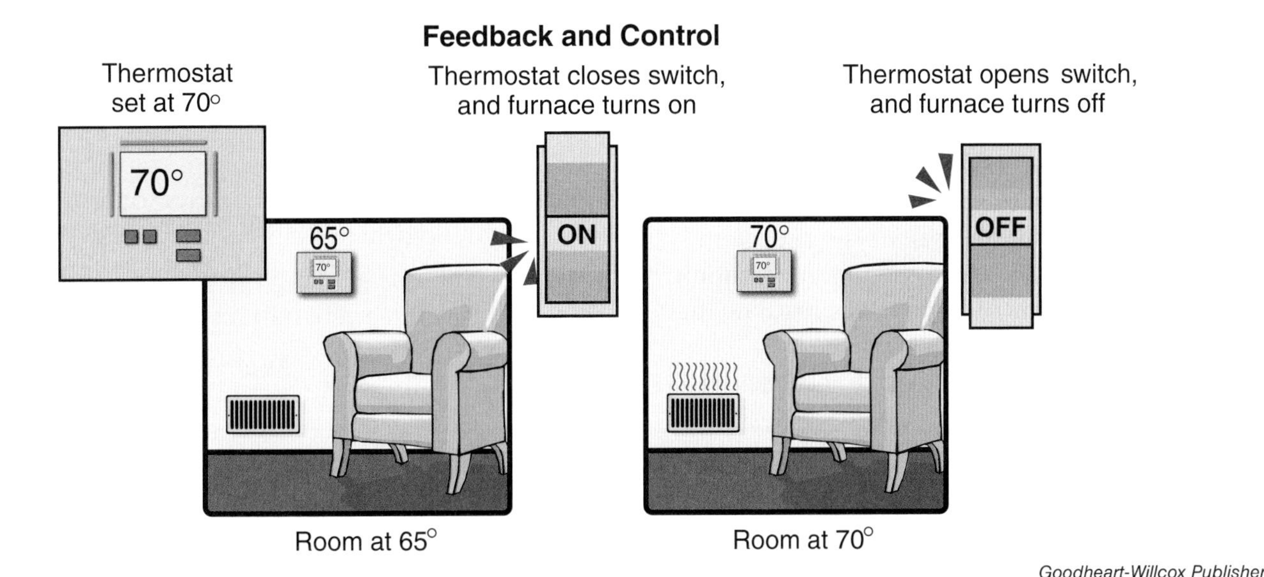

Figure 9-16. A thermostat regulates a heating system through feedback and control.

Summary

- ✔ Systems are any entity or object that consists of parts, each of which has a relationship with all other parts and to the entity as a whole.
- ✔ Technological systems are a group of components designed by people to function together to complete a desired task.
- ✔ Systems thinking is the process of understanding and mentally exploring systems in multiple ways.
- ✔ The components in technological systems are goals, inputs, processes, outputs, and feedback and control.
- ✔ Humans develop technological systems with a goal in mind.
- ✔ Inputs are the resources that go into the system and are used by the system.
- ✔ Inputs include people, materials, tools and machines, energy, information, finances, and time.
- ✔ Technological systems require that a series of tasks be completed.
- ✔ The steps needed to complete these tasks are called processes.
- ✔ Processes used in technological systems are problem-solving and engineering design processes, production processes, and management processes.
- ✔ Production processes are actions completed to perform the function of the technological system.
- ✔ Management processes include planning, organizing, actuating, and controlling.
- ✔ All technological systems are designed to produce specific outputs, or results, good or bad, of the operation of any type of system.
- ✔ Feedback is the process of using information about the outputs of a system to regulate the inputs.
- ✔ Control is the feedback loop that causes management and production activities to change through evaluation, feedback, and corrective action.

Check Your Engineering IQ

Now that you have finished this chapter, see what you learned by taking the chapter posttest.
www.g-wlearning.com/technologyeducation/

Test Your Knowledge

Answer the following end-of-chapter questions using the information provided in this chapter.

1. In what way or ways is technology a system?
2. _____ is the process of understanding and mentally exploring systems in multiple ways.
3. List the five major components of a technological system.
4. *True or False?* Each technological system is developed to meet one goal.
5. List the seven inputs to a technological system.
6. Materials can be classified as _____.
   A. natural
   B. composite
   C. synthetic
   D. All of the above.
7. The ability to do work is known as _____.
8. Explain the difference between data and information.
9. List the six steps in the engineering design process.
10. Actions completed to perform the function of a technological system are called _____.
11. What are the four functions of management activities?
12. List three potential undesired outputs of a technological system.
13. The process of using information about the outputs of a system to regulate the system is called _____.
14. What is control?

Critical Thinking

1. Select an example of a technological system and describe how its development has affected society.
2. Select another example of a technological system and describe how its development has impacted natural systems.
3. Describe how systems thinking influences the engineering design process.
4. Explain the importance of feedback and control in technological systems.

STEM Applications

1. Select a technological device you use. List its inputs, processes, and desired and undesired outputs. Organize your answer using the following headings:
   Device:
   Major inputs:
   Production processes used to make the device:
   Desired outputs:
   Undesired outputs:
2. Apply Newton's second law of motion to calculate the answers to the following questions. Here is an example of how to answer the questions:

   How much force is needed to accelerate a 1400 kg car 2 meters per second per second? Write the formula:

   $$F = ma$$

   Fill in given numbers and units. Remember that force is often measured in newtons, acceleration is often measured in m/s^2, and mass is often measured in kilograms.

   $F = 1400\ kg \times 2$ meters per second per second

   Solve for the unknown:
   2800 kg-meters/second/second or 2800 N

   A. A net force of 16 N causes a mass to accelerate at a rate of 5 m/s^2. What is the mass of the object?
   B. What acceleration will result when a 10 N net force is applied to a 4 kg object? a 6 kg object?
   C. What is the force on a 2000 kg elevator that is falling freely at 9.8 m/s^2?
   D. How much force is needed to accelerate a 65 kg ice skater 2 m/s^2?

Engineering Design Challenge

Newtonian System Design

Background

Systems can be thought of as an organized group of related components that function as a whole. Systems can be found in nature or can be man-made. Man-made systems, also known as technological systems, are designed by people to perform some type of task or achieve desired goals.

Sir Isaac Newton's work led him to establish the three laws of motion. Understanding these laws is important when designing any type of technological system. These laws are:

- An object at rest will remain at rest and an object in motion will remain in motion at a constant velocity, unless acted on by an external force.
- The force of an object is equal to its mass multiplied by its acceleration.
- For every action by a force, there is an equal and opposite reaction by another force.

In this challenge, you will create a complex system to perform a simple task while focusing on demonstrating Newton's three laws of motion in action. You may have heard complex systems used for simple tasks referred to as Rube Goldberg machines. Rube Goldberg was a famous engineer and cartoonist who created drawings of elaborate machines designed to perform some type of simple task, such as using a napkin, scratching one's back, or opening an umbrella.

Situation

You have recently been hired by the Newton Museum of Science to design and create an elaborate and outlandish system to complete the simple task of raising a flag. The purpose of your Rube Goldberg-type system is not only to raise a flag but to teach others about Newton's three laws of motion by demonstrating the laws in action.

The system should perform at least six different steps to raise the flag. The operator of the system may touch the system only one time, to set the system in motion. Once the system has been set in motion, the operator should not touch any component of the system. When designing the system, you must plan that two steps represent Newton's first law of motion, two steps represent Newton's second law of motion, and two steps represent Newton's third law of motion. To help others understand how each of the six steps represents each law of motion, create a postcard to display alongside each component of your system. Each postcard should contain a full explanation of how that step in the system illustrates the law of motion.

Desired Outcomes

- A working system that performs six different steps to achieve the desired goal of raising a small flag.
- A system that has at least two different actions that represent each of the laws of motion.
- A postcard for each system step that provides a full explanation of how that action demonstrates one of Newton's three laws of motion.
- A five-minute presentation to the class that demonstrates the functioning of your system and explains how Newton's three laws of motion are represented in the actions of your system.

Materials

- A variety of materials can be used, such as marbles, string, balsa wood, toy cars, dominos, pulleys, or electric motors.
- Postcards
- Small flag
- Mechanical fasteners
- Tape
- Glue

Beginning the Process

- Remember, the engineering design process is an iterative approach to solving problems, which means you can and should go back and forth between the different steps.
- Determine your engineering design team based on the students' individual skill sets. Work with your team to formulate a problem statement based on the background and situation provided above. A good problem statement should address a single problem and answer "who, what, where, when, why?" However, the problem statement should not imply a single solution to the problem. Your class may create different problem statements based on their own interpretations of the situation and the teams' prior interests and experiences. Your engineering team will use your defined problem statement as the starting point to creating a valid solution.
- Apply the knowledge the team has obtained and employ the engineering design process in order to propose new or improved solutions to the problem defined from the situation provided. This will be a natural progression from brainstorming, to generating ideas, to producing prototypes/models, to evaluating success, and to iteratively refining the solution.

CHAPTER 10

Inputs to Technological Systems

Check Your Engineering IQ 

Before you read this chapter, assess your current understanding of the chapter content by taking the chapter pretest.
www.g-wlearning.com/technologyeducation/

Learning Objectives

After studying this chapter, you should be able to:

- ✔ Describe the inputs of technological systems.
- ✔ Summarize the types of skills and knowledge various groups of people bring to technological systems.
- ✔ Explain the types of tools and machines used as inputs to technological systems.
- ✔ Categorize materials based on their properties.
- ✔ Contrast the different types of information that are inputs to technological systems.
- ✔ Explain the major types and sources of energy used as inputs to technological systems.
- ✔ Identify different sources of finances used as inputs to technological systems.
- ✔ Summarize the importance of time with regard to technological systems.

Technical Terms 

biotechnology
composite materials
consumers
corporation
data
debt financing
distance multipliers
entrepreneurs
equity financing
exhaustible resources
first-class lever
force multipliers
fulcrum
gases
genetic materials
hand tools
inclined planes
inexhaustible resource
information
inorganic materials
knowledge
lever
lever arm
liquids
machines
managers
mechanics
natural materials
natural resources
organic materials
partnership
production workers
pulleys
renewable resource
screw
second-class lever
sole proprietorship
solids
support staff
synthetic materials
technicians
third-class lever
tools
wedge
wheel and axle

Close Reading

As you read the chapter, outline the hierarchy of the groups of people involved in technological systems. Think about where you fit within this hierarchy.

Humans have lived on the earth for about 2 1/2 million years. This might seem to be a long time, but it is relatively short, considering the earth is about 5 billion years old. In the short span of human history, people have learned how to build and use tools, machines, systems, and materials to change their environment. These technological artifacts are developed through problem-solving processes and engineering practices. This ability to develop technology has led to many kinds of technological systems. All of these systems are made up of general components, including inputs, processes, outputs, feedback and control, and goals. This chapter explores in more depth the inputs that are common to all technological systems.

As you recall, all inputs can be grouped into seven major categories, **Figure 10-1**:

- People.
- Tools and machines.
- Materials.
- Information.
- Energy.
- Finances.
- Time.

Inputs are resources used to operate technological systems. They are the elements that are changed by technological processes or are used by technology to change other inputs.

People

People are a key input to technological systems. People produce these systems. Their minds create and design the systems and their outputs. People use their skills to make and operate the systems. Their management abilities make systems operate efficiently. The systems satisfy people's needs and wants. People are fundamental to technology.

Different people bring different knowledge, skills, and abilities to technological systems. For example, scientists generally have knowledge needed to help create products and processes, **Figure 10-2**. Fields of knowledge scientists commonly study include physics, biology, materials science, geology, and chemistry.

Engineers apply scientific and technological knowledge to design products, structures, and systems. See **Figure 10-3**. They determine appropriate materials and processes needed to produce products and systems or perform services. For example, civil engineers determine the correct structure for a bridge to carry vehicles across a river and electrical/computer engineers design circuits for computers.

Production workers process materials, build structures, operate transportation vehicles, service products, or produce and deliver communication products.

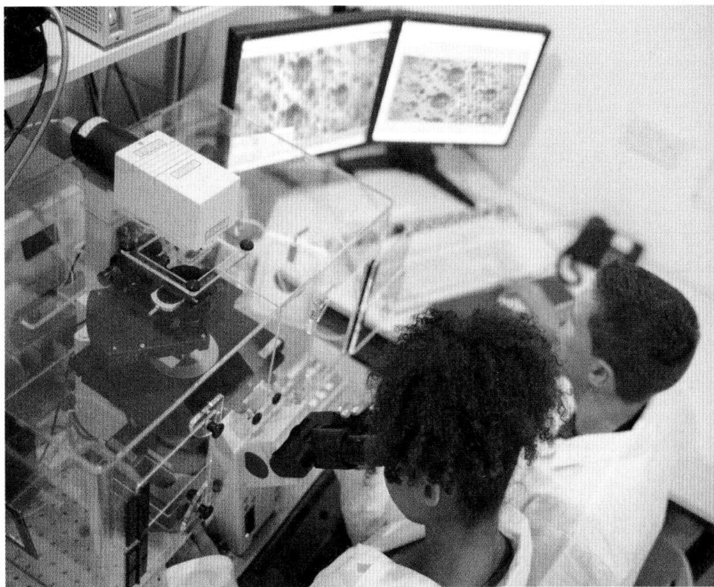

©iStock/Thinkstock.com/Wavebreakmedia

Figure 10-2. Scientists are involved in various fields of knowledge when creating technological products and processes.

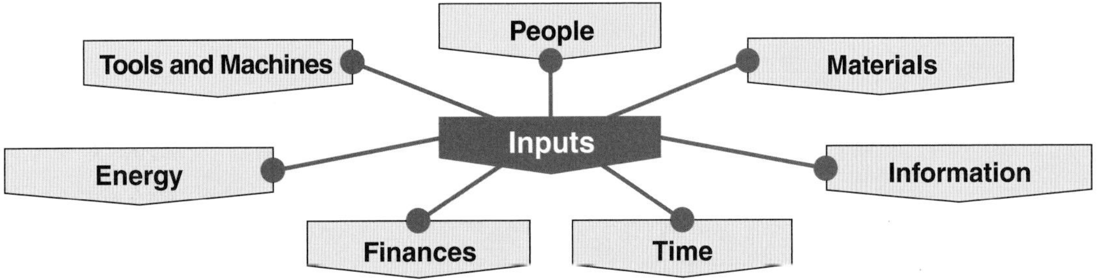

Goodheart-Willcox Publisher

Figure 10-1. Inputs to technological systems can be grouped into seven major categories.

Lev Kropotov/Shutterstock.com

Figure 10-3. This engineer is reviewing plans at a new construction site.

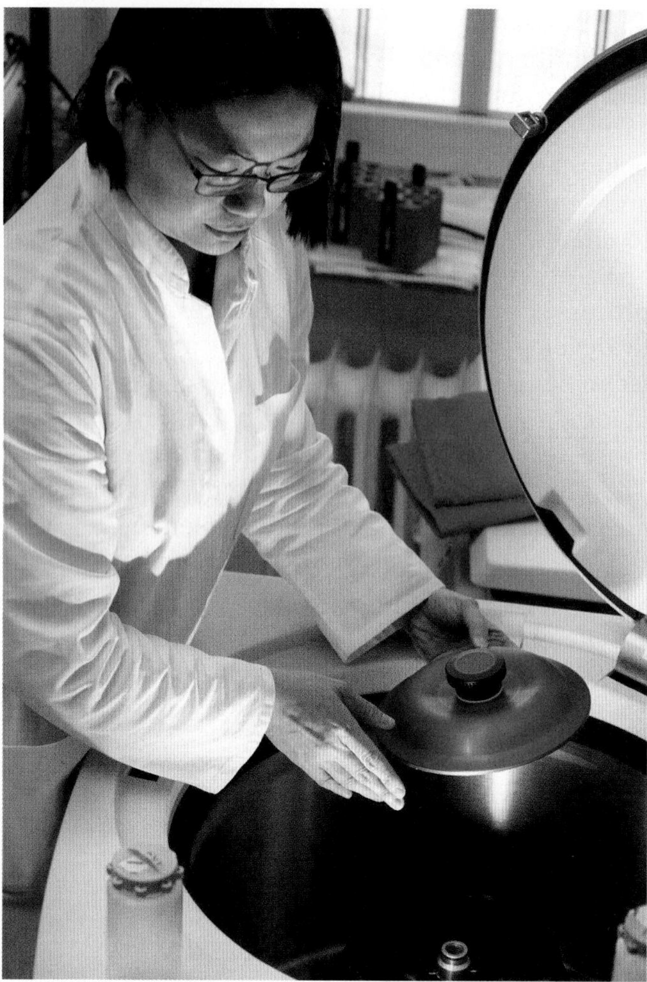

anyaivanova/Shutterstock.com

Figure 10-4. This technician is involved with medical technology.

Production workers in manufacturing and construction are often categorized as unskilled, semiskilled, or skilled workers. The designation is based on training and work experience. Skilled workers in laboratories and product-testing facilities are known as ***technicians***, **Figure 10-4**. Skilled workers in service operations are often called ***mechanics***.

Other groups of people establish or help direct businesses. Those who create businesses are called entrepreneurs. ***Entrepreneurs*** have a vision of what can be done and are willing to take risks to see it happen. ***Managers*** organize and direct the work of others in these businesses. They set goals, structure tasks to be completed, assign work, and monitor results. Nonmanagerial ***support staff*** carries out tasks such as keeping financial records, maintaining sales documents, and developing personnel systems.

The final group involved in technological systems is consumers, **Figure 10-5**. ***Consumers*** are people who financially support technological

dotshock/Shutterstock.com

Figure 10-5. Consumers are fundamental to all technological systems.

Career Connection Engineering Manager

An engineering manager is responsible for supervising a team of engineers in the development, creation, evaluation, and improvement of solutions to technical problems.

Students interested in engineering management careers can get a jump-start in high school by taking a variety of STEM courses, specifically technology courses focused on manufacturing and digital electronics. Mathematics courses related to trigonometry and calculus and advanced science courses in physics are also important. A minimum of a bachelor's degree in engineering along with a master's degree in either engineering management or technology management is required for most engineering manager positions.

Individuals pursuing an engineering management position must be able to work in a collaborative environment, make decisions, establish expectations, and lead groups of people.

Engineering managers are needed in a variety of industries. Typically, these individuals start working as an engineer in a specialized area. As they develop their skills and expand their education, they then have opportunities to be hired as an engineering manager to supervise teams of engineers and handle complex projects.

Examples of specific jobs for potential engineering managers include:

- Directing and coordinating operations in an engineering industry or enterprise.
- Hiring engineers, technicians, and support staff to complete engineering projects.
- Evaluating employees and writing performance reviews.
- Establishing and monitoring project goals, plans, resources, and time lines.
- Overseeing the preparation of budgets and contracts.
- Reviewing project and design proposals.

Engineers are typically in high demand, which means there will continue to be a need for engineering managers. Professional organizations for engineering managers include American Society for Engineering Management.

systems by spending money on products or services. Consumers are the reason for the system in the first place. Their attitudes about styling, price, and service directly affect the systems' outputs. Consumers' money is spent on the products or services. Therefore, consumers financially support the system.

One element common to the development of technology within all these groups of people is creativity. Creativity is the ability to see a need or way to make life easier, and design systems and products to meet the need or desire. Creative people have this ability. In some instances, they develop ideas for a product and decide what size, shape, and color the product will be, **Figure 10-6**. Their decisions add beauty to the world through well-designed products and structures. Creative people's innovations can lead to improvements in existing technology.

One well-known example of a creative person is Arthur Fry, a chemist who is now retired from the 3M Company. Fry was aware that another scientist at 3M was attempting to invent a stronger adhesive for some of the company's tapes by experimenting with various molecules. This scientist, instead, discovered a new adhesive that adhered to paper, but was not as sticky as tape. He was not sure what to do with this discovery, but he did mention it to other employees at 3M. Mr. Fry, a member of his church's choir, typically marked his place in the songbook with little pieces of paper that invariably fell out of the book. He suddenly thought that a bookmark made with the new adhesive would stick to the page, but could also be removed.

Peter Bernik/Shutterstock.com

Figure 10-6. This woman uses her creative abilities to design clothing that is aesthetically appealing.

Rawpixel/Shutterstock.com

Figure 10-7. A variety of people contributed to the development of Post-it® Notes.

He prepared samples of his idea and distributed them to people at 3M to use and critique. In that process, he and his manager realized they had something more than a bookmark. They had a new way of communicating and organizing information: self-attaching notes, called Post-it® Notes, **Figure 10-7**. Other 3M employees, with the support of management, joined the effort to design and produce this new item. After testing various forms and solving a series of manufacturing problems, they came up with a product that many people still use.

We can see from this example that various groups of people contributed to this invention in different ways. Some products create their own

Think Green Organic Cotton

Traditional cotton farming uses more pesticides than farming any other types of crops. Traditional cotton farming can cause harm to the natural systems of the environment, and could potentially hurt the people who work in these fields or who use products made from that cotton. The chemicals in the pesticides get into the ground, the water, and the air, causing contamination. Even wildlife near these cotton fields may be affected. Cotton is a common material found in everyday items such as clothing, towels, and sheets.

Organic cotton production systems are designed to have a low impact on the environment by maintaining soil fertility while reducing risks involved with toxic fertilizers and pesticides. Organic cotton is grown without the use of pesticides or fertilizers by using scientific knowledge to naturally support the growth of cotton. These methods include planting crops at the proper times based on meteorological data analysis, intercropping (planting different crops together that support each other), crop rotation, using specific mulches that repel insects, and introducing natural predators for pests. These production systems pose little health risks to farmers and reduce impact on the land, the water supply, and the surrounding wildlife. In recent years, the number of organic cotton crops has increased. As a result, major manufacturers have started using organic cotton in their products, and the products can be found in several stores. As you can see, changing one system of cotton production can have far-reaching effects on a variety of technological and natural systems.

demand. In this example, we can see that some changes in society lead to even more needs and demands. In this case, 3M has expanded its product line in the Post-it area to include such items as flip charts, tape flags, and room-decorator kits.

Tools and Machines

Humans are the only species on the earth that can develop and use technology because humans are tool builders and users. ***Tools*** are the artifacts that expand what humans are able to do. People use tools to increase their ability to do work. Tools include such diverse items as milking machines used on dairy farms, machine tools used in factories, hammers and saws used on construction sites, medical equipment used in hospitals, and automobiles used to transport people and cargo.

There are several ways to categorize tools. One way is by their area of activity, **Figure 10-8**. For example, microscopes and telescopes are used in scientific activities. Tennis rackets and pitching machines are used in recreational activities. Another way to group tools is by the major technological system in which they are used. For example, some people in agriculture use plows, some people in communication use radio transmitters, and some workers in the field of transportation use trucks. The most common way to classify tools, however, is as either hand tools or machines. Each category is described in more depth below.

Hand Tools

Almost every technological product or system uses a common set of hand tools to produce, maintain, and service products and equipment. ***Hand tools*** are simple, handheld artifacts requiring human-muscle power, air, or electric power to make them work. These tools can be classified by their purpose, **Figure 10-9**:

- Measuring tools are used to determine the size and shape of materials and parts.
- Cutting tools are used to separate materials into two or more pieces.
- Drilling tools are used to produce holes in materials.

Dan76/Shutterstock.com

Mediaphotos/Shutterstock.com

Deyan Georgiev/Shutterstock.com

Dmitry Yashkin/Shutterstock.com

Figure 10-8. Tools are used in many types of activities.

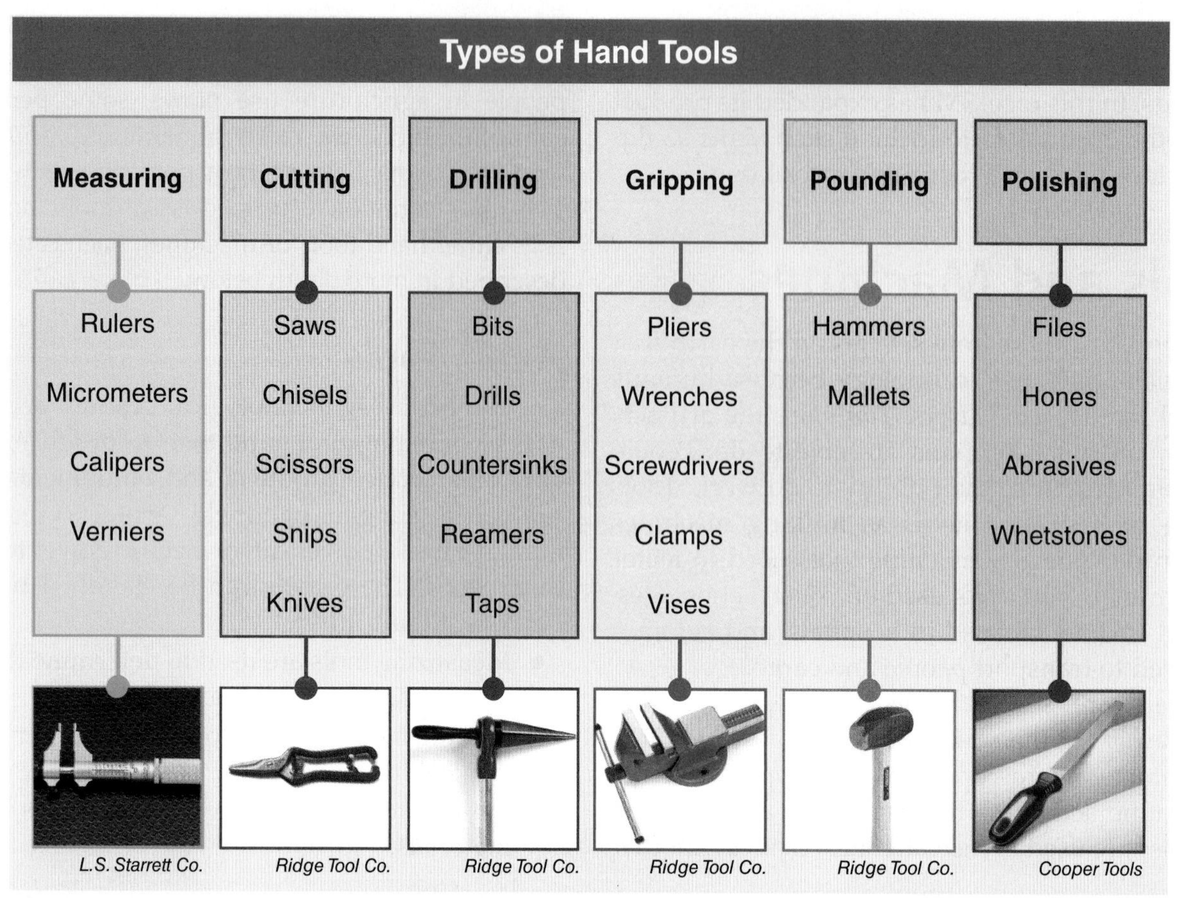

Goodheart-Willcox Publisher

Figure 10-9. Hand tools can be grouped by the actions they perform.

- Gripping tools are used to grasp and, in many cases, turn parts and fasteners.
- Pounding tools are used to strike materials, parts, and fasteners.
- Polishing tools are used to abrade and smooth surfaces.

Machines

Machines are artifacts that transmit or change the application of power, force, or motion. They can be simple or complex. Complex machines are made up of more than one simple machine. Simple machines work on two basic principles: the principle of the lever and the principle of the inclined plane, **Figure 10-10**. There are six categories of simple machines:

- Lever.
- Wheel and axle.
- Pulley.
- Inclined plane.
- Wedge.
- Screw.

The lever, the wheel and axle, and the pulley operate on the principle of the lever. The inclined plane, the wedge, and the screw operate on the principle of the inclined plane.

Levers

If you have ever pried open a crate with a crowbar or pulled a nail with a claw hammer, you have used a lever. A ***lever*** is a simple machine that multiplies the force applied to it and changes the direction of a linear force. The ***lever arm*** is a rod or bar that rests and turns on a support. This support is called a ***fulcrum***, **Figure 10-11**. You apply a force at one end of the rod, or bar, to lift a load at the other end. The purpose is to help lift weight more easily.

Levers are grouped into three categories: first class, second class, and third class, **Figure 10-12**. Each class of lever applies force differently to move the load. In ***first-class levers***, the fulcrum is between the load and the effort. A pry bar is an example of a first-class lever. In ***second-class levers***, the load

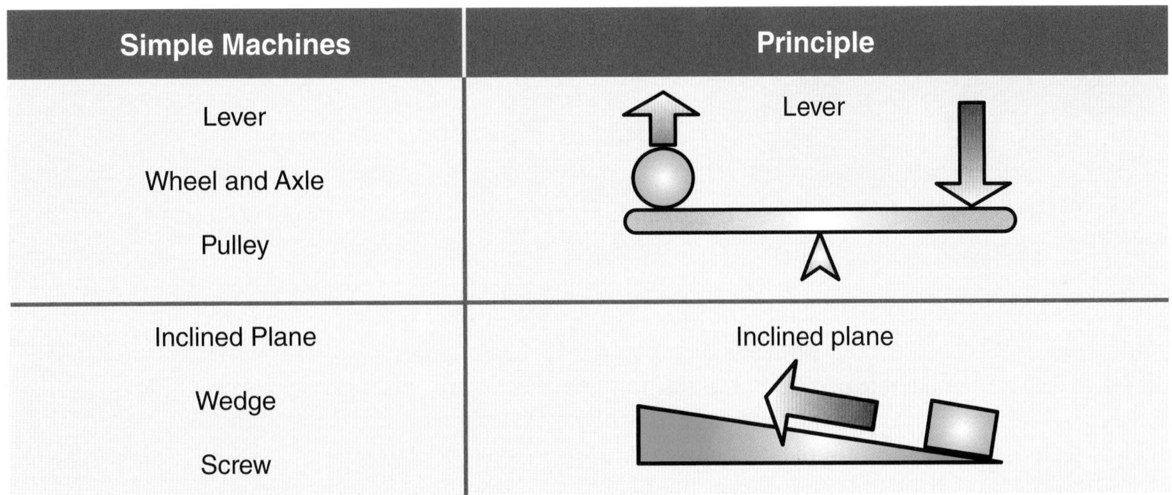

Simple Machines	Principle
Lever Wheel and Axle Pulley	Lever
Inclined Plane Wedge Screw	Inclined plane

Goodheart-Willcox Publisher

Figure 10-10. The six types of machines work on two basic principles.

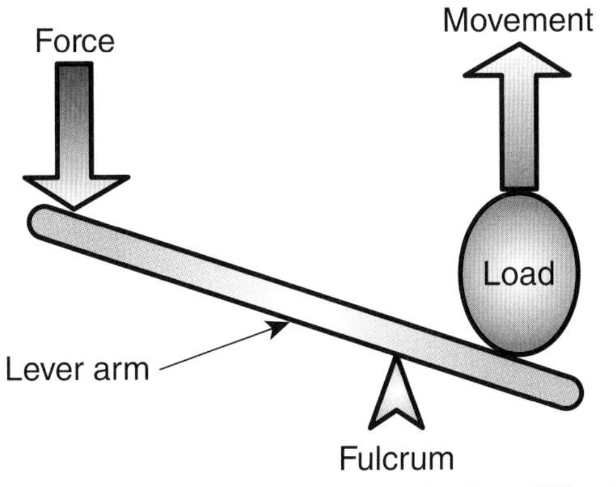

Goodheart-Willcox Publisher

Figure 10-11. Levers use a lever arm and a fulcrum, to which a force is applied and a load is moved. The crowbar uses the principle of the lever.

is between the effort and the fulcrum. The wheelbarrow uses the principle of a second-class lever. In ***third-class levers***, the effort is placed between the load and the fulcrum. A person moving dirt with a shovel illustrates the principle of a third-class lever.

These simple machines can be either force multipliers or distance multipliers. When levers increase the force applied to the work at hand, they are ***force multipliers***. The fulcrum is close to the load, and the force is applied at the other end, allowing a heavy load to be lifted with a light force.

A ***distance multiplier*** increases the amount of movement applied to the work at hand. On a lever that is a distance multiplier, the fulcrum is close to the force, and the load is at the other end. The load moves a greater distance than the force, but a large

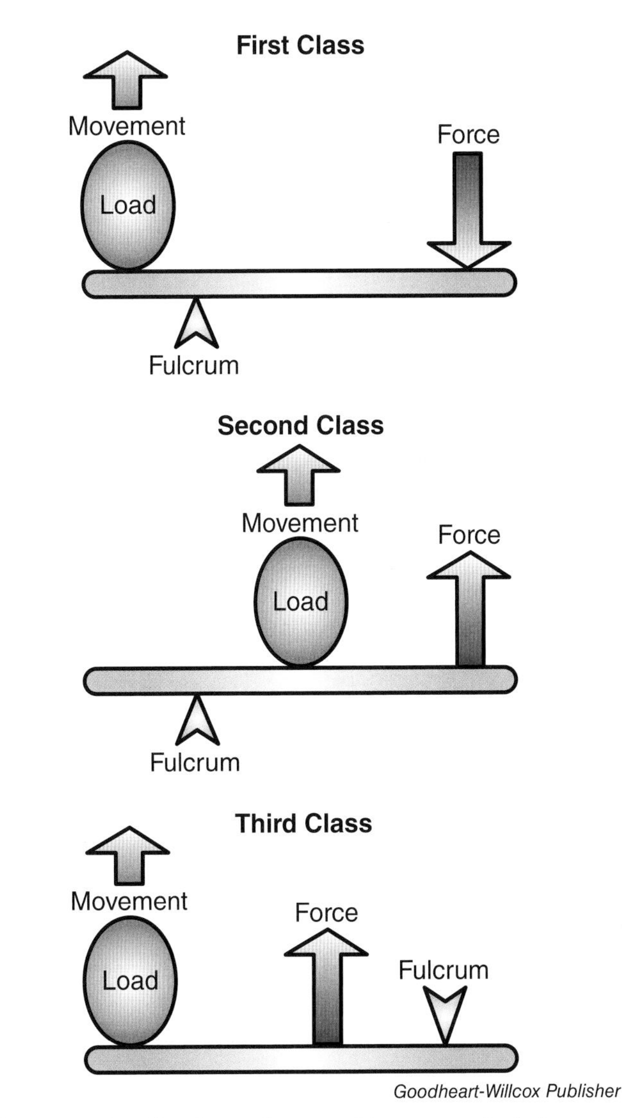

Goodheart-Willcox Publisher

Figure 10-12. The three classes of levers are first class, second class, and third class.

Academic Connection: Argumentative Essay

Nuclear Power

Argumentative writing is a type of writing that requires people to investigate a topic, collect and evaluate data related to the topic, and establish and present a position on the topic based on evidence. A valid argument provides facts and data for any opposing views to the position presented. Scientists and engineers often face issues requiring the need to take a particular stance on the topic. In doing so, they must must be able to defend their positions based on logic and fact. A common practice among engineers and scientists is to engage in arguments based on information that they have evaluated and communicate the resulting conclusions effectively.

The production of electricity through nuclear fission has been a controversial topic for decades. Supporters of nuclear energy argue that nuclear power is a cleaner alternative to burning fossil fuels. Therefore, it can reduce carbon emissions and dependency on imported fuel supplies. Opponents to nuclear power, however, argue that it poses a great threat to people and the environment. If nuclear energy production is not handled properly, it can result in a nuclear meltdown that can have devastating effects. For example, the Chernobyl power plant explosion in the Ukraine released a large amount of radioactive contaminates into the atmosphere that impacted the health of people for years. It is estimated that up to 4000 cancer deaths were related to this incident. More recently, in 2011, the Fukushima Daiichi nuclear power plant was hit by a tsunami, causing three of the plant's six nuclear reactors to melt down. This resulted in the release of radioactive material into the surrounding environment.

Research the topic of nuclear energy production and write an argumentative essay on whether or not this practice should continue. Support your argument with appropriate evidence.

An argumentative essay can be written using the following format:

I. Introduction
  a. Introduce the topic
  b. Present your position
II. Body
  a. Topic sentence for reason 1
    i. Supporting evidence
  b. Topic sentence for reason 2
    i. Supporting evidence
  c. Topic sentence for reason 3
    i. Supporting evidence
III. Conclusion
  a. Restate your position
  b. Counter arguments
  c. General conclusions

force is required to move a light load. These two applications of levers are shown in **Figure 10-13**.

Wheel and Axles

A *wheel and axle* is a shaft attached to a disk. This simple machine acts as a second-class lever. The shaft, or axle, acts as the fulcrum. The circumference of the disk acts as the lever arm. If the load is applied to the shaft, the wheel and axle becomes a force multiplier, **Figure 10-14**. Automotive steering wheels use this principle. A 15″ wheel attached to a 1/4″ shaft multiplies the force 60 times.

Force Multiplier

100 lbs.
x
4x
25 lbs.
25 pounds of force move a 100-pound load.
Force
Load
Fulcrum

Distance Multiplier

Moving the force end one foot lifts the load four feet.
4x
Load
Force
x
Fulcrum

Goodheart-Willcox Publisher

Figure 10-13. Levers can be used to multiply force or distance.

Multiply Force

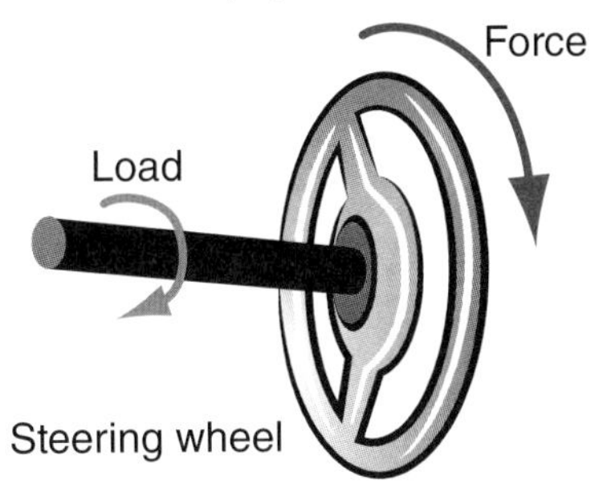

Multiply Distance

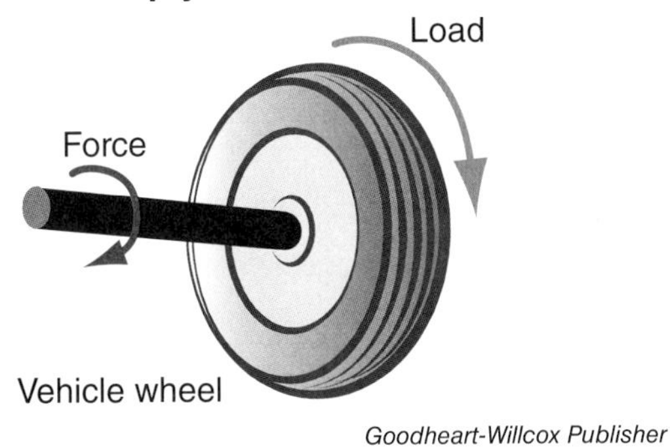

Goodheart-Willcox Publisher

Figure 10-14. A wheel and axle can multiply force or distance.

If the load is applied to the disk, the wheel and axle becomes a distance multiplier. Automobile transaxles use this type of wheel and axle. One revolution of the axle causes the wheel to revolve one time. The circumference of the wheel is many times that of an axle. Therefore, the vehicle moves a considerable distance down the road for each revolution. For example, a 20″ wheel attached to a 1/2″ shaft multiplies the distance for each revolution of the shaft 40 times.

Pulleys

Pulleys are grooved wheels attached to an axle. They also act as second-class levers. Pulleys can be used for three major purposes, **Figure 10-15**. A single pulley can be used to change the direction of a force. Two or more pulleys can be used to multiply force or to multiply distance. The number and diameters of the pulleys used determine the mechanical advantage (force multiplication) of a pulley system.

Inclined Planes

Inclined planes are sloped surfaces used to make a job easier to do. The principle of the three simple machines in this category (inclined plane, wedge, and screw) is that it is easier to move up a slope than up a vertical surface, **Figure 10-16**. The simplest application of this principle is the inclined plane. The inclined plane is used to roll or drag a load from one elevation to another. Common examples of inclined planes are roadways in mountains and ramps to load trucks.

Wedges

A second application of the inclined plane principle is the ***wedge***. This device is used to split and separate materials and to grip parts. A wood chisel, a firewood-splitting wedge, and a doorstop are examples of this simple machine.

Screws

The ***screw*** is the third simple machine using an inclined plane. This simple machine is actually an inclined plane wrapped around a shaft. The screw is a force multiplier. Each revolution of the screw moves the screw into the work only a short distance. For example, a 1/2″ × 12 machine screw is 1/2″ in diameter and has 12 threads per inch. With one revolution

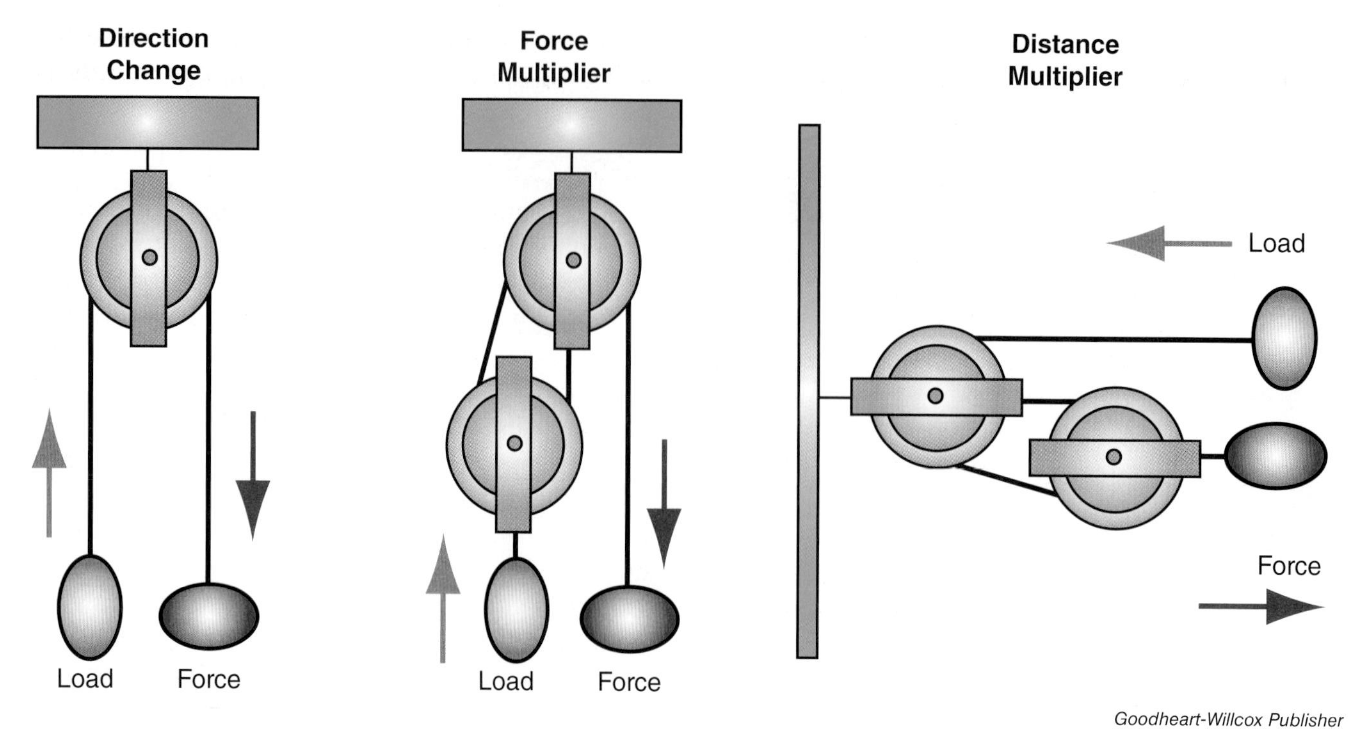

Figure 10-15. Pulleys can be used to change the direction of a force, multiply force, or multiply distance.

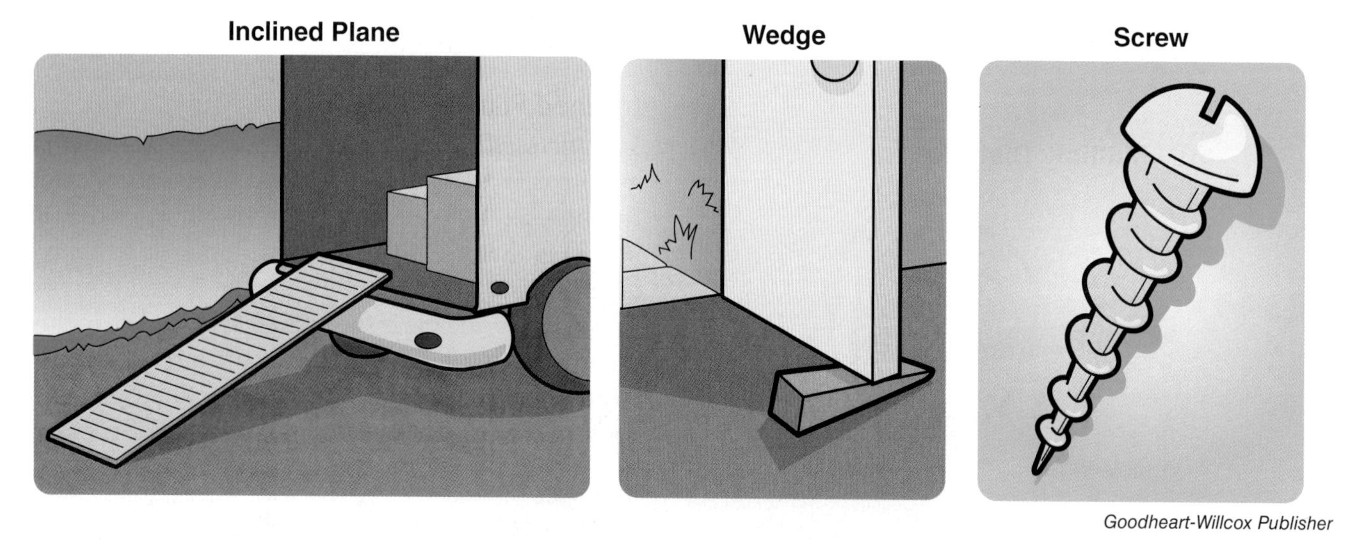

Figure 10-16. The three types of inclined plane machines are the inclined plane, the wedge, and the screw.

of the screw, the circumference moves about 1 1/2″, but the screw moves into the work only 1/12″. In this example, the force is multiplied about 18 times.

Materials

We are living in a material world. Everywhere you look, you see materials. They come in all sizes, shapes, and types. Materials possess a number of specific properties. All material is made up of one or more of the elements occurring naturally on the earth. These natural elements, which number less than 100, combine to produce literally thousands of compounds. To understand materials and to use them effectively, a person must know about the types of materials and the properties materials exhibit.

Types of Materials

In Chapter 9 you learned that materials can be classified as natural, synthetic, or composite. They can also be classified by their origin, their ability to be regenerated, and their physical state, **Figure 10-17**.

Natural, Synthetic, or Composite

Many materials occur naturally on the earth. These materials are often called ***natural resources***. Iron, carbon, petroleum, and silica are examples of ***natural materials***. They can be refined and combined to make products.

Other materials are manufactured, or ***synthetic materials***. The most common synthetic materials are plastics. They are developed and produced from cellulose (vegetable fibers), natural gas, and petroleum.

Composite materials are combinations of natural and synthetic materials that are mixed to create items with other desirable properties. For example, concrete is a mixture of water, cement, sand, and gravel. Concrete molds to almost any shape and then hardens into a long-lasting material requiring little care.

Origin

Materials that come from living organisms are called ***organic materials***. For example, wood, cotton, and flax are products of plant fibers. Wool and leather are products of animals. Petroleum, coal, and natural gas are the products of decayed and fossilized organic materials. Materials that do not come from living organisms are called ***inorganic materials***. For example, metals and ceramic materials are inorganic.

Ability to Be Regenerated

Some materials naturally occur on the earth in limited amounts. Human action or nature cannot replace these materials. The quantity of these materials is finite. Once they are used up, there will be no more. They are called ***exhaustible resources***. Metal ores, coal, petroleum, and natural gas fall into this category of materials.

Other materials have life cycles and can be regenerated. Human action or nature can produce them. They are called ***genetic materials*** because living things produce them. These materials are the results of farming, forestry, and fishing activities, among others. Wood, meat, wool, cotton, and leather are all genetic materials. The technology associated with producing these materials is called ***biotechnology***.

Still other materials are the products of natural reactions. For example, carbon dioxide and oxygen are the by-products of natural processes. Water is purified through natural processes in lakes, wetlands, and rivers. Natural evaporation produces salt around saltwater.

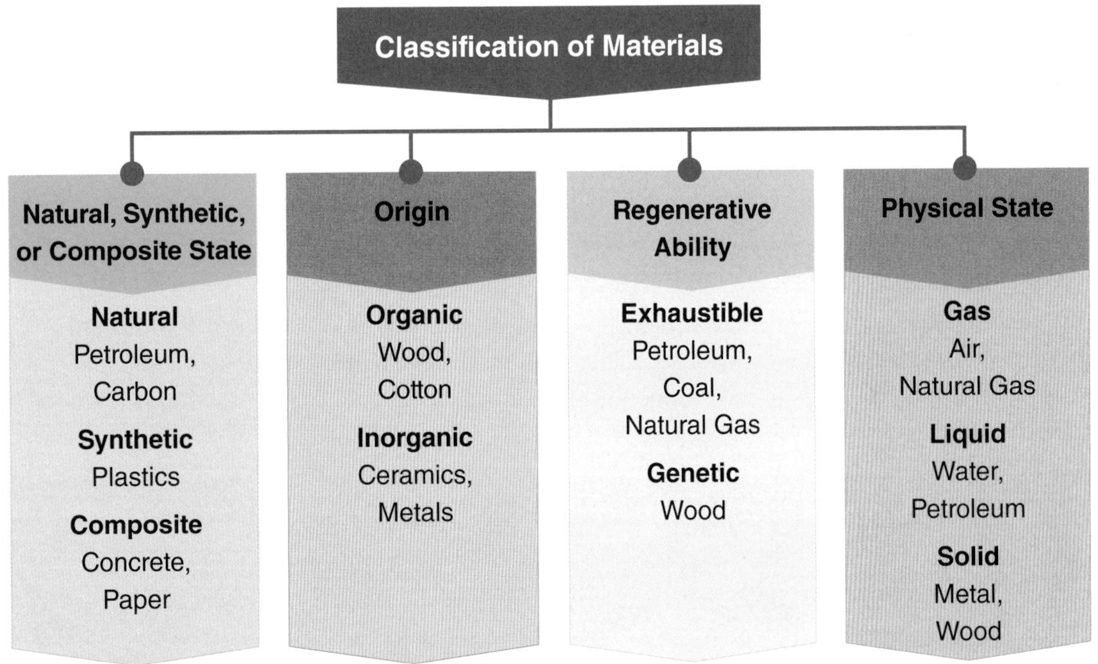

Figure 10-17. Materials can be classified as natural, synthetic, or composite, and by their origin, ability to be regenerated, or physical state.

Physical State

Another important way of grouping materials is by their physical states—that is, as gases, liquids, or solids. ***Gases*** are materials that easily disperse and expand to fill any space. They have no physical shape, but they do occupy space and have volume. Gases can be compressed and put into containers. Examples of gases are the oxygen we breathe, the fuel used for rockets, the carbonation added to beverages, and the compressed air used to inflate tires.

Liquids are visible, fluid materials that will not normally hold their sizes and shapes. They cannot be easily compressed. Common liquids include drinking water, fuels used for transportation vehicles, and coolants for industrial processes.

Solids are materials that hold their sizes and shapes. They have internal structures that cause them to be rigid. These materials can support loads without losing their shapes. Solids can be divided into four categories, **Figure 10-18**:

- Metallic materials (metals) are inorganic substances that have crystalline structures—that is, their molecules are arranged in box-like frameworks called *crystals*. They are the most widely used of all engineering materials. Metals are generally used as alloys, or mixtures of a base metal and other metals or nonmetallic materials. For example, steel is primarily an iron-carbon alloy, and brass is a copper-zinc alloy.
- Polymeric materials (plastics) are synthetic materials containing complex chains of hydrogen-carbon (hydrocarbon) molecules. These materials are either thermoplastic (softened when heated) or thermosetting (made rigid by heat).
- Ceramics are nonmetallic, mostly inorganic crystalline materials, such as clay, cement, plaster, glass, abrasives, or refractory material.
- Composites are a combination of two or more kinds of materials. One material forms the matrix, or structure. The other material fills the structure. Fiberglass is a composite with a glass fiber structure filled with a plastic resin. Wood is a natural composite with a cellulose-fiber

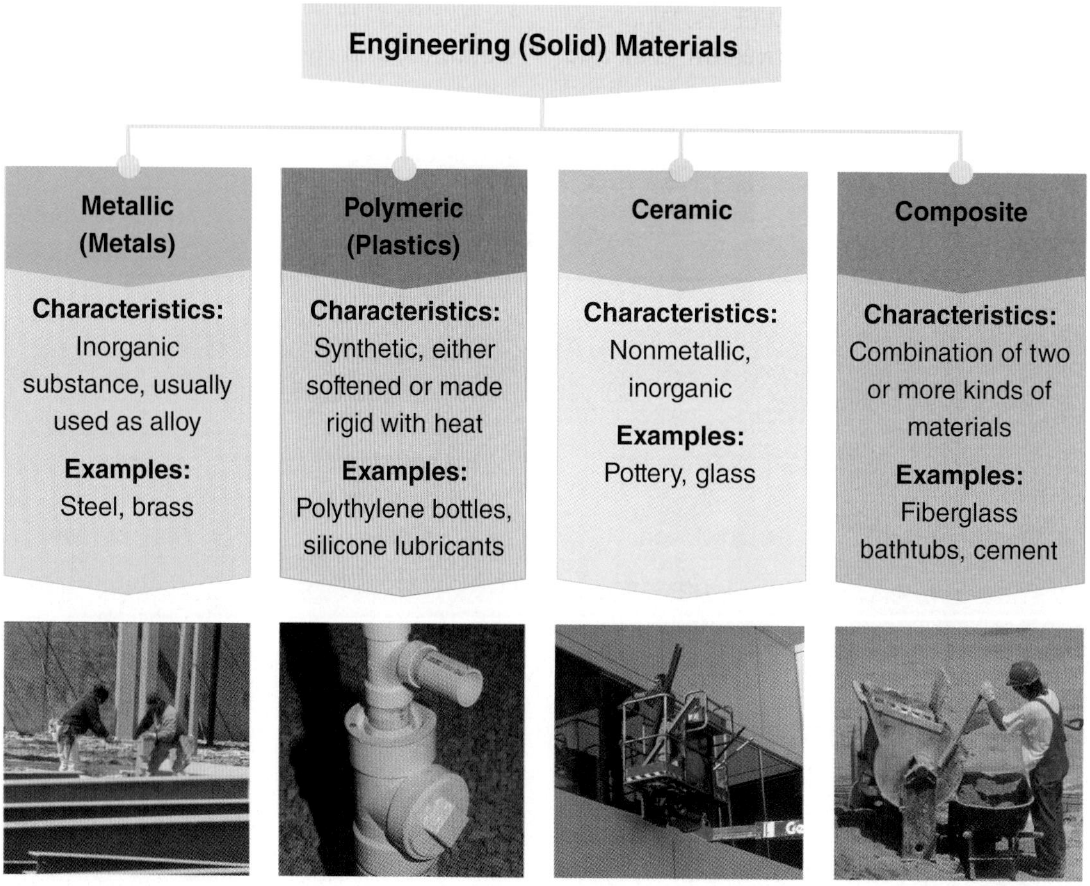

Engineering (Solid) Materials

Metallic (Metals)	Polymeric (Plastics)	Ceramic	Composite
Characteristics: Inorganic substance, usually used as alloy	**Characteristics:** Synthetic, either softened or made rigid with heat	**Characteristics:** Nonmetallic, inorganic	**Characteristics:** Combination of two or more kinds of materials
Examples: Steel, brass	**Examples:** Polythylene bottles, silicone lubricants	**Examples:** Pottery, glass	**Examples:** Fiberglass bathtubs, cement

Goodheart-Willcox Publisher

Figure 10-18. Solids can be grouped in one of four categories: metallic, polymeric, ceramic, or composite.

STEM Connection: Mathematics

The Law of Equilibrium

A Greek mathematician who lived more than 2000 years ago proved the law behind the workings of the lever. Archimedes had observed how a small force can move a great weight. From this observation, the law of equilibrium was created. A lever is in equilibrium when the product of the weight (w_1) and distance (d_1) on one side of the fulcrum (the center of gravity) is equal to the product of the weight (w_2) and distance (d_2) on the other side of the fulcrum. The mathematical formula is $w_1 \times d_1 = w_2 \times d_2$.

Thus, if Roberto, who weighs 150 pounds, is 2′ from the fulcrum of a seesaw, how far from the fulcrum would Becky, who is 60 pounds, have to sit to achieve balance? Using the above formula, where w_1 is 150, d_1 is 2, and w_2 is 60, we can calculate the distance (d_2) as follows:

$$60 \times d_2 = 150 \times 2$$
$$60 \times d_2 = 300$$
$$d_2 = 300 \div 60$$
$$d_2 = 5$$

d_1 d_2 w_1 w_2

Goodheart-Willcox Publisher

Thus, Becky would have to sit 5′ from the fulcrum. If Isaac, who weighs 120 pounds, is 3′ from the fulcrum, how far from the fulcrum would Josie, who is 90 pounds, have to sit to achieve balance?

structure filled with lignin, a natural glue, which bonds the structure together.

- Advanced materials are the result of advances in materials science. These advances have improved traditional materials such as ceramic, metallic, polymer, composite, or semiconductors at an atomic level, creating materials that have entirely different and even adaptable characteristics.

Properties of Materials

All materials exhibit a specific set of properties. For example, the properties of iron are different from those of oak. These properties are considered when materials are selected for specific uses. Seven common properties, **Figure 10-19,** are:

- Acoustical properties are the reaction of a material to sound waves. Acoustical transmission (the ability to conduct sound) and acoustical reflectivity (the ability to reflect sound) are measures of acoustical properties.
- Electrical and magnetic properties are the reaction of a material to electrical and magnetic forces. These properties are described in terms of electrical conductivity (the ability to conduct electrical current) and magnetic permeability (the ability to retain magnetic forces).

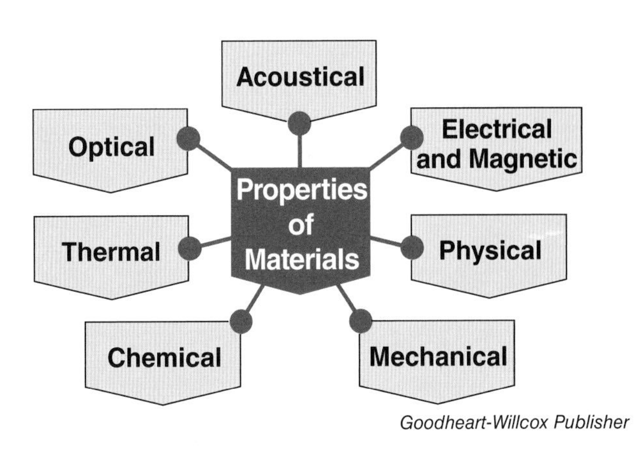

Goodheart-Willcox Publisher

Figure 10-19. Properties of materials can be grouped under seven categories.

- Physical properties are characteristics based on the structure of a material, including size, shape, density, moisture content, and porosity.
- Mechanical properties are the reaction of a material to a force or load. This property affects the material's strength (ability to withstand stress), plasticity (ability to flow under pressure), elasticity (ability to stretch and return to the original shape), ductility (ability to be bent), and hardness (ability to withstand scratching and denting).
- Chemical properties are the reaction of a material to one or more chemicals in the outside environment. These properties are often described in terms of chemical activity (the degree to which the material enters into a chemical action) and corrosion resistance (the ability to resist attack from other chemicals).
- Thermal properties are the reaction of a material to heating and cooling. This property is expressed as thermal conductivity (the ability to conduct heat), thermal shock resistance (the ability to withstand fracture from rapid changes in temperature), and thermal expansion (the change in size due to temperature change).
- Optical properties are the reaction to visible light waves. Optical properties include color (reflected waves), optical transmission (the ability to pass light waves), and optical reflectivity (the ability to reflect light waves).

Information

In contrast to other living species, humans have the ability to think, reason, and speak. They can observe what is happening around them, make judgments about those observations, and explain those judgments to other people. This unique human ability requires knowledge, which is derived from data and information, **Figure 10-20**.

Rawpixel.com/Shutterstock.com

Andrey_Papov/Shutterstock.com

Ingram Publishing/Thinkstock.com

Figure 10-20. Data are processed into information used by people to gain knowledge.

Data are all the raw facts and figures people and machines collect. ***Information*** is data that has been sorted and categorized for human use. Data processing involves collecting, categorizing, and presenting data so humans can interpret it. Information is grouped into three areas:

- **Scientific information**. This information is organized data about the laws and natural phenomena in the universe. Scientific information describes the natural world.
- **Technological information**. This information is organized data about the design, production, operation, maintenance, and service of human-made products and structures. Technological information describes the human-made world.
- **Humanities information**. This information is organized data about the values and actions of individuals and society. Humanities information describes how people interact with society and the values individuals and groups of people hold.

Information that people learn and apply is called ***knowledge***. Knowledge is the result of reasoned human action. This result guides people as they determine which course of action to take. Although knowledge can be described as being derived from scientific, technological, or humanities information, in reality, all these types of knowledge must be considered for any problem needing a solution. Knowledge derived from science might provide a theoretical base for the solution. Knowledge derived from technology is used to implement the solution. Knowledge derived from the humanities tells us if the solution is acceptable to society.

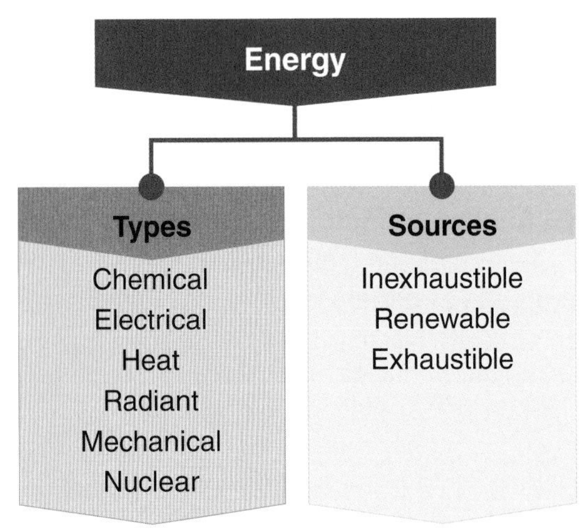

Goodheart-Willcox Publisher

Figure 10-21. The two main factors in understanding energy are the type of energy and the source from which it comes.

Energy

Energy is key to our survival, but as the law of conservation of energy states, it cannot be created or destroyed. Energy can only be converted from one form to another. After it is converted, energy powers factories, heats and lights homes, cooks foods, propels vehicles, drives communication systems, powers farm machinery, and supports construction activities.

Various types of energy exist. For example, everyone uses human energy to complete tasks. Human energy falls short of meeting all our needs, however. We have a limited supply, and we do not want to use it all on work. Also, some tasks cannot be done with human energy alone. For example, a house cannot be heated with human energy because the heat radiated from a human body is not enough to warm the house. Other sources of energy are necessary. Energy can be viewed from two vantage points: type and source, **Figure 10-21**.

Types of Energy

Energy can be grouped into six major types:

- **Chemical energy**. This energy is stored in a substance and released by chemical reactions.
- **Electrical energy**. Moving electrons create electrical energy.
- **Thermal (heat) energy**. Heat energy comes from the increased molecular action heat causes.
- **Radiant (light) energy**. The sun, fire, and other matter, including light, radio waves, X-rays, and UV and infrared waves, produce radiant energy.
- **Mechanical energy**. Moving water, animals, people, and machines produce mechanical energy.
- **Nuclear energy**. Splitting atoms or uniting atomic matter produces nuclear energy.

Sources of Energy

Energy is available from three major sources. These sources are grouped in terms of the supply. The first source of energy is an ***inexhaustible resource*** because it cannot be entirely used up or consumed. The sun is the fundamental source of inexhaustible energy for our solar system. The sun's actions warm the earth, create wind and weather, generate lightning, and, indirectly with the moon, create the ocean tides. Solar energy has always been available and will continue to be so.

A second source of energy comes from living matter. Human- and animal-muscle power can provide valuable energy. Wood and other plant matter can be burned as a fuel. This type of energy is a ***renewable resource***. It can be used up, but can also be replaced within the normal life cycle of the energy source.

The third source of energy is exhaustible. The quantity found on the earth limits these sources. Coal, petroleum, and natural gas are exhaustible energy sources. Neither the sun nor human action can create additional supplies of these sources as they take millions of years to form. The goal of wise energy utilization is to maximize the use of inexhaustible sources, recycle the renewable sources, and use minimal exhaustible energy.

Technology Explained Power Station

Power station: A technological system that extracts energy from different types of fuels and transmits the energy as electricity to consumers.

Power stations typically burn fossil fuels such as coal, oil, or natural gas as a means to generate electricity. Other power stations produce energy by splitting apart atoms of certain elements such as uranium and plutonium, which is called nuclear fission.

When power stations burn fuels or split atoms, the type of energy that is produced is mostly thermal energy. The thermal energy is used to heat water, converting it into high-pressure steam. The steam is sent to turn a turbine, which is a spinning wheel with blades similar to a windmill. When the steam blows past the blades, the turbine spins and transforms the thermal energy into mechanical energy. The turbine is connected to an electricity generator that converts the mechanical energy of the spinning blades into electrical energy by forcing the movement of electric charges (electrons) present in copper wire, much like a water pump pushes water through a pipe. The generator forces the movement of electrons through the wire by moving a magnet near the wire. The turbine is what moves the magnet. When the turbine spins, it also spins a magnet in the middle of a series of copper wire spools. The magnetic charge of the magnet is what forces the electrons to move, creating electricity.

The electricity travels out of the generator through electrical cables into a step-up transformer. The transformer boosts the electricity generated to a high voltage, enabling it to travel long distances along wire cables without losing as much energy. Once the electricity reaches its destination (a building), it enters a step-down transformer, which converts the electricity to a lower voltage that is safe to use in the building. The electricity then travels through cables into the building where appliances and electronics are plugged in.

In summary, see **Figure A**. Electricity used to power televisions or charge cell phones starts off as a shipment of fuel (oil, gas, coal) which is then burned in a furnace, or as radioactive elements (uranium and plutonium) whose atoms are split. Either method heats water into steam. The steam then moves a turbine, which spins a magnet inside a generator that produces electrical energy. That electrical energy is then transmitted through a series of cables to your home.

Cleaner ways of turning turbines and generating electricity include water and wind. Water is used to turn turbines in hydroelectric dams. Wind is used to turn large wind turbines. Ocean currents and waves are used to generate electricity as well as geothermal energy under the earth's surface.

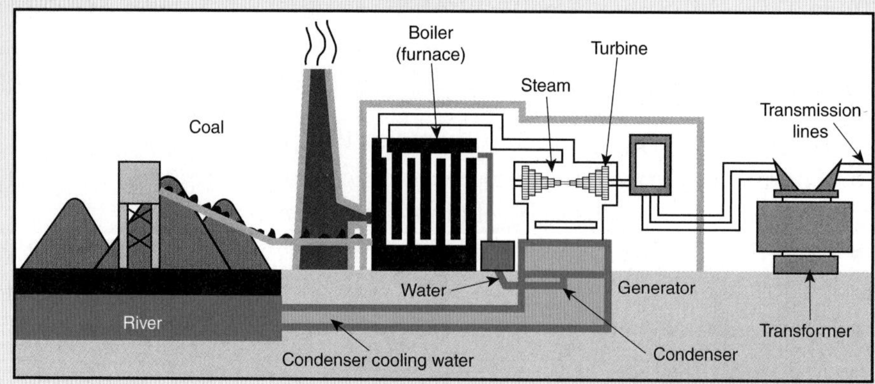

Figure A. A coal-fired power plant.

Finances

Technological systems require people, tools and machines, materials, and energy. These resources are generally purchased. For example, we pay people wages, or salaries, for their labor and knowledge. We buy or lease machines and purchase materials and energy. These actions require money, which provides the financial foundation of the technological activity.

Money to develop and operate technological systems can be obtained through equity financing or debt financing, **Figure 10-22**. In ***equity financing***, money is raised by selling partial ownership in a technological system or company. If one person owns the company, the operation is called a ***sole proprietorship***. If two or more people own it, they might form a ***partnership***, in which each person owns a portion of the company. In another circumstance, people might form a corporation. A ***corporation*** is a legal entity formed by people to own a company. They sell shares, which are certificates of ownership in the corporation. In all three cases, the person or group of people raises the money by selling equity—a portion of the company. Raising money in this way is called equity financing.

People also raise money by borrowing from other sources. Banks, insurance companies, or investment groups might loan money to support the activities of a company. This loan constitutes a debt that must be repaid. This type of financing is called ***debt financing***.

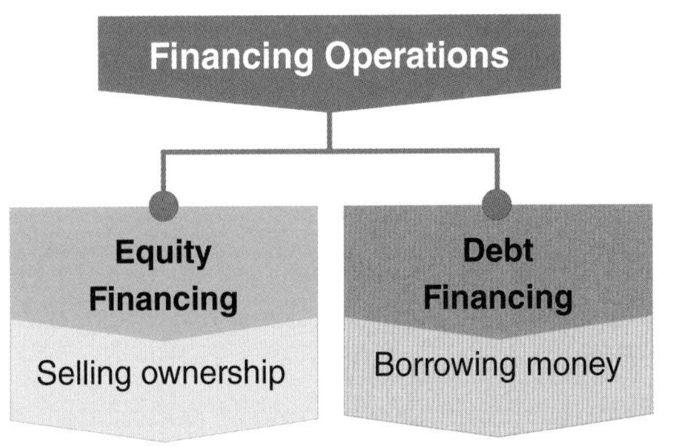

Goodheart-Willcox Publisher

Figure 10-22. Financing the operations of technological systems and companies is achieved through equity and debt financing.

Time

Humans have always been aware of time, but our measurement, perception, and use of it have changed in a variety of ways. See **Figure 10-23**. For example, early people measured time by the rising and setting of the sun. They knew they had only so much daylight in which to hunt and gather food, and they acted accordingly. Later, farmers were very aware of particular months of the year because certain periods were more conducive to growing healthy crops.

Now, although time in and of itself has not changed, technology has accelerated its use and changed our standards of measurement. We allocate machine time, computer time, and sales response time, for example. At one point, we measured time only in years, months, and days. Hours and seconds were then observed. Now, engineers worry about nanoseconds (billionths of a second) in computer processing. Time is becoming an even more valuable resource for technological systems as the rate of development increases rapidly. New technologies build on old technologies at a rapid rate, and the resulting products are made and disseminated quickly.

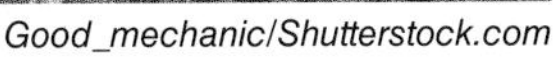
Good_mechanic/Shutterstock.com Monkey Business Images/Shutterstock.com US Department of Commerce

Figure 10-23. Time has been measured in different ways throughout human history. Early Egyptians used sundials that measured the time of day by the sun's shadow. We currently use various types of clocks and watches. The NIST F1 at the National Institute of Standards and Technology keeps time with an uncertainty of about 3×10^{-16}, meaning it will neither gain nor lose a second in more than 100 million years.

Summary

- ✔ All technological systems involve inputs that are processed into outputs.
- ✔ People contribute to technological systems through their varying knowledge, skills, and abilities.
- ✔ People use tools and machines to increase their ability to do work.
- ✔ Tools are typically categorized as either hand tools or machines.
- ✔ Simple machines that work on the principle of the lever include the lever, wheel and axle, and pulley.
- ✔ Simple machines that work on the principle of the inclined plane are the inclined plane, wedge, and screw.
- ✔ Types of materials can be classified by origin, ability to regenerate, or physical state.
- ✔ Common properties of materials are acoustical, electrical and magnetic, physical, mechanical, chemical, thermal, and optical.
- ✔ The ability to communicate requires knowledge, which is derived from data and information.
- ✔ Information that people learn and apply is called knowledge.
- ✔ Various types of energy include chemical, electrical, thermal, radiant, mechanical, and nuclear.
- ✔ Sources of energy can be inexhaustible, renewable, or exhaustible.
- ✔ The goal of wise energy utilization is to maximize the use of inexhaustible resources, recycle the renewable resources, and use minimal exhaustible resources.
- ✔ We use finances to purchase other inputs.
- ✔ Money to develop and operate technological systems can be obtained through equity financing or debt financing.
- ✔ Time affects the amount and kinds of technology we produce.

Check Your Engineering IQ

Now that you have finished this chapter, see what you learned by taking the chapter posttest.
www.g-wlearning.com/technologyeducation/

Test Your Knowledge

Answer the following end-of-chapter questions using the information provided in this chapter.

1. The resources used to make a technological system operate are called _____.
2. Explain the statement, "Consumers are the reason for technological systems."
3. *True or False?* A drilling tool is used to abrade and smooth surfaces.
4. Name three types of simple machines that use the principle of the lever.
5. This drawing shows a(n) _____-class lever.

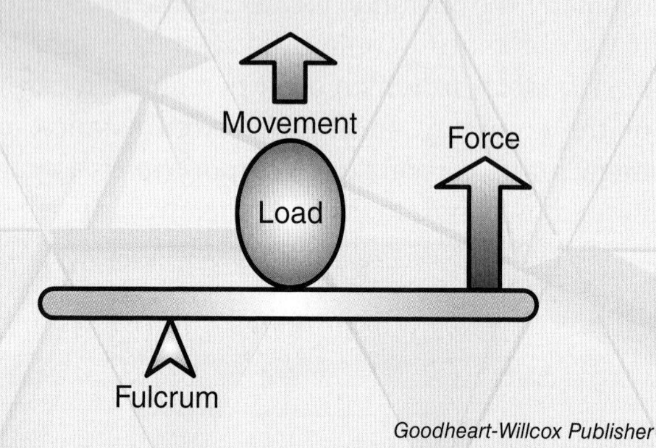

Goodheart-Willcox Publisher

6. The three types of simple machines that use the principle of the inclined plane are the inclined plane, the screw, and the _____.
7. Give an example of a genetic material.
8. In their physical state, materials can be classified as solids, gases, and _____.
9. *True or False?* The mechanical properties of a material deal with the material's reaction to a force or load.
10. Information can be classified into three areas: technological, humanities, and _____.
11. *True or False?* Energy that comes from the increased molecular action heat causes is called *thermal energy*.
12. Give an example of an exhaustible source of energy.
13. Name two ways to raise money to develop and operate technological systems.
14. *True or False?* Our perception of time has remained constant throughout history.

Critical Thinking

1. Look around the room you are in. List the human abilities that were used to design, construct, and decorate the room and the products within it, and list three ways each ability was used.
2. Select a product made from exhaustible materials or a task that uses an exhaustible energy source. Describe how that product could be made or how the task could be completed using renewable materials or renewable energy sources.
3. Design and sketch a simple device that uses at least three of the six simple machines to perform a specific task.

STEM Applications

1. Create a social network profile for a specific element. Use the periodic table of elements to describe the element's characteristics (properties), relatives (groups and periods), as well as occupation (what is it used for?). Do not forget to include a profile photo.

Simple Machine Science Kit

Background

Technology uses tools to process materials to create products. The tools often use one or more of the six basic machines. These machines are the lever, the wheel and axle, the pulley, the inclined plane, the wedge, and the screw.

Situation

The Science Experiments Company markets instructional kits for use in elementary and middle school science programs. This company has hired you as a designer for its physics kits. Design a kit containing common materials, such as tongue depressors, mousetraps, rubber bands, thread spools, small wooden blocks, and string. Using these materials, develop a technological system using three or more of the simple machines to lift a tennis ball at least 6″ off a table. You must also include an explanation of how each simple mechanism is demonstrated in the system, using the materials that you selected.

Desired Outcomes

- A working system that uses three or more of the six basic machines to achieve the desired goal of raising a tennis ball 6″ off a table.
- An explanation of how each simple machine was created and how it demonstrates the principles of the machine in the system.
- An explanation of how each stage in the system demonstrates one of Newton's three laws of motion.
- A five-minute presentation to the class that demonstrates the functioning of your system and explains how the different simple machines are represented in the actions of your system.

MilanB/Shutterstock.com

This compound pulley uses more than one pulley to move a load.

CHAPTER 11

Technological Processes

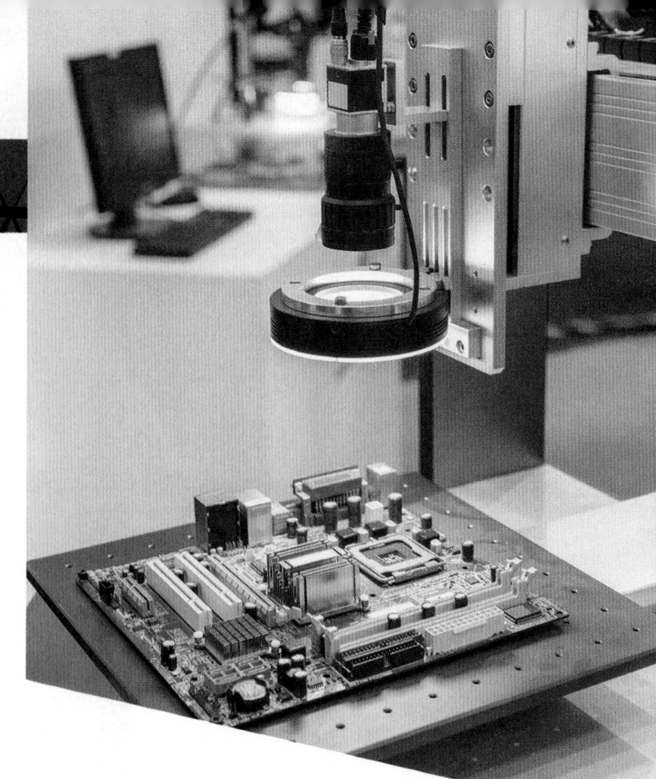

Check Your Engineering IQ

Before you read this chapter, assess your current understanding of the chapter content by taking the chapter pretest. www.g-wlearning.com/technologyeducation/

Learning Objectives

After studying this chapter, you should be able to:

- ✔ Explain the three major types of technological processes and their relationship to one another.
- ✔ Describe the steps used in the engineering design process.
- ✔ Analyze the major processes or activities used in agriculture and biotechnology, communication and information, construction, energy and power, manufacturing, medicine, and transportation.
- ✔ Explain the steps involved in management processes.

Technical Terms

assembling
buildings
casting and molding
civil engineering structures
commercial structures
conditioning
conversion and processing
decoding
diagnosis
drilling
encoding
engineering structures
finishing
forming
foundation
graphic communications
growth
guidance
harvesting
heavy engineering structures
ideation
industrial structures
mining
pathways
planning
power
prevention
primary processing
propagation
propulsion
receive
residential structures
retrieved
secondary processes
sensor
separating
servicing
site preparation
stored
structure
superstructure
support systems
suspension
telecommunication
telecommunications technology
terminals
transmitting
treatment
ultrasonic sensor
utilities
vehicular systems

Close Reading

Before you read the chapter, think of an everyday piece of technology. As you read, think of how the various engineering, technological, and managerial processes were used in creating that technology.

A technological system has inputs, processes, and outputs. Inputs are the resources that go into the system. Processes are what happen within a system. They change inputs into outputs. Processes are used to create products, structures, energy-conversion systems, communication messages, and transportation systems. They are also used to:

- Grow and process crops.
- Produce products and structures.
- Treat medical conditions and illnesses.
- Convert and apply energy.
- Communicate information and ideas.
- Transport people and cargo.

In addition, managerial processes ensure that the technological system runs efficiently and produces quality products and few unwanted outputs.

All of the above processes can be classified under three major headings, **Figure 11-1**. These headings are problem-solving and engineering design processes, production processes, and management processes. Both the problem-solving and engineering design process and the management process can be seen as generic in that the same basic process can be used, no matter what the technology, **Figure 11-2**. The steps in the production process, however, vary according to the technological system under consideration.

Problem-Solving and the Engineering Design Process

All technology has been developed to meet human needs and wants. Each device or system is designed to solve a problem by extending human capabilities. See **Figure 11-3**. Early technology creation often focused on designing tools to make work easier. New housing technology made living more comfortable. New agriculture tools made farming easier. New transportation devices helped humans move loads from place to place. New communication methods made information exchange faster.

Many modern technological devices are designed to help solve a problem or to solve a problem better than a previous solution. For example, spacecraft are designed to solve the problem of how to travel through space to desired locations to study the universe. New spacecraft will also be designed to solve the same problem but do so in a manner that enables faster travel. Some products are designed to make money. How many people really need an electric toothbrush? Do standard toothbrushes fail to meet our needs? The electric model was most

Technological Processes

Problem-Solving and Engineering Design

Andresr/Shutterstock.com

Production

Matva/Shutterstock.com

Management

Monkey Business Images/Shutterstock.com

Figure 11-1. The engineering design, production, and management processes are used to develop and operate technological systems.

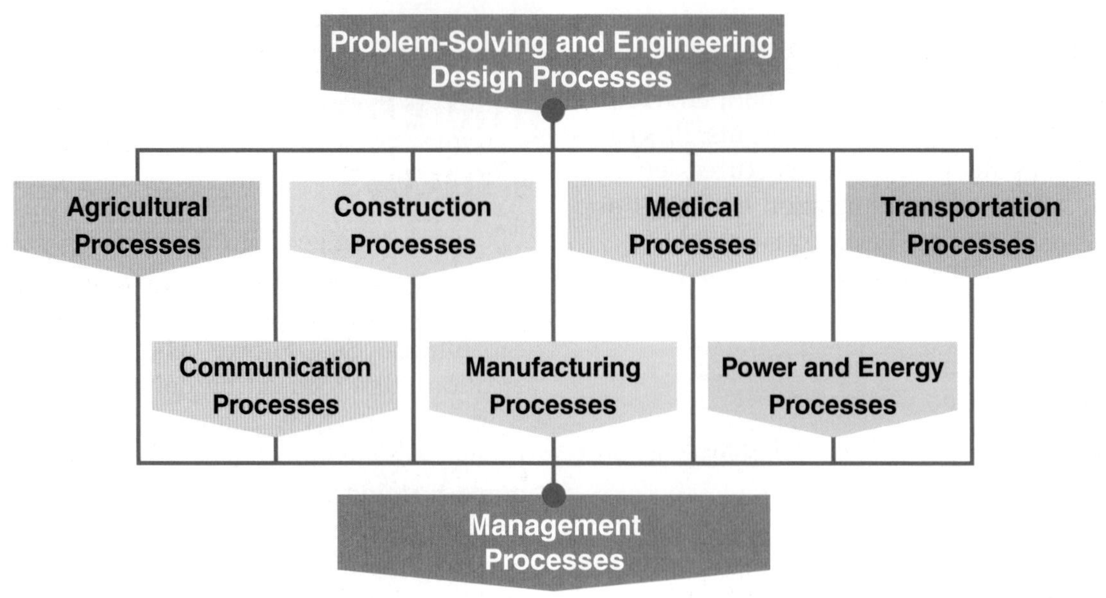

Goodheart-Willcox Publisher

Figure 11-2. The same engineering design and management processes are used for all seven technological areas.

SKatzenberger/Shutterstock.com

Figure 11-3. Technology is designed to solve problems. Drones are used to monitor irrigation systems on farms, helping in disaster relief, and deliver packages and covering breaking news stories.

likely designed to give inventors and manufacturers an opportunity to profit from their idea. There is nothing wrong with this motive. In fact, the drive to make money has brought us many things we now take for granted. The change simply suggests that our civilization has progressed a great deal. Now, we have time to apply technology to luxuries. We are beyond the survival mode of our ancestors.

No matter what the motive, people tend to follow a common procedure to solve problems. This is typically a trial-and-error method of developing a solution that includes the following actions:

1. Develop an understanding of the problem through observation and investigation.
2. Devise a plan for solving the problem.
3. Implement the plan.
4. Evaluate the plan.

Engineers, designers, and other technical professionals, however, try to avoid costly trial-and-error approaches to solving problems by employing a more informed and purposeful problem-solving procedure called the engineering design process. These professionals follow steps similar to the trial-and-error approach. The key difference, however, is that they make informed decisions about developing a solution before actually producing the solution. The solution is then produced through thorough three-dimensional modeling and predictive analysis. Predictive analysis is the practice of extracting information from purposefully collected data to determine patterns that aid in predicting future outcomes. The steps in the engineering design process, **Figure 11-4**, are:

1. Identify and define a problem.
2. Gather information.
3. Develop a design solution.
4. Model and make a solution.
5. Evaluate the solution.
6. Communicate the final solution.

Each step builds on the others. However, they are not always done in order and are often revisited throughout the problem-solving process. Recall

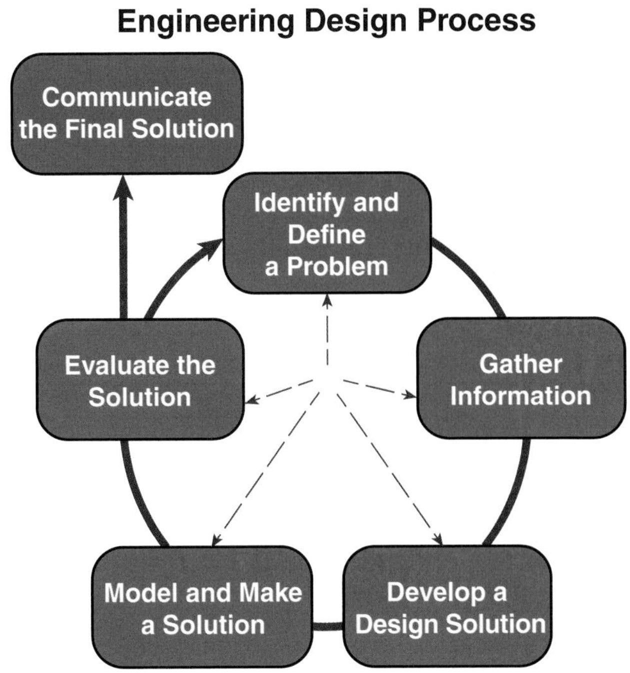

Engineering Design Process

Communicate the Final Solution

Identify and Define a Problem

Gather Information

Develop a Design Solution

Model and Make a Solution

Evaluate the Solution

Goodheart-Willcox Publisher

Figure 11-4. The engineering design process has six steps. Each step affects the others.

from Chapter 9 that all technological processes have feedback. The results from one step might cause the designers to retrace their steps. For example, a prototype built in the one step might show major problems with a design. The solution might have to be modified by changing it and selecting a new "best" solution. The following sections explain each of these steps separately.

Identify and Define a Problem

Technology development starts with a problem. These problems, however, are seldom clearly seen or defined. The first step in the design process identifies and describes the task being undertaken.

This step involves defining the problem and listing limitations or requirements for a solution. These requirements include the criteria and constraints regarding such characteristics as function, appearance, size, operations, manufacturing, marketing, and finances. Problems that might result from various environmental, political, ethical, and social issues might also be identified. These criteria and constraints communicate the expectations for the solution.

Gather Information

Once the problem is defined, designers seek solutions by gathering all potentially valuable information. This information is then carefully reviewed, in order to understand the necessary concepts to design solutions. Other tasks in this step include conducting research to understand unknown phenomena and examining current or prior solutions to similar problems through patent searches.

Develop a Design Solution

Engineers create many possible solution designs by letting their minds create ideas (a process called ***ideation***) and by brainstorming with others. They continually check and critique their answers, redefining their solutions and improving the designs. Their sketches show ways to meet the challenge. Engineers first make rough sketches to capture thoughts. They refine the sketches to mold thoughts into more specific solutions, **Figure 11-5**. Engineers then choose the most promising ideas and refine their work. They analyze ideas to select the optimal solution, realizing that criteria and constraints sometimes compete with one another and that trade-offs must occur. Engineers know that this best solution might not meet all the criteria and the constraints perfectly. They do understand, however, that the solution must be safe, efficient,

sayhmog/Shutterstock.com

Figure 11-5. Designers use sketches as they explore possible solutions for design problems.

Career Connection Process Engineer

Product manufacture requires certain processes be done to transform input resources into a desired output product. The people responsible for devising, developing, studying, and improving these processes are process engineers. Process engineers work to design and evaluate systems involved in a wide range of manufacturing operations, such as pharmaceutical, petrochemical, and mineral processing industries. Process engineers determine ways to enhance the efficiency, quality, and safety of the process specific to those operations.

Coursework for those interested in this career includes advanced courses in statistics, mathematics, physics, and chemistry at the high school level. Students should also explore technology and engineering coursework, especially in control systems, manufacturing, and electronics. Process engineering requires a bachelor's degree in engineering or another technical field.

Process engineers should be able to problem solve, troubleshoot, analyze data, think in terms of systems, and use a variety of data analysis and design software and tools.

Process engineers can work as self-employed contractors or as salaried employees in manufacturing industry companies. With additional education and experience, individuals within a company can advance to become senior process or manufacturing engineers, as well as engineering managers.

Potential projects for process engineers include:

- Evaluating and troubleshooting manufacturing and control systems and equipment.
- Implementing process enhancements to improve the cost effectiveness of system operations.
- Analyzing how system alterations impact the overall production process.
- Managing projects designed to improve the economical value of manufacturing operations.
- Developing written documents to support the development and operation of technological systems.

The job outlook for process engineers is steady. The average salary for a process engineer is relatively high compared to most careers in the United States. Popular industries for employment are petrochemicals, computer technologies, and pharmaceuticals. Professional organizations include the Society for the Advancement of Material and Process Engineering.

and functional (work properly) and that it must be producible and marketable within cost limits. Once the solution is chosen, it must be detailed. Specifications and general characteristics of the final solution must be established and modeled using computer design and simulation software.

Model and Make a Solution

Frequently, engineers produce some type of model or prototype of the expected solution, **Figure 11-6**. A model or prototype allows designers and managers to review the solution's performance. Models might be physical, graphic, or mathematical. Often, designers use a computer to develop graphic models based on mathematics. They then refine these models to optimize the solution. Finally, they construct physical (working or appearance) models or prototypes.

Evaluate the Solution

Throughout the design process, engineers are constantly evaluating. They will often test and

RAGMA IMAGES/Shutterstock.com

brett_land/Shutterstock.com

Figure 11-6. Models are used to test product and structure designs. They can be created on computers or made physically.

evaluate different ideas or parts of a design to determine whether these elements will perform as desired. At this stage in the engineering design process, engineers want to validate that the selected design solution will meet all criteria and constraints once the solution is fully implemented in the environment in which it will perform. The results from evaluating and validating the solution design can then be used to revise the design solution before the project is completed.

Communicate the Final Solution

The final solution must be carefully specified for production. Engineers develop graphics showing size, shape, and component arrangement. They formulate a material list or bill of materials and produce specifications for the materials to be used. These documents communicate the characteristics of the product, structure, media, or system. Engineers also prepare written and oral reports to gain approval for the solution from decision makers. Once the engineering design process ends, the design solution moves into the production process.

Production Processes

Production processes are actions that create physical solutions to problems. Examples of production processes include growing and harvesting crops, constructing structures, and generating communication messages. As noted earlier, the engineering design process is used across various production processes. This process generates the description of the solution. The same processes or sequence of events can be used to design agricultural, food-processing, manufacturing, construction, energy and power, medical, communication, information-processing, and transportation systems.

Each of the technological systems has its own unique production processes. These processes transform the solution designs into tangible solutions.

Agricultural and Biotechnological Processes

Since our early history, people have raised crops and domesticated animals. This activity is called farming, or agriculture. Agriculture is the art of cultivating the soil, growing crops, and raising livestock. Broadly speaking, this area includes farming, fishing, and forestry. Each of these areas involves a crop with a biological cycle including birth, growth, maturity, and death.

Farming is growing plants and animals for a specific use. See **Figure 11-7**. Typical crops are fruits, vegetables, grains, and forage for animals. Individuals apply farming processes as they grow plants for landscapes and home gardens.

Fishing is harvesting fish from lakes and oceans for commercial use. Fish can also be raised and harvested in controlled areas called *fish farms*. Individuals also use fishing processes for recreational purposes and food.

Forestry is growing trees for commercial use. The trees might be used for lumber and veneer (thin sheets of wood), paper, or other products. Agricultural practices involve the following major steps, **Figure 11-8**:

Zigzag Mountain Art/Shutterstock.com

Figure 11-7. Trees grown on this farm will typically be sold to garden centers and landscaping companies.

1. Propagation.
2. Growth.
3. Harvesting.
4. Conversion and processing.

Agriculture produces food and fiber for people to use. This activity starts with the birth of crops by planting seeds, rooting cuttings from plants, or allowing or enabling animals to breed. This step is called ***propagation***. Propagation allows biological organisms to reproduce.

After new animals are born or plants appear, growth occurs. Agriculture provides a proper environment for this to happen. ***Growth*** involves providing feed and water for animals or cultivating and watering (irrigating) crops. When the animal or plant has reached maturity, harvesting occurs. ***Harvesting*** is the process of gathering. In the case of agriculture, it is the process of gathering crops or butchering animals to process into consumable products.

In the final step, food products undergo ***conversion and processing*** to create foodstuffs. For example, wheat is ground into flour. Meat is cut into steaks and roasts.

Agricultural practices are enhanced through the application of biotechnological processes. For example, through combining biological processes with physical technology, we have genetically altered bacteria. These bacteria are used to protect crops from insects. Agricultural and biotechnological processes are discussed in depth in Chapters 25 and 27.

Communication and Information Processes

The exchange of information and ideas is called communication. The simplest communication processes involve spoken language. Spoken language is not technology, however, because no technology is used in the process. Technology became involved as civilization grew and people developed additional techniques to help them communicate better. For example, we now use technology to produce printed, graphic, and photographic media to share information and express ideas. We group these techniques under the general heading called graphic communications. We have also developed ***telecommunications technology***, which allows us to communicate using electromagnetic waves. Another type of communication, information processing, uses computers to exchange information and ideas.

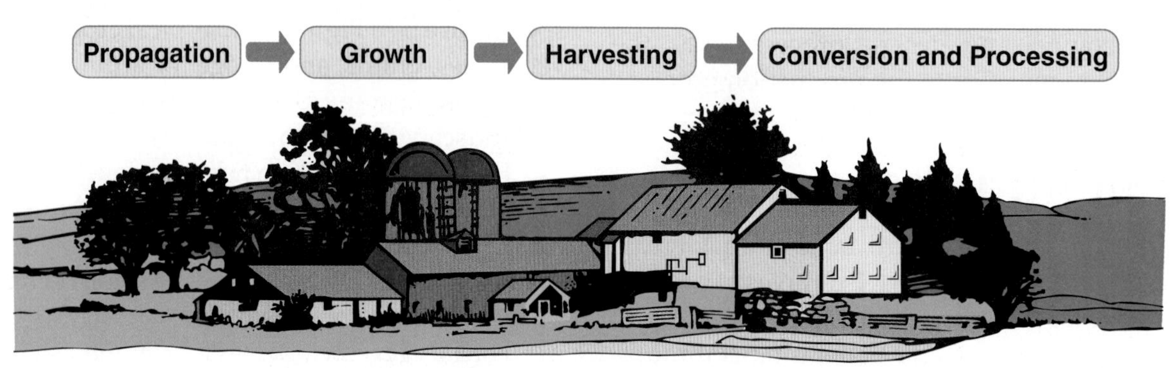

Goodheart-Willcox Publisher

Figure 11-8. Agricultural practices using technology involve four major steps.

Academic Connection: History

The Tennessee Valley Authority (TVA)

During the Great Depression of the 1930s, the United States, along with many other countries in the world, went through a period of high unemployment, declining sales, falling prices, and rising debt. One result of these conditions was a new willingness on the part of the American people to allow the federal government to become involved in improving social conditions. The administration of Franklin Delano Roosevelt proposed a way to use technology to help some of the poorest areas of the country, while putting many people back to work.

In 1933, Congress authorized the administration to create a federal program called the *Tennessee Valley Authority* (*TVA*). Under this program, the federal government became involved in the production of electric power. The TVA covered seven states: Tennessee, Kentucky, Alabama, Mississippi, Georgia, Virginia, and North Carolina. Some residents of these states did not even have electricity in their homes yet. The TVA built dams and built and operated hydroelectric plants to improve transportation and control floods. Under the TVA, the residents of this region were able to purchase cheap power. Although the Depression created distressing conditions, some people's lives actually improved in some ways because of the application of technology. The TVA is still operating today. Conduct research on the Internet to determine what areas the TVA is involved in now.

Graphic Communications

Graphic communications is a process in which messages are visual and have two dimensions. This category includes the printed messages commonly found in books, magazines, owners' and service manuals, and promotional flyers. Graphic communications media also include technical graphic messages, such as technical illustrations and engineering drawings. The final graphic communications medium is photographic communication. This group includes the film and print media coming from photographic processes.

Telecommunications

Telecommunications is the transmission of information over a distance for the purpose of communication. Telecommunications processes often depend on electromagnetic waves to carry their messages. Telecommunications include audio, video, and Internet communications services. Telecommunications techniques include broadcast (television and radio), hardwired (telephone and telegraph), and surveillance (radar and sonar) systems.

Information Processing

Information processing can be described as manipulating data to produce useful information. This processing involves changing (processing) information so the information is in a useful form. Information processing includes the actions of obtaining, recording, organizing, storing, retrieving, displaying, and sharing information. This processing generally involves the use of computer systems to locate and obtain the information, manipulate it into a useful form, and send it out to a client or customer. The outputting and sharing of the information is the communication part of the process.

Steps in the Communication Process

All communication technologies involve five major steps, **Figure 11-9**:

1. Encoding.
2. Transmitting.
3. Receiving.
4. Decoding.
5. Storing and retrieving.

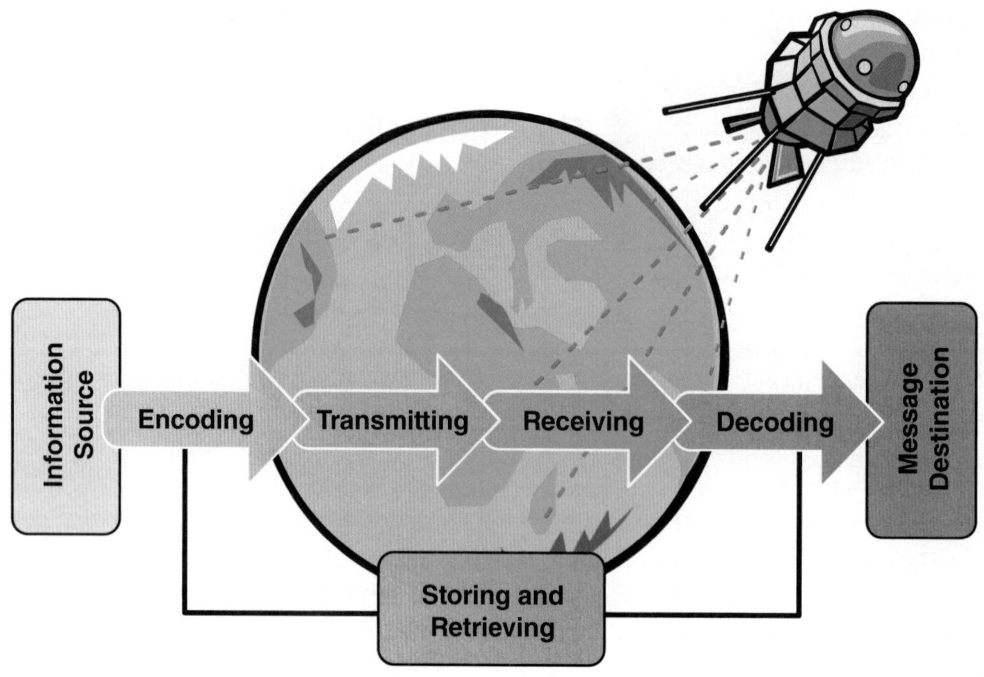

Figure 11-9. All communication technologies involve five major steps.

Communication technology organizes information so people can receive the information. The communication process starts with an information source. Most often, this source is the human brain. In some cases, machines are used. The first step in the communication process is encoding the information. ***Encoding*** involves changing information into a format or pattern that can be transmitted. This process might involve producing images on film, pulses on a light wave, electrical charges on a tape, or graphics on paper.

Transmitting the coded message from the sender to the receiver is the next step. Transmission can include moving printed materials, broadcasting radio and television programs, transmitting telephone messages along wires or fiber-optic strands, and using pulses of light to send messages between ships at sea. Transmitting a message is not enough. Someone or something must ***receive*** the message. Receiving the message requires recognizing and accepting the information. A radio must recognize the coded radio waves. The telephone receiver must recognize the pulses of electricity.

The received message must then undergo decoding. ***Decoding*** means the coded information must be changed back into a recognizable form. The decoded information is presented in an audio or visual format people can understand. The radio receiver changes radio waves into sound waves. The telephone changes electrical impulses to sound.

Throughout the communication process, information undergoes storage and retrieval. Storage processes allow the information to be ***stored***, or retained for later use. Books are shelved in libraries. Recorded music is stored on your smartphone. Data can be stored on remote servers. Pictures can be stored in file folders. Later, this information can be ***retrieved***, or brought back. The information can be selected and delivered back into the communication process. Storage and retrieval can happen at any time in the communication system. Information is commonly stored at the source or at the destination.

Construction Processes

Humans first used construction technology to produce shelter. This effort allowed early humans to move out of caves and into crude homes. These humble beginnings took the form of huts and tents. Now there are many types of constructed buildings. Construction technology also includes many types of structures that support other activities, including:

- Roads, canals, and runways for transportation systems.
- Factories and warehouses for manufacturing activities.

- Studios, transmission towers, and telephone lines for communication.
- Dams and power lines for electricity-generation and distribution systems.

Constructed works are everywhere. These works can be grouped into two major categories: buildings and civil, or heavy, engineering structures, **Figure 11-10**. ***Buildings*** are structures erected to protect people and machines from the outside environment. They are used for three major purposes. They are used as ***residential structures***, that is, places where people live. These structures can be homes, townhouses, condominiums, and apartment buildings. Buildings are also used as ***commercial structures***. These structures are the stores and offices used to conduct business. Government buildings, such as schools, city halls, and state capitols, are also in this category. Finally, buildings can be ***industrial structures***. These structures house machinery used to make products. They include power plants, factories, and transportation terminals.

The second type of constructed work is ***civil engineering structures***, which are also called ***heavy engineering structures***. These structures are primarily designed by civil engineers. Common civil structures are roads, dams, communication towers, railroad tracks, pipelines, airport runways, irrigation systems, canals, and electricity-transmission lines.

Most construction projects include several steps, **Figure 11-11**. These steps will be presented in more depth in Chapters 17 and 18. They include the following:

1. Preparing the site.
2. Setting the foundation.
3. Building the superstructure.
4. Installing the utilities.
5. Enclosing the superstructure.
6. Finishing the structure.
7. Completing the site.
8. Servicing the structure.

Not all construction projects use all the steps listed. Most projects, however, start with ***site preparation***. This task includes removing existing buildings, structures, brush, or trees that interfere with locating the new structure. Workers roughly grade the site and establish the desired slope. Surveyors then locate the exact spot for the new structure.

Setting the base, or ***foundation***, for the structure generally involves digging pits or trenches. Concrete or rock is then placed in the holes. If the hole extends to solid rock, additional foundations might not be needed. Foundations provide a stable surface on which to construct buildings.

The ***superstructure*** of a project is constructed on a foundation. Superstructures include the framework of the building or tower. They also include the pipes for pipelines, surfaces for roads and airport runways, and tracks for railroads.

Most constructed structures include utilities. ***Utilities*** are a system of a structure that provides water, electricity, heating, cooling, or communications. Waste pipes, electrical and communication wire, and heating and cooling ducts are all examples of utilities. Many runways, railroads, and highways also require lighting and communication systems.

After the utilities are installed, the superstructure must be enclosed. Walls need interior and

Buildings

Paul Matthew Photography/Shutterstock.com

Civil Engineering Structures

Shackleford Photography/Shutterstock.com

Figure 11-10. Construction technology is used to erect buildings and civil structures.

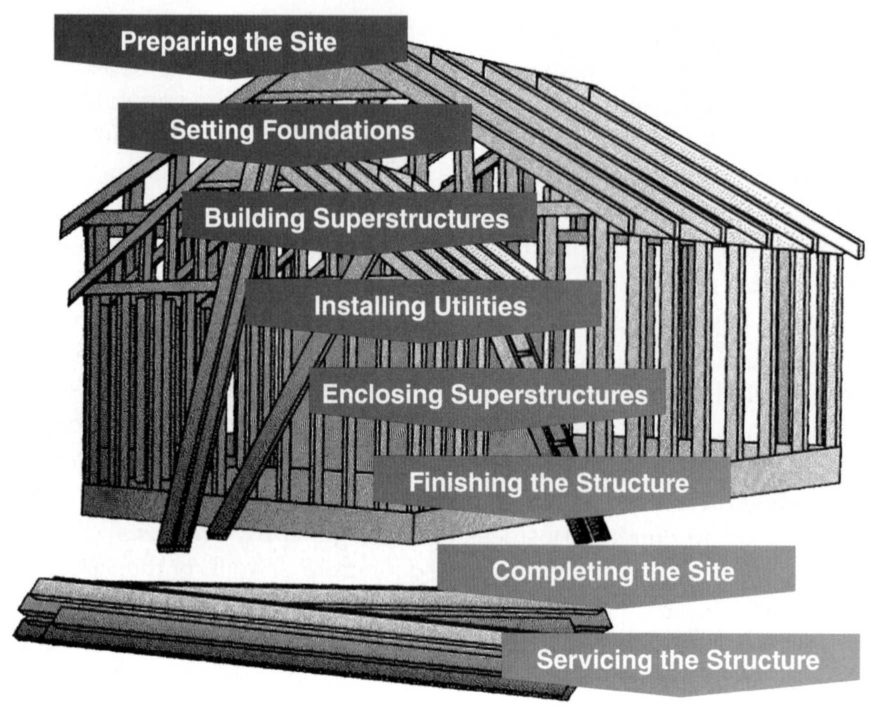

Goodheart-Willcox Publisher

Figure 11-11. Most construction projects move through these eight steps.

exterior skins. Roofs, ceilings, and floors must be covered. Doors and windows must be installed.

Once all this work is done, the structure must be completed. Walls, ceilings, doors, windows, and building trim must be painted. Likewise, communication towers must be coated to protect them from natural elements. Runways and roadways must have traffic stripes painted on them and signs installed.

The area around the structure must also be completed. The site must be graded to reduce erosion. Shrubs, grass, and trees are added to landscape the site and keep the soil in place. Hardscape, such as driveways, sidewalks, and retaining walls, is completed.

The result of this sequence of steps is a new constructed structure. During its life, the structure will require servicing. ***Servicing*** is the maintenance, repair, and reconditioning that structures require to maintain their integrity. Examples of servicing include painting walls, repairing roofs, and replacing utility-systems components. Servicing keeps the structure in good working order.

Energy and Power Processes

Energy is the ability to do work. All technological systems require energy. ***Power*** (which is sometimes confused with energy) is the rate at which work is accomplished or energy is changed from one form to another. This rate is calculated by dividing the work done by the time taken to do it.

Energy and power processes are covered in detail in Chapter 19. For now, however, note that six basic forms of energy exist and that technology aids in transforming and applying energy. The six basic forms of energy are:

- **Mechanical energy**. The energy found in moving objects.
- **Chemical energy**. The energy contained in molecules of a substance.
- **Radiant (light) energy**. Energy in the form of electromagnetic waves.
- **Thermal (heat) energy**. The energy associated with the movement of atoms and molecules.
- **Electrical energy**. The energy associated with moving electrons.
- **Nuclear energy**. The energy released when atoms are split or united.

The process of transforming and applying energy involves gathering or collecting the resources, converting the energy into a new form, transmitting the energy, and applying the energy to do work, **Figure 11-12**.

Gathering Resources

Energy resources include resources such as petroleum, coal, wind, sunlight, and moving water.

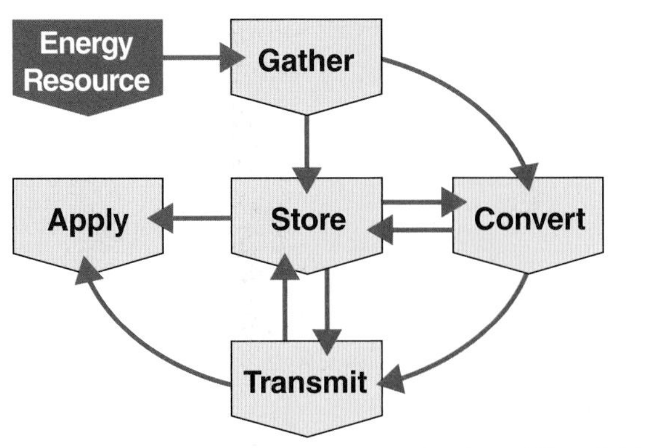

Goodheart-Willcox Publisher

Figure 11-12. The process of transforming energy involves four major steps.

The type of gathering techniques used depends on the resource to be used. Coal is gathered through mining. Natural gas and petroleum are gathered through drilling. Sunlight is captured with solar collectors. Water is collected using dams. Wind energy is captured with turbines.

Converting Energy

Energy resources are converted from one form to another through a variety of processes. Internal combustion engines convert fuels (gasoline, diesel fuel, and natural gas, for example) into mechanical energy. Hydroelectric plants use turbines and generators to change the mechanical energy in flowing water into electricity. Home furnaces change the chemical energy in fuels, such as coal, fuel oil, and natural gas, into heat (thermal energy). Solar panels convert the radiant energy of sunlight into electricity. Wind turbines convert the mechanical energy of blowing wind into electricity. Hundreds of energy converters exist that are designed to perform specific tasks.

Transmitting Energy

Energy must often be moved from one place to another. For example, the energy from an engine must be transmitted to the wheels driving a car. Heat energy must be moved from a furnace to the appropriate rooms in a house. Electrical energy must be moved from generating stations to homes, businesses, and factories.

Applying Energy

Energy, in itself, is of little use to people until it is applied to a task. The energy must power a device or power a process. For example, energy is useful when it causes a vehicle to move, a machine to perform a task, or a device to light a space, **Figure 11-13**.

Not all energy is used immediately after it is gathered, but is stored for later use. The reservoir behind a dam stores the energy in moving water. A battery stores the chemical energy of its parts. Tank farms store petroleum for later energy applications.

Manufacturing Processes

In everything you do, you use manufactured products. You ride to school in a manufactured car or bus. You look out of the building through a manufactured window. You wear manufactured clothing. You are reading a manufactured book and sitting on a manufactured chair in a room lighted with manufactured fixtures. It is difficult to imagine a world without manufactured products. No matter what the product, however, all manufacturing activities involve three stages, **Figure 11-14**. These stages are obtaining resources, producing industrial materials through primary processing, and creating finished products through secondary processing.

Obtaining Resources

Manufacturing activities start by obtaining material resources. This effort involves searching for or growing materials that can be harvested or extracted from the earth. For example, farmers and foresters grow the trees, plants, and animals needed

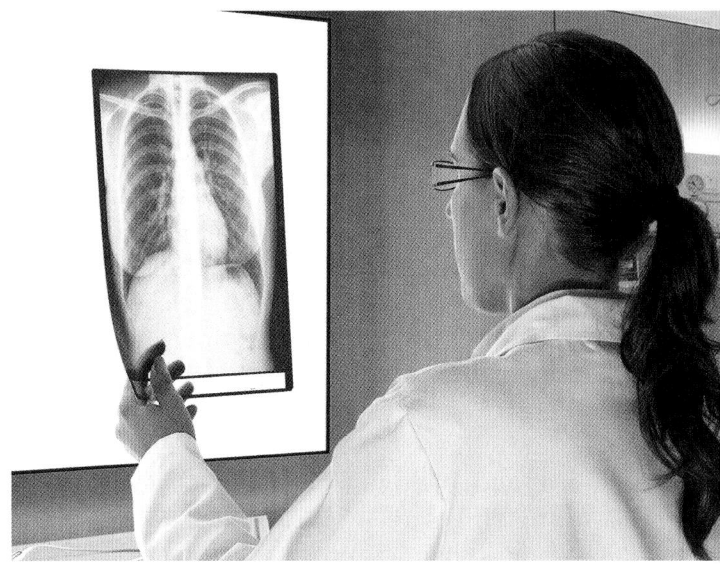

Fernando Madeira/Shutterstock.com

Figure 11-13. Energy is used to do work. Here, light enables a physician to examine an x-ray.

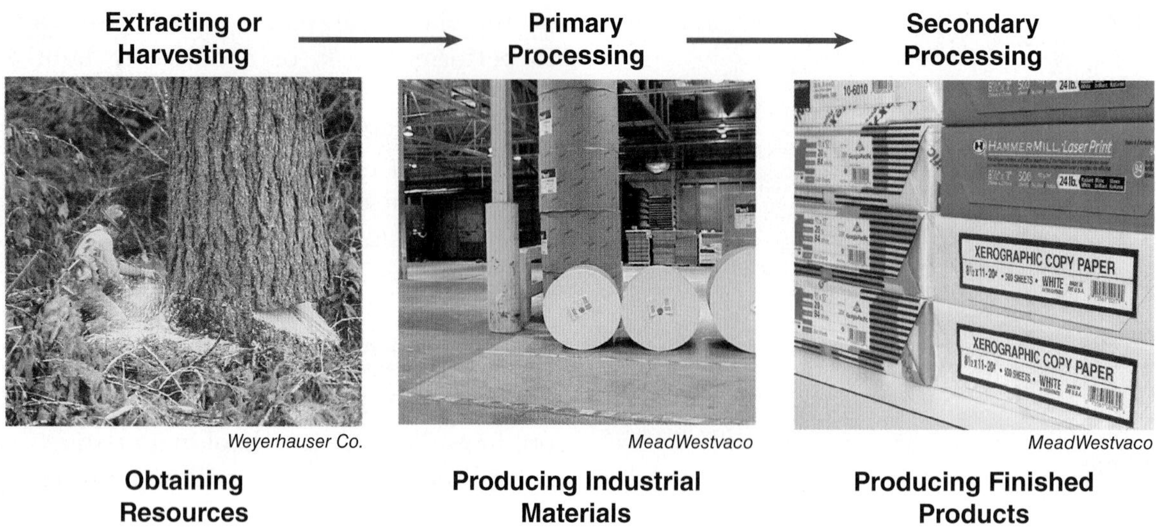

Figure 11-14. The manufacturing process begins by obtaining material resources. These resources are then changed into industrial materials. The industrial materials are used to make products.

to support the manufacturing processes. Exploration companies locate petroleum, coal, and metal-ore reserves. Once resources are grown or found, they are gathered or obtained using one of three processes:

- Harvesting. Recall that harvesting is the process of gathering. In this case, genetic (living) materials are gathered from the earth or bodies of water at the proper stage of their life cycles.
- *Mining*. The process of obtaining materials from the earth through shafts or pits.
- *Drilling*. The process of obtaining materials from the earth by pumping them through holes drilled into the earth.

Producing Industrial Materials

Primary processing is the step in which material resources are converted to industrial materials. For example, in steel mills, iron ore, limestone, coke, and other materials are converted into steel sheets and bars. In forest-products, trees are converted into paper, lumber, plywood, particleboard, and hardboard. In smelters, aluminum and copper ore are converted into usable metal sheets, bars, and rods.

Typically, these materials are refined using heat (smelting and melting, for example), mechanical action (crushing, screening, sawing, and slicing, for example), and chemical processes (oxidation, reduction, and polymerization, for example). The result is an industrial material or a standard stock. Examples include a sheet of plywood, glass or steel, a plastic pellet, and metal rods and bars.

Creating Finished Products

Most industrial materials have little value to the consumer. What would you do with a sheet of plywood, a bar of copper, or a bag of plastic pellets? These materials become valuable when they are made into products. They are the material inputs to the ***secondary processes*** of manufacturing. These processes change industrial materials into industrial equipment and consumer products, **Figure 11-15**.

Secondary processing changes materials in six basic ways. Casting and molding processes, forming processes, and separating processes give materials size and shape. Materials also undergo conditioning, finishing, and assembling.

- ***Casting and molding*** introduce a liquid material into a mold cavity where it solidifies into the proper size and shape.
- ***Forming*** uses force applied from a die or roll to reshape materials.
- ***Separating*** uses tools to shear or machine away unwanted material. This activity shapes and sizes parts and products.
- ***Conditioning*** uses heat, chemicals, or mechanical forces to change the internal structure of material. The result is a material with new, desirable properties.
- ***Finishing*** coats or modifies the surfaces of parts of products to protect them, make them more appealing to consumers, or both.

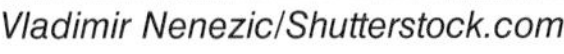

Vladimir Nenezic/Shutterstock.com

LDprod/Shutterstock.com

Figure 11-15. Secondary manufacturing produces industrial equipment and consumer products.

- *Assembling* brings materials and parts together to make a finished product. They are bonded or fastened together to make a functional device.

The results of obtaining resources, producing industrial materials, and manufacturing products are the devices we use daily. More information about these manufacturing activities is in Chapters 14 and 15.

Medical Processes

People in industrialized nations are living longer. This longer life expectancy is partly the result of better medical care. Technological and scientific advances have given health-care professionals new tools to treat patients, **Figure 11-16**. These tools are used in three major areas:

- *Prevention*. This area is using knowledge, technological devices, and other means to help people maintain their health. Prevention focuses on such health concerns as nutrition, exercise, and immunizations.
- *Diagnosis*. This area is using knowledge, technological devices, and other means to determine the causes of abnormal body conditions. Diagnosis matches symptoms to possible causes. For example, a person who experiences back pains might have vertebrae damage, muscle strains, or cancer of the spine. By using various tests, health-care professionals can determine the actual cause of the pain.

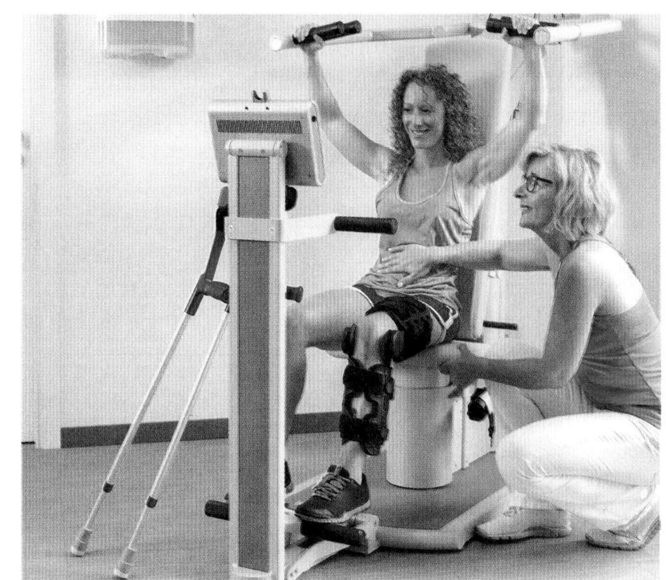

belushi/Shutterstock.com

Figure 11-16. Medical technologies help people stay well through disease prevention, diagnosis, and treatment.

- *Treatment*. This area is using knowledge, technological devices, and other means to fight disease, correct body malfunctions, or reduce the impact of a physical condition. Treatment includes such things as surgery, drug therapy, physical therapy, and other medical actions. You will learn more about medical technology in Chapters 25 and 26.

Transportation Processes

Humans have always had the desire to move around. They frequently move themselves and their

Technology Explained Hybrid Vehicles

Hybrid vehicle: A vehicle combining two or more sources of power.

With the high cost of gasoline, people are looking for more energy-efficient automobiles. One alternative to the standard car is the hybrid vehicle. A hybrid car is any vehicle combining two or more sources of power.

Hybrid vehicles are not a new concept. Diesel-electric railroad locomotives have been in use for decades, **Figure A**. They combine an internal combustion engine and electric generators and motors. Many submarines are also hybrids. They combine nuclear power with electric generators and motors. Hybrid commuter buses use diesel engines for parts of their routes and overhead electric lines to power motors for other parts of their routes.

Konstantin Menshikov/Shutterstock.com

Figure A. A common hybrid vehicle is the diesel-electric locomotive.

Hybrid cars are a combination of a gasoline-powered car with an electric car. In a standard American car, an engine uses gasoline as fuel. The gasoline engine turns a transmission that drives the wheels. In an electric car, batteries are the power source (fuel). The electricity powers a motor. The motor drives the wheels through a transmission.

A hybrid vehicle uses both types of power, **Figure B**. The gasoline engine is smaller than one in a traditional car. The engine might be as small as 1.0 liter and have as few as three cylinders. This gasoline engine uses advanced technologies to increase its efficiency and reduce its emissions. The engine can provide power to the wheels or to the motor or generator to recharge the batteries.

The vehicle also has an advanced electric motor. This power source has special electronic controls allowing it to be both a motor and a generator. At times, it draws energy from the batteries to accelerate the car. At other times, it functions as a generator. The motor recharges the batteries by regenerative braking. During this phase, the energy from forward momentum is captured during braking. Therefore, no external power supply is needed for recharging the batteries.

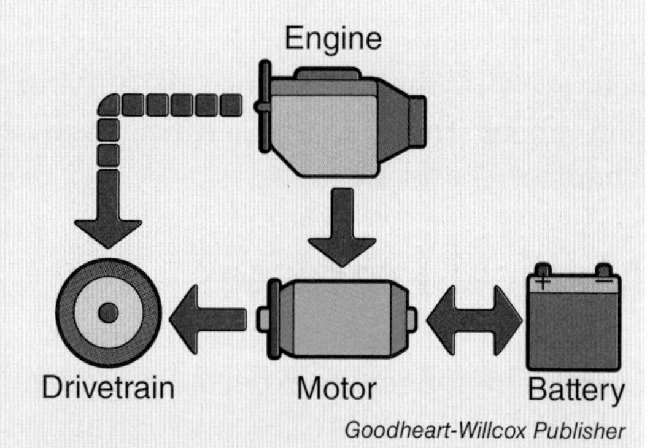

Goodheart-Willcox Publisher

Figure B. The major parts of a hybrid vehicle power system.

The operation of the vehicle changes under different driving conditions. During initial acceleration, the electric motor is the primary source of power. The gas engine starts up under heavy acceleration or to turn the generator. The generator, in turn, charges the battery.

During city driving, the electronic control system controls both power sources. The gas engine and electric motor are used equally. The engine starts and stops, depending on the situation. At high speeds, the gas engine is the primary source of power. The electric motor provides some power.

possessions from one place to another. This movement is called transportation. One way to understand transportation is to look at the media in which the various transportation systems have been developed. The three basic media are land, water, and air. A fourth one, space, is on the horizon, **Figure 11-17**. Another way to understand transportation is to look at its subsystems. The subsystems are vehicular systems and support systems.

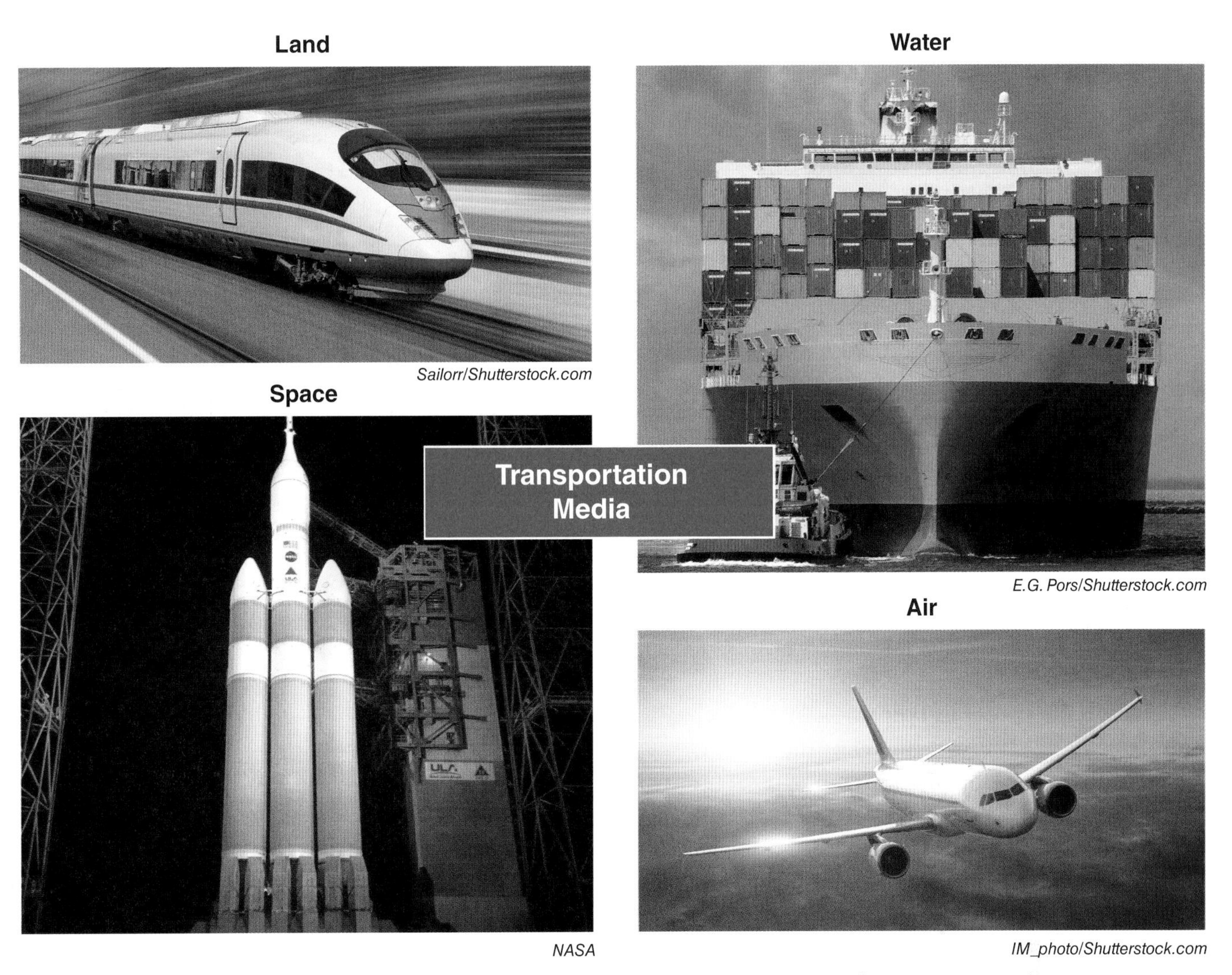

Figure 11-17. Transportation can be conducted on land, in water, and through the air. Space transportation systems may be used routinely in years to come.

Transportation Media

The earliest movement of people occurred on land. First they walked the land. They then tamed animals to help them move from one place to another. Finally, they developed various land vehicles from horse-drawn wagons prior to the invention of the automobile to modern magnetic-levitation trains.

The second transportation medium is water. Water transportation also has a long history. Transportation in this area has ranged from dugout log canoes to nuclear submarines.

The third major transportation medium is air. This medium is fairly modern. Air transportation began with balloon travel and now extends to supersonic airplanes.

We are starting to develop a fourth transportation medium—space. Today, we explore only the near reaches of space. The future holds the promise of space transportation systems that will take humans back to the moon, to asteroids, and to other planets.

Transportation Subsystems

Most transportation systems contain two major subsystems: vehicular systems and support systems. ***Vehicular systems*** are onboard technical systems that operate a vehicle. See **Figure 11-18**. Most vehicular systems are made up of five systems:

- *Structure*. This system provides space for devices. The structure, or framework, includes passenger, cargo, and power-system compartments.
- *Propulsion*. This system generates motion through energy conversion and transmission.
- *Guidance*. This system gathers and displays information so the vehicle can be kept on course.

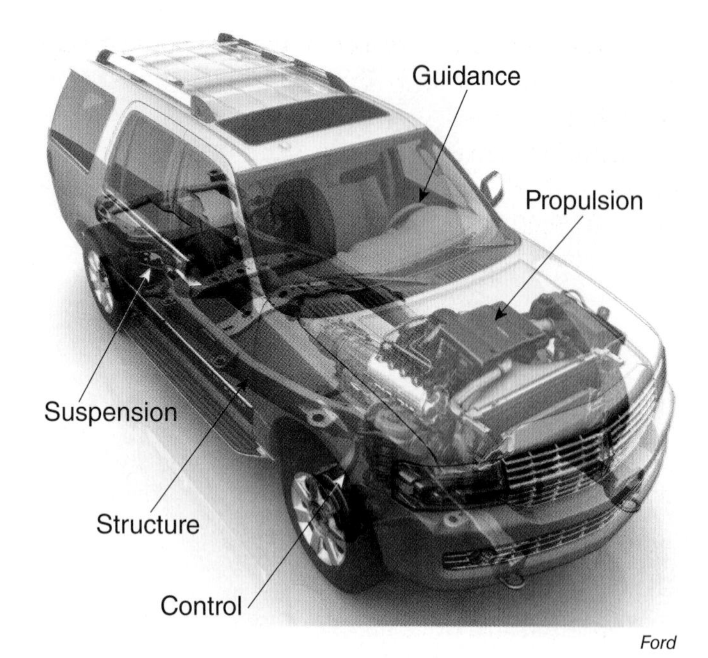

Ford

Figure 11-18. Most vehicular systems contain five systems.

- *Control*. This system makes it possible to change the speed and direction of a vehicle.
- *Suspension*. This system keeps the vehicle held in or onto the medium being used (land, water, air, or space).

Support systems, **Figure 11-19**, are external operations and facilities that maintain transportation systems. Support systems include pathways and terminals:

- *Pathways*. This system contains the structures that vehicles travel. Pathways include roads, railways, waterways, and flight paths.
- *Terminals*. This system contains structures that house passenger and cargo storage, and loading facilities.

Juan Novakosky/Shutterstock.com

Figure 11-19. Transportation support systems include pathways such as railway terminals.

A complete discussion of transportation technology is found in Chapters 23 and 24.

Management Processes

The third major type of process used in technology is management. Management processes are designed to guide and direct other processes. Management provides the vision for an activity. Management includes four steps, **Figure 11-20**. The first step is planning. ***Planning*** involves developing goals and objectives. The goals can be broad or specific and focus on such areas as production, finance, and marketing.

Once plans are developed, the activity must be structured. Procedures to reach the goal must be established. Lines and levels of authority within the group or enterprise must also be drawn. These actions are called *organizing*.

After the activity is organized, actual work must be started. This is called *actuating*. Workers must be assigned to tasks. They must be motivated to complete the tasks accurately and efficiently.

Finally, the outputs must be checked against the plan. This is called *controlling*. Control is the feedback loop that leads to the adjustment of management activities.

Sometimes, we think of management only in terms of companies. In reality, every human activity is managed. Goals are set. A procedure for

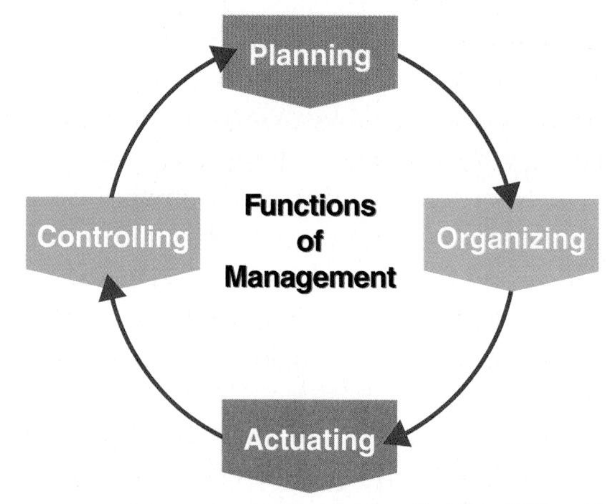

Goodheart-Willcox Publisher

Figure 11-20. Management involves setting goals (planning), establishing a structure to meet them (organizing), assigning and supervising work (actuating), and monitoring the results (controlling).

finishing the task is established. Work is started and completed. Results are evaluated. For example, a coach plans, directs, and evaluates the plays a football team makes. Homeowners develop plans to mow their lawns. Even something as simple as getting an after-school snack includes a plan.

Group activities are also managed. Some group activities take place within social, religious, and educational organizations. Others occur in technological enterprises. The management of technological activities and companies is explored in more detail in Section 5 of this book.

STEM Connection: Science and Mathematics

Ultrasonic Sensors

A ***sensor*** is a device that detects or measures a physical quantity and responds to it by recording information and initiating some other function. Sensors allow technological systems to respond to various inputs. This is similar to the way the human body responds to external stimuli. For example, when your hand touches something hot, signals are sent through your nervous system to your brain. These signals tell you to remove your hand so you do not get burned. A sensor detects signals as various forms of energy and converts those signals into another form of energy that can be measured to determine a physical quantity. Much like human senses, different sensor types enable technological systems to measure a variety of physical quantities (pressure, light, sound, distance) to determine the proper system responses. Analyzing and interpreting resulting data requires mathematics and scientific knowledge. To explore this mathematics and science connection, we can examine how an ultrasonic sensor collects and interprets physical quantities.

An ***ultrasonic sensor*** in a system functions like a human eye. It can detect distances, shapes, sizes, speed, and direction. Ultrasonic sensors determine these physical quantities in much the same way that bats use echolocation to locate objects in the dark. Bats produce an ultrasonic sound (pitch higher than the human hearing range) that travels through the air, bounces off objects, and goes back to the bat's ears. The bat's brain then determines the distance and shape of objects based on the amount of time that the sound takes to echo back to the bat's ear. The farther away an object is, the longer the sound takes to return to the bat's ears.

To mimic echolocation, an ultrasonic sensor has two parts: a transmitter and receiver. Like the bat's mouth, the transmitter sends out an ultrasonic sound signal and like the bat's ears, the receiver receives the signal once it has echoed off any nearby objects. This is where mathematics is important. The signals then need to be analyzed to compute the physical quantities. The sensor must measure the distance to an object by computing the time it takes for the sound to travel to an object and back, **Figure A**. To determine the object's left to right location, the sensor system has to be designed using principles of algebra, geometry, and trigonometry. Ultrasonic sensors can use two receivers (bats also have two ears) to compute information based on right triangles to determine an object's position left to right. The signal from the transmitter, to the object, and back to the two different receivers will have two different distances. This information is used to calculate the position using basic algebra and trigonometry.

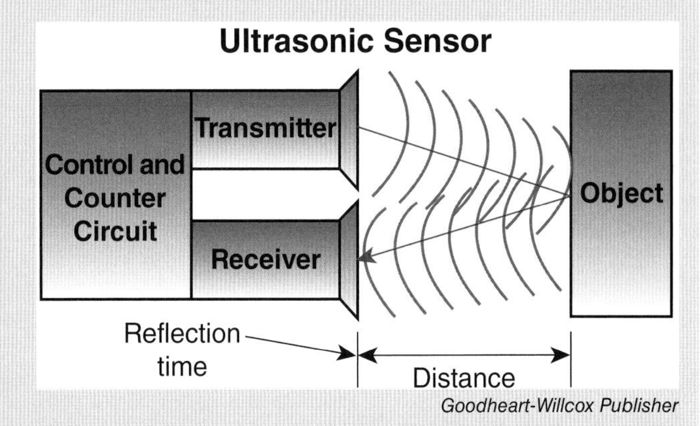

Goodheart-Willcox Publisher

Figure A. Ultrasonic sensors operation.

Summary

- ✔ All technological systems have processes.
- ✔ Processes are what happen within a system, such as creating products, structures, and communication messages.
- ✔ Processes can be classified under three headings: problem-solving and the engineering design process, production processes, and management processes.
- ✔ Processes often use problem-solving methods. These methods guide the creative activities of the field.
- ✔ The engineering design process uses informed and purposeful problem-solving procedures in which a solution is developed before it is actually produced.
- ✔ Each step of the procedure builds on the other steps.
- ✔ Once the system and outputs are designed, the production processes are used.
- ✔ Production processes are actions that create physical solutions to problems.
- ✔ The engineering design process is used across various production processes.
- ✔ Each technological system has a unique production process.
- ✔ Technological systems include agriculture and biotechnology, construction, energy and power, manufacturing, medical, and transportation.
- ✔ All technological systems are managed. They are planned, organized, actuated, and controlled.

Check Your Engineering IQ

Now that you have finished this chapter, see what you learned by taking the chapter posttest.
www.g-wlearning.com/technologyeducation/

Test Your Knowledge

Answer the following end-of-chapter questions using the information provided in this chapter.

1. Name the three major types of technological processes.
2. The design process and the management process can be seen as _____, in that the same basic process can be used, no matter what the technology.
3. *True or False?* The steps in the production process vary according to the technological system under consideration.
4. *True or False?* The best solution to a problem resolves all the criteria and constraint requirements.
5. _____ processes are actions that create physical solutions to problems.
6. Name the steps in the agricultural production process.
7. _____ is a process in which messages are visual and have two dimensions.
   A. Telecommunications
   B. Information processing
   C. Graphic communications
   D. None of the above.
8. Organize the following steps that occur in the communication process.
   A. Decode
   B. Encode
   C. Store and retrieve
   D. Transmit
   E. Receive
9. Give one example of a civil engineering structure.
10. _____ energy is energy found in moving objects.
    A. Thermal
    B. Chemical
    C. Mechanical
    D. Electrical

11. List the three major processes used to gather or obtain manufacturing materials.
12. *True or False?* Secondary processing changes material resources into industrial materials.
13. _____ uses tools to shear or machine away unwanted material.
    A. Forming
    B. Separating
    C. Conditioning
    D. Finishing
14. The three areas in which health-care professionals use technological tools are prevention, diagnosis, and _____.
15. Transportation media include water, air, space, and _____.
16. List the five vehicular systems common to most vehicles.
17. _____ involves developing goals and objectives.

Critical Thinking

1. Select a product or structure that is familiar to you. Develop a chart similar to the one below. List five considerations used in each of the technological processes used to produce the item.

Product or structure name:	
Process	**Considerations**
Engineering design and problem solving	
Production	
Management	

2. Select a current world problem that can be partially solved through technology. List the role each process would play in solving the challenge.
3. List and describe how you have used the production processes in one of the technological areas (agriculture and related biotechnology, communication and information, construction, energy and power, manufacturing, medicine, or transportation) to meet a need. Use a chart similar to the one shown.

Product or structure name:	
Technology	**What you did and the need that was met**
Agriculture and related biotechnology	
Communication and information	
Construction	
Energy and power	
Manufacturing	
Medicine	
Transportation	

STEM Applications

1. As you have learned, agriculture processes produce food and fiber for people to use. This process, called propagation, enables the creation of new plants from a variety of sources, such as seeds, cuttings, bulbs, and other plant parts. To control and enhance propagation processes, extensive knowledge of plant science is required. In this activity, we will explore the science behind these propagation methods. Two common methods are using cuttings and micropropagation. Research these methods. Then, obtain the resources and materials necessary to propagate a new plant from cuttings of a live plant and conduct the micropropagation of a plant.
2. Prepare a presentation that explains the science behind the propagation process methods and describes how science led to the improved methods.

CHAPTER 12

Outputs, Feedback, and Control

Check Your Engineering IQ 

Before you read this chapter, assess your current understanding of the chapter content by taking the chapter pretest.
www.g-wlearning.com/technologyeducation/

Learning Objectives

After studying this chapter, you should be able to:

- ✔ Describe the major types of outputs of various technology systems.
- ✔ Compare desirable and undesirable outputs of technological systems.
- ✔ Compare intended and unintended outputs of technological systems.
- ✔ Compare immediate and delayed outputs of technological systems.
- ✔ Recall the definition of feedback, as used in technology systems.
- ✔ Contrast the two major types of internal control systems and their functions.
- ✔ Analyze the major components of internal control systems and their functions.
- ✔ Differentiate between manual and automatic control in internal control systems.
- ✔ Evaluate external controls in technology systems.

Technical Terms 

adjusting devices
analytical systems
automatic control systems
closed-loop control system
data-comparing devices
delayed outputs
desirable outputs
electrical and electronic controllers
electrical or electronic sensors
electromechanical controllers
fluidic controllers
immediate outputs
intended outputs
inventory control
judgmental systems
magnetic (electromagnetic) sensors
manual control systems
mechanical controllers
mechanical sensors
monitoring device
open-loop control system
optical sensors
thermal sensors
undesirable outputs
unintended outputs
wage control

Close Reading

As you read this chapter, think of an example technology, either current or no longer in use. Use the three output categories given in this chapter to describe this technology's outputs.

All technological systems are designed to meet specific needs and wants. A direct relationship exists between the need (the reason for the system) and the output (the satisfying of the need). These outputs can be categorized in three ways, **Figure 12-1**:

- Desirable or undesirable.
- Intended or unintended.
- Immediate or delayed.

The people designing and using technology should strive to maximize the desirable results. But undesirable outputs can never be totally eliminated. They need to be held, however, to a minimum. To achieve this goal, people have created control systems that compare the results of the outputs with the goals. As seen in Chapter 9, these systems use feedback to regulate inputs. Outputs and feedback and control systems are essential to understanding and improving technological systems.

Outputs

Recall that resources (inputs) are transformed (processed) into things people want or need (outputs). These outputs can be seen as positive (desirable) or negative (undesirable). They can also be described as planned (intended) or unplanned (unintentional) outputs. Finally, outputs can happen right now (immediately) or appear at a later date (delayed). In this section, you will explore each of these types of outputs.

Desirable and Undesirable Outputs

The needs for and outputs of technology are countless. Each of the technological systems we have already classified (agriculture and biotechnology, communication and information technology, construction technology, energy and power technology, manufacturing technology, medical technology, and transportation technology) has a general type of output, **Figure 12-2**. For example, agricultural and related biotechnology can help people satisfy their needs for nourishment. Biotechnology's common outputs are plants and animals that can be processed into food. Communication technology helps satisfy the need for information. This technology's outputs are often media messages. Construction technology satisfies the need for shelter and support structures for other activities. This technology's outputs are buildings and civil structures. Energy and power technology helps satisfy the needs for heat, light, and motion. The output is mechanical, thermal, radiant, chemical, electrical, or nuclear energy. Manufacturing technology satisfies the need for tangible goods. This technology's outputs are consumer and industrial products. Medical technology helps satisfy our need for health and well-being. This technology's outputs are devices and treatments for medical ailments and for promotion of healthy bodies. Transportation technology satisfies a desire to move humans and things. This technology's output is the movement of people and goods (cargo). These outputs benefit people and so are called ***desirable outputs***.

Undesirable outputs are those outputs that are not wanted or planned for, **Figure 12-3**. For example, some manufacturing activities produce fumes and toxic chemical by-products. If improperly treated, these outputs poison the air and water around us. Poorly planned agricultural and construction projects can cause soil erosion. The result can be the loss of valuable topsoil and increased flooding. Communication technology can use billboards. Some people

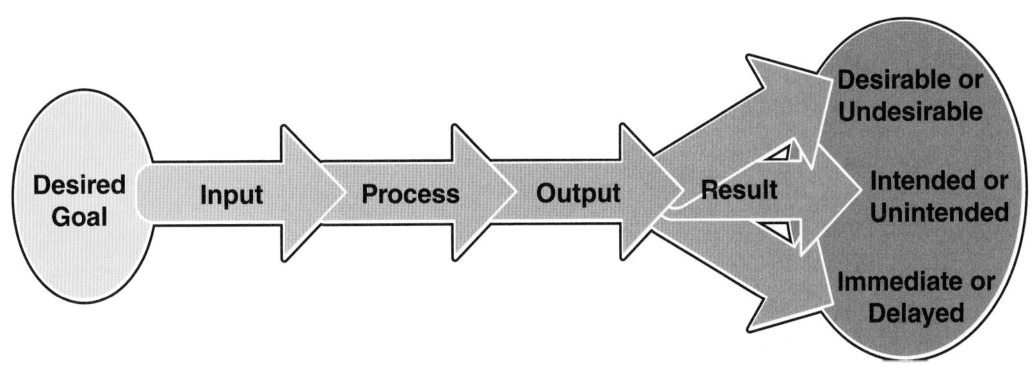

Figure 12-1. The outputs of a technological system can be categorized in three ways.

Figure 12-2. Each major type of technological system has a general type of output.

Desirable Output

Terrance Emerson/Shutterstock.com

Undesirable Output

Alex Tihonovs/Shutterstock.com

Figure 12-3. Technological system outputs can be desirable or undesirable.

find billboards unsightly. Also, people differ on the value of some communication messages. Transportation systems create exhaust fumes that pollute the air. Airport and highway noise affect nearby residential areas. Newly built rail lines and roads can divide established neighborhoods or farms. These are all examples of the undesirable outputs of technology.

Intended and Unintended Outputs

Intended outputs are products or services designed and produced with specific goals in mind. These outputs vary from refrigerators to magazines to municipal bus transportation. There are thousands of products and services that are intended outputs.

Depending on your point of view, intended outputs may be desirable or undesirable. For example, suppose you plan to make 10″-diameter plywood disks that must be cut from rectangular sheets of plywood. Cutting circles from rectangular material wastes material. You can plan for the waste, consider it in pricing your product, and reduce it to a minimum. The waste, however, cannot be totally eliminated. Both the disks and the waste are intended. They are in the plan for the system. The disks are the intended, desirable output. The waste is the intended, undesirable output.

Technological systems can also produce ***unintended outputs***, that is, outputs that were not anticipated when the system was designed. For example, some heating systems produced from the 1940s through the 1960s used asbestos as insulation for pipes. The material was considered an excellent insulator but was later determined to cause lung cancer in people who worked with it. This was an undesirable, unintended output of the technological system.

Immediate and Delayed Outputs

Technological systems are commonly designed to produce products or services for use now. These are known as ***immediate outputs***. For example, steel is not produced for use in 2071, it is produced for use this year. This practice reduces inventory costs and loss due to theft and spoilage. The production of steel is an example of an immediate, intended, and desirable output.

Technological systems can also create ***delayed outputs***, or results that occur at a later date than expected or planned. For example, chemical propellants used in aerosol (spray) cans and as refrigerants have affected the ozone layer around the earth. The accumulation of this matter has taken decades to reach a dangerous level. The same is true of the sulfur dioxide coal-fired power plants produce. The resulting acid rain kills forests in the United States and Canada. These examples show delayed, unintended, and undesirable outputs of technology. Thus, although technological systems are designed to produce specific outputs, many kinds of outputs can occur. The outputs of various types of technological systems are shown in **Figure 12-4**.

Feedback and Control

As noted earlier, humans have created control systems so they can reach desired goals. Most of these systems use feedback. Feedback uses

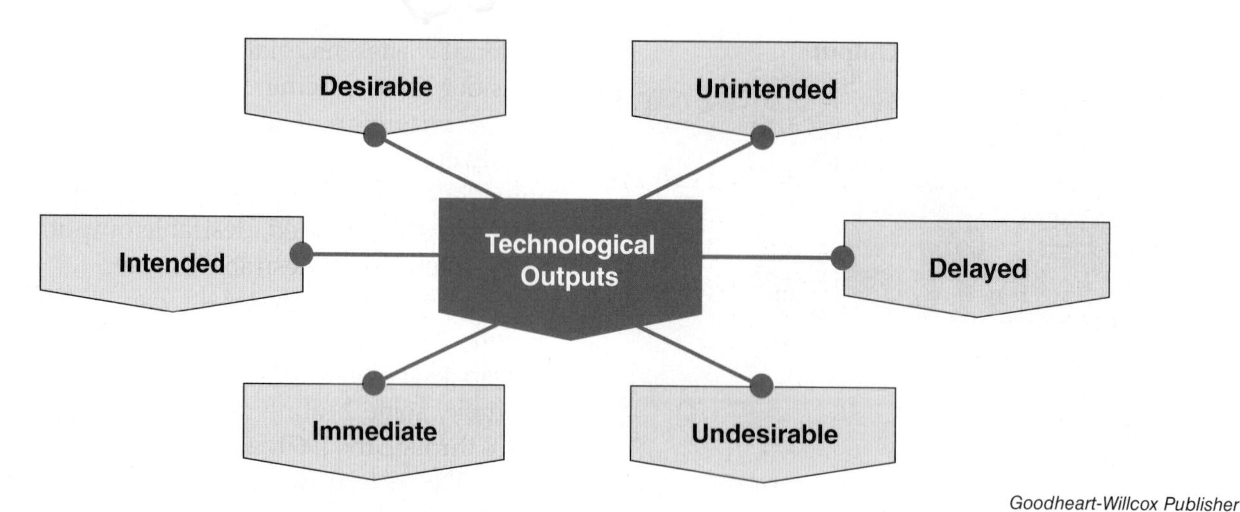

Goodheart-Willcox Publisher

Figure 12-4. The outputs of various types of technological systems.

information from the output of a system to control, or regulate, the system. We often view control systems as internal. That is, we look at how operating processes are controlled so outputs meet specifications. To understand technological system control completely, we also need to consider controls that are external to the system. For example, political actions and personal values also exercise control. We explore both internal and external areas of control in the rest of this chapter.

Internal Controls

Internal control systems can be viewed from three different angles. Systems can be explored according to their type: open loop or closed loop. Systems can also be explored based on the components found in many control systems—monitoring, data-comparing, and adjusting devices. Finally, systems can be explored according to control system operation—manual versus automatic.

Types of Control Systems

Technology control systems include open-loop systems and the closed-loop systems, **Figure 12-5**.

In an ***open-loop control system,*** output information is not used to adjust the process. This system uses no feedback to compare the results with the goal. The system is set and then operates without the benefit of output information. Common examples in a home include a vacuum cleaner or a lightbulb. A vacuum cleaner works the same way whether a floor is dirty or clean. A lightbulb will stay on until it is turned off.

Open Loop

FMC Corp.

Closed Loop

American Electric Power

Figure 12-5. Technology uses open-loop and closed-loop control systems. This food-processing equipment runs at a set speed (open loop). These power generators automatically adjust to increased demand for electricity (closed loop).

Career Connection Control Engineer

Mark Agnor/Shutterstock.com

Control engineers design and improve systems to better achieve the desired system behaviors or outcomes. These professionals apply control theory to design automated systems and processes. ***Control theory*** is an engineering and mathematical way of thinking that focuses on monitoring and correcting systems' behaviors using output performance measures as feedback to control the system. This involves understanding and use of sensors to collect the output performance of a system as feedback for correcting the system's performance.

High school students interested in this field should gain a solid foundation in statistics in order to understand the collection and analysis of data. Students should also take classes in physics, calculus, and technology and engineering.

A bachelor's degree in engineering is required for a career as a control engineer. Control engineering is often part of the electrical, electronic, mechatronics, or mechanical engineering programs at universities and colleges.

Successful control engineers must be able to work in a multidisciplinary environment. They must be able to work closely with coworkers in all areas of an organization, requiring excellent communication skills and abilities. Control engineers must also have the ability to troubleshoot various electrical and mechanical issues.

Control engineers work in a variety of areas, but typically work in the production and manufacturing industry. Positions as senior engineers or engineering managers typically have additional education and comprehensive experience in the field.

The types of projects control engineers work on can include:

- Designing, developing, and using computerized control systems for retrieving and interpreting data.
- Testing and troubleshooting instrumentation and control systems.
- Ensuring performance output compliance with the organization's goals and industrial standards.
- Providing technical guidance and consultancy support for technicians and administration.

Joining a professional organization is a good way to keep up on trends in the field and to network with other control engineers. The Control Systems Society of the Institute of Electrical and Electronics Engineers is the world's largest technical professional society for control engineers.

Most technological systems use feedback in their control systems. Using feedback means output information is used to adjust the processing actions. Feedback is used to control various stages or factors within a system or the entire system. Systems using feedback are called ***closed-loop control systems***.

Manufacturing processes commonly use closed-loop systems, **Figure 12-6**. After using market research to gather and analyze customers' positive and negative reactions, manufacturers evaluate present and future product designs. Using a process called quality control, manufacturers compare parts, assemblies, and finished products with engineering standards, **Figure 12-7**. Quality control ensures that the product performs within an acceptable range. Using ***inventory control***, manufacturers compare

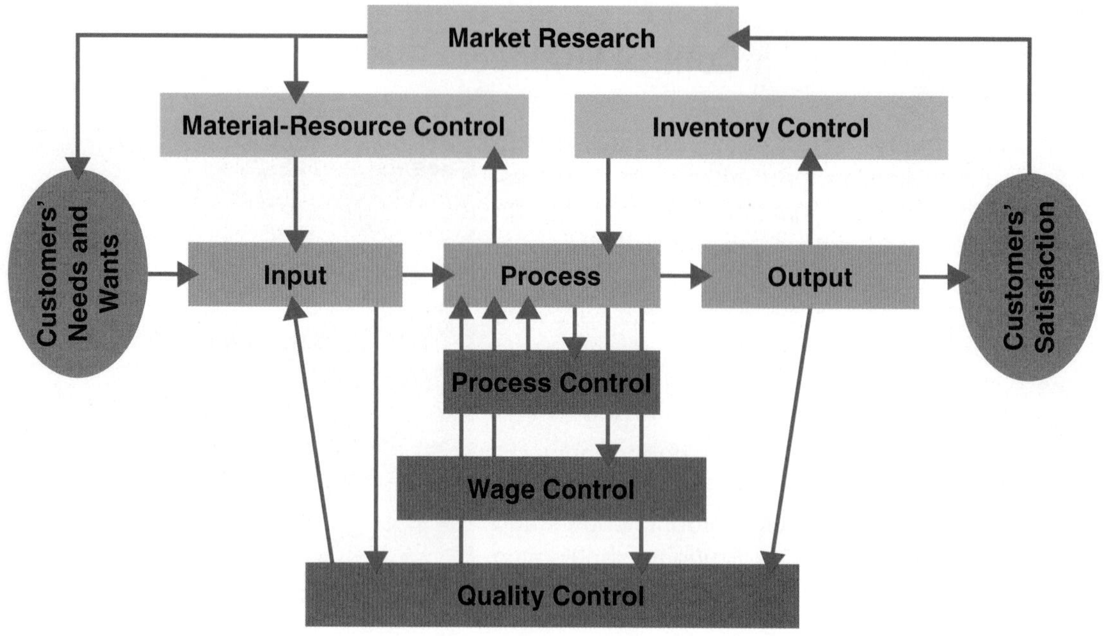

Goodheart-Willcox Publisher

Figure 12-6. With feedback from market research, manufacturers use various types of control to meet customers' needs and wants. Information enters the control system, is processed, and is then fed back into the system.

Akimov Igor/Shutterstock.com

Figure 12-7. This worker is using a caliper to verify the accuracy of a manufactured steel part.

Reynolds Metals Co.

Figure 12-8. This worker is checking the inventory of tread plates used in making truck bodies.

finished goods in warehouses and products in process with sales projections, **Figure 12-8**. The goal is to match, as closely as possible, product production with product demand. Process control allows manufacturers to closely monitor the operation of machines and equipment. The goal is to see that each manufacturing process produces the proper outputs. Using material-resource control (or planning), manufacturers monitor the need for materials and the quantity on order. The goal of this activity is to reduce the raw material inventory. Finally, using a process called ***wage control***, manufacturers monitor the number of hours worked by employees who produce products. The goal of this activity is to keep the labor costs of a product within planned limits.

Control is not limited to manufacturing, of course. Automobiles have emission control systems. Engines have governors to control their operating

Think Green Waste Heat Recovery

Waste heat can be an undesirable output of all production processes. When machines and people use energy to perform some type of work, heat is released into the surrounding environment. It is considered waste because it is typically energy that is transformed into mostly unusable thermal energy. The more waste heat that is produced as a by-product, the less efficient a process may be. However, waste heat is not avoidable. As you may recall, energy cannot be created nor destroyed, it can only be transferred to another form. When machines use mechanical energy, some of that energy is transformed to thermal energy and often wasted.

Waste heat recovery uses the thermal energy by-product of production processes for other purposes. The heat can be directed elsewhere in a production facility to aid in some other process, reducing the need for energy consumption during that process. For example, the heat produced by large facility air conditioners during the day can be directed into a thermal energy storage unit. The heat is then pumped back into the facility during the night to maintain a comfortable temperature. Waste heat can even be shared with other facilities to minimize energy consumption.

speeds. Control systems monitor the ink being applied to paper during the printing process. Thermostats control the temperatures in buildings, ovens, and kilns. Fuses and circuit breakers control the maximum amount of electric current in a circuit. In other words, control is everywhere.

Components of Control Systems

Technological services, products, information, and structures must meet the needs of their users. To ensure that this happens, control systems are designed with three major components, **Figure 12-9**:

- Monitoring devices or sensors.
- Data-comparing devices.
- Adjusting devices.

Monitoring devices

A ***monitoring device*** (or sensor) gathers information about the action being controlled. It monitors inputs, processes, outputs, or reactions. Many types

Monitor

©iStockphoto.com/kickstand

Compare

Renishaw, Inc.

Adjust

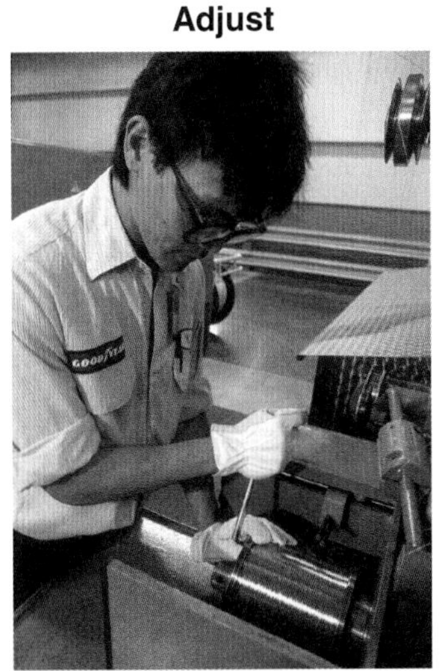

Goodyear Tire and Rubber Co.

Figure 12-9. Control systems monitor performance, compare it against standards, and adjust the system to ensure the output meets the goal.

STEM Connection: Science

Chlorofluorocarbons (CFCs)

MaxyM/Shutterstock.com

As this chapter has discussed, new technologies sometimes produce undesirable consequences. This was the case with a group of synthetic organic compounds called chlorofluorocarbons, or CFCs. CFCs contain carbon, chlorine, and fluorine, and often include hydrogen. They change easily from liquid to gas or from gas to liquid and were used as refrigerants.

The development of CFCs in the late 1920s was originally considered to be a miracle find. Until that time, refrigerators contained toxic gases and fatalities sometimes occurred when the gases leaked from the refrigerators. CFC refrigerant seemed to solve the problem. This refrigerant is odorless, colorless, nonflammable, and noncorrosive. Within a few years, CFCs became the standard refrigerant for refrigerators and air conditioners.

As time went on, however, scientists discovered that CFCs broke down ozone molecules in the upper atmosphere of the earth. In 1978, the United States started to ban the use of some CFCs. An international treaty, called the Montréal Protocol, followed this action in 1987. All but one of the United Nations member nations ratified this treaty. The treaty banned the production of ozone-depleting substances, such as CFCs, in industrial countries by 1996, while developing countries were allowed to continue to produce these substances until 2010. What is now used as refrigerant in most air conditioners and refrigerators?

of information are gathered, including material size and shape, temperatures of enclosures, speed of vehicles, and impacts of advertising. The information can also vary, in that it can be very specific and mathematical or more general. In addition, information can be gathered and recorded in short or long intervals.

The type of information gathered during the monitoring step depends on how the data will be used. Process-performance data is very narrow and focused. Data about the impacts of technological systems is much more general. For example, the data needed to determine the emission levels of a motorcycle engine are more specific than the data needed to determine the impact of the air-transportation system on business. Process-operating information can be gathered using several types of monitoring devices:

- *Mechanical sensors* can be used to determine the position of components, force applied, or parts movement.
- *Thermal sensors* can be used to determine changes in temperature.
- *Optical sensors* can be used to determine light level or changes in light intensity, **Figure 12-10**.
- *Electrical or electronic sensors* can be used to determine the frequency of or changes in electric current or electromagnetic waves. See **Figure 12-11**.

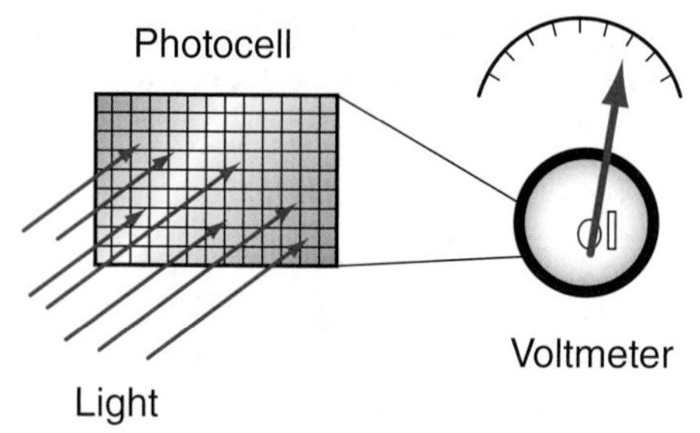

Goodheart-Willcox Publisher

Figure 12-10. One commonly used optical sensor is a photographic light meter. Light striking the photocell generates electricity. A voltmeter measures the electricity. Strong light produces more voltage than weak light.

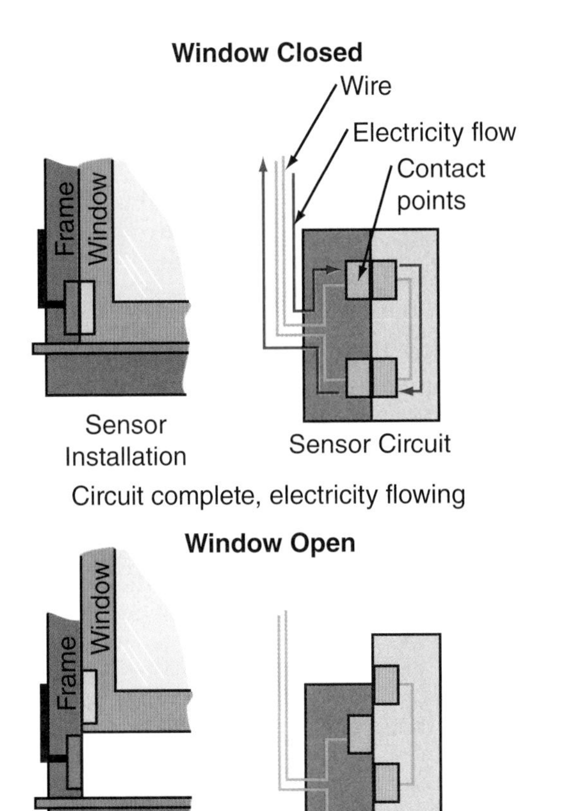

Figure 12-11. This electrical sensor, part of a home security system, can detect open windows.

- ***Magnetic (electromagnetic) sensors*** can be used to determine whether or not changes are occurring in the amount of current flowing in a circuit.

Data-comparing devices

The data gathered by monitoring devices is matched against expectations. The results are compared to the intent of the system through the use of ***data-comparing devices***.

Comparisons can be analytical or judgmental. ***Analytical systems*** mathematically or scientifically make the comparison. Statistics might be used to see whether or not the output meets the goal. For example, the temperature of an oven can be compared with the setting on the knob. The size of a part can be checked with lasers and compared with data in a computer file. The position of a ship can be compared with the signals from a satellite. The analytical type of comparison is commonly used in closed-loop control systems.

Analytical systems can also be used for managerial control systems. For example, the number of products produced might be directly compared with the production schedule. The number of passengers buying bus tickets might be compared with sales projections. The number of people listening to a radio station might be compared with total listeners for all stations. Analytical systems try to remove human opinion from the evaluation process. The information is reduced to percentages, averages, or deviations from expectations.

Judgmental systems allow human opinions and values to enter into the control process. Open-loop control systems require human judgment. For example, the speed at which you drive a car is based on your judgment. You do not apply mathematical formulas to decide when to accelerate and when to brake as you drive on a mountain road.

Likewise, human opinion is used in determining the value of system outputs. Judgments are used in the design phase of technological systems. It is almost impossible to select a design by analytical means. For example, people decide the styling, color, and shape of most products and buildings. They select the content of magazines and telecommunications programs. After the output is available, analytical means might be used to compare the results with expectations. After a television program is produced, the number of viewers watching the program can be measured. The decision to produce the show, however, is a judgment.

Adjusting devices

A comparison of outputs to expectations might indicate that a system is not operating correctly. The output might not meet specifications, so the system might need modification. Controllers do modifications through the use of ***adjusting devices***. These devices might speed up motors, close valves, increase the volume of burning gases, or perform a host of other actions. The goal is to cause the system to change and, therefore, produce better outputs. Several types of adjusting devices, or system controllers, exist, **Figure 12-12**. They include:

- ***Mechanical controllers***. These controllers use cams, levers, and other types of linkages to adjust machines or other devices.

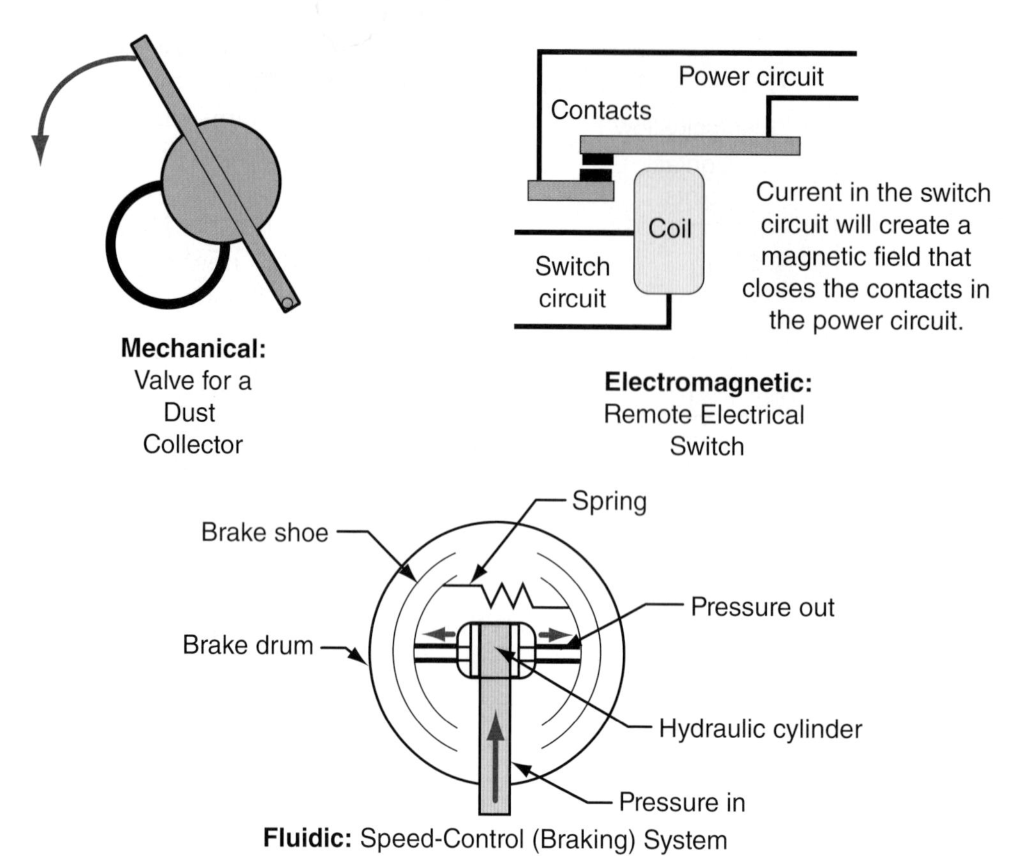

Goodheart-Willcox Publisher

Figure 12-12. System controllers, or adjusting devices, modify the system to produce better outputs.

- ***Electromechanical controllers***. These controllers use electromagnetic coils and forces to move control linkages and operate switches to adjust machines or other devices.
- ***Electrical and electronic controllers***. These controllers use electrical devices (switches, relays, and motors, for example) and electronic devices (diodes, transistors, and ICs, for example) to adjust machines or other devices.
- ***Fluidic controllers***. These controllers use fluids to adjust machines or other devices. The fluids include oil (in hydraulic controllers) and air (in pneumatic controllers).

The Operation of Control Systems

Technological control systems can also be classified as manual or automatic. ***Manual control systems*** require humans to adjust the processes. ***Automatic control systems*** can monitor, compare, and adjust systems without human interference.

Manual control

A good example of a manual control system is the automobile. The driver uses various speed and direction controls to guide the vehicle to the destination (goal). As the vehicle enters a city, the driver eases up on the accelerator to reduce the speed to the new speed limit. Brakes might be applied to meet the new conditions. The steering wheel is turned as the road curves. In this example, the human eye gathers information. The eye sees a new speed limit sign. To determine the speed of the vehicle, the eye also scans the speedometer. The brain compares the posted speed limit with the speedometer reading. The brain commands the foot to change its position to allow the speed to be adjusted until it matches the speed limit. In this example, the human eye monitors the system. The brain compares the system's actual and desired performances. The foot physically adjusts the system.

Manual control systems are everywhere. We use them in such simple actions as riding a bicycle, adjusting the temperature of a gas stove when cooking, setting the focus on a camera, and checking a fence post for plumb. Also, we use them in performing more complicated actions, such as landing an airplane, **Figure 12-13**.

Technology Explained Integrated Circuit (IC)

Integrated circuit: A piece of semiconducting material, in which a large number of electronic components are formed.

The information age began with the invention of the transistor. A transistor is a tiny electronic device that allowed engineers to eliminate bulky electronic devices, such as the vacuum tube. Technology has made even greater advancements in the field of electronics, expanding transistor technology to create the integrated circuit (IC), or microchip.

An IC is a thin piece of pure semiconductor material, usually silicon, **Figure A**. Tens of thousands of electronic devices and their interconnections can be produced in one IC. ICs must be manufactured in a sterile environment, **Figure B**. Individual electronic devices on the chip are typically very small. ICs are very complex and require careful design. Building them requires a high level of skill and can be very time-consuming. ICs are typically manufactured using the following steps:

1. Oxidizing the surface. This prevents impurities from entering the silicon.
2. Coating the surface with a photoresist, a light-sensitive chemical. When developed, the photoresist prevents the oxidized layer of the IC from being etched away.
3. Placing a pattern over the photoresist and exposing it to light. The unexposed areas are then washed away.
4. Etching the IC with acid. This creates unprotected areas on the silicon.
5. Introducing impurities into the silicon (doping). These impurities create the tiny transistors, diodes, capacitors, and resistors in the IC, giving the chip its electronic properties.
6. Connecting the components with small aluminum leads.
7. Placing the IC in a protective casing.

Audrius Merfeldas/Shutterstock.com

Figure A. One integrated circuit can house thousands of tiny electronic devices.

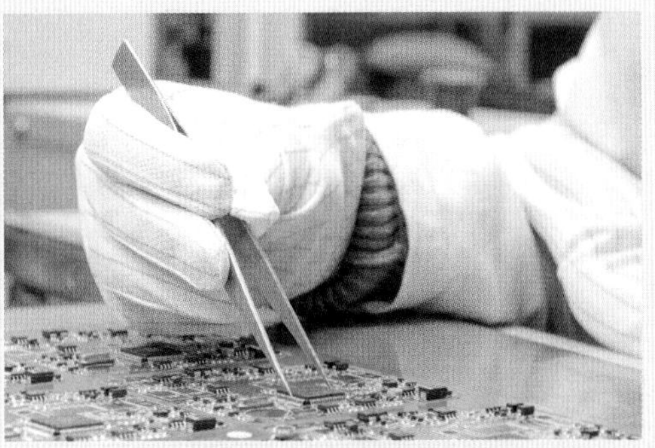

servickuz/Shutterstock.com

Figure B. ICs must be manufactured carefully in a sterile environment.

Regien Paassen/Shutterstock.com

Figure 12-13. This pilot uses a manual control system to land an airplane. He monitors the approach to the runway while comparing the actual flight path with the desired one. His hands and feet change control surfaces and speed to land the plane safely.

Automatic control

A simple example of automatic control is the thermostat in a heating system, **Figure 12-14**. The device measures the room temperature. The thermostat compares the temperature of the room with the desired temperature. As long as the temperatures are within a preset range, no action is taken. When the room temperature drops below the set temperature range, a switch is activated. This switch turns on the heating unit. The unit operates until the upper limit of the temperature range is reached. At this point, the thermostat turns off the unit. Automatic doors in stores, electronic fuel-injection systems in vehicles, and automatic emergency braking systems are other examples of automatic control systems. More sophisticated automatic control systems help pilots land the new generation of jet airliners, assist workers in controlling the output of electric generating plants, and help contractors lay flat concrete roads.

External Controls

We have looked at control primarily as an internal system component. As we said earlier, control also comes from outside the system, **Figure 12-15**. For example, environmental responsibility is a societal goal that technological systems must address. Systems, therefore, are designed to reduce air, water, and noise pollution.

Public opinion is also a control factor. The opinions of many different groups and associations affect public policies. Our news media reports on issues controlling technological systems. The debate between environmentalists and forest-products companies over timber cutting is one example. Another example is the dialogue between the pronuclear and antinuclear groups. The controversy over the types of nets that should be used in fishing is still another example of external control.

This discussion is not designed to list all the important public issues that exert some control on technology. Looking only at the physical controls built into technological systems, however, gives an incomplete view. We need to see that, primarily through feedback, both internal and external control systems affect technology by creating new inputs, processes, and outputs, **Figure 12-16**.

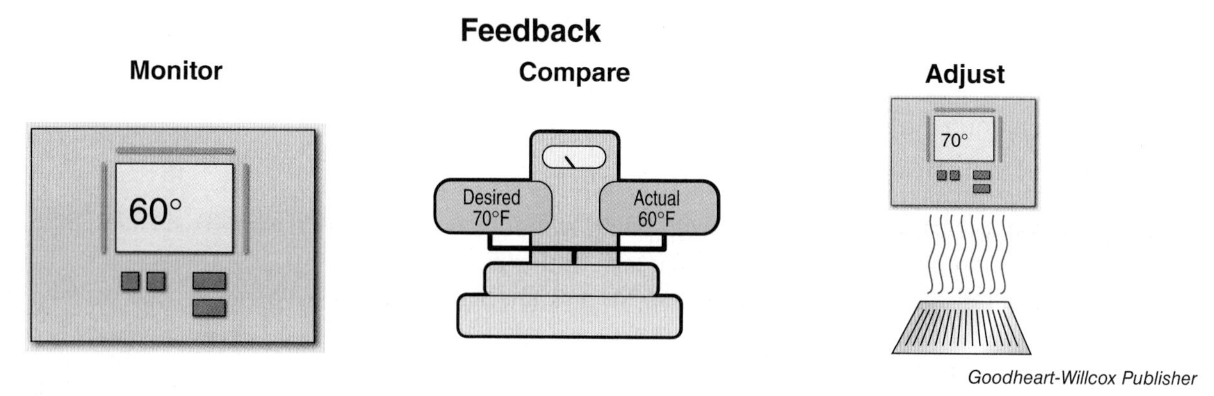

Goodheart-Willcox Publisher

Figure 12-14. A thermostat is a common automatic control system found in many homes and buildings.

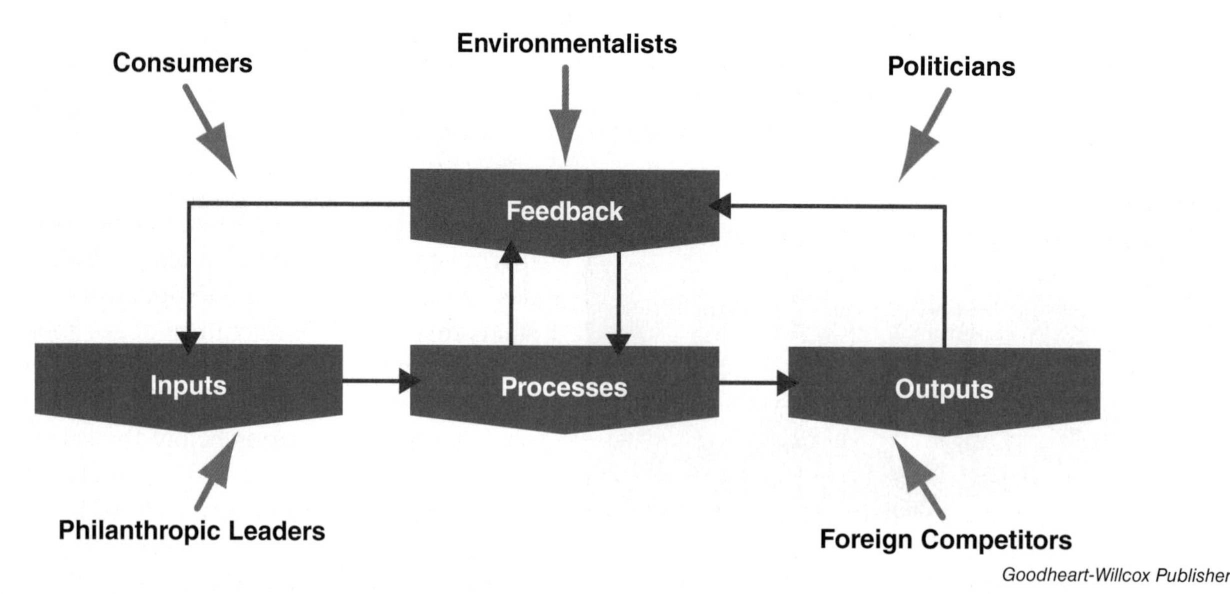

Goodheart-Willcox Publisher

Figure 12-15. External forces exercise control on technological systems.

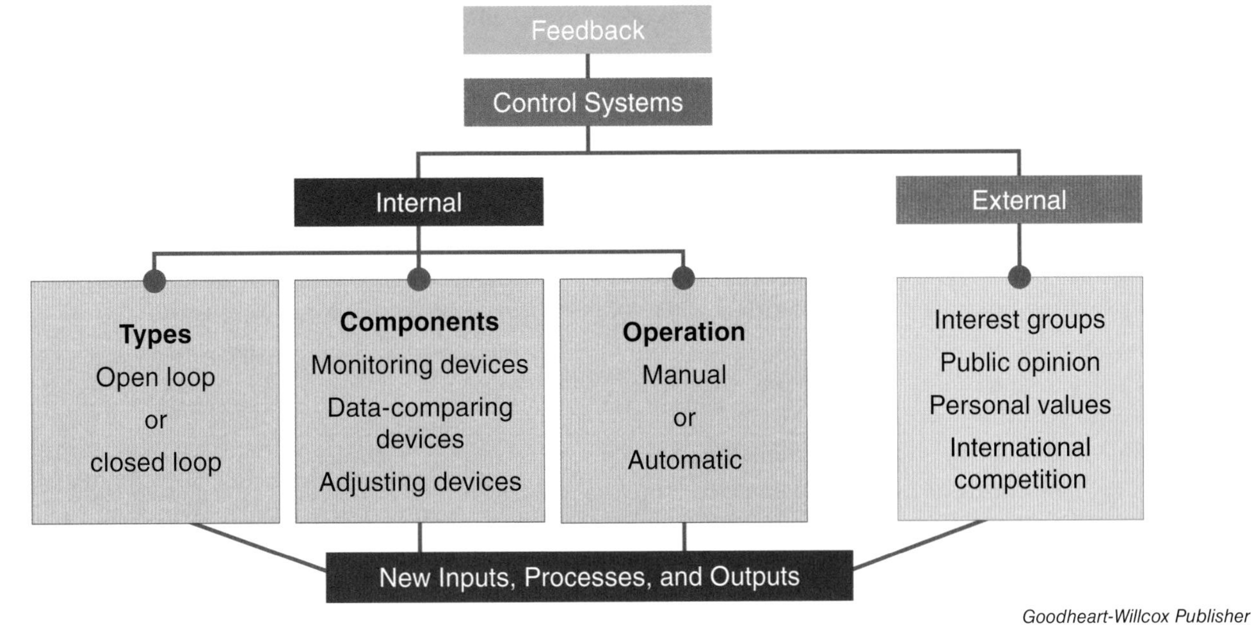

Figure 12-16. Primarily through feedback, both internal and external control systems lead to new inputs, processes, and outputs.

Academic Connection: Literacy

Reading Comprehension Strategies

Literacy is the ability to read and write. Effective reading and writing skills and practices also enable people to develop competency in their fields of study or work. When reading technical and scientific subjects, following a few reading strategies can help you understand and retain the content. These strategies include:

- Determine the meaning of key terms, phrases, and symbols used.
- Analyze the structure of relationships between key terms and concepts.
- Evaluate the author's purpose by providing a constructed explanation of a scientific or technical topic or idea.
- Be accurate when citing specific details about a scientific or technical analysis.

This chapter is about the outputs, feedback, and control of technological systems. To aid in your understanding of these concepts, employ these important reading strategies by completing the following tasks:

1. Document all chapter key terms or phrases along with their meaning in a scientific or engineering journal or notebook.
2. Create a visual aid or diagram that depicts the relationship between desirable, undesirable, intended, and unintended outputs.
3. Read an article from a scientific journal about the outcomes of a specific technological system. Based on your reading, develop five statements that summarize the author's intent for the article.
4. Identify a technological system that has had undesirable outputs and cite specific evidence that supports that the outputs were unintended.

Completing these tasks will help you develop competency in engineering and better understand complex and abstract concepts, such as systems, enabling you to explore and prepare for a variety of outputs when designing a system or product.

Summary

- ✔ All technological systems are designed to meet specific needs and wants.
- ✔ A direct relationship exists between the need (the reason for the system) and the output (the satisfying of the need).
- ✔ The outputs of the systems can be desirable or undesirable, intended or unintended, and immediate or delayed.
- ✔ Most control systems use feedback to increase the probability that outputs will be desirable.
- ✔ Internal control systems can be categorized by the type of control system, the components of the control system, or the operation of the control system.
- ✔ Types of control systems include open-loop systems and the closed-loop systems.
- ✔ Control system components include monitoring devices or sensors, data-comparing devices, and adjusting devices.
- ✔ Control systems can also be classified as manual or automatic.
- ✔ External controls are those controls that come from outside a system.

Check Your Engineering IQ

Now that you have finished this chapter, see what you learned by taking the chapter posttest.
www.g-wlearning.com/technologyeducation/

Test Your Knowledge

Answer the following end-of-chapter questions using the information provided in this chapter.

1. Name one major desirable output for each of the classified technology systems.
2. Give one example of an undesirable output.
3. Products or services designed and produced with specific goals in mind are _____. Outputs that were not anticipated when the system was designed are _____.
4. The production of steel is an example of a(n) _____ output.
   A. immediate
   B. intended
   C. desirable
   D. All of the above.
5. Feedback can be defined as using information from the _____ of a system to control, or regulate, the system or its inputs.
6. What is the major difference between open-loop control systems and closed-loop control systems?
7. List the three major types of devices found in most internal-control systems.
8. _____ sensors are a type of monitoring device that can be used to determine the level of light or changes in the intensity of light.
   A. Mechanical
   B. Thermal
   C. Optical
   D. Electrical
9. *True or False?* Comparisons made by data-comparing devices can be analytical or judgmental.
10. _____ can speed up motors, close valves, increase the volume of burning gases, or perform a host of other actions.
11. _____ controllers are a type of adjusting device using cams, levers, and other types of linkages to adjust machines.
    A. Electromechanical
    B. Mechanical
    C. Electrical
    D. Fluidic
12. Give one example of an automatic control system.
13. *True or False?* Public opinion is an external control factor.

Critical Thinking

1. Select a technology you use every day. Prepare and complete a chart similar to the following.

Outputs	
Desirable	Undesirable
Intended	Unintended
Immediate	Delayed

2. For the same technology, diagram the control system used. List the ways information is gathered, comparisons are made, and adjustment techniques are used.
3. For the same technology, list the outside influences that helped determine its design and helped control its operation.
4. With two or more of your classmates, complete the following challenge:
   A. Identify a technological system or device you all use, such as bicycles or desk lighting.
   B. List a control system needed for the device or system.
   C. Brainstorm ways to provide this control.
   D. Sketch your solution.

STEM Applications

1. A commonly used technological device is a motion-activated light or electrical switch. Design and conduct an experiment that determines the slowest motion this device will detect and the farthest angle to each side that the device can recognize motion. Present a brief summary of your research, with appropriate sketches, or drawings.

Engineering Design Challenge

Automated Control System Challenge

Background

Throughout Section 3, you have been learning about technological systems and ways that inputs can be processed into desired outputs. You have explored ways that different mechanisms are used to transform resources into a variety of products. Additionally, you have analyzed how systems are designed and operated to solve problems in a variety of technological fields and industries. In this chapter, you learned more about how outputs and feedback of a system can be used to inform the processes that happen within that system. Data collected from a system's output can be monitored by control system's and used as feedback for controlling the processes of the system. These control systems enable technological operations to be automated.

Situation

A major university has hired your engineering team to develop a solution to their campus parking and commuting problem. The university has expanded into three campuses across the town in which it is located. Students now must travel between campuses on a daily basis. The town's transportation infrastructure is not capable of supporting the traffic on the roadways or parking between campuses. Additionally, the geography of the area does not enable existing roadways to be expanded or additional parking to be added. Therefore, a rapid transportation system must be designed to carry students between campuses. This system must operate on a track above the roadways in a manner that does not impact the current transportation infrastructure. This system must be fully automated and operate twenty-four hours a day, seven days a week. Your team must design and test a model of an automated rapid transportation system using various control components, such as monitoring, data-comparison, and adjusting devices.

Desired Outcomes

For this project, your team will create a system that contains only those processes and controls needed to manage a rapid transportation system. The simulation will consist of a transport that must stop at three stations, Campus 1, 2, and 3. Based upon the input of three switches at each campus, the transport must move directly from the current position to the requested campus station. The transport can start at whichever campus you prefer. However, once you start the system, the transport must travel directly to the campus that calls for the transport. You must also include a safety light on the transport that operates while the transport is moving, **Figure A**.

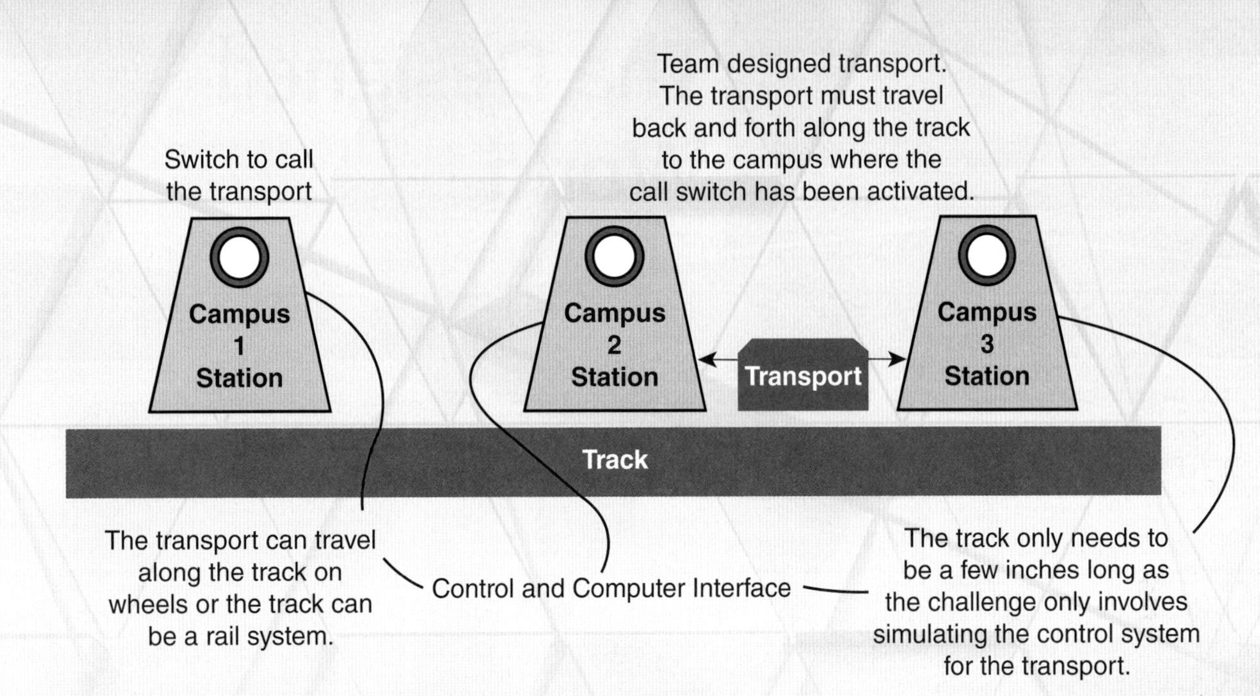

Figure A. A diagram of the simulated track for the automated transport system.

Materials

This project requires robotics equipment and software, such as Lego Mindstorms, VEX robotics, Tetrix, or Fischertechniks.

Beginning the Process

- Remember that the engineering design process is an iterative approach to solving problems, meaning you can and should move back and forth between the different steps.
- Determine your engineering design team based upon the class's individual skill sets. Have a discussion with your team to create a problem statement based on the background and situation provided above. A good problem statement should address a single problem and answer who, what, where, when, and why. However, the problem statement should not suggest there is a single solution to the problem. Your classmates will create their own problem statements based on their own interpretations of the situation and the teams' prior interests and experiences. Your engineering team will use your defined problem statement as the starting point to create a valid solution.
- Using the knowledge your team has obtained and employing the engineering design process, propose new or improved solutions to the problem defined from the situation provided. This will follow a natural progression from brainstorming, to generating ideas, to producing prototypes/models, to evaluating success, and to iteratively refining the solution.

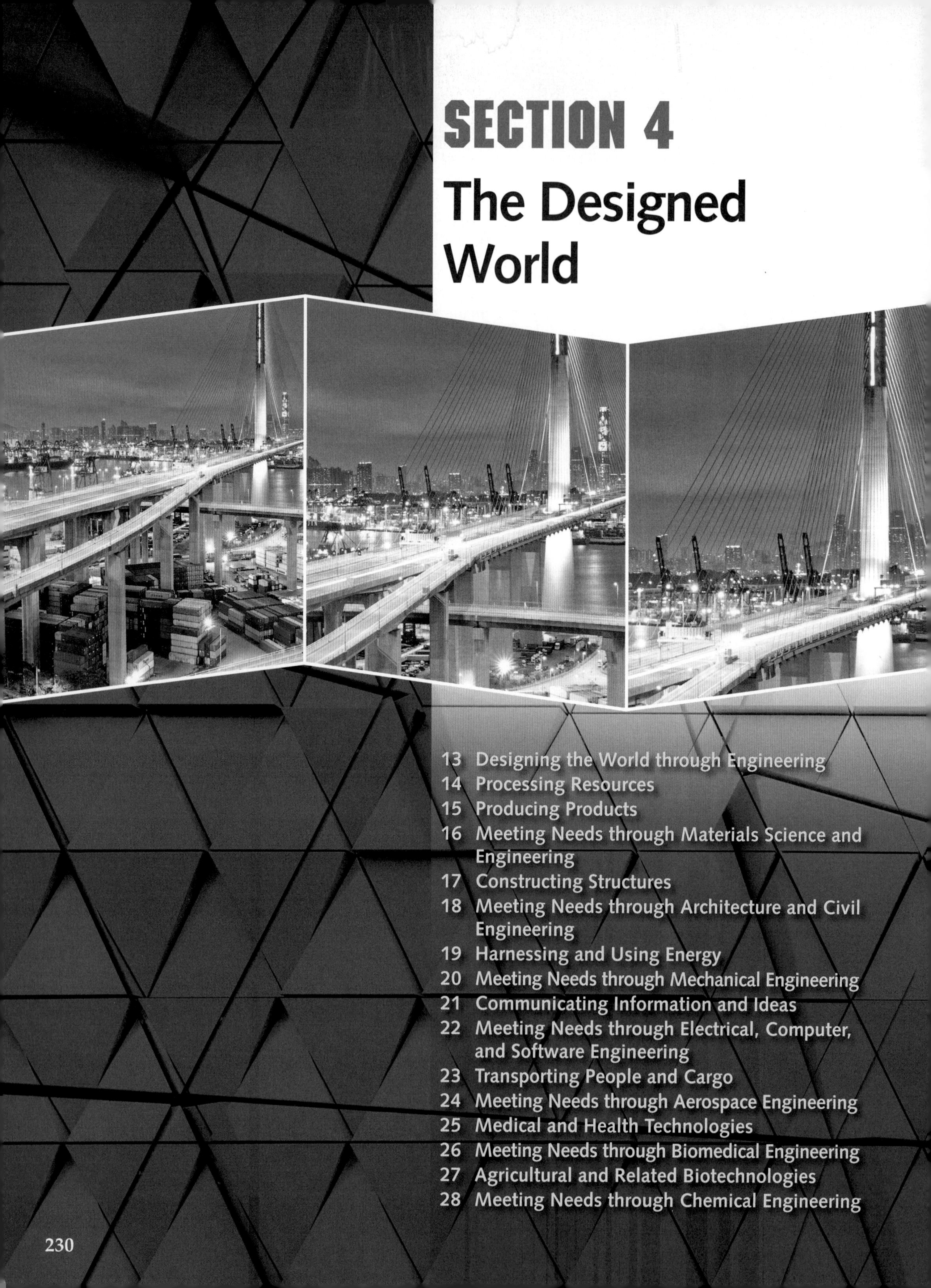

SECTION 4
The Designed World

Tomorrow's Technology Today

Smart Materials

What makes a smart material "smart"? Smart materials are extraordinary because they have the ability to dramatically change their original state when external forces are applied. Smart materials react differently to these forces, or stimuli. No two smart materials are the same.

How are smart materials different from "normal" materials? A common material, such as wood, can be exposed to external stimuli, such as fire. As wood burns, its properties change, and it is eventually reduced to ash. Extinguish the fire, or remove the stimulus, and the ash remains—the wood will never again return to its original state. Compare this to a smart material called *magneto-rheostatic*. Under normal circumstances, magneto-rheostatic is a liquid. However, if placed in a magnetic field, the liquid transforms into a solid—a change that occurs in less than a second. Remove the magneto-rheostatic from the magnetic field, and it is a liquid once more.

While smart materials may seem like the wave of the future, they actually have many common applications. Piezoelectric smart materials are used with the airbag sensor in your car. The force of an impact creates an electric pulse that deploys the airbag. The wire used in dental braces is a shape memory alloy whose form changes with temperature. The wire change follows a predetermined pattern designed for each patient's teeth. Over time, the shape memory alloy slowly corrects imperfect teeth. When designing new products, smart materials can be more easily manipulated to serve a specific purpose than normal materials.

For example, shape memory alloys change their shape when exposed to extreme temperature change. Some shape memory alloys also possess pseudo-elasticity, a quality that enables shape change through the application of force or pressure. Pseudo-elastic shape memory alloys have been used in eyeglass design for many years. Special frames are marketed as "bendable"—good for accident-prone people whose normal frames break easily. When these special frames are bent, or force is applied to them, they adapt to the pressure. Once the force is removed, they return to their original shape. What seems like a magic pair of glasses is really the application of a smart material.

As our understanding of smart materials grows, scientists will discover even more impressive applications. It is hard to imagine what revolutionary new products they will discover in the future!

While studying this chapter, look for the activity icon to:

- **Assess** your knowledge with self-check pretest and posttests.
- **Practice** technical terms with e-flash cards, matching activities, and vocabulary games.
- **Reinforce** what you learn by submitting end-of-chapter questions.

www.g-wlearning.com/technologyeducation/

CHAPTER 13

Designing the World through Engineering

Check Your Engineering IQ 

Before you read this chapter, assess your current understanding of the chapter content by taking the chapter pretest.
www.g-wlearning.com/technologyeducation/

Learning Objectives

After studying this chapter, you should be able to:

- ✔ Distinguish between the natural world and the designed world.
- ✔ Explain the purpose of designing through engineering.
- ✔ Identify seven categories of technologies.
- ✔ Describe the role of engineering design in each of the categories of technologies.

Technical Terms 

agriculture
biotechnology
construction technologies
energy
green buildings
information and communication technologies
manufacturing technologies
medical technologies
power
transportation technologies

Close Reading

Chapter 13 distinguishes between the designed world and the natural world and provides an overview of the various areas of technology that are created through engineering. As you read this chapter, consider the role engineering plays in creating the designed world that you interact with every day.

Each day, people interact with countless technologies that are designed through engineering. These technologies, which may be products, systems, or processes, influence our quality of life. The home we live in, the car we drive, the cell phone we use to communicate, and medical technologies all impact our ability to live enjoyable and healthy lives. At times our input might be solicited to inform the design of new technologies, **Figure 13-1**. However, in most instances, others create products that we then purchase, use, and evaluate to determine if they are effective, high-quality, and safe for us and the environment. Therefore, understanding the designed world is vital in order to be an informed citizen.

The Designed World

One distinction between humans and other living organisms is our ability to create new technologies, driven by our desire to fulfill needs and wants. Some animals have been observed to use tools, but humans are unique in that they design and make tools and technologies that do not currently exist.

Designing can broadly be described as taking what does not currently exist in the natural world and creating it. This has resulted in the *designed world*, which encompasses anything created by humans that is not represented in the *natural world*.

Rawpixel.com/Shutterstock.com

Figure 13-1. A focus group is a group of people tasked to participate in a discussion about a particular product or topic in order to provide valuable feedback on the product or topic.

The natural world includes rocks, plants, trees, and animals, **Figure 13-2**. Without humans designing or intervening, the natural world would remain as it is.

Humans have been creating the designed world for thousands of years. Initially, they created technologies largely for survival, **Figure 13-3**. These technologies included rudimentary shelters to survive the natural elements to tools used for hunting, gathering, and protection. By examining different technological artifacts, one can understand what life was like

Andrew Mayovskyy/Shutterstock.com

Figure 13-2. The natural world.

Andrey Burmakin/Shutterstock.com

Figure 13-3. Early technologies were created mostly to enhance survivability.

during that time period. Even inventions associated with discoveries of early humans, such as the wheel, were designed to achieve a specific goal.

Although the designed world is often synonymous with electronic artifacts such as computers, TVs, camcorders, and cell phones, it is also represented in our tools, modes of transportation, information and communication systems, and biotechnologies. Because technological advancements and changes define the time period, they also affect our perceptions of what technology actually is. For example, the pencil has become such a staple of everyday life that it is not considered technology or a component of the designed world. However, two hundred years ago, pencils and paper were considered the most advanced form of the designed world.

scyther5/Shutterstock.com

Figure 13-5. Artist creating a new design.

Designing through Engineering

People create the designed world for many reasons. The foremost reason is to meet basic needs for food, safety from danger, and shelter from the elements. Also, people attempt to satisfy wants or desires for a high quality of life by designing through engineering.

New products and processes are often conceptualized through designing. See **Figure 13-4**. Designing can take many forms to address human needs and wants. For example, artistic design, **Figure 13-5**, focuses on creativity and aesthetics.

Monkey Business Images/Shutterstock.com

Figure 13-4. An engineer using predictive analysis software to conceptualize and optimize a design.

Engineering design, on the other hand, is associated with achieving specific goals and is associated with the application of mathematical and scientific knowledge. The engineering design process, as presented in Chapter 4, provides a general framework for how engineers work through designing.

People's wants and needs are constantly changing. Therefore, the designed world is continually evolving and undergoing constant improvements and modifications. Therefore, designing may be based on problems requiring a solution, but there is never one best solution. That is why products continue to be refined and improved.

Engineering design is an important part of many technologies and industries. These are discussed in the following sections.

Technologies and Industries

Although different industries have varying purposes, goals, processes, and outcomes, they are often interrelated and overlap. See **Figure 13-6**. For example, manufacturing processes and technologies affect construction technologies. Products such as construction materials can be manufactured more efficiently, influencing how to design for construction demands. A good example is modular construction, the process of manufacturing homes and buildings in a factory, which blurs the line of construction and manufacturing. Technologies can be separated into the following categories:

Zapp2Photo/Shutterstock.com

Figure 13-6. The goals of engineering design vary among industries, but the technologies are often interrelated.

- Manufacturing technologies.
- Construction technologies.
- Transportation technologies.
- Energy and power technologies.
- Information and communication technologies.
- Medical technologies.
- Biotechnology and agricultural technologies.

Manufacturing Technologies

Manufacturing technologies cover a wide range of processes, systems, and procedures to create the products people need or want. Consider products that you encountered on your way to school today. Each product, including your toothbrush, backpack, textbook, and car or bus, was created through manufacturing technologies. See **Figure 13-7**.

Phonix_a Pk.sarote/Shutterstock.com

Wasant/Shutterstock.com

eggeegg/Shutterstock.com

Figure 13-7. Each product you encounter on a daily basis is a result of manufacturing technologies and processes.

Technological advancements within manufacturing as well as other technological areas have changed manufacturing processes over time. For example, products were once primarily manufactured item by item and custom designed by an expert. However, within the last one hundred years, manufacturing technologies have become much more efficient through assembly lines, quality control, computers, robotics, and standardization of parts and job descriptions. These processes will be further discussed in Chapters 14 and 15.

Engineers are challenged to design and develop manufacturing technologies in an ever-changing world. Global economies and ongoing developments require designing through engineering to constantly react to new requests. One challenge is to produce parts that have less of an impact on our environment. Engineers must develop a sound final product that minimally impacts the environment during the manufacture process or procurement of natural materials. For example, plastic bags can cause damage to ecosystems, take a long time to degrade, are toxic, and are highly dependent on oil to produce. As engineers design new products to replace plastic bags, they must not only consider usefulness for the consumer, but also the materials and processes used to create the product.

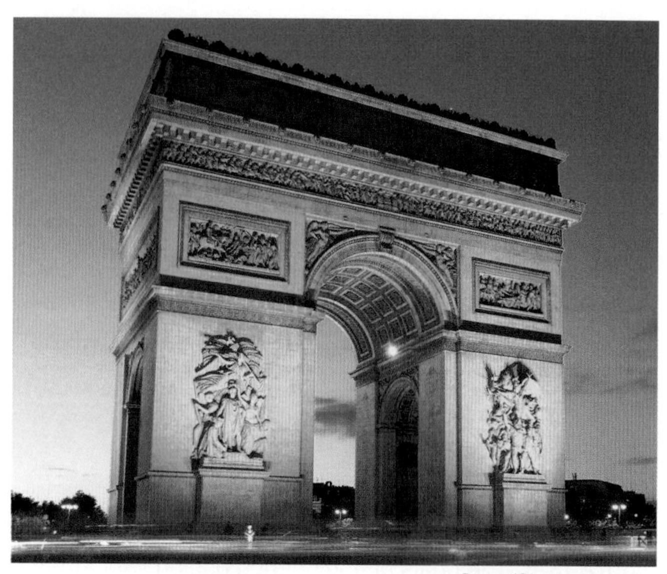

Samot/Shutterstock.com

Figure 13-8. Many structures are designed for shelter, while others serve different purposes. The Arc de Triomphe is a monument in Paris.

Construction Technologies

Construction technologies encompass the design, assembly, and maintenance of structures. This technology is largely related to civil engineering and architecture. Examples of structures are homes, schools, office buildings, museums, and monuments. See **Figure 13-8**. Construction technologies also include structures such as bridges, dams, roads, and temporary storage facilities. See **Figure 13-9**.

Andrew Zarivney/Shutterstock.com

Figure 13-9. Construction technologies also include structures such as bridges, dams, and roads.

When designing new structures, engineers must consider many factors, including the quality of building materials, environmental forces such as weather conditions, and the demands of users. They must also understand and apply mathematical and scientific concepts such as material strength, loads, and compressive and tensile forces. See **Figure 13-10**.

Engineers in the construction technologies face challenges when designing new products and systems. An aging infrastructure is one such challenge. Many highways and bridges across the United States are outdated and require attention. Some need to be rebuilt, while others need fresh ideas to be more sustainable, which means it is capable of lasting for a long period of time. Designers use their knowledge of manufactured products that are stronger and more efficient than before.

Another challenge is the current demand and regulations for construction technologies to address ***green buildings***, which are structures that are

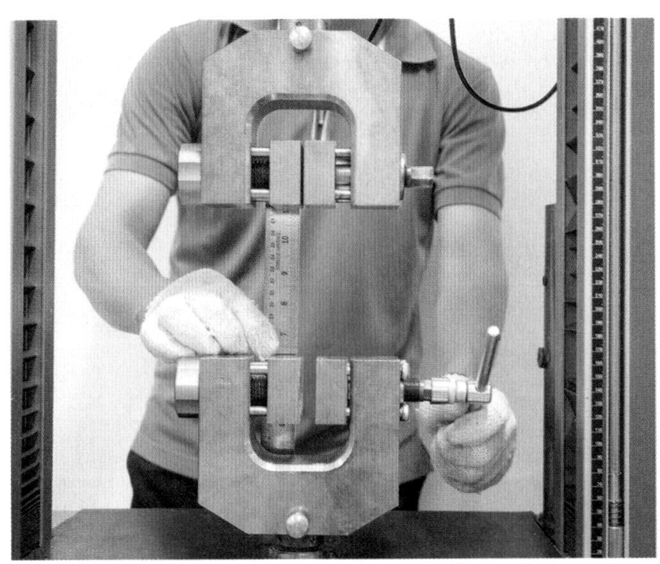

Mrs_ya/Shutterstock.com

Figure 13-10. When designing new structures, engineers must understand and apply mathematical and scientific concepts such as material strength, loads, and compressive and tensile forces.

environmentally responsible, sustainable, and resource efficient. See **Figure 13-11**. This requires a designer to consider a building's life cycle, which covers the building's proposal, erection, operation, upkeep, renewal, and demolition. See **Figure 13-12**.

Moosician/Shutterstock.com

Figure 13-11. Green buildings are structures that are environmentally responsible, sustainable, and resource efficient. Some green buildings feature vertical gardens.

Transportation Technologies

The designed world also includes changes to the natural world to accommodate the desire to move freely and conveniently throughout the world. ***Transportation technologies*** move people and products from one location to another through land, water, air, and space travel. Transportation technologies make possible travel to and from work or school, movement of goods from one geographic location to another, and recreational travel.

Aerospace engineering largely focuses on designing new transportation technologies, such as more efficient materials, structures, and fuel systems, related to space travel. Space travel has become possible as oversight from NASA has largely become the responsibility of the private sector. Organizations such as SpaceX and SpaceDev now compete regularly for missions to transport goods to the *International Space Station*, **Figure 13-13**. Additionally, travel to and colonization of Mars is being explored.

Designing transportation technologies must be done in coordination with manufacturing and construction technologies. For example, advancements in transportation technologies, such as solar or electric vehicles, affect how the manufacture of those vehicles and their components will occur. Construction technologies and the manner in which roadways are designed and built also affect the design of transportation technologies.

When designing transportation technologies, engineers respond to new needs and wants to create better products. For example, understanding local traffic patterns and traffic jams allows an engineer to design a new road that will reduce bottlenecks and divert vehicles. When designing cars, engineers also respond to varying needs. One such need is for safer transportation. As a result, vehicles have been designed with driver override systems, which disregard the driver's input. For example, brake override systems use sensors that recognize issues that may be irregular for a driver's normal input. The car's computer system can then provide the appropriate feedback to make corrections. Another safety feature is biometric vehicle access, which uses facial or fingerprint technologies to access and use a vehicle.

Proposal

Rawpixel.com/Shutterstock.com

Construction

Unkas Photo/Shutterstock.com

Operation

Artens/Shutterstock.com

Upkeep

David Spates/Shutterstock.com

Renewal

ingehogenbijl/Shutterstock.com

Demolition

Monika Gruszewicz/Shutterstock.com

Figure 13-12. A building's life cycle must be considered when the structure is designed.

Another need that engineers address is the desire for comfort. For example, heated seats, remote start, and multimedia options increase the comfort of traveling. Vehicles continue to be designed with new technologies such as satellite radio, screen-based directions, and GPS. See **Figure 13-14**.

Transportation technologies continually face challenges that engineers must consider in order to design efficient vehicles. These challenges include the need to reduce dependence on exhaustible resources, demand for electric powered vehicles, and increasing populations.

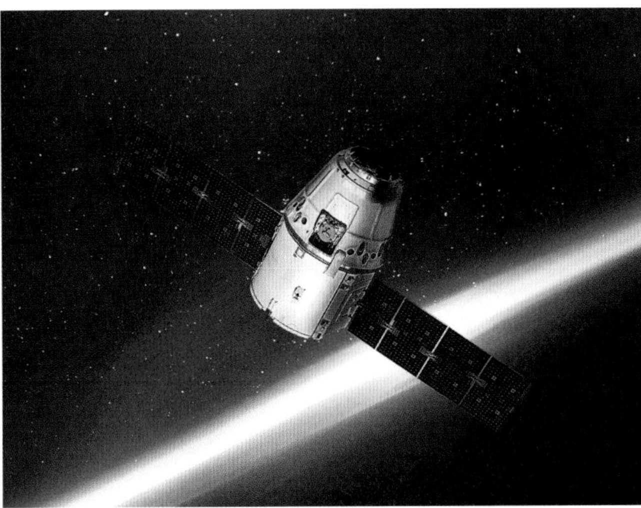

Dima Zel/Shutterstock.com

Figure 13-13. Organizations such as SpaceX and SpaceDev compete for missions to transport goods to the International Space Station.

ambrozinio/Shutterstock.com

Figure 13-14. Vehicles are continuously improved and equipped with new technologies to make travel safer and more comfortable. This head-up display provides a driver with important information such as speed or directions within their viewpoints. This enables the driver to see important information without looking away from the roadway.

Energy and Power Technologies

Energy and power are two distinct but related concepts. ***Energy*** is defined as the ability to do work. There are different forms of energy, which will be discussed in Chapter 19. ***Power*** can be defined as the amount of consumed energy of a technological system over time.

Energy and power technologies have considerable impact on an individual's life, affect other areas of technologies, and can impact the environment and availability of natural resources for future generations. When designing energy and power technologies, engineers consider the inputs, processes, outputs, and feedback of a system. They must anticipate the type of energy to be used; the processes used to convert, transmit, or store it; and the overall outputs required.

Engineers face increasing energy and power challenges when designing new technologies. Such engineers are well prepared and may even apply their mechanical engineering background to design technologies that consume less energy and power. Events have pressured designers to overcome future challenges. One example is the 2016 gas shortage on the East Coast, as a result of a leak in a critical pipeline. Since then, engineers have continued to work on designing more fuel-efficient cars, as well as electric cars. Additionally, engineers play a significant role in designing energy and power technologies through smart grids, which are networks of generation, transmission, distribution, and power consumers of energy and power. See **Figure 13-15**.

Information and Communication Technologies

Information and communication technologies are technologies that provide the ability to record, store, manipulate, analyze, and transmit data across

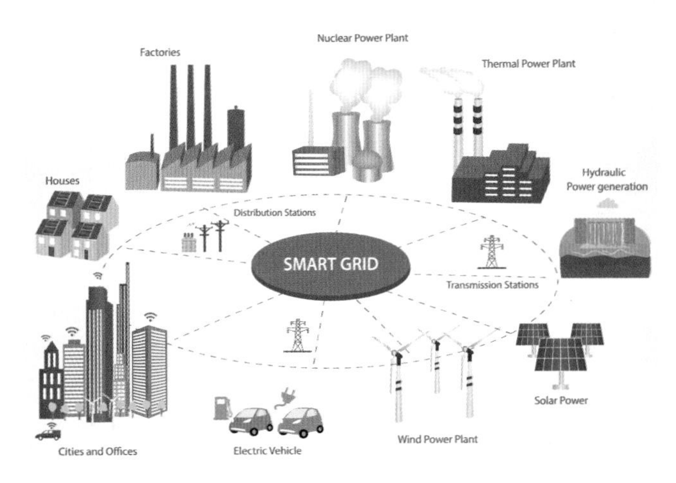

monicaodo/Shutterstock.com

Figure 13-15. Engineers play a significant role in designing energy and power technologies through smart grids.

Career Connection Mechatronics Engineer

Mechatronics is a field that combines mechanical and electrical technology with information, computer, and control systems. A mechatronics engineer unites the principles of electronics, mechanics, and computing to develop, improve, and operate automation technologies. Examples of these technologies are industrial robots and automated guided vehicles used in industries like computer integrated manufacturing.

science photo/Shutterstock.com

High school students interested in a mechatronics engineering career should take a variety of STEM courses. Such courses include technology courses related to digital electronics and engineering design; mathematics courses, such as trigonometry and calculus; and physics.

Many postsecondary institutions offer degrees directly related to mechatronics engineering and technology. Programs typically include multiple levels of calculus and physics as the foundation. Further courses in statics, material properties, circuits, fluid mechanics, and automated control systems are included in the program. Programs also focus on engineering design to meet the specific objectives of applications requiring new automated control systems. Many mechatronics engineering jobs require a graduate degree in science.

Individuals pursuing a degree in mechatronics should be able to think analytically and critically. They should be comfortable working in a collaborative environment. Experience in robotics and automation and a solid foundation in mathematics and science are required.

Mechatronics engineers are hired in many industries, including automotive, aerospace, manufacturing, mining, forestry, and defense. However, many mechatronics engineers have created their own companies. Other jobs that can be obtained with a degree in mechatronics include control systems engineer, manufacturing engineer, mechanical engineer, electrical engineer, and automation engineer. Because industries seek to automate and improve their processes and services by integrating the latest hardware and software developments, these engineers are in high demand.

A mechatronics engineer may work on the following types of projects:

- Enhancing manufacturing production systems using electronic and mechanical processes combined with new computer technology.
- Developing new technologies and products to extend the ability to automate processes, such as extracting raw materials from the earth.
- Building and testing new robotics technology to solve industrial problems.

Organizations related to this career path include the American Society of Mechanical Engineers, Society of Manufacturing Engineers, and Institute of Electrical and Electronics Engineers.

various modes. These technologies encompass a wide range of products and processes, which continually change based on other technological advancements. See **Figure 13-16**.

Engineers design and use information and communication technologies concurrently to create new products. They use these technologies not only to share ideas with others, but also to design more efficient, effective, and innovative methods to transfer information between humans and machines. Whereas communication was once an explicit process, often human to human, or through means such as smoke signals, books, and cave paintings, it has increasingly become a process that humans do not see. Information is almost instantaneously sent across computers digitally, represented as 1s and 0s.

Engineers with computer and electrical engineering degrees face many challenges in designing information and communication technologies. Finding new ways to teach students is one of these challenges. Engineers have opportunities to create new technologies that can communicate and transfer information to a wide range of individuals. They have developed new technologies such as Massive Open Online Courses (MOOCs). See **Figure 13-17**. MOOCs are online courses that allow a large number of participants to attend and participate in a course via the Internet.

GaudiLab/Shutterstock.com

Figure 13-16. Information and communication technologies are becoming increasingly interactive, slimmer in design, and flexible in use.

megaflopp/Shutterstock.com

Figure 13-17. Online learning is growing in popularity.

Another engineering challenge is ensuring the security of cyberspace. For example, a traditional solution is creating firewalls, or perimeter defenses on networks, to block access from external hackers. Finding new ways to protect personal privacy and national security has become increasingly important. Many businesses have been the victims of cyber attacks, resulting in the release of confidential information and leakage of e-mails.

Medical Technologies

The ability to live healthier and longer lives can be attributed to advancements in medical technologies. ***Medical technologies*** create the tools used in the prevention, diagnosis, monitoring, and treatment of illness, as well as repair of injury. Examples of tools developed to treat illness and injury include diagnostic tools, such as CT scans and MRIs; preventative measures, such as vaccines; monitoring equipment, such as heart monitors; surgical tools and equipment, such as surgical robots; and devices used in repair of the human body, such as prosthetic limbs. See **Figure 13-18**. Because people encounter medical technologies throughout their lives and must make decisions directly related to living a healthy lifestyle, understanding medical technologies is essential.

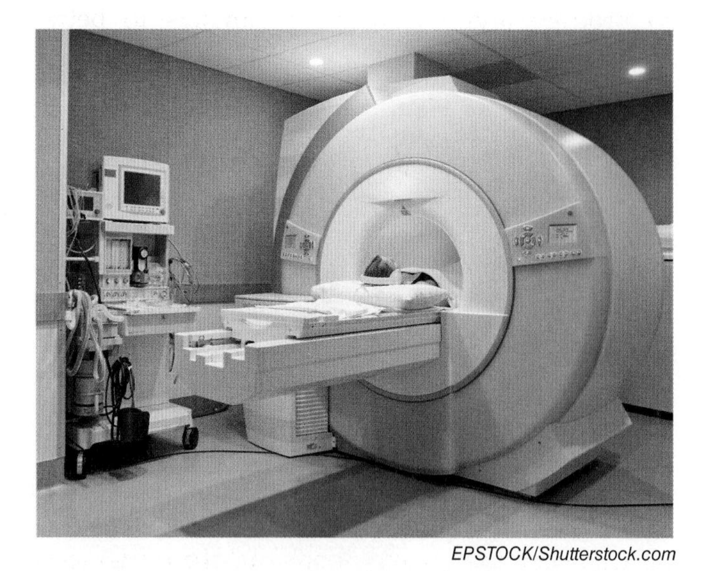
EPSTOCK/Shutterstock.com

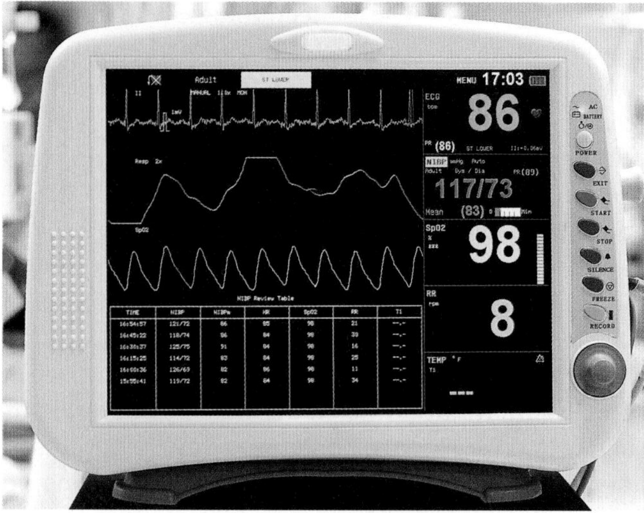

Tewan Banditrukkanka/Shutterstock.com

Figure 13-18. Medical devices such CT scanners and heart monitors are the result of advancements in medical technologies.

Biomedical engineers apply principles of engineering, such as identifying and researching the problem and developing, testing, and evaluating solutions, to design new technologies. They must understand the use of medical technological tools and understand the human body in order to design new devices and solutions to medical problems.

One challenge faced by biomedical engineers is the engineering of better medicines. There is a need for new drugs that are adjusted to an individual's specific genetic makeup. Since each person's DNA is unique and may influence his or her susceptibility to certain diseases, engineers and scientists seek to create personalized medicine tailored just for the body chemistry of individuals. See **Figure 13-19**.

Think Green Green Building

It is becoming increasingly important to design structures in a way that reduces their impact on the environment. This is known as *green building*. Engineers are challenged to design buildings that are sustainable, which refers to having minimal negative long-term environmental impacts throughout its entire life cycle. Factors to be considered include the materials used in construction, the effect of the structure on the local environment, and the energy required to run the structure once complete. A team of different engineers with highly specialized knowledge is required to design a green structure.

A set of rating systems called the Leadership in Energy and Environmental Design (LEED), developed by the US Green Building Council, is used to determine how a green building can be designed and operated. There are four LEED certification levels for buildings, which are certified, silver, gold, and platinum. These certification levels are awarded to a building based on a point system that ranks the structure's potential environmental impact and benefits for humans.

One example of a green building is the health clinic built at Arizona State University. This LEED-Platinum building reduced the previous facility's carbon footprint by 20%, which resulted in the preservation of over 5,000 square feet of natural space. Additionally, engineers used more thermal barriers, mechanical systems that were of higher-efficiency, and better designed lighting systems. These choices resulted in an almost 50% reduction in energy usage below a standard for buildings of similar size. Green projects are becoming more widespread to ensure future generations continue to have access to the natural world.

Mopic/Shutterstock.com

Figure 13-19. Each person's DNA is unique to them and may influence their susceptibility to certain diseases.

Biotechnology and Agricultural Technologies

Biotechnology and agricultural technologies are fields concerned with the use and application of living organisms to meet humans' needs and wants. ***Agriculture*** is the growing of plants and animals to be used for food or other technological products. ***Biotechnology*** is the application of living organisms, in whole or in part, to create technology. See **Figure 13-20**. Both of these fields have existed for thousands of years and make it possible for humans to produce and use living organisms to better their lives.

The engineering design of biotechnology and agricultural technologies requires knowledge of ecosystems. Ecosystems are interconnected groups of organisms living together and interacting with their environment. In addition, engineers in this field must understand how living organisms function, the use of fertilizations and plant control mechanisms, and methods for analyzing the effect intervention may have on humans.

In comparison to other areas of technology, engineers designing biotechnology and agricultural technologies spend a significant amount of time researching and testing new technologies prior to development. Every contribution that these engineers make has a significant effect on humans and the environment. Many new technologies in this area raise social and ethical concerns. Thus, it is important for individuals to be informed about how these technologies are created and the trade-offs that come with them.

One of the greatest engineering challenges is to design solutions for feeding world hunger. Biological and chemical engineering degrees provide the background necessary to design solutions. Genetically modified organisms (GMO) were developed to increase food surplus, allow food to last longer, and increase the size of crops. However, engineers are challenged to address safety questions regarding potential negative impacts of growing and consumption of GMO foods on health and ecosystems.

Wayhomestudio/Shutterstock.com

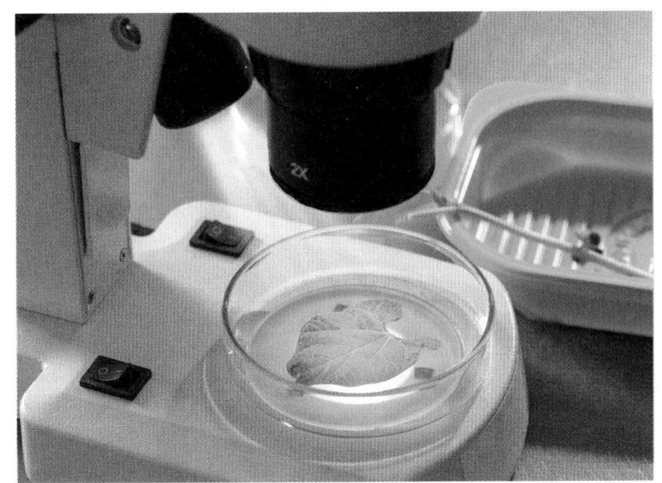

HUANGSHENG ZU/Shutterstock.com

Figure 13-20. Biotechnology is the application of living organisms, in whole or in part, to create technology.

Summary

- ✔ The natural world includes rocks, plants, trees, and animals. The designed world encompasses anything created by humans that is not represented in the natural world.
- ✔ Engineering design involves the application of mathematical and scientific knowledge to meet human needs and wants.
- ✔ The benefits and drawbacks of new technology are often evident only after some time has passed.
- ✔ Technologies can be separated into seven areas—manufacturing, construction, transportation, energy and power, information and communication, medical, biotechnology and agricultural. These areas often overlap.
- ✔ Manufacturing technologies cover a wide range of processes, systems, and procedures to create the products people need or want.
- ✔ Construction technologies encompass the design, assembly, and maintenance of structures.
- ✔ Transportation technologies move people and products from one location to another through land, water, air, and space travel.
- ✔ When designing energy and power technologies, engineers consider the type of energy to be used; the processes used to convert, transmit, or store it; and the overall outputs required.
- ✔ Information and communication technologies are technologies that provide the ability to record, store, manipulate, analyze, and transmit data across various modes.
- ✔ Medical technologies create the tools used in the prevention, diagnosis, monitoring, and treatment of illness, as well as repair of injury.
- ✔ Biotechnology and agricultural technologies are concerned with the use and application of living organisms to meet human's needs and wants.

Check Your Engineering IQ

Now that you have finished this chapter, see what you learned by taking the chapter posttest.
www.g-wlearning.com/technologyeducation/

Test Your Knowledge

Answer the following end-of-chapter questions using the information provided in this chapter.

1. What is meant by the term *designed world*?
2. Identify seven categories of technologies.
3. *True or False?* The negative consequences of newly designed systems are immediately evident.
4. Manufacturing technologies are used to create the products that people _____ or _____.
5. Structures that are environmentally responsible, sustainable, and resource efficient are called _____ buildings.
6. Transportation technologies move people and products through what four types of travel?
7. What is the difference between energy and power?
8. Information and communication technologies provide the ability to record, store, _____, analyze, and transmit data across various modes.
9. New medical technologies are designed by _____ engineers.
10. The application of living organisms, in whole or in part, to create technology is called _____.
    A. an ecosystem
    B. biotechnology
    C. green design
    D. agricultural technology

Critical Thinking

1. What do you view as technology? Has this chapter changed your perspective for what is included in the designed world?
2. Why might it be difficult to anticipate the negative effects of a new technology? Name one new technology for which this is a concern.
3. Research MOOCs and list several benefits and drawbacks of these online courses.
4. How do artistic designing and engineering designing differ? How do they overlap?

STEM Applications

1. Choose one of the seven categories of technologies and research a challenge that engineers face in developing a particular technology within that field. Prepare a research report on the pros and cons of this technology.
2. Create a presentation that shows your viewpoint on whether the technology you researched in Question #1 should or should not be further developed.

CHAPTER 14

Processing Resources

Check Your Engineering IQ 

Before you read this chapter, assess your current understanding of the chapter content by taking the chapter pretest. www.g-wlearning.com/technologyeducation/

Learning Objectives

After studying this chapter, you should be able to:

- ✔ Summarize the evolution of manufacturing.
- ✔ Describe the equipment and systems used in modern manufacturing.
- ✔ Differentiate between primary manufacturing and secondary manufacturing.
- ✔ Differentiate among the three types of natural resources.
- ✔ Describe the processes used to obtain raw materials and transform them into industrial materials.

Technical Terms 

additive manufacturing
artificial intelligence (AI)
automated guided vehicles
automatic storage and retrieval systems
automation
clear-cutting
coal
computer-integrated manufacturing (CIM)
computer numerical control (CNC)
conveyor systems
engineered wood products
extrusion
fluid mining
form utility
fossil fuels
genetic materials
hardwood lumber
horizontal drilling
hydraulic fracturing
just-in-time (JIT) manufacturing
Kaizen
lean manufacturing
lumber
mass customization
mass production
mechanical processes
minerals
natural gas
petroleum
primary manufacturing
programmable logic controller (PLC)
rapid prototyping
robots
secondary manufacturing
seed-tree cutting
seismographic study
selective cutting
softwood lumber
standard stock
subtractive manufacturing
surface mining
thermal processes
timber cruising
underground mining
yarding

Close Reading

Engineers often read back and forth between text and visual diagrams to develop a deep understanding of concepts. As you read through this chapter and examine the visual diagrams, think about how manufacturing processes and technologies have evolved over time. Consider the reasons for these changes and contemplate the impact they have had on society, culture, politics, and the environment.

Each of the products you use every day is the result of a technological production system. Production systems involve the processes used to take raw materials from the earth and transform them into usable commodities.

Some of the earliest technologies were developed to create production systems. These systems create ***form utility***, which is the change in the form of materials to make them more valuable. For example, which is more valuable to people—lumber or furniture built from the lumber? Wood in the form of a final furniture product is more useful than the lumber itself. Therefore, the actions that change a raw material into a useful product provide form utility.

Manufacturing and Construction

The goal of a production system is to produce a product or a structure. The production activities that create products are called *manufacturing*, and the production activities that produce structures are called *construction*. See **Figure 14-1**. Construction is often described as producing a structure on the site where it will be used. A manufactured product is most likely produced in one location, such as a factory, and used by a consumer in a different location.

However, in some cases the boundaries between construction and manufacturing are blurred. For example, modular construction consists of manufacturing components of a structure in a warehouse. The components are then shipped to the site where the structure is assembled.

Thousands of different production activities fall under the categories of manufacturing and construction. This chapter discusses manufacturing and the primary processes that produce the industrial materials used to manufacture products.

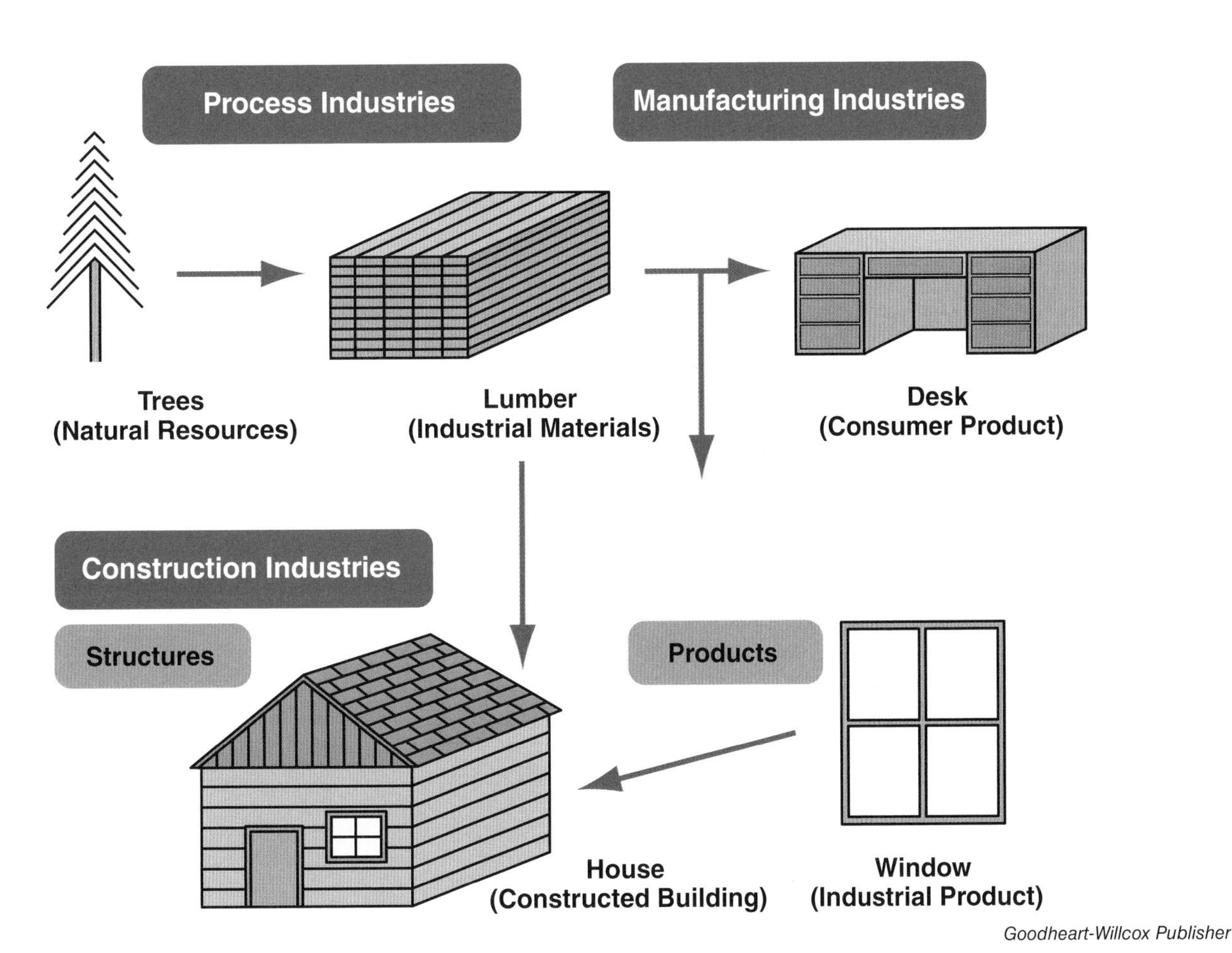

Goodheart-Willcox Publisher

Figure 14-1. Production systems convert natural resources into industrial materials. These materials can be manufactured into industrial or consumer products or used in construction.

The Evolution of Manufacturing

Manufacturing is the process of producing industrial materials with desirable properties and finished consumer products. Manufacturing technologies are developed to effectively and efficiently facilitate the conversion of raw materials into more valuable industrial materials and finished products.

Throughout history, people have been manufacturing products to meet their needs and extend their human capabilities. Primitive forms of manufacturing involved the production of tools for hunting and gathering, weapons for protection and military dominance, and construction materials for building shelters. These products enabled people to survive longer. Increased life expectancies afforded them the time necessary to continuously refine and improve their manufacturing processes and technologies.

Early manufacturing involved skilled artisans, or skilled craft workers, who produced handmade products using specialty tools and materials. See **Figure 14-2**. These handmade products were unique and could be produced only one at a time. Therefore, this early process of manufacturing was time consuming and resource exhaustive.

In the late 1700s, an American inventor named Eli Whitney, **Figure 14-3**, changed the way people approached manufacturing. Whitney championed the idea of using interchangeable parts in his musket production company. Using interchangeable parts enabled unskilled workers to quickly produce a large quantity of goods at a lower cost because all of the product parts were identical, or standardized.

Everett Historical/Shutterstock.com

Figure 14-3. Eli Whitney.

Goodheart-Willcox Publisher

Goodheart-Willcox Publisher

Figure 14-2. These people are practicing early production activities.

Career Connection Petroleum and Natural Gas Engineer

Christian Lagerek/Shutterstock.com

Petroleum and natural gas engineers design and use systems that aid in the discovery, extraction, processing, and transportation of petroleum and gas. Using the principles of geosciences and engineering, these engineers develop, improve, and operate technologies involved in oil and natural gas well drilling and stimulation, resource storage and handling, and environmental remediation.

High school students interested in a petroleum and natural gas engineering career should take a variety of STEM courses. Courses focused on technology and engineering will help prepare students for this career. Additionally, mathematics courses related to trigonometry and calculus and advanced science courses in chemistry and geology will be beneficial.

Several postsecondary institutions that are located in geographical areas of shale gas and oil expansion offer degrees directly related to petroleum and natural gas engineering. Programs typically include multiple levels of calculus, chemistry, physics, geology, and macroeconomics as their foundation. These programs offer further courses in fluid mechanics, thermodynamics, statics, well stimulation, and reservoir design. Programs also focus on engineering design to meet specific objectives in oil and natural gas well production and oil and natural gas storage.

Individuals pursuing a petroleum and natural gas engineering pathway should have a multidisciplinary skill set, the ability to think critically, the ability to work in a collaborative environment, an interest in economics, the ability to conduct scientific investigations, and a solid foundation in earth and physical sciences.

Petroleum and natural gas engineers may be hired in the oil and gas industries or by government and environmental agencies. These engineers are in high demand due to the increased focus on extracting natural gas and oil from the large shale rock formations found throughout the United States.

A petroleum and natural gas engineer may work on the following types of projects:

- Enhancing oil and natural gas drilling operations.
- Developing new technologies and techniques for efficient and economical extraction of oil and natural gas.
- Improving processes of oil and natural gas handling and storage to reduce negative environmental impacts.
- Researching and evaluating the impact of oil and natural gas extraction operations on the surrounding environment.
- Providing professional advice to government agencies.

The professional organization for this field is the Society of Petroleum Engineers (SPE). This organization offers both professional and student memberships.

As a result of part standardization, the labor required to manufacture products could be distributed among an unskilled workforce. This concept also led to the introduction of precision tools and machines, such as the milling machine invented later by Whitney, to help improve the efficiency and quality of production.

The standardization of parts also made part replacement easier when repairs were needed. Prior to the interchangeability of parts, a product would have to be repaired by the specialist who created it.

In the early 1900s, Henry Ford, founder of the Ford Motor Company, further revolutionized the manufacturing industry by further developing the moving assembly line technique of mass production. ***Mass production*** is the process of creating a large number of standardized products efficiently. The assembly line technique increased the speed at which products, such as automobiles, could be produced by separating the assembling process into small and easily managed steps. See **Figure 14-4**. These steps could then be handled by unskilled laborers at a lower cost. This enabled the Ford Motor Company to mass produce automobiles at a price affordable to many middle-class Americans.

Everett Collection/Shutterstock.com

Figure 14-4. Automobiles being mass produced on an assembly line.

However, Henry Ford faced problems with his workers due to the mundane daily tasks of performing the same specific operation on an assembly line, dangerous work environments, and low wages. As a result, Ford worked to stabilize the labor force by raising wages and establishing what is now known as *Fordism*. Fordism is a model of economic expansion and technological progress that focused on the mass production of inexpensive and standardized goods through assembly lines consisting of specialized tools and equipment operated by unskilled laborers. These laborers were then paid higher wages, which provided them the means to afford the goods they were producing. Ford also worked with the workers' unions to create the three work shifts that are still in use today, enabling around-the-clock factory operation.

Another entrepreneur in the automotive industry, Kiichiro Toyoda, worked to advance manufacturing processes in the early 1900s. Toyoda, the founder of the Toyota Motor Company, and engineer Taiichi Ohno, developed and perfected the strategy of ***just-in-time (JIT) manufacturing***. JIT is a production strategy in which companies purchase the parts for their goods only as they are needed. This drastically reduces the costs associated with handling and storing a large inventory of parts. As a result of this strategy, Toyota Motor Company and others raised their productivity significantly by minimizing waste, such as storage cost and space, involved with maintaining a large inventory of parts.

Minimization of waste also led to the management philosophy of the Toyota Production System called lean manufacturing. ***Lean manufacturing*** is a systematic strategy to reduce operation costs and improve the overall consumer value of products by eliminating wastes and inefficiencies throughout the manufacturing process. Toyota Motor Company identified seven types of wastes in the manufacturing process that could be minimized. These are the transportation of products, the inventory of parts, the motion of products that can result in damage, products in waiting, overprocessing of products, over-production of products, and part defects.

Lastly, following World War II, the Toyota Motor Company introduced the continuous improvement philosophy called *Kaizen*, which is Japanese for *good change*. ***Kaizen*** promotes constant change in manufacturing operations based on employee input in a way that humanized the workplace. Employees are empowered to experiment with their work in order to spot areas where waste could be reduced or methods could be developed to increase efficiency. Employees are rewarded for ideas that result in process improvement.

As the twentieth century progressed, additional advancements and technological breakthroughs transformed the manufacturing industry. For example, by the late 1900s, Jervis B. Webb Co. developed electronically controlled conveyor systems, a critical component of modern day manufacturing systems. ***Conveyor systems*** move materials, parts, or products in a fixed path along a production line. See **Figure 14-5**. The first electronically controlled conveyor system created by this company allowed processes in manufacturing to begin to be automated, or coordinated by computer control systems. See **Figure 14-6**.

Olga Serduk/iStock/Thinkstock.com

Figure 14-6. A modern assembly line.

Modern Manufacturing

Highly efficient manufacturing techniques are important because manufacturing is central to a nation's economy. Economic success of a nation is ultimately tied to global trade, which is based on the goods produced through manufacturing industries. Additionally, the manufacturing industry creates jobs. The majority of jobs today are directly or indirectly dependent on manufacturing. These jobs may be related to the actual production of goods or related to the service industries tied to those goods, such as retail, marketing, and repair. Therefore, there is a constant competition between nations to develop new products and faster and more cost-efficient production methods.

Marcin Balcerzak/Shutterstock.com

Figure 14-5. A conveyor belt system.

The development of more advanced manufacturing processes is based on new and improved tools and techniques. These tools and techniques are founded in scientific understanding and engineering practices. Increased scientific knowledge and engineering abilities can help to optimize manufacturing processes, enhance the quality of final products, and increase the availability of sophisticated computer controlled systems that allow faster and cheaper production. Additionally, computer-aided manufacturing processes allow for ***mass customization***, which is the ability of manufacturing systems to produce a large output of customized products at a low cost.

Machines controlled by computer systems have replaced many factory workers. The term ***automation*** refers to the operation and control of equipment and processes by automatic systems.

Think Green Green Materials

Green materials are materials that are considered environmentally friendly. These materials must meet specific criteria. For a material to be considered green, it must not tax exhaustible natural resources, it must not cause carbon dioxide emissions when processed, and it must either be recyclable or biodegradable. Manufacturers have made strides to not only make their products recyclable, but also to make the products out of recycled materials while emitting less carbon dioxide in processing. Carbon dioxide is considered a greenhouse gas because its concentration in Earth's atmosphere can contribute to global warming. One major way carbon dioxide is released into the atmosphere is through the burning of fossil fuels.

To mitigate these impacts, manufacturers have implemented new practices, such as using plant fiber, rather than petroleum to create plastic. These bioplastics are made of renewable resources, are cleaner to process, are recyclable, and are biodegradable. Another example is the use of bamboo in the construction industry as a substitute for some hardwood applications. Bamboo is a highly sustainable crop that grows quickly in a wide variety of conditions, unlike most hardwoods.

Computer-Integrated Manufacturing

Computer-integrated manufacturing (CIM) is the approach of using computers to control the entire production process. This integration relies on closed-loop control systems, which were discussed in Chapter 12. Remember that closed-loop control systems are coordinated by feedback from real-time sensors. These systems enable individual manufacturing processes to exchange information with each other and initiate corresponding manufacturing actions. The advantage of this approach is a more accurate and much faster automated manufacturing process. Additionally, computer-integrated manufacturing is a safer alternative to having employees perform potentially hazardous tasks.

Computer-integrated manufacturing may use robots, programmable logic controllers (PLCs), computer numerical control machines, automatic storage and retrieval systems, automated guided vehicles, and artificial intelligence (AI) systems. These components are discussed in the following sections.

Robotics

Robots are programmable part-handling or work-performing devices, often used to replace human labor in industrial settings, and can perform tasks automatically or with varying degrees of direct human control. The Czech author Karel Capek first coined the word *robot* in the play *Rossum's Universal Robots*. He derived the word *robot* from the Czech word *robota*, which means *work*. The first robots were designed in the 1940s to handle radioactive materials. The first industrial robot was developed in 1962. The industrial robot's functions were limited to picking up an object and setting it down in a new location. This simple type of robot is called a *pick-and-place robot*. In manufacturing today, industrial robots, **Figure 14-7,** are often used to manipulate

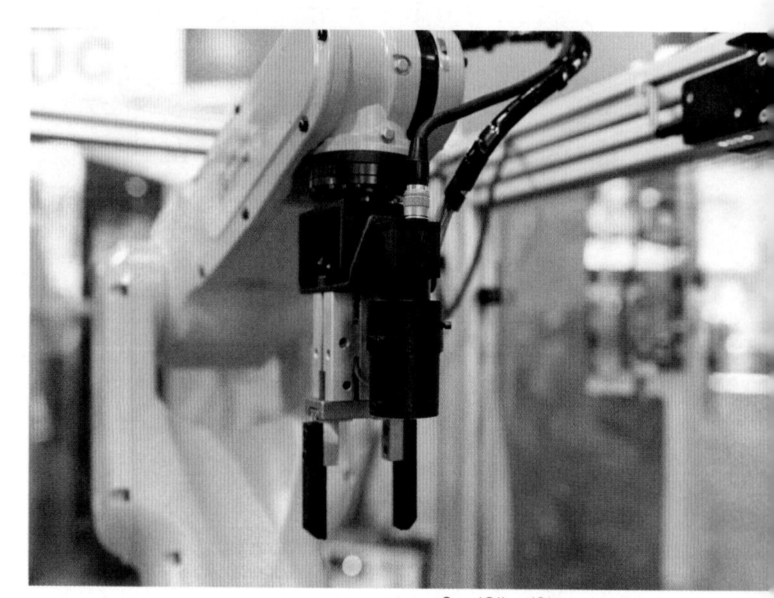

SvedOliver/Shutterstock.com

Figure 14-7. One type of industrial robot.

materials by welding, painting, assembling, moving materials and parts, and inspecting products.

A typical industrial robot has five main parts—an arm, an end effector, a drive, a controller, and a sensor. See **Figure 14-8**. Most robot arms resemble a human body in that they have a body, a shoulder, an elbow, a forearm, and a wrist. These parts of the arm allow the robot to move in three axes, or straight lines, in respect to a coordinate system. The motions along these axes can be left and right, up and down, and in and out. The base allows the robot to move left and right. The shoulder and elbow create the up-and-down and in-and-out motions. The wrist rotates to create the final positioning for the tool or material. The end effector is the hand connected to the wrist. The structure of this "hand" varies based on its function. For example, an end effector can be a gripping device or a paint spray nozzle.

The drive is composed of hydraulic, pneumatic, or electric devices. The drive system moves the arm and end effector to a desired position. The controller is the brain of the robot. This brain controls the arm movements created by the motors or hydraulic systems. Computer commands direct the motions. Sensors provide feedback that allows the robot to know where its end effector is and determine the moves needed to move the effector into a new position.

Today, there are many different types of robots that are used in a wide variety of applications outside of manufacturing. Robots can mow a lawn, vacuum a house, gather samples on the moon, clean up toxic waste, disarm explosives, and search the ocean floor.

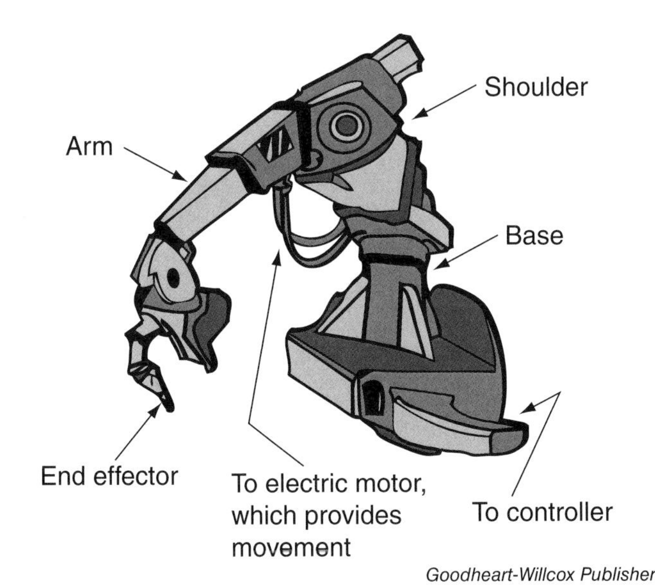

Goodheart-Willcox Publisher

Figure 14-8. Parts of a robot.

Computer Numerical Control

Computer numerical control (CNC) is the automatic control of a machine by means of a computer program that controls the machine's motions. See **Figure 14-9**. A CNC machine produces automatic, precise motion of the machine's tools or the material being processed. This motion is in two or more directions along specified axes. The tool or workpiece can be automatically and precisely positioned along the length of travel on any axis. For example, a person turns a handwheel that is attached to a lead screw (a screw used to translate rotating motion into linear motion) to position the worktable on a milling machine. In CNC milling machines, a motor responding to computer-programmed commands rotates the lead screw to position the worktable.

Programmable Logic Controllers (PLCs)

A ***programmable logic controller (PLC)***, or programmable controller, is a device that uses microprocessors to control machines or processes. See **Figure 14-10**. A technician on the factory floor generally creates the program for the controller.

genkur/Shutterstock.com

Figure 14-9. A CNC machining center can be programmed to do many different tasks.

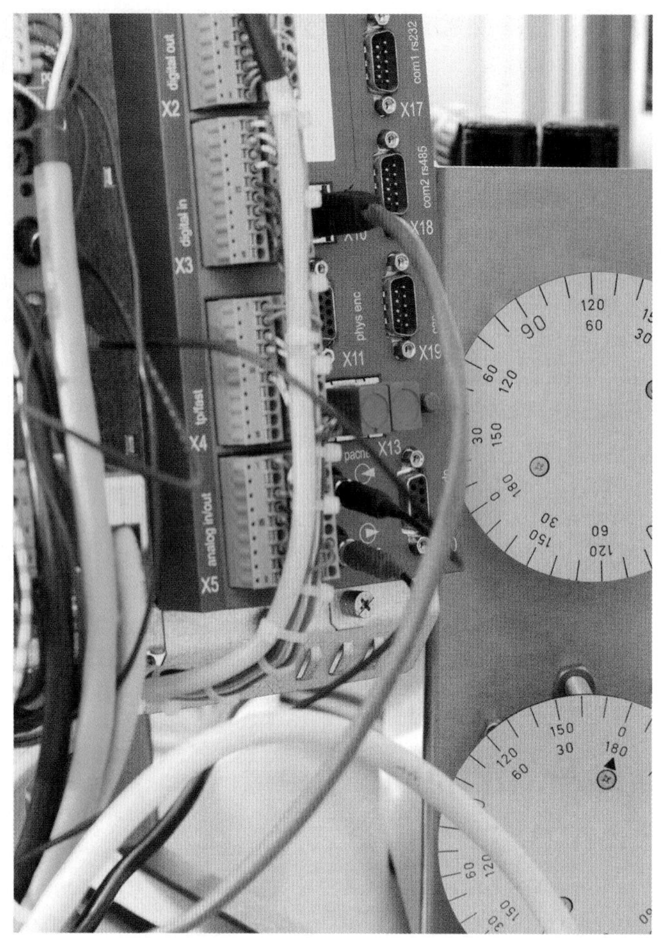

Gianni Furlan/iStock/Thinkstock.com

Figure 14-10. A programmable logic controller (PLC).

PLCs have three basic functions—input, control, and output. A PLC works by analyzing its inputs and, depending on their states, turning on or off its outputs. For example, the controller might receive input such as temperature, liquid levels, or shaft speed. Based on this input and the logic written into the controller, it activates appropriate outputs. A flame might be reduced to lower the temperature, a valve might be opened to fill a tank, or the speed of a motor might be changed to improve a machine's operation.

Automatic Storage and Retrieval Systems

Automatic storage and retrieval systems consist of a series of computer-controlled conveyors and vehicles that transport and store materials, parts, or final products. These systems require minimal human input. They are used in manufacturing operations in which large volumes of materials or products are transported in and out of storage and where the potential for damage to products or materials through transportation errors is high. Automatic storage and retrieval systems reduce the labor needed for transporting and storing materials, parts, or products and minimize the space required for storage and retrieval. These systems also enhance the accuracy of tracking materials.

Automated Guided Vehicles

Automated guided vehicles are computer-controlled devices that use sensors to navigate from one place to another. These vehicles are used in manufacturing industries to transport materials, parts, and final products throughout the facility or warehouse. Automated guided vehicles can be a component of automatic storage and retrieval systems. These vehicles help automate a manufacturing process to increase efficiency and reduce costs.

Artificial Intelligence

Artificial intelligence (AI) is the intelligence exhibited by a manufactured device or system. AI is becoming more widely used in manufacturing. Most AI systems involve a computer to act as the brain for the system. These systems often use a type of AI called *machine vision*. The purpose of a machine vision system is to program a computer to "see and be aware of" a scene using digital cameras or other sensors. The system then acts on this understanding to inspect a part, locate a cutter, or pick up a randomly positioned part. Other uses for AI include handwriting recognition for handheld communication devices, optical character recognition (OCR) programs for scanners, speech-recognition systems for automated customer call centers, and face recognition for security systems. See **Figure 14-11**.

Rapid Prototyping

Rapid prototyping is the fabrication of a scaled part or model of a product using three-dimensional computer-aided design (CAD) data. Rapid prototyping is often used to visualize and test products before establishing a full production process. However, with the wider availability of rapid prototyping techniques and tools, such as 3-D printers, this process has enabled computer-automated solid manufacturing. This process is used to manufacture production-quality goods.

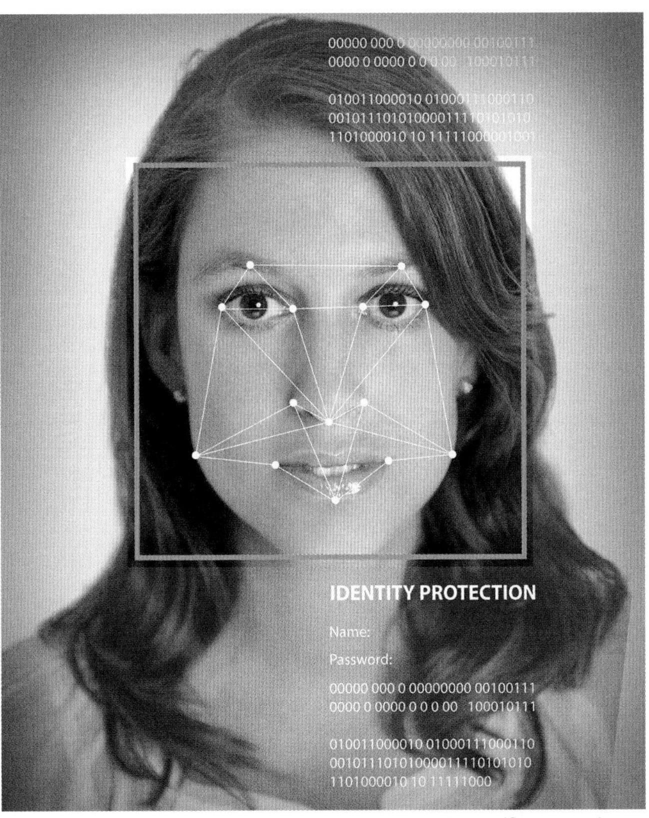

Franck Boston/Shutterstock.com

Figure 14-11. Facial recognition systems are used to identify and verify a person's identity from a video or image using biometrics. Biometrics is the measurement of human characteristics such as a person's iris, retina, face geometry, and fingerprints.

CNC machines follow a ***subtractive manufacturing*** process that uses computer control to remove material from raw stock until a desired shape and size is reached. Conversely, the ***additive manufacturing*** process of rapid prototyping begins with the creation of 3-D geometric data through solid modeling CAD software. These data are then processed into X, Y, and Z coordinates, which are used to direct the rapid prototyping machine or 3-D printer to place successive layers of materials to create the predesigned three-dimensional object. See **Figure 14-12**. Innovation in rapid prototyping and additive manufacturing continues to revolutionize the manufacturing industry.

Primary and Secondary Manufacturing

Manufacturing can be grouped into two major categories. ***Primary manufacturing*** includes the resource processing systems of manufacturing.

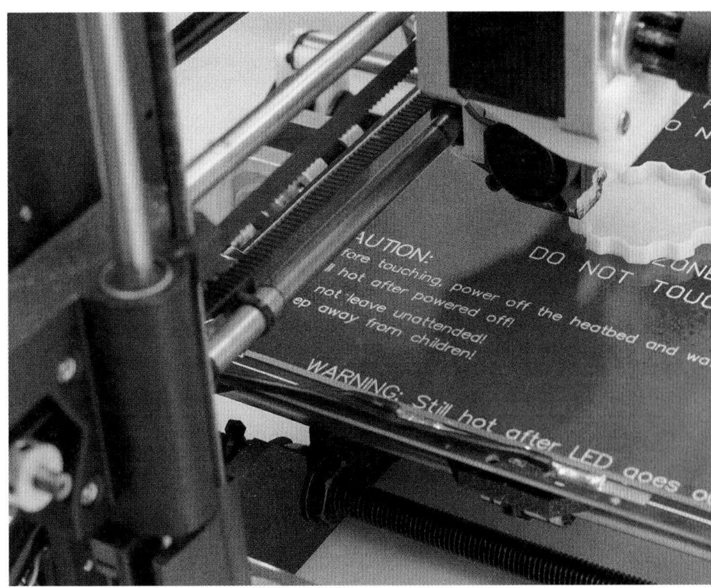

hopsalka/Shutterstock.com

Figure 14-12. A three-dimensional printer creating a product using an additive manufacturing process.

Secondary manufacturing encompasses product-specific manufacturing systems. Each of these activities plays a unique role in converting a material resource into a product to meet human needs and desires. See **Figure 14-13**. This chapter focuses on primary manufacturing, describing the processes used to obtain raw materials from the earth and transform them into industrial materials. Secondary manufacturing is examined in Chapter 15.

All technological objects are made of materials that can be traced back to one or more natural resources. For example, plastic materials are made from plant fiber (cellulose) or natural gas. Glass is made from silica sand and soda ash. Cotton is grown and harvested from plants. Plywood comes from trees in the forest. Steel is made from iron ore, limestone, and coal. However, few materials occur in nature in a usable state. Generally, materials must be converted into new forms before products or structures can be made from them. This conversion of materials is considered *primary manufacturing*.

Primary manufacturing typically involves two actions. The materials must first be located and obtained and then processed to turn them into an industrial material. Obtaining materials involves extracting or gathering them. Trees, crops, and domesticated animals are grown and harvested. Minerals and hydrocarbons (petroleum and coal) are searched for and extracted from the earth by drilling or mining.

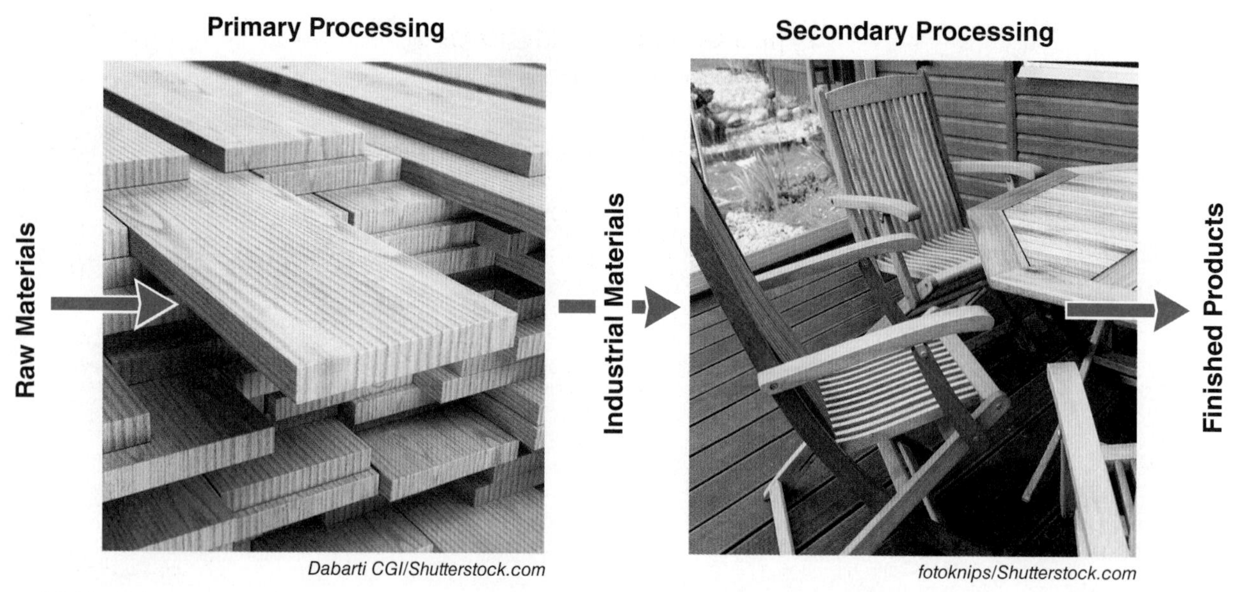

Figure 14-13. Primary manufacturing converts natural resources into industrial materials. These materials can then be manufactured into industrial or consumer products through secondary manufacturing processes.

Once the natural resources have been obtained, they are transported to a processing location where they are changed into industrial materials. These materials then become the inputs to secondary manufacturing activities. An example is the changing of iron ore, limestone, and coke into steel. See **Figure 14-14**. The steel might then be processed into bars, rods, sheets, or pipes that end up as a part of an automobile, bridge, cell tower, or other product structure made for industry and consumers.

Figure 14-14. Primary processing activities change natural resources into industrial goods, such as these steel rods.

Types of Material Resources

Without material resources, production is not possible. Production technology uses materials and energy as inputs and makes products and structures as outputs. Three types of natural resources can become the inputs to production systems—genetic materials, fossil fuel materials, and minerals. See **Figure 14-15**.

Genetic Materials

Many resources come from living things. ***Genetic materials*** are obtained during the normal life cycles of plants or animals through farming, fishing, and forestry. See **Figure 14-16**.

US Department of the Interior

Figure 14-15. Genetic, fossil fuel, and mineral materials are used as inputs to production systems.

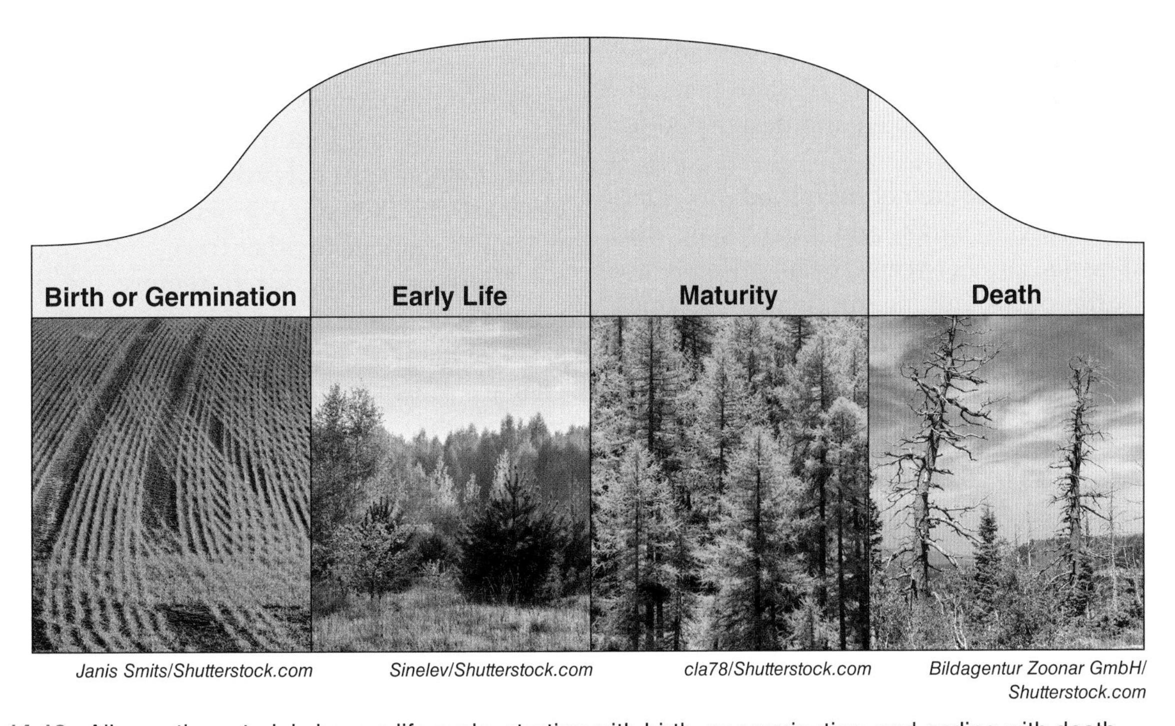

Janis Smits/Shutterstock.com Sinelev/Shutterstock.com cla78/Shutterstock.com Bildagentur Zoonar GmbH/Shutterstock.com

Figure 14-16. All genetic materials have a life cycle, starting with birth, or germination, and ending with death.

Typical genetic materials used in production systems are grains (such as wheat, oats, barley, and corn), vegetable fibers (such as wood, flax, and cotton), and animals (such as meat, hides, and wool). The origin of all genetic materials is in birth or germination. The appearance of animal life is called *birth.* Plant life generally starts with germination, which is the growth from seeds or spores. Young plants and animals grow rapidly early in their life cycles. Their growth generally slows down as they reach older age. This period is called *maturity.* The organism is still healthy at maturity, but it stays about the same size. The length of this maturity stage varies widely. The maturity stage of mushrooms is measured in days, while redwood trees have a maturity stage lasting centuries. All organic life ends at some point, however. The plant or animal dies from old age, disease, natural disaster, or human interference.

Most genetic materials are easy to find. Trees and farm crops are grown on plots of land. However, those who seek to harvest wild plants and animals seek genetic resources that are sometimes hard to find.

Harvesting genetic resources should be done at the proper stage of growth. This stage varies with the growth cycle and growing habits of the organism. Many trees are harvested during their mature phase. However, since some young trees do not grow in the shade of older trees, all trees in a single plot may need to be harvested at one time. New trees can then be planted, which will grow in the cleared area. Likewise, most farm crops are planted at one time in fields. The plants are fertilized and irrigated to stimulate their growth and most of the crop matures at the same time. Therefore, the entire crop can be harvested at one time.

Logging

Harvesting trees requires forest management. Private forests are generally designed and managed to maximize the amount of harvested material. Each stand of trees is evaluated and designated for a specific use.

Roads and logging are not permitted in wilderness areas. National parks, and many state and provincial parks, protect scenic beauty. Roads are allowed in the parks, but logging is not. National, state, and provincial forests generally are multiuse lands. Lakes, hiking trails, and camping areas are set aside for recreational use, but logging is permitted in selected areas to harvest mature trees.

The logging process, explained later in this section, requires both planning and action. Planning for the removal of trees involves several steps. First, the forest is studied to determine whether or not it is ready for harvesting. This study involves a process called ***timber cruising*** in which teams of two or three foresters measure the diameters and heights of the trees. Their task is to find stands of trees that can be economically harvested. These teams also prepare topographical maps showing the locations and elevations of the features on the potential logging site.

Forest engineers then plan the proper way to harvest the trees. They plan logging roads and loading sites. These engineers also select the type of logging to match the terrain and the type of forest. Three logging methods exist:

- ***Clear-cutting***. All trees, regardless of species or size, are removed from a plot of land that is generally less than 1,000 acres. See **Figure 14-17**. This process allows for replanting the area with trees that cannot grow in competition with mature trees. Also, the number of tree species can be controlled.
- ***Seed-tree cutting***. All trees, regardless of species, are removed from a large area, except for three or four per acre. These trees are used to reseed the area. The type of seed trees left controls the number of reseeded species.
- ***Selective cutting***. Mature trees of a desired species are selected and cut from a plot of land. This technique is used in many pine forests, where tree density is limited.

These steps are the prelude for the main activity called *logging*. See **Figure 14-18**. Workers move equipment into the forest to remove the trees. Loggers called *fellers* use a chain saw to cut down (fell) the appropriate trees. A machine that shears the trees off at ground level can harvest smaller trees. Loggers are careful to drop the trees into a clear area so they are not damaged and so they do not cause damage to other trees.

Workers called *buckers* remove the limbs and tops. These parts of the tree are called *slash*. The slash is piled for later burning or chipping so nutrients are returned to the soil. The trunk is then cut into lengths called *logs*.

chris kolaczan/Shutterstock.com

Figure 14-17. Clear-cutting removes all trees from a certain area, allowing for replanting of the whole area.

Figure 14-18. Logging starts with selecting and cutting trees. The felled trees are collected using yarding. The trees are then hauled to a mill for storage and further processing.

Workers gather the logs in a central location called the *landing*. This gathering process is called ***yarding*** and can be accomplished in several ways. First, workers use chokers (cables) to bind the logs into bundles. They can then use cables to drag logs using high-lead and skyline yarding. See **Figure 14-19**. These systems use a metal spar (pole), cables, and an engine. High-lead yarding drags the logs to the landing. Skyline yarding lifts and carries the logs over rough or broken terrain.

On gentle terrain, ground yarding is used. Tractors and an implement called an *arch* are used to drag the logs, bound together with a choker, to the landing. Very steep terrain might require helicopter yarding.

Once logs arrive at the landing, workers load them onto trucks and move them to the processing plant. This plant might be a lumber, plywood, particleboard, hardboard, or paper mill. The logs are often stored at the mill in ponds or stacked up and sprayed with water to prevent cracking and insect damage.

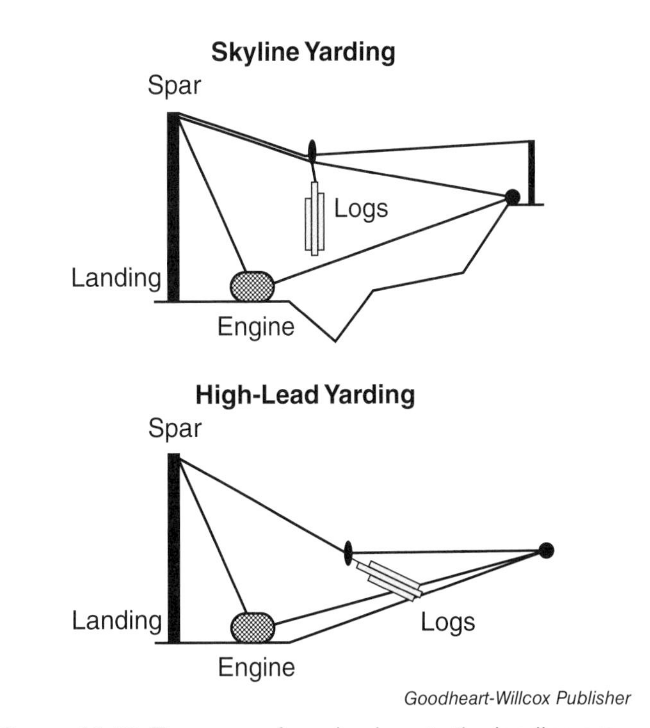

Figure 14-19. Two ways of moving logs to the landing are skyline yarding and high-lead yarding.

Fossil Fuel Materials

Fossil fuels, formed from the remains of prehistoric living organisms, are mixtures of carbon and hydrogen called *hydrocarbons*. Fossil fuels are found in a vast number of products, from fuels to medicines.

Types of Fossil Fuels

Most hydrocarbon products are made from three fossil fuel resources:

- Petroleum.
- Natural gas.
- Coal.

Petroleum is an oily, flammable mixture of different solid and liquid hydrocarbons. The composition of petroleum depends on where the petroleum is found in the earth. Petroleum is the principal source of fuels, such as gasoline, diesel fuel, and heating oil. Lighter hydrocarbon products that come from petroleum are gases, which include methane, ethane, butane, and propane. These gases are widely used in producing plastic resins, such as polypropylene and polyethylene.

Natural gas is a combustible gas occurring in porous rock. This gas is composed of light hydrocarbons. Typically, natural gas is about 85% methane, and the rest is made up of propane and butane. Natural gas is used as a fuel for homes and industry. Compared with other fossil fuels, natural gas burns more cleanly. Complex pipeline networks distribute the gas to potential users. Natural gas is also used to make plastics, chemicals, and fertilizers. In some cases, it is compressed and used as a fuel for vehicles. In this form, it is called *compressed natural gas (CNG).*

Coal is a combustible solid composed mostly of carbon. Coal does not burn cleanly, and its sulfur content is a source of chemicals that make acid rain. Also, its bulk makes it costly to ship.

The principal types of coal are the following:

- **Lignite.** This soft and porous material is made from peat that has been pressed by natural action. Lignite gives off more heat when burned than peat. It is primarily burned in power plants to generate electricity.
- **Bituminous coal.** This most commonly found coal is harder than lignite. Bituminous coal is sometimes called *soft coal* because it can be easily broken into various sizes. See **Figure 14-20**. This coal is widely used for power generation and for heating to produce coke for steelmaking. Bituminous coal can also be used for coal gasification (the process of converting coal to a fuel gas using high temperatures while avoiding the combustion of the coal) and other chemical processes.
- **Anthracite coal.** This shiny black coal is the hardest coal. It has the highest carbon content of all the types of coal. Anthracite coal burns without smoke. It is often used for heating, steel production, and other smokeless fuel needs, such as coal-fired foods.

Locating and Obtaining Petroleum and Natural Gas

Most fossil fuel resources are buried under the surface of the earth. Locating and extracting these resources is challenging. The techniques for

Jeffboat Shipyard

Figure 14-20. Bituminous coal is sometimes called *soft coal.*

BP

Figure 14-21. This geologist is seeking deposits of petroleum.

extracting natural gas and petroleum are different from those used for coal.

Finding rock formations that might contain deposits of oil and gas is the job of geologists and geophysicists. See **Figure 14-21**. Petroleum comes from decayed plant and animal matter that have been covered by layers of sediment from rivers. The layers built up and created great pressure and heat. Over millions of years, the pressure and heat turned the organic matter into oil.

Oil and gas are generally found in porous rock under a layer of impervious (dense) rock. The oil and gas collect under the dense rock. These deposits can be under oceans, mountains, deserts, or swamps. They can be near the surface, as they are in the Middle East, or several miles beneath the land or sea.

One of the most accurate ways to search for petroleum and natural gas is a ***seismographic study***. See **Figure 14-22**. In this technique, a small explosive charge is detonated in a shallow hole. The shock waves from the explosion travel into the earth. When the waves hit a rock layer, they reflect back to the surface. Seismographic equipment uses two

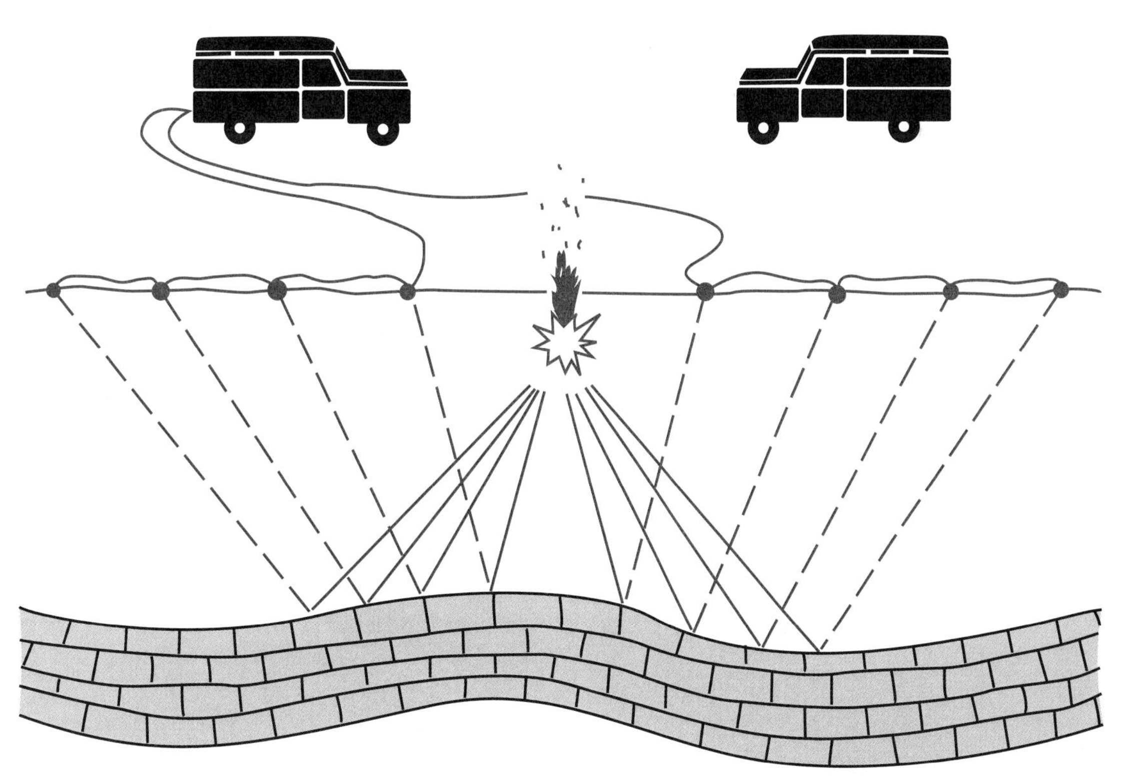

Shell Oil Co.

Figure 14-22. In a seismographic survey, an explosive charge is detonated, creating a shock wave that travels down into the earth. The boundaries between different types of rock reflect the shock waves back to the detectors on the surface. Geologists measure the time it takes for the waves to reach the detectors. The timing information gives the geologists an idea of the shape of a formation.

listening posts to measure the shock waves bouncing off various rock layers. Measuring the time it takes the waves to go down to the layers and reflect back allows the geologist to construct a map of the rock formations.

Other methods used to search for oil or gas are geological mapping, fossil study, and core samples from drilling to find deposits. Geological mapping measures the strength of magnetic forces. The geological study aids in the selection of a good site for exploration. If an area has never produced oil or gas, it is called a *potential field*. Producing fields are called *proven reserves*.

A drilling rig is brought to promising sites. This rig can be either a land rig or an offshore drilling platform. See **Figure 14-23**. Drilling involves rotating a drill bit on the end of a drill pipe. Lengths of pipe are added to the drill pipe as the hole gets deeper. See **Figure 14-24**. During drilling, a mixture of water, clay, and chemicals is pumped down the drill pipe. This mixture is called *mud*. The mud flows out through holes in the drill bit to cool and lubricate. This mixture also picks up the ground-up rock and carries it to the surface. Finally, it seals off porous rock and maintains pressure on the rock. Pressure is maintained to prevent a blowout, which can occur when oil surges out of the well. Blowouts are dangerous and waste large quantities of oil and gas.

Early oil wells were drilled straight down or at a specific angle. Modern techniques, such as ***horizontal drilling***, allow the well to be drilled along a curve to reach deposits that cannot otherwise be tapped. See **Figure 14-25**. Horizontal drilling is now widely done through shale rock formations to extract previously unreachable natural gas deposits. This process is combined with ***hydraulic fracturing***, which involves pumping a high volume of water underground. The pressure fractures the shale rock formations to release trapped pockets of natural gas. See **Figure 14-26**.

Once an oil or gas deposit is found, the drilling rig is replaced with a system of valves and pumps. See **Figure 14-27**. The recovered resource flows through pipes into storage tanks. From the well, the petroleum is transported to refineries. Natural gas is compressed and sent to petrochemical plants. Pipeline companies sell natural gas to home-heating and electric-power customers.

Locating and Obtaining Coal

Coal started as plant matter thousands of years ago. In moist areas, the plant matter did not decay easily and layered up to make peat, a brownish-black plant matter that looks similar to decayed

Land Rig

BP

Offshore Drilling Platform

Gulf Oil Co.

Figure 14-23. Oil and gas rigs allow people to drill wells on land and under lakes and seas.

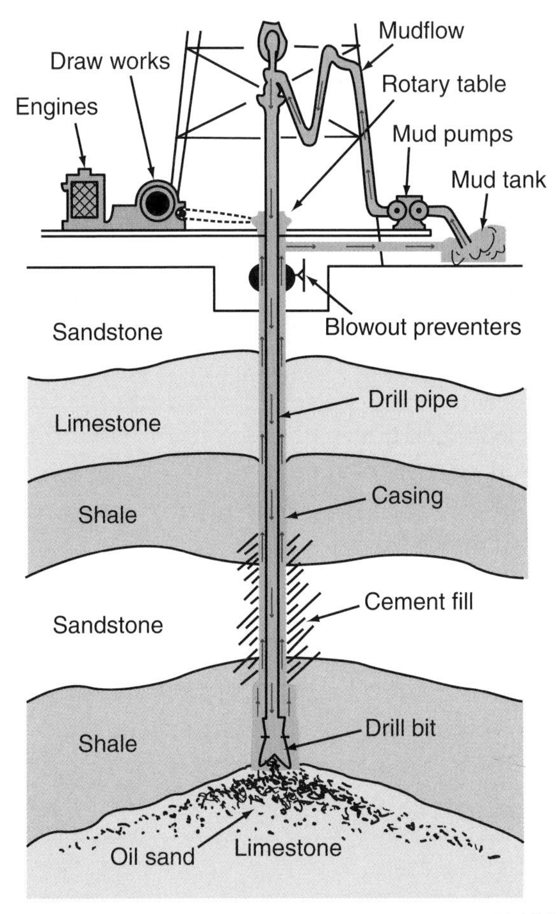

Mobil Oil Co.

Figure 14-24. Drilling involves rotating a drill bit on the end of a drill pipe.

BP

Figure 14-25. Modern techniques allow wells to be drilled along curved lines to tap difficult-to-reach deposits of oil and gas.

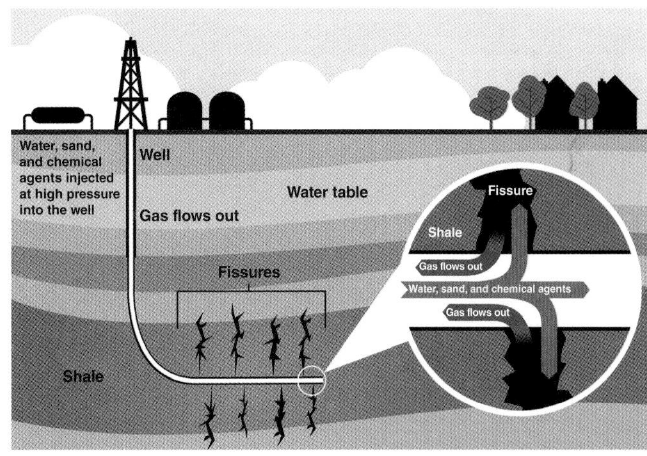

jaddingt/Shutterstock.com

Figure 14-26. The process of hydraulic fracturing to obtain shale gas.

AMAX

Figure 14-27. A valve system, called a *Christmas tree*, is set on top of an oil well.

wood. Dry peat burns, but it gives off a great deal of smoke. Peat that was buried by sediment thousands of years ago was subjected to pressure and heat, which changed the peat into coal.

Coal is the most abundant fossil fuel and is found on every continent. Most of the known reserves are in the Northern Hemisphere. These reserves are generally recovered through mining operations. The types of mining used to extract coal are surface mining and underground mining.

Surface mining, also called open-pit mining, is used when the coal vein is not very deep underground. This type of mining generally involves four steps:

1. The surface layers of soil and rock are stripped from above the coal. This material is called *overburden*. The overburden is saved for later use in reclaiming the land.

Academic Connection: Argumentative Essay

Locating and Obtaining Natural Gas

Argumentative writing is a type of writing that requires people to investigate a topic, collect and evaluate data related to the topic, and establish and present a position on the topic based on evidence. A valid argument also provides facts and data pertaining to any opposing views to the position presented. Scientists and engineers are commonly required to take a particular stance on an issue and defend their position effectively based on logic and fact.

The United States is continuing to expand its efforts to locate and obtain natural gas from shale rock formations. It is believed that increased extraction of natural gas in the United States will help reduce dependency on foreign energy resources. Additionally, the natural gas extracted from these rock formations is believed to be a cleaner source of energy than burning coal. However, obtaining these resources remains a controversial topic because of the process used to extract it from the earth. The hydraulic fracturing process, often called *fracking*, involves drilling deep underground through shale rock formations. Large amounts of water are injected into the formations to fracture the shale rock and release the trapped natural gas. The United States continues to drill more natural gas wells throughout the country.

Research the topic of hydraulic fracturing and write an argumentative essay on whether or not this practice should continue. Support your argument with appropriate evidence. Write your essay in the following format:

- I. Introduction
  - A. Introduce the topic
  - B. Present your position
- II. Body
  - A. Topic sentence for reason 1
    - i. Supporting evidence
  - B. Topic sentence for reason 2
    - i. Supporting evidence
  - C. Topic sentence for reason 3
    - i. Supporting evidence
- III. Conclusion
  - A. Restate your position
  - B. Counterarguments
  - C. General conclusions

2. The coal is dug up with giant shovels, **Figure 14-28**.
3. The coal is loaded on trucks or railcars to be transported to a processing plant.
4. The site is reclaimed by replacing the topsoil and replanting the area. This step is important to help reduce the impacts of the mining activity on the environment. In some instances, the reclaimed sites are converted into usable land for specific needs, such as sports activities.

Underground mining requires shafts in the earth to reach the coal deposits. The three types of underground mining are shaft mining, slope mining, and drift mining. See **Figure 14-29**. Shaft mining requires a vertical shaft to reach the coal deposit. Workers then dig horizontal shafts to remove the coal, **Figure 14-30**. Slope mining is used when the vein is not too deep under the ground. A sloping shaft is dug to reach the coal. The miners then dig a horizontal shaft to follow and remove the coal vein. Drift mining is used when the coal vein

American Electric Power

Figure 14-28. In surface mining, giant shovels are used to dig up and load the natural resource.

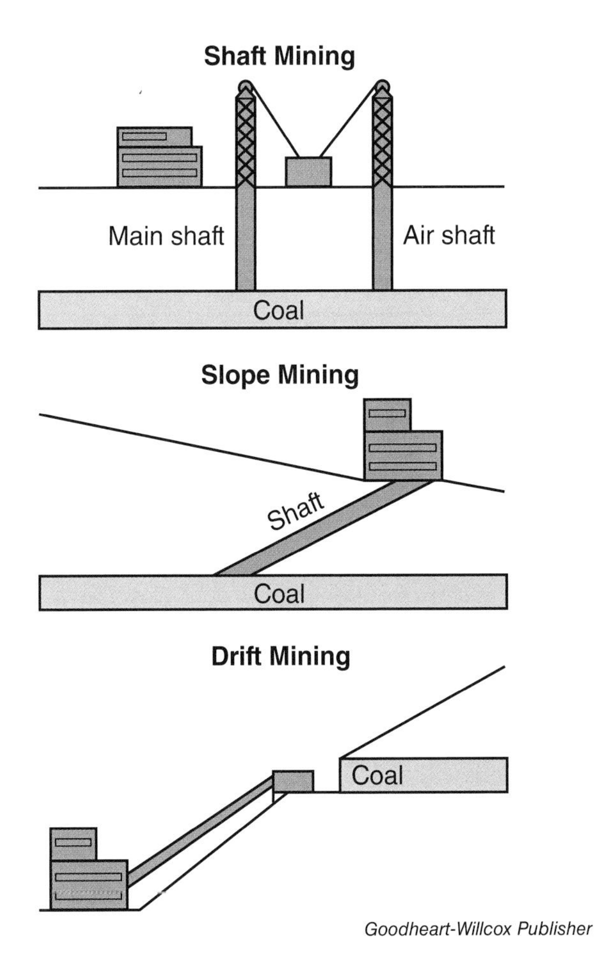

Goodheart-Willcox Publisher

Figure 14-29. The three types of underground mining.

FMC Corp.

Figure 14-30. This long-wall mining machine is recovering coal from an underground mine.

extends to the surface of the earth. The miners dig a horizontal shaft directly into the vein.

Miners involved with underground mining use elevators to remove the mined coal. The same elevators move people and equipment in and out of the mine. Those working with slope and drift mines often use coal cars or conveyors to remove the coal. See **Figure 14-31**.

Minerals

Minerals are substances with a specific chemical composition that occurs naturally. They are different from fossil fuel resources, which are all chemical mixtures. Typical minerals are iron ore, bauxite (aluminum ore), and sulfur.

Minerals are classified in various ways. They can be grouped by their chemical compositions. This grouping includes native elements (elements that occur naturally in a pure form), oxides, sulfides, nitrates, carbonates, borates, and phosphates. However, a more useful classification groups minerals that have economic value into families with similar features. These groups include the following:

- **Ores.** These naturally occurring solid minerals comprise metals chemically combined with other elements. Ores can be processed to separate the metal from the other elements. An example is hematite.

Tractor

AMAX

Conveyor

AMAX

Figure 14-31. Coal is removed from slope and drift mines on tractors or conveyors.

- **Nonmetallic minerals.** These substances do not have metallic qualities. An example is sulfur.
- **Ceramic minerals.** These fine-grained minerals are formable when wet and become hard when dried or fired. These minerals include kaolinite, aluminum oxide, and silicon carbide.
- **Gems.** These stones are cut, polished, and prized for their beauty and hardness. Some examples of gemstones are diamond, sapphire, and ruby.

Over the years, valuable minerals have been found by one of two methods. Early prospectors looked for interesting rock formations or mineral deposits in streams. Prospectors staked claims on promising locations and then either worked the claims or sold them to mining companies.

Modern prospectors are usually trained geologists who study earth formations to discover deposits of a specific mineral. The geologists first identify an area that might contain the mineral based on their knowledge of how ores are formed and where they occur. The selected area is then reduced in size through geological investigations that can include testing of the surface and subsurface geology (the earth's physical structure). This investigation might include studying variations in gravity, magnetism, and other variables. The end result is locating a deposit of ore that can be extracted.

Extracting minerals from the earth is commonly done by mining. Open-pit mines for minerals are generally much deeper than those for coal, extending several thousand feet into the earth. They appear as giant inverted cones with ridges around the edges. The spiral ridge is the road used to move equipment into the mine and minerals out of it. See **Figure 14-32.**

Fluid mining is also used to extract minerals. The ***fluid mining*** technique uses two wells extending into the mineral deposit. Hot water is pumped down one of the wells. The water dissolves the mineral and is forced up the other well. This mining process is often called the *Frasch process.* The Frasch process is widely used to mine sulfur found in the limestone rocks covering salt domes.

Brush-Wellman

Figure 14-32. Open-pit mineral mines can extend several thousand feet into the earth.

Technology Explained 3-D Printer

3-D printer: A device that produces three-dimensional solid objects from binary data on a digital file created using computer-aided design software.

3-D printers are changing the way products are produced. 3-D printing is an additive manufacturing process used to produce almost any three-dimensional product by adding layer by layer of material to create the product.

To print a three-dimensional product, a three-dimensional blueprint of the product is first made using computer-aided design (CAD) software. This blueprint can then be sent to the 3-D printer to print the object. The most common material used to print three-dimensional objects is plastic. However, materials such as glass, wax, metal, and even paper can be used.

©iStock/Thinkstock.com/S?bastien Bonaim?

Figure A. A three-dimensional sphere being printed.

Some 3-D printers have a spool of plastic material in a string-like form. When the printer receives the CAD data, the plastic is pulled through the printer head, where it is melted. The plastic is deposited on a plate, where it instantly cools. The printer head then adds layer on layer of plastic until the final product is formed. See **Figure A**.

Some 3-D printers use two separate materials that bind together to make the individual layers of the object. When these printers receive the CAD data, the printer nozzle first lays out a fine powder and then applies a binder agent or liquid glue to the powder. This powder can be composed of different types of materials and is available in various colors. Some 3-D printers print support material that helps to hold the object together as it is printed. This support material is then removed by bathing the object in water or a chemical solution to dissolve it.

As 3-D printing technology advances, 3-D printers are becoming more affordable. New applications are being continuously developed. 3-D printers are now used in the food industry to create elaborate food products. In the medical field, genetic materials are being used to print human tissue for body parts. Additionally, large 3-D printers have been created to print entire homes in a short amount of time.

Minerals can be found in groundwater, lakes, and oceans. A common way to extract minerals from water is through evaporation. Seawater or water from salt lakes is pumped into basins. Solar energy is then used to cause the water to evaporate, leaving the mineral resource behind. A number of minerals have been recovered from the Great Salt Lake in Utah in this manner.

Material Processing

The goal of primary processing is to convert material resources into industrial materials, which are often referred to as ***standard stock***. For example, primary processing converts wheat into flour; aluminum ore into aluminum sheets, bars, and rods; logs into lumber, plywood, particleboard, hardboard, and paper; natural gas into plastic pellets, film, and sheets; and silica sand and soda ash into glass. These industrial materials are used as inputs in further manufacturing or construction activities.

Primary processing production activities can be grouped by the type of energy used. See **Figure 14-33**. This grouping includes the following processes:

- Mechanical processes.
- Thermal (heat) processes.
- Chemical and electrochemical processes.

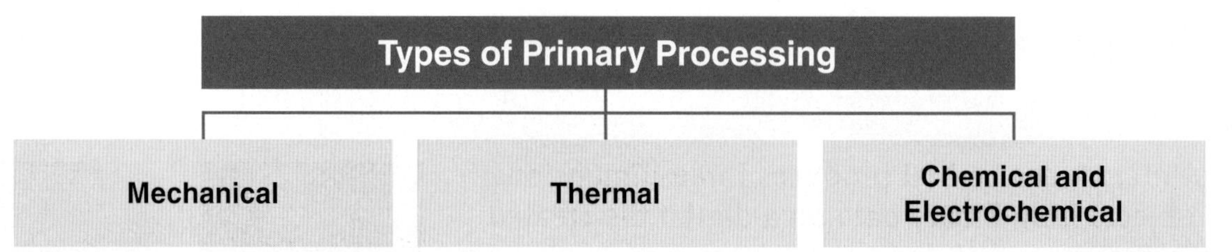

Goodheart-Willcox Publisher

Figure 14-33. The three types of primary processes are mechanical, thermal, and chemical and electrochemical.

More than one type of process may be involved in producing a material. For example, steel is made from iron ore, coke, and limestone using a thermal process. Some steel that is produced is formed into bars, rods, and sheets using mechanical processes. Some sheets of steel are coated with zinc to produce galvanized steel, which is resistant to rusting. The process for producing galvanized steel is an electrochemical process.

Mechanical Processes

Mechanical processes use mechanical forces to change the forms of natural resources. Compression (pressure) crushes the material to reduce its size or change its texture. Shearing forces cut and fracture the material. Material can be run over screens to sort it by size.

A number of natural resources are first processed by mechanical means. The production of forest products from trees is a common example. Wood can be cut or sheared into new shapes. These mechanical actions are used to produce lumber, plywood, and particleboard.

Producing Lumber

Trees were one of the first natural resources used by humans. Trees provided the raw material for shelters and crude tools, as well as fuel for cooking and heating. Even today, wood is a major energy source in developing countries around the world. In ancient Egypt, wood became a basic material for carpentry and boat-building. This use of wood, a natural resource, continues in modern civilization.

One widely used form of wood is lumber. A piece of ***lumber*** is a flat strip, or slab, of wood. Lumber is available in two types:

- ***Softwood lumber*** is produced from needle-bearing trees, such as pine, cedar, and fir. This lumber is used for construction purposes, for shipping containers and crates, and for railroad ties. Softwood lumber is produced in specific sizes called *nominal sizes*, which are used for the purpose of identification and not to describe the real size of the object. Typical sizes range from 1 × 4 (3/4″ × 3 1/2″ actual size) and the common 2 × 4 (1 1/2″ × 3 1/2″ actual size) to as large as 4 × 12 (3 1/2″ × 11 1/2″ actual size). Softwood lumber is available in standard lengths in 1′ increments. Generally, it is available in lengths from 6′ to 16′.
- ***Hardwood lumber*** is produced from deciduous (leaf-bearing) trees that lose their leaves at the end of each growing season. Examples are poplar, maple, oak, cherry, and walnut trees. Hardwoods are widely used for cabinetmaking and furniture making, for making shipping pallets, and for manufacturing household decorations and utensils. Hardwood lumber standard thicknesses range from 5/8″ thick to 1″ thick rough boards, known as 4/4 (pronounced "four-quarter") to as large as 4″ thick rough boards, or 16/4. The boards are available in random widths and lengths. Hardwoods are usually not cut to specific widths and lengths.

The largest amount of lumber is produced from softwood trees. See **Figure 14-34**. At the mill, logs are stored in ponds to prevent checking (cracking) and to protect them from insect damage. These logs are the material input for lumber manufacturing. See **Figure 14-35**. Changing logs from a natural resource into an industrial material involves the following steps:

1. The log is removed from the pond and cut to a standard length. This length is established to give the mill a uniform input and maximum yield from the log.
2. The log is debarked either with mechanical trimmers or high-pressure water jets. The bark is a by-product that can be used as fuel for the mill or sold as landscaping mulch.

Boise Cascade

Figure 14-34. Logs are shipped to the mill on trucks or floated down rivers. Lumber production starts when the logs arrive at the mill.

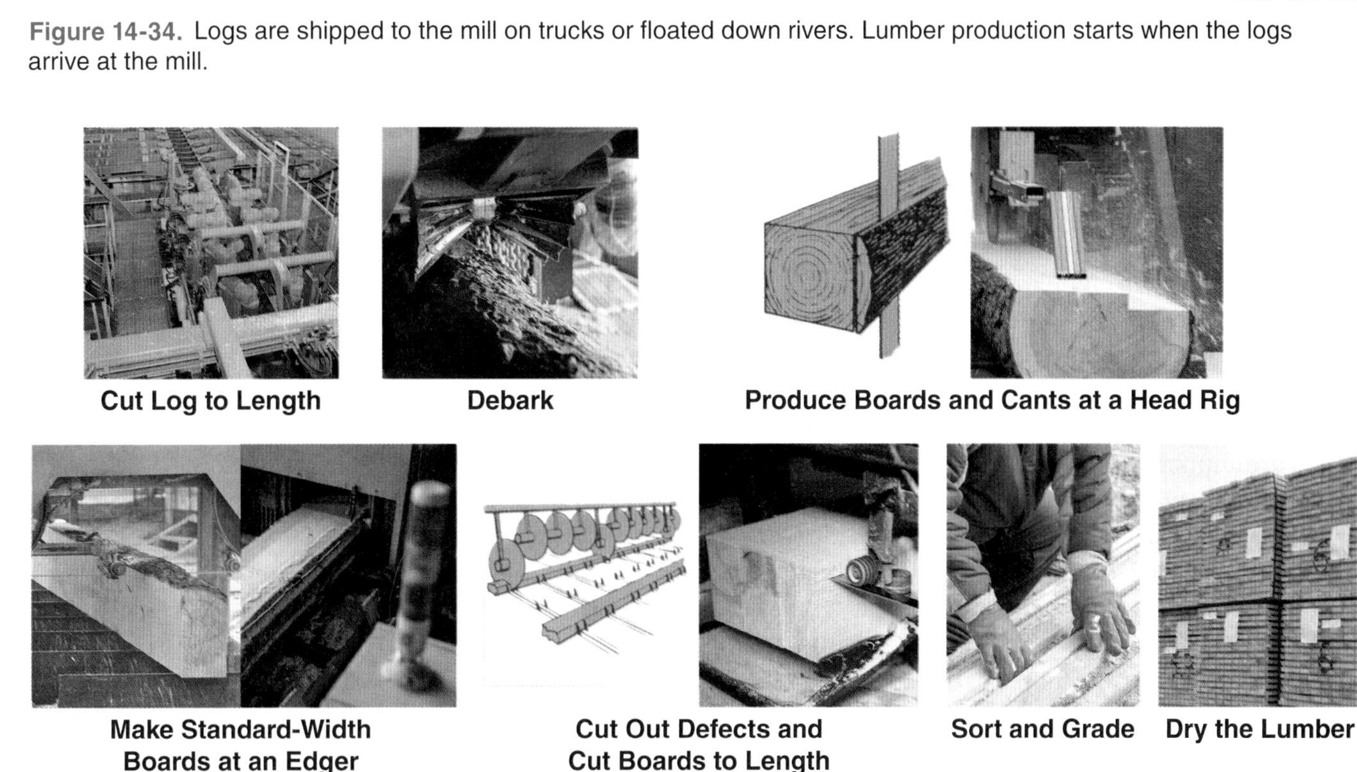

Clockwise from top left: Dalibor Sevaljevic/Shutterstock.com; Rustam Shigapov/Shutterstock.com; Goodheart-Willcox Publisher; genkur/Shutterstock.com; al7/Shutterstock.com; Guryanov Andrey/Shutterstock.com; Goodheart-Willcox Publisher; Ciprian Nasalean/Shutterstock.com; Olinchuk/Shutterstock.com; geno4ka01/Shutterstock.com

Figure 14-35. The steps followed to change logs into lumber.

3. The log is cut into boards and cants at the head rig. A head rig is a huge band saw that cuts narrow slabs from the log. When the square center section (called the *cant*) remains, the cant either remains at the head rig to be cut into thick boards or moves to the next step.
4. The cant is cut into thin boards at a resaw. This machine is a group of circular or scroll saw-type blades evenly spaced to cut many boards at once. Small logs often move directly from the debarker to the gang saw, bypassing the head rig.
5. The boards are cut to standard widths at an edger saw. This machine has a number of blades on a shaft. The blades can be adjusted at various locations to produce standard widths from 2″ to more than 12″.
6. The edged boards are cut to standard lengths at a trim saw. This machine has a series of blades spaced 2′ apart. The operator can actuate any or all the blades. This allows for cutting out defects and producing standard lengths of lumber. The boards can be 6′, 8′, 10′, 12′, 14′, or

16′ long. All the blades cut low-quality boards. This cuts the board into 2′-long scrap pieces. The scrap from the edgers and trim saws is used as fuel for the sawmill or becomes the raw material for board products and paper.

The processed lumber then moves onto the green chain, where the boards move down and are inspected and sorted by quality. The lumber is air or kiln (oven) dried to make it a more stable product. The dried lumber is shipped as an industrial material.

Some lumber receives special processing. See **Figure 14-36**. Short boards can be processed into longer boards using finger joints cut in the ends of the boards. The boards are then glued end to end to form a continuous ribbon of lumber. The ribbon is cut into standard lengths as it leaves a glue-curing machine.

The leftover wood scraps, fibers, and chips produced as waste throughout the lumber production process can be used to create other engineered wood products. ***Engineered wood products*** combine the leftover waste with adhesives to create composite materials (materials made from two or more different materials that, when combined, have different properties from the individual materials), such as oriented strand board (OSB), particleboard, fiberboard, or plywood. This is one example of how waste items are repurposed into useful products. In many cases, these engineered wood products are stronger than natural lumber. For example, glued laminated lumber and laminated veneer lumber are made by gluing several layers of wood components together to form larger pieces of lumber. These products are made stronger by alternating the direction of the wood grain of each layer during the gluing process.

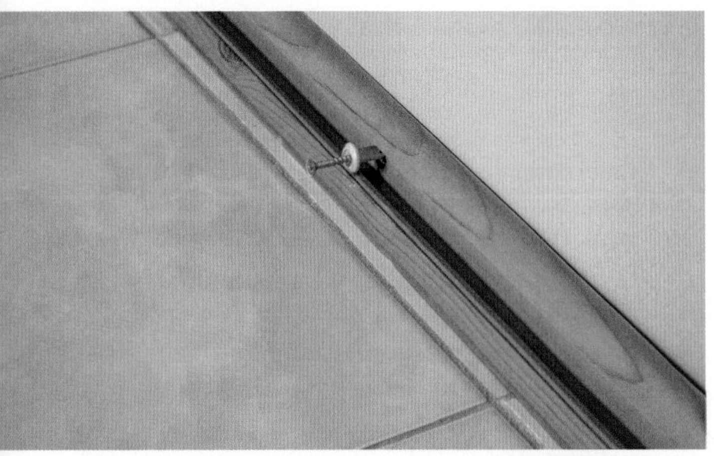

ESBuka/Shutterstock.com

Figure 14-36. Wood molding used in homes receives special processing.

Another common forest product is plywood. Plywood is a composite material made up of several layers of wood. See **Figure 14-37**. Plywood is more stable than solid lumber because cross-grained layers in the plywood reduce warping and expansion.

The outside layers are called *faces*. Between the faces are layers called *cross bands*. The grain of the cross bands is at a right angle (90°) to the face grain. The layer in the center is called the *core*. The core's grain is parallel with the face grain. Plywood with only three layers does not have cross bands. Three-layer plywood has a core with its grain running at a right angle to the face layers.

Three types of cores are used for plywood. See **Figure 14-38**. The most common is *veneer-core plywood*. A veneer is a thin sheet of wood sliced, sawed, or peeled from a log. Plywood used for cabinetwork and furniture usually has a lumber or a particleboard core. Lumber-core plywood has a core made from pieces of solid lumber that have been glued up to form a sheet. Particleboard-core plywood has a core made of particleboard. Particleboard is made up of wood chips that are glued together under heat

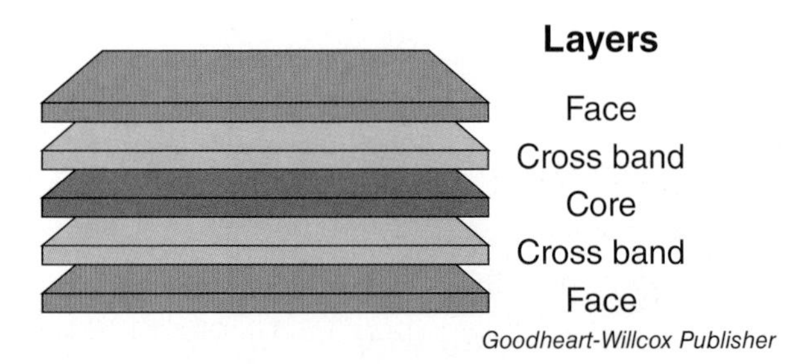

Goodheart-Willcox Publisher

Figure 14-37. Plywood is made of layers of wood glued together under pressure. The number and the thickness of the layers are varied to make different types of plywood.

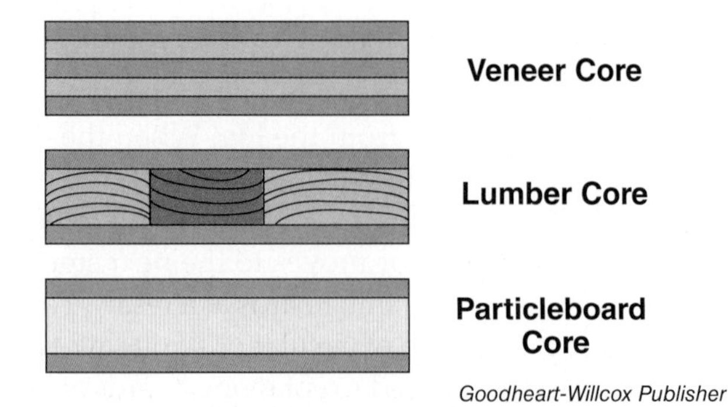

Goodheart-Willcox Publisher

Figure 14-38. Plywood is available with three types of cores—veneer, lumber, and particleboard.

and pressure. Plywood is typically available in 4′ × 8′ sheets. Thicknesses from 1/8″ to 3/4″ are available.

Veneer-core plywood is produced in two stages. The first stage makes the veneer. The veneer is sliced or peeled from the log and moves through a dryer. The dried veneer is sheared into workable-size pieces. Defects are cut out or are patched.

Now the veneer is ready for the second stage—plywood production. Glue is applied between the layers. The layers of veneer are stacked up. The sheet is then placed in a heated press. The press is closed, and the glue cures under heat and pressure.

After pressing, the sheet is removed, trimmed to size, and sanded. The completed sheets are inspected and loaded for shipment.

Thermal Processes

Thermal processes use heat to melt and reform a natural resource. Smelting is a thermal process widely used to extract metals from their ores. The extracted metals are often heated and pushed through a die to produce long continuous shapes for specific purposes in a process called ***extrusion***. This process often creates a long piece of metal with a uniform cross section, such as a hollow tube, round bar, U-channel, or L-channel. Extrusion can also be used for other materials, such as plastics or ceramics. Other thermal processes are used to make glass and cement.

Three common thermal processing activities are steelmaking, glassmaking, and petroleum refining. These use a combination of thermal and chemical processes. Thermal energy melts the materials. During the melting process, chemical reactions take place to produce a new material. The new material is then shaped into standard stock. The shaping process uses mechanical techniques. The material is cast, drawn, rolled, or squeezed into new sizes and shapes.

Chemical and Electrochemical Processes

Chemical processes and electrochemical processes break down or build up materials by changing their chemical compositions. Chemical and electrochemical processes are used to produce synthetic fibers, pharmaceuticals, plastics, and other valuable products. These processes are also used to chemically combine different materials, such as zinc and steel, to prevent corrosion or rusting.

STEM Connection: Science and Engineering

Synthetic Fuels

Asking questions and defining problems regarding global issues or concerns are common scientific and engineering practices. Scientists and engineers must ask questions or define problems in order to construct explanations and design solutions for those issues or concerns. They often analyze global concerns to develop ideas and parameters for a solution to meet societal needs and wants.

The earth's supply of fossil fuels is limited and will one day be exhausted. Scientists and other researchers have been working on creating synthetic fuels, also called *synfuels*. Synthetic fuels result from the actions of various chemical processes on such materials as coal and biomass (organic matter).

For example, coal can be turned into gas through a process called *gasification*. In this method, mined coal is combined with steam and oxygen and becomes a mixture of carbon monoxide, hydrogen, and methane. This mixture can then be used in place of natural gas. In another process, organic matter is fermented to produce ethyl alcohol. This product is then mixed with gasoline and becomes a fuel called *gasohol*.

Although synthetic fuels might eventually be the answer to our limited supply of fossil fuels, many drawbacks still exist. For example, the process of creating synfuels from coal requires a lot of coal, thus reducing the supply. Producing synfuels is also extremely costly at this time. Scientists continue to work on these and other problems associated with synfuels. Can you name a synfuel currently being investigated and a problem being addressed in the research?

Summary

- ✔ Early manufacturing involved skilled craft workers who produced handmade products one at a time.
- ✔ In the late 1700s, interchangeable, standardized parts were introduced, enabling unskilled workers to quickly produce more goods at a lower cost.
- ✔ In the early 1900s, Henry Ford developed the moving assembly line technique of mass production.
- ✔ Manufacturing strategies, such as just-in-time (JIT) manufacturing and lean manufacturing, have reduced manufacturing costs and minimized waste and inefficiencies.
- ✔ Computer-integrated manufacturing (CIM) is the approach of using computers to control the entire production process.
- ✔ Modern manufacturing makes use of robots, programmable logic controllers (PLCs), computer numerical control machines, automatic storage and retrieval systems, automated guided vehicles, and artificial intelligence (AI) systems.
- ✔ A typical industrial robot has five main parts—an arm, an end effector, a drive, a controller, and a sensor.
- ✔ Computer numerical control (CNC) is the automatic control of a machine by means of a computer program that controls the machine's motions.
- ✔ In a subtractive manufacturing process, material is removed from raw stock until a desired shape and size is reached. In an additive manufacturing process, successive layers of materials are placed to create a predesigned three-dimensional object.
- ✔ Primary manufacturing includes the resource processing systems of manufacturing in which raw materials are obtained and then transformed into industrial materials.
- ✔ Primary processing production activities include mechanical processes, thermal processes, and chemical and electrochemical processes.
- ✔ Mechanical processes are used to produce lumber, plywood, and particleboard.

Check Your Engineering IQ

Now that you have finished this chapter, see what you learned by taking the chapter posttest.
www.g-wlearning.com/technologyeducation/

Test Your Knowledge

Answer the following end-of-chapter questions using the information provided in this chapter.

1. The term _____ refers to production activities that create products.
2. The manufacturing industry was revolutionized in the early 1900s by the development of _____, which made use of an assembly line.
3. List the five main parts of a robot.
4. *True or False?* 3-D printers use an additive manufacturing approach.
5. Primary manufacturing converts natural resources into _____.
6. The three types of natural resources that are inputs to production systems are genetic materials, fossil fuel materials, and _____.
7. Genetic materials are obtained through farming, _____, and forestry.
8. Genetic resources must be _____ at the proper stage of growth.
9. *True or False?* Fossil fuels are mixtures of carbon and nitrogen.
10. List the three main fossil fuel resources.
11. Releasing trapped pockets of natural gas by pumping water underground to produce enough pressure to fracture shale rock formations is called _____.
    A. hydraulic fracturing
    B. fluid mining
    C. ground yarding
    D. core sampling
12. In _____ mining, miners dig a horizontal shaft directly into a coal vein that extends to the surface of the earth.
13. *True or False?* Minerals are substances with a specific chemical composition that occurs naturally.
14. List the three types of primary processes.
15. Which type of primary processing is used to produce lumber?

Critical Thinking

1. Choose a particular manufacturing industry and explain its importance to the nation's economy.
2. Write a report on a controversial resource-recovery issue that applies to your area of the country. The report can deal with a number of issues, including protecting the environment, wildlife, or the supply of the resource. State your opinion as to how you think the issue can best be resolved.

STEM Applications

1. Select a material not discussed in this chapter. Examples are paper, hardboard, nylon, and polyethylene. Research the processes used to produce the material you choose. Report your results in a chart similar to the one shown below.

Material:
Primary natural resource used to make the material:
Primary production process (steps):
Uses for the material:

2. Select a resource from the categories of genetic materials, fossil fuels, and minerals. Research how it is located (exploration) and recovered from the earth (extraction). Use a chart similar to the one below to record your findings.

Resource:
Exploration processes:
Extraction processes:

3. Build a scale model to show a selected mining or drilling technique.

Engineering Design Challenge

Lumber Processing System

Background

Some states in the United States are rich in natural resources. These states have traditionally located and extracted their natural resources and then shipped them to other states or countries to be processed into industrial materials. However, resource-rich states can generate more value by using primary manufacturing processes to create industrial materials locally. With advances in technology, primary-manufacturing processes can take place in the same area where the resources are obtained, enhancing the economy of the region.

The development of new systems and processes leads to an increased ability to produce industrial materials and consumer goods at a higher quality and faster rate. However, these new types of processes require a different workforce than traditionally used in the resource extraction industries, such as mining and drilling. These manufacturing industries require a workforce skilled in engineering and technology, with a strong knowledge of mathematics and science.

Situation

You have recently been hired as part of a team of manufacturing and mechatronics engineers by a rural logging company. This company is interested in developing their own process for transforming their products into more valuable lumber. The company has received grant funding to create an automated system to produce quality lumber materials for their local construction industry.

Your team has been tasked to design an automated factory to transform the raw material into finished lumber. The process should start with the raw material and move throughout the process of creating finished lumber with little human intervention. This process needs to be quick, efficient, and accurate. Your team will design and create a working model of the automated process to present to the president of the logging company.

Desired Outcomes

- A working model of a simulated automated lumber processing system in which the simulated raw logs can be manually loaded into the automated system.
- Simulation of the following processes using sensors, motors, and servos:
  - Cutting the logs to length.
  - Debarking and squaring the logs.
  - Resawing the logs into boards.
  - Rough sanding the boards.
  - Depositing the lumber into storage containers.
- A system that produces the most lumber in the least amount of time. Your team should record the amount of time it takes their system to process a log and use this data to calculate the efficiency of the system. The results should be used to improve their system.
- A five- to ten-minute presentation to the class that introduces the designed solution, outlines the production of the system, and identifies any economic, social, or environmental impacts of the work. The presentation should explain how the lumber production waste will be used to ensure that the team's system has minimal impact on the environment.

Materials and Equipment

- Robotics equipment, such as Lego® robotics, Vex Robotics, or FischerTechnik.
  - Conveyor belts.
  - Motors.
  - Servos.
  - Light sensors.
  - Ultrasonic sensors.
  - Limit switches.
- Simulated logs (Lincoln Logs®)

Procedure

- Determine your engineering design team based on the class's individual skill sets.
- Remember the engineering design process is an iterative approach to solving problems, which means your team can and should go back and forth between the different steps.
- As a team, formulate a problem statement based on the background and situation provided. A good problem statement should address a single problem and answer the questions *who*, *what*, *where*, *when*, and *why*. However, the problem statement should not imply a single solution to the problem. Your team may create different problem statements based on their own interpretations of the situation and prior interests and experiences. Use the defined problem statement as the starting point to creating a valid solution.
- Be sure to document your process by keeping an engineering notebook!
- Apply the knowledge your team has obtained and employ the engineering design process in order to propose new or improved solutions to the defined problem. Progress from brainstorming, to generating ideas, to producing prototypes/models, to evaluating success, and to iteratively refining the solution.

This activity develops the skills used in TSA's Computer-Integrated Manufacturing Production event.

TSA Modular Activity

This activity develops the skills used in TSA's Computer-Integrated Manufacturing Production event.

Computer Numerical Control (CNC) Production

Activity Overview

In this activity, you will design, fabricate, and use a CNC machine to produce parts needed to create the furthest traveling mousetrap car using a CNC machine.

Materials

Participants will use one mousetrap to power their vehicle. The vehicle itself will be fabricated from:

- A single 12″ × 12″ square of 1/4″ material.
- The material must be flat stock.
- Plastic, wood, or metal may be used, but the entire car must be cut from one type of material.

Background Information

- **General.** In this activity, you will use critical-thinking skills to select appropriate applications of CNC programming, select appropriate materials and fabrication techniques for your solution, and demonstrate the application of your solution.
- **CNC machined parts.** You will create the following parts:
  - Three or four wheels, as determined by the design.
  - Frame rails as needed.
  - Lever to increase the force applied to the drive wheel.
  - Other parts required by the design.

Guidelines

- Document the entire process for designing and producing the parts.
- Submit the CNC programs used to produce the mousetrap car parts.
- Assemble the mousetrap car using parts fabricated only with a CNC machine.
- Operate the mousetrap car to demonstrate the distance the car can travel.

Evaluation Criteria

Projects will be evaluated using the following criteria:

- Thorough documentation of the project design and CNC programming.
- Quality in the fabrication of the project components.
- The distance that the mousetrap car can travel.

AKK photo/Shutterstock.com

CNC machines are controlled by programs entered into the console.

CHAPTER 15

Producing Products

Check Your Engineering IQ

Before you read this chapter, assess your current understanding of the chapter content by taking the chapter pretest.
www.g-wlearning.com/technologyeducation/

Learning Objectives

After studying this chapter, you should be able to:

- ✔ Define *secondary manufacturing*.
- ✔ Describe the six different types of manufacturing processes.
- ✔ Compare and contrast different types of production systems.
- ✔ Describe the influence of social, economic, environmental, and manufacturability factors on product design and development.
- ✔ Explain the trade-offs that are made in the design of products.
- ✔ Describe procedures used for quality control.
- ✔ Explain what is meant by a *manufacturing enterprise*.

Technical Terms

batch production
bonding
casting and molding processes
chemical conditioning
conditioning processes
continuous production
customized production
durable products
elastic range
finishing processes
fixture
flowchart
forming processes
fracture point
heat treating
jig
machining
manufacturing enterprise
mechanical conditioning
mechanical fastening
nondurable products
plastic range
separating processes
shearing
thermal conditioning
yield point

Close Reading

Engineers create questions based on facts and data about what they are observing. As you read this chapter, develop a list of questions that you would need to answer if you were planning to design and produce a novel product.

Secondary Manufacturing

Primary manufacturing produces industrial materials that have little worth to the average person. How is a sheet of steel, a 2 × 4 stud, a pound of polypropylene pellets, or an ingot of pure aluminum useful in daily life? The term *secondary manufacturing processes* describes the actions used to change industrial materials into products. See **Figure 15-1**.

Product-manufacturing systems change industrial materials into products by causing them to acquire a desired shape and size. These changes occur through a variety of manufacturing processes, such as forming, separating, casting, and molding. The materials are then assembled into products by processes such as welding, fastening, or gluing.

The outputs of secondary manufacturing systems are either industrial products or consumer products. Industrial products are items that companies use to conduct business. Consumer products are developed for end users, such as homeowners, athletes, or students.

Types of Manufacturing Processes

Thousands of manufacturing processes are used to change the size and shape of materials, to fasten materials together, to give materials desired properties, and to coat the surfaces of products. See **Figure 15-2**. These secondary manufacturing processes can be classified into six groups:

- Casting and molding.
- Forming.
- Separating.
- Conditioning.
- Assembling.
- Finishing.

Each of these six processes has actions or concepts common to all the other processes in its group. However, within each group are specific techniques and actions that vary from others in the

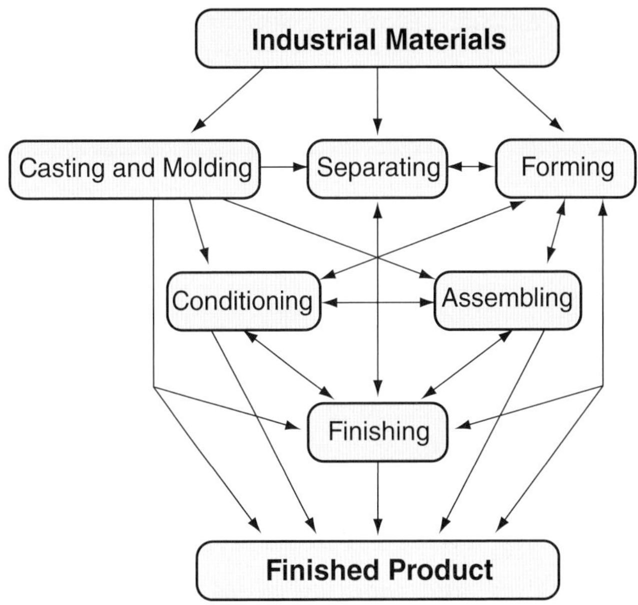

Goodheart-Willcox Publisher

Figure 15-2. The six groups of manufacturing processes.

Fedor Selivanov/Shutterstock.com

borisow/Shutterstock.com

Figure 15-1. Secondary manufacturing processes turn industrial materials such as lumber into useful products, such as chairs and tables.

Academic Connection: Explanatory Essay

Manufacturing and the Economy

Explanatory writing is a type of writing that requires you to examine and convey complex ideas and concepts clearly and accurately. Explanatory writing is a form of technical reporting that enables you to share important information to an invested audience. Constructing explanations is a common practice of scientists and engineers. As a scientist or engineer, you will need to construct, communicate, and revise an explanation often in response to questions. The explanation needs to be based on reliable and valid data collected from a variety of sources.

One task in explanatory writing involves a cause-and-effect relationship. This task is to show why certain conditions exist and to show the influence of one event on another. Using valid and reliable data, write an explanatory essay describing why manufacturing is central to a nation's economy. This essay should include an introduction that defines the purpose of the essay, a body including all of the supportive information for the writing prompt, and a conclusion.

same group. In other words, each group is similar to a family. They do many things alike and look alike, but each member is unique.

The six processes are described in the following sections. The first three—casting and molding, forming, and separating—give size and shape to pieces of material.

Casting and Molding Processes

Casting and molding processes give materials shape by introducing a liquid material into a mold. The mold has a cavity of the desired size and shape. Liquid material is poured or forced into the mold, where it is allowed to solidify before being removed. In this discussion, we refer to molten and fluid materials as *liquids*. The term *molten* refers to materials heated to a fluid state. These materials are normally solid at room temperature. Fluid materials, such as water and casting plastics, are liquid at room temperature.

All casting and molding processes involve five basic steps, **Figure 15-3**. These steps are as follows:

1. Producing a mold of the proper size and shape.
2. Preparing the material.

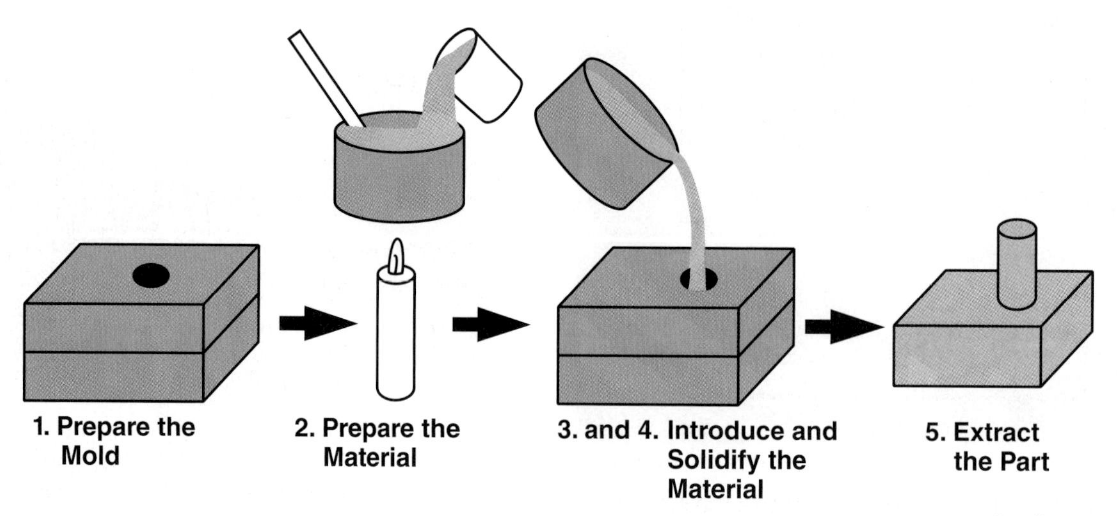

Goodheart-Willcox Publisher

Figure 15-3. All casting and molding processes produce parts and products using five basic steps.

3. Introducing the material into the mold.
4. Solidifying the material.
5. Extracting the product from the mold.

Forming Processes

All materials react to outside forces, **Figure 15-4**. ***Forming processes*** apply force through a forming device to cause the material to change shape. Small forces cause a material to flex (bend). When the force is removed, the material returns to its original shape. When the force is increased to the point at which the material does not return to its original shape, the material has reached its ***yield point***. The range between rest (no force applied) and the yield point is called the material's ***elastic range***.

In the forming process, the force applied is in a specific range that is above the material's yield point and below the material's ***fracture point*** (the point at which a material breaks due to the amount of force applied to it). Above the yield point, the material is permanently deformed. The greater the force applied, the more the material will be stretched, compressed, or bent. The range to which a material can be stretched, compressed, or bent without being destroyed is called the material's ***plastic range***. All forming processes operate in the plastic range of the material.

Separating Processes

Separating processes remove excess material to make an object of the correct size and shape. Separating processes are essential for subtractive manufacturing activities, which involve the removal of undesired materials to achieve a desired form. Whereas casting and forming processes change the shape and size of materials without any removal, separating processes remove material by either machining or shearing. See **Figure 15-5**.

Machining, which removes excess material in small pieces, is based on the motion of a tool against a workpiece to remove material. This process cuts material away using three methods—chip removal, flame cutting, and nontraditional machining.

Shearing uses opposing edges of blades, knives, or dies to fracture the unwanted material away from the work. Material is cut to create the desired size and shape. Shearing processes can be used to cut material to length, produce an external shape, or generate an internal feature. Cutting to length is generally accomplished with opposing blades. The upper blade moves downward to deform and fracture the material where it contacts the lower blade. Internal and external shapes are often made with a punch and a die. The die has a cavity of the desired shape. The punch fits into this cavity. The material is placed on the die. When the punch moves downward, the material is sheared into the shape of the die cavity. Punches and dies are used to produce holes, slots, and notches.

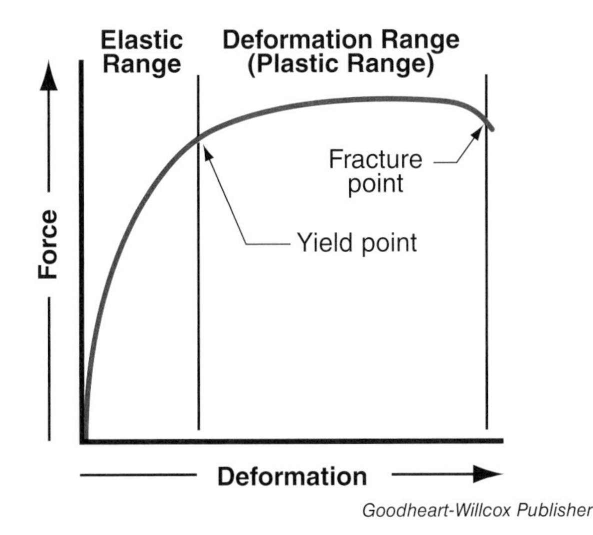

Goodheart-Willcox Publisher

Figure 15-4. This stress-strain chart shows the elastic and plastic ranges for a material. Note the yield and fracture points.

Conditioning Processes

To change the internal properties of a material, ***conditioning processes*** are used. The material might need to be harder, softer, stronger, or more

Types of Separating Processes

Machining
- Chip removal
- Flame cutting
- Nontraditional

Shearing

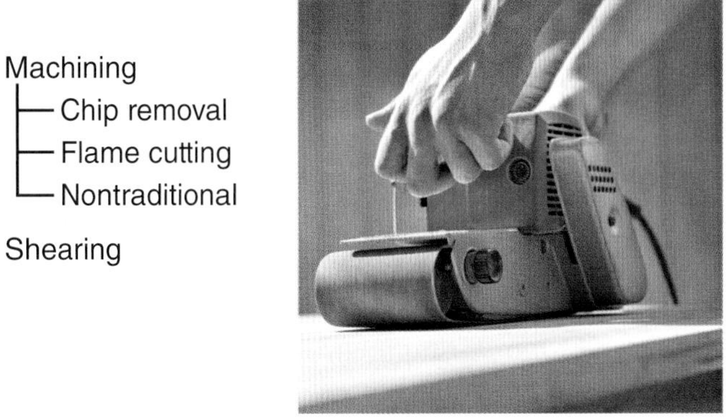

Goodheart Willcox Publisher; Fotokostic/Shutterstock.com

Figure 15-5. A sanding machine uses chip removal to remove unwanted material.

easily worked. The three types of conditioning processes are mechanical, chemical, and thermal.

Mechanical conditioning is the use of mechanical forces to change the internal structure of a material. For example, most metals become harder as they are squeezed, stretched, pounded, or bent. In ***chemical conditioning***, chemical actions change the properties of a material. For example, the lenses

STEM Connection: Science

Forming and Conditioning Materials

The science of the molecular structure of materials must be considered when materials are formed to produce a product. The molecular structure of a substance determines its physical properties. For example, a substance that is made up of long chains of molecules is usually flexible but durable. All materials have a temperature at which their molecular structure changes, which results in a change in their physical properties.

Metals can be formed cold or hot. Ceramics are formed cold. Plastics and glass are formed hot. When a metal is cold formed, internal stresses are built up and cause the material to become brittle. This is called *work hardening*. Work hardening is relieved by heat treating the metal. ***Heat treating*** involves heating the metal and allowing it to cool slowly.

Bethlehem Steel Co.

Figure A. These hot parts are leaving a heat-treating furnace. They will be quenched to harden them.

When a metal is hot formed, it must be heated above its recrystallization point. The recrystallization point is the temperature at which the grains of a material's structure realign into a new structure or state. Hot forming takes place above this point, while cold forming takes place below this point. Hot forming prevents work hardening of the metal. As the material cools, it forms a normal structure. The material shaped by hot forming is stress free. The minimum temperature for hot forming is different for each material.

Metals are conditioned for certain products using heat. *Heat treating* is a term used to describe thermal conditioning processes used on metals. These processes include the following:

- ***Hardening*** increases the hardness of a material. See **Figure A**. The part is heated to a specific temperature and allowed to soak. Soaking means that the metal is kept at the desired temperature until the part is evenly heated throughout. The part is then rapidly cooled in a tank of oil or water (quenched). Quenching reduces the crystallization of the internal structure of the metal that takes place when the metal is cooled slowly. This process increases the hardness of the metal.
- ***Annealing*** softens a part and removes internal stress. The part is heated to a specific temperature, allowed to soak, and then removed from the oven and cooled slowly to room temperature.
- ***Tempering*** relieves internal stress in a part. Hardening often creates internal stress, which causes a part to crack under use. Heating the part to a specific temperature and allowing it to slowly cool removes this stress. The tempering temperature is much lower than the annealing temperature. Tempering is used for many metal parts and most glass products.

of many safety glasses are chemically treated to make them shatterproof.

Thermal conditioning, the most common conditioning process, uses heat. Thermal conditioning processes include heat treating, firing, and drying. ***Heat treating*** is a term used to describe the thermal conditioning processes used on metals, in which the metal is heated and allowed to cool slowly. Firing is a thermal conditioning process used on ceramic products. Most ceramics are made from clay materials that are plastic when wet. After drying, the clay can be heated to a high temperature. The water is driven out of the clay particles, and the grains bond together to make a solid structure. Drying, a common thermal conditioning process, involves removal of excess moisture from materials. For example, ceramic materials and wood products must be dried before they are useful. Drying can be allowed to occur naturally or be helped by adding heat.

Assembling Processes

Look around you. How many products with one part do you see? You might notice such items as paper clips or straight pins. Most items, however, are assembled (fit together) from two or more parts. Through assembling processes, a simple product, such as a lead pencil, is created from five parts. The barrel is two pieces of wood glued around the graphite "lead." The eraser is held onto the barrel with a metal band. In fact, a lead pencil uses the two methods by which products are assembled—bonding and mechanical fastening.

Bonding holds plastic, metal, ceramic, and composite parts to each other by cohesive or adhesive forces. Cohesive forces hold the molecules of one material together. Adhesive forces occur between different kinds of molecules.

Mechanical fastening uses mechanical forces, such as friction, to hold parts together. For example, a part might be pressed or driven into a hole slightly smaller than the part. The friction between the parts causes the parts to remain together. This type of fit is called a *press fit*. Press fits can be used to hold bearings in place on a shaft. In other cases, the parts might be bent and interlocked to hold the parts together. This type of joint is called a *seam*. Many sheet metal parts are held together using seams. The most widely used method to hold parts together is mechanical fasteners, **Figure 15-6**. Examples of fasteners are staples, rivets, screws, nails, pins, bolts, and nuts.

Finishing Processes

Finishing processes, the last of the secondary processes most products go through, are techniques that protect products and enhance their appearance. Some finishing processes change the surface of the product, while others involve the application of a coating.

Most metals corrode if they are not protected in some way. Metals are easy to protect by changing the surface chemically. For example, anodizing converts the surface of aluminum products to aluminum oxide. This type of finish is called a *converted surface finish*.

Coating involves applying a film of finishing material, such as paints or varnishes, to a product or base material. Coatings protect the surface and can add color, **Figure 15-7**. Finishing processes involve cleaning the surface, selecting the finish, and applying the finish.

Product Production

Materials require a variety of secondary manufacturing processes to produce industrial and consumer products. These products are produced through different types of manufacturing systems

Paul Vinten/Shutterstock.com

Figure 15-6. The lug nuts on this wheel are mechanical fasteners.

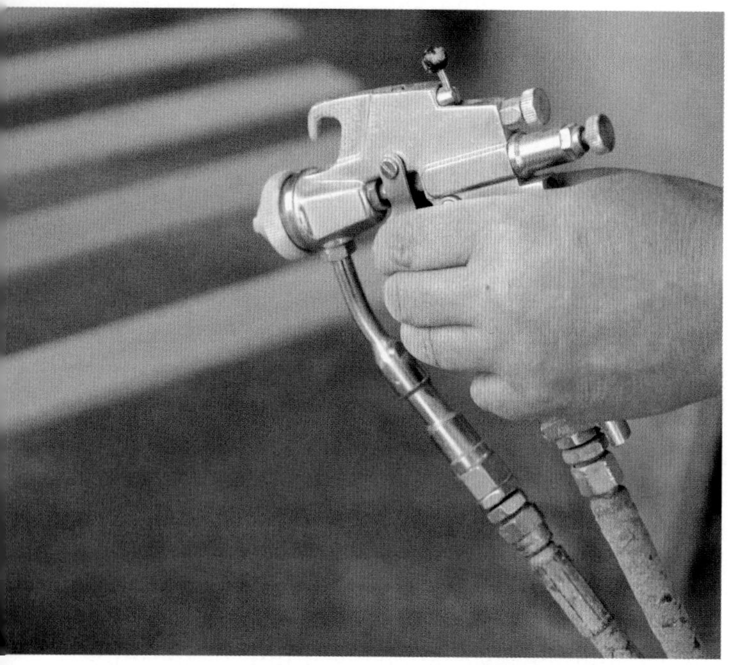

©iStockphoto.com/OwenPrice

Figure 15-7. Paint is being applied to a metal part using a hand-operated spray gun.

ChristinLola/Thinkstock.com

Figure 15-8. A worker is building a customized guitar.

that are classified as customized production, continuous production, and batch production.

Customized production systems are used to individually produce unique products that are tailored to meet the specific needs of the customer. Custom products are often more expensive and valuable than mass-produced products because more time and resources are needed to ensure the products meet the customer specifications. Additionally, these products are usually held to higher quality standards because they are produced with a close relationship to the end user. See **Figure 15-8**. ***Continuous production*** systems, on the other hand, are used to produce a large quantity of similar products in the least amount of time possible without interruption. See **Figure 15-9**.

Batch production systems manufacture products in groups or batches instead of individually or continuously. This type of system is necessary for manufacturing companies that want to produce many similar products with different variants. For example, a shoe manufacturer may want to produce the same model of shoe but in various batches based on shoe size. See **Figure 15-10**.

Due to the development of innovative manufacturing technologies, a new type of production system has emerged. Mass customization production systems use computer-integrated systems and rapid prototyping technologies to combine the flexibility of individual customization with the low costs of continuous production.

juananbarros/Thinkstock.com

Figure 15-9. A continuous production system that produces a large quantity of olive oil.

Think Green Product Life Cycle and Reduction

In the manufacturing industry, *product life cycle* refers to the stages a product goes through from its design through its production to its eventual disposal. Examining this entire process helps manufacturers determine ways to increase the efficiency of the production while minimizing wastes that are costly and harmful to the environment. Finding ways to minimize waste throughout a product's life cycle is called *reduction*. One example of reduction is when manufacturers change the size and shape of a product design to use mostly materials in their standard sizes. This practice reduces the processing of the materials, minimizing waste and reducing waste disposal costs. For instance, a standard sheet of plywood is typically 4′ × 8′. Therefore, if a product is designed to use 2 full sheets of plywood in its production versus using only partial sheets, the amount of energy for separating or cutting the sheets is minimized and the amount of waste to be disposed of is reduced.

Now think about ways you can reduce waste in common activities. Examples are making two-sided photocopies, using cloth instead of paper items, and turning off lights when you are not using them. What other examples can you think of?

You may be surprised at how small adjustments in your everyday life can help the environment.

Guryanov Andrey/Shutterstock.com

Figure 15-10. A batch production system used for producing various types of leather boots.

Product Design and Development

The engineering design process is used to devise and refine a product that is marketable and profitable. This process is also used to continuously improve products and to drive change in a product over time to better meet the ever-shifting demands of customers. Think about how the engineering design process has been responsible for the evolution of products over the years, such as the telephone. Telephones have evolved from the original landline-based apparatus for transmitting sounds telegraphically to today's smartphones. This is largely the effect of the continual evolution of technology.

A variety of concerns prompt changes and improvements to products over time, including customer, societal, economic, and environmental concerns. For example, customers may demand that a product be more user-friendly or have enhanced performance. Product designers employ the engineering design process to develop new products or improve existing products in order to address these concerns. See **Figure 15-11**.

ESB Professional/Shutterstock.com

Figure 15-11. A product designer contemplating design factors.

Socioeconomic Design Considerations

Product designers and manufacturers make product development decisions that are influenced by societal shifts and trends. Societal trends such as increased interest in healthy living, changes in fashions, environmental concerns, and increased mobility have influenced product development and have led to the development of such items as fitness trackers and smart watches. Product developers who obtain feedback from target customers and adapt their products to anticipate trends that are more likely to produce more useful and desirable products.

Product development is also greatly influenced by economic shifts related to the specific industry. For example, the price of crude oil is driven by the supply and demand of the world market. If the demand for crude oil rises or if the supply of crude oil is lessened by natural disasters or political

STEM Connection: Mathematics

Consumer Data

Product design and development decisions involve prediction based on consumer data. Mathematics in the form of statistics, which is the study of the collection, analysis, and interpretation of data, provides a tool for examining consumer data to make informed decisions about product development.

Data is collected and analyzed to interpret patterns or changes in the customer base so manufacturers can adapt to consumer trends. Consumer data is collected through the examination of sales reports and market research techniques, such as sample surveys. Using statistical experiments, the product designer can make inferences about sales projections and which products will be the most successful. See **Figure A**.

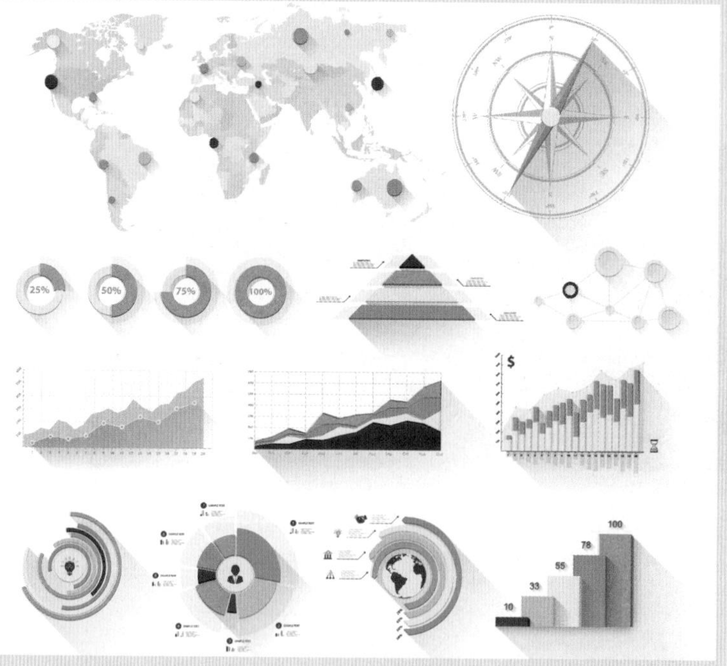

vasabii/Shutterstock.com

Figure A. Consumer data to be used for product development.

events, the price of crude oil increases. If the price continues to increase, petroleum-based products or products reliant on petroleum will be affected by this economic shift. As a result, product developers may look to cheaper alternatives to crude oil for petroleum-based products. Products dependent on crude oil or gasoline, such as the automobile, may be redesigned to rely on a totally different resource. Examples are cars that are powered by electricity, solar power, or natural gas. The cost of resources is a major factor in product design and development.

Just as product design and development are impacted by societal and economic factors, product production impacts the society and economy. Some of these impacts can be intended or unintended.

The automobile is an example of how product development can drive societal and economic change. The introduction of the automobile, **Figure 15-12**, sparked tremendous amounts of intended and unintended societal and economic change. The automobile was intended to solve the problem of getting people and cargo from point A to point B in less time. Because people were able to get from one place to another faster, they were then able to travel farther to work. They were then able to live outside of major cities and commute to work. As a result, a suburban infrastructure was created. As the automobile became a requirement for more individuals, the demand for fuel increased. This increased demand impacted the economy of countries all over the world.

Everett Historical/Shutterstock.com

Figure 15-12. The introduction of the automobile had a great impact on society.

The increased use of the automobile even changed home design. For example, Frank Lloyd Wright, a famous architect, recognized that the expanded use of vehicles would result in people walking less on sidewalks, as well as increased amounts of traffic on roadways. In response, construction of homes was planned farther away from roadways. Not only did automobiles impact the construction of new homes, they also changed the world's economy as they created a market for other industries, such as fast food restaurants.

Environmental Design Considerations

Product designers and manufacturers must understand environmental impacts when developing products. The obtaining and conversion of natural resources into industrial and consumer products can take a major toll on the environment. For example, building a hydroelectric dam to generate electricity can alter the ecosystem for aquatic life; extraction of natural gas from subterranean rock formations could potentially contaminate water sources; wind turbine farms could impact the migration routes of birds; and burning of fossil fuels such as coal can contribute to increased levels of greenhouse gases in the atmosphere. See **Figure 15-13**. Therefore, it is important that product designers and manufacturers minimize negative environmental impacts to help safeguard the health of people and wildlife.

Because product design and development impact the environment, the governments of certain countries establish and enforce regulations that define how materials are obtained and used in product development and how manufacturing processes should be conducted to minimize negative impacts. Additionally, government agencies may even establish tax benefits to manufacturers that are more environmentally friendly.

The Starbucks® Coffee Company's introduction of the EarthSleeve™ is an example of how environmental factors influence product design and development.

John Carnemolla/Shutterstock.com

Figure 15-13. A mine used to extract and obtain uranium.

The EarthSleeve™ was a new version of their hot-cup sleeve, which protects the customer's hands from the hot temperature of the beverage inside. See **Figure 15-14**. The sleeve was redesigned to require fewer natural materials to make while also increasing the amount of material that can be recycled.

Goodheart-Willcox Publisher

Figure 15-14. A hot-cup sleeve is a product designed to protect a customer's hands from the hot temperature of a beverage.

Manufacturability Design Considerations

The design of products is affected by factors that are specific to the ability to effectively manufacture the product. Materials used to make a product, the production process, and the product itself must all be safe.

In addition, the proper materials should be selected to ensure that the product could be produced at a cost that enables the product to be sold at a competitive price while making a profit. The cost of the materials, as well as the cost of shipping and transporting the materials, must be considered. The designer and manufacturer must understand the mass properties of the materials used in production. Mass properties such as the weight, volume, density, and surface area determine needed storage space, transportation costs, the machines needed to process the materials, and the cost of finishing the materials.

Materials should also be selected to guarantee that a product will be functional for a determined amount of time. Products may be designed to be durable or nondurable. ***Durable products*** are designed to operate for a long period of time. Examples include household appliances, cars, and furniture. ***Nondurable products*** are designed to be used only for a short period of time. These products include plastic cups, paper products, cosmetics, and food.

Some products are produced with a built-in obsolescence, which means the manufacturer has predetermined that the product will be out-of-date or useless within a known period of time. These types of products include lightbulbs, computers, and cell phones. These products are designed this way to require customers to purchase the product more than once.

Designing for manufacturability of a product also involves trade-offs. A trade-off is a situation where an action diminishes one aspect of something in order to enhance another aspect. Trade-offs in the design of products generally involve the quality of the materials used for the product versus the consumer cost of the product. For example, a higher quality and more expensive material could be used to make a product and improve its functionality or durability. However, doing so would increase the cost of the final product. The increased cost may then make the product unaffordable to many of the targeted costumers. Therefore, a trade-off needs to be made and in this instance, it may be using a lower quality material that reduces the durability of the product but maintains its affordability.

Career Connection Manufacturing Engineer

Monkey Business Images/Shutterstock.com

Manufacturing engineering is a field that involves planning, designing, and optimizing the processes used to transform raw materials into finished products. Manufacturing engineers work in a variety of areas, including food and medicine production. Professionals in this field use creative and analytical thinking and problem-solving skills to improve production processes and ultimately enhance industry profits. These engineers also strive to minimize the negative impacts of manufacturing on the environment.

High school students interested in manufacturing engineering careers should take a variety of STEM courses, specifically technology courses focused on computer-integrated manufacturing (CIM) and digital electronics. Mathematics courses related to trigonometry and calculus and advanced science courses in physics are also important.

Many postsecondary institutions currently offer degrees in industrial or manufacturing engineering. Programs typically include multiple levels of calculus and physics as the foundation. Courses in thermal dynamics, material properties, electric circuit analysis, fluid mechanics, and automated control systems are also included.

Individuals pursuing a degree in manufacturing engineering should have a solid foundation in mathematics and science. They should have a multidisciplinary skill set, good problem-solving skills, the ability to work in a collaborative environment, experience in robotics and automation, and the ability to perform cost-benefit analyses.

Manufacturing engineers work in a variety of production industries. They also work in the areas of quality control and CIM. Additionally, manufacturing engineers can become plant or operations managers.

Manufacturing engineers are in increasingly high demand. Factors that affect demand include the increasing globalization of production and the efforts of the United States to revitalize the country's manufacturing industry.

A mechanical engineer may work on the following types of projects:

- Streamlining an industry's production process.
- Improving the quality of finished products.
- Designing systems that have a minimal impact on the environment.
- Developing new technologies and products to extend automation.

There are a variety of professional organizations for each engineering specialization. The Society of Manufacturing Engineers is a professional organization that specifically works to advance manufacturing technology and produce a skilled manufacturing workforce.

Trade-offs also involve the choice of materials or production processes versus the impact that these materials and processes may have on the environment or human health and safety. Trade-offs require manufacturers to make ethical and moral decisions, such as determining certain safety features of an automobile based on production costs. Furthermore, consumers must make trade-off decisions when they are buying these products. For example, a consumer may choose to purchase a more affordable automobile with fewer safety features over an expensive vehicle with all available safety features.

Designing the Production Process

Product development involves designing the production process as well as the product. The secondary manufacturing processes required to produce the product and the facilities needed to do so must be identified.

Designing the manufacturing process involves planning all of the procedures involved with transforming the materials into an end product. A flowchart is often used for this purpose. A ***flowchart***, **Figure 15-15**, is a diagram that uses symbols to represent a sequence of actions or operations of a complex system such as a manufacturing facility. A flowchart can be used to help determine the most efficient layout of the facility work floor, which includes the machines, materials, and equipment needed to produce the products. The production process and facilities should be continually refined to minimize the waste of resources, which increases production efficiency and enhances profits.

Quality Control

Quality control is the process by which manufacturers check that their products are being produced to meet the desired specifications. See **Figure 15-16**.

FLOW PROCESS CHART

Product:	Desk Clock	Part Name:	Leg	Part No.:	2
Prepared by:	Employee	Position:	Technician		
Approved by:		Position:	Supervisor		
Flow Begins:	2.1	Flow Ends:	2.17	Date:	4/3/16

Process symbols and number used: ● Operations 10; ■ Inspections 1; ➡ Transportations 4; D Delays 0; ▽ Storage 1

Task No.	Process Symbols	Description of Task	Machine Required	Tooling Required
1	● □ ⇨ D ▽	Crosscut Better End	Miter Saw	
2	● ■ ⇨ D ▽	Crosscut to Finish Length of 8 ½" and Inspect Length	Miter Saw	Inspection Gauge
3	○ □ ➡ D ▽	Transport to Table Saw		
4	● □ ⇨ D ▽	Cut Top Grooves ¼"x3/4"	Table Saw	
5	● □ ⇨ D ▽	Cut Bottom Grooves ¼" x 1"	Table Saw	
6	○ □ ➡ D ▽	Transport to Router Table		
7	● □ ⇨ D ▽	Cut Clock Face Groove 1/8"	Router	
8	● □ ⇨ D ▽	Cut Glass Panel Groove 1/8"	Router	
9	● □ ⇨ D ▽	Cut Back Panel Groove 1/8"	Router	
10	○ □ ➡ D ▽	Transport to Wirth Machine		
11	● □ ⇨ D ▽	Drill ¼" holes (2) for Pins	Wirth Machine	
12	○ □ ➡ D ▽	Transport to Work Tables		
13	● □ ⇨ D ▽	Sanding	Orbital Disc Sander	
14	● □ ⇨ D ▽	Finishing		
15	○ □ ⇨ D ▼	Storage till Assembly		

Goodheart-Willcox Publisher

Figure 15-15. Manufacturing flowchart for one part of a desk clock.

bikeriderlondon/Shutterstock.com

Figure 15-16. A quality control specialist visually inspecting the quality of products.

Quality control is achieved in several ways. First, fixtures or jigs can be used to help ensure that secondary manufacturing processes, such as separating, assembling, and finishing, are conducted in a precise manner. A ***fixture*** is a device that secures a workpiece to a machine while machining operations are performed in order to ensure they are performed in the correct location each time. A ***jig*** is a device that guides the machines or tools performing the operations to help ensure the operations are continuously repeated in the same manner to produce the same results each time. See **Figure 15-17**.

Additionally, the quality of a workpiece can be examined at quality control checkpoints. Go/no-go gauges can be used at these checkpoints to determine whether a product meets the desired specification. If the product meets the specification, then it is a *go* and can move to the next step of the production process. If the product does not meet the specification, then it is a *no go*. This means it must be removed from production or modified to meet the specifications. These go/no-go gauges are designed to check specific qualities such as the size and shape of the product at different stages throughout the manufacturing process.

Quality control procedures can be simple or complex. Some manufacturers use hand-operated machinery and visually inspect every product for quality. Other manufacturers use precision equipment, such as CNC machines, cameras, and sensors in computer-integrated manufacturing systems, to

Goodheart-Willcox Publisher

Figure 15-17. Fixtures and jigs are devices that help ensure the quality production of products. This jig is used to cut the board to the precise length each time.

automatically detect errors in products. Using control systems, cameras, and sensors in combination with various machines in the manufacturing process enables a computer system to inspect each product visually for quality and to detect errors. The ability for production systems to visually inspect, or "to see," using imaging and analytic technologies can be called *machine vision*.

Manufacturing Enterprises

Designing products requires numerous iterations of refinements and critical decisions about design trade-offs. Additionally, production requires a precise and extensive coordination of efforts and resources. Many individuals communicating and working together as a team are needed to develop and enact the proper procedures to produce quality products. A ***manufacturing enterprise*** is an organization that contains all of the components needed to operate and maintain a facility to convert materials into finished products with the goal of generating a profit.

Therefore, the enterprise encompasses both the making and selling of products for a profit. A customer-centered approach to a manufacturing enterprise provides a direction for all production activities and helps establish a role for all of the team members within the manufacturing organization. These roles include management, research and development, production, quality control, marketing and sales, purchasing, shipping and receiving, customer support, and maintenance.

Summary

- ✔ Secondary manufacturing involves the actions used to change industrial materials into products.
- ✔ Casting and molding processes give materials shape by introducing a liquid material into a mold.
- ✔ Forming processes apply force through a forming device to cause the material to change shape.
- ✔ Separating processes remove excess material to make an object of the correct size and shape. Machining and shearing are two types of separating processes.
- ✔ Conditioning processes, which change the internal properties of a material, include mechanical and chemical conditioning.
- ✔ Assembling processes, which fit parts together, include bonding and mechanical fastening.
- ✔ Finishing processes protect products and enhance their appearance. Some change the surface of the product, while others involve the application of a coating.
- ✔ Secondary manufacturing processes include customized production, continuous production, batch production, and mass customization.
- ✔ Customer, societal, economic, and environmental concerns influence product design and development. Materials and processes needed to manufacture the product are also considerations.
- ✔ Manufacturers use quality control systems to check that their products are being produced to meet the desired specifications.
- ✔ Quality control is the process by which manufacturers check that their products are being produced to meet the desired specifications.
- ✔ Trade-offs in the design of products include the quality of the materials used for the product versus the consumer cost of the product and the choice of materials or production processes versus their impact on the environment or human health and safety.
- ✔ A manufacturing enterprise is an organization that contains all of the components needed to operate and maintain a facility to convert materials into finished products with the goal of generating a profit.

Check Your Engineering IQ

Now that you have finished this chapter, see what you learned by taking the chapter posttest.
www.g-wlearning.com/technologyeducation/

Test Your Knowledge

Answer the following end-of-chapter questions using the information provided in this chapter.

1. Secondary manufacturing processes change industrial materials into _____.
2. Casting and molding processes give materials shape by introducing a(n) _____ material into a mold.
3. In forming processes, the force applied is in a range that is _____.
   A. above the material's plastic range and below the material's yield point
   B. above the material's yield point and below the material's elastic range
   C. above the material's yield point and below the material's fracture point
   D. above the material's fracture point and below the material's elastic range
4. The two kinds of separating processes are machining and _____.
5. What are the three types of conditioning processes?
6. Bonding holds parts together by cohesive or _____ forces.
7. List three examples of mechanical fasteners.
8. Finishing processes change the surface of the product or involve the application of a(n) _____.
9. _____ production systems produce a large quantity of similar products in the least amount of time possible without interruption.
   A. Batch
   B. Continuous
   C. Customized
   D. Primary
10. *True or False?* Product development decisions are influenced by societal shifts and trends.
11. List four properties of production materials that affect storage and transportation costs.

12. *True or False?* Product development includes designing the product and the production process.
13. An example of a(n) _____ in product development is the choice of materials or production processes versus the impact that these materials and processes may have on the environment.
14. _____ gauges can be placed at quality control checkpoints to determine whether a product meets the desired specification.
15. What is a manufacturing enterprise?

Critical Thinking

1. Choose a manufacturing industry and write a report that analyzes the impact that industry has on the environment.
2. Choose any type of product and describe the potential trade-offs the product designer had to consider when developing the product.
3. How is the engineering design process used in product development?

STEM Applications

1. Select a simple product made from more than one part and do the following:
   A. List the parts the product is made of.
   B. Select one part. List the steps you think were used to manufacture it.
   C. Complete a form similar to the following for one of the steps.

Product:
Part name:
Production process:
Step needed to complete the process:

2. Investigate the concept of market research to explore the development of market surveys. Create a survey to determine consumer preferences for a specific product and collect responses to the survey from your school. Research how to analyze the results of the survey by calculating the mean, median, and mode of the collected data. Present the results to your class.
3. Build a simple product, such as a kite, using secondary manufacturing processes. List each step and the type of process used.
4. Select a product and build a go/no-go gauge that you would use for quality control in the production of that product.

Engineering Design Challenge

Manufacturing Enterprise

Background

Manufacturing systems encompass all of the processes involved in transforming raw materials into more valuable industrial materials or consumer products. To operate a manufacturing enterprise, many individuals work collaboratively to design, develop, make, and service products. These enterprises use different types of production systems to meet their specific needs of producing something of value to a consumer.

Advancements in design and manufacturing technologies have generated new approaches to product development. One of these approaches is *mass customization*. Mass customization production systems combine the flexibility of individual product customization with the low costs of continuous production. This type of production process enables customers to specify their own product design features. They can purchase the exact product they want at a reasonable price.

Situation

Fashion trends affect what clothes people wear, the shoes they buy, and the backpacks they use. Because of these trends, there are always new products being developed and produced. You have decided to take advantage of changing trends and mass customization production strategies to develop a product and establish a manufacturing enterprise. Therefore, you have challenged yourself to design and produce a new product based on a current fashion trend while applying a mass customization approach. You must apply creativity to develop a product that customers can tailor to their own specific needs and wants.

Desired Outcomes

- A customizable product.
- Manufacturing enterprise business plan.
- Product production flowchart and procedure.

Procedure

- Conduct market research in relationship to your product.
- Design your product using reliable and valid consumer data.
- Develop a business plan for selling your product.
- Establish a procedure for producing the basic product.
- Determine points in the process in which different product features can be adapted to meet the specific needs of your customers.
- Create a flowchart to graphically represent production procedures.
- Assign individuals to specific roles within the manufacturing enterprise.
- Establish a customer base and collect their individual criteria for your customizable product.
- Produce your product based on the criteria provided by your customers.

panpote/Shutterstock.com

These sneakers are mass produced in a garment production factory.

CHAPTER 16

Meeting Needs through Materials Science and Engineering

Check Your Engineering IQ

Before you read this chapter, assess your current understanding of the chapter content by taking the chapter pretest. www.g-wlearning.com/technologyeducation/

Learning Objectives

After studying this chapter, you should be able to:

- ✔ Summarize the evolution of materials science.
- ✔ Identify the methods of classifying materials.
- ✔ Describe the atomic structure of materials.
- ✔ Identify the different types of material properties.
- ✔ Identify the classifications of materials based on their atomic structure and characteristics.
- ✔ Summarize the role of materials scientists and engineers.
- ✔ Describe innovative methods used by materials scientists and engineers to create materials.

Technical Terms

acoustical properties
advanced materials
alloy
atomic number
biomimicry
ceramics
chemical bonding
chemical properties
composites
conductor
ductility
elastomers
electrical and magnetic properties
electron orbitals
elements
flammability
hardness
heat of combustion
insulator
ion
materials science and engineering
mechanical properties
metals
molecules
nanotechnology
optical properties
physical properties
polymers
semiconductors
smart materials
smelting
tensile strength
thermal conductivity
thermal fatigue
thermal properties
thermoplastics
thermosets
valence electrons

Close Reading

As you read this chapter, think critically about the differences that exist between materials. Look for patterns among the different classes of materials while visualizing the characteristics that distinguish them from one another. Make predictions about future improvements in materials and the developments needed to modify and create new materials.

In Chapters 14 and 15, you read about primary and secondary processing of materials. Both processes manipulate materials at the macro, or observable, level. The study of solid materials at the atomic level is called ***materials science and engineering***. This knowledge informs the design of new products and the selection of manufacturing systems.

Materials Science and Engineering

Materials science and engineering classifies solid materials by their atomic and molecular characteristics and differences in properties. Physical, chemical, thermal, electrical, and magnetic properties are used to distinguish materials from one another. These properties assist scientists and engineers in determining the suitability of a material for a particular application.

While once primarily a scientific field, materials science and engineering is now an interdisciplinary field, which means it requires collaboration across disciplines. Discoveries and new knowledge through the scientific method inform and drive the engineering design process. Technological advancements have closely aligned these two fields. Knowledge of the science and properties of materials is necessary in order to recognize why and how certain materials are used.

Evolution of Materials Science

Although materials are now processed with knowledge of their atomic and molecular structure, materials were once understood and created through a trial-and-error approach. The modification of materials dates back thousands of years. Often, time periods are identified by the materials most commonly manufactured. The primary time periods include the following:

1. During the Stone Age, prior to 10,000 BCE and through 4000 BCE, humans manipulated natural materials such as rocks, animal bones and skins, clay sediment, and glass. They created tools by shaping materials. For example, stone was chipped away to make cutting tools, **Figure 16-1**. Such tools were used to create items that were needed to survive.
2. The Bronze Age ended the Stone Age with the introduction of ***smelting***, a process of extracting metal from an ore through heating and cooling. This method resulted in the creation of bronze from the ore *cassiterite*, a mineral of tin oxide, **Figure 16-2**. From approximately 4000 BCE to 1000 BCE, bronze was produced through smelting. Bronze is

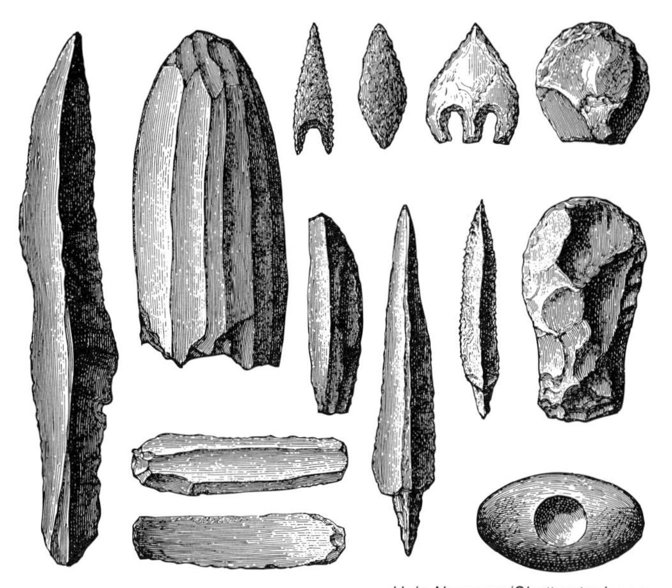

Hein Nouwens/Shutterstock.com

Figure 16-1. Early cutting tools chipped away from stone.

andreevarf/Shutterstock.com

Figure 16-2. A piece of cassiterite mined from the earth.

John A Davis/Shutterstock.com

Figure 16-3. Casting process comparable to what would have occurred during the Bronze Age.

an ***alloy*** (a metal comprised of two or more metallic elements) of tin and copper. The alloy was stronger than the two materials alone. Tools were created during the Bronze Age through alloyed casting and forging. See **Figure 16-3**. These tools helped humans to hunt, defend themselves, and farm.

3. The Iron and Steel Age began around 1000 BCE, marking the increased use of iron. Through iron smelting, metal was obtained by extracting iron from ores. This extraction required high temperatures and furnaces able to withstand the heat. This period contrasted with the Bronze Age in that blacksmiths increasingly shaped materials instead of using casting methods. See **Figure 16-4**. The process of creating materials during this time (often smelting) produced variations of iron and steel. The technology available at the time, such as bloomery furnaces, made it difficult to produce steel. See **Figure 16-5**. It was much easier and more economical to produce wrought iron.
4. The Nonferrous and Polymer Age is traced back to 1907 with the introduction of the first synthetic plastic, Bakelite. See **Figure 16-6**. The word *ferrous* is derived from the Latin word *ferrum*, meaning *iron*. The Nonferrous and Polymer Age represented a time when materials that did not contain iron were being introduced. This time period sparked a scientific laboratory focus for materials science. Since that time, polymers such as synthetic plastics, have become widespread. They are used in a variety of technological products such as nylon, polyethylene, and PVC. See **Figure 16-7**. Scientific knowledge continues to contribute to advancements in polymer products.

Nejron Photo/Shutterstock.com

Figure 16-4. A blacksmith shaping materials.

Elena11/Shutterstock.com

Figure 16-5. A bloomery furnace, once used for smelting iron from oxides.

Leonardo Emilliozzi/Shutterstock.com

Figure 16-6. Bakelite was one of the first synthetic plastics.

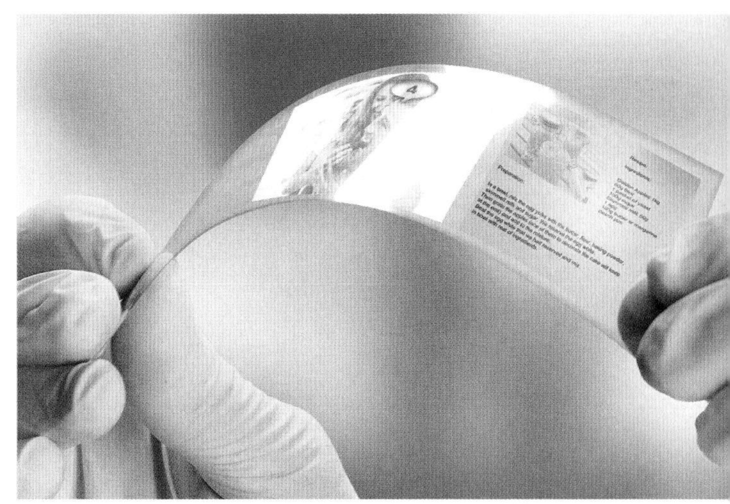

BONNINSTUDIO/Shutterstock.com

Figure 16-8. Graphene application. One micron (μm) is one millionth of a meter.

katatonia82/Shutterstock.com

Figure 16-7. Samples of synthetic grass used for sporting events.

5. The current era in materials science could be referred to as the advanced materials age. This age is producing scientific breakthroughs and technological advancements that are resulting in the creation of new materials that are stronger and more efficient. One such material, graphene, comprises extremely thin, atom-thick carbon sheets that are stronger than steel and very low weight. See **Figure 16-8**. One application of graphene is very thin and highly flexible cell phone and television screens. New materials are being created through manipulation at the atomic level through nanotechnology, which is discussed later in this chapter.

Although time periods are often identified by the creation and use of a new product, these time periods are not always the same around the world. Societies advanced at different rates, some faster than others.

Classifying Materials

Materials scientists and engineers have categorized materials based on differences in their characteristic properties. Such characteristics include atomic structure, material properties, and the intended engineering application. This has created a class system that separates materials into four divisions—metals, polymers, ceramics, and composites. Other classification systems also include semiconductors and advanced materials. As subdivisions further separate materials, distinctions in categories can overlap—material classification does not fully separate categories without blurring boundaries.

Atomic Structure of Materials

All materials—solid, liquid, and gas—are comprised of tiny units called atoms. Within each atom are smaller particles—protons, neutrons, and electrons. The nucleus of an atom contains one or more protons and a similar number of neutrons. One or more electrons orbit around the nucleus and are bound by an electrical charge. Electrons have a

STEM Connection: Mathematics

Calculating Heat Flow

Materials scientists and engineers use mathematical formulas and calculations to determine the appropriateness of a material for a specific engineering application. Heat flow is one example of a mathematical calculation that is often used.

Thermal conductivity varies among materials. In other words, different materials conduct or transfer heat at different rates. This measure of thermal conductivity is denoted by the letter *k*. The k values of materials have already been established in laboratories and are closely related to the R-values placed on building materials. The amount of heat (Q) that is transferred is represented as change from the hot (T_{hot}) section to a cold (T_{cold}) section of a material. This change is denoted as delta (Δ). The following formula is used to determine heat flow:

$Q = kA\,(\Delta T)/t$

Calculations for determining heat loss of a home due to the materials selected are shown in the following example. If the outside temperature is 75°F and the inside is 32°F, and the walls made of pine are 3″ thick with a surface of 8′ × 8′, the heat loss would be calculated as follows:

k of pine = .075
Δ T = 43 (i.e., 75°F – 32°F)
Thickness (t) = 3″ = 3/12′ = 1/4′ = .25′

Area = 5 × 64 = 320 ft^2 (We use 5 since heat loss through the floor is not counted. 5 represents the amount of sides. 64 is the square footage of each wall and was found by multiplying 8 × 8).

Solved: Q = .075 × 320 × 42 /.25 = 40,320 BTU/hour

negative electrical charge, protons have a positive electrical charge, and neutrons are neutral. See **Figure 16-9**.

The Bohr atomic model illustrates the electron configuration of an atom, **Figure 16-10**. As the electrons orbit the nucleus, they are distributed into shells. Each of the shells can hold only a fixed number of electrons. As the distance from the nucleus increases, so does the energy. Shells are ordered from 1 to 2 to 3 to 4, and so on. The maximum number of electrons each shell can contain is represented by the formula $Ne = 2n^2$, in which *Ne* represents the electron capacity of each shell and *n* represents the number of the shell. According to this formula, shell 1 can contain up to two electrons; shell 2 can contain up to eight electrons, shell 3 can contain up to 18 electrons, and shell 4 can contain up to 32 electrons.

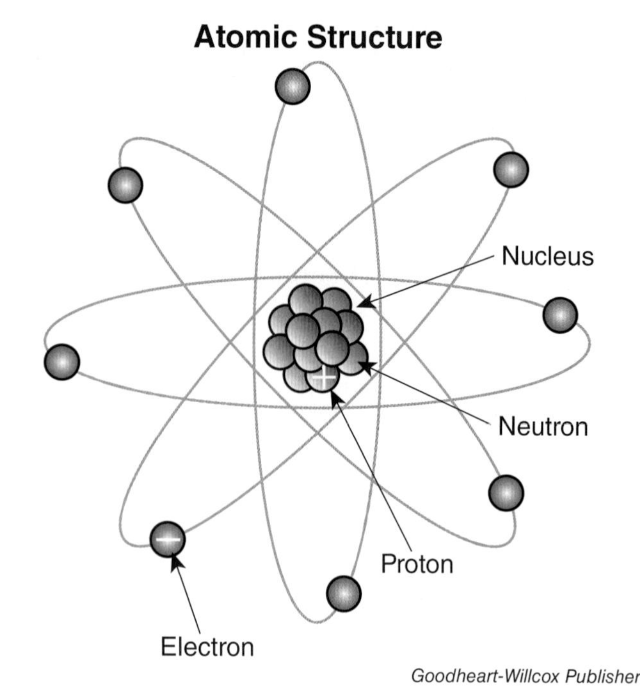

Figure 16-9. The components of an atom.

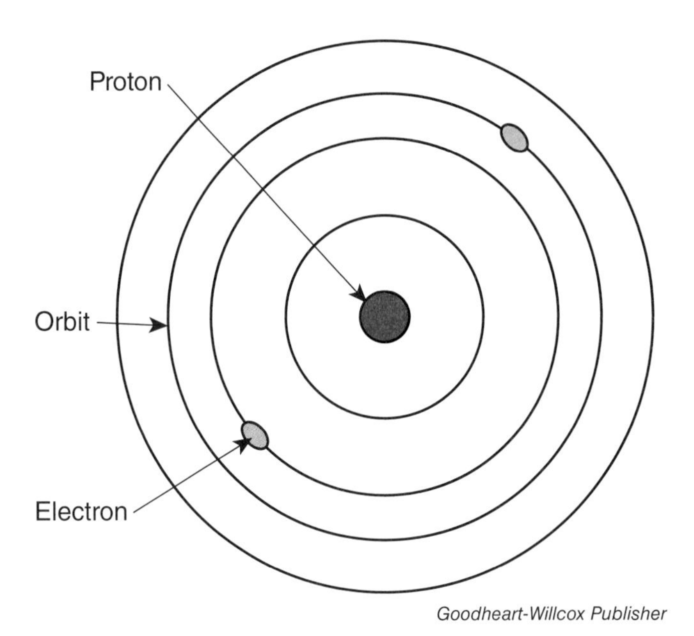

Figure 16-10. Bohr atomic model representation.

The distribution of the electrons into shells affects how the atom functions in nature. Though the electrons furthest away from the nucleus have the highest energy, they are also less bound to the nucleus compared to electrons that are closer. Electrons in the outermost shell are called ***valence electrons***. When a valence electron gains enough energy, it can break away from its atom. When an atom gains or loses a valence electron, it becomes unbalanced and is called an ***ion***.

The configuration of valence electrons determines the electrical conduction ability of a material. For example, a ***conductor*** is a material that contains atoms with only one valence electron that is loosely bound and can become easily free to conduct electrical current. Gold, copper, and silver are excellent conductors. An ***insulator*** is a material that does not conduct electrical current under typical conditions, due to valence electrons that are tightly bound to an atom. Rubber, glass, and plastics are examples of insulators. See **Figure 16-11**.

Materials that are comprised of only one type of atom are called *elements*. ***Elements*** are substances that cannot be broken down any further through chemical reactions. Every atom that makes up an element is unique to that element. The elements, such as hydrogen, gold, and silicon, are classified in the *periodic table*, **Figure 16-12**.

The periodic table orders elements horizontally by ***atomic number***, or the number of protons in the nucleus of their atoms, which is the same as the number of electrons that are distributed through electron orbitals. ***Electron orbitals*** are areas within an atom where electrons have a high potential of being located. All atoms of an element

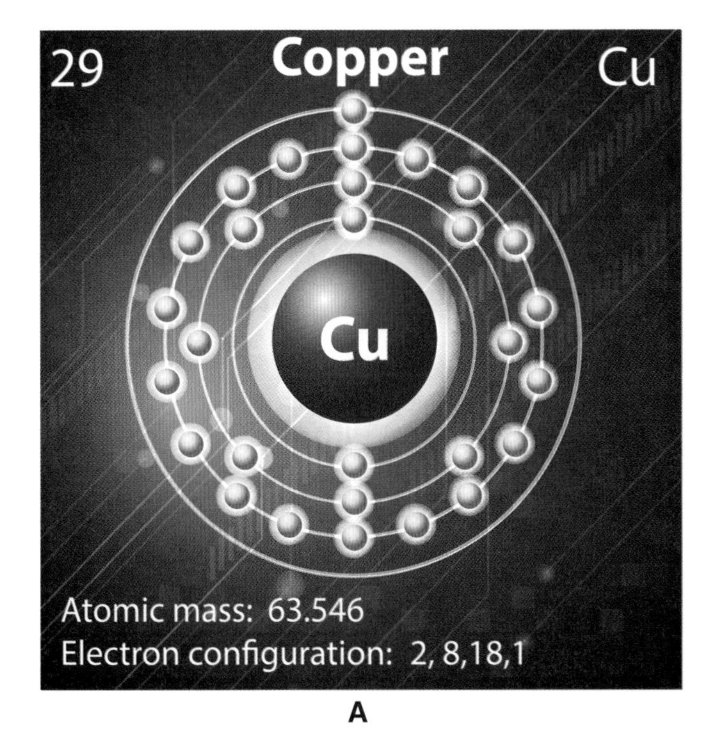

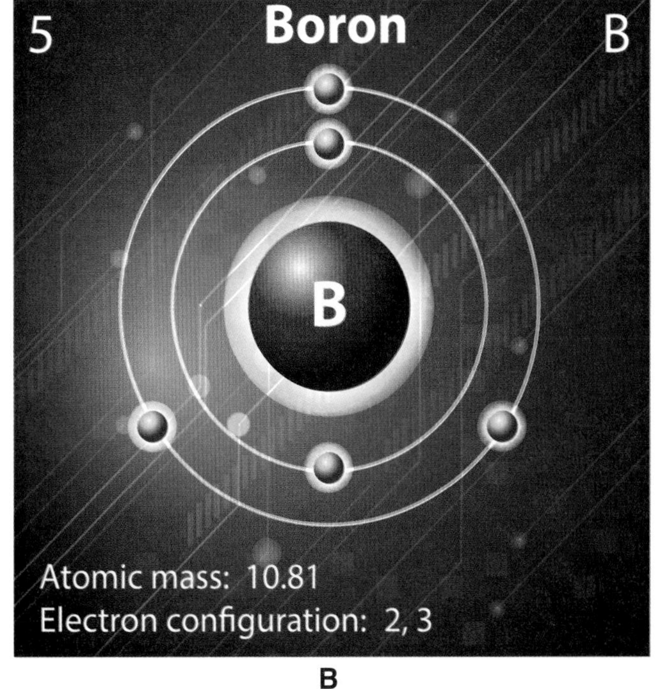

Figure 16-11. Electron configuration comparison between conductors and insulators. A—Copper (a conductor). B—Boron (an insulator).

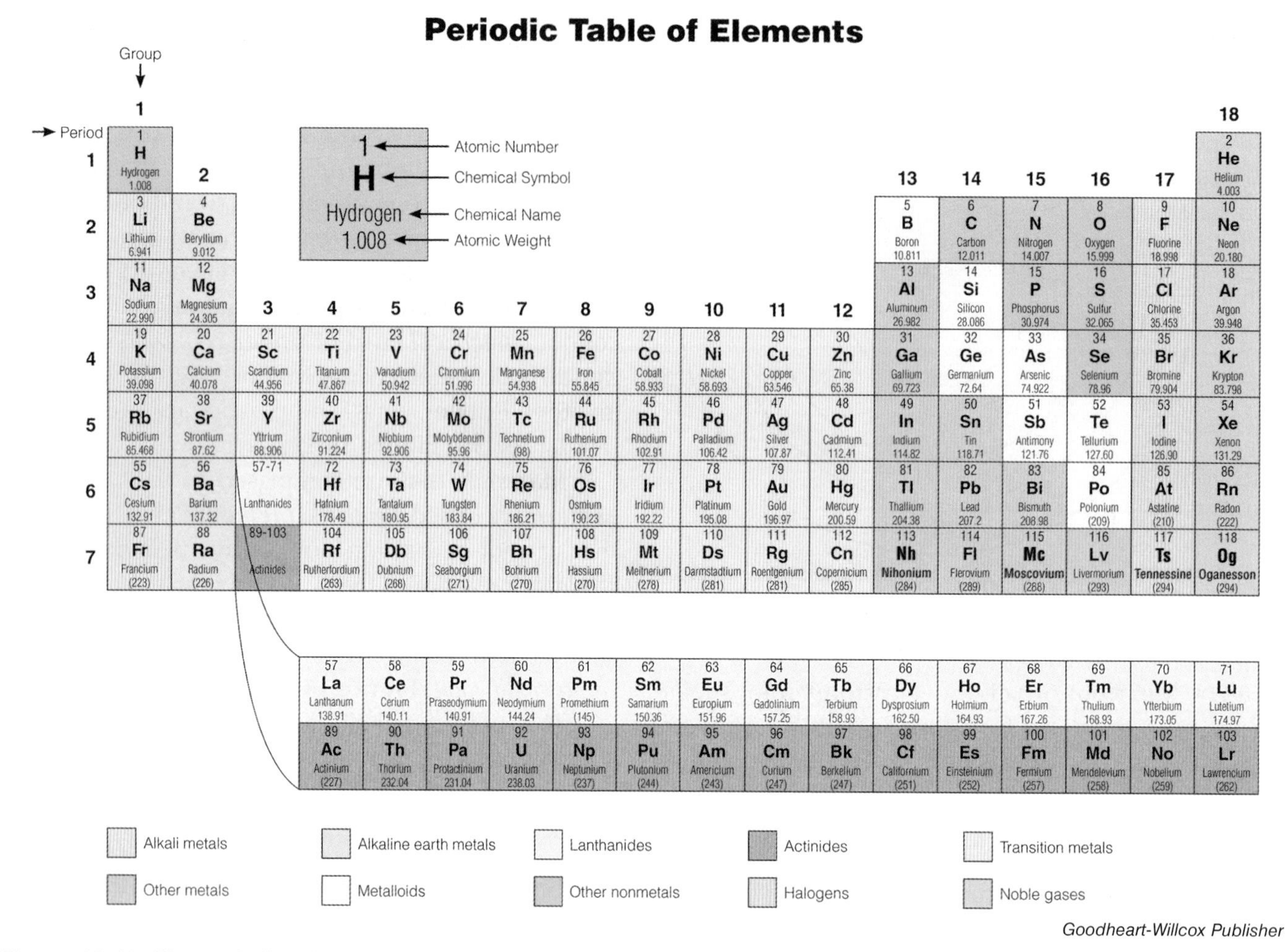

Figure 16-12. The periodic table of the elements.

Think Green Embodied Energy

Materials scientists and engineers are challenged to create products that are stronger, lighter, and more efficient than ever before. They must also consider the costs to make the product and the equipment necessary to produce it. These costs are often referred to as the *cradle-to-grave* costs.

Materials scientists and engineers must also address energy concerns when creating new materials. The Environmental Protection Agency (EPA) and other organizations advocate for energy use to be reduced, especially in the construction of new buildings. One factor that materials scientists and engineers must consider is embodied energy.

Embodied energy is the total quantity of energy used to obtain raw materials, manufacture, construct, and transport the materials. Embodied energy is often expressed as MJ/kg and is calculated in terms of measurements and material used. An MJ (mega joule) is equal to approximately 948 BTUs, or roughly the heat produced as a result of lighting 948 matches.

Architects and engineers are constantly searching for materials with less embodied energy. Materials scientists and engineers are challenged to develop such materials. For example, to air-dry wood requires significantly less energy than kiln drying.

are balanced, or neutral, meaning that they have the same number of protons as electrons when they are in their normal state.

The elements on the periodic table are arranged in columns based on similar chemical properties, which determine which element group they belong to. These groups include metals, nonmetals, transition metals, metalloids, halogens, and noble gases. Understanding the information on the periodic table allows you to determine which elements will react or combine with other elements to form substances with different chemical and physical properties.

Chemical Bonding

The electron configuration of atoms affects whether electrons can be shared or transferred. ***Chemical bonding*** occurs when the valence electrons of atoms are attracted to one another and are transferred or shared between the atoms. There are two main types of chemical bonding. Covalent bonding occurs when atoms share valence electrons. Ionic bonding occurs when one of the atoms accepts or gives the other atom valence electrons. See **Figure 16-13**. In chemistry, a compound is created as these atoms bond together.

The grouping, or bonding, of two or more atoms creates ***molecules***. Water (H_2O) is a common example of a molecule containing the two elements, hydrogen and oxygen. The structure and connections between molecules and atoms is called *molecular structure*. See **Figure 16-14**. Scientists and engineers must understand the atomic structure, electron configuration, and molecular structure of materials in order to create new technological products.

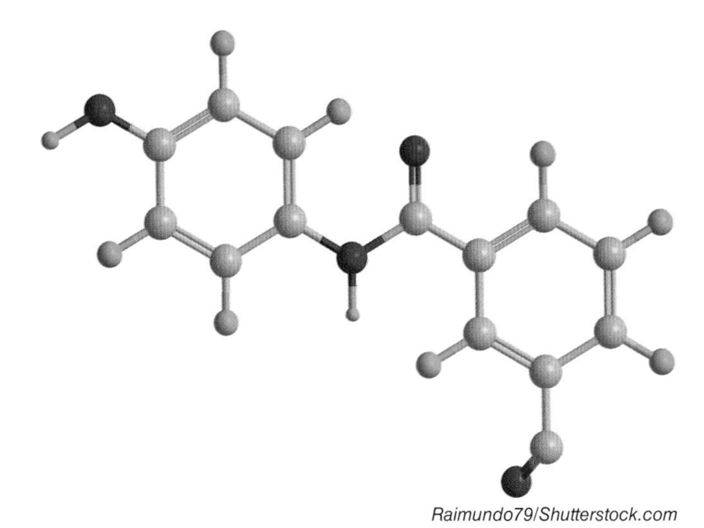

Raimundo79/Shutterstock.com

Figure 16-14. The molecular configuration of Kevlar®.

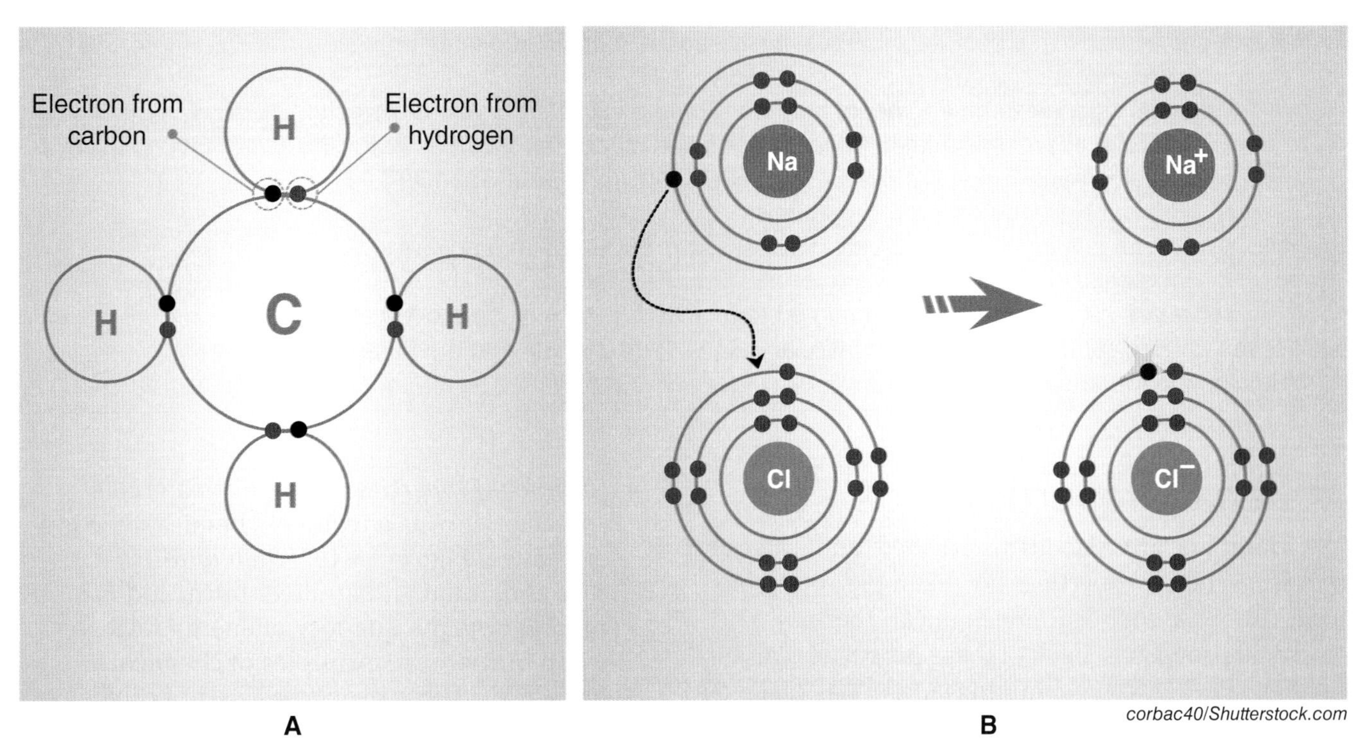

corbac40/Shutterstock.com

Figure 16-13. There are two different types of bonding. A—Covalent bonding. B—Ionic bonding.

Career Connection Materials Scientists and Engineers

Materials scientists and engineers conceptualize, test, process, and create materials that are used in the development of almost every created product. They work with a variety of materials including metals, polymers, ceramics, composites, semiconductors, and advanced materials. They understand the atomic and molecular structure of materials and the thermal, electrical, and mechanical properties they exhibit. Their knowledge requires interdisciplinary understanding including chemistry, mechanical engineering, and biology. This knowledge affords them the opportunity to create more advanced and innovative materials.

High school students interested in materials science and engineering careers should take a variety of STEM courses. These courses include mathematics courses (algebra, trigonometry, and calculus) and science courses (biology, chemistry, and physics).

Many college programs have bachelor's degrees directly related to Materials Science and Engineering. Some programs directly address this field; others involve the study of specific classes of materials, such as polymers, ceramics, or metallics. Programs typically include mathematical and scientific focus as the foundation, with further courses describing the properties, structure, processing, and performance of all classes of materials. Programs also focus on engineering design to meet specific objectives of applications requiring new materials. However, few graduate programs are specifically related to materials science and engineering.

Individuals who do well in the materials science and engineering field have good critical thinking skills. They work well collaboratively and have a broad understanding of multiple fields. They are able to use the scientific method to carry out investigations and have an understanding of physical and earth sciences.

Materials scientists and engineers are hired in a variety of industries. They are employed in manufacturing in both primary and secondary processing. They also work in service industries, conducting research, consulting, and providing engineering services.

Materials scientists and engineers may work on the following types of projects:

- Improve operations related to natural gas and oil drilling.
- Develop materials and processes to reduce environmental impacts.
- Research and develop new materials.
- Serve as a consultant, providing professional expertise and advice to industries and government groups.

There are several organizations related to this field. They include the Materials Research Society (MRS); ASM International; ASTM International; and the Minerals, Metals, and Materials Society (TMS).

Material Properties

Materials are also distinguished and categorized by their properties. For example, materials have different chemical, mechanical, thermal, and electrical properties. Each of these properties can be further subdivided, containing additional distinctive properties. Engineers must consider material properties in order to make informed design decisions based on the intended engineering application. Material properties include the following:

- ***Chemical properties*** describe a material's characteristics as a result of chemical reaction or change, **Figure 16-15**. ***Flammability***, the ease a material will ignite or burn, and ***heat of combustion***, the quantity of energy released when burned, are examples of chemical properties. These properties are not directly observed until the structure of the chemical is changed. Other chemical properties are toxicity, oxidation states, and pH.

Robert cicchetti/Shutterstock.com

Figure 16-15. A leaf changing colors is an example of physical and chemical changes.

- *Mechanical properties* describe a material's reaction to physical forces. Specifications and limits are determined through testing. *Tensile strength* is the maximum pulling force, or tension, a material can withstand prior to failure. See **Figure 16-16**. *Hardness* is a measurement describing the resistance of a material through indentation, scratch, and rebound tests. *Ductility* defines the ability of a material to be stretched or lengthened into wire before failure occurs. Other examples of mechanical properties are wear resistance, impact strength, corrosion resistance, and density.

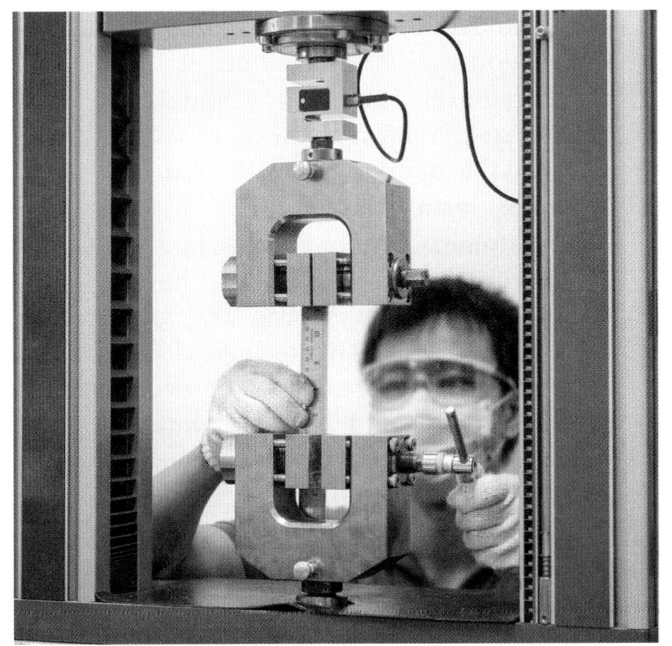

Mrs_ya/Shutterstock.com

Figure 16-16. A material undergoing a tensile strength test.

- *Physical properties* are the characteristics due to the structure of a material, including size, shape, density, moisture content, and porosity.
- *Thermal properties* determine the effect temperature has on a material. The extent a material can conduct, or transfer, heat is ***thermal conductivity***. The amount of stress a material can withstand from repeated changes in heating and cooling is referred to as ***thermal fatigue***. Other thermal properties include thermal expansion, thermal resistance, and specific heat.
- *Electrical and magnetic properties* are characteristics that describe a material's ability to conduct electrical current by allowing electrical charge to transfer from atom to atom, **Figure 16-17**, as well as its magnetic permeability (ability to retain magnetic forces). Electric current is conducted efficiently by conductors but not by insulators. Other electrical properties include resistivity and dielectric strength.
- *Acoustical properties* describe how a material reacts to sound waves. These properties include acoustical transmission (the ability to conduct sound) and acoustical reflectivity (the ability to reflect sound).
- *Optical properties* describe how a material reacts to visible light waves. Optical properties include color (reflected waves), optical transmission (the ability to pass light waves), and optical reflectivity (the ability to reflect light waves).

Classification and Characteristics of Materials

Based on their atomic structure and characteristics, materials can be classified into the following categories:

- **Metals.** Most of the elements on the periodic table are metals. Metals containing a single type of atom are called *pure metals*. Metals containing multiple atoms or a mixture of metals are called *alloys*. See **Figure 16-18**.

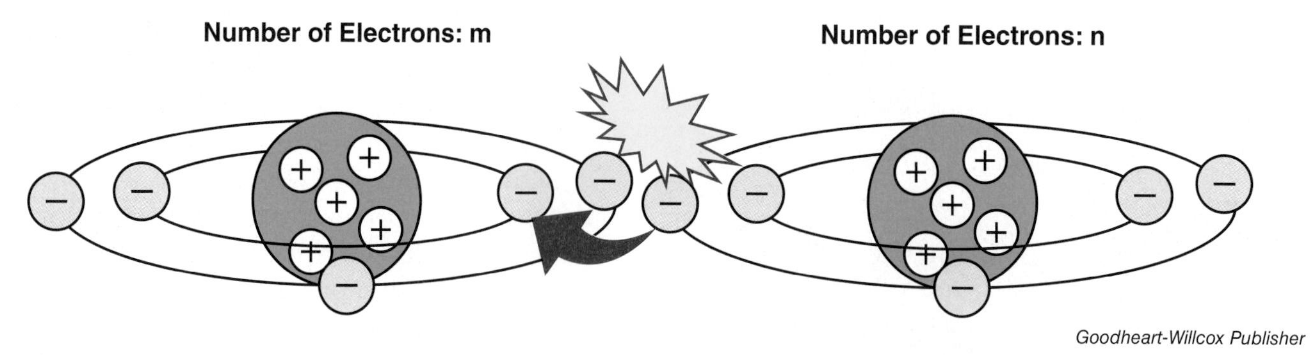

Goodheart-Willcox Publisher

Figure 16-17. An electrical charge being transferred from atom to atom.

Aksenenko Olga/Shutterstock.com

Vladimiroquai/Shutterstock.com

Figure 16-18. Metals are pure metals or alloys. A—Copper is an example of a pure metal. B—Car rims are an example of an alloy. A car rim is commonly made up of steel, an alloy of iron and carbon.

Metals have strong electrical and thermal conductivity; are solid at normal temperatures; are shiny, hard, malleable, and ductile; can be difficult to burn; and have a high density. Examples of pure metals include gold, silver, aluminum, and copper. Brass, steel, and bronze are examples of alloys.

- **Polymers.** Materials containing many repeating molecules are called ***polymers***. Polymers can be either natural or synthetic. See **Figure 16-19**. Naturally occurring polymers include cotton, DNA, starch, and proteins. Polymers can be flexible or rigid. They have low melting points, are nonconductive, and can be flammable. Four classes of polymers, **Figure 16-20**, are as follows:
  - **Thermoplastics.** *Thermoplastics* do not share electrons between atoms, and thus have no covalent bonding. Thermoplastics can be softened when heated to be remolded. Examples are acrylic, nylon, polypropylene, and Teflon™.
  - **Thermosets.** *Thermosets* cannot be reheated to form a new mold once already cured. Examples are vulcanized rubber, epoxy resin, and Bakelite.
  - **Elastomers.** *Elastomers* can be thermoplastic or thermoset. They are elastic polymers that can stretch, but return to their original shape when released. Examples of elastomers are silicone rubber, chloroprene, and fluoroelastomers.
  - **Fibers.** Fibers are extruded polymers that have been pulled through a die to form small threads. Examples are nylon and polyesters.
- **Ceramics.** Ceramics, the oldest material created by humans, can be characterized by specific properties. At room temperatures, ***ceramics*** have high brittleness. Consider

Prakan wongsupa/Shutterstock.com preecha2531/Shutterstock.com

A B

Figure 16-19. Polymers can be natural or synthetic. A—Wool is a natural polymer. B—Teflon™ pans are synthetic polymers.

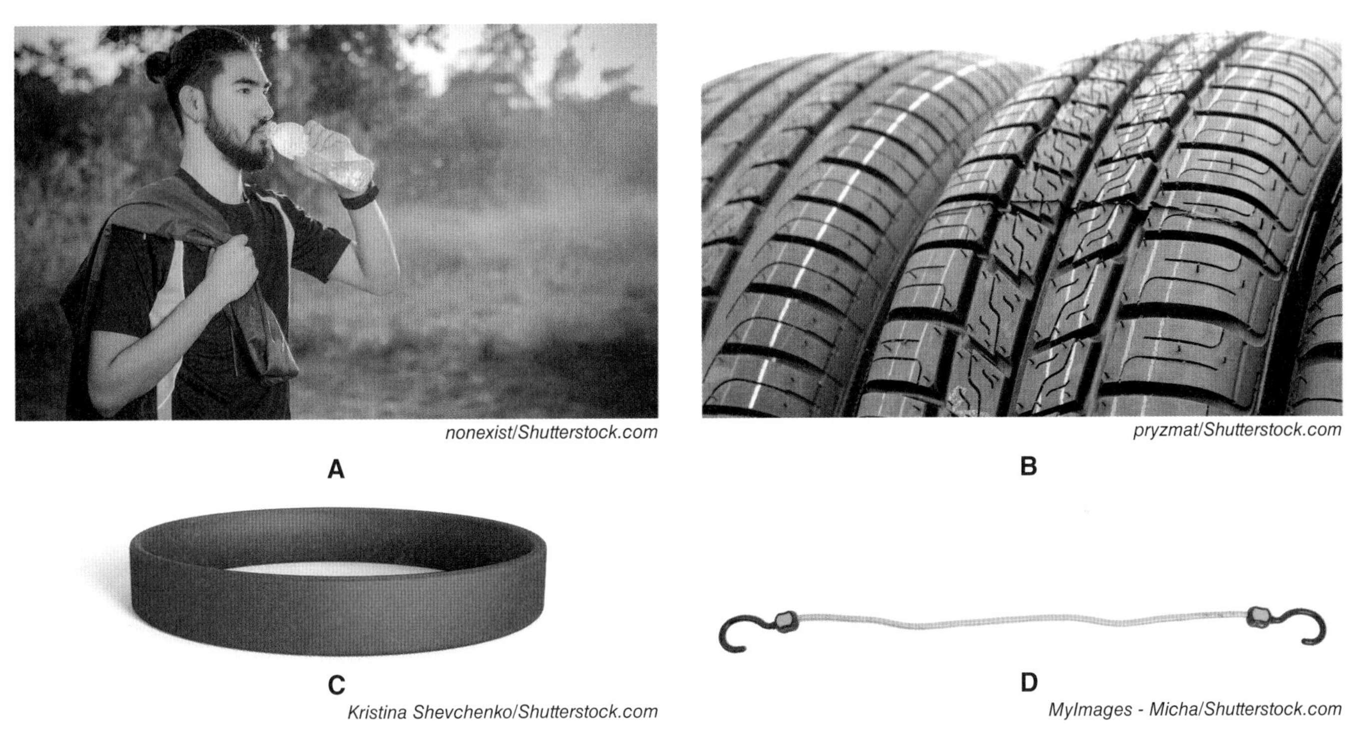

nonexist/Shutterstock.com pryzmat/Shutterstock.com

A B

C D

Kristina Shevchenko/Shutterstock.com MyImages - Micha/Shutterstock.com

Figure 16-20. Polymers. A—Drinking bottles are made from polyethylene, a thermoplastic. B—Vulcanized rubber is a thermoset polymer. C—Silicone rubber is an elastomer. D—Nylon fiber is commonly used in the designed world.

how a ceramic plate shatters when dropped. Ceramics also are poor electrical and thermal conductors, have very high melting points, and are nonflammable. Examples of traditional ceramics include clay pottery, plates and bowls for dining, and bricks. Today, ceramics are used in technological products such as cell phones and computers, as well as in aviation and electrical components.

- **Composites.** *Composites* are comprised of multiple materials with different chemical or physical properties. When composites are created, the component materials retain their individual characteristics. See **Figure 16-21**. Composites are made to benefit from the individual properties of each material to create an overall more preferable material. For example, a composite might retain the strength of the original materials but also be lighter and have a higher flexibility than each material by itself. Composites are created to be strong without increasing costs or weight. Composites can be polymers, metallic, ceramic, or combinations. Building construction frequently uses composites. Examples are engineered wood,

TLaoPhotography/Shutterstock.com

Figure 16-21. Carbon fiber Kevlar® is an example of a composite.

such as plywood and laminated veneer lumber, fiberglass, and Trex® decking. See **Figure 16-22**.

- **Semiconductors.** ***Semiconductors*** are materials that are between conductors and insulators in terms of their ability to conduct electrical current. A semiconductor is neither a poor nor a good conductor or insulator. Due to their electron configuration, semiconductors are frequently used by the electronics industry and are often found in circuits. Many elements found in the periodic table group 14, such as silicon and germanium, serve as the basis of electronic components. See **Figure 16-23**. Semiconductors typically contain four valence electrons in their outermost shell.
- **Advanced materials.** Materials science and engineering are creating innovative materials to meet the needs of high-tech applications. Some of these ***advanced materials*** improve traditional materials, such as ceramic, metallic, polymer, composite, or semiconductors. Others have entirely different characteristics from traditional materials. Examples are smart materials, and nanostructured materials. ***Smart materials*** have properties that can be modified in a controlled manner by external forces. For example, self-healing materials are smart materials that are able to repair damage as a result of typical usage, allowing the material to last longer. Increasingly, advancements in nanostructured materials are allowing new smart materials to be created with changes at the atomic level, resulting in unique properties compared to other materials.

Anna Bolotnikova/Shutterstock.com

Figure 16-22. Laminated veneer lumber used in home construction.

STEM Connection: Science

Creating New Elements

The periodic table is an arrangement of all of the chemical elements known to man. Currently, 118 elements exist and are represented on the periodic table. The first 98 elements on the table occur naturally. Elements 99 to 118 are synthetic (man-made) elements, created as the result of laboratory work.

Scientists strive to create element 119, the heaviest element known. Since this element will decay in a fraction of a second, the fusion of two lighter elements is an extremely quick process. By using a linear particle accelerator, the scientists will merge titanium and berkelium to make element 119. This accelerator shoots a particle of titanium at over 67 million miles per hour through a 400-foot tube. As this particle smashed into a target of berkelium atoms, scientists collected data through specially designed devices. This data provides the scientists with evidence that the new element does indeed exist. Current models predict that the heaviest element has at most 126 protons. Scientists have further research and studies to conduct in order to determine what other elements might exist.

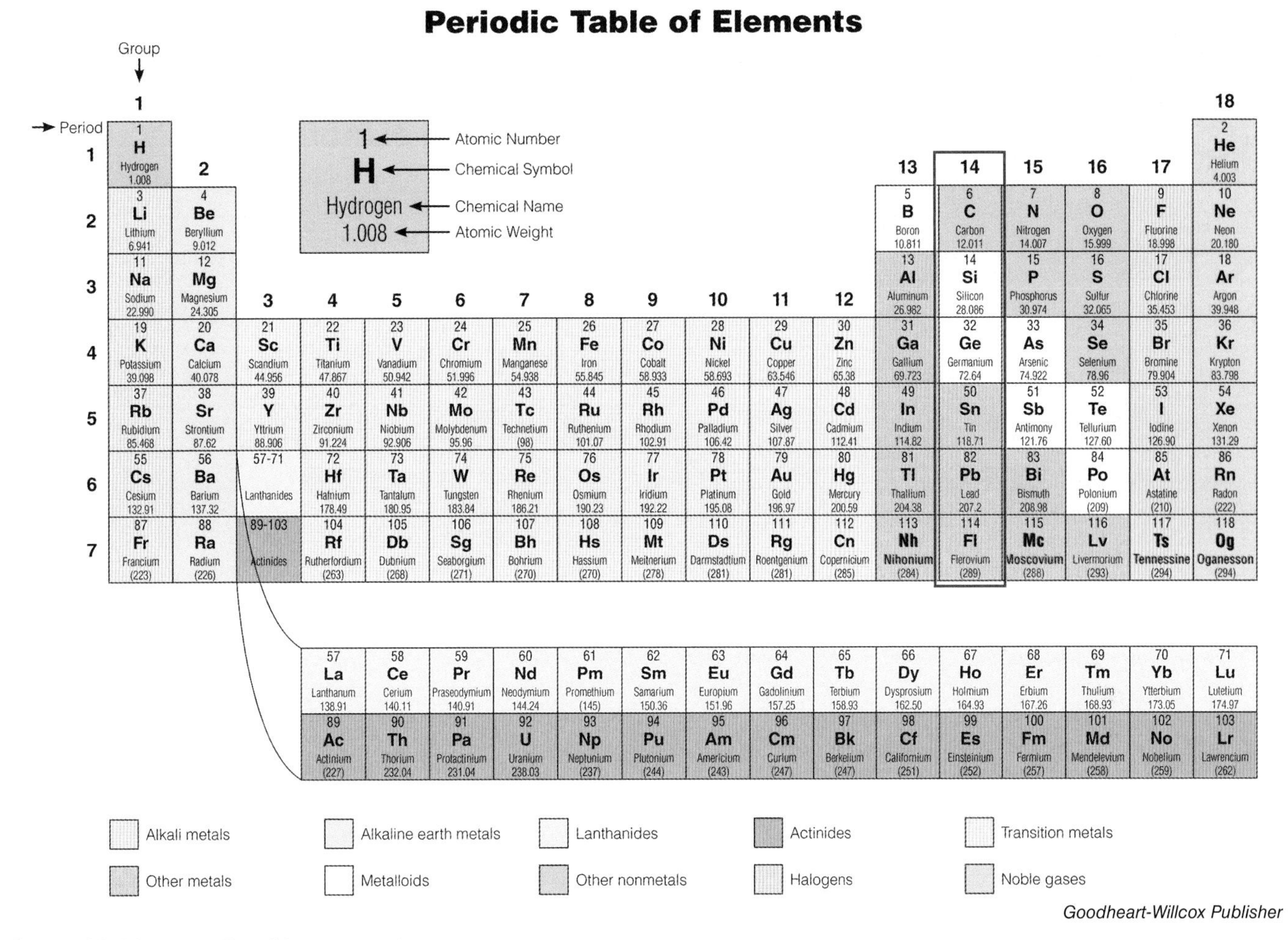

Figure 16-23. Periodic table group 14.

Advancements in Materials

Attention to advanced materials has resulted in a materials revolution. Classes of materials once had specific purposes for particular applications. Now, however, advanced knowledge of materials has allowed for the creation of materials that can meet multiple, competing demands. For example, a product once requiring transparency would often use glass as an appropriate material. Advancements in new materials have made polymers or ceramics available as a transparent alternative. Alternatives have also been created for high strength materials used in building construction. Instead of steel or concrete, stronger composites can be used that have the same compressive or tensile strength but are lighter and cheaper to produce.

Technological advancement has increased demand for the modification and manufacture of new materials. Materials scientists and engineers have an increased role and impact on the creation of innovative technological products. More time is spent in laboratories creating materials that need to meet very specific needs.

For example, the space industry consistently requires materials that are stronger and more durable. Reinforced carbon-carbon (RCC) was developed to allow space shuttles to withstand temperatures up to 2700°F (1482°C). This composite material was designed with the specific goal of being able to withstand ascent from and reentry into the earth's atmosphere.

Additionally, scientists have begun to create metamaterials. Metamaterials are materials created by scientists to exhibit characteristics that cannot be found in nature. For example, as you will read in Chapter 20, scientists are working toward creating an invisibility cloak, which requires the structure of the metamaterial to be smaller than the wavelength of light—blocking light ray access to the material covered by the cloak. Researchers have created a successful prototype of this type of invisibility cloak,

Academic Connection: Argumentative Essay

Advantages, Disadvantages, and Ethical Concerns of Nanotechnology

Argumentative writing is a type of writing that requires people to investigate a topic, collect and evaluate data related to the topic, and establish and present a position on the topic based on evidence. A valid argument also provides facts and data pertaining to any opposing views to the position presented. Scientists and engineers are commonly required to take a particular stance on an issue and defend their position effectively based on logic and fact.

Materials scientists and engineers are using nanotechnology to create new products that range from better tennis balls to breakthroughs in medicines. The use of nanotechnology for the manufacture of new products has given rise to concerns such as environmental impacts, health issues, and ethical questions.

Research the topic of nanotechnology and identify recent advancements in nanomaterials. Choose a new or emerging nanomaterial and write an argumentative essay on whether this material should or should not be pursued in the advancement of new materials. Support your argument with appropriate evidence and empirical evidence while considering other points of view.

Write your essay in the following format:

I. Introduction
   A. Introduce the topic
   B. Present your position
II. Body
   A. Topic sentence for reason 1
      i. Supporting evidence
   B. Topic sentence for reason 2
      i. Supporting evidence
   C. Topic sentence for reason 3
      i. Supporting evidence
III. Conclusion
   A. Restate your position
   B. Counterarguments
   C. General conclusions

but it is extremely small, measuring only ten micrometers wide. That is equivalent to .00001 meters!

Environmental concerns also require materials scientists and engineers to respond to needs for new materials. Efforts to reduce the use of plastic bags have resulted in researchers creating products that are more sustainable and environmentally friendly. One such project has been to make disposable bags out of biodegradable crops. Materials scientists and engineers are using components of crops to make molecules that can be developed into polymers. This is blurring the lines between the natural world and the technological world. These polymers can then be processed into manufactured goods to replace traditional plastic bags. This material can then be broken down through ultraviolet light into its original building blocks with minimal effect on the environment.

Biomimicry

Materials scientists and engineers are challenged to create innovative materials that are designed to be stronger, more flexible, sustainable, less costly, and more environmentally friendly. To create these materials, scientists and engineers often turn to the natural world for inspiration. Through ***biomimicry***, the study of processes and designs that are in nature, characteristics of nature are then imitated in new materials.

Velcro® is one of the best known examples of a product created through biomimicry. Scientists examined the microscopic structure of burdock burrs and saw features that could be modeled into a new material. See **Figure 16-24**.

Other examples of biomimicry include super adhesive materials that are based on the gecko's ability to stick to almost any surface. The understanding of how geckos are able to climb walls and how their body exhibits chemical reactions is quickly being transferred into new adhesive materials.

Nanotechnology

Nanotechnology, the manipulation of individual atoms and molecules of materials, is increasingly changing the role of materials scientists and engineers in creating materials. At the nanoscale, around 1 to 100 nanometers, a scanning tunneling microscope (STM) is required to see individual atoms. To gain perspective, one human hair is roughly 75,000 nanometers thick. See **Figure 16-25**.

sebastienlemyre/Shutterstock.com

Figure 16-24. Burdock burrs.

Two types of nanomanufacturing processes are top-down nanomanufacturing and bottom-up nanomanufacturing. In top-down nanomanufacturing, materials are shaped or reduced down to the nanoscale. Bottom-up nanomanufacturing is the addition of atoms and molecules to build up a material.

Nanotechnology benefits reach across many sectors of industry and society. New processes and products allow for materials to be stronger and more efficient, improving everyday products. Additionally, advances in nanotechnology improve the environment by reducing pollution, since materials can be created from the ground up rather than machining products, which creates additional waste. Nanotechnology has many applications in health and medicine, including the early detection of diseases and in tools doctors use during surgery.

Nanotechnology has raised ethical concerns related to environmental and health impacts. One concern is that nanoparticles might be toxic. The scale of the particles may allow them to cross through the brain membrane, resulting in harmful chemicals entering the bloodstream. As clothing and materials begin to be coated by nanoparticles, more research is needed to determine the effect these may have on health and the environment.

Selection of Materials

Understanding the atomic and molecular structure of materials and the properties they exhibit informs the selection of materials and the creation of new products. This knowledge is used to find the ideal combination of characteristics in a material and ensure high quality at minimal cost. Material selection is based on fabrication requirements, such as the material's machinability, castability, and weldability. Additional factors that influence material selection include intended final shape, required mechanical properties, service necessities, processing costs, tolerances, material availability, and the cost of obtaining the material.

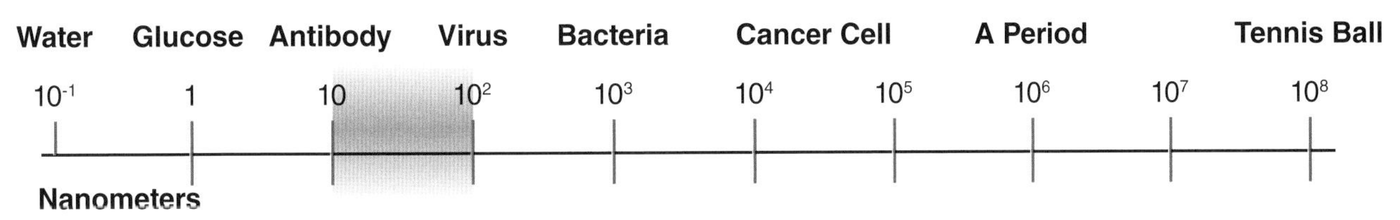

Goodheart-Willcox Publisher

Figure 16-25. A representation of nanoscale.

Summary

- ✔ Materials science and engineering classifies solid materials by their atomic and molecular characteristics and differences in properties. Scientists and engineers examine these properties to determine the suitability of a material for a particular application.
- ✔ Time periods are identified by the materials most commonly manufactured. The primary time periods include the Stone Age, the Bronze Age, the Iron and Steel Age, the Nonferrous and Polymer Age, and the current advanced materials age.
- ✔ Scientists and engineers must understand the atomic structure, electron configuration, and molecular structure of materials in order to create new technological products.
- ✔ The configuration of valence electrons determines the conduction ability of a material.
- ✔ Categories of material properties include chemical properties, mechanical properties, physical properties, thermal properties, electrical and magnetic properties, acoustical properties, and optical properties.
- ✔ Based on their atomic structure and characteristics, materials can be categorized as metals, polymers, ceramics, composites, semiconductors, and advanced materials.
- ✔ The four classes of polymers are thermoplastics, thermosets, elastomers, and fibers.
- ✔ Materials scientists and engineers have an increasing role and impact on the creation of innovative technological products. They create materials to meet very specific needs.
- ✔ Materials scientists and engineers are challenged to create innovative materials that are designed to be stronger, more flexible, sustainable, less costly, and more environmentally friendly.
- ✔ Scientists and engineers use biomimicry to develop materials. They study nature's processes and designs and imitate the characteristics of nature in new materials.
- ✔ Nanotechnology, the manipulation of individual atoms and molecules of materials, is changing the role of materials scientists and engineers in creating materials.

Check Your Engineering IQ 

Now that you have finished this chapter, see what you learned by taking the chapter posttest.
www.g-wlearning.com/technologyeducation/

Test Your Knowledge 

Answer the following end-of-chapter questions using the information provided in this chapter.

1. What is materials science and engineering?
2. The introduction of smelting, a process of extracting metal from an ore through heating and cooling, marked the beginning of the _____ Age.
   A. Iron and Steel
   B. Stone
   C. Nonferrous and Polymer
   D. Bronze
3. The four main types of materials are metals, ceramics, polymers, and _____.
4. *True or False?* Each of the shells of an atom can hold the same maximum number of electrons.
5. Materials that do not conduct electrical current under typical conditions are called _____.
6. Chemical bonding occurs when the _____ of atoms are attracted to one another and are transferred or shared between the atoms.
7. _____ properties describe a material's reaction to physical forces.
8. *True or False?* Metals have weak electrical and thermal conductivity.
9. *True or False?* Thermoplastics can be softened when heated and remolded.
10. Which type of material has high brittleness?
11. *True or False?* Composites are made of multiple materials with different chemical or physical properties.
12. What are smart materials?
13. *True or False?* Technological advancement and the materials revolution have reduced demand for the modification and manufacture of new materials.
14. Biomimicry involves the imitation of processes and designs found in _____ to develop new materials.
15. The manipulation of individual atoms and molecules of materials is called _____.

Critical Thinking

1. Consider materials currently in use and how they might define the time period you are living in. What materials do you foresee as remaining in use? Which materials might be replaced or modified in the future?
2. When choosing materials, engineers do not consider only one property of a material. For example, which material is the best conductor of electrical current—gold or copper? Why is this material not frequently used in wiring? What material is most commonly used in residential wiring and why?

STEM Applications

1. Select an element from the periodic table and do the following:
   A. Using the atomic number, identify the number of electrons, neutrons, and protons in the atom.
   B. Create a 2-D drawing illustrating the location of the protons, neutrons, electrons, and shells of the atom.
   C. Build a 3-D model accurately depicting the drawing.
2. Consider a common product that you may use on a regular basis.
   A. Reflect on the chemical, mechanical, thermal, and electrical properties, and how you can improve those to improve the product.
   B. Justify the changes you would make in the specific properties of the materials used in the product.
   C. Create a sketch, 2-D drawing, and 3-D drawing depicting what changes you would make to improve the product.
   D. Create a prototype of the improved product.

Engineering Design Challenge

Advanced Material Creation

Background

Materials scientists and engineers select the most appropriate materials for engineering applications. To make informed decisions, they must know the atomic and molecular structure of materials. They must also understand the thermal, mechanical, electrical, and chemical properties of materials. One example of replacing a material with a better material is the replacement of metal components with polymers. Small ball bearings made out of polymers, such as Teflon™, have replaced metal ball bearings in many applications. The polymer bearings are corrosion-resistant, are inexpensive to manufacture, and do not require lubrication like metal bearings do.

Situation

In an effort to be more efficient and capture thermal and electrical energy, car manufacturers require a new material that can meet multiple criteria. The battery of the car must collect energy as a result of friction during the braking process. The materials engineer is challenged to create a new material that the automotive engineers can integrate into their new technological process. Your goal is to research and design a conceptual model of such a material that can be created in the laboratory. The material must meet the following criteria:

- **Mechanical properties.** Hardness, corrosion resistance, and impact strength.
- **Thermal properties.** High thermal conductivity, low thermal fatigue.
- **Electrical properties.** Low resistance, high electrical conductivity.

In addition to these criteria, consider other factors when designing the material. These factors are raw material costs and processing costs, machinability, casting properties, and the tooling required.

Desired Outcomes

- A model or prototype of a new advanced material depicting a combination of materials that can be used to create a new material.
- A report describing the benefits of each submaterial and the anticipated costs of production.

Materials and Equipment

- Examples of various types of materials.
- Testing equipment or fixtures to test the mechanical properties, thermal properties, and electrical properties of common materials.
- Computer simulation software.

Procedure

- Identify the problem and formulate it from your own perspective. This is an important step in the engineering design process.
- Use your knowledge of engineering design to guide you as you define the problem and create a viable solution to it.
- Research and test the different properties of materials to determine which ones meet your needs. Document the materials and properties in a table that you can refer to when making a later decision.
- Identify the materials that meet each of the criteria referenced. Consider where these materials would be best placed. Would they serve as the core of the material or as a coating?
- Create a mock-up of your product.

Dima Moroz/Shutterstock.com

Praiseaeng/Shutterstock.com

Maik Kirsten/Shutterstock.com

Teflon™, Velcro®, and Bakelite were all created through materials engineering.

CHAPTER 17

Constructing Structures

Check Your Engineering IQ

Before you read this chapter, assess your current understanding of the chapter content by taking the chapter pretest.
www.g-wlearning.com/technologyeducation/

Learning Objectives

After studying this chapter, you should be able to:

- ✔ Explain the two major types of construction.
- ✔ Summarize the characteristics of the major types of constructed buildings.
- ✔ Describe the steps involved in constructing a structure.
- ✔ Summarize the characteristics of the frameworks used in buildings.
- ✔ Explain how buildings are enclosed.
- ✔ List the types of utility systems used in buildings.
- ✔ Discuss the characteristics of materials used in finishing buildings.
- ✔ Summarize the characteristics of different heavy engineering structures.

Technical Terms

arch bridges
arched dam
beam bridges
bearing surface
buttress dam
cantilever bridges
ceiling joists
ceilings
commercial buildings
drywall
fascia board
floor joists
footing
forced-air heating systems
geothermal systems
gravity dam
headers
heat pump
hot water heating
industrial buildings
landscaping
manufactured home
pile foundations
potable water
rafters
reinforced concrete
residential buildings
sheathing
sill
slab foundations
soffit
sole plate
spread foundation
studs
subfloor
suspension bridges
top plate
trap
truss
truss bridges
wastewater

Close Reading

As you read the following chapter, make a list of the different types of buildings mentioned. Then think of an example in your city of each of these buildings and analyze how these buildings were constructed.

Human beings have three basic needs: food, clothing, and shelter. Each of these needs can be satisfied using technology. Agriculture and related biotechnologies help us grow, harvest, and process food. Manufacturing helps us to produce natural and synthetic fibers. These fibers are the inputs for clothing and fabric manufacture. Materials and manufactured goods can also be fabricated into dwellings and buildings using construction technologies. Construction uses technological actions to erect a structure on the site where the structure will be used.

Construction technology consists of two types of structures: buildings and heavy engineering structures, **Figure 17-1**. Buildings protect people, materials, and equipment from the elements. They also provide security for people and their belongings. Heavy engineering structures help our economy function effectively by providing the infrastructure to support transportation, production, and communication systems and industries.

B Brown/Shutterstock.com

ThamKC/Shutterstock.com

Figure 17-1. Construction erects buildings and heavy engineering structures.

Buildings

Buildings are grouped into three types: residential, commercial, and industrial. See **Figure 17-2**. These groupings are based on how the buildings are used, but other types of buildings do exist. These special buildings follow the same construction steps as the other buildings. This section describes each of the main types, provides an overview of the other types, and discusses the general steps involved in constructing buildings.

Types of Buildings

Residential

Artazum/Shutterstock.com

Commercial

Paul Matthew Photography/Shutterstock.com

Industrial

FUN FUN PHOTO/Shutterstock.com

Figure 17-2. Construction is used to build residential, commercial, and industrial buildings.

Types of Buildings

Different building types have different uses. We live in buildings, buy products in buildings, and work in buildings, to name just a few activities done in buildings. To help you understand the types of buildings people design, build, and use, we will explore three major types of buildings and then briefly review some more unique structures.

Residential Buildings

Residential buildings are structures in which people live. These buildings can be single-family or multiple-unit dwellings. The multiple-unit dwellings include apartments, town houses, and condominiums.

A residential building can be either owner occupied or rented from the owner. The owner of a dwelling is responsible for the dwelling's maintenance. In some types of dwellings, such as condominiums, maintenance costs are shared among owners. Each owner belongs to and pays fees to an association. This group elects officers who manage the maintenance of common areas such as entryways, garages, parking areas, and lawns. The association is also responsible for exterior repairs and insurance on the building. The individual owners maintain their own living quarters and insure their personal belongings against fire and theft.

Commercial Buildings

Commercial buildings are used for business purposes. These buildings can be publicly or privately owned. Commercial buildings are built in a range of sizes. Retail stores, offices, and warehouses are commercial buildings.

Industrial Buildings

Industrial buildings house manufacturing processes. These buildings are used to protect machinery, materials, and workers from the weather. The building supports the machines and supplies the utility needs of the manufacturing process. Many industrial buildings are specially built for one manufacturing process.

Other Building Types

Commercial, industrial, and residential buildings are common in most locations. Looking around your town or city, however, you might see other types of buildings, **Figure 17-3**, including:

- **Monuments** honor the accomplishments or sacrifices of people or groups.

dibrova/Shutterstock.com

Richard Cavalleri/Shutterstock.com

Alan L Meakin/Shutterstock.com

RomanSlavik.com/Shutterstock.com

robert cicchetti/Shutterstock.com

Figure 17-3. Buildings are constructed for a variety of uses.

- **Cultural buildings** house theaters, galleries, libraries, performance halls, and museums. They host musical, dramatic, and dance performances; literary activities; and art exhibits.
- **Government buildings** house governmental bodies. Examples include city halls, post offices, police stations, firehouses, state capitols, and courthouses.
- **Transportation terminals** are used to load and unload passengers and cargo from transportation vehicles. Examples are airports, train and bus stations, freight terminals, and seaports.
- **Sports arenas and exhibition centers** are used for sporting events, concerts, trade shows, and conventions.
- **Agricultural buildings** include barns and storage buildings used to house livestock, shelter machinery, and protect farm products (grain and hay, for example).

As noted earlier, these special buildings are built using the same construction steps used for a single-family home.

One special type of building is the ***manufactured home***. Recall that manufacturing produces products in a factory. The completed product is transported to its place of use. This is exactly how manufactured homes are produced, **Figure 17-4**. Most of the structure is built in a factory. This type of home is usually built in two halves. The floors, walls, and roof are erected. The plumbing and electrical systems are then installed. The structure's interior and exterior are enclosed and finished. This step includes installing flooring, painting walls, setting cabinets and plumbing fixtures, and installing appliances and electrical fixtures.

The two halves of the structure are transported to the site, where a foundation is already in place. Each half is lifted from its transporter and placed on the foundation. The two halves are then bolted together. The final trim that connects the halves is installed. The utilities are hooked up, and the home is ready for the homeowner. Similar techniques are used to produce temporary classrooms, construction offices, and modular units that can be assembled into motels or nursing homes.

Constructing Buildings

Constructed structures start with architectural and engineering plans that were produced using the engineering design process discussed in Section 2. The owner's needs and budget are the most important constraints on these plans. But three other factors also put constraints on the plans:

Figure 17-4. Building a manufactured home.

- **Zoning laws.** These laws are government regulations restricting how a piece of land can be used.
- **Building codes.** These codes are regulations controlling the design and construction of a structure to provide for human safety and welfare.
- **Best (professional) practices.** These practices are the accepted methods or processes the profession considers to be the most appropriate ways to complete an activity or build a structure.

Within the constraints the design provides, most construction projects follow the same basic steps, **Figure 17-5**, including:

1. Preparing the site.
2. Setting foundations.
3. Building the framework.
4. Enclosing the structure.
5. Installing utilities.
6. Finishing the exterior and interior.
7. Completing the site.

Each type of structure has specific actions taken during each step. This helps complete the structure on time. We will look at the steps used to construct a small single-family home. Other construction activities are discussed later in the chapter.

Single-family homes are designed to meet the needs of the owners. These needs include comfort, security, and protection, **Figure 17-6**. To meet these needs, a home must be properly designed and constructed. The construction process starts with locating, buying, and preparing a site.

Preparing the Site

A home location needs to be carefully selected. The location should meet the needs of the people who will live there. For example, a family with children might investigate the schools serving the area. Parents might also consider the distances to work, shopping, recreation, and cultural facilities. The condition of other homes in the neighborhood and building codes are other factors to consider.

Once the site is chosen, it is purchased from the current owner. Next, the site is cleared to make room for the structure. The location of the new building is marked out. The area is cleared of obstacles. Whenever possible, locate buildings to save existing trees and other plant life. The site might require grading to level the site, **Figure 17-7**. Grading prepares areas for sidewalks and landscaping and helps water drain from the site. These preparations are needed for the next step, setting foundations.

Setting Foundations

The foundation acts as the feet of a building. Try to stand on just your heels. You will be unstable and wobble. Likewise, a building without a proper

Preparing the Site

Setting Foundations

Building the Framework

Enclosing the Structure

Installing Utilities

Finishing the Exterior and Interior

Completing the Site

Gehl Co.

Figure 17-5. Most construction projects follow the same basic steps.

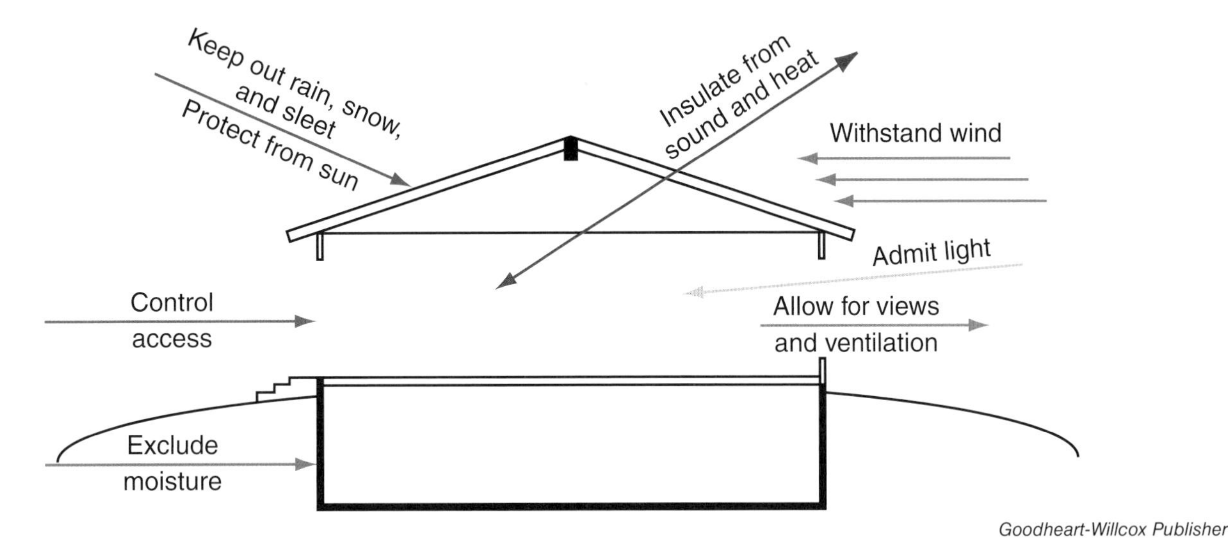

Goodheart-Willcox Publisher

Figure 17-6. A home must meet many needs for its owner.

muratart/Shutterstock.com

Figure 17-7. A site must be cleared and graded before building construction can begin.

foundation settles unevenly into the ground. Such a building leans, becomes unstable, and might collapse. The Leaning Tower of Pisa in Italy is an example of a building that has a poor foundation. Over time, the tower has settled and noticeably leans to one side.

A complete foundation has two parts: the footing and the foundation wall, **Figure 17-8**. The footing spreads the load over the bearing surface. The ***bearing surface*** is the ground on which the foundation and building will rest. This can be rock, sand, gravel, or marsh. Each type of soil offers unique challenges for the construction project.

The type of foundation used is selected to match the soil of the site, **Figure 17-9**, including:

- ***Spread foundations***. This type of foundation is used on rock and in hard soils, such as clay. The foundation walls sit on a low, flat pad called a ***footing***. On wide buildings, posts support the upper floor between the foundation walls. These posts also rest on footings.
- ***Slab foundations***. These types of foundations are used for buildings built on soft soils. They are sometimes called *floating slabs*. The foundation becomes the floor of the building. Such foundations allow the weight of the building to be spread over a wide area. This type of foundation is used in earthquake-prone areas because it can withstand vibration.
- ***Pile foundations***. These types of foundations are used on wet, marshy, or sandy soils. Piles are driven into the ground until they contact solid soil or rock. They are large poles made of steel, wood, or concrete. Piles are widely used for high-rise buildings, marine docks, and homes in areas that flood easily.

Each type of foundation is built in a unique way. Spread foundation construction begins with a site survey to locate the foundation placement, **Figure 17-10**. The site is then excavated in preparation for the footings and the walls. Buildings with basements require deep excavations. Buildings with crawl space require some excavation, but not as deep as for basements. After excavation, footing

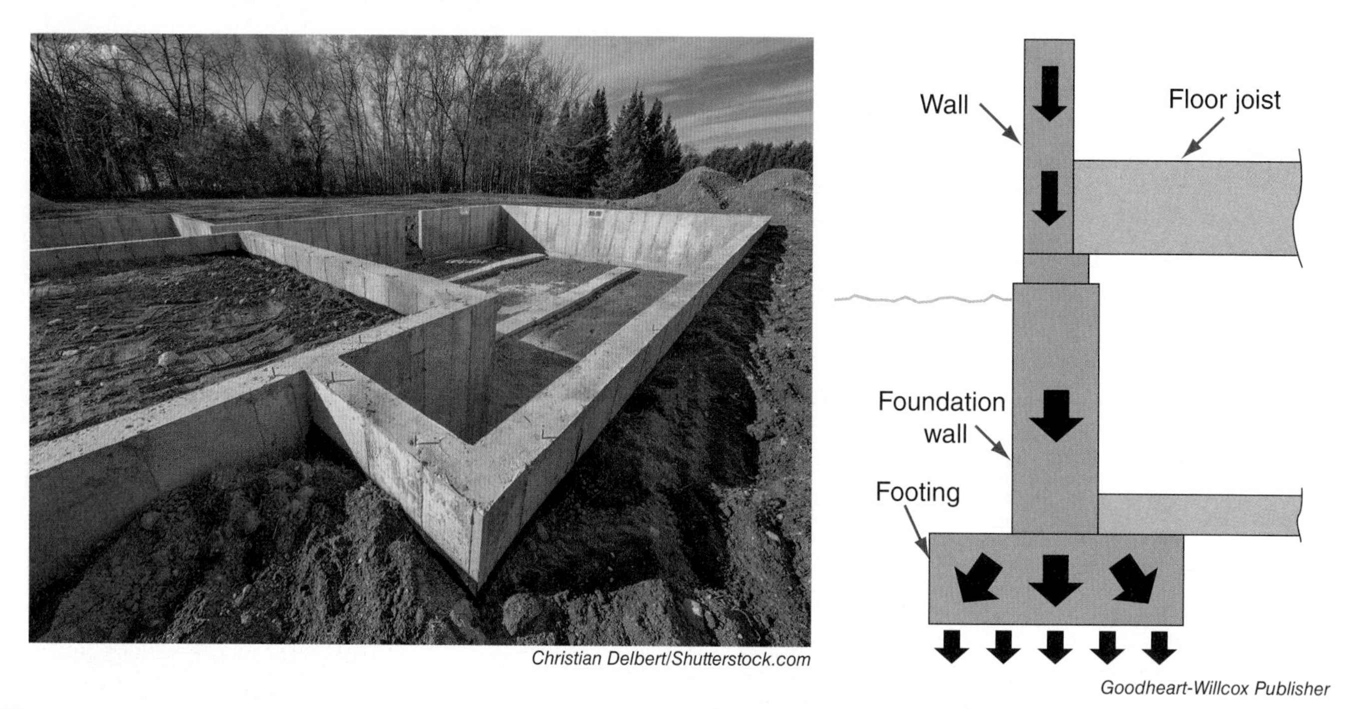

Christian Delbert/Shutterstock.com

Goodheart-Willcox Publisher

Figure 17-8. The foundation wall and footing spread the building's weight onto the bearing surface.

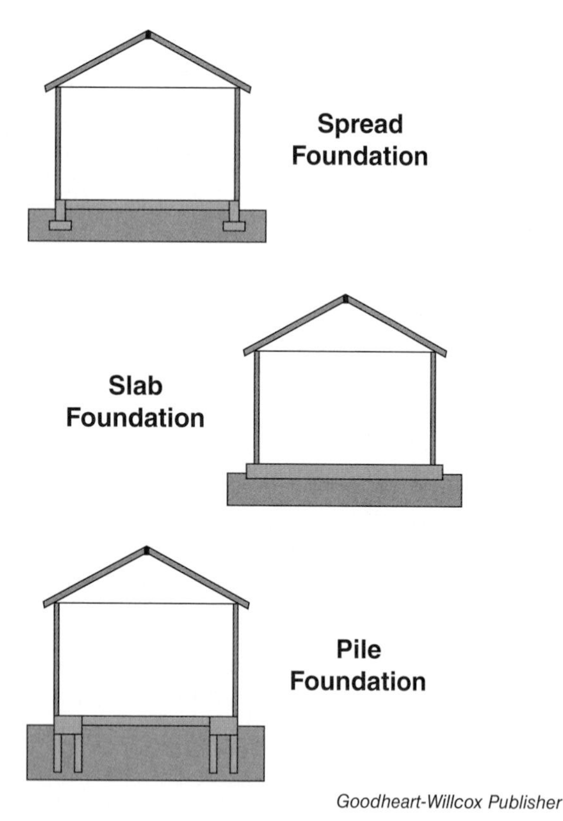

Goodheart-Willcox Publisher

Figure 17-9. Three types of foundations used for buildings are spread, slab, and pile.

Inland Steel Co.

Figure 17-10. This worker is surveying a site for a new building. The survey determines foundation location.

forms are set up next. Forms are lumber frames that hold the wet concrete until the concrete cures (hardens). They give the footings or slabs height and shape. Concrete is poured and leveled off. When the concrete is cured, the forms are removed. Walls of poured concrete or concrete block are built on top of footings. Slabs are ready for aboveground superstructures. Wooden foundations use no concrete for either footings or walls.

Think Green Alternative Construction Methods

Alternative construction methods can be used to create more affordable and efficient housing. These methods often use natural, locally available, and energy-efficient renewable resources. These resources help reduce the cost of construction and the building operations. Several popular types of alternative construction methods include cordwood masonry, straw-bale construction, and earthship construction.

Cordwood masonry, **Figure A**, uses cordwood, which are small debarked logs, and cob, a natural building material made mostly of subsoil and water. The cordwood and cob are used to build the walls of the structure, similar to the way bricks and mortar are used. The logs are laid out and the cob holds them in place. The exposed log ends on each side of the wall are visually appealing. The walls are often insulated during construction by creating a hollow cavity in the wall that is filled with natural insulating materials such as saw dust. The result of this alternative construction method is an energy-efficient building made of natural and locally available materials that is energy efficient in regards to heating and cooling because it is able to store, distribute, or block exterior temperatures.

Straw-bale construction, **Figure B**, involves stacking straw bales in rows on a foundation. The bales are tied together using materials such as wood, rebar, or wire mesh. The straw bales are then covered with some type of earth, clay, or cement-based mixture. Straw bales are excellent renewable construction materials. Compared to the length of time needed to grow trees, straw grows rapidly. Straw bales also have excellent thermal properties, resulting in reduced heating and cooling costs.

Earthship construction, **Figure C**, uses locally available materials to create passive solar homes. Passive solar homes are built to maximize the sun's energy. Walls, floors, and roof are located to effectively distribute, collect, store, or block heat. Earthship construction uses dense thermal masses for the outer walls, mostly in the form of rammed-earth tires, used tires, which are filled and compacted with dirt. The compacted dirt naturally regulates internal temperatures of the structure by storing heat, reducing energy needs. Although these sustainable structures are made of dirt and recycled materials, they still offer the same comforts and amenities as any traditional modern home.

David Majestic/Shutterstock.com

Figure A. Cordwood masonry is economical and energy efficient.

ushi/Shutterstock.com

Figure B. This home is being built using straw bales.

Erik Wannee

Figure C. Earthship buildings are passive solar buildings made of natural materials.

Building the Framework

The foundation becomes the base for the next part of the building, the framework. Erecting the framework gives the building its size and shape. The framework includes the floors, interior and exterior walls, ceilings, and roof. Door and window locations are also determined at this time.

The framework can be built out of three different materials, **Figure 17-11**. Small and low-cost buildings have frameworks made from lumber. Most industrial and commercial buildings have either steel or reinforced concrete frameworks. ***Reinforced concrete*** is concrete that has wire mesh or steel bars embedded into it to increase its tensile strength. Building the framework involves three steps.

Floors

Floors are constructed first, **Figure 17-12**. Homes with slab foundations use the surface of the slab as the floor. Those with basements or crawl spaces use lumber floors.

Lumber floors start with a wood ***sill*** bolted to the foundation. Floor joists are then placed on the sill. ***Floor joists*** are horizontal support pieces that extend across the structure and carry the weight of the floor. The span (distance between outside walls) and the load on the floor determine the size and spacing of the joists. A ***subfloor*** is installed on top of the joists. It is usually made from plywood or particleboard and acts as a base for finished flooring material such as carpet, tile, or wood.

Walls

The wall frames are placed on top of the floor. These frames support both exterior and interior walls. Wall framing is often made of 2×4 or 2×6 construction-grade lumber, **Figure 17-13**. A framed wall has a strip at the bottom called the ***sole plate***. Nailed to the sole plate are upright framing members called ***studs***. The length of the studs is determined by ceiling height. At the top of the wall, the studs are nailed to double ribbons of 2×4s called a ***top plate***, or wall plate. Door and window openings require headers above them. ***Headers*** carry the weight from the roof and ceiling across the door and window openings. Shorter studs, called *trimmer studs*, hold up the headers.

Technology...

Slab Floor

Brandon Bourdages/Shutterstock.com

Lumber Floor

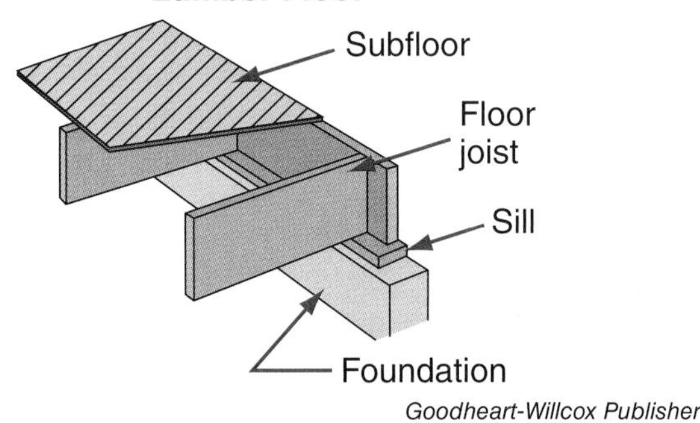

Goodheart-Willcox Publisher

Figure 17-12. The floors in single-family homes are either concrete slabs or lumber.

Lumber

Flas100/ Shutterstock.com

Steel

Lev Kropotov/Shutterstock.com

Reinforced Concrete

Russia Today/Shutterstock.com

Figure 17-11. The materials used for framework are lumber, steel, and reinforced concrete.

Lev Kropotov/Shutterstock.com

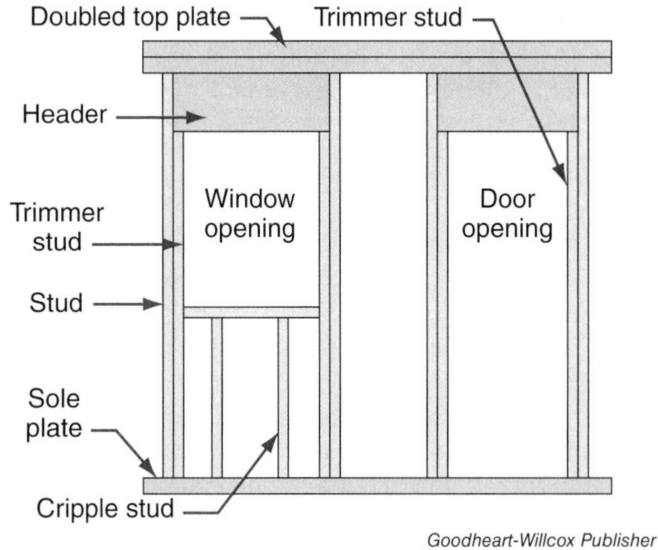

Goodheart-Willcox Publisher

Figure 17-13. Some of the parts of a wood-framed wall.

Goodheart-Willcox Publisher

Figure 17-14. Roof and wall frame construction.

Ceilings

The walls support the ceiling and roof, **Figure 17-14**. The ***ceiling*** is the inside surface at the top of a room. The roof is the top of the structure that protects the house from the weather.

Ceiling joists support the ceiling. These joists rest on the outside walls and some interior walls. Interior walls that help support the weight of the ceiling and roof are called load-bearing walls or bearing walls.

Roofs

The roof forms the top of the building. There are many types of roofs, including gable, flat, hip, gambrel, and shed, **Figure 17-15**. The type of roof is

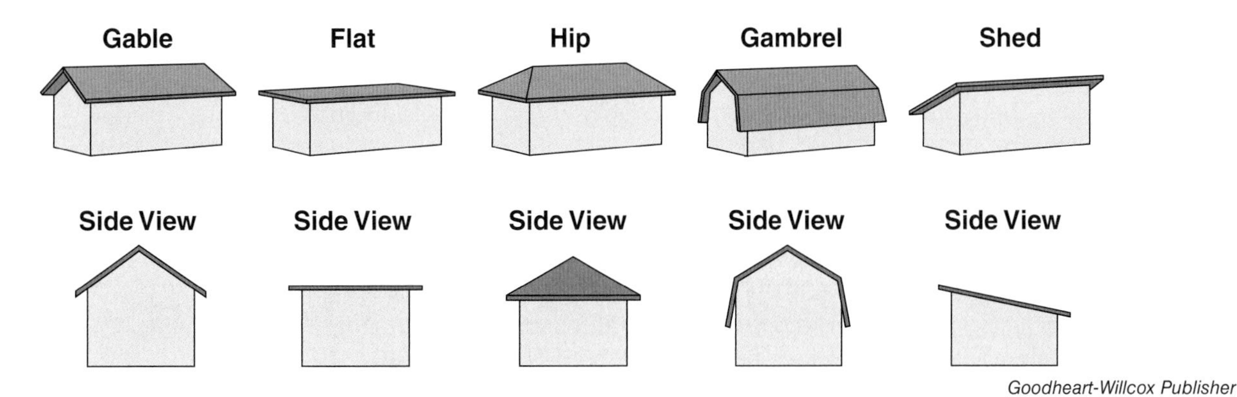

Goodheart-Willcox Publisher

Figure 17-15. Some popular types of roofs used on homes.

Career Connection Construction Manager

A construction manager directs and coordinates the construction and maintenance activities of structures and facilities from project development to completion. This includes overseeing the project's organization, budgeting, and scheduling. To achieve project goals, construction managers must also supervise personnel in a variety of fields, such as carpentry, plumbing, electrical, and heating and cooling.

goodluz/Shutterstock.com

Many high schools offer coursework in technology and engineering. Such coursework introduces students to a variety of skills and concepts needed to understand construction management. Additionally, students interested in construction management should explore the options available at their career and technical centers. Many centers offer career programs in construction, architecture, engineering, and carpentry, which can be beneficial to a future construction manager.

There are numerous two-year colleges offering associate degrees in construction technology or construction management. These degrees combined with construction experience can qualify a student to work as a construction manager on small projects. However, many companies require their construction managers to have bachelor's degrees in construction management or construction science. Many universities and colleges offer bachelor-degree programs in these areas. The programs often require coursework in project management, design, construction methods, building codes, and legal regulations, as well as coursework in mathematics, statistics, and cost control management.

People interested in becoming construction managers should have the ability to work collaboratively with a variety of individuals. Construction managers work with engineers, architects, skilled tradespeople, and a variety of subcontractors to coordinate efforts to achieve project goals. In doing so, construction managers need to be skilled in directing people and resolving conflicts among personnel. This requires problem-solving skills and the ability to make critical on-the-spot decisions.

Construction activity continues to expand due to population increases, aging infrastructures, and the drive to make current structures more efficient and sustainable. Therefore, the need for skilled construction personnel and managers is projected to grow for the near future.

Potential projects undertaken by construction managers include:

- Developing project timelines and worktables.
- Preparing project budgets and cost estimates.
- Creating and interpreting construction contract agreements.
- Analyzing legal requirements and training construction personnel to comply with all regulations and building codes.
- Preparing progress reports and delivering them to project stakeholders and customers.

chosen for its appearance and ability to withstand the weather. For example, a hip roof looks out of place on a Spanish-type home. And flat roofs are a poor choice in areas that receive heavy snow because the roof cannot support the weight of deep snow.

The first step in roof construction is building the roof frame with rafters. ***Rafters*** are angled boards resting on the top plate of the exterior walls. Often, a special structure called a truss is used. A ***truss*** is a triangle-shaped structure that contains both rafters and ceiling joists in one unit. Trusses are manufactured in a factory and then shipped to the building site. The rafters or trusses are covered with plywood or particleboard sheathing. This step completes the erection of the frame.

More recently, technological development has enabled buildings to be constructed automatically from engineering plans using large 3-D printers. This is changing the way people think about constructing structures. Essentially, engineers and architects can design a computer model of a structure and 3-D print it directly on the desired building site using a large 3-D printer that is assembled over the construction site. The large 3-D printer creates the structure by automatically pouring concrete into the desired geometric shapes based on the engineering plans. More development in 3-D printers for construction purposes could potentially decrease the amount of resources needed to produce affordable housing.

Enclosing the Structure

After the framework is complete, the structure must be enclosed. The roof and wall surfaces need to be covered. This process involves enclosing the walls and installing the roof. All homes have both interior and exterior wall coverings. These coverings improve the looks of the building and keep out the elements (rain, snow, wind, sunlight).

Enclosing the exterior walls involves covering all the exterior surfaces, **Figure 17-16**. Plywood, fiberboard, or rigid foam sheets called ***sheathing*** are used to cover the walls. Most foam sheets have a reflective backing to improve the insulation value of the sheet. Most homes built today have a layer of plastic over the sheathing to prevent heat loss.

Normally, the roof is put in place before the utilities are installed. See **Figure 17-17**. The actual roof surface has two parts. Plywood or waferboard sheathing is applied over the rafters. Builder's felt is often applied over the roof sheathing. Wood or fiberglass shingles, clay tiles, or metal roofing are then installed over the sheathing and felt. Flat and shed roofs often use a built-up roof. A built-up roof starts with laying down sheets of insulation. Roofing felt is laid down, followed by a coat of tar, which is covered with gravel.

On many structures, the overhang of the roof is also finished. A ***fascia board*** is used to finish the ends of the rafters and the overhang. The ***soffit*** is installed to enclose the underside of the overhang. Soffits can be made of aluminum, vinyl, or plywood. They must have ventilation holes or vents to prevent moisture and heat buildup in the attic.

Once the sheathing and roof are installed, the openings for doors and windows are cut out. The

Goodheart-Willcox Publisher

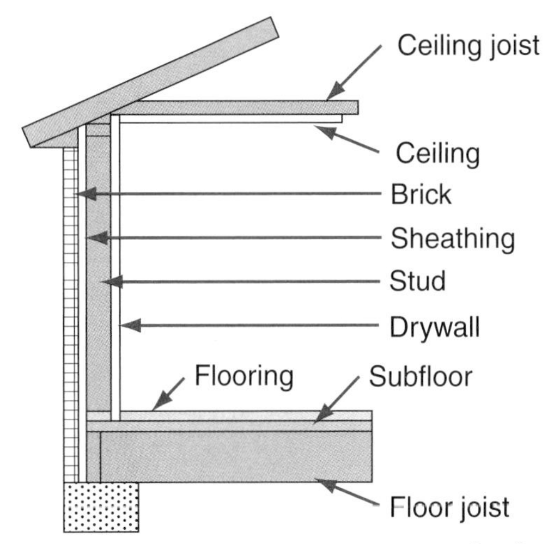

Goodheart-Willcox Publisher

Figure 17-16. A cross-sectional diagram of a finished wall. The photo shows brick being applied as a siding material.

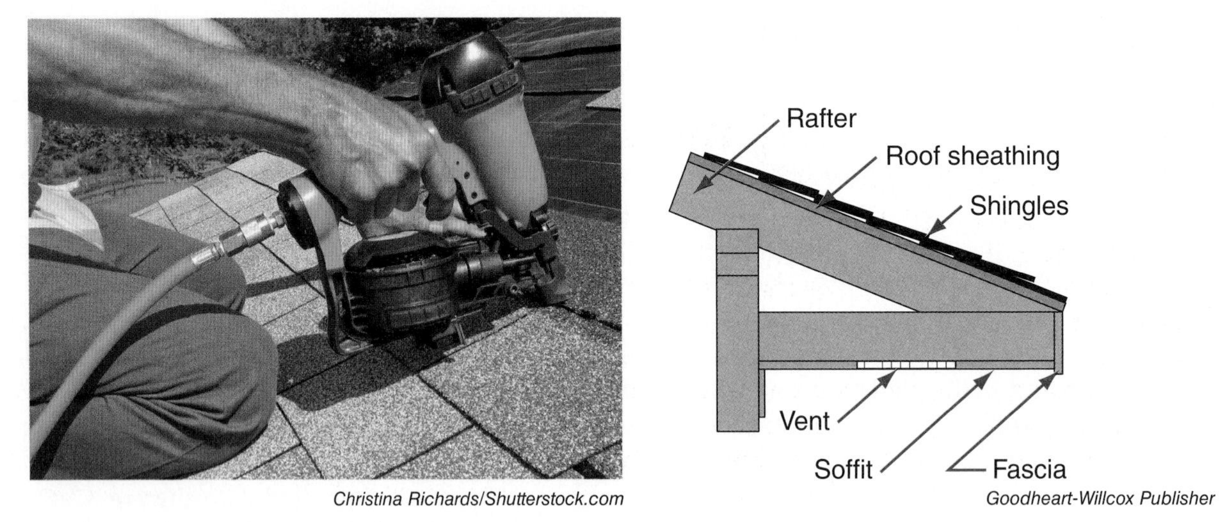

Christina Richards/Shutterstock.com

Goodheart-Willcox Publisher

Figure 17-17. A partially finished roof. The photo shows asphalt shingles being installed on a new roof.

doors and windows are then set in place. Now the house is secure and weather tight.

Installing Utilities

Utilities are usually installed after the building has been enclosed. This prevents theft and weather damage. Some parts of the utilities are installed earlier, such as large plumbing lines. The utility system includes four major systems:

- Electrical.
- Plumbing.
- Climate control.
- Communications.

Electrical systems

The electrical system delivers electrical power to the house. The power is brought into the house through wires to a meter and distribution panel. This panel splits the power into 110-volt and 220-volt circuits. Each circuit has a circuit breaker to protect against current overloads.

Appliances such as clothes dryers, electric ranges, water heaters, and air conditioners require 220-volt power. Circuits for smaller appliances, lighting, and wall outlets use 110 volts. Outlets might have power fed to them at all times. Switches can also control outlets. **Figure 17-18** shows a 110-volt

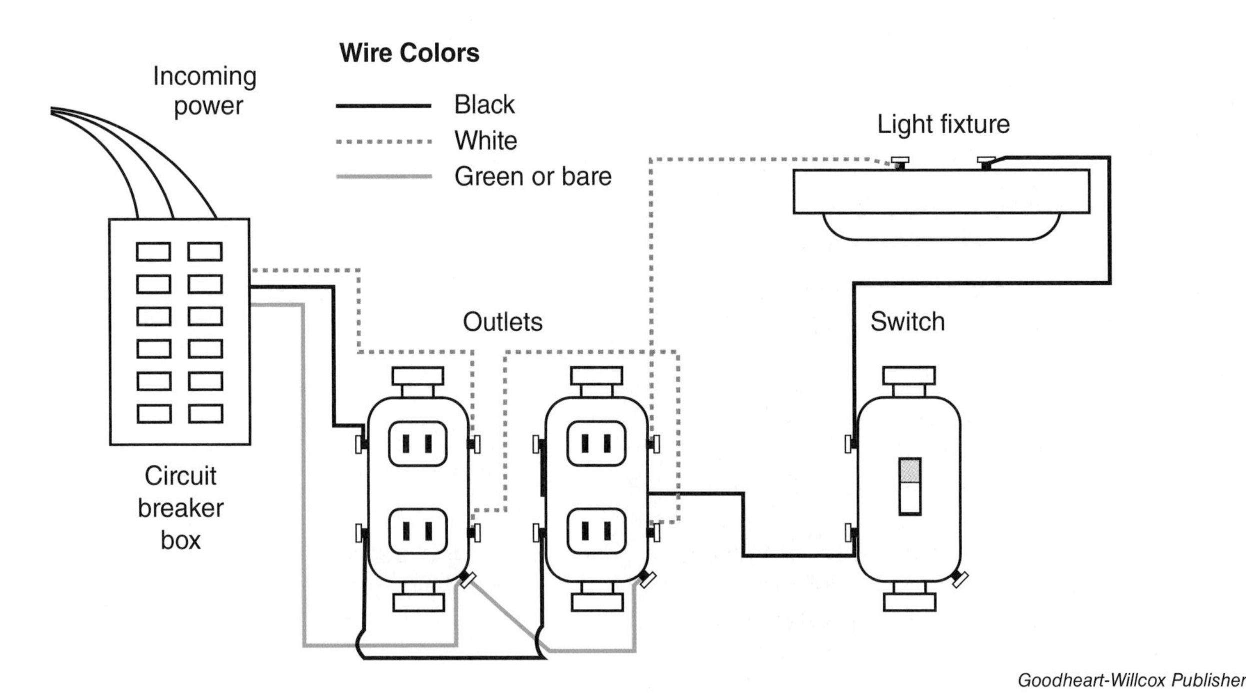

Goodheart-Willcox Publisher

Figure 17-18. A 110-volt electric circuit. Note how the switch controls the light, but not the outlets.

Concrete Calculations

The footing of a home spreads the load of the structure over the ground where the building will rest, **Figure A**. The foundation wall, typically made from poured concrete or concrete block, is then built on top of the footings. Therefore, concrete must be purchased, which requires calculations of how much is needed. If too much is ordered, the extra concrete will need to be disposed of, resulting in higher costs for the construction project. If not enough is ordered, then additional challenges are created. Therefore, it is important to do the calculations to order the correct amount.

To begin, we will calculate the materials needed for a simple foundation of 24′ × 24′ with a footing 12″ thick and 16″ wide and a foundation wall 8″ thick and 8′ tall. This includes ordering cubic yards of concrete for pouring the foundation as well as the rebar used to reinforce it.

First, determine the perimeter of the foundation by adding the sides. In the case of a rectangle, the formula would be:

$$P = \text{side 1} + \text{side 2} + \text{side 3} + \text{side 4}$$

Or:

$$24' + 24' + 24' + 24' = 96'$$

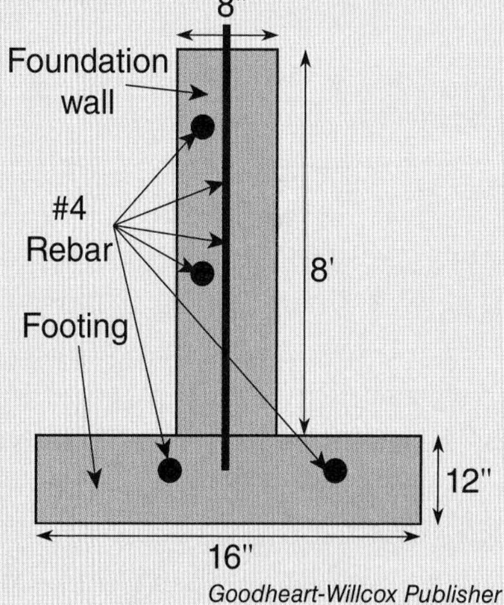

Goodheart-Willcox Publisher

Figure A. Foundation cross-section.

Next, find the volume of the footing by multiplying the product of the thickness and width by the perimeter. However, be sure to make sure all the units are the same.

Change inches to feet:

$$16'' \times 1'/12''$$

$$16'/12 = 1.33 \text{ ft (rounded to the nearest hundredth)}$$

Now calculate the volume:

$$1' \times 1.33' \times 96' = 127.68 \text{ cubic feet}$$

Next, find the volume of the foundation wall by multiplying the product of the thickness and width by the perimeter. Again, make sure all units are the same.

Change inches to feet:

$$8'' \times 1'/12''$$

$$8'/12 = 0.67 \text{ ft (rounded to the nearest hundredth)}$$

Now calculate the volume:

$$0.67' \times 8' \times 96' = 514.56 \text{ cubic feet}$$

Cubic feet must be converted to cubic yards because this is the way concrete is ordered. There are 27 cubic feet in 1 cubic yard.

$$514.56 \text{ ft}^3 \times 1 \text{ yd}^3/27 \text{ ft}^3 = 19.06 \text{ yd}^3$$

You would probably want to order about 20 cubic yards of concrete.

Now look at the amount of rebar needed for the foundation. Rebar is used to increase the tensile strength of the concrete. For typical homes, size #4 rebar is suspended vertically every 3′ in the concrete of the foundation wall. Each piece is typically 10′ long, which allows it to extend into the footing for a better connection between both sections. Additionally, two more pieces of rebar are suspended horizontally in the foundation wall and two more horizontally in the footing, **Figure A**. How many 10′ pieces of #4 rebar are needed for the foundation?

circuit with wall (duplex) outlets and a ceiling light. The outlets always have power. The circuit to the light has a switch, however.

Most 110-volt circuits are limited to 15 or 20 amps. Therefore, a number of different circuits are required to supply various parts of the home. A kitchen might have one or two circuits based on the number of appliances used there. One circuit might feed two bedrooms because there are few appliances in these rooms.

Plumbing systems

Plumbing systems have two parts. One part supplies potable water, **Figure 17-19**. ***Potable water*** is safe for drinking. The other part of the system carries away wastewater. Plumbing fixtures and systems are designed to prevent mixing of potable water and wastewater and to stop sewer gas from leaking into the dwelling.

The potable-water system starts with a city water supply or a well for the house. The water enters the house through a shutoff valve and might pass through a water conditioner to remove impurities, such as iron and calcium. The waterline is split into two branches. One line feeds the water heater. The other line feeds the cold-water system. Separate hot and cold waterlines feed fixtures in the kitchen, bathrooms, and utility room. Toilets receive only cold water. Most waterlines have shutoff valves before they reach the fixture. For example, the waterlines under a sink should have a shutoff valve. The valve allows repairs to be made without stopping the flow of water to the rest of the house.

The second part of the plumbing system is the ***wastewater*** system. This system carries used water away from sinks, showers, tubs, toilets, and washing machines. The wastewater is routed to a city sewer line or to a septic system. At each fixture and appliance, a device called a ***trap*** is provided. A trap is a *U*-shaped piece of pipe that remains full of water. The water in the line stops gases in the sewer system from leaking into the home. Wastewater systems have a network of vents to prevent the water from being drawn out of the traps. The vents also allow sewer gases to escape above the roof without causing any harm.

Homes that use natural gas have a third type of plumbing. Gas lines carry natural gas to furnaces, stoves, water heaters, and other appliances. Shutoff valves are installed at the entrance and at each major appliance.

Climate-control systems

In many homes, the climate-control system is used to heat the building in winter and cool it in summer. This can be done with a single unit or with separate heating and cooling units. Heating systems can directly or indirectly heat the home.

In a direct heating system, the fuel is used in the room to be heated. Direct heating might use a stove or a fireplace that burns wood or coal. Other direct heating methods use electrical power. These systems use resistance heaters installed in the walls or along the baseboards. Also, ceiling radiant wires or panels are also used to supply heat.

Indirect systems heat a conduction medium, such as air or water, which carries the heat from

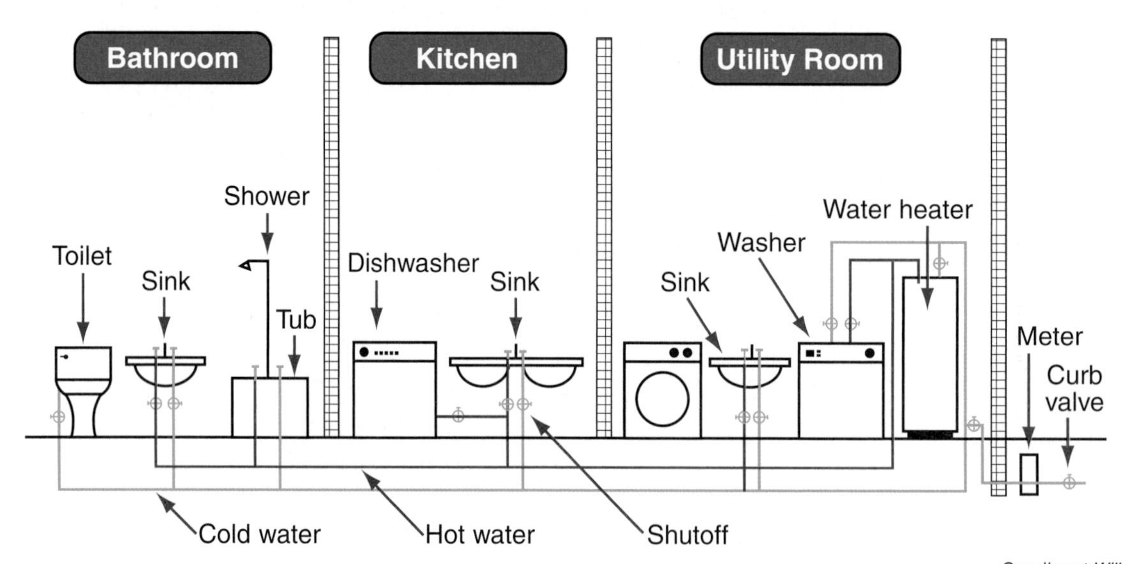

Goodheart-Willcox Publisher

Figure 17-19. The potable-water system for a home. Each fixture has a shutoff valve.

furnace to rooms. The heat is then given off to the air in the room. See **Figure 17-20**. The energy sources for these systems are electricity, coal, oil, wood, natural gas, or propane.

Furnaces that heat air as a conduction medium are called ***forced-air heating systems***. Forced-air furnaces draw air from the room. This air is heated as it moves through the furnace. A fan delivers the heated air through ducts to various rooms.

Hot water heating uses water to carry the heat. The water is heated in the furnace, pumped to various rooms, and passes through room heating units that have metal fins surrounding the water pipe. The fins pass the heat into the room.

Some homes are heated with active or passive solar-energy systems. Passive solar systems use no mechanical means to collect and store heat from the sun. Active solar systems use pumps or fans to move a liquid or air to collect solar heat. After the heat is collected, the liquid or air is moved to a storage device. Solar heating systems are effective in areas that have ample sunshine.

Many buildings have cooling systems to cool the air during the warm months of the year. Cooling systems use compressors, evaporators, and condensers. The system has a fan that draws the air from the room and passes it over a cold evaporator. This is similar to a forced-air furnace. Instead of the air being heated, however, it is cooled before it is returned to the room.

Another system used in climate control is a unit called a ***heat pump***. A heat pump works as a cooling and heating system by capturing heat in the atmosphere. This pump can be operated in two directions. Operating in one direction, the heat pump acts similar to an air conditioner. The heat pump takes warmth from inside the house and discharges it outside. In the winter, the heat pump works in reverse. The pump takes warmth from the outside and brings it into the house. Heat pumps can use air, water, or the ground as a heat source. Those that use air work best in areas that do not get very cold, such as the southern United States. Groundwater heat pumps use well water as the heat source. The water is pumped from the well to the heat pump, has heat removed or receives excess heat, and then returns to the well. ***Geothermal systems*** use coils buried a few feet down in the soil where the temperature remains at a constant 55°F regardless of the season. Air or water is pumped through the coils to take on or give off heat as needed. Groundwater and geothermal systems can be used in colder climates. However, a small furnace or other auxiliary heat source may be needed as a backup.

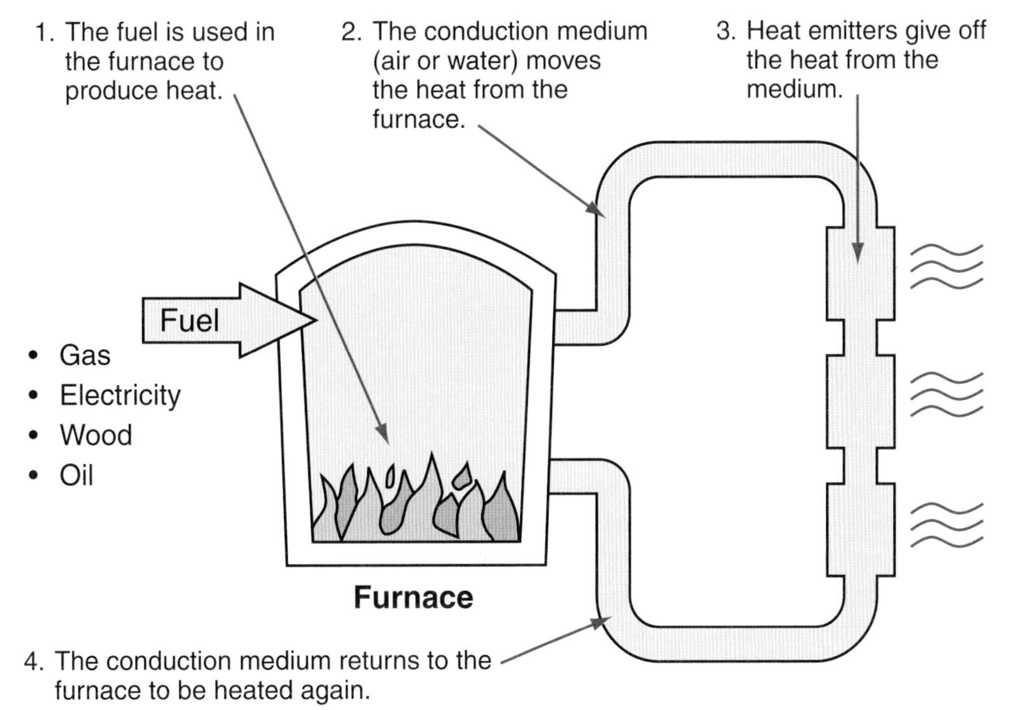

Goodheart-Willcox Publisher

Figure 17-20. Indirect heating systems heat a conduction medium. The resulting heat is then put into the air in the room.

Academic Connection: Communications

Word Origins

Many people would be pleased if their names become part of everyday language as a result of their inventions. John McAdam, a Scottish engineer who experimented with road construction, might be doubly pleased. His work with road construction resulted in two words in common use today.

As mentioned in this chapter, John McAdam developed the crushed-stone road. This new type of road had three layers of crushed rock compacted into a solid mass. The road was also made slightly convex. McAdam's design improved roads tremendously because it spread the traffic load and forced rainwater off the surface. This type of roadway is commonly known as *macadam*.

The other word, even more familiar, is also related to roads. In an effort to improve roads even more, people used tar to bind the crushed rock together. This process was called *tarmacadam* or, as it is called when used on runways, *tarmac*. See **Figures A** and **B**.

Modern roads are still built using John McAdam's principles. Can you find another common word we use today that is based on someone's name and invention?

©iStockphoto.com/resonants

Figure A. An asphalt road, a blacktop road, or a tarmac road.

©iStockphoto.com/Pesky Monkey

Figure B. Fresh asphalt (tarmac).

Communication systems

Most homes have communication systems, such as telephone, Internet, and television, which require special wiring. Telephone wiring and television cables are normally installed during the construction of the building. Installation after a building is complete requires considerable work to feed the wires through attics, under floors, and inside walls.

Finishing the Exterior and Interior

The final step in building construction is finishing the exterior and interior. Siding is the finish covering used on a wood building. Wood shingles and boards, plywood, hardboard, brick, stone, aluminum, vinyl, steel, and stucco can be used as siding. Trim is strips of wood covering the joints between window and door frames and the siding.

The interior walls are finished next. Insulation is placed between the studs and around the windows and doors of all exterior walls. This reduces heat loss on cold days and heat gain on hot days. The most common type of insulation is fiberglass blankets or batts. A vapor barrier of polyethylene film is attached to the studs over the insulation. This barrier prevents moisture from building up in the insulation.

Once insulation and utilities are in place, the interior wall surfaces can be covered. The most widely used interior wall covering is gypsum wallboard, commonly known as ***drywall***. Drywall is a sheet material made of gypsum bonded between layers of paper. The sheets of drywall are nailed or screwed onto the studs and ceiling joists. The fastener heads and drywall seams are then covered with a coating called joint compound. The compound is applied in several thin coats. This is done to smooth the surfaces and joints between the sheets of drywall.

The inside and outside of the house are now ready for the finishing touches. Interior wood trim is installed around the doors and windows. Kitchen, bathroom, and utility cabinets are set in place. Floor coverings, such as ceramic tile, wood flooring, carpet, or linoleum, are installed over the subflooring. Baseboards are installed around the perimeters of all the rooms. The exterior siding and wood trim are painted. Interior trim is painted or stained. The walls are painted or covered with wallpaper or wood paneling. Lighting fixtures, switch and outlet covers, towel racks, and other accessories are installed. The floors and windows are cleaned. Now the home is finished and ready to be occupied.

Completing the Site

Completing the building is the most important part of the project. Other work remains to be done, however. The site must be finished. Earth must be moved to fill in areas around the foundation. Sidewalks and driveways need to be installed. The yard area needs to be landscaped.

Landscaping is trees, shrubs, and grass that are planted to help prevent erosion and improve the appearance of the site. These plants can divide the lot into areas for recreation and gardening. Landscaping can be used to screen areas for privacy, direct foot traffic, and shield the home from wind, sunlight, and storms, **Figure 17-21**. Notice how trees and lawn improve the appearance of a building site and how a sidewalk can be used to guide foot traffic through the yard.

ppa/Shutterstock.com

Figure 17-21. A completed building site.

Heavy Engineering Structures

Construction activities do not always produce buildings. They can be used to produce civil structures, or heavy engineering structures. These structures include highways, rail lines, canals, pipelines, power-transmission and communication towers, hydroelectric and flood-control dams, and airports. They provide the paths for the movement of water, people, goods, information, and electric power. These projects can be grouped in various ways. For this discussion, we group them into transportation, communication, and production structures.

Transportation Structures

Transportation systems include railroad lines, highways and streets, waterways, and airport runways. Other constructed works help vehicles cross uneven terrain and rivers. These structures include bridges and tunnels. Pipelines are land-transportation structures used to move liquids or gases over long distances.

Roadways

The Romans built the first engineered roads more than 2000 years ago. Their influence remained until the 1700s, when modern road building started. Today's roads have their roots in the work of the Scottish engineer John McAdam. He developed a crushed-stone road built of three layers of crushed rock, laid in a 10″ (25 cm) thick ribbon. Later, this roadbed was covered with an asphalt-gravel mix that is very common today. A more recent development is the concrete roadway.

Road building starts with selecting and surveying the route. The route is then cleared of obstacles such as trees, rocks, and brush. The roadway is graded so it will drain. Proper drainage prevents roadways from being damaged by freezing and thawing. Also, a dry roadbed withstands heavy traffic better than a wet, marshy one. Another reason for grading is to keep the road's slope gentle. Elevation changes are described using the term *grade*. Grades are expressed in percentages. A road with a 5% grade gains or loses 5′ of height for every 100′ of distance. Most grades are kept below 7%.

Once the roadbed is established, the layers of the road are built, **Figure 17-22**. The graded dirt is compacted, and a layer of coarse gravel is laid. This is followed with finer gravel that is leveled

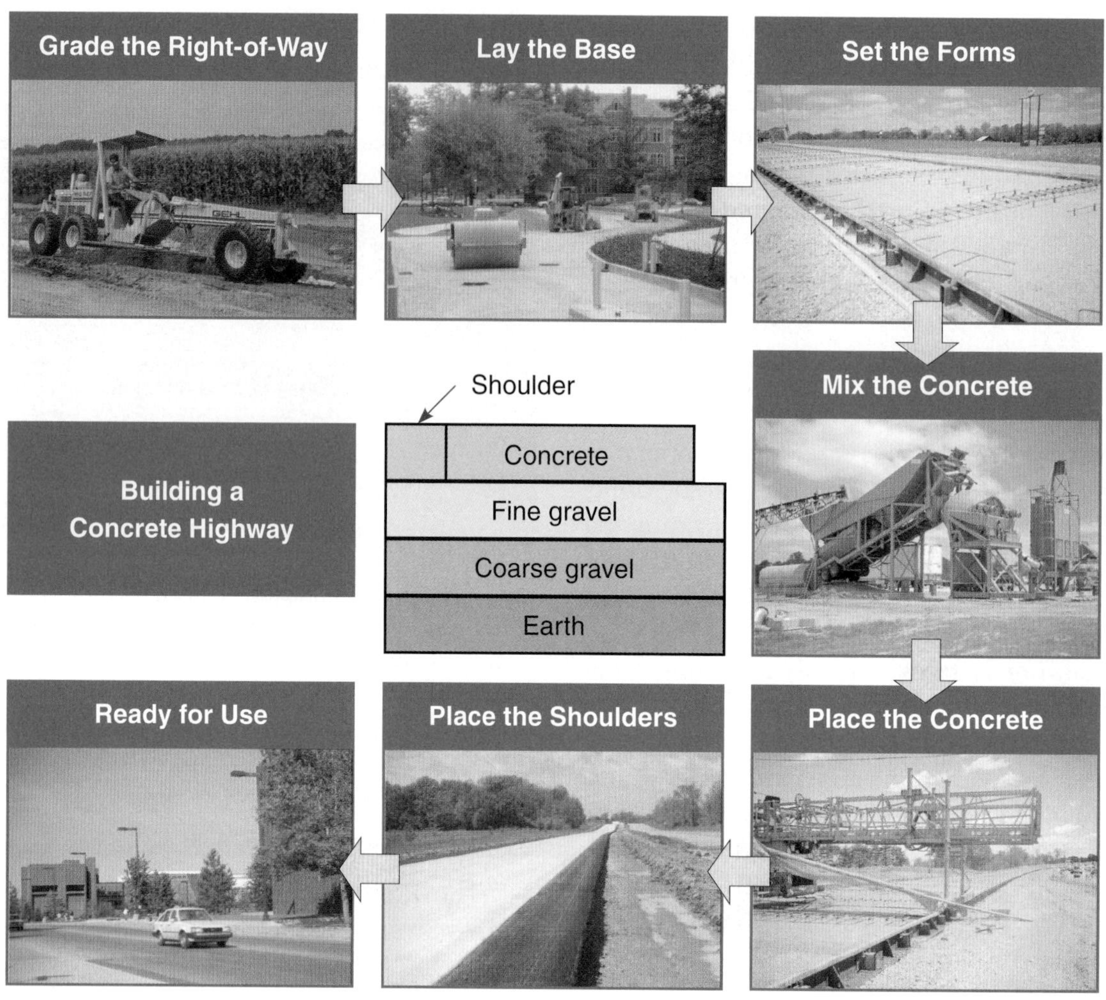

Figure 17-22. Road building is a step-by-step process.

and compacted. Next, the concrete or asphalt top layer is applied. Concrete roads are laid in one layer. Asphalt is generally applied in two layers: a coarse undercoat and a finer topcoat. Finally, the shoulders, or edges, of the road are prepared. The shoulders can be gravel or asphalt.

Bridges

Another constructed structure vital for transportation is the bridge. Bridges provide a path for vehicles to move over obstacles. These obstacles include marshy areas, ravines, other roads, and bodies of water. Bridges can carry a number of transportation systems. These systems include highways, railroads, canals, pipelines, and footpaths.

Generally, there are fixed and movable bridges. A fixed bridge does not move. Once the bridge is set in place, it stays there. Movable bridges can change their positions to accommodate traffic below them. This type of bridge is used to span ship channels and rivers. The bridge is drawn up or swung out of the way so ships can pass.

Bridges have two parts, **Figure 17-23**. The substructure spreads the load of the bridge to the soil. The abutments and the piers are parts of the substructure. The superstructure carries the loads of the deck to the substructure. The deck is the part used for the movement of vehicles and people across the bridge.

The superstructure a bridge has indicates the type of bridge it is. The most common types of bridges are beam, truss, arch, cantilever, and suspension, **Figure 17-24**.

Beam bridges use concrete or steel beams to support the deck. This type of bridge is widely used when one road crosses another one. Beam bridges are very common on the interstate highway system.

Truss bridges use small parts arranged in triangles to support the deck. These bridges can carry heavier loads over longer spans than beam bridges. Many railroad bridges are truss bridges.

Tanya Puntti/Shutterstock.com

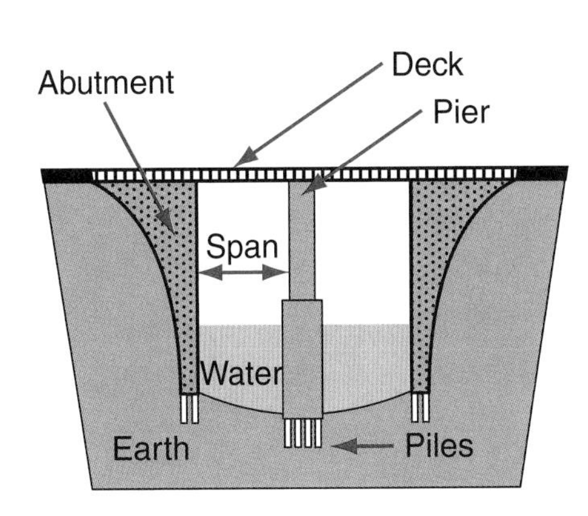

Goodheart-Willcox Publisher

Figure 17-23. The parts of a bridge. An arch bridge is shown in the photo.

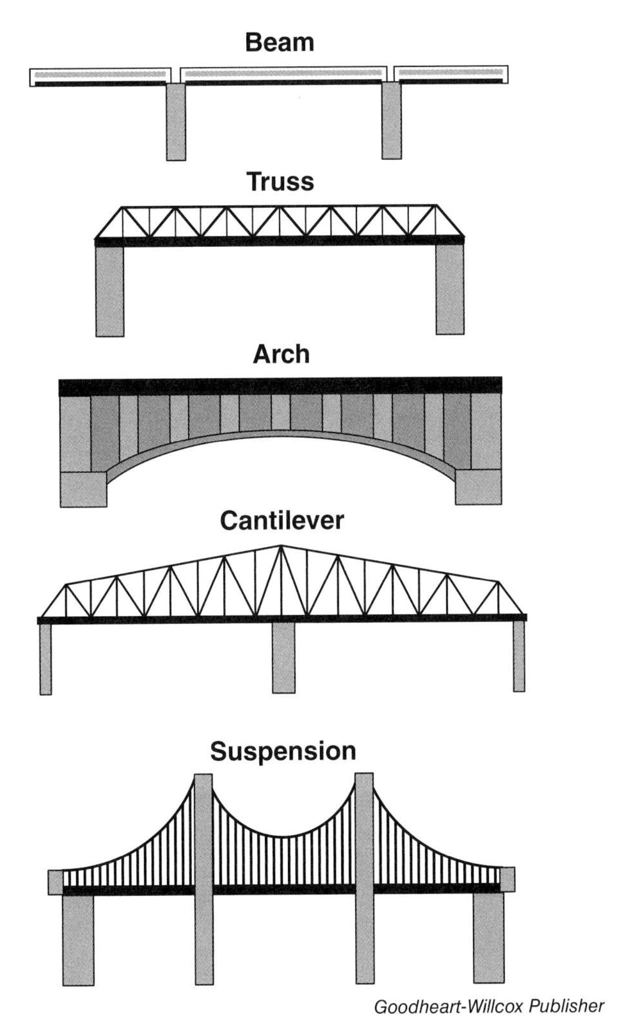

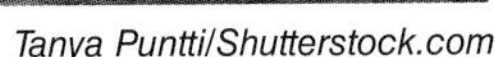

Goodheart-Willcox Publisher

Figure 17-24. Five common bridge types are the beam, the truss, the arch, the cantilever, and the suspension.

Arch bridges use curved members to support the deck. The arch can be above or below the deck. Arch bridges are used for longer spans. One of the longest arch bridges spans more than 1650′ (503 m).

Cantilever bridges use trusses extending out in both directions from the support beams, similar to arms. The ends of the arms can intersect with the road leading up to the bridge or hook up to another truss unit to form a longer span. The arms transmit the load to the center.

Suspension bridges use cables to carry the loads. A large cable is suspended from towers. From the large cable, smaller cables drop down to support the deck. Suspension bridges can span distances as great as 4000′ (1220 m) and longer.

Communication Structures

Most telecommunication technology relies on constructed towers to support antennas. These towers are usually placed on a concrete foundation. A steel tower is built on top of the foundation. Once the tower is complete, the signal wiring can be installed. Similar techniques are used to construct towers for power-transmission lines. See **Figure 17-25**.

Production Structures

Some structures used for production activities are not buildings. For example, petroleum refineries are a mix of machinery and pipelines. Irrigation systems are constructed to bring water to farms in dry areas. Evaporation basins are built to recover salt and other minerals from seawater.

Another important production structure is the dam. Dams are used to control flooding, supply water, create lakes, or generate electricity. Several types of dams exist. One type is called a ***gravity dam***. This dam's upstream side is vertical, whereas

kanvag/Shutterstock.com

Figure 17-25. This mobile phone base station transmits telephone messages.

the downstream side slopes outward. The sheer weight of the concrete the dam is made from holds the water back. See **Figure 17-26**.

Two more types of dams are the rock dam and the earth dam. An earth dam is shown in **Figure 17-27**. A rock dam looks similar to two gravity dams placed back-to-back. Both sides slope outward. Rock and earth dams must be covered with a waterproof material to prevent seepage. Clay is often used for this covering.

A ***buttress dam*** has a solid upstream side and is supported on the downstream side with a series of supports, or buttresses.

An ***arched dam*** spreads the pressure onto the walls of the canyon where the dam is built. The arched shape increases the strength of the dam.

Zack Frank/Shutterstock.com

Figure 17-27. Earth dams can be economical to build, making them attractive in some locations.

Matt Ragen/Shutterstock.com

Figure 17-26. A gravity dam uses the weight of the material to resist the pressure of the water against it.

STEM Connection: Science

Bridge Design and Natural Forces

All structures are designed to withstand certain natural forces specific to their region. Therefore, it is necessary for engineers and architects to have a thorough knowledge of the environment and weather patterns at a construction site. Using this knowledge, they can design structures that ensure minimal damage is caused during natural extremes, while maintaining the cost efficiency of construction. However, if proper scientific knowledge of nature and physics is not taken into account when designing structures, the results can be disastrous. One notable engineering disaster involved natural forces acting in an unexpected way on a suspension bridge in Tacoma, Washington.

National Park Service

Figure A. Tacoma Narrows Bridge moving in the wind.

The Tacoma Narrows Bridge was designed to withstand wind forces up to 120 miles per hour (mph), which was well above the normal wind speeds in that location. However, in 1940 the bridge collapsed in winds of only 40 mph. So what happened? When it was designed, the bridge designers did not take aerodynamics into account. Aerodynamics is the study of the way in which the air passing over a solid body experiences the forces of lift and drag. The bridge materials may have been strong enough to withstand the force of the wind, but the way the wind passed over the shape of the bridge caused forces other than the wind itself to act upon the structure. In this case, the wind caused the bridge to experience lift force. Lift force opposes the weight of an object, causing it to rise. In the case of an airplane, wing shape forces air to pass underneath it, pushing the plane upward.

In this situation, the bridge shape made it experience lift as the wind blew past it. The lift caused the bridge to vibrate. The vibrations moved the bridge in the form of waves. On windy days, the bridge would wave up and down so much that the bridge was nicknamed Galloping Gertie, **Figure A**. However, what caused the bridge to collapse was the perfect resonance force from a 40 mph wind. Resonance occurs when an external force, such as wind, forces an object to continuously vibrate at a greater and greater frequency until the object breaks. This is essentially what happens when a singer's voice breaks a glass. The pitch of the voice breaks the glass. The 40 mph wind was the perfect speed to cause the vibrations to resonate and ultimately reach a wave frequency that destroyed the bridge.

Using this scientific knowledge, engineers now design bridges with dampeners that interrupt resonant waves of motion, preventing them from increasing to levels that will cause a bridge collapse.

Summary

- ✔ Construction can be divided into two kinds of projects: buildings and heavy engineering structures.
- ✔ Different building types have different uses.
- ✔ Three commonly used types of buildings include residential buildings, commercial buildings, and industrial buildings.
- ✔ Residential buildings are constructed for people to live in and can be single-family or multiple-unit dwellings.
- ✔ Commercial buildings are used for business purposes and are built in a range of sizes.
- ✔ Industrial buildings house manufacturing businesses. They protect and support the machinery, materials, and workers needed for manufacturing processes.
- ✔ An owner's needs and budget are the most important constraints on a constructed structure.
- ✔ Other constraints include zoning laws, building codes, and best practices.
- ✔ Preparing a site includes choosing a site, and then clearing and grading the site.
- ✔ Foundations support buildings. A poorly built foundation will cause a building to settle unevenly and become unstable.
- ✔ Three common foundations are spread, slab, and pile.
- ✔ Erecting the framework gives the building its size and shape.
- ✔ Frameworks can be made from lumber, steel, or concrete.
- ✔ Enclosing a structure involves enclosing the walls and installing the roof.
- ✔ Utilities are usually installed after the building has been enclosed.
- ✔ Common utilities include electrical, plumbing, climate-control, and communications systems.
- ✔ Typical exterior finishing materials on buildings include siding, shingles, plywood, brick, stone, and stucco.
- ✔ Typical interior finishing materials on buildings include insulation, drywall, and wood trim.
- ✔ Final grading and installing sidewalks, driveways, and landscaping complete a building site.
- ✔ Heavy engineering structures include transportation structures, communication structures, and production structures.

Check Your Engineering IQ

Now that you have finished this chapter, see what you learned by taking the chapter posttest.
www.g-wlearning.com/technologyeducation/

Test Your Knowledge

Answer the following end-of-chapter questions using the information provided in this chapter.

1. Name two kinds of constructed works.
2. *True or False?* The steps in constructing buildings vary based on what type of building is being constructed.
3. Post offices and firehouses are examples of _____ buildings.
4. What is a manufactured home?
5. *True or False?* Spread foundations are used on wet or sandy soils.
6. The two types of floors used in homes are concrete slab or _____.
7. What is a fascia board used for?
8. Name the two types of water systems that are part of a plumbing system.
9. _____ is trees, shrubs, and grass that are planted to help prevent erosion and improve the appearance of the site.

Matching: Match each description with the correct construction step. Letters can be used more than once.

10. Sheathing the walls
11. Installing drywall
12. Grading
13. Installing footings
14. Landscaping
15. Installing a heat pump
16. Installing a subfloor
17. Installing the roof
18. Driving in piles
19. Marking the building site
20. Placing floor joists
21. Adding baseboards
22. Installing a sidewalk

A. Preparing the site
B. Setting foundations
C. Building the framework
D. Enclosing the structure
E. Installing utilities
F. Finishing the exterior and interior
G. Completing the site

23. What does the term *grade* mean, as used in this chapter?
24. On bridges, the _____ spreads the load of the bridge to the soil.
    A. substructure
    B. superstructure
    C. deck
    D. None of the above.
25. Tall dams holding back large quantities of water are called _____.

Critical Thinking

1. Use a chart similar to the one shown here to list and describe a few of the constructed structures you see as you travel from your home to school.

Structure	Type of Construction	Description and Use

2. Select one structure you saw in completing the previous assignment and make a drawing of the structure. Then label the major parts and summarize their characteristics.
3. Develop a preventive maintenance schedule for your home. List the items or portions of the structure that need maintenance and the type of maintenance each needs.
4. Explain how various natural forces may impact your home.
5. Describe ways in which alternative construction methods help reduce the use of natural resources.

STEM Applications

1. Research different shed designs and select one that you would like to install on your property. Next, research the soil in your area and use this information along with the research of shed designs to determine what type of foundation (spread, slab, or pile) would be best for installing the shed that you selected. Then, create a drawing of the foundation for your shed and calculate the amount of concrete needed for the foundation. Lastly, create a presentation of your results.
2. Create an accurate scale model of the shed from the previous question based on the wood framing techniques that you learned in this chapter. Label each component of the shed frame.
3. Locate a bridge in your area. Research how it was designed to withstand the natural forces that occur at its site. Then create a scale model of the bridge to provide a demonstration as to how the natural forces affect the structure.

Engineering Design Challenge

Load Bearing Heavy Engineering Structure

Background

In this chapter you learned about heavy engineering structures, including transportation assemblies such as bridges. There are a variety of different bridge types (beam, truss, arch, cantilever, and suspension) used for different situations. All of these bridges consist of two parts, the substructure and the deck. The substructure is the part of the bridge that carries the load of the deck and spreads that load out to the ground. The deck is the section that supports the load of the items moving across the bridge. In this challenge, you will research different types of bridges and design and produce a model bridge to hold the most weight in the most efficient manner possible.

Situation

Your team has been asked to use their knowledge of bridges and engineering to build the strongest and most efficient model bridge, using only 1/8″ × 1/8″ strips of balsa wood, wood glue, and 2-lb test fishing line. The bridge must span a gap of 12″. The bridge will be tested to failure by placing weights on the middle of the bridge. The efficiency of the bridge will then be calculated and compared to other model bridges.

Desired Outcomes

Your model bridge must meet the following requirements:

- The bridge must span a gap of 12″.
- The bridge must be freestanding.
- The bridge deck must be at least 2″ wide.

Additionally, your team must also develop a technical report that describes how the bridge was designed using engineering principles.

To test the model bridges, your team should record the following structural data to calculate the efficiency compared to other model bridges:

- Bridge weight
- Total weight capacity of the bridge

After collecting this information, your team can perform the following calculations to determine the structure's efficiency points:

Calculate efficiency points determined by the following formula:

$$\text{Bridge efficiency} = \frac{\text{maximum load (g)}}{\text{mass of structure (g)}}$$

The structure with the highest efficiency will score 20 points. The remaining scores will be based on percentage of the top score. For example:

First place bridge efficiency = 1550

Second place bridge efficiency = 1250

$$\frac{1250}{1550} = 81\%$$

This percentage will be multiplied by the 20 points possible, which equals 16.2 points.

Materials and Equipment

This project is restricted to the following materials:

- 1/8″ × 1/8″ strips of balsa wood
- Wood glue
- 2-lb test fishing line

Testing requires the following materials:

- Two tables spaced 12″ apart on which each end of the bridge will rest
- Weights to place on the bridge
- A scale for weighing the bridge

Beginning the Process

- Use your knowledge of engineering design to guide you as you define the problem and create a viable solution to it.
- Remember the engineering design process is an iterative approach to solving problems, meaning you can and should go back and forth between the different steps.

- Determine your engineering design team based on students' individual skills. Talk with your team to formulate a problem statement based on the background and situation provided. A good problem statement should address a single problem and answer the questions "who, what, where, when, and why?" The problem statement should not suggest there is a single solution to the problem. Your class may create different problem statements based on each team's interpretation of the situation and their prior interests and experiences. Your engineering team will use your defined problem statement as the starting point to creating a valid solution.
- Apply the knowledge you have obtained and use the engineering design process to propose new or improved solutions to the problem defined from the situation provided. This will be a natural progression from brainstorming, to generating ideas, to producing prototypes/models, to evaluating success, and to iteratively refining the solution.

CHAPTER 18

Meeting Needs through Architecture and Civil Engineering

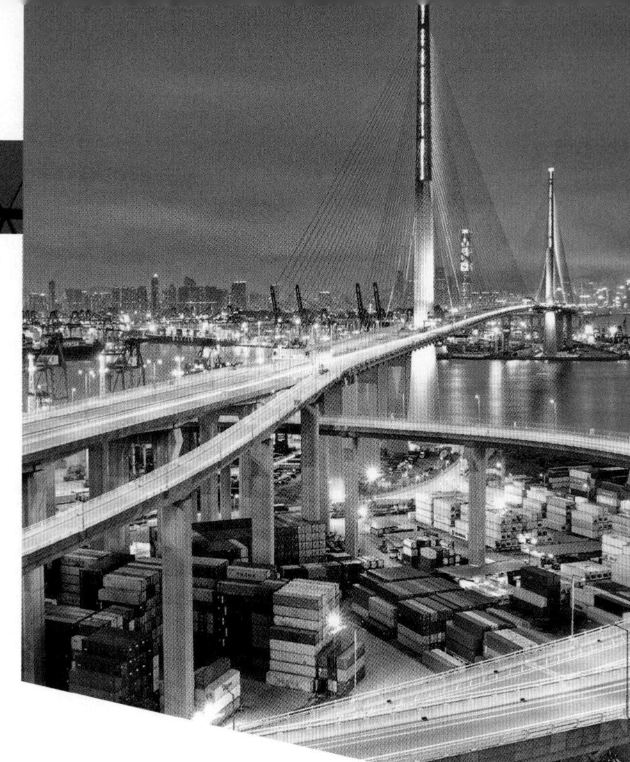

Check Your Engineering IQ

Before you read this chapter, assess your current understanding of the chapter content by taking the chapter pretest.
www.g-wlearning.com/technologyeducation/

Learning Objectives

After studying this chapter, you should be able to:

- ✔ Explain how civil engineering and architecture have shaped modern civilization.
- ✔ Define architecture and civil engineering.
- ✔ Differentiate between the work of an architect and a civil engineer.
- ✔ Explain civil engineering specializations.
- ✔ Demonstrate how architectural drawing conventions are used to convey structure designs.
- ✔ Explain how the forces of tension, compression, torsion, and shearing affect structural members.
- ✔ Describe how static and dynamic loads apply stress to a structure.
- ✔ Evaluate the strength of structural components.
- ✔ Apply knowledge of mathematics, science, and engineering to analyze the forces applied to structures.

Technical Terms

abutment
aesthetics
arch
architects
architecture
bending moment
bending stress
bubble plans
buckling
cantilever
compression
critical value
dead loads
dynamic loads
elevations
environmental engineering
force
free body diagram
geotechnical engineering
live loads
loads
moment
plan view
point load
post and beam construction
reaction forces
shear
shear force diagram
solid mechanics
static loads
strain
stress
structural engineering
tension
thrust
torsion
transportation engineering
water resource engineering

Close Reading

As you read the following chapter, create your own diagrams that depict how different structures and their components withstand the loads and forces applied to them.

Architecture and civil engineering provide us with structures that satisfy the basic human need for shelter and also enable us to work and play. As you learned in Chapter 17, the construction of these structures begins with development of architectural and engineering plans. The individuals who create these plans are architects and civil engineers who have a variety of specializations. These professionals use their knowledge of mathematics and science, combined with design skills, to plan structures that meet the criteria and constraints of the project. Architects and civil engineers each have distinct methods for developing plans for constructing structures. This chapter will explain the architecture and civil engineering fields, as well as explore the practices and processes of both.

Stefano Ember/Shutterstock.com

Figure 18-1. An ancient Roman bridge in the Swiss Alps.

The Development of Architecture and Civil Engineering

Early humans roamed the earth, hunting and gathering food. People were nomadic, meaning they moved from place to place, following food and water sources. These early people did not need permanent shelters. They used shelters such as caves as they hunted moving animals. This practice changed as people moved to an agricultural society. This advancement in civilization provided people with a more constant and consistent source of food, enabling them to live in one place. Families created permanent shelters. They built structures that protected them from the climate of their region.

As people created permanent homes, they saw the advantages of living closer to other people. The more people nearby, the more they could work together to achieve a variety of tasks or goals. This resulted in the development of villages and towns. As villages or towns expanded, people began to specialize in specific occupations, knowing that they could call on others in the community who had other skills and abilities.

As communities grew, there was a need to create larger buildings, such as churches and civic houses, where people could meet to discuss community issues. As more towns were settled, there was a need for transportation structures such as roads and bridges, **Figure 18-1**. People were then needed to design and plan these structures. This also included designing the appearance of the structure, which was based on the societal, cultural, and religious trends of the time. These actions evolved into the professional practices of architecture and civil engineering.

Architecture and Civil Engineering

As civilization evolved, people developed new materials such as bricks, concrete, and steel, as well as better construction methods. These advancements meant building bigger, better, and more complex structures for a variety of purposes. These advancements and complexities also required more scientific and technical knowledge, resulting in collaboration between professionals in architecture and engineering.

Architecture

Look at the structures around you. Many of them serve similar purposes, whether it is residential, commercial, or industrial. There are infinite variations in shape, form, and size to meet specific desires and needs. See **Figure 18-2**. Some of these variations are based on space and functional requirements. Other variations are based on the art or beauty of a structure, based on specific taste.

Flat Style Vector Art/Shutterstock.com

Figure 18-2. Homes are built in a variety of shapes, forms, and sizes to meet specific desires and needs of owners.

Architecture is the art and science of designing and building structures. This includes ***aesthetics*** (the beauty of a structure) and its spatial and functional layout. ***Architects***, **Figure 18-3**, are the professionals who design structures. Architects must be able to create structures that are both aesthetically pleasing and functional. Architects must be creative. They must also understand construction methods, material properties, and structural design principles. This knowledge ensures their creative ideas can be brought to life. An architect can create an elaborately designed building, but it is worthless if it is impossible to build. Architects must also understand how people's surroundings make them feel, and how they impact their thought processes and actions.

Phovoir/Shutterstock.com

Figure 18-3. Architects designing a floor plan.

Civil Engineering

Architects focus on the aesthetics and functional and spatial layout of buildings. Civil engineers focus on the structural elements of design, ensuring that buildings can withstand all types of conditions. Civil engineers understand physical and natural sciences, as well as mathematics, modeling, and design, **Figure 18-4**. A civil engineer provides the plans that enable a structure to stand and support itself. Many states have laws that require construction plans and specifications for certain types of structures be prepared by a registered civil engineer.

Civil engineering involves designing the infrastructure necessary to support the functions of society, improve communities, and enhance people's lives. Civil engineers design and plan the construction of residential, industrial, and commercial buildings as well as heavy engineering structures, such

bikeriderlondon/Shutterstock.com

Figure 18-4. These civil engineers are showing plans to workers at a construction site.

as roads, bridges, tunnels, communication towers, and energy production and manufacturing facilities. This work mostly involves conducting structural analyses to develop and optimize structure designs, as well as working in the field to monitor and assess the progress of construction.

Because civil engineering is such as broad field, there are several areas of specialization, **Figure 18-5**. These specializations ensure that all aspects of the construction and operation of a structure meet the needs of society and the environment.

Structural Engineering

Structural engineering is the study of the framework of structures: designing, analyzing, and constructing structural components or assemblies to resist the stresses and strains of loads and forces that affect them. Using their knowledge of solid mechanics, structural engineers plan the assembly of elements to create a structure that is able to stand and support itself. ***Solid mechanics*** is the study of how stresses are distributed as they are applied to certain materials and the resulting strains on the materials. This knowledge is used to ensure the structural integrity of buildings or heavy engineering structures in regards to their natural environment.

Geotechnical Engineering

Geotechnical engineering is the study of the way soil behaves under the stresses and strains created by soil or rock at proposed sites. Geotechnical engineers understand the ways structures interact with their surrounding environment. Geotechnical engineers design the foundations of structures, plan the excavation of build sites, select the route for roads and highways, and minimize the negative impacts that structures have on the environment.

Water Resource Engineering

Water resource engineering focuses on ways to provide safe and reliable access to water for a

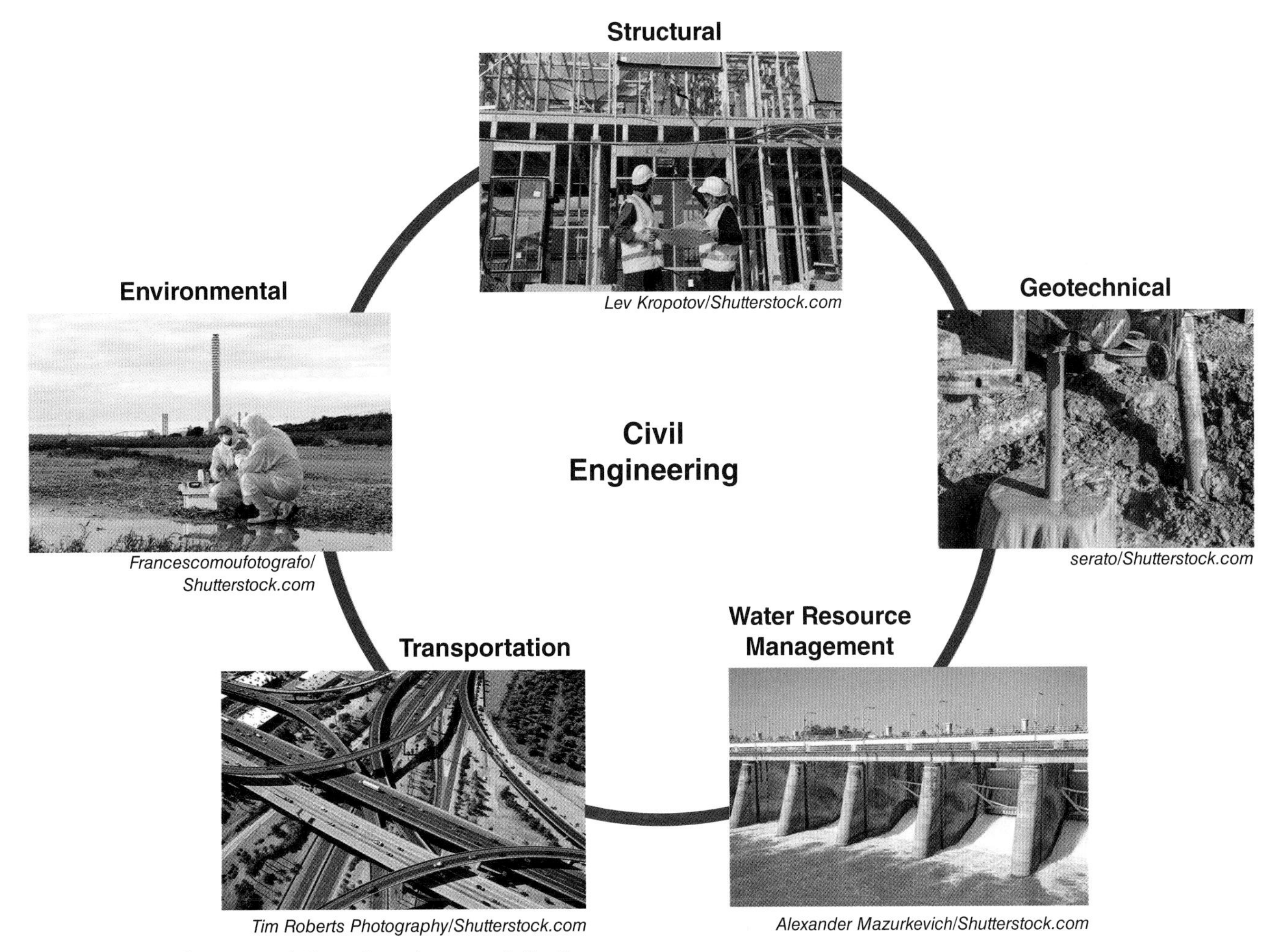

Figure 18-5. Common civil engineering specializations.

variety of uses, to remove and treat wastewater, and to avoid the damages that can result from excess water. Water resource engineers design and optimize systems that control, move, and store water. Water resource engineers also oversee the construction and maintenance of these systems.

Transportation Engineering

Transportation engineering involves the design and optimization of transportation infrastructure. Transportation engineers design and oversee the construction of safe and efficient systems that support the movement of people and goods from one place to another, such as highways, subways, airports, and railroads.

Environmental Engineering

Environmental engineering focuses on managing the use of natural resources to minimize the negative impacts that human activity can have on the environment. Their work also includes developing new and better ways to dispose of waste and to clean up pollution.

Architectural Design and Drawing Fundamentals

Understanding basic architectural design and drawing principles and practices is necessary to develop and read any architectural or engineering building plans. Technical drawing and design techniques and conventions have been established in the architecture and engineering fields. These techniques and conventions allow architects and civil engineers to understand the design and plan for a structure.

As you learned in Chapter 5, orthographic projections depict three-dimensional objects in two dimensions. Projection creates a multiview drawing consisting of three orthographic views of the object. These views fully describe all of the object's features. These views are all aligned with one another, **Figure 18-6**. For mechanical products and systems, these views are usually the top, right side, and front of an object.

In architectural design, architects use drawings to depict their structure designs that are similar to the orthographic projections and multiview drawings discussed in Chapter 5. However, the drawings are called by different names. For example, the top view of a building is considered the plan view. The ***plan view*** includes cross-sectional views of the different levels of the building, **Figure 18-7**. The plan view is the basis for projecting the other views of the building, such as a foundation plan, multiple floor plans, and a roof plan. The front, side, and back views of a building are shown on an architectural drawing. They are called ***elevations***.

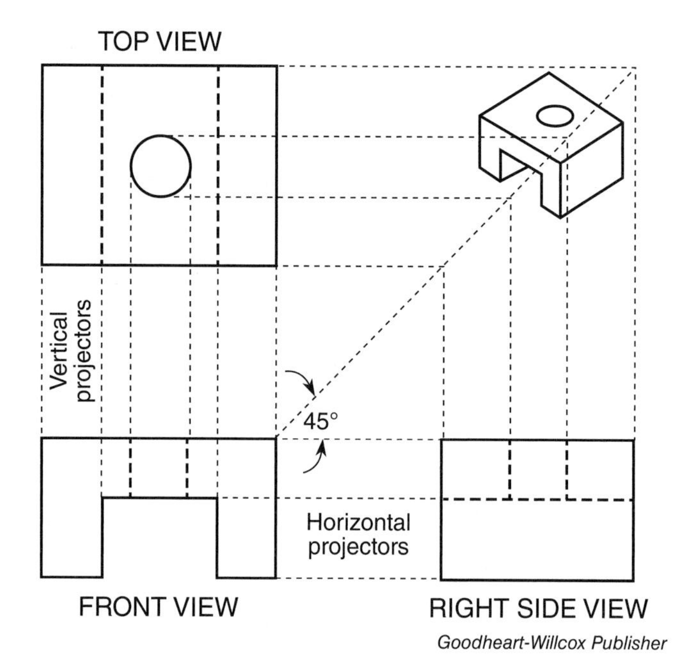

Figure 18-6. Orthographic projection of an object.

Plan views are the basis for the construction process. However, these plans can sometimes look overwhelming. This is why drawing conventions and standard symbols and sizes have been established. They have been established for many items, including walls, windows, doors, and appliances, **Figure 18-8**. There are also several types of lines used to describe features in a floor plan, **Figure 18-9**.

When developing architectural plans, architects must decide on the best use of space. They sometimes begin by sketching out bubble plans. ***Bubble plans*** are rough drawings that assist in brainstorming a structure's layout. Bubble plans can then be refined into a more practical sketch for creating a real floor plan, **Figure 18-10**. Once the plan is refined, computer-aided design software is used to create the architectural plans. These plans include the size and location dimensions of every element, **Figure 18-11**. The plans must match the elevations of the structure.

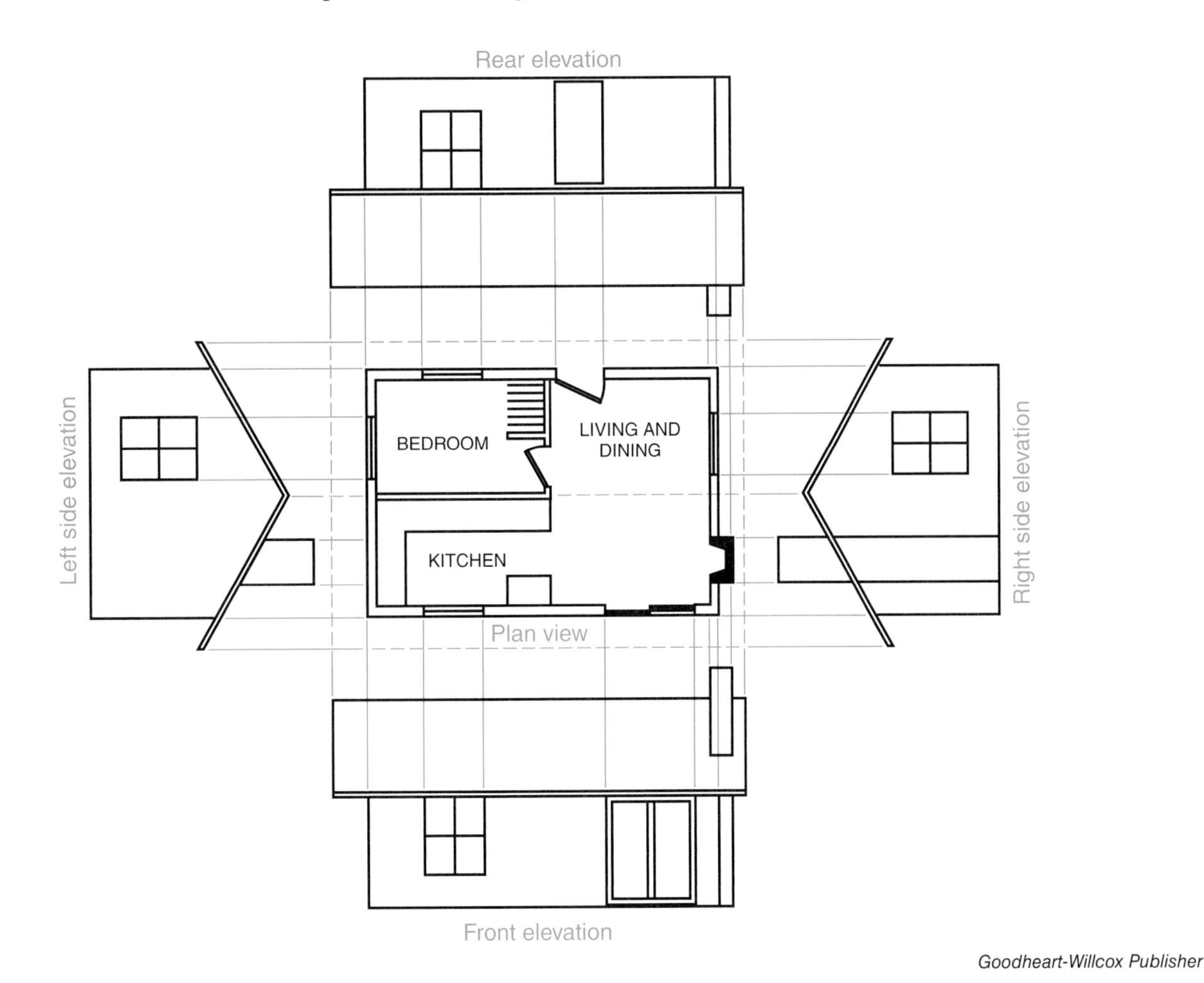

Goodheart-Willcox Publisher

Figure 18-7. Projection of elevations from the plan view.

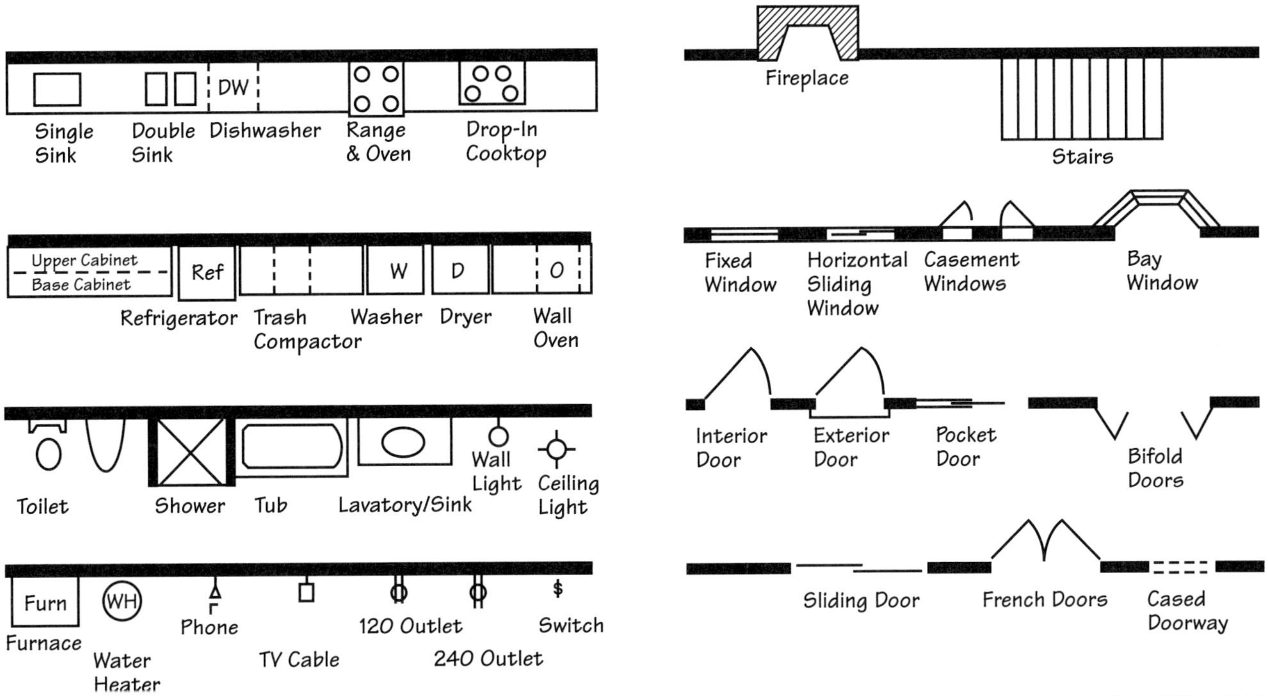

Single Sink
Double Sink
DW
Dishwasher
Range & Oven
Drop-In Cooktop
Fireplace
Stairs
Upper Cabinet
Base Cabinet
Ref
Refrigerator
Trash Compactor
W
Washer
D
Dryer
O
Wall Oven
Fixed Window
Horizontal Sliding Window
Casement Windows
Bay Window
Toilet
Shower
Tub
Lavatory/Sink
Wall Light
Ceiling Light
Interior Door
Exterior Door
Pocket Door
Bifold Doors
Furn
Furnace
WH
Water Heater
Phone
TV Cable
120 Outlet
240 Outlet
$
Switch
Sliding Door
French Doors
Cased Doorway

Goodheart-Willcox Publisher

Figure 18-8. Architectural drawing symbols.

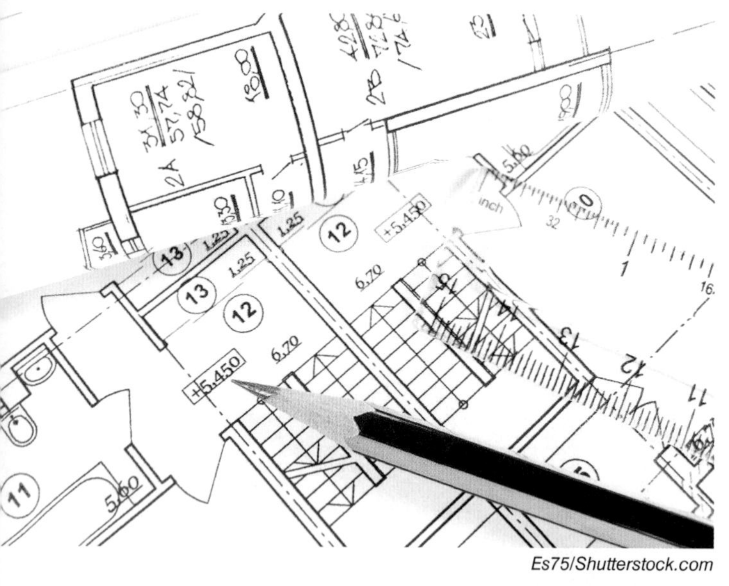

Es75/Shutterstock.com

Figure 18-9. Different lines used for creating a floor plan.

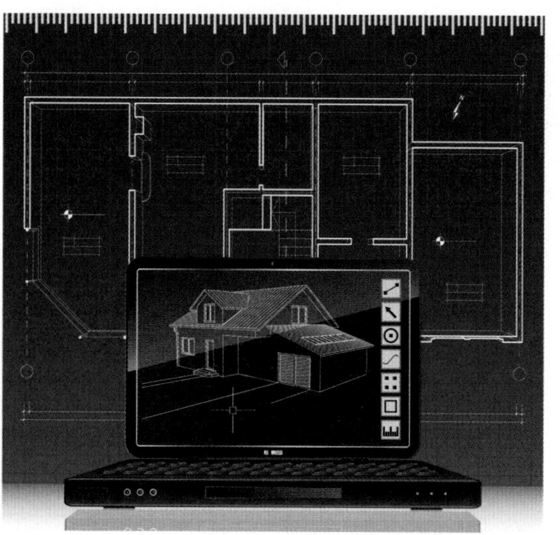

AiVectors/Shutterstock.com

Figure 18-11. A floor plan and elevation created using CAD software.

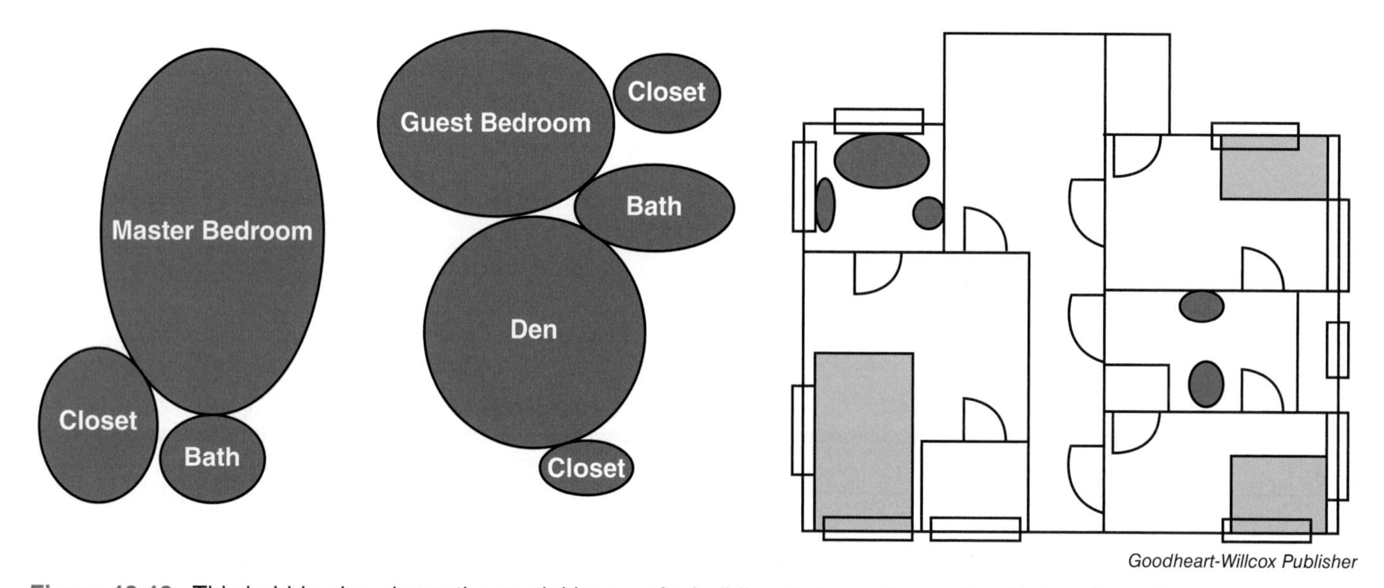

Goodheart-Willcox Publisher

Figure 18-10. This bubble plan shows the spatial layout of a building that was then refined into a floor plan.

Civil Engineering Foundations

In order to design and plan structures that function properly, engineers must understand how materials and structural components react to a variety of applied forces and loads. There are several basic structural components such as columns, beams, arches, and trusses that are placed together strategically. This placement allows components to resist the natural forces and the actions of people that affect structures.

Forces and Loads

Structures are constantly under some type of strain or stress due to a variety of forces applied to them. A ***force*** is a pushing or pulling applied to an object as a result of an interaction with another object. ***Strain*** is the relative change in size or shape of an object due to stress caused by these forces. ***Stress*** is the force per unit of area or unit of length of an object.

Tension and Compression

The two key types of forces that affect structures are tension and compression. As mentioned, forces either pull or push on an object, resulting in

Academic Connection: History

Ancient Architecture

Civilizations throughout history are often defined by their structures. This is because a civilization's cultural traditions and religions, elements and resources of certain geographical regions, and technological development influenced the practices and processes for planning and producing their structures. These influences shaped the way structures look and the methods in which they were built during many different time periods at diverse locations around the world. For this reason, an ancient Egyptian structure, **Figure A**, is very different from an ancient Grecian-Roman structure, **Figure B**.

Early Egyptian societies believed in many gods and carried out a variety of rituals in respect to these deities. The design of structures was often under the direction of a priest who would plan the shape and form of structures to facilitate the rituals and to honor their gods. On the other hand, early Greek and Roman architecture followed their societies' focus on civic life. They built with newly discovered materials such as marble and concrete. The structures were planned to accommodate municipal and governmental activities. The use of different materials between the two societies makes it possible to distinguish between their architecture.

Nestor Noci/Shutterstock.com

Figure A. Ancient Egyptian structure.

eFesenko/Shutterstock.com

Figure B. Ancient Grecian-Roman structure.

strain. A ***tension*** force pulls material apart, much like two people pulling on a rope. A ***compression*** force pushes on a material and squeezes it, **Figure 18-12**. Materials vary in their ability to withstand tension and compression forces. For example, concrete is very strong in regards to compression, making it ideal for columns or blocks that support a large amount of weight. However, concrete does not hold up to tensile forces very well, making it less than ideal for cables or long beams. Steel, however, is resistant to stretching from tensile forces, so it is often used for cables on bridges.

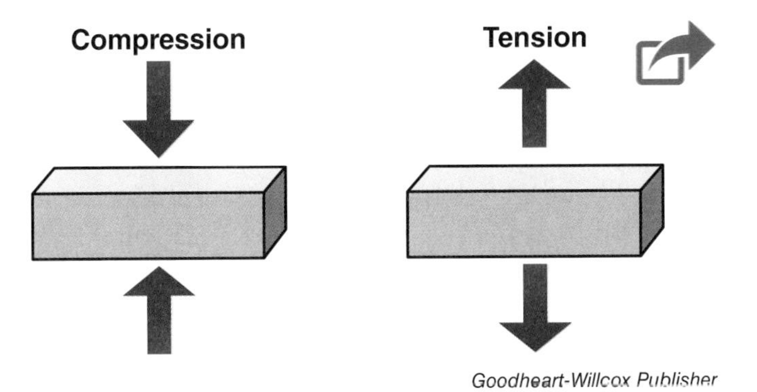

Goodheart-Willcox Publisher

Figure 18-12. A compression force pushes on a material and squeezes it. A tension force pulls a material apart.

Bending Stress

In many instances, a force applied to a structure causes the materials to bend. ***Bending stress*** causes materials to experience both tension and compression. For example, if a beam was laid across two pillars and a weight was placed on the middle of the beam, the beam would bend. When the beam bends, the top of the beam would squeeze together

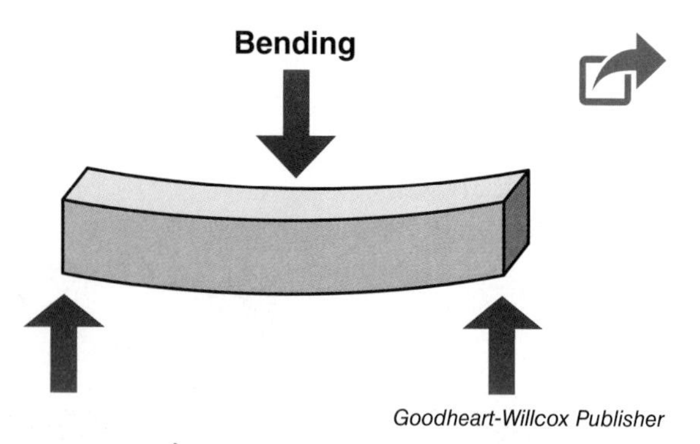

Goodheart-Willcox Publisher

Figure 18-13. A load applied to a beam will cause the beam to bend. The bending beam experiences bending stress, meaning the top of the beam experiences compression while the bottom experiences tension.

under compression, while the bottom of the beam would be pulled apart under tension, **Figure 18-13**. To counteract this, materials can be combined to increase their strength against different forces. For example, steel rebar is added to concrete used for bridges to withstand both compression and tension.

Torsion and Shear Forces

There are many other forces that affect the way structures function. Two more important forces that affect structural design are torsion and shear.

Torsion is an applied rotational force that causes an object to twist, **Figure 18-14**. ***Shear*** is the result of unaligned forces pushing two segments of an object in opposite directions, causing them to slide over each other, **Figure 18-15**. Strong shear forces can cause structural materials to tear in half, leading to devastating structural failures. Therefore, civil engineers use a variety of complex mathematical models and calculations to predict how structures will withstand the different forces applied to them.

Static and Dynamic Loads

Loads originate from a variety of sources, including wind, settling ground, earthquakes, gravitational field pull, changing temperatures,

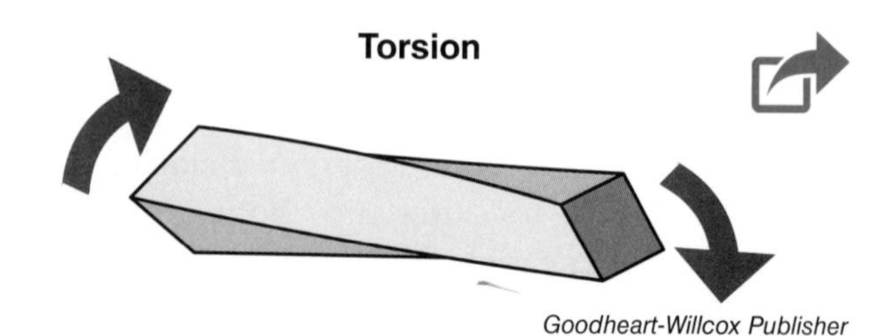

Goodheart-Willcox Publisher

Figure 18-14. Torsion is an applied rotational force that causes an object to twist.

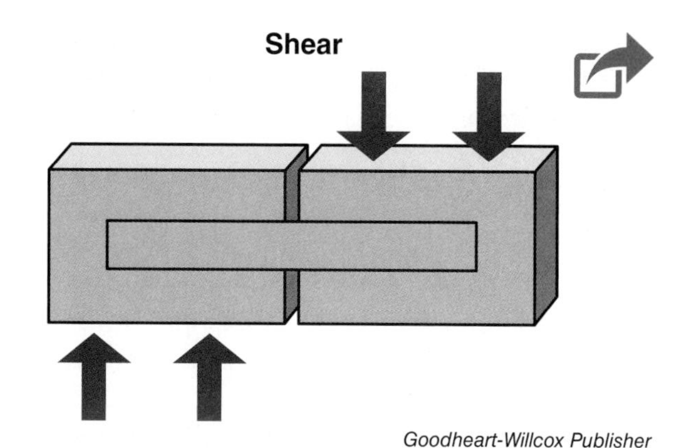

Goodheart-Willcox Publisher

Figure 18-15. Shear results from unaligned forces pushing two segments of an object in opposite directions, causing them to slide over each other.

and heavy snowfalls. Addressing loads and weight distribution is a complex and ongoing issue when designing structures. Civil engineers must determine the loads that will act on a structure and how extreme those loads might be. Civil engineers have a thorough understanding of physical and natural sciences, as well as the mathematical skills necessary to calculate design optimizations. The two main types of loads are static and dynamic.

Static loads are the weight of the structure itself (dead load) and weight added to the structure under normal use (live load). They are unchanging or slowly changing forces applied to a structure that must be accounted for when designing it. ***Dead loads*** include any permanent component of the building such as the floors, walls, roof, or beams. The dead load of a structure is enduring and does not change without altering the structure itself. It is a force that continually acts on a structure due to the earth's gravitational pull.

Live loads are another type of static load. These are the loads that change slowly. They are not permanent forces acting on a structure. ***Live loads*** involve the weight of things such as furniture, appliances, people, and equipment. These items can be moved around and can fluctuate at times. For example, you may purchase a new couch for your house and then have 12 people over for a party. The new couch and the additional people are now part of the live load on the structure. However, after the party, the 12 people go home, resulting in a lower live load. As you can see, the live load fluctuates. But it is important to remember that it doesn't happen abruptly, so it does not cause a sudden stress on the structure. Nevertheless, live loads are important to account for because of the potential safety issues involved with changing loads. The end user of the structure should not use the structure in a manner that was not approved by the engineer. This is the reason for building codes. They are predetermined and followed by engineers when designing buildings.

Dynamic loads are rapidly or abruptly changing, impermanent forces acting on a structure. These loads are exerted by forces such as a wind gust or earthquakes. Because these forces occur abruptly, they are often more damaging than the same force exerted gradually. There are five major types of dynamic loads that civil engineers must assess when designing structures.

Wind loads can exert a large amount of force on a structure. Strong gusts of wind can occur suddenly, greatly stressing the structural components of a building. The impact of wind pressure on a structure also increases with the height of a structure because wind speed increases at higher elevations. Engineers must use their knowledge of science to determine potential wind extremes at proposed building sites. Using this information, engineers can determine the amount of wind bracing needed to protect the structure. The wind bracing adds weight to the structure and increases costs. Consequently, the dead load of the added wind bracing must be accounted for in the design of the structure. Trade-offs must be made between the amount of bracing needed versus the amount of wind the building must withstand. It is not cost effective to build a structure that has enough wind bracing to withstand wind levels that have historically occurred rarely at the building site. The correct amount of wind bracing ensures the safety of people in one of these extreme cases and results in the least amount of damage to the structure.

Snow loads are created by heavy snowfalls, **Figure 18-16**. They increase stress on a structure, and can cause damage. One foot of wet snow on the roof of a building can add 21 pounds per square foot of pressure to the structure. Additionally, if the snow suddenly slides off the roof, like a miniature avalanche, the structure receives another sudden stress from the weight being removed quickly. The sudden removal of the weight causes the materials of the structure to spring back, similar to the spring action of a diving board. This movement can also damage the structure. Engineers must determine the snowfall extremes of a region and design their structures accordingly.

Thomas Riggins/Shutterstock.com

Figure 18-16. Heavy snow loads are a danger to structures.

Seismic loads, or earthquakes, are created by the abrupt movement of the earth's crust. Earthquakes create extreme strain on structures. When earthquakes occur, the bottom of a structure is forced to move with the ground. However, based on Newton's first law of motion, the top of the building tries to maintain its original state of rest, or inertia. The top of the building is pushed and pulled by the movement of the bottom of the building. This action causes the top of the building to sway back and forth as the earth continues to shake, **Figure 18-17**. The top

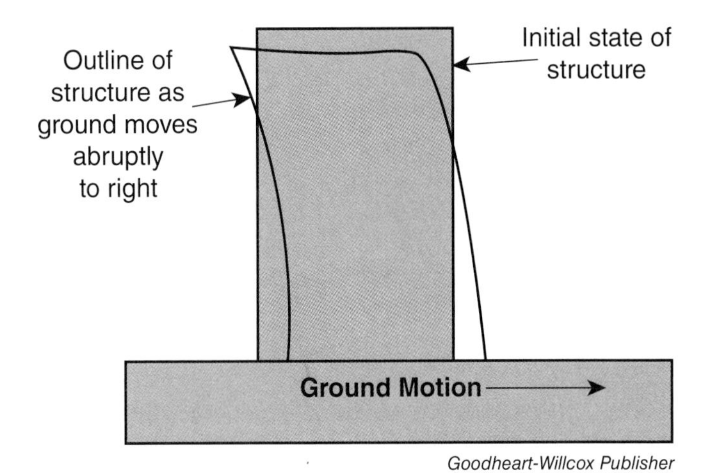

Goodheart-Willcox Publisher

Figure 18-17. The swaying of a structure due to the movement of the ground during an earthquake.

of the building will sway back and forth at a greater distance than the bottom of the building. This movement can cause devastating damage to the building. Therefore, engineers must design structures to bend and sway with the movement of the earth.

Settlement load occurs when the soil that a structure rests on settles unevenly. This can be due to either natural changes or the weight of the structure. When the soil settles, structures experience stress from settlement load. The settling soil causes the building to sink or change shape, resulting in its components experiencing tensile, compression, torsion, or shearing forces. The Tower of Pisa, commonly known as the Leaning Tower of Pisa, **Figure 18-18**, leans because it was improperly constructed on soft soil. The soft soils were unable to support the weight of the tower and the tower sank unevenly into the ground.

When considering a site for construction, engineers analyze soil to determine potential settlement loads. This information will indicate how a future structure will withstand the resulting forces. Using this information, engineers can design structures accordingly. For example, some structures in certain areas will require the construction of heavy concrete pillars deep into the ground. These pillars are called *deep piles.* The deep piles are placed through soft soils into the solid earth underneath, which is strong enough to support the weight of the structure. The structure then rests on the piles instead of the soft soil, **Figure 18-19**.

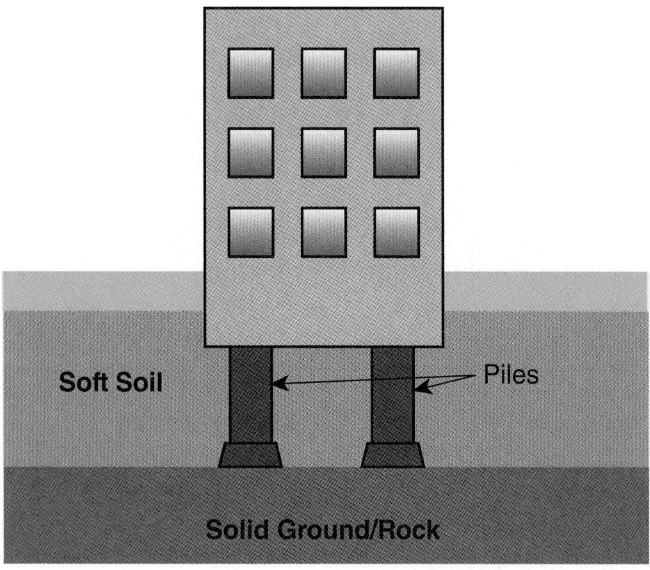

Goodheart-Willcox Publisher

Figure 18-19. Concrete piles are used to support the foundation of a structure on soft soil by connecting it to the solid earth/rock below.

Steven Bostock/Shutterstock.com

Figure 18-18. The Tower of Pisa in Italy was not designed to lean. The soil could not bear the weight of the tower, causing it to sink unevenly into the ground.

Thermal loads are created by the materials used to build structures as they expand and contract with temperature changes. As the materials change shape, parts of the structure will begin to pull or push on other parts of the structure, causing the structure to experience a thermal load. To account for these loads, engineers design structures with precisely placed expansion and roller joints. These joints give parts of the structure the freedom to expand or contract as temperatures change, **Figure 18-20**.

Structural Members, Shapes, Materials, and Strength

A structure's strength is its ability to withstand the strain or stress caused by the forces that a load applies to it. Therefore, the strength of a structure is

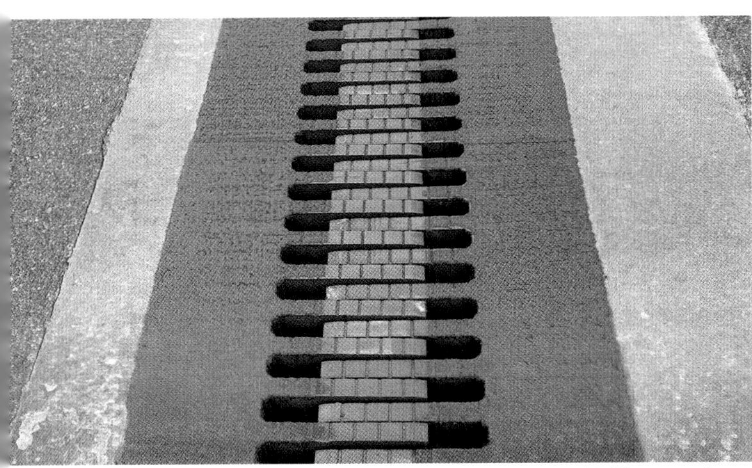

Jerry Lin/Shutterstock.com

Figure 18-20. Expansion joints on bridges allow for expansion and contraction of materials caused by temperature changes.

a function of the size, shape, and properties of the materials used to make its members. The combination of different shapes, sizes, and materials are considered when building all sorts of structures.

The size of the structural members affects the amount of weight it can hold. A beam that is 2″ thick will support less weight than a beam that is 20″ thick. It is important to remember, however, that increasing the size of structural members also increases the cost of the structure and can often limit its functionality. One way to improve a structure's strength without increasing its size is to change the shape of its components. Different shapes are used to make basic structural components such as beams, arches, and trusses. Changing material shape can increase its strength.

Think about a piece of paper. If you hold the paper horizontally, **Figure 18-21A**, it would collapse under its own weight. However, if you slightly curve the piece of paper, **Figure 18-21B**, it is able to hold its own weight plus the weight of a pencil. The material's size or properties did not change, only its shape. The same thing will happen if you fold the paper back and forth, **Figure 18-22**. Place that paper across two books and it will hold its own weight, plus more.

Another example of shape increasing strength is an egg. An eggshell is very thin and breaks easily. But it is nearly impossible to break an egg by squeezing it in the vertical position with one hand, **Figure 18-23**. This is because the curved shape of the eggshell equalizes the forces that you apply to it.

As you can see, curving and folding materials can enhance their strength. This is because arches and triangles are stronger than basic rectangular shapes, **Figure 18-24**. A load applied to a rectangle can cause the top member to bend and break. However, when a load is applied to the top of a curve or arch, the forces are spread outward to the ground, **Figure 18-25**. The ground pushes back on the arch, causing it to be under pure compression.

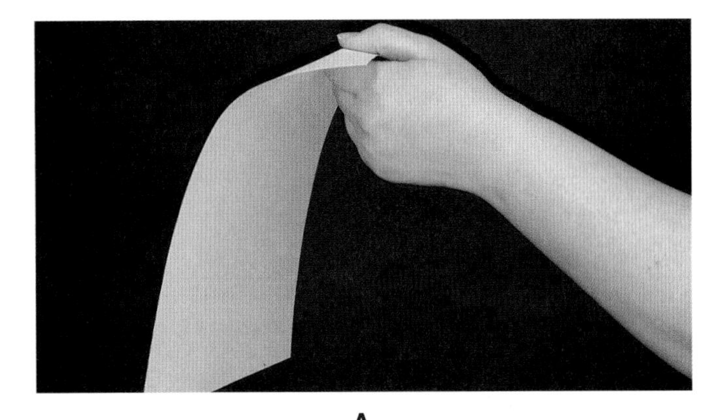

A

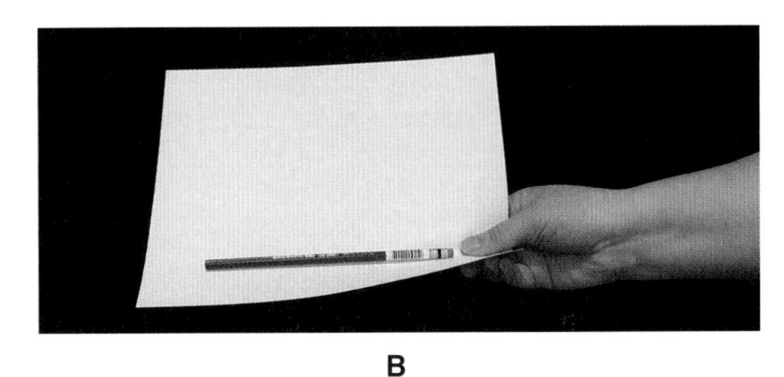

B

Goodheart-Willcox Publisher

Figure 18-21. Material strength can be improved by changing its shape. A—One sheet of paper cannot even support its own weight. B—Curve the paper, however, and it can hold its own weight plus the weight of a pencil.

Goodheart-Willcox Publisher

Figure 18-22. By folding this sheet of paper into an accordion pleat, it is able to support the weight of a book.

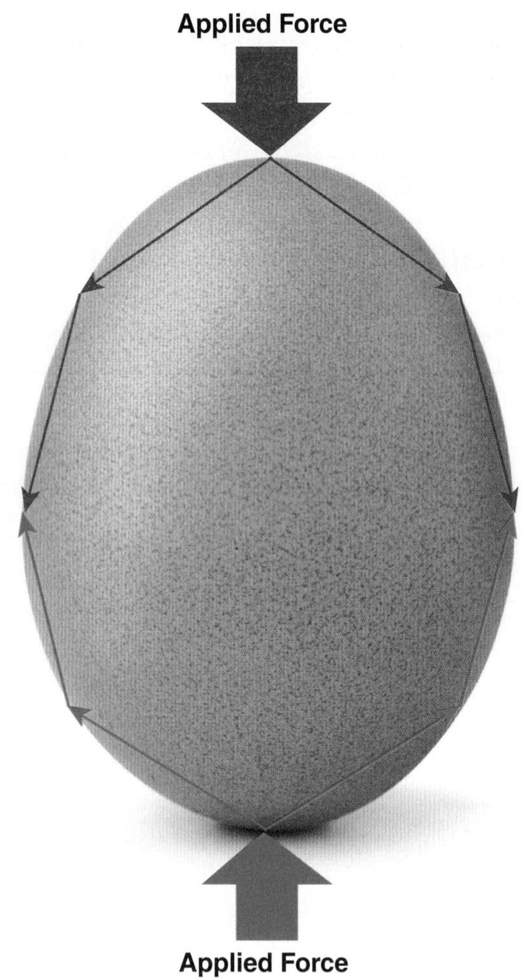

Coprid/Shutterstock.com

Figure 18-23. The curved shape of an eggshell distributes applied force evenly to the entire shell. Many structures use similarly curved elements to enhance their strength.

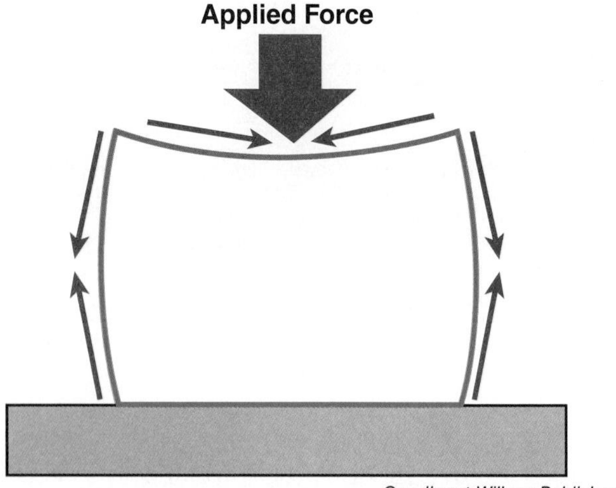

Goodheart-Willcox Publisher

Figure 18-24. Applying a force to the top of a rectangle. The top experiences bending stress (tension and compression) and the sides experience compression. As a result of too much force, the top will bend and the sides will buckle, causing the rectangle to collapse.

Triangles are the strongest shape for supporting a load. This is because when a load is placed at the top of a triangle, the forces are displaced in a way that counteracts the deformation of the triangle. The bottom of the triangle balances the weight by pulling on the sides of the triangle. A triangle cannot change shape without changing the lengths of its sides or breaking them, **Figure 18-26.**

Different shapes are used to create the basic types of structural components. These components are placed strategically to resist natural forces and actions of people that affect structures. ***Post and beam construction*** is essentially a rectangle.

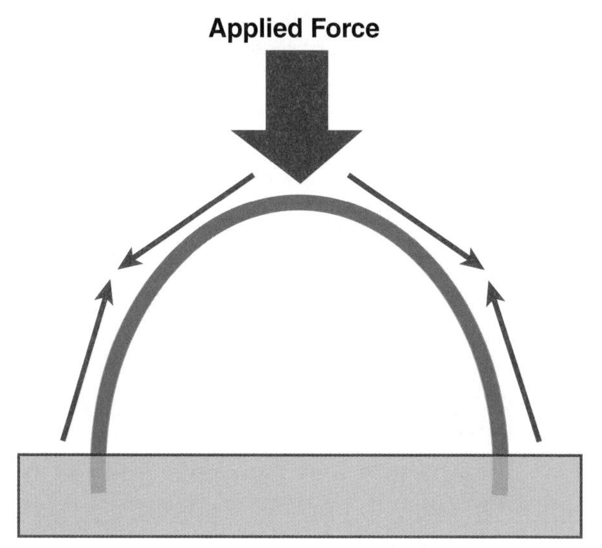

Goodheart-Willcox Publisher

Figure 18-25. Applying a force to the top of an arch. The curve distributes the force outward to the ground.

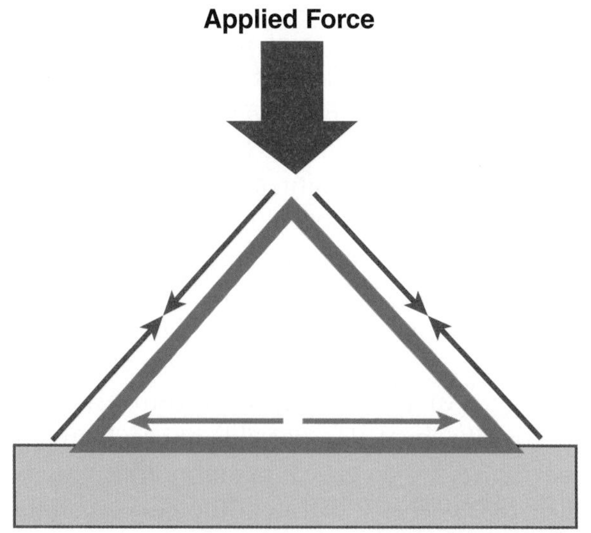

Goodheart-Willcox Publisher

Figure 18-26. Applying a force to the top of a triangle. This is the strongest shape for supporting a load.

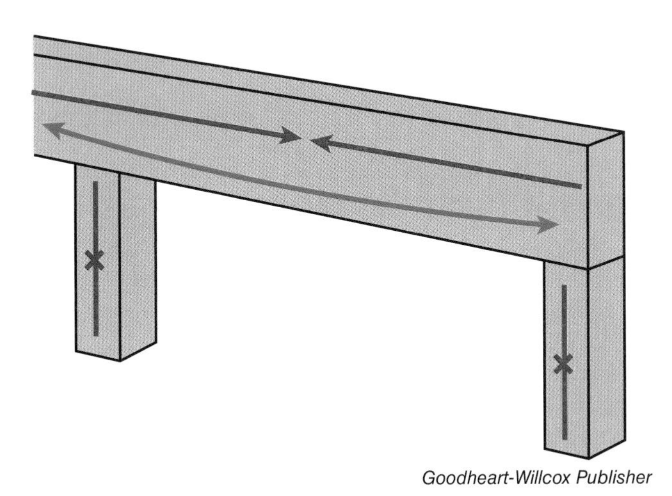

Goodheart-Willcox Publisher

Figure 18-27. In post and beam construction, the top of the beam experiences compression, the bottom of the beam experiences tension, and the posts experience compression.

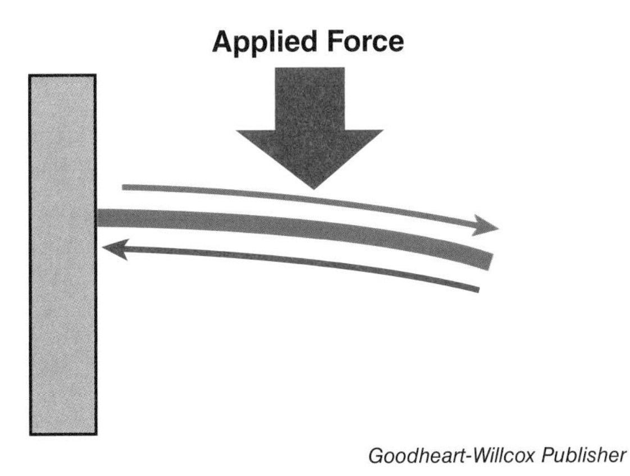

Goodheart-Willcox Publisher

Figure 18-28. When a force is applied to the unfixed end of the cantilever beam, the top of the beam will experience tension while the bottom will experience compression.

It consists of a horizontal element, or beam, that spans a gap and rests on two columns or posts, **Figure 18-27**. When supporting a load, the beam bends, causing it to experience both tension and compression. The posts experience compression as they support the weight of the beam and the load.

A ***cantilever*** is often used to create balconies or bridges. A cantilever is a long, projecting beam that is supported on only one end, **Figure 18-28**. The load applied to a cantilever is transferred to the fixed end of the beam and then transmitted to a larger mass that can help absorb it.

An ***arch*** is a curved structural component used to support a load while spanning a space. As you read earlier, the curvature allows the force applied by the load to be carried outward to the ground. The ends of the arch then push outward at the base. The pushing is called ***thrust***. In order for the arch to remain standing, the thrust must be stopped by an element of the arch called an ***abutment***. The abutment restrains the outward push of each end of the arch by tying it to the ground. Because the ends of the arch cannot move, the force from the load that was spread out to the ground is then transmitted back to the center of the arch, causing the load of one side of the arch to support the other side. This means that the entire arch is under compression, **Figure 18-29**.

A truss contains straight members connected to form triangular units to support a load while spanning a space. You can see trusses in bridges,

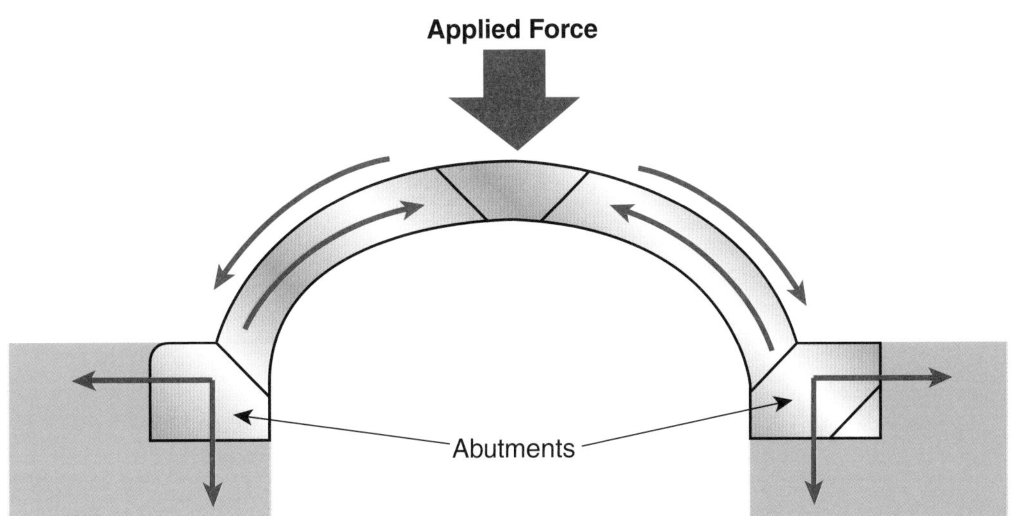

Goodheart-Willcox Publisher

Figure 18-29. Abutments restrain the outward push of each end of the arch. The force from the load that was spread out to the ground is transmitted back to the center of the arch, causing the load on one side of the arch to support the other side.

towers, roofs, and cranes. The strength of the series of triangles supports the loads. The members that form the triangular units will experience either pure tension or compression. In **Figure 18-30**, note that when weight is applied to the top of the truss structure that the different members experience either tension or compression. To support different loads and to achieve different functions, a variety of truss configurations are used.

The properties of the materials used for each structural component or member are important for enhancing the strength of a structure. As you learned in Chapter 16, different materials have different chemical, mechanical, thermal, and electrical properties that make them suitable for different purposes. Civil engineers, prior to selecting the materials of the structural components, study established material properties in order to make informed design decisions based on the intended engineering applications.

Structural Analysis

The major task of civil engineers is conducting structural analyses to make informed decisions about how structures should be designed. They must perform the proper calculations to determine whether or not various structural members will be able to support the forces applied to them. Therefore, it is important to understand the formulas used for calculating stress and the concepts or methods used for analyzing how forces affect structural members.

Tensile and Compressive Stress

Calculating tensile and compressive stress is important for conducting structural analysis. Objects that are under pure tensile or compressive stress are thought to be under normal stress. Normal stress is expressed as the applied normal stress (either tension or compression) over the area that it is applied.

$$\sigma = \frac{Fn}{A}$$

σ = normal stress

Fn = force applied

A = area

For example, this formula can be used to calculate the compressive stress that a load applies to a post. If a 200 lb beam is placed on top of a 5″ × 5″ post, then the compressive stress would be calculated as:

$$\sigma = \frac{200 \text{ lb}}{5'' \times 5''}$$

Therefore, the post would experience 8 lb per square inch (psi) of stress.

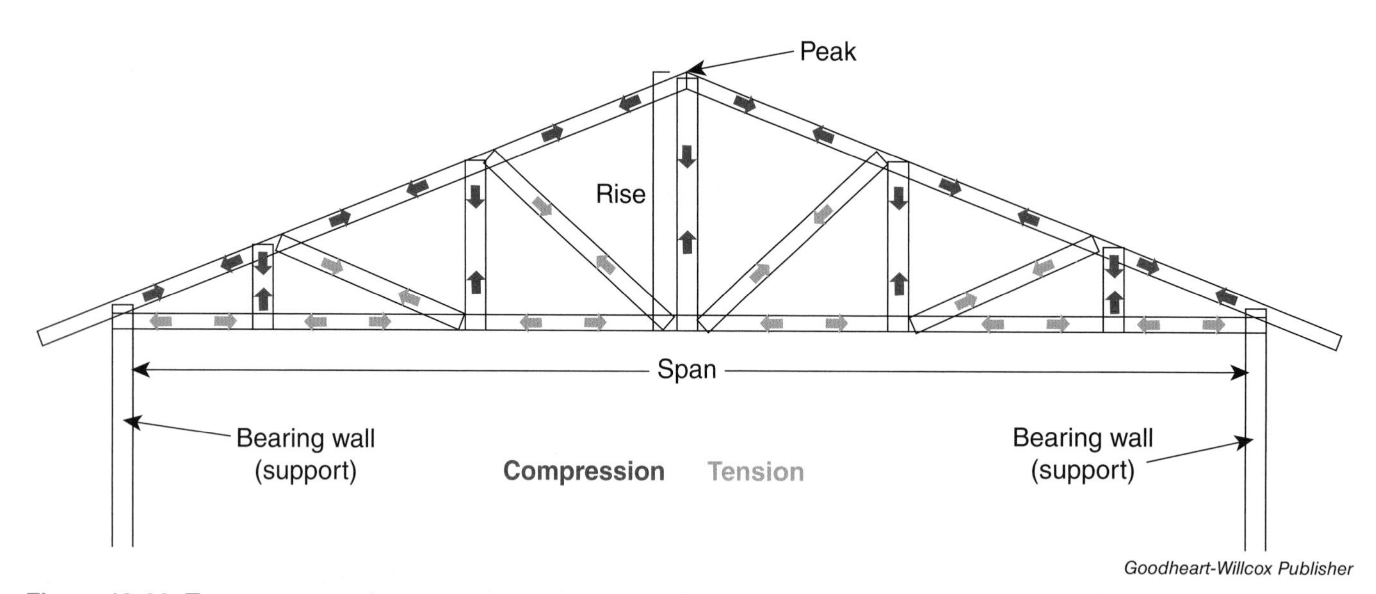

Figure 18-30. Trusses are used to create the roof structure of most homes. Blue represents the members of the truss that are under tension. Red represents compression.

Buckling

One cause of structural failure is ***buckling***. Buckling occurs when a straight structural element, such as a post, bends under compression, causing structural failure. Buckling can be seen by placing one end of a thin ruler on a table and applying pressure to the other end. As you increase pressure, the ruler will, at some point, suddenly bend outward or buckle. The amount of pressure that causes this buckling is the element's ***critical value***. When a structural element reaches this value, it is unstable. Once it buckles, the force applied then amplifies because the bend creates a lever action, **Figure 18-31**. The force will eventually cause the element to fail or break.

Horizontal structural elements can also buckle. This happens when an element under a load follows an easier bending pathway in a lateral direction, **Figure 18-32**, as opposed to the normal deflection or bending. This creates a twisting motion, causing the element to fail.

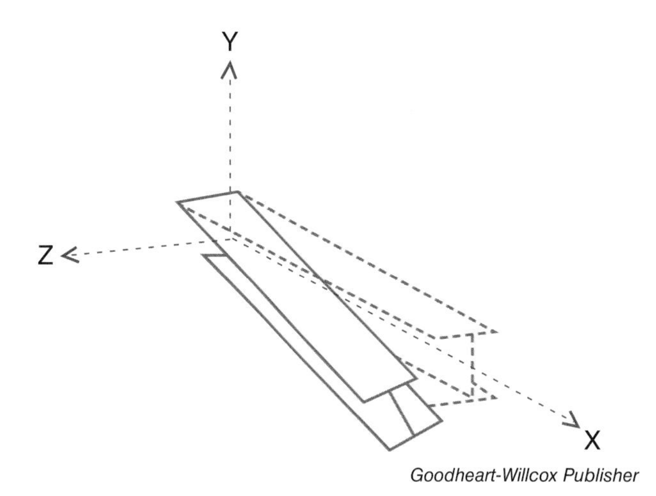

Goodheart-Willcox Publisher

Figure 18-32. Compressive forces can buckle a horizontal member in the lateral direction.

Strain

Strain is the deformation of an object due to stress. Calculating it is very important when designing structures. Strain can be expressed as the change of length of an object divided by the initial length of the object experiencing stress:

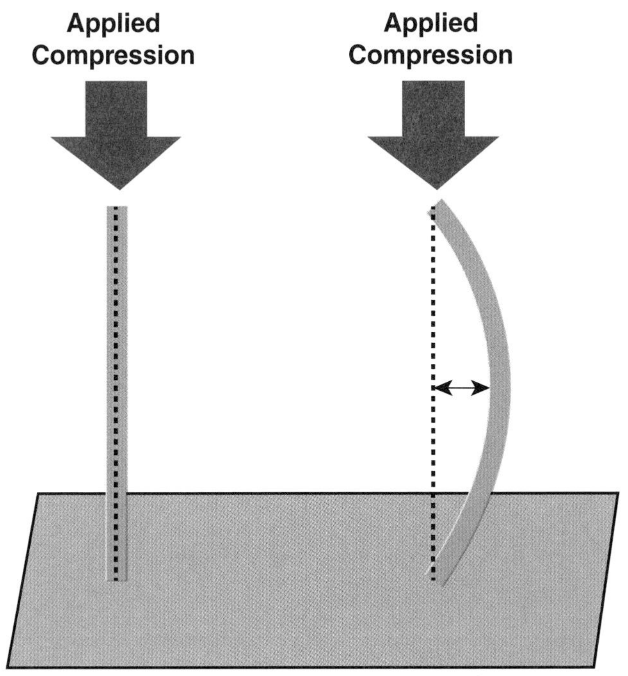

Goodheart-Willcox Publisher

Figure 18-31. A column buckling under axial compression.

$$\varepsilon = \frac{dl}{l_0}$$

ε = strain

dl = change in length

l_0 = initial length

Calculating strain can predict the elongation or compression of structural members under stress. This is key to making informed design decisions. This prediction can be made using Young's Modulus of Elasticity, which is a measure of stiffness for a variety of materials. This measurement is actually the ratio of stress to strain for a material:

$$E = \frac{\text{stress}}{\text{Strain}}$$

$$E = \frac{F \div A}{dl \div l_0}$$

Where:

F = force applied

A = area

dl = change in length

l_0 = initial length

E = Young's Modulus

Using this information, we can insert the predetermined elastic moduli into the equation and calculate the change in length of a structural member based on different forces applied to it.

Shear and Moment Diagrams

Another important structural analysis skill is determining the shearing force that occurs when a load is applied to a horizontal structural element, such as a beam. As described earlier, shear is the force that causes two parts of an object to slide over one another. You can think of shearing force affecting a beam as a knife cutting through a cucumber. When pressure is applied to the knife, it will shear the cucumber. When a load is applied to a point on a beam, it will act like the knife on the cucumber. It is important to determine how these shear forces occur at different points on a beam to determine where shearing may be the greatest. This is done by creating a ***shear force diagram***.

To create a shear force diagram for a ***point load***, or a load that is applied in one spot on a beam, you must first create a free body diagram of the beam. The ***free body diagram*** shows where the forces are applied. This is necessary for calculating the ***reaction forces*** that resist the applied load at the beam supports. For example, a free body diagram of a beam, **Figure 18-33**, shows a beam that has a point load of 10 kilonewtons (kN). It also shows two different beam supports (pin and roller) that are spaced four meters apart. When calculating the reaction forces at supports A and B, remember that the beam is static (not moving). Therefore, the sum of all the vertical forces in the Y direction, which includes the two reaction forces (F_{Ay} and F_{By}) and the applied force, must equal zero or $\Sigma Fy = 0$. Also, because the beam is static, it is not rotating. The rotation of a beam about an axis as a result of an applied force is called ***moment***. To calculate moments, multiply the force by the distance from a point. Once again, because the beam is static, the sum of the moments about any point must also equal zero or $\Sigma Mp = 0$.

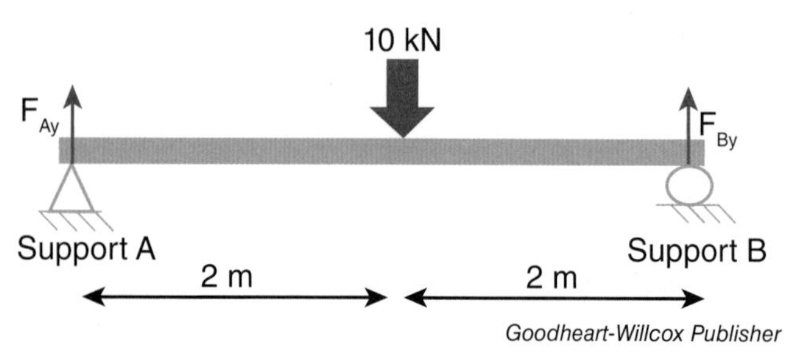

Figure 18-33. A free body diagram for a beam experiencing a point load.

Using this information, we can calculate the reaction forces at the supports. Refer again to **Figure 18-33**. The two reaction forces are F_{Ay} and F_{By} and 10 kN of force is applied two meters from each support. We can calculate one of the reaction forces using this information as well as the equation that the sum of all moments about any point must equal zero. Start with support B. The 10 kN of force is applied two meters from support B and the reaction force F_{Ay} is four meters from support B. Therefore, the sum of the product of 10 kN and two meters, and the product of F_{Ay} and four meters equals zero:

$$\Sigma Mp = 0$$

$$\Sigma Mp = 0 = (10 \text{ kN})(2\text{m}) + (-F_{Ay})(4\text{m})$$

Using this equation, we can solve for the reaction force F_{Ay}, which is 5 kN. Knowing the magnitude of the F_{Ay} reaction force, use the equation that the sum of all vertical forces equals zero to calculate the F_{By} reaction force. The sum of all vertical forces (reaction forces and applied loads) equals zero. This means the sum of the reaction forces F_{Ay} and F_{By}, as well as the applied load of –10 kN, is equal to zero as in the following equation:

$$\Sigma Fy = 0$$

$$\Sigma Fy = 0 = F_{Ay} + F_{By} - 10 \text{ kN}$$

We can now plug in our value for F_{Ay} to solve for F_{By}.

$$\Sigma Fy = 0 = 5 \text{ kN} + F_{By} - 10 \text{ kN}$$

$$F_{By} = 5 \text{ kN}$$

When working with forces, the positive and negative signs for the force magnitudes refer to the direction of the force. In this example, all upward forces are positive and downward forces are negative. Therefore, F_{By} equals a positive 5 kN.

Now that the reaction forces have been calculated, we can create our shear force diagram toanalyze this structural member. Refer to **Figure 18-34**. Begin the shear force diagram by

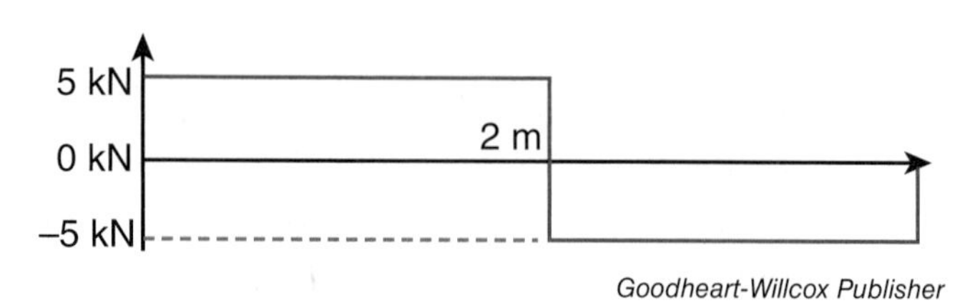

Figure 18-34. A shear force diagram for a beam experiencing a point load.

STEM Connection: Mathematics

Calculating the Weight of a Structure

The weight of a structure can be determined using simple calculations. The dimensions of a structure's elements are used to determine the volume of each element. The calculated volume of the element can then be multiplied by the specified weight of the material used to make the element. For example, if the element has a volume of 23 m^3 and is made of steel, then the volume can be multiplied by the weight of 1 m^3 of steel to find the weight of the entire element. Then each element's weight can be added to determine the weight of the structure. The challenge, however, is determining whether a structural element can support its own weight, as well as withstand forces exerted on it by other loads. This is because the dimensions of the element must be known for the calculations, and these dimensions are ultimately dependent on the element's weight. Therefore, it can be challenging to calculate the loads that a structural element can support. As a result, it is necessary to rely on knowledge of structural testing and analysis to perform the proper calculations.

The following drawing illustrates the steps for calculating the dead load of three 20 meter long steel I beams, using the dimensions provided.

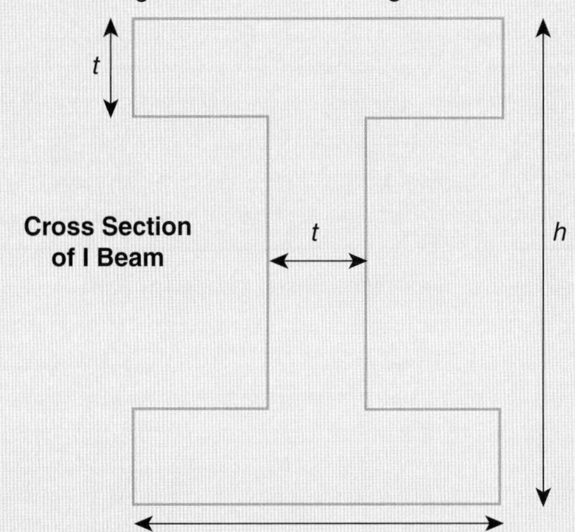

h	42.23 cm
w	20.447 cm
t	2.032 cm
Weight	7850 kg

Goodheart-Willcox Publisher

Step 1—Determine the sectional area of the I beam using the measurements provided. Determine the area of the three rectangles that make up the cross section of the I beam. Then add the area of each rectangle to determine the sectional area of the beam:

$41.55\ cm^2 + 41.55\ cm^2 + 77.55\ cm^2 = 160.65\ cm^2$

Step 2—Multiply the sectional area of the beam by the desired length to determine the beam volume. Convert centimeters squared to meters squared:

$0.016065\ m^2 \times 20\ m = 0.3213\ m^3$

Step 3—Multiply the beam volume by the weight of one cubic meter of steel:

$0.3213\ m^3 \times 7850\ kg = 2522.205\ kg$

Step 4—Multiply the weight of one beam by the total number of beams:

$2522.205\ kg \times 3 = 7566.615\ kg$

The dead load of the three beams is 7566.615/kg.

graphing the reaction force for support A, which is a positive 5 kN. The reaction force for support A is plotted on the y axis at 5. The rest of the forces are plotted moving across the beam at each location where there is a load acting on the beam. Therefore, the next force is a negative 10 kN. This force is plotted at the distance along the x axis where it is applied and at –5 on the y axis. The next force on the diagram is the reaction force at support B, which will end up on zero in the y direction on the diagram because the magnitude of the force is positive 5 kN. **Figure 18-34** can be used to further analyze the beam by calculating the bending moment. Recall that a moment is rotational force that occurs when a load is applied perpendicularly to a point at a specific distance from that point. Therefore, ***bending moment*** is bending that occurs in a beam because of moment. Every point where a shear force diagram crosses the x axis will be a minimum or maximum bending moment.

To calculate bending moment on a beam, create a moment diagram, **Figure 18-35**. The moment diagram displays the magnitude of the moment at every point along a beam. To create the moment diagram, start at the x axis origin and move along the beam to stop and calculate the moment before and after each force in the y direction. Remember, moment is equal to the force times the distance. Therefore, immediately to the right of support A, the beam is only experiencing the F_{Ay} reaction force. So, directly to the right of support A, the beam is experiencing little to no bending. However, as you move farther from this point along the beam, the moment will increase along with the distance.

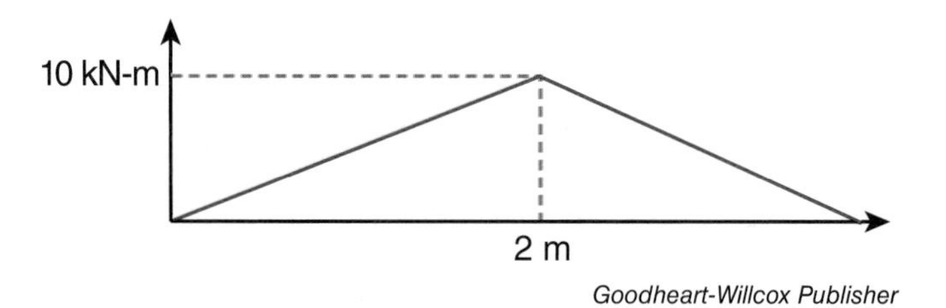

Figure 18-35. A moment diagram for a beam experiencing a point load.

If you look at the point on the beam immediately before the applied force of –10 kN, which is two meters away from support A, the moment would be equal to 5 kN times two meters, or 10 kilonewton-meters. At the moment immediately to the right of the applied load, two forces in the y direction are interacting with the beam. These two forces are the 5 kN F_{Ay} reaction force and the –10 kN. As a result, the moment at two meters across the beam will be (5 kN × 2 m) – (10 kN × 0 m), which would also equal 10 kilonewton-meters. However, as you move closer to support B, the distance for the applied force increases. At the point on the beam immediately to the left of support B, the moment would be equal to (5 kN × 4 m) – (10 kN × 2 m), which equals zero. Remember that the sum of the moment in a static beam should be equal to zero. A beam with a point load will have a moment diagram that looks like a triangle. A beam that has a uniformly distributed load across the beam will have a parabolic shape. Samples of free body, shear, and moment diagrams for a beam with a uniformly distributed load across a beam can be seen in **Figure 18-36**.

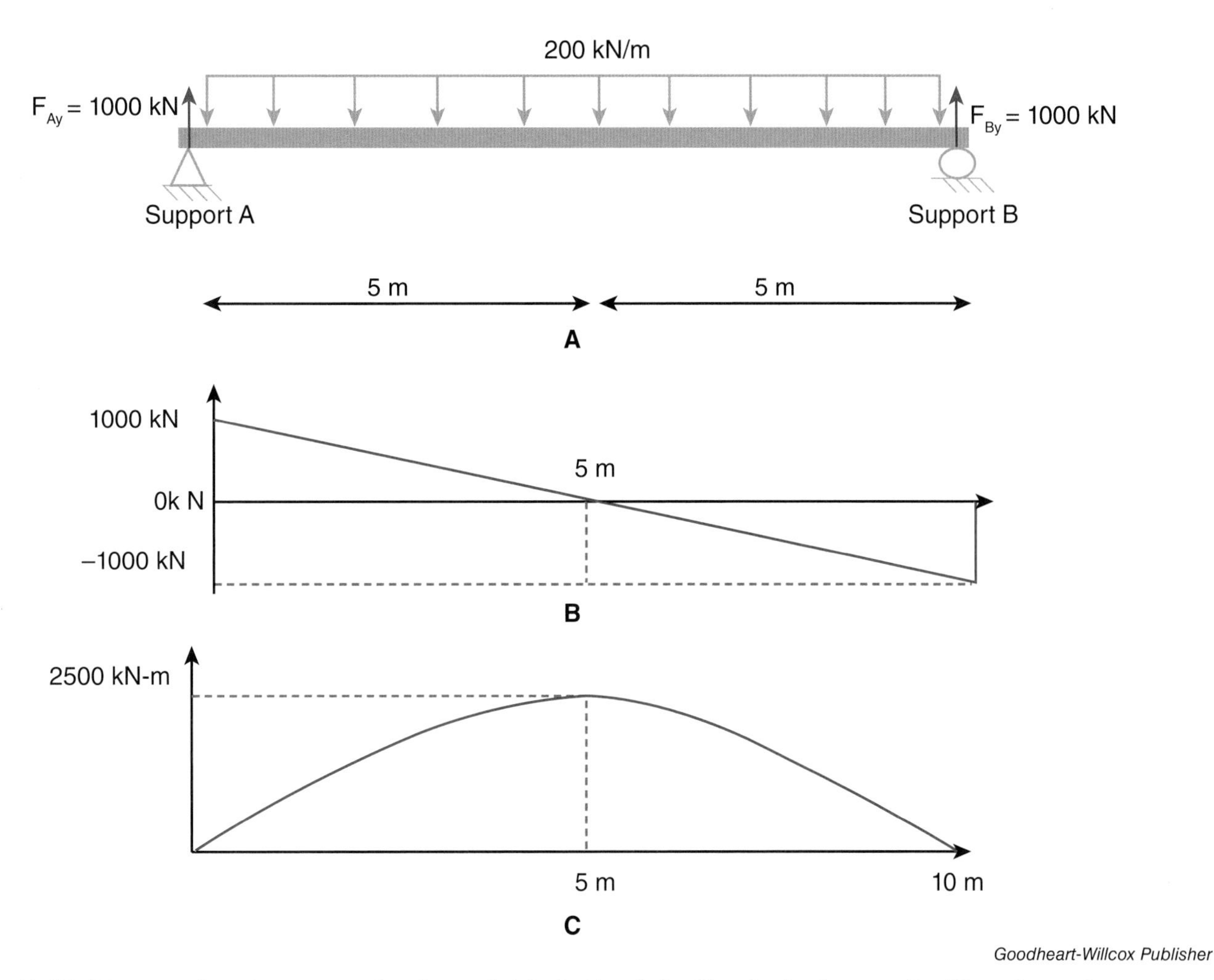

Figure 18-36. Diagrams for a beam experiencing a distributed load. A—Free body diagram. B—Shear force diagram. C—Moment diagram.

Summary

- ✔ The design and planning of structures in early communities evolved into the professional practices of architecture and civil engineering.
- ✔ Architecture is the art and science of designing and building structures.
- ✔ Architects are the professionals who design structures.
- ✔ Civil engineers design and plan the construction of residential, industrial, and commercial buildings as well as heavy engineering structures, such as roads, bridges, tunnels, communication towers, and energy production and manufacturing facilities.
- ✔ Civil engineers focus on the structural elements of design, ensuring that buildings can withstand all types of conditions.
- ✔ Specializations within the civil engineering field include structural, geotechnical, water resource, transportation, and environmental engineering.
- ✔ Technical drawing and design techniques and conventions have been established in the architecture and engineering fields.
- ✔ These techniques and conventions allow architects and civil engineers to understand the design and plan for a structure.
- ✔ Orthographic projections depict three-dimensional objects in two dimensions.
- ✔ Multiview drawings consist of three orthographic views of the object. These views fully describe all of the object's features.
- ✔ The plan view is the basis for projecting the other views of the building.
- ✔ The front, side, and back views of a building shown on architectural drawings are called elevations.
- ✔ Drawing conventions and standard symbols have been established for many items, including walls, windows, doors, and appliances. There are also several types of lines used to describe features in a floor plan.
- ✔ In order to design and plan structures that function properly, engineers must understand how materials and structural components react to a variety of applied forces and loads.
- ✔ The placement of basic structural components allows those components to resist the natural forces and actions of people that affect structures.
- ✔ Structures are constantly under some type of strain or stress due to a variety of forces applied to them.
- ✔ A tension force pulls material apart. A compression force pushes on a material and squeezes it.
- ✔ Bending stress causes materials to experience both tension and compression.
- ✔ Torsion is an applied rotational force that causes an object to twist.
- ✔ Shear is the result of unaligned forces pushing two segments of an object in opposite directions, causing them to slide over each other.
- ✔ Loads are the forces that pull or push on buildings or heavy engineering structures, causing stress.
- ✔ Static loads are the weight of the structure itself (dead load) and weight added to the structure under normal use (live load).
- ✔ Dynamic loads are rapidly or abruptly changing, impermanent forces acting on a structure.
- ✔ The strength of a structure is a function of the size, shape, and properties of the materials used to make its members.
- ✔ Changing material shape can increase its strength.
- ✔ The major task of civil engineers is conducting structural analyses to make informed decisions about how structures should be designed.
- ✔ Civil engineers use a variety of formulas, concepts, and methods to calculate stress and analyze the way forces affect structural members.

Check Your Engineering IQ 

Now that you have finished this chapter, see what you learned by taking the chapter posttest.
www.g-wlearning.com/technologyeducation/

Test Your Knowledge 

Answer the following end-of-chapter questions using the information provided in this chapter.

1. Define architecture.
2. Define civil engineering.

Matching: Match each of the following civil engineering disciplines with the correct definition.

A. Water resource engineering
B. Environmental engineering
C. Structural engineering
D. Transportation engineering
E. Geotechnical engineering

3. The study of the framework of structures: designing, analyzing, and constructing structural components or assemblies to resist the stresses and strains of loads and forces that affect them.
4. Focuses on the behavior of soils under the stresses and strains created by soil or rock at proposed sites.
5. Focuses on ways to provide safe and reliable access to water for a variety of uses, to remove and treat wastewater, and to avoid the damages that can result from excess water.
6. Involves designing and overseeing the construction of safe and efficient systems that support the movement of people and goods from one place to another.
7. Focuses on managing the use of natural resources to minimize the negative impacts that human activity can have on the environment.

8. In architectural drawing, the front, side, and back views of a building are called _____.
9. What is the purpose of drawing conventions in the architectural and engineering fields?
10. *True or False?* Bubble plans are refined plans that include the size and location dimensions of every element.
11. What are force, strain, and stress?
12. A(n) _____ force pulls material apart, much like two people pulling on a rope. A(n) _____ force pushes on a material and squeezes it.
13. Explain bending stress.
14. _____ is a twisting force and _____ is a force that causes two parts of an object to slide over one another.
15. Name two types of static loads.
16. List five different dynamic loads that can affect a structure.
17. One way to improve a structure's strength without increasing its size is to change the shape of its _____.
18. *True or False?* Triangles are the weakest shape for supporting a load.
19. A(n) _____ is a long projecting beam that is supported on only one end.
    A. arch
    B. truss
    C. cantilever
    D. None of the above.
20. The major task of civil engineers is conducting _____ to make informed decisions about how structures should be designed.
21. *True or False?* When a structural element reaches critical value, it is unstable.
22. What is a shear force diagram?
23. The rotation of a beam about an axis as a result of an applied force is called _____.

Critical Thinking

1. Explain how civil engineering and architecture have impacted civilization over time.
2. Describe the differences and similarities between the work of an architect and a civil engineer.
3. How can changing the shape of a material increase its strength?
4. Explain how the forces of tension, compression, torsion, and shearing affect a roof truss.

STEM Applications

1. Demonstrate how architectural drawing conventions are used to convey structure designs by sketching out the floor plans and elevation views of your home.
2. Apply your knowledge of mathematics, science, and engineering to analyze the forces applied to the beam in the two free body diagrams shown here. First, calculate the reaction forces at the supports, and then create a shear and moment diagram for each beam.

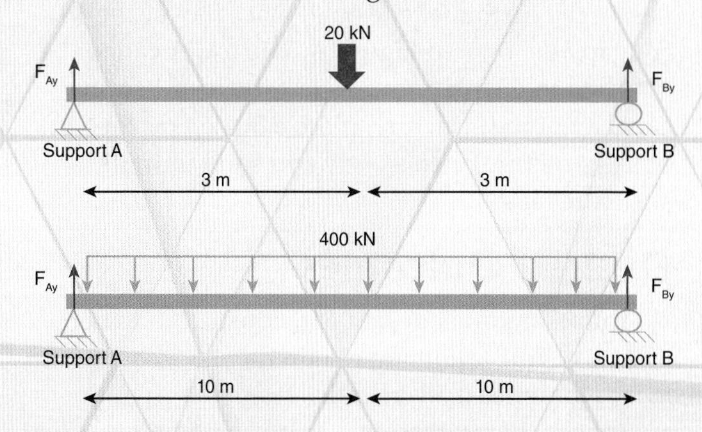

Engineering Design Challenge

Vertical Structure Challenge

Background

You have learned about the forces that structures must be able to support and withstand. An assortment of structural members of different shapes and sizes is used to support the structure and withstand these forces. Determining the best design for the structure requires an understanding of structural analysis concepts in order to compute the load capacity of structural components. Using this knowledge, your team will design and build a model of an efficient vertical load-bearing structure, using limited materials.

Situation

Your team will use their civil engineering knowledge and skills to build the tallest, strongest, and most efficient tower using a maximum of 24 sheets of standard copy paper. The paper will be used to create the structural members of the tower. You may use 3/4″ cellophane tape to join the paper members. The tower must be at least 50 cm tall and span a 10 cm gap, **Figure A**. Your tower must support at least 2.5 kg and withstand an earthquake test, using an earthquake-testing platform (a movable platform connected to a rigid frame by rubber bands or springs). Once the tower passes the earthquake testing, place it over the 10 cm gap as shown in **Figure A**. The tower cannot be attached to the platform. To test the load-bearing capacity of the structure, place a bucket on top of the structure, then slowly add sand until the structure fails.

Desired Outcomes

Your team must design and construct a tower that meets the following requirements:

- Stands at least 50 cm tall
- Spans a 10 cm gap
- Composed of no more than 24 sheets of paper
- Supports at least 2.5 kg during earthquake testing

Your team must also develop a technical report that describes how the tower was designed using engineering principles. The report should also include the calculations for determining load capacity and concepts applied to withstand seismic loads.

To test your tower, you need to collect the following structural data for calculating the tower's performance and efficiency compared to other towers:

- Tower height
- Tower mass
- Total weight capacity of tower

After collecting the structural data, perform the following calculations to determine the structure's performance based on loading capacity and its efficiency:

- Performance points
  - Up to 50 points for the strongest structure in the class:

$$O = \frac{\text{total height (cm)} \times \text{total weight (g)} \times 50}{\text{performance points of highest in class}}$$

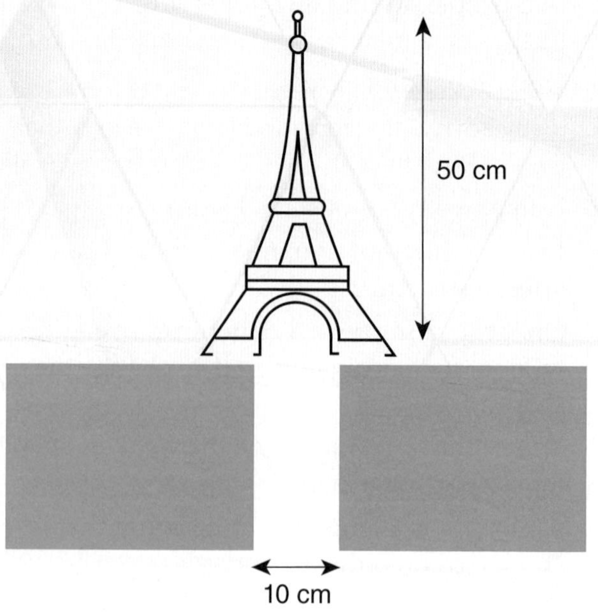

Figure A. Efficient tower.

- Efficiency points:
    - Qualifying structures will be awarded efficiency points, determined using the following formula:

      $$\text{tower efficiency} = \frac{\text{maximum load (kg)}}{\text{mass of structure (kg)}}$$

    - The structure with the highest efficiency will score 20 points. The remaining scores will be based on percentage of the top score. For example:

      Top tower efficiency = 1550

      Second place tower efficiency = 1250

      $$\frac{1250}{1550} = 81\%$$

    - This percentage will be multiplied by the 20 points possible, which equals 16.2 points.

Materials

This project is restricted to the following materials:

- 24 sheets of standard copy paper
- 3/4″ cellophane tape
- Scissors

Testing requires the following materials:

- Earthquake-testing platform
- Bucket
- Sand
- Spring scale

Beginning the Process

- Use your knowledge of engineering design to guide you through the process of defining the problem and creating a viable solution to it.
- Remember, the engineering design process is an iterative approach to solving problems, meaning you can and should go back and forth between the different steps.
- Choose your engineering design teams based on the individual skill sets. Talk with your teammates to formulate a problem statement based on the background and situation provided. A good problem statement should address a single problem and answer "who, what, where, when, why?" However, the problem statement should not imply a single solution to the problem. Different teams will create different problem statements, based on their own interpretations of the situation and their interests and experiences. Your engineering team will use your defined problem statement as the starting point to creating a valid solution.
- Apply the knowledge your team has obtained and employ the engineering design process in order to propose new or improved solutions to the problem defined from the situation provided. This will be a natural progression from brainstorming, to generating ideas, to producing prototypes/models, to evaluating success, and to refining the solution.

TSA Modular Activity

Structural Engineering

Activity Overview

In this activity, you will create a balsa-wood bridge and determine its failure weight (the load at which the bridge breaks).

Materials

- Grid paper
- 20′ of 1/8″ × 1/8″ balsa wood
- 3″ × 5″ note card
- Glue

Background Information

General. There are several types of bridges. The length of the span and available materials generally determine the type of bridge used in a particular situation. For this activity, a truss design is considered the most efficient.

Truss bridges. The truss bridge design is based on the assumption that the structural members carry loads along their axes in compression or tension. The members along the bottom of the bridge carry a tensile load. Those along the top of the truss carry a compressive load. The members connecting the top and bottom chords (members) can be in tension or compression.

Gussets. Gussets are plates connected to members at joints to add strength. The structural steel members are welded or bolted to the gusset. When designing your bridge, include a gusset at each joint, if possible.

Wood properties. Due to its molecular structure, wood can normally carry a greater load in tension than it can in compression. Also, a shorter member can carry a greater compressive load than a longer member.

Guidelines

- Create a scale sketch of the bridge before building.
- Two pieces of balsa wood can be glued together along lengthwise surfaces. No more than two pieces of balsa can be glued together. Do not use excessive amounts of glue.
- Gussets cut from the 3″ × 5″ card can be no larger than the diameter of a US quarter. One gusset cannot touch another gusset. This plate cannot be sandwiched between two pieces of balsa wood.
- The bridge design must take into account the loading device. Your instructor will provide specific guidelines for bridge length, bridge width, and the required details for attachment of a loading device.
- Your bridge will be weighed before being loaded.

Evaluation Criteria

Your project will be evaluated using the following criteria:

- Accuracy of sketch, compared to completed bridge.
- Conformance to guidelines.
- Efficiency (failure weight ÷ bridge weight).

Razvan Sera/Shutterstock.com

Architects and civil engineers are responsible for designing and planning structures such as buildings and bridges.

CHAPTER 19

Harnessing and Using Energy

Check Your Engineering IQ

Before you read this chapter, assess your current understanding of the chapter content by taking the chapter pretest. www.g-wlearning.com/technologyeducation/

Learning Objectives

After studying this chapter, you should be able to:

- ✔ Differentiate between energy, work, and power.
- ✔ Define potential energy and kinetic energy.
- ✔ Describe the six major forms of energy and how they are applied to do work.
- ✔ Identify exhaustible, renewable, and inexhaustible energy sources.
- ✔ Describe the different methods of inexhaustible-energy conversion.
- ✔ Describe the different methods of renewable-energy conversion.
- ✔ Identify common ways to heat homes and buildings.
- ✔ Recall the major parts of an electric-energy generation-and-conversion system and their functions.

Technical Terms

active collectors
biochemical conversion
biofuels
biogas
chemical energy
conduction
convection
direct active solar system
direct-gain solar system
electrical energy
energy converter
indirect active solar system
indirect-gain solar system
insolation
kinetic energy
mechanical energy
nuclear energy
passive collectors
potential energy
prime mover
radiant energy
radiation
stroke
thermal energy
Trombe wall
water turbine
work

Close Reading

Before you read this chapter, consider your daily routine. List the different types of energy associated with your actions and any energy converters that you are familiar with. As you read, add to your list if necessary. Also, make an outline of the different types of converters. Include at least one example of each type.

The word *energy* is used in different ways. For instance, you might say, "I don't have the energy to mow the lawn." Topics such as energy dependence on foreign petroleum and energy conservation are discussed. However, not everyone using the word knows its exact meaning.

The word *energy* comes from the Greek word *energeia,* which means *work.* As time passed, the word came to describe the force that makes things move. Today, *energy* is defined as *the ability to do work.* This ability includes a broad spectrum of acts. Energy is used in simple human tasks, such as walking, running, and exercising. The ability to do work can be obtained from petroleum and then used to power a ship across the ocean. See **Figure 19-1**. Energy is used to provide motion in vehicles and machines and produce heat and light. Energy is fundamental to communication technologies and is used in manufacturing products and constructing structures. Energy is everywhere and is used by all of us.

Energy, Work, and Power

It takes energy to do work. For example, you must eat well if you plan to run a marathon or build a structure. When you do one of these things, you might say, "I worked really hard!" In scientific terms, ***work*** is applying a force that moves a mass a distance in the direction of the applied force. Lifting boards off the floor and placing them on a table is work. The boards have mass, and they were moved some distance. If you are aware of the power of a machine, then you can recognize how quickly or slowly it will function. Whereas work involves force and distance, power involves work and time. For example, one machine may force 1000 gallons of water through a 50′ pipe in 60 seconds. Another machine may force the same amount of water, the same distance, in 120 seconds. The first machine is twice as powerful as the second machine. Power is calculated as work divided by time. Common units for power include joules per second (J/sec), foot-pounds per second (ft-lb/sec), and newton-meters per second (N-m/sec). Power can also be defined by each of the energy systems in which it is being considered.

Types of Energy

Energy can be associated with a force doing work or with a force that has the capability of doing work. ***Kinetic energy*** is energy in motion; it is involved in moving something. A hammer striking a nail is an example of kinetic energy. Other examples of kinetic energy are a sail capturing the wind to power a boat and a river carrying a boat or turning a waterwheel.

Energy Is Used for...

Human Activities

Goodheart-Willcox Publisher

Technological Systems

Akif Oztoprak/Shutterstock.com

Figure 19-1. Energy is used in all actions, from human movement to powering complex technological devices.

Not all energy is in use at any given time. Energy stored for later use is called ***potential energy***. See **Figure 19-2**. Energy in this condition has the capability, or potential, of doing work when it is needed. For example, water stored behind a hydroelectric dam possesses energy. The water releases this energy to turn a turbine when it flows through the power-generating plant. A flashlight battery and a gallon of gasoline are other examples of potential energy.

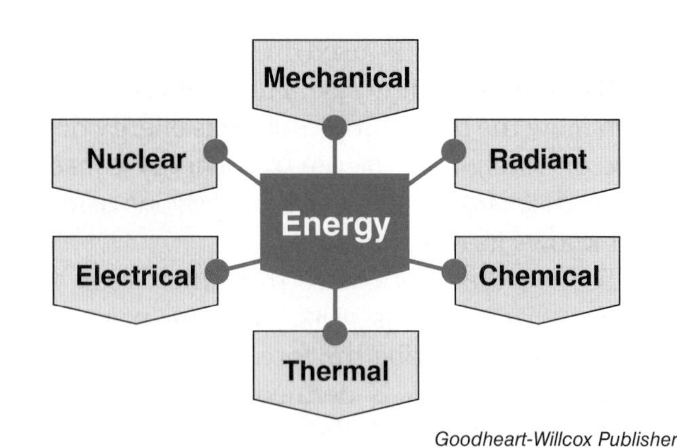

Goodheart-Willcox Publisher

Figure 19-3. Energy can be grouped into six major forms.

Forms of Energy

Energy is everywhere. The ability to do work is in the fires that burn wood in a fireplace. Energy is in sunlight, wind, and moving water. The hundreds of examples of energy can be grouped into six major forms. See **Figure 19-3**. The following are the six primary forms of energy:

- Mechanical.
- Radiant.
- Chemical.
- Thermal.
- Electrical.
- Nuclear.

These energy sources do work that ends up as motion, light, or heat. They are used to power manufacturing machines, light buildings, propel vehicles, and produce communication messages.

AMAX

Figure 19-2. The coal being mined has potential energy. The energy will be released when the coal is burned.

Mechanical Energy

Mechanical energy is energy that an object possesses in relation to its motion or due to its position. Often, it is produced by motion of technological devices, such as machines, but there are other types of mechanical energy. Wind and moving water have motion and, thus, are also mechanical energy sources. See **Figure 19-4**.

Kevin George/Shutterstock.com

Figure 19-4. This waterfall is a natural source of mechanical energy.

Radiant Energy

Radiant energy is energy in the form of electromagnetic waves, which are discussed in Chapter 21. These waves range from ELF radio waves (long waves) to gamma rays (short waves). Cool objects give off longer waves than hot objects do. Low-frequency waves contain less energy than high-frequency waves of the same amplitude.

The main source of radiant energy is the sun. Radiant energy is also emitted by objects heated with a flame or from a lightbulb in a lamp. Sometimes, radiant energy is called *light energy*. This is not completely correct because many waves with wavelengths longer or shorter than the wavelengths of light possess radiant energy. Examples of radiant energy include radio waves; microwaves; infrared, visible, and UV light; x-rays; and gamma rays.

ProstoSvet/Shutterstock.com

Figure 19-5. Energy to run this cordless drill is stored inside the drill's battery.

Chemical Energy

Chemical energy is energy stored within a chemical substance. Chemical energy is released when a substance is put through a chemical reaction. A chemical reaction occurs when bonds are broken. This can be done by rapid oxidation (burning) or other chemical actions, such as digestion in our bodies as acids break down food.

Typical sources of chemical energy are fuels used to power technological machines. See **Figure 19-5**. The most common fuels are petroleum, natural gas, and coal. Wood, grains (such as corn), and biomass (organic garbage) are also sources of chemical energy.

©iStockphoto.com/DJClaassen

Figure 19-6. This log burning in a fireplace is giving off both radiant and thermal energy.

Thermal Energy

Thermal energy (*heat energy*) is energy created by the internal movement of atoms in a substance. These particles are always in motion. If the atoms move or vibrate rapidly, they emit heat, or thermal energy. The faster the atoms move, the more heat they give off. See **Figure 19-6**.

Thermal energy cannot be seen directly. You may be able to see its effects, however, such as the heated airwaves above a road on a very hot day. Thermal energy is usually felt. The energy strikes a surface, such as your skin, and elevates its temperature.

Thermal energy is widely used in technological devices. This energy provides the energy for our heating systems and some electrical generating plants.

Electrical Energy

Electrical energy is created by electrons moving along a conductor. The conductor can be a wire in a man-made electrical system. The conductor can also be the air, as with lightning when there is a very high electric voltage. Lightning is a natural source of electrical energy. Electrical energy is used as a basic source for other forms of energy. This energy

Academic Connection: History

The Origin of Horsepower

After developing the first truly practical steam engine, James Watt needed to describe its output in a way that potential customers could understand. In those days, horses drove much of the machinery. Watt decided the best way to describe the output of an engine was in terms of the number of horses it could replace.

©iStockphoto.com/trueblu

In 1782, Watt began working on the formula that would eventually lead to the unit of power known as *horsepower*. After consulting with experts who designed horse-driven machinery, Watt determined that an average horse can perform 22,000 ft-lb of work per minute. Since the rotary-motion steam engine was new technology, potential customers were skeptical. Watt was afraid that, if factory or mine machinery failed because the engine used to power it was too weak, the public would blame the steam engine technology and not the miscalculation of the purchaser. Watt decided it would be better to understate the power of his engines.

If a customer bought a two-horsepower engine, Watt wanted that engine to be able to do the same amount of work as three actual horses. In order to accomplish this, Watt added 50% to his estimation of the amount of work a horse could do over a given time. This caused 1 horsepower to be equivalent to 33,000 ft-lb of work per minute, which is the value still used today.

As technology improves, the amount of horsepower available to the average person increases. Research the horsepower rating for several types of automobiles. How many actual horses would be required to produce the same amount of power? (Remember, a horse can produce only 0.67 horsepower.)

is often converted into heat energy (for example, to warm buildings) and into light energy (for example, to illuminate our homes). See **Figure 19-7**.

©iStockphoto.com/iahulbak

Figure 19-7. This worker is using an electric arc to generate high heat to weld (melt) two steel parts together.

Nuclear Energy

Nuclear energy is energy produced from the splitting of atoms. When the internal bonds of atoms are split, they release vast quantities of energy. This process is called *fission*. Combining two atoms into a new, larger atom also releases large amounts of energy. This process is called *fusion*.

Interrelationship of Energy Forms

All forms of energy are related to one another. For example, radiant energy can be used to produce heat. It can be harnessed by solar collectors and then distribute heat. A fire causes fuel to undergo a chemical action. For example, coal can be changed into carbon dioxide and water. In the process of this chemical action, heat is given off. The mechanical motion of an electrical generator causes magnetic lines of force to cut across any nearby conductor. This process induces an electrical current in the conductor.

Sources of Energy

Energy is a basic input to all technological systems. The three basic types of energy resources, **Figure 19-8**, are as follows:

- Exhaustible energy resources.
- Renewable energy resources.
- Inexhaustible energy resources.

Exhaustible Energy Resources

Exhaustible energy resources are materials that cannot be replaced. Once they are used up, we will no longer have that energy source. Exhaustible resources include petroleum, natural gas, and coal. These fossil fuels originated from living matter. Millions of years ago, plant and animal matter was buried under the earth. Over time, this matter was subjected to pressure, and it decayed. This resulted in deposits of solid fuels (coal and peat), liquid fuels (petroleum), and gaseous fuels (natural gas). These deposits have been found in many locations on the earth. Chapter 14 describes how the deposits are located and the fuels are extracted.

Uranium is another exhaustible energy source. This energy source is an element that developed when the solar system came into being. Uranium is a radioactive mineral used in nuclear power plants.

Renewable Energy Resources

Renewable energy resources are biological materials that can be grown and harvested. Human propagation, growing, and harvesting activities directly affect the supply of these resources. Common renewable energy resources are wood and grains. They can be burned directly to generate thermal energy. Corn is often converted to alcohol (ethanol), which then can be used as a fuel. See **Figure 19-9**.

Organic matter, such as garbage, sewage, straw, animal waste, and other waste, can be an energy resource. This matter is often referred to as *biomass resources*. The prefix *bio-* means "having a biological, or living, origin." The resources can be traced back to plant or animal matter. These organic materials can be burned directly to produce fuels called ***biofuels***. Organic wastes can also be broken down by bacteria in the absence of oxygen to produce ***biogas***. Biogas is predominantly methane, a highly flammable gas. Biogases can replace some exhaustible fuel resources.

Inexhaustible Energy Resources

Inexhaustible energy resources are sources of energy that are not expected to run out. They are part of the earth's solar weather system. This natural

Exhaustible

©iStockphoto.com/iahulbak

Renewable

blphoto1/Shutterstock.com

Inexhaustible

nares mongkol/Shutterstock.com

Figure 19-8. The three types of energy resources.

©iStockphoto.com/photosbyjim

Figure 19-9. The corn in the foreground can be converted into ethanol at the plant in the background.

cycle starts with solar energy. About one-third of the solar energy reaching the earth's atmosphere is reflected back into space. The other two-thirds enter the atmosphere. The atmosphere absorbs much of this solar energy.

About one-fourth of this absorbed energy powers what can be called a *water cycle*. See **Figure 19-10**. A small portion of the earth's water is in rivers and lakes. The majority of the water is contained in oceans that cover much of the globe. Solar energy causes the water in the oceans to heat and evaporate. The warm water vapor rises into the atmosphere and forms clouds. The clouds rise and are carried inland by the wind. As the clouds travel

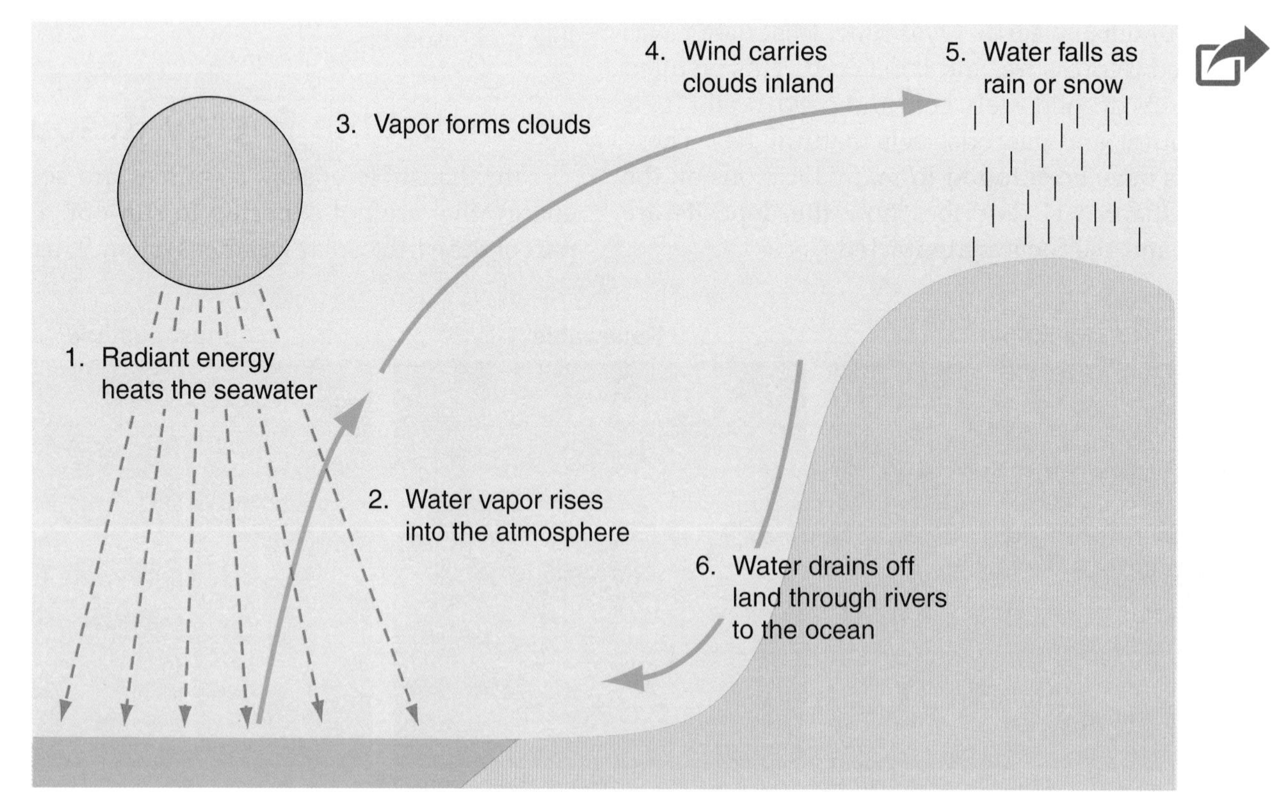

Goodheart-Willcox Publisher

Figure 19-10. The Sun's radiant energy powers the earth's water cycle. The water cycle provides an inexhaustible energy source.

Think Green Efficient Energy Use

Energy efficiency is a kind of reduction. An energy-efficient appliance is something that still works in the same way as other appliances while using less energy. There are several ways of achieving a more efficient use of energy. Examples of products with energy-efficient alternatives include insulation, lamps, automobiles, and household appliances.

Energy efficiency is important for various reasons. Two of the most important reasons are that it helps reduce depletion of fossil fuels, and it reduces the emission of carbon dioxide into the atmosphere.

Brian A Jackson/Shutterstock.com

Figure A. Smart thermostats and apps for mobile devices help homeowners control their energy remotely.

A well-insulated home helps regulate the temperature inside. Therefore, the use of a furnace or air conditioner is reduced. Automobile bodies are being redesigned to minimize drag, helping reduce fuel usage. Incandescent lamps are being replaced with LED lamps, which use less energy to give off light. Smartphone apps and smart thermostats, as seen in **Figure A,** allow homeowners to control their house and smart technologies remotely.

upward, the water vapor cools. This cooling effect condenses the water vapor into droplets, which fall to earth in the form of rain or snow. Much of the water runs off the land and collects in rivers. From there, the water flows into the ocean, where it begins the cycle once again.

Not all the water follows this exact pattern. Plants use some quantity of water in their respiratory cycles. Some water ends up in lakes, from which it evaporates into the air to join clouds. Other portions flow underground into rivers and oceans.

Solar energy also heats the land, but the heating effects on the oceans and the land are different. Temperature differences are created because different amounts of solar energy strike various areas of the globe. Warmer air rises and is replaced by cooler air. This air movement is called *wind*. The water cycle, winds, and direct solar energy become inexhaustible energy resources. They produce energy through hydroelectric power generators (water), wind generators, and solar converters.

Another inexhaustible energy resource is geothermal energy. Geysers—hot springs that shoot out intermittent jets of hot water and steam—are examples of this energy resource. See **Figure 19-11**. This energy source comes up from the Earth (*geo-*) in the form of heat (*thermal*). Geothermal energy uses water heated by the hot core of the Earth. This energy is usually tapped by wells and used to heat buildings or to power electrical generators.

CE Photography/Shutterstock.com

Figure 19-11. Geothermal energy produces this geyser in Yellowstone National Park.

Energy Conversion

Science tells us that energy can be neither created nor destroyed. However, it can be converted into other forms and applied to do work. For example, we burn fuels to change water into steam. Steam contains energy in the form of heat. The steam might then be passed through devices used to warm rooms or dry lumber. Energy powers most devices that process materials and information.

Energy converters should be viewed as part of a larger system. An ***energy converter*** is a unique device that has energy as its input and its output. Mechanical energy is the input to a turbine in a hydroelectric generator. Electricity (electrical energy) is the output. This same electricity can be the input to several other energy converters. Incandescent lamps convert electrical energy into light (radiant energy). Motors convert electrical energy into rotary motion (mechanical energy). Resistance heaters change electrical energy into heat (thermal energy).

Your body is an energy converter. This converter converts food (your fuel) into energy. Energy moves muscles, allowing you to walk, talk, and see. Likewise, an automobile engine is an energy converter. The engine converts the potential energy in gasoline into heat energy to produce mechanical motion. Energy converters power our factories, propel our transportation vehicles, heat and light our homes, and help produce our communication messages. See **Figure 19-12**.

Humans have developed hundreds of energy converters that process energy in various ways. The following are four broad categories of energy-conversion systems:

- Inexhaustible-energy converters.
- Renewable-energy converters.
- Thermal-energy converters.
- Electrical-energy converters.

Powering a Factory

Hamik/Shutterstock.com

Propelling a Vehicle

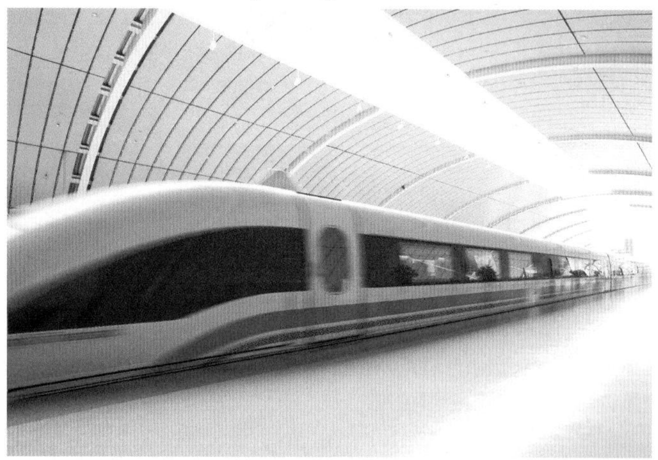

06photo/Shutterstock.com

Figure 19-12. Energy is the foundation for technology. This foundation powers our machines and carries us across long distances.

Inexhaustible-Energy Converters

The earliest energy-conversion technologies were designed to power simple devices. These devices fall into a category that mechanical engineers call prime movers. A ***prime mover*** is any device that changes a natural source of energy into mechanical power.

Most early prime movers used wind power and waterpower. Wind power and waterpower are inexhaustible energy sources. Almost all societies used energy converters in transportation. Wind and flowing water helped move their boats.

Several uses for energy converters emerged, however. On land, windmills and waterwheels became important technological devices developed to harness the forces of wind and water. See **Figure 19-13**. These two devices convert natural mechanical energy (flowing air or running water) into controlled mechanical energy. For example, they can produce the motion needed to power a water pump or an electric generator. Other important converters use solar, geothermal, and ocean energy. These converters can be used to produce energy needed to heat and light homes or power other technological devices.

©iStockphoto.com/Enjoylife2

Figure 19-13. Windmills were early energy converters.

©iStockphoto.com/rcp

Figure 19-14. Sails were one of the earliest technological devices to harness wind power.

Wind-Energy Conversion

The sun is the original source of most of the energy on Earth. This energy is stored in growing plants and animals and in decayed organic matter. Decayed organic matter includes peat, coal, natural gas, and petroleum.

The sun causes winds to blow all over Earth. An unequal heating of Earth's surface produces these air currents. Each day, the sun's rays heat the landmasses and water they touch. Not all areas are touched at the same time or with equal energy, however. Polar areas receive less solar energy than do areas near the equator. Areas under cloud cover receive less solar heating than do areas in direct sunlight. The heat from the land and water warms the air above it. The warm air rises. Cooler air moves in to replace it. This movement produces air currents we call *wind*.

The air above the hot areas near the equator is always rising. The cooler polar air moves toward the equator. In addition, air above water heats and cools more rapidly than air over land. Thus, during the morning hours, the air above the water warms quickly and rises. Cool air from the land moves in to replace it. In the evenings, the air above the land stays warmer longer than the air above the water. The cooler sea air moves inland to replace the rising land air.

Early humans designed technological devices to use air currents. An early use of wind power was a ship's sail. See **Figure 19-14**. This device was developed in Egypt around 12,000 years ago. Sails remained the primary power for ships until the development of the steam engine in the late 1700s.

Wind power, the principle of the sail, was adapted to land applications with the development of the windmill. The windmill's first use was probably in the Middle East around 200 BCE. See **Figure 19-15**. These mills were used for grinding grains into flour.

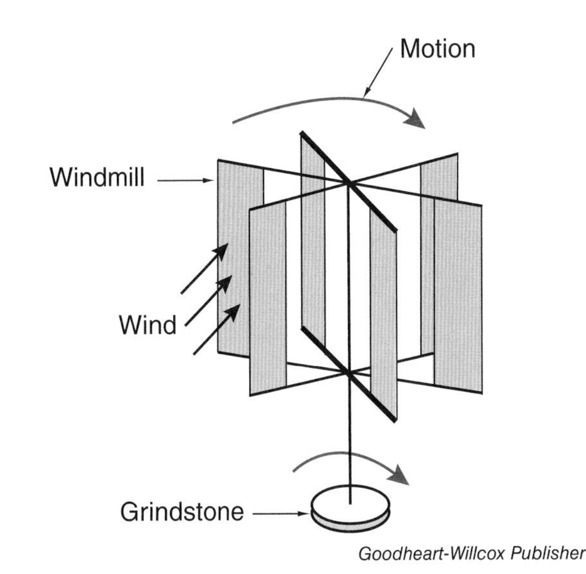

Goodheart-Willcox Publisher

Figure 19-15. A design thought to be used in ancient windmills. The windmill harnesses the wind for ease in grinding.

This early wind-conversion technology eventually led to today's windmills and turbines. Modern windmills are primarily used to pump water for livestock on large western cattle ranches. Wind turbines are used to power electric generators.

Windmill and wind turbine designs can be grouped into two classes—horizontal axis and vertical axis. See **Figure 19-16**. The horizontal-axis design has one or more blades connected to a horizontal shaft. The wind flows over the blades, causing them to turn. To see this action, place a household cooling fan in front of a blast of air. The blades turn, even though the fan's power is off.

Vertical-shaft designs have blades arranged around the shaft. As the wind blows past the blades, torque (turning force) is generated, rotating the shaft. The most common vertical-shaft devices are the Darrieus and Savonius wind turbines.

A great deal of experimentation is being done to develop and improve the efficiency of existing wind turbines to increase the capability of electric generators. Large numbers of these devices are grouped together in wind farms located in various parts of the western United States. See **Figure 19-17**.

Water-Energy Conversion

A moving gas (air) powers windmills. Likewise, moving liquids can provide the energy to power technological devices. One of the earliest devices used to capture this energy was the waterwheel. This device is essentially a series of paddles

©iStockphoto.com/GregC

Figure 19-17. A windmill farm in California.

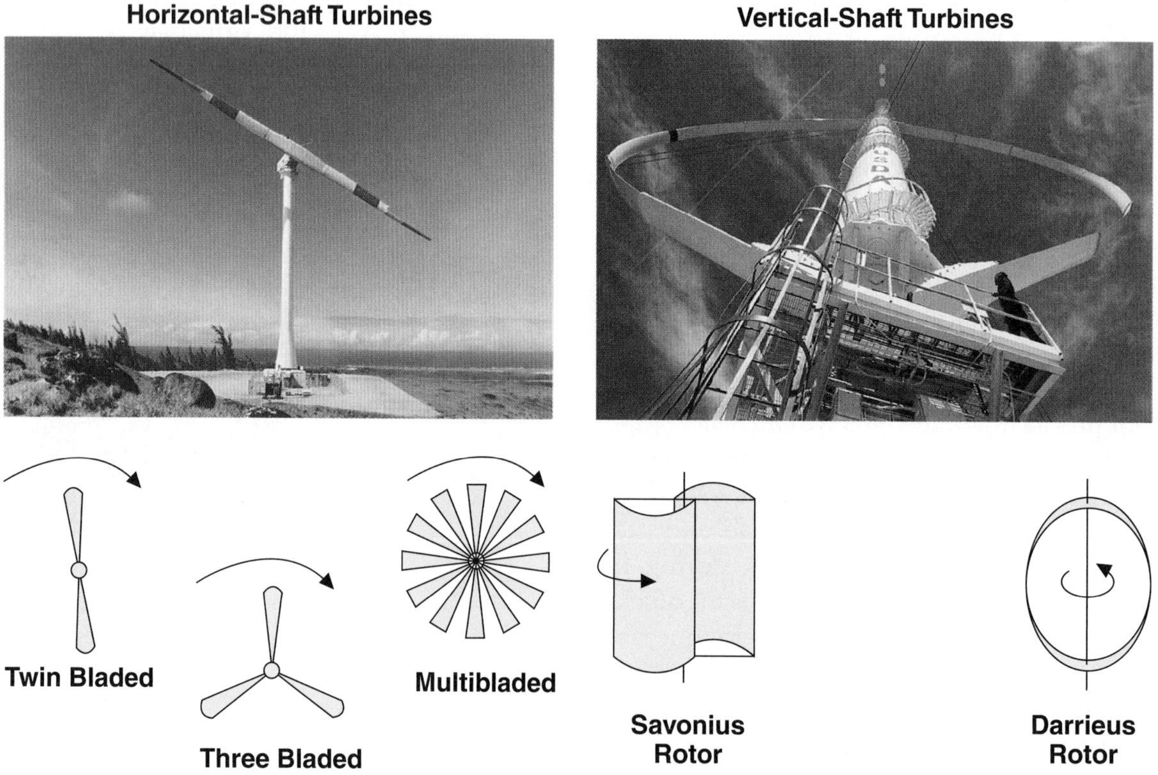

Horizontal-Shaft Turbines

Vertical-Shaft Turbines

Twin Bladed

Three Bladed

Multibladed

Savonius Rotor

Darrieus Rotor

US Department of Energy

Figure 19-16. The two classes of wind turbines. On the left are examples of wind turbines with their axes parallel with the airflow (horizontal shafts). On the right are examples with their axes at right angles to the wind (vertical shafts).

extending outward from a shaft. The flowing water drives the paddles, causing the wheel to rotate.

Waterwheels powered the first factories of the Industrial Revolution. The wheels were produced in two basic designs—undershot and overshot. See **Figure 19-18**. Water rushing under the waterwheel powers an undershot waterwheel. The overshot wheel is powered by water falling onto it from an overhead trough or pipe.

A modification of the waterwheel is the water turbine. The ***water turbine*** is a series of blades arranged around a shaft. As water passes through the turbine at a high speed, the blades spin the shaft. Water turbines are used widely to power electric generators in hydroelectric power plants. A steam turbine is a similar device. This device uses steam (hot water vapor), however, to drive the turbine. Coal, oil, natural gas, and nuclear power plants' generators are driven by steam turbines.

Solar-Energy Conversion

The need to conserve exhaustible energy resources has brought solar energy into consideration as a replacement resource. A solar converter uses the constant energy source of the sun.

Solar energy is very intense before it reaches the earth. The energy can produce temperatures approaching 10,800°F (5982°C). As it reaches the earth's protective ozone layer, however, much of this energy is absorbed and heats the atmosphere. Water vapor in the air absorbs additional energy. Some solar energy reaches the earth and is available as an inexhaustible energy source.

The amount of this available energy depends on the inclination (angle to the horizon) of the sun and the atmospheric conditions (cloud cover) over the earth. The term ***insolation*** is used to describe the solar energy available in a specific location at any given time. Insolation varies with the seasons and the weather. The maximum insolation on a clear, summer day is about 1000 megawatts (MW) per square kilometer (0.38 square mile).

Most solar energy–conversion systems have two major parts—a collection system and a storage system. Solar collectors depend on the principle that black surfaces absorb most of the solar energy that strikes them. This causes black surfaces to gather heat when they are exposed to sunlight. Solar collectors can be grouped into two categories—passive collectors and active collectors.

Passive Collectors

Passive collectors directly collect, store, and distribute the heat they convert from solar energy. In a passive solar-collection system, an entire house is actually a solar collector. The building sits quietly in position as the sun heats it. This can be done in three ways. See **Figure 19-19**.

A ***direct-gain solar system*** allows the radiant energy to enter the home through windows, heating inside surfaces. An ***indirect-gain solar system*** uses a black concrete or masonry wall (***Trombe wall***) that has glass panels in front of it. The wall has openings

Waterwheels

Undershot

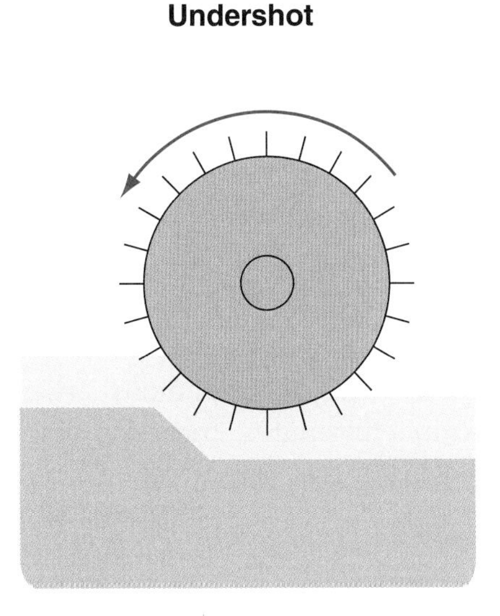

Overshot

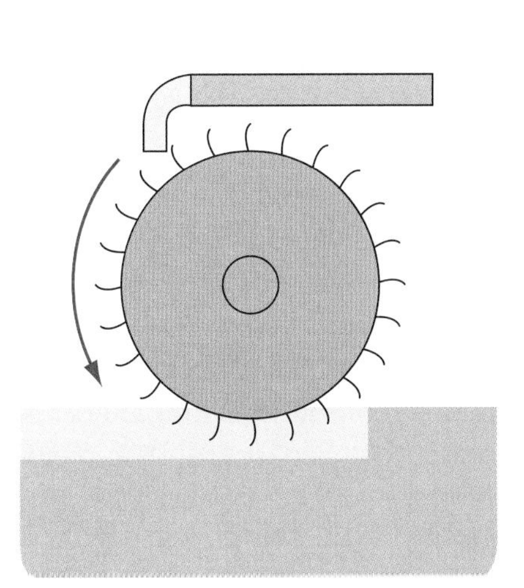

©iStockphoto.com/esemelwe

Figure 19-18. The two common types of waterwheels.

Direct Gain

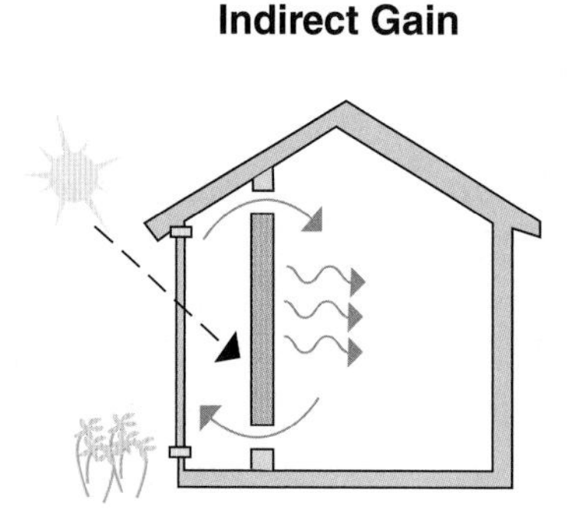

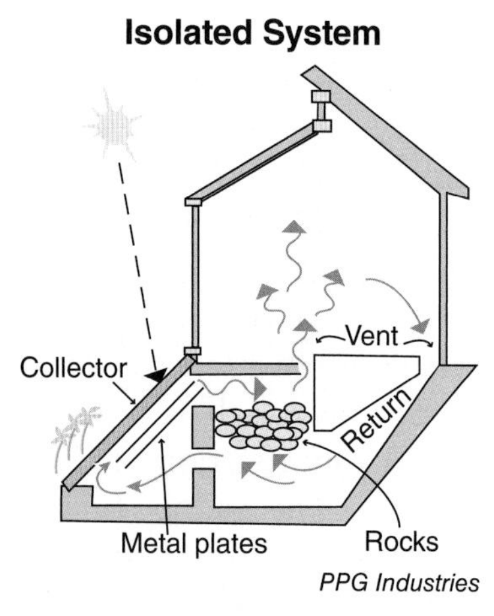

PPG Industries

Figure 19-19. Three types of passive solar systems can be used to heat a home. Direct gain is the simplest system. The indirect system uses a Trombe wall. One type of isolated system uses collectors separate from the house.

at its bottom and top. As sunlight strikes its surface, the wall is heated. In turn, the air between the wall and the glass panels becomes heated by energy radiating from the wall. The warm air rises and flows into the building through the openings at the top of the wall. This creates natural convection currents that draw cooler, heavier air into the openings at the bottom. This new air, in turn, is heated and rises. The Trombe wall retains a great deal of heat. Consequently, after the sun sets, the wall continues to radiate the heat and warms the air between the wall and the glass panels.

Adobe homes in the southwestern part of the United States use a similar solar-heating principle. The solar energy striking the adobe brick warms the surface. During the day, the energy slowly penetrates the thick wall. This penetration takes about 12 hours. In the evening, the heat finally reaches the inside of the dwelling and provides warmth during the cool nights. By morning, the wall is cool. The wall insulates the rooms from the daytime heat. The 12-hour lag between heating and cooling makes an effective daytime cooling and nighttime heating system for areas that have hot days and cool nights.

An isolated solar system uses solar collectors or greenhouses separate from the house. These collectors are built below the level of the house. The heat generated in the collector can be channeled directly to heat the home. Additional heat is stored in a thermal mass (rock bed) for night and cloudy-day heating.

Active Collectors

Active collectors use pumps to circulate water. As they do this, they collect, store, and distribute the heat that was converted from solar energy. These systems are used in many areas to provide hot water and room heat (space heating). Two major types of active-collector systems are *indirect* and *direct*.

A typical ***indirect active solar system*** has a series of collectors. Each collector has a black surface to absorb solar energy. Above or below this surface is a network of tubes, or pipes. Water is circulated through these channels. As the fluid passes a warm black surface, it absorbs heat. The warm water is then pumped to a heat exchanger. The heat exchanger can heat water for domestic use or provide thermal energy for a heating system. See **Figure 19-20**.

A ***direct active solar system*** does not have a heat exchanger. The water circulated in the system is used as domestic hot water. This water flows directly to such areas as household faucets, washing machines, and showers.

Active solar converters are used in electric-power generation. The converters produce steam. The steam drives the turbines in the generation plant. See **Figure 19-21**.

Another type of solar converter is the ***photovoltaic cell*** (solar cell), which converts light energy into electricity. The cells are made of certain semiconductor materials, such as crystalline silicon. Small bundles of light energy, called *photons*, impact the cell. These photons strike the cell, causing electrons to dislodge from the silicon wafer. The electrons move in one direction across the wafer. A wire attached to the wafer provides a path for the electrons to enter an electrical circuit. A second wire allows electrons from the circuit to return to the wafer. This current flow supplies the power for each device.

Solar radiation
Collector
Hot water or space heating
Pump
Heat exchanger

US Department of Energy

Goodheart-Willcox Publisher

Figure 19-20. The solar panels on the building power a system similar to the one in the diagram.

Peteri/Shutterstock.com

Figure 19-21. This large bank of solar collectors collects solar energy to produce electricity near Pilsen, Czech Republic, European Union.

©iStockphoto.com/kozmoat98

Figure 19-22. This portable highway warning sign is powered by a series of solar cells.

The cell can be a small device powering a pocket calculator or part of an array of units providing electricity for such devices as satellites, solar-powered vehicles, and portable signs. See **Figure 19-22**.

Geothermal-Energy Conversion

Geothermal energy is heat originating in the earth's molten core. This energy can be found at great depths all over the planet. At certain locations, however, it reaches the surface and appears as volcanoes, hot springs, and geysers. This energy is tapped in a number of ways. Electricity is readily produced using geothermal energy to form steam that drives a steam turbine–powered generator. Geothermal energy is used in industry for applications such as food dehydration. See **Figure 19-23**.

Geothermal energy is also used for direct heating. Geothermal heat pumps use the constant 55°F (12.8°C) temperature of groundwater to heat homes. Often, the water is pumped from a well extending into an underground aquifer. The water enters the heat pump. The pump removes heat from the water and transfers it to the dwelling. The cool water is returned to the aquifer through a second well.

Ocean-Energy Conversion

Ocean energy is an inexhaustible source that has only recently been considered a major source of energy for the coming generations. The oceans cover more than 70% of the globe. They contain

N.Minton/Shutterstock.com

Figure 19-23. Piping at a geothermal power station.

two important sources of energy. These sources are thermal and mechanical (the motion of waves and tides).

Ocean Thermal–Energy Conversion (OTEC) Systems

Ocean thermal–energy conversion (OTEC) systems use the differences in temperature between the various depths of the ocean. The basic system has three steps. First, warm ocean water is used to evaporate a working fluid. Second, the vapors are fed into a turbine turning an electrical generator. Finally, cold ocean water is used to condense the vapors to complete the energy-conversion cycle. The process requires water with at least a 36°F (20°C) difference. This occurs only at the equator.

Ocean Mechanical–Energy Conversion Systems

Ocean mechanical–energy conversion systems use the mechanical energy in the oceans to generate power. Two sources of mechanical energy are used—wave energy and tidal energy. See **Figure 19-24**.

Several wave-energy convertor (WEC) devices are currently being used, and newer models that are more efficient are being developed. One of the first offshore wave machines to be able to generate electricity to a grid system was the Pelamis Wave Energy Converter. This technology converted the motion of the ocean surface waves to generate electricity. Another device, a mechanical surface follower, is a simple design. The surface follower is used as a navigational aid. Inside of a buoy floating in the water is a mechanism that uses the up-and-down movement of the buoy to ring a bell or blow a whistle. A second type of buoy used for navigational aid is a pressure-activated device. The device uses the bobbing action the waves create to compress air in a cylinder. As the water rises, it compresses the air. When the buoy falls, the compressed air is released, powering a small generator. The resulting electricity can power a navigational light.

E.O./Shutterstock.com

Figure 19-24. Ocean tides and waves contain tremendous energy that can be tapped to meet human needs.

Tidal-energy devices use the difference between the height of the ocean at high tide and the height at low tide to generate power. As the ocean rises, water is allowed to flow over a dam into a basin. As the tide recedes, the water flows back through turbines, generating electricity. See **Figure 19-25**. A

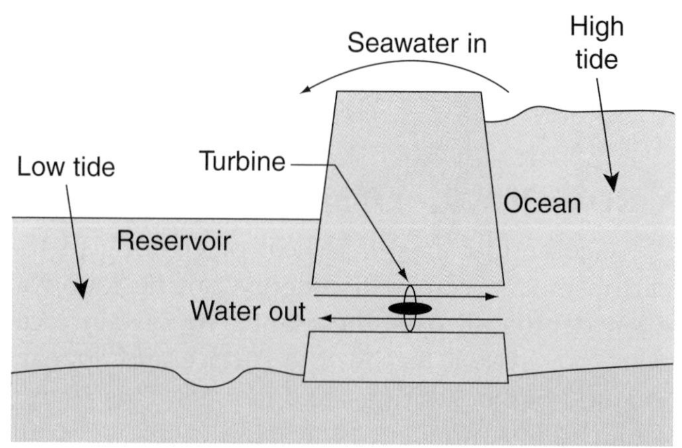

Goodheart-Willcox Publisher

Figure 19-25. A tidal-power ocean energy–conversion system.

system in France creates power from water flow in both directions. The rising tide drives the turbine in one direction. Later, the falling tide powers the turbine in the other direction.

Renewable-Energy Converters

Renewable-energy converters convert inexhaustible energy into usable sources. Our early ancestors depended heavily on renewable energy resources. They burned wood and cattle dung to heat their homes and cook their food. In 1850, these energy sources provided 90% of energy needs. In many parts of the world, these resources are still very important. In the United States today, however, they supply just a small percentage of our energy needs.

Biomass resources are vegetable wastes and animal wastes generated through biological actions. Biomass resources are obtained from the following sources:

- **Forest-products industry.** Resources include sawdust, bark, logging slash (waste), wood shavings, scrap lumber, and paper, for example. See **Figure 19-26**.
- **Agriculture and food processing.** Resources include corncobs, nutshells, fruit pits, grain hulls, sugarcane bagasse (pulp), and manure.
- **Municipal waste.** Resources include sewage and solid waste (garbage).

Bioenergy conversion takes place through thermochemical and biochemical means. These processes are discussed in the following sections.

Thermochemical Conversion

Thermochemical conversion produces a chemical reaction through the application of heat. The most common method is direct combustion, in which the biofuel is burned to produce heat for buildings or to produce steam to power electric generation plants. This system is widely used in the forest-products industry. Burning mill waste produces the steam that heats buildings, dries lumber, and operates the processing equipment.

In a thermochemical process called pyrolysis, the material is heated in the absence of oxygen. The heat causes the biofuel to form liquids, solids, or gases. The solids are carbon and ash. The liquids are similar to petroleum and require further processing. The gases are flammable hydrocarbons. All these materials can be directly or indirectly used as fuels. Other names used for this process are *liquidification* and *gasification*.

Another thermochemical conversion process is *liquefaction*. In this process, the biofuel is heated at moderate temperatures under high pressure. During heating, steam and carbon monoxide, or hydrogen and carbon monoxide, are present. A chemical action takes place, converting the material into an oil that has more oxygen than petroleum. This oil requires extensive refining to develop usable fuels.

Wood from Forest Thinning

©iStockphoto.com/LyaC

Logging Waste

©iStockphoto.com/LyaC

Figure 19-26. Wood from forest thinning and logging waste can provide material for biomass conversion.

STEM Connection: Science

Laws of Gases

The expansion of gas produced by burning fuel powers the internal combustion engine. This artifact was created and can be successfully used because of various scientific discoveries regarding the properties of gases.

For example, the scientist Robert Boyle studied various gases and saw that a relationship exists between the pressure and volume of a gas. Boyle's law states that, when a temperature is held constant, the volume of a fixed mass of gas varies inversely with the pressure. That is, when the pressure increases, the volume of gas decreases. The pressure doubles when the gas is compressed to half its volume. In the internal combustion engine, the pressure increases against the cylinder walls and piston.

Another scientist, Jacques Charles, also studied the behavior of gases. His law states that the volume of a fixed mass of gas at a constant pressure varies directly with absolute temperature. That is, if pressure is kept constant, volume and temperature are directly related. As the temperature increases, the volume increases. How does this law apply to the internal combustion engine?

©iStockphoto.com/wolv

Biochemical Conversion

Biochemical conversion uses chemical reactions caused by fungi, enzymes, or other microorganisms. The two common biochemical conversion processes are anaerobic digestion and fermentation. ***Anaerobic digestion*** is a controlled decaying process that takes place without oxygen. The material used is agriculture waste, manure, algae, seaweed (kelp), municipal solid waste, and paper. The reaction produces methane (a flammable gas) as the biomaterials decay. See **Figure 19-27**.

Fermentation is a very old process using yeast, a living organism, to decompose the material. The yeast changes carbohydrates into ethyl alcohol (ethanol). Grain, particularly corn, is often used for this process. Ethanol can be directly burned or can be mixed with gasoline as an automobile fuel.

Juergen Faelchle/Shutterstock.com

Figure 19-27. A biogas converter in a cornfield.

Thermal-Energy Converters

The Industrial Revolution was greatly dependent on heat engines. Electrical motors have replaced these engines in most industrial applications. Transportation is one exception. We still depend on thermal energy, however. The comfort of your home depends on heating and cooling systems. Many industrial processes use heat to cook, cure, or dry materials and products. The following sections describe two major applications of thermal energy—heat engines and space heating.

Heat Engines

Today, most transportation systems are based on fossil fuel–powered engines. These technological devices burn fuel to produce heat. In turn, the heat is converted into mechanical energy. All heat engines can be classified as either internal combustion engines or external combustion engines. These classifications are based on the location of the thermal energy source. Internal combustion engines burn the fuel within the engine. External combustion engines burn the fuel away from the engine.

Internal Combustion Engines

Gasoline is widely used in land- and water-transportation vehicles, **Figure 19-28**. Common internal combustion engines are gasoline and diesel engines. Jet and rocket engines are also internal combustion engines. To create power, these engines use expanding gases produced by burning fuel. They change heat energy into mechanical motion.

Gasoline and diesel engines change heat energy into mechanical energy. They drive a reciprocating piston by igniting a fuel. A crankshaft changes the piston's reciprocating motion into rotary motion. See **Figure 19-29**.

Gasoline engines can be either two stroke–cycle engines or four stroke–cycle engines. A ***stroke*** is the movement of a piston from one end of a cylinder to another. A cycle is a complete set of motions needed to produce a surge of power. A two-stroke engine moves the piston up and back once (two strokes—one up and one down) to produce a power stroke. A four-stroke engine moves the piston up and back

risteski goce/Shutterstock.com

Figure 19-28. This internal combustion engine is used in race cars.

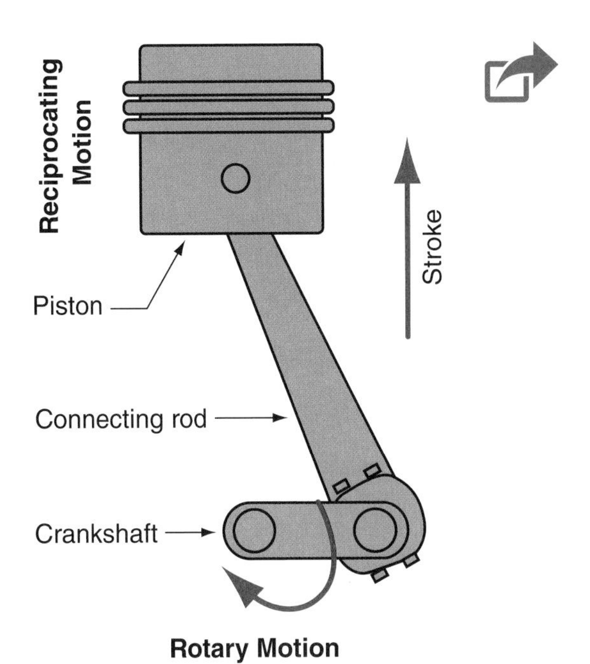

Goodheart-Willcox Publisher

Figure 19-29. The reciprocating piston in an internal combustion engine turns a crankshaft.

twice (four strokes) to produce a power stroke. The four stroke–cycle engine is the most common type. See **Figure 19-30**. During the four strokes, the following actions take place:

1. **Intake stroke.** The piston moves downward to create a partial vacuum. Atmospheric pressure forces a fuel-and-air mixture into the cylinder.
2. **Compression stroke.** The intake valve closes, and the piston moves upward. As the piston moves up, the fuel-air mixture is compressed in the small cavity at the top of the cylinder. The area forms a combustion chamber for the power stroke.
3. **Power stroke.** The spark plug produces an electrical spark that ignites the compressed fuel-air mixture. The fuel-air mixture expands. The resulting gases force the piston downward in a powerful movement.
4. **Exhaust stroke.** The piston moves upward to force the exhaust gases and water vapor from the cylinder. At the end of this stroke, the engine is ready to repeat the four-stroke cycle.

Single-cylinder engines are common for low-horsepower applications, such as lawn mowers, cement mixers, and portable conveyors. For more demanding applications, several cylinders are combined into one engine. The cycle of each cylinder is started at a different point in time so the engine

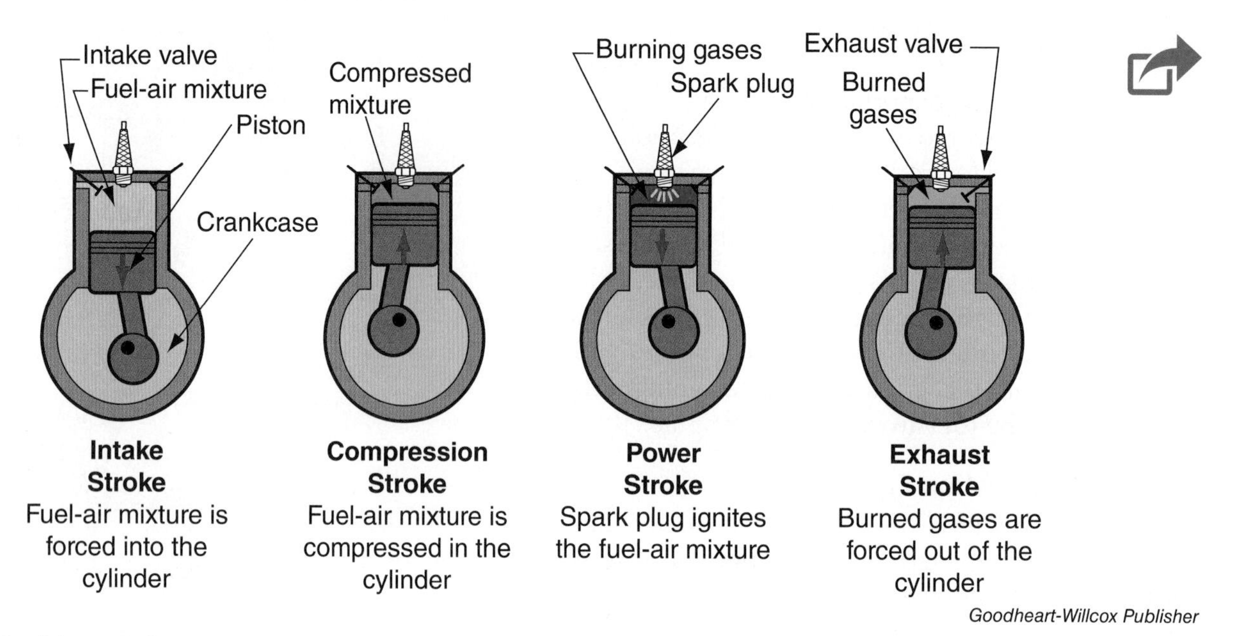

Goodheart-Willcox Publisher

Figure 19-30. A four-stroke engine moves the piston up and down twice to produce a power stroke.

has a series of closely spaced power strokes. Four-cylinder, six-cylinder, and eight-cylinder engines are common.

External Combustion Engines

Most external combustion engines are steam engines. The steam engine uses the principle that steam occupies more space than the water from which it came. In fact, one cubic centimeter (1 cc) of water produces 1700 cc of steam.

The operation of the steam engine is simple, **Figure 19-31**. Water is heated in a boiler until it changes into steam. This high-pressure steam is introduced into a closed cylinder that contains a free-moving piston. The steam forces the piston down. Next, cold water is introduced into the cylinder, condensing the steam. The resulting water takes up only 1/1700th as much space as the steam did, so a vacuum is formed in the cylinder. This causes the piston to be drawn up. At the top of the piston's stroke, a fresh supply of high-pressure steam is introduced into the cylinder. The engine repeats its cycle.

A flywheel changes the reciprocating (up-and-down) motion of the engine to rotary motion, or the motion of an object as it spins or rotates around an axis, such as a wheel. This rotary motion can be used to power any number of technological devices. In past times, steam engines powered ships, locomotives, cars, and many machines in factories.

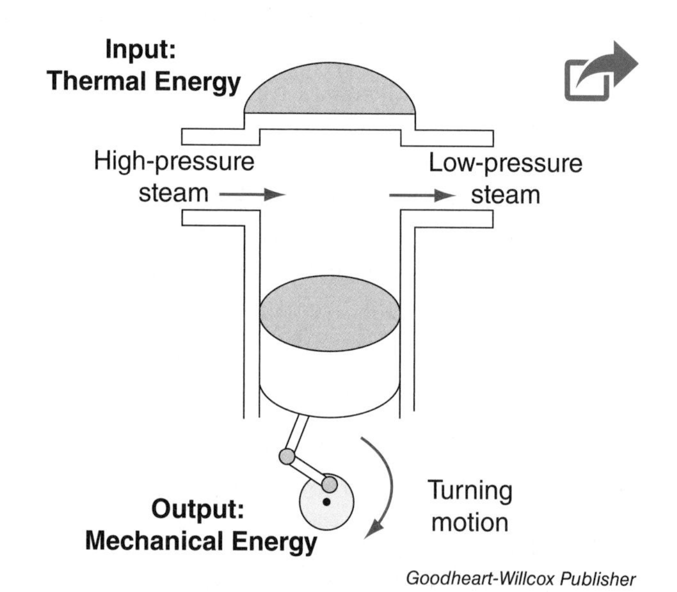

Goodheart-Willcox Publisher

Figure 19-31. The operation of a simple steam engine.

Space Heating

Thermal energy is used to heat buildings and other enclosed spaces. Three basic types of heat transfer are conduction, convection, and radiation. See **Figure 19-32**.

Conduction

Conduction is the movement of heat along a solid material or between two solid materials touching each other. This movement takes place without any flow of matter. The movement of

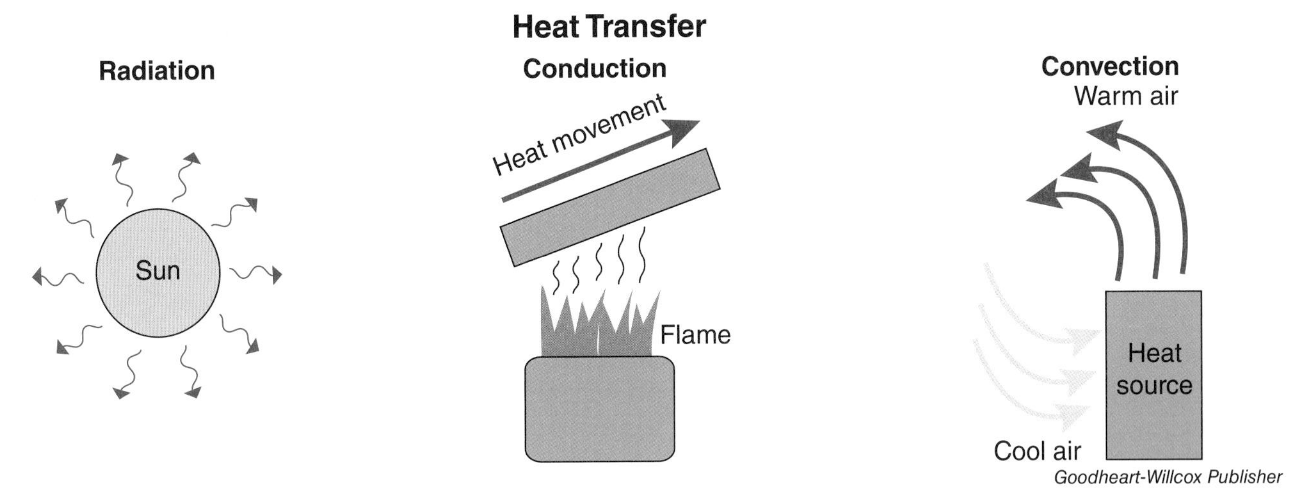

Figure 19-32. Examples of radiation, conduction, and convection. Radiation, conduction, and convection are the three important heating processes.

energy is from the area with a higher temperature to the area with a lower temperature. Conduction heats a pan on an electric heating plate.

Convection

Convection is the transfer of heat between or within fluids (liquids or gases). This heat transfer involves the actual movement of the substance. The process uses currents between colder areas and warmer areas within the material. Convection can occur through natural action or through the use of technological devices. The wind is an example of natural convection action. Forced hot–air heating systems are technological devices that warm homes with convection currents.

Radiation

Radiation is heat transfer using electromagnetic waves. The strength of the radiation is directly related to the temperature of the radiating medium. Hot objects radiate more heat than cooler objects do. The heat people feel on a bright, sunny day is from solar radiation. Also, if you bring your hand close to a hot metal bar, you can feel the heat radiate from it.

This heat transfer heats only the solid objects it strikes. Radiation does not heat the air it travels through. Radiant heaters in warehouses keep workers warm in a building, but the air still feels cold. See **Figure 19-33**.

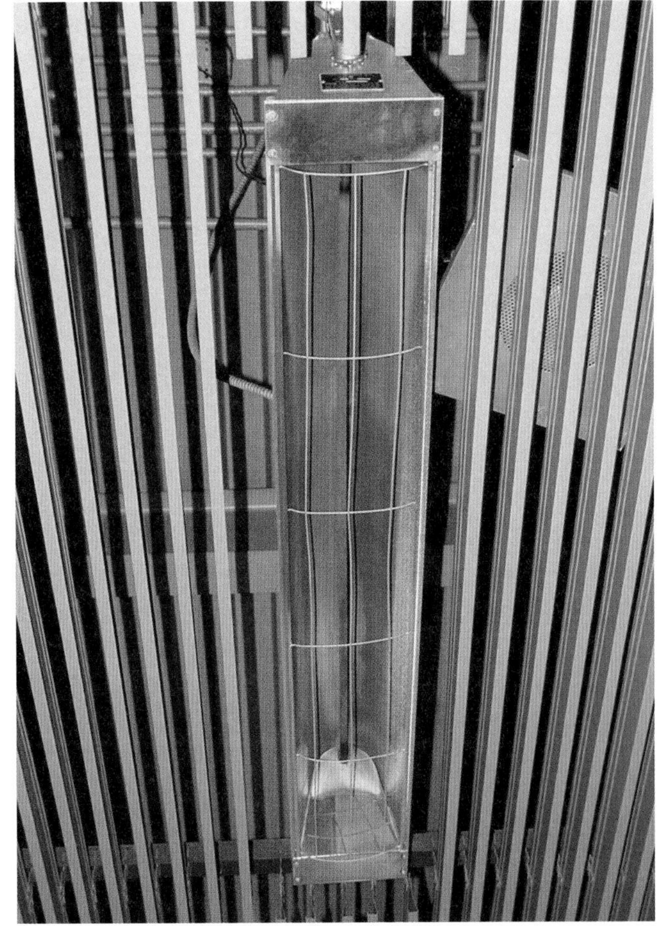
©iStockphoto.com/Acerebel

Figure 19-33. This radiant heater is used to heat an industrial space.

Career Connection Power Plant Operator

Power plant operators control and monitor boilers, turbines, and generators in electrical power–generating plants. They monitor instruments to maintain voltage and regulate electricity flow from the plant. These operators start or stop generators and connect or disconnect them from circuits to meet changing power requirements. To maintain round-the-clock operations, power plant operators usually work 8- or 12-hour shifts. Employers seek high school graduates with strong mathematics and science skills for entry-level power plant–operator positions. Most power plant operators receive extensive on-the-job and classroom instruction. Several years of training and experience are required to become a fully qualified operator.

Lefteris Papaulakis/Shutterstock.com

Power plant operators may be required to take the Plant Operator (POSS) and Power Plant Maintenance (MASS) exams. Both of these exams are offered by the Edison Electric Institute (EEI) and include questions concerning reading comprehension skills, mechanical assembly, mathematics, and mechanical concepts. Additional questions include an applicant's ability to interpret tables and graphs. However, nuclear reactor power plant operators need licenses, and these professionals must pass exams to become licensed through the US Nuclear Regulatory Commission (NRC).

Individuals who have strong communication, organizational, and problem-solving skills do well in this field. Students interested in pursuing a degree and career as a power plant operator should take vocational classes such as drafting, AutoCAD, HVAC, technology and engineering education, and welding. Many companies may expect an applicant to have at least an associate's degree and on-the-job training. Project management and leadership training programs are also vital.

Power plant operators usually work at plants that produce electricity through such means as generators, water turbines, or nuclear reactors. They monitor pieces of equipment, document energy production rates, and check gauge readings periodically. Potential projects for power plant operators include:

- Controlling power-generating equipment, which may use any one type of fuel, such as coal, nuclear fuel, or natural gas.
- Reading charts, meters, and gauges to monitor voltage and electricity flow.
- Checking equipment and indicators to detect evidence of operating problems.
- Adjusting controls to regulate the flow of power.
- Starting or stopping generators, turbines, and other equipment as necessary.

Heat Production

The conversion of energy into heat has been a goal of humans since before recorded history. People living in colder climates have always been challenged to heat their living spaces. A number of different methods are currently used to produce thermal energy to heat materials and buildings. These include burning fuels, capturing heat from the surroundings, and converting electrical energy.

Fuel Conversion

Fuel converters include fossil-fuel furnaces, wood-burning stoves, and fireplaces. A furnace has a firebox, a heat exchanger, and a means of heat distribution. See **Figure 19-34**. The fuel is burned in the firebox to generate thermal energy. Convection currents pass through the cells of the heat exchanger and raise its temperature. This thermal energy is transferred in the heat-distribution chamber to a heating medium (water or air). The medium is then passed over or through the heat exchanger.

In some systems, water is heated or turned to steam. The fluid is then piped to radiators in various locations. These radiators use convection and radiation currents to heat the room. Other systems blow air through ducts to areas needing heat. Convection currents circulate the warm air within the enclosure.

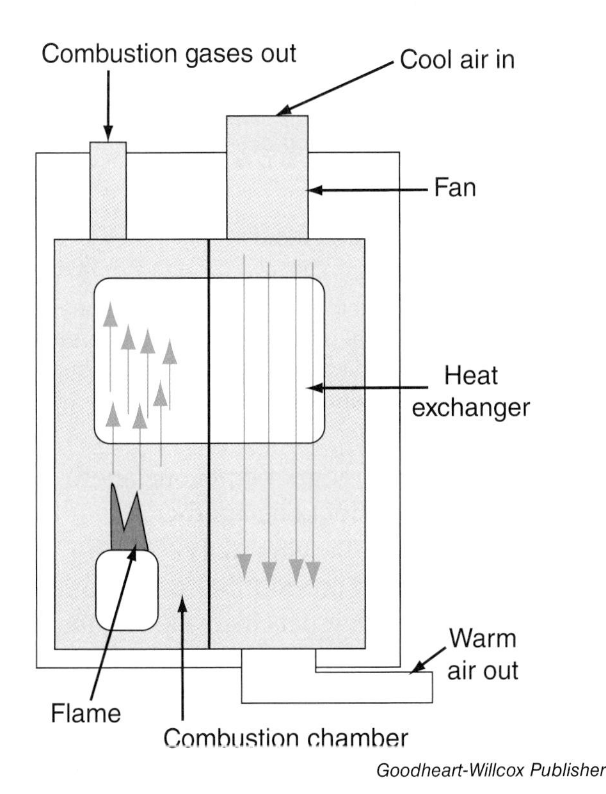

Figure 19-34. The airflow through a gas-fired, hot air furnace.

Atmospheric Heat

The atmosphere has heat available, no matter how cold the day seems. The standard device used to capture this heat is called a *heat pump*. This pump is actually a refrigeration unit that can be run in two directions. In one direction, the pump removes heat from the room and releases it into the atmosphere. This is part of what an air-conditioning (cooling) unit does. When a heat pump is operated in the opposite direction, the pump takes heat from the outside air and releases the heat inside a building.

Heat pumps work on the principle that a liquid absorbs heat when it vaporizes and releases heat when it is compressed. The system consists of a compressor, cooling or condenser coils, evaporator coils, and a refrigerant. See **Figure 19-35**.

In a heat pump, a heat-transfer medium, such as ammonia, is allowed to vaporize in the evaporator coils. The heat needed to complete this task is drawn from the material around the coils. This material might be air (in an atmospheric heat pump) or water from a well (in a groundwater heat pump).

The refrigerant gas is then compressed. This action causes the material to give off heat through the condenser coils. The heat can be used to warm air or water. The air or water is then transferred to the rooms needing heat.

The system can be reversed to produce cooling for air conditioning. The heat for the evaporation is drawn from within the building. The heat from compressing the gas is expelled into the outside atmosphere.

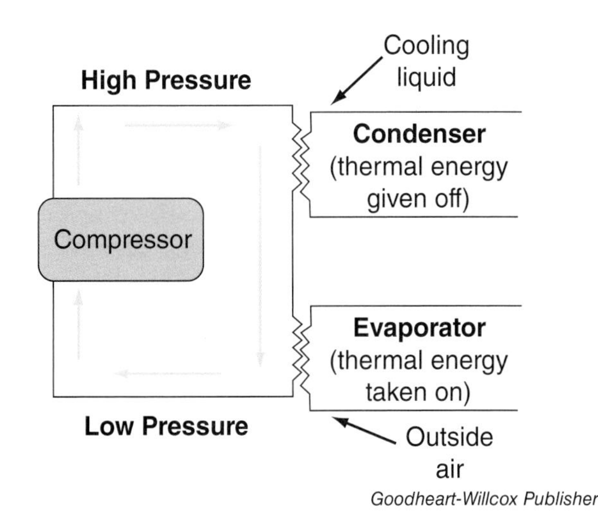

Figure 19-35. The operation of a simple heat pump.

Electric Heat

A heat pump is one way electricity is used to heat a building. Electricity powers the compressor drawing heat from the air. Another method using electricity is a furnace that uses an electric resistance heater.

One common method of heating uses electric resistance heaters in each room. These heaters have special wires that have a high resistance to electrical current. The wires become very hot when electricity passes through them. The hot wires warm the air around them. Convection currents transfer the heat to all parts of the room.

Another type of electric heating heats with radiation (radiant heating). This system uses high-resistance wires installed in the ceiling. When electricity passes through them, they become warm. This warmth radiates into the room similar to the way heat radiates from a hot bar of steel. The electromagnetic waves emitted from the system warm objects in the room.

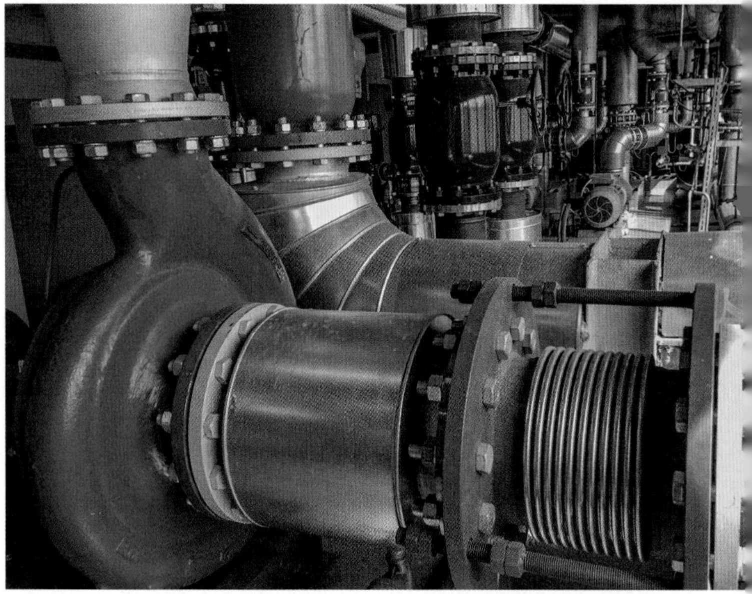

DyziO/Shutterstock.com

Figure 19-36. Water pumps with large electric motors.

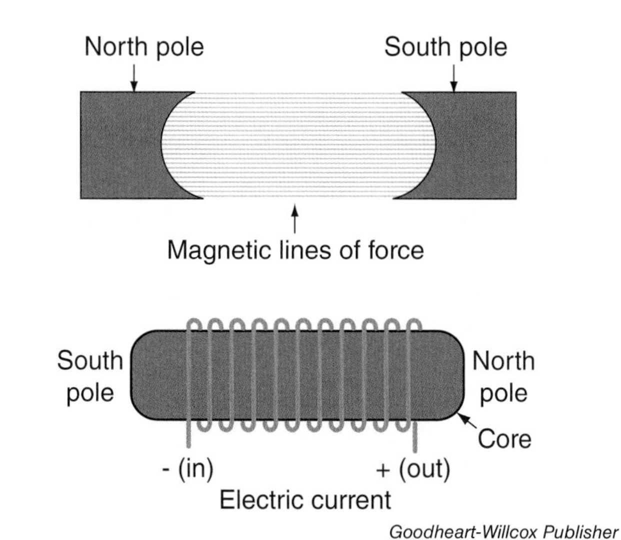

Goodheart-Willcox Publisher

Figure 19-37. In an electric motor, magnetic lines of force travel between opposite poles of a magnet. Current flowing through wires wrapped around an iron core produces magnetic poles at the ends of the core.

Electrical-Energy Converters

Electricity is important in all areas of life. Without electricity, stores would close, food would spoil, and many homes would become dark and cold. The electric motor is a universally used source of power. Electricity is generated and distributed for a vast array of applications. Electric motors, electricity generation, and electricity distribution are discussed in the following sections.

Electric Motors

The average home has more than 40 motors in it. Motors can be found in clocks, refrigerators and freezers, DVD players, furnaces and air conditioners, clothes washers and dryers, shavers, hair dryers, and many other appliances. Electric motors are also found on construction sites and in factories. See **Figure 19-36**. They play a vital part in most agricultural, transportation, and communication systems.

Electric motors convert electrical energy into mechanical energy. The electric motor is based on the laws of magnetism and electromagnetism. See **Figure 19-37**. These laws state the following:

- Like poles of a magnet repel one another, and unlike poles attract one another.
- Electric current flowing in a conductor induces a magnetic field around the conductor.

There are two magnets in an electric motor. The outer magnet is stationary and is called the field magnet. The inner magnet is an electromagnet that can rotate and is called the armature. When electric current passes through the armature, it induces magnetism in the armature core. A north pole is on one end, and a south pole is on the other. The direction the wire wraps around the core and the direction of the current determine which end is the north pole. See **Figure 19-38**.

The north field pole and the north armature pole are next to each other. These are like poles, which repel each other. Therefore, the armature spins one-quarter turn. Now, the unlike poles attract each other, and the armature continues rotating another one-quarter turn. The direction of the current is reversed. This action occurs mechanically in direct current (dc) motors. Alternating current (ac) motors use the existing direction changes in the line current. The current direction change reverses the poles on the armatures. Now, the entire action is repeated. Like poles repel and turn the armature another one-quarter of a turn. Unlike poles then attract, turning the armature the final one-quarter turn to complete a full revolution of the armature.

Electricity Generation

Electricity may be generated by engines, wind, water, fossil fuels (such as coal), and nuclear power. See **Figure 19-39**. Electricity generation uses the principles of electromechanical-energy conversion, meaning the conversion of electric energy into mechanical energy or vice versa.

In most commercial systems, water or steam is used to turn a turbine to produce electricity. This type of electricity is called *hydroelectricity*. A water-powered plant, called a *hydroelectric generating plant*, uses a dam to develop a water reservoir. See **Figure 19-40**. The water is channeled through large pipes into the turbines in the generating plant. Steam and water turbines have a series of blades attached to a shaft. As the water or steam strikes the turbine blades, the shaft turns. This shaft is attached to a generator. The generator changes the mechanical energy into electrical energy.

A steam-powered electrical plant uses fossil fuels or nuclear energy to produce the steam that drives the generator's turbines. Nuclear power stations and fossil-fuel power stations both use thermal energy to create the mechanical energy used to produce electricity. These methods of generating electricity are commonly used to power cities. Fossil-fuel electrical plants burn coal, natural gas, or fuel oils to produce thermal energy. A nuclear plant uses atomic reactions to heat water in a primary system. The heated water is used to produce steam in a secondary system. The steam then drives the generator's turbine. See **Figure 19-41**. Keeping the primary and secondary systems separate prevents the water in the reactor from entering the steam turbines. This reduces the hazards for workers in the plant and for people living near the plant.

Wind can also be used to turn turbines. One wind turbine does not generate enough power for large-scale power distribution. Wind turbines are typically grouped into wind farms.

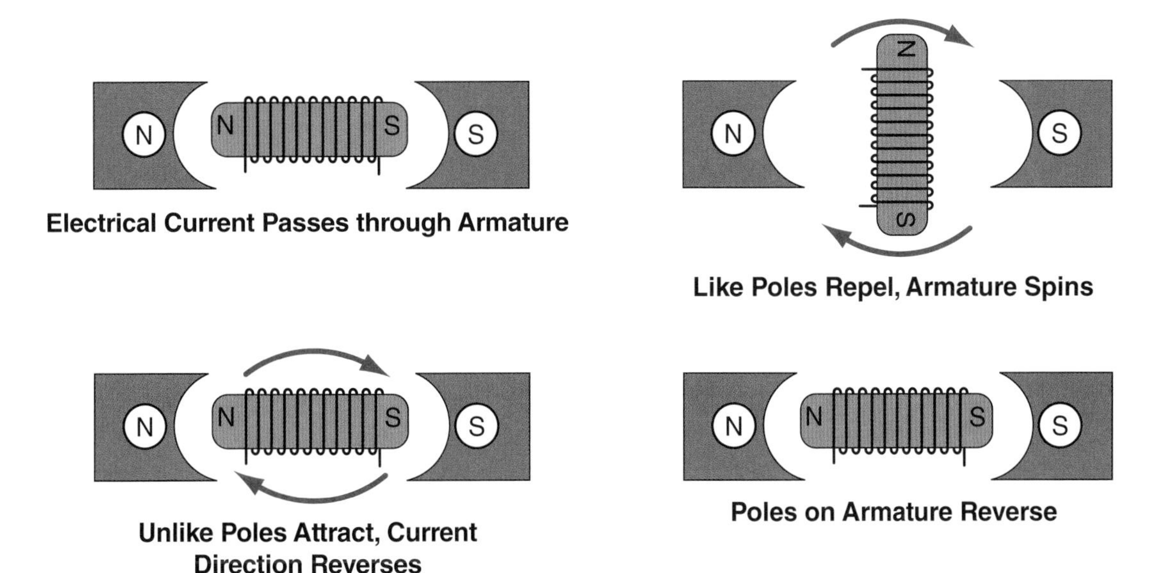

Goodheart-Willcox Publisher

Figure 19-38. Electric motors operate by using a stationary magnet and an armature that switches its magnetic poles.

Zeljko Radojko/Shutterstock.com

Iko/Shutterstock.com

Vaclav Volrab/Shutterstock.com

adziohiciek/Shutterstock.com

Figure 19-39. Electrical energy is generated by different types of devices.

Shasta Dam

US Department of Energy

Inside a Hydroelectric Plant

US Department of Energy

Figure 19-40. Hydroelectric power. Shasta Dam is in northern California. Note the water-delivery pipes that lead to the power plant in the lower-right portion of the left-hand picture. In the right-hand picture are the generators above the water turbines.

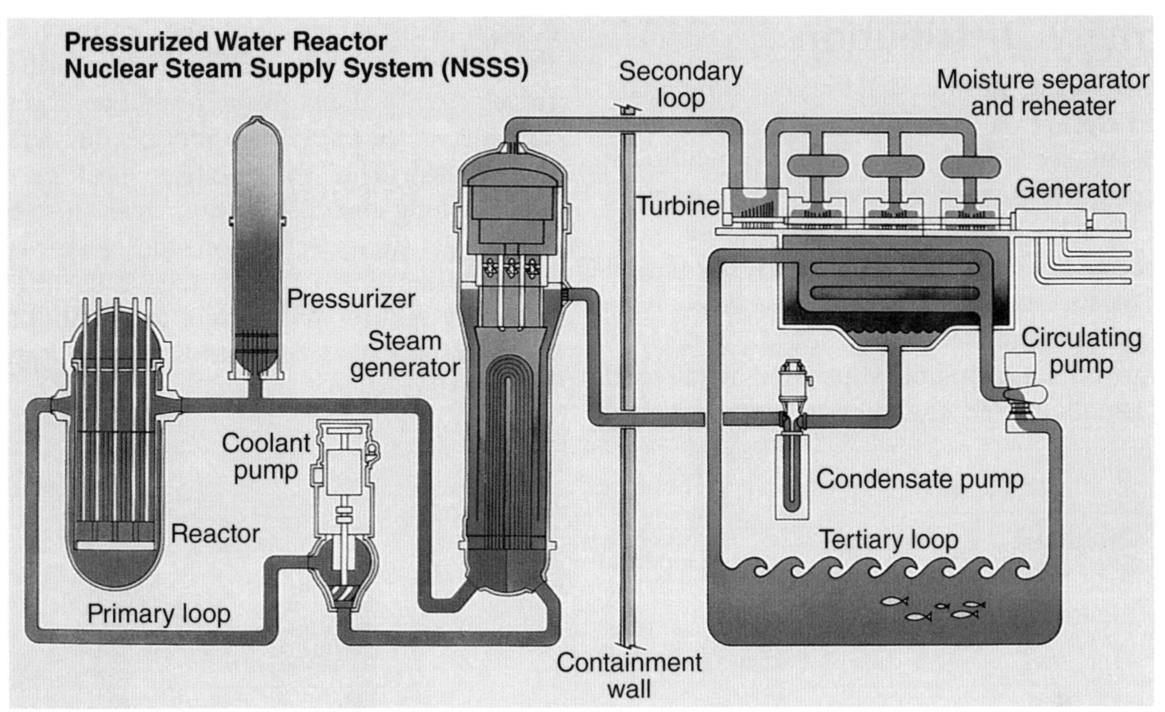

Westinghouse Electric Corp.

Figure 19-41. An NSSS. Notice how the water in the primary loop is used to heat the water in the secondary system. The contents of the two loops remain separate.

Electric Generator

An electric generator is a device that converts mechanical energy into electrical energy. Different types of electric generators are produced for different applications. Large power-generating facilities are needed to power buildings. Smaller electrical generators may be portable or used for automobiles. Alternators are a type of electrical generator used in automobiles. An alternator's energy comes from a belt being turned on a pulley. This belt is turned by the engine. Portable electric generators may be used to temporarily power a house. These generators are similar to alternators in that a gas-powered engine is used to turn the turbine.

An electric generator has two magnets, **Figure 19-42**. Similar to a motor, the outside magnet of the generator is a stationary electromagnet called the *field magnet*. The inside magnet is a series of wires wound on a core. This part is called the *armature* and is able to rotate on its axis.

An electrical current is allowed to flow in the coils of the field magnet. This action produces an electromagnetic field around the field magnet that cuts through the armature. When the water or steam turbine spins the armature, the wires on the armature cut through the magnetic lines of force around the field magnet. This induces a current in the armature. Once current has been induced, the mechanical energy is converted to electrical energy. After the electrical energy has been produced, it is transmitted through a conductor away from the electrical generator to be distributed.

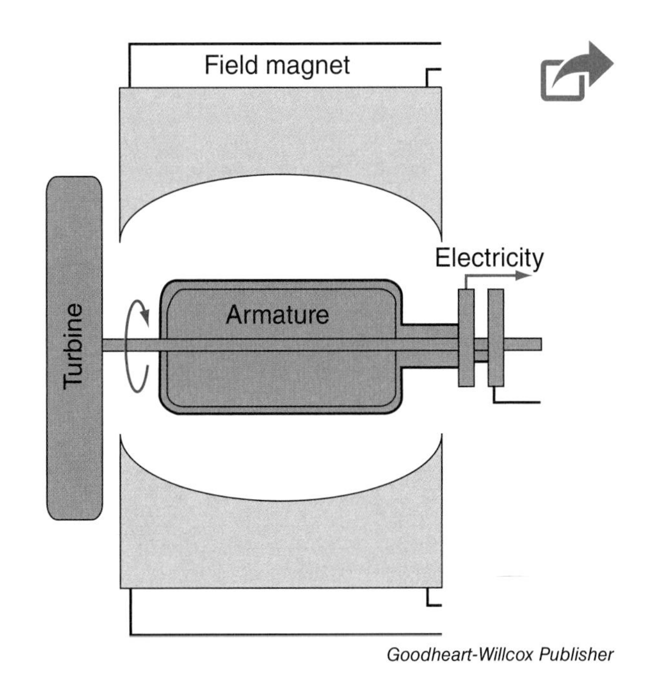

Goodheart-Willcox Publisher

Figure 19-42. An electric generator.

Electricity Distribution

The electricity produced in the generating plant is passed through a step-up transformer. A transformer is an electrical device with the capability of transferring electrical energy between two or more circuits. This transformer is called a *step-up transformer* because it steps up, or increases, the output voltage of the electrical current. The very high voltage reduces power losses in transmission.

Large transmission lines supported on tall steel towers usually carry the high-voltage electrical current to distant locations. See **Figure 19-43**. When the current reaches the area in which it will be used, the electricity flows through another transformer. This *step-down transformer* reduces, or steps down, the voltage. The lower-voltage electrical current moves along the distribution lines. Just before it reaches its final destination, the electricity enters another step-down transformer. This transformer generally reduces the current to 110 and 220 volts for residential use. Some industrial applications use 440 or 880 volts.

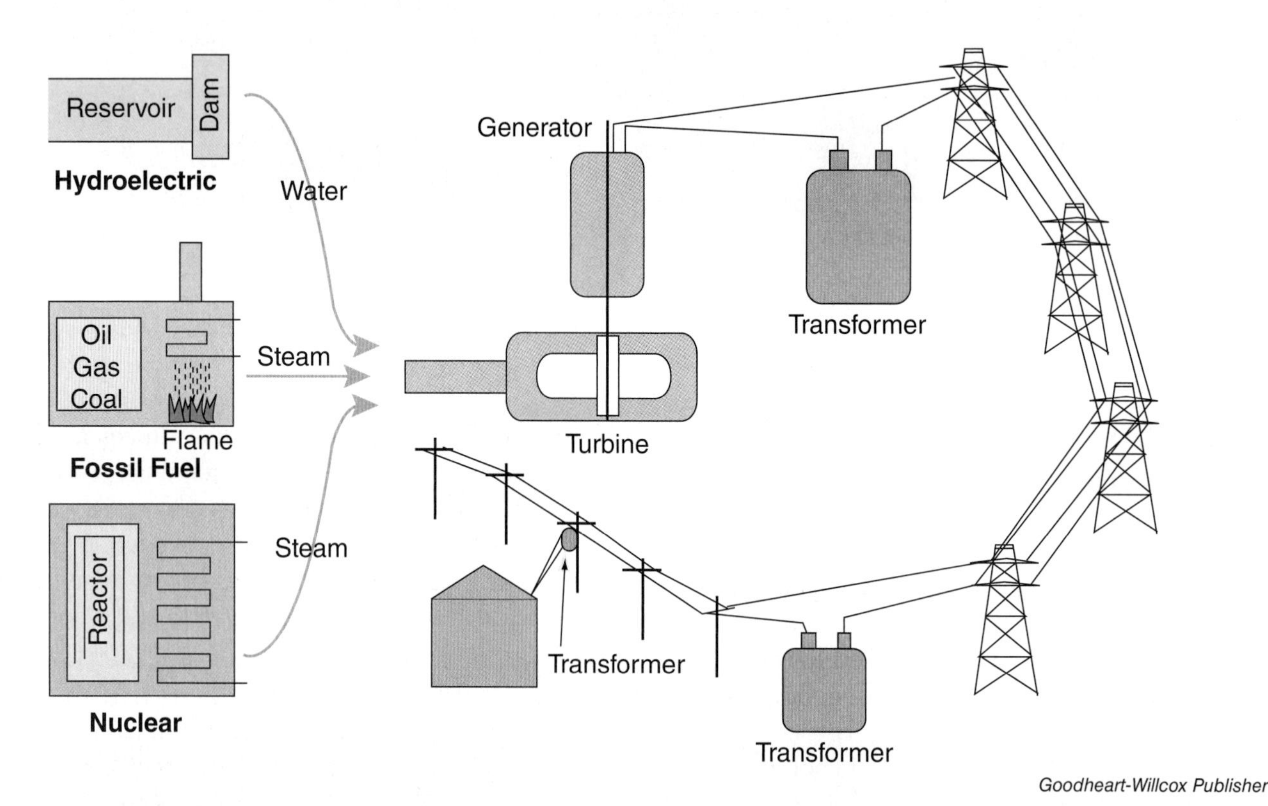

Figure 19-43. Electrical systems generate, transform, and distribute electricity. A nuclear power plant generates the electricity. A giant transformer station steps up and distributes electricity. High-voltage power lines take electric power over long distances.

irisphoto1/Shutterstock.com

The Hoover dam creates power for 1.3 million people in Arizona, California, and Nevada.

Summary

- ✔ *Energy* is defined as the ability to do work.
- ✔ Kinetic energy is energy in motion. Potential energy is energy that is stored for later use.
- ✔ The six primary forms of energy are mechanical, radiant, chemical, thermal, electrical, and nuclear.
- ✔ The three basic types of energy resources are exhaustible, renewable, and inexhaustible.
- ✔ An energy converter is a device that has energy as its input and its output.
- ✔ Inexhaustible energy converters change wind energy, water energy, and solar energy into other types of energy.
- ✔ Most solar energy–conversion systems have two major parts—a collection system and a storage system.
- ✔ Passive solar collectors directly collect, store, and distribute the heat they convert from solar energy. Active collectors use pumps to circulate the water collecting, storing, and distributing the heat they convert from solar energy.
- ✔ Geothermal energy (heat originating in the earth's molten core) is used to drive steam turbine–powered generators and for direct heating.
- ✔ Oceans contain two important sources of energy—thermal (wave) and mechanical (tide) motion.
- ✔ Biomass resources are vegetable wastes and animal wastes generated through biological actions. Bioenergy conversion takes place through thermochemical conversion and biochemical conversion.
- ✔ Two major applications of thermal energy are heat engines, such as internal combustion engines in vehicles, and heating of buildings and other enclosed spaces.
- ✔ Electricity generation uses the principles of electromechanical-energy conversion. In a steam-powered electrical plant, fossil fuels or nuclear energy is used to produce the steam that drives the generator's turbines.

Check Your Engineering IQ

Now that you have finished this chapter, see what you learned by taking the chapter posttest.
www.g-wlearning.com/technologyeducation/

Test Your Knowledge

Answer the following end-of-chapter questions using the information provided in this chapter.

1. Define *energy.*
2. Energy that is involved in moving something is called _____ energy.
3. _____ can be defined as "applying a force that moves a mass a distance in the direction of the applied force."
4. *True or False?* Wind is a mechanical energy source.
5. The main source of radiant energy is _____.
6. *True or False?* Chemical energy is also known as *heat energy.*
7. Two processes associated with nuclear energy are fission and _____.

Matching questions: For Questions 8 through 16, match each energy source with its type. Answers can be used more than once.

8. Corn
9. Falling water
10. Wood
11. Wind
12. Natural gas
13. Petroleum
14. Sunshine
15. Biomass materials
16. Coal

A. Exhaustible
B. Renewable
C. Inexhaustible

17. *True or False?* An energy converter has energy as both its input and its output.
18. A Trombe wall is used in a(n) _____.
    A. indirect active solar system
    B. direct-gain solar system
    C. direct active solar system
    D. indirect-gain solar system
19. Where does geothermal energy originate?
20. Two sources of mechanical energy in oceans are _____ energy and _____ energy.

21. The thermochemical conversion process called _____ involves heating a biofuel at moderate temperatures under high pressure.
    A. pyrolysis
    B. liquefaction
    C. gasification
    D. liquidification
22. Name two common biochemical conversion processes.
23. Internal combustion engines use expanding gases produced by burning fuel to change _____ energy into _____ motion.
24. Radiation is _____.
    A. the movement of heat between two solid materials touching each other
    B. the release of heat from a compressed liquid
    C. heat transfer using electromagnetic waves
    D. the transfer of heat between or within fluids
25. Fossil fuels or _____ are used in a steam-powered electrical plant to produce the steam that drives the generator's turbines.

Critical Thinking

1. Consider two technologies you encountered today. Depict how energy, work, and power are components of these two technologies.
2. Consider the region you reside in. What type of solar conversion system would you create and why?

STEM Applications

1. Construct a simple device to change wind energy into rotating mechanical motion.
2. Design and construct a passive solar device that heats the air in a cardboard box. Prepare a sketch of the device and an explanation of how it works.

CHAPTER 20
Meeting Needs through Mechanical Engineering

Check Your Engineering IQ

Before you read this chapter, assess your current understanding of the chapter content by taking the chapter pretest. www.g-wlearning.com/technologyeducation/

Learning Objectives

After studying this chapter, you should be able to:

- ✔ Explain the role and responsibilities of a mechanical engineer.
- ✔ Summarize the evolution of mechanical engineering.
- ✔ Explain how forces and motion result in work.
- ✔ Describe mechanical methods used to transfer energy.
- ✔ Describe the two basic categories of systems used to change the type, direction, or speed of a force.
- ✔ Perform calculations for work, torque, power, gear speed, and heat transfer.
- ✔ Identify mechanical engineering specializations and industries that employ mechanical engineers.

Technical Terms

bottom up manufacturing
cam
conductive heat transfer
fatigue failure
flexible belt drive
fluid mechanics
foot-pounds (ft-lb)
gears
horsepower
hydraulics
hydrodynamics
hydrostatics
joules (J)
kilowatt-hour
linear motion
mechanical engineering
mechatronics
metamaterials
power transmission
reciprocating motion
rotary motion
solid and structural analysis
solid mechanics
spur gear
static structural failure
thermodynamics
torque
watt

Close Reading

Before you read this chapter, consider how the concepts discussed in Chapter 19, including energy, power, and work, can be used to solve real world problems. Then, as you read, recognize how scientific and mathematical principles are applied to designing, creating, and improving mechanical systems.

Nearly every technological device you encounter daily has been created through engineering. Each technology is designed by applying a specialized body of knowledge. For example, in order to design an automotive engine, **Figure 20-1**, an understanding of energy, work, and power is required. In addition, knowledge of thermodynamics, hydraulics, pneumatics, mechatronics, robotics, drafting, 3-D modeling, forces, torque, and strengths of materials is needed. After the design is tested, the results are analyzed, and corrections or improvements are made.

In the case of a system driven by an engine, consider the following scenario. After identifying that the horsepower was insufficient in the first iteration, the engineer also recognized that the size and weight of the new engine would need to be reduced. Overheating also needed to be reduced. Engineers applied principles of mathematics, science, and engineering to solve these problems. They considered alternative materials, beginning with the investigation of properties such as tensile and compressive strength, while taking into consideration thermal conductivity to prevent overheating. They examined mechanical components to determine if improved designs, such as better gears or pulleys, would impact horsepower output. Additionally, they considered the resources needed to create and manufacture the engine.

David Hurtley Creative/Shutterstock.com

Figure 20-1. The development of an automotive engine involves mechanical engineering.

The field that is most often responsible for creating products such as engines and evaluating criteria and constraints for design is mechanical engineering. ***Mechanical engineering*** is an engineering discipline that works with mechanical systems, or machines that cause or involve movement.

Evolution of Mechanical Engineering

Mechanical engineering is one of the oldest disciplines of engineering. It is also considered one of the broadest and most comprehensive, overlapping with many other disciplines. Mechanical engineering involves knowledge and skills from a number of subject areas, including science and mathematics.

Early Mechanical Engineering

Mechanical engineering dates back thousands of years. Ancient China, Greece, and Egypt are examples of early civilizations that experimented with principles of mechanical engineering. During these early times, mechanical products, such as the chariot and screws, were created. See **Figure 20-2**. These artifacts were rudimentary and were often crafted out of natural materials, including wood, stones, and bones. During the Middle Ages, however, more complicated mechanical devices were designed and produced. These devices included camshafts, crankshafts, and chain drives.

3DMI/Shutterstock.com

Figure 20-2. A Roman racing chariot is an example of early mechanical engineering.

Career Connection Mechanical Engineer

Mechanical engineers design, build, test, and create mechanical, fluid, robotics, and thermal systems. They invent new products and innovate existing solutions to prior problems. These engineers create tools, machines, engines, air-conditioning units, and building systems, such as elevators. They also oversee the design of processes used to manufacture these products.

Herrndorff/Shutterstock.com

High school students interested in mechanical engineering should take mathematics courses, such as algebra, trigonometry, and calculus. They should also take science courses, including biology, chemistry, and physics. Other beneficial courses include technology and engineering courses that apply mathematics and scientific principles.

Many colleges and universities offer mechanical engineering programs. A bachelor's degree is required to enter the mechanical engineering field. Students take specifically related mathematics and science courses, as well as engineering and design classes. Programs are typically four years, but some include five- or six-year degrees. These longer programs include a master's degree aligned with an internship.

Individuals who do well in the mechanical engineering field have good mathematical and mechanical skills. They also have good creative, critical thinking, and listening skills.

Due to the broadness of mechanical engineering, individuals can be expected to be employed in many fields. These fields include, but are not limited to, architectural, manufacturing, agricultural, computers and electronics, and transportation.

The mechanical engineering field is expected to grow five percent through 2024. Individuals who stay informed regarding technological advances would benefit from advancements in their field.

A mechanical engineer may work on the following types of projects:

- Designing and creating a more efficient air-conditioning system.
- Improving the efficiency and safety of elevator or escalator systems.
- Developing material handling systems, such as conveyor belts and automated transfer systems.
- Improving engines or turbines for an aviation system.

The American Society of Mechanical Engineers (ASME) and the Society of Automotive Engineers (SAE) are two professional associations for mechanical engineers.

In this early phase of mechanical engineering, much of the designing was the result of tinkering. Mathematical equations and scientific principles were minimally involved. It was not until the seventeenth century that mathematician and physicist Sir Isaac Newton articulated his three laws of motion (detailed later in this chapter) and developed calculus. These developments resulted in Newtonian mechanics, which described the relationship among force, mass, and motion in the natural world. Newton's methods still serve as a mathematical foundation for physics and mechanical engineering problems to this day.

During Newton's time, mechanical engineering was still primarily within the larger discipline of engineering. However, in the nineteenth century, mechanical engineering began to separate from the general engineering field. This separation resulted from the increased development of machines and tools, requiring individuals with specialized knowledge to design and fix them.

Mechanical Engineering Today

Mechanical engineers today create products that do not include obvious moving, or *mechanical*, parts. Also, present-day mechanical engineering does not always result in products that are entirely new. One example of this is the bladeless wind turbine.

Although windmills have been around for hundreds of years, the bladeless wind turbine, **Figure 20-3**, is an innovative conversion process. Historical wind turbines translate the kinetic energy of the wind into mechanical energy that is acquired from the motion of large blades that are rotated. The mechanical energy is then converted to electrical energy. The bladeless wind turbine allows the wind to move electrically charged droplets against the direction of an electric field. This increases the potential energy of the particle. In turn, the excess energy is harvested by a collection system.

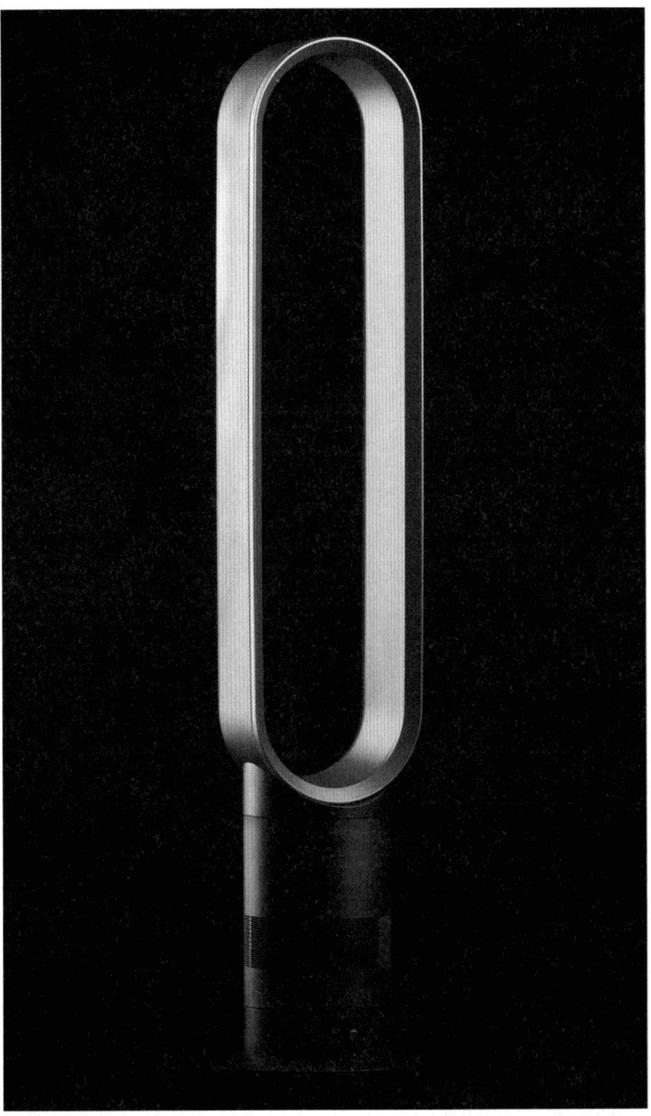

LeStudio/Shutterstock.com

Figure 20-3. A prototype of a bladeless wind turbine.

Compared to windmills with blades, the windless turbine lacks large moving parts. Noise and danger to animals are reduced, resulting in less disturbance to the environment.

Mechanical engineering has undergone a change in the scale of moving parts. For example, nanotechnology allows mechanical engineers to make parts and machines that are built at the atomic and molecular level. This is the result of ***bottom up manufacturing***, the manufacture of products by building them up from the atomic and molecular level. For example, a nanomotor is less than one square micrometer in size and can fit inside a human cell, yet can rotate at speeds of 18,000 rpm (revolutions per minute) for 15 continuous hours. See **Figure 20-4**.

An area of specialization fundamental to mechanical engineering, yet embedded across all other engineering fields, is detailing design solutions through drafting and computer-aided design. This is discussed in depth in Chapter 5. Although designing still involves aspects of drafting (**Figure 20-5**), it also involves computer-aided design (**Figure 20-6**). Mechanical engineers use computer-aided drafting (CAD) to draft ideas, analyze designs, recognize flaws in a design, and communicate those ideas to potential or current customers. Mechanical engineers can use CAD software to generate 2-D and 3-D models or create a simulation of a model/prototype to simulate stresses and expected performance.

Scientific and Mathematical Principles

Mechanical engineers work with information and concepts that have been established within the scientific and mathematical community and are employed across all engineering branches. For example, all engineers apply understanding and

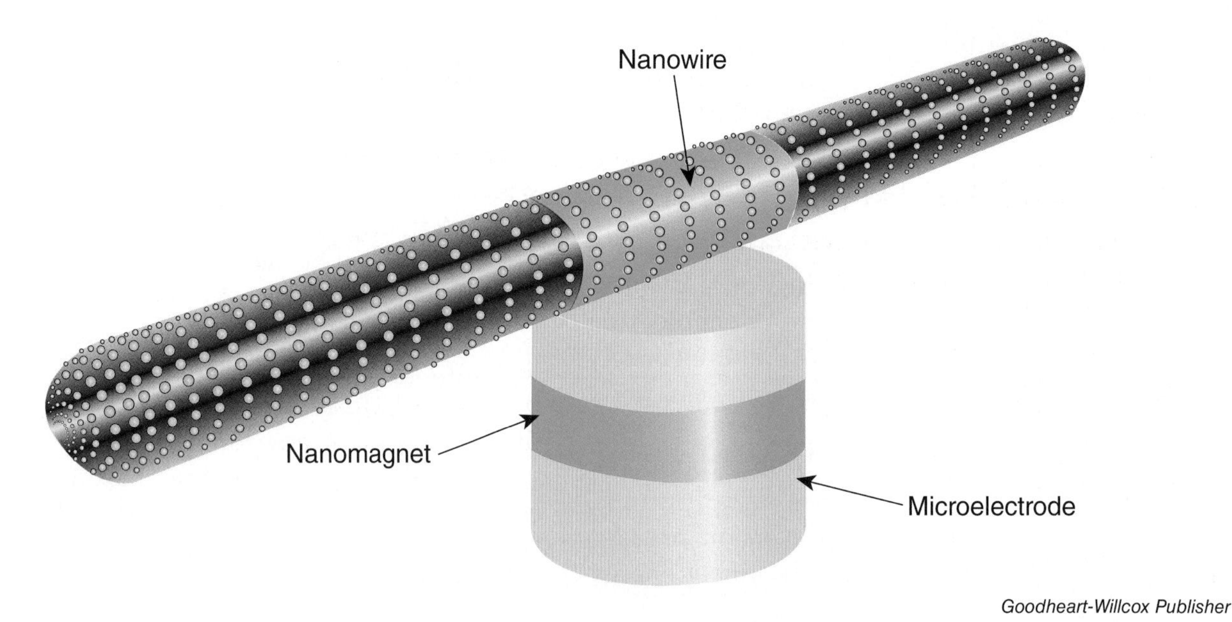

Goodheart-Willcox Publisher

Figure 20-4. Nanomotors operate on a tiny scale. This nanomotor has three primary components—a nanowire, nanomagnet, and microelectrode.

Africa Studio/Shutterstock.com

Figure 20-5. Drafting is still used in industry and by mechanical engineers to develop design solutions.

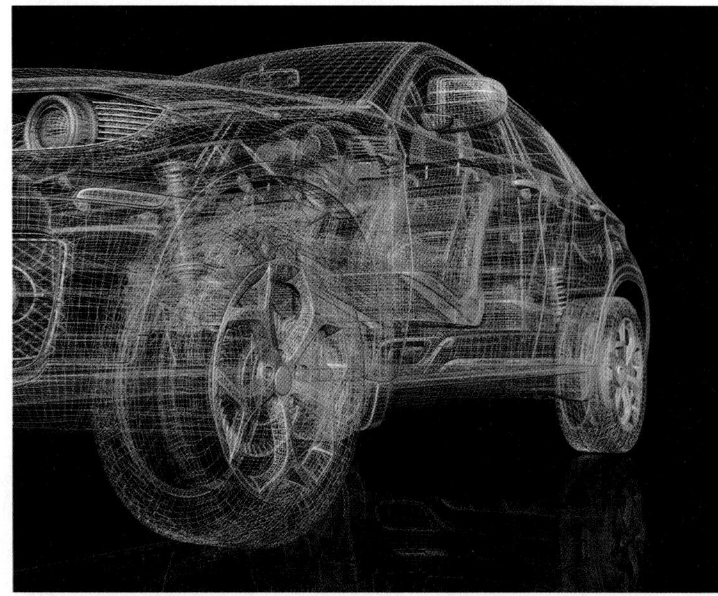

Master3D/Shutterstock.com

Figure 20-6. Computer-aided drafting is a quick and efficient way to develop design solutions.

knowledge of the concepts of energy, power, and work. However, a specific type of engineer may be needed in a specific context. Civil engineers use structural analysis to examine and predict the loads on different materials within the broad context of an entire structure, whereas mechanical engineers would apply those same concepts to mechanical systems such as elevators or escalators.

Mechanical engineers need a strong foundation in scientific principles and mathematics, including physics, calculus, differential equations, and linear algebra. The following sections include examples of scientific principles and mathematical equations used by mechanical engineers.

Technology Explained Invisibility Cloaks

Invisibility cloak: Device that uses optical camouflage to create the illusion of invisibility.

An invisibility cloak might seem like a magical device found only in stories and movies, but it is currently in research and development and may soon become reality. Scientists have explored two options for achieving invisibility.

Optical camouflage creates the illusion of invisibility by blending the intended object into its background. Optical camouflage employs technology similar to a weather forecaster's blue screen. A camera records the background image that appears behind the newscaster. The image will simultaneously be projected onto the person, who is also wearing a special silver coat that enhances, or improves, the resolution and the quality of the projection. Optical camouflage makes a person or object "invisible" from only one vantage point. In reality, it masks an object rather than making it invisible.

Another method currently being developed is a covering, or cloak, that diverts light rays around the desired object. When we *see* an object, we actually see light that has struck the object and reflected off it into our eyes. If light did not strike the object, but instead flowed around it, we could not see the object.

In order to construct a material that can deflect light, scientists have turned to the technology of metamaterials. ***Metamaterials*** are materials created by scientists to exhibit characteristics that cannot be found in nature. In order to achieve invisibility, the structure of the metamaterial must be smaller than the wavelength of light—blocking light ray access to the material covered by the cloak. Researchers have created a successful prototype of this type of invisibility cloak, but it is extremely small, measuring only ten micrometers (.00001 meters) wide.

The small size is not the only reason this invisibility cloak has restricted applications. The cloak can only cover still objects, so it could not actually be worn like a cloak. However, it could make a building or motionless army tank invisible. However, following current design and construction, an invisibility cloak large enough to cover even a bicycle would be incredibly heavy.

While a true invisibility cloak has yet to be created, the technology is in the works. Because light rays and sound waves work in a similar way, scientists are also attempting to create a sound barrier that would serve much like an invisibility cloak for the ears.

Forces and Motion

Energy sources do work that ends up as motion, light, or heat. Energy powers manufacturing machines, lights buildings, propels vehicles, and produces communication messages. Many of the problems mechanical engineers solve require an understanding of how forces and motion result in work.

Forces

Forces can be described as a *push* or *pull* that results from the interaction between two objects. Consider a ball at rest. For the ball to move, you kick, pull, or push on it. The greater you kick, pull, or push it, the greater the force on the object. This is called an increase in the *magnitude* of the force that was applied. Another element used to describe a force is *direction*. If you kick the ball at rest to the right, it will move in that direction. See **Figure 20-7**.

Forces can be quantified. To describe force, mechanical engineers often work with the SI unit newton (N). One newton is the force necessary to accelerate a one-kilogram mass at a rate of 1 m/s^2. The equation is written as follows:

$$1 \text{ newton} = (1 \text{ kg})(1 \text{ m/s}^2) \text{ or } 1 \text{ N} = 1 \text{ kg} \times \text{m/s}^2.$$

Forces are not always the result of direct contact between objects. For example, forces may be magnetic or gravitational.

Motion

Newton's laws of motion are the foundation of mechanics. They explain why objects do or do not move. The three laws are as follows:

1. **Newton's First Law of Motion.** This law, also called the *law of inertia*, states that an object at rest will remain at rest, and an object in motion will stay in motion at the same speed and direction, unless an unbalanced force acts on it. An unbalanced force is one that has more force on one side. For example, if you push a wall, the forces are typically balanced and it will not topple over. This is because the wall pushes back with an equal, opposite force.
2. **Newton's Second Law of Motion.** This law describes the applied force on an object in relationship to the object's mass and acceleration. This law is expressed in the following equation:

   $$\vec{F} = m\vec{a}$$

Both force and acceleration are notated as vectors, with an arrow symbol displayed above them. In simplest form, a vector quantifies magnitude and direction. One key point of Newton's Second Law is the direction of the *net* force, or overall force acting on an object, is the same as the direction of the acceleration. As you kick, or accelerate, the soccer ball to the right, then it will move to the right.

3. **Newton's Third Law of Motion.** This law states that for every action there is an equal and opposite reaction. Newton's Third Law of Motion can be seen as you sit in a chair. As you sit, your body applies a downward force on the chair, while the chair exerts an upward force on your body.

Fotokostic/Shutterstock.com

Figure 20-7. A soccer player can adjust the magnitude of the kick and the direction toward or away from the goal.

Types of motion

There are three types of motion. These types are as follows:

- ***Linear motion*** is movement in a straight line. Objects exhibiting linear motion move in a straight line indefinitely. Linear motion can be measured in two ways—speed and direction. These two measurements make up what is called *velocity vector*. Drawing a line along a ruler is an example of linear motion.
- ***Rotary motion*** is spinning around an axis, or motion in a circle. The spinning tires or the movement of the steering wheel of a car are examples of rotary motion. Rotary motion can be measured by the number of degrees turned during a set period of time or by the revolutions completed in one minute (rpm). In addition, the direction of motion can also be described. Typically, rotary

motion is described as either clockwise or counterclockwise.

- ***Reciprocating motion*** is moving back and forth. It is the motion exhibited by an up-and-down or back-and-forth movement of an object. Reciprocating motion is usually repeated over and over again. The movement of a piston of an internal combustion engine is an example of reciprocating motion. There are two ways to measure reciprocating motion. *Throw* describes the distance between the two extremes of the motion. The *period* is the length of time required for each cycle to return to its original position.

The three types of motion are depicted in **Figure 20-8**. A band saw uses linear motion, a circular saw uses rotating motion, and a scroll saw uses reciprocating motion.

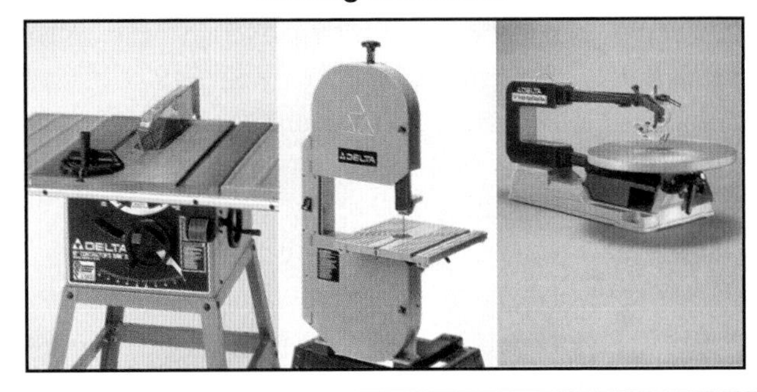

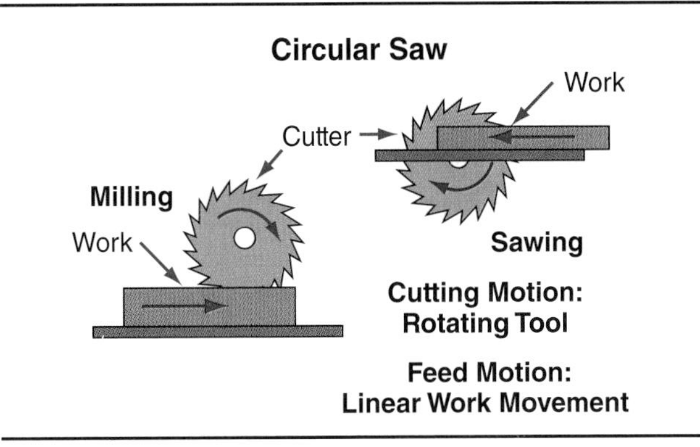

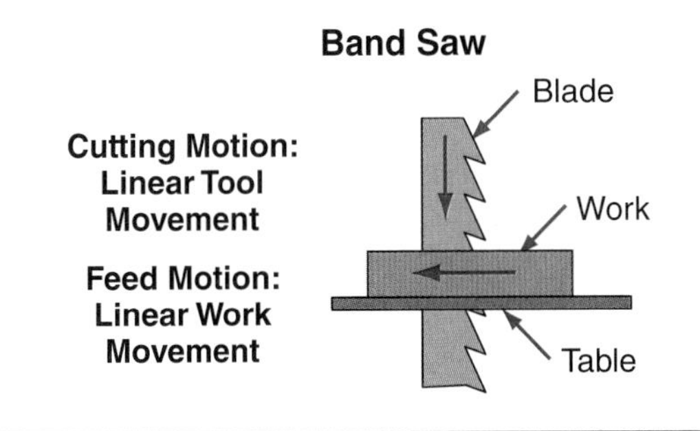

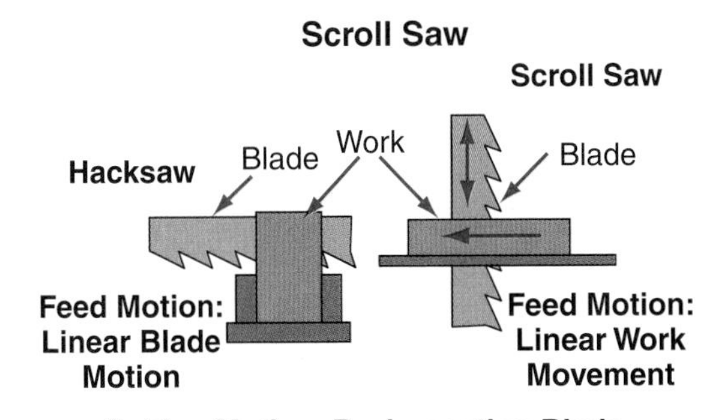

Delta; Goodheart-Willcox Publisher

Figure 20-8. Each type of saw illustrates a different type of motion.

Power Transmission

A main activity in energy conversion involves changing the type or direction of a load's motion. This action is called ***power transmission***. Power transmission takes the energy generated by a converter and changes it into motion. An example is the reciprocating motion of a piston in the cylinder of an internal combustion engine. The crankshaft changes this motion to rotary motion. The end of the crankshaft is attached to a pulley. The pulley drives the V-belt. The belt travels in a linear and rotary motion around a second pulley. The second pulley changes the linear movement back into rotary motion. See **Figure 20-9**.

Applying motion to perform work often requires changing both the type of motion and the motion's direction. Two basic categories of systems are used to change the type, direction, or speed of a force. These are mechanical-power (solid mechanics), and fluid-power (fluid mechanics) systems.

Mechanical-Power Systems

Solid mechanics involves the analysis of the behavior of solid materials or systems when they are subjected to stresses, loads, and other external forces. Mechanical engineers use that knowledge and information to design, improve, and create technological products.

Mechanical systems use moving parts to transfer motion. See **Figure 20-10**. This is the oldest method of transferring energy. Various mechanical methods are used in technological devices. Common mechanical methods used to transfer energy are described in the following sections. See **Figure 20-11**.

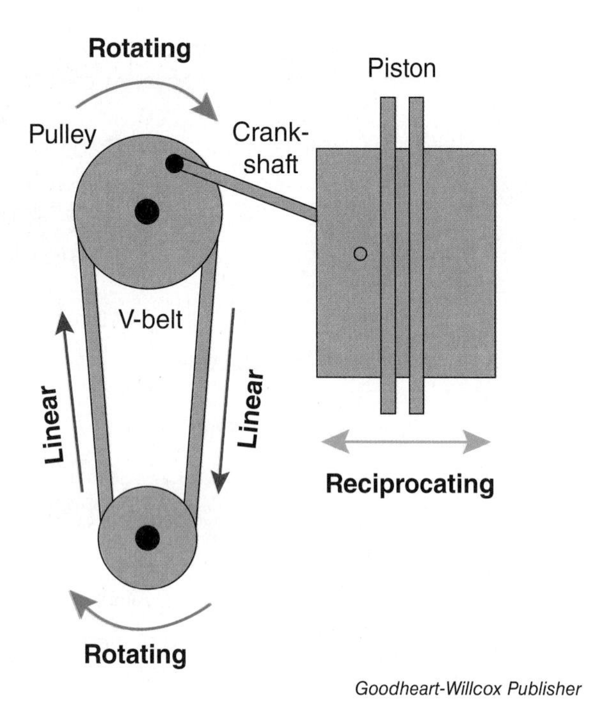

Goodheart-Willcox Publisher

Figure 20-9. Mechanisms that change the type and direction of motion. This system can change rotating motion into reciprocating motion, or vice versa.

American Electric Power

Figure 20-10. This crane uses a mechanical means to lift a load.

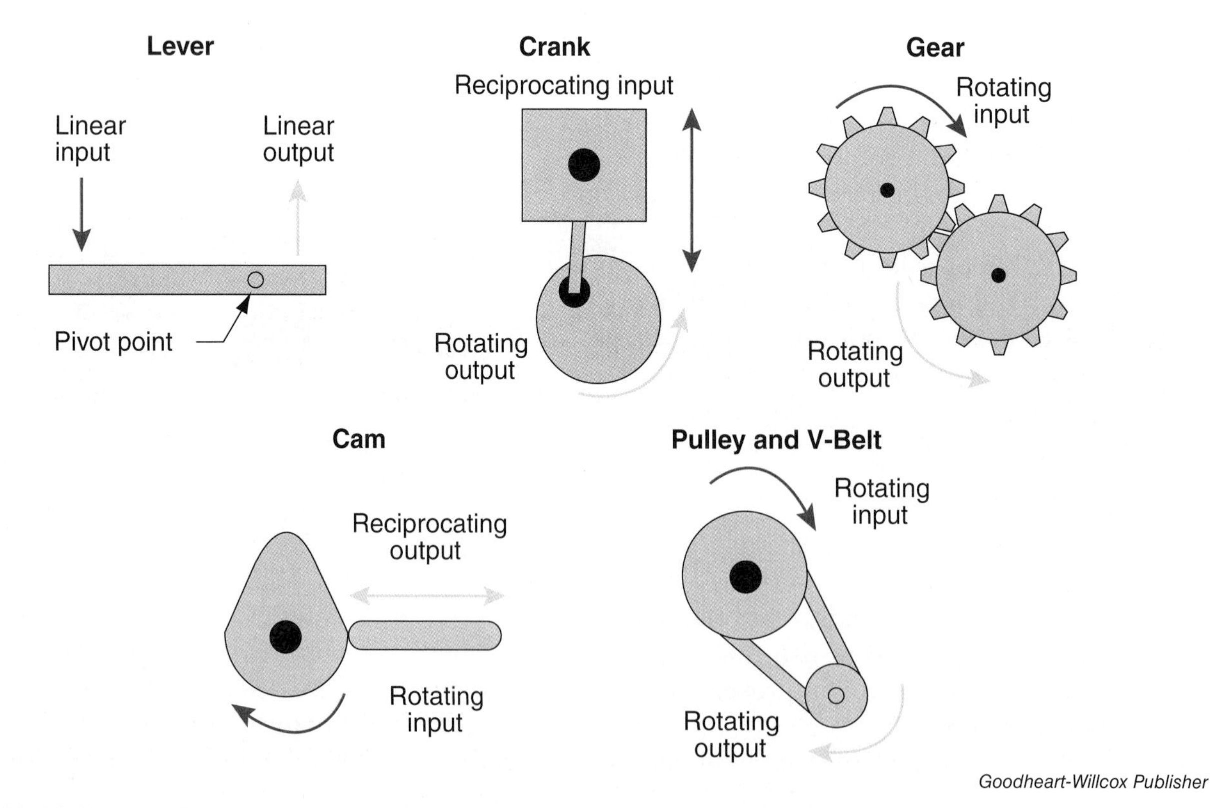

Goodheart-Willcox Publisher

Figure 20-11. Multiple mechanical techniques change the type, direction, or speed of a moving force.

Levers

Levers change the direction or intensity of a linear force. A downward force applied to one end of a lever causes the lever arm to pivot on its fulcrum. The opposite end moves in an upward direction. The location of the fulcrum determines whether the device multiplies the amount of the output force or the distance it moves. Many door-handle mechanisms in automobiles transfer motion with levers. The three categories of levers—first class, second class, and third class—were discussed in Chapter 10. Each class of lever applies force differently to move the load. Refer to the images of levers in Chapter 10.

Cranks

A crank is another component used in mechanical systems. It is an arm attached to a shaft that rotates. As it rotates, the reciprocating motion is transferred or received from the shaft. See **Figure 20-12**. A pivot pin near the outside edge of a wheel or disk changes reciprocating motion into rotating motion. The diameter of the swing of the crank determines whether the amount of the force or the distance of the force is multiplied. An internal combustion engine transfers power from the piston to the transmission using this type of process.

Gears

Torque, or the force to rotate an object around an axis, is affected by the drive and driven gears. See **Figure 20-13**. Two or more wheels with teeth on their circumferences that work together to change the direction of a rotating force are called ***gears***. The diameters of the drive (input) gears and driven (output) gears affect the overall performance of the system. Specifically, gear sizes will determine whether the system is a force multiplier (the *driven* gear rotates faster) or a distance multiplier (the *driven* gear turns over a greater area). The following gear size examples are in relation to the other gears in the system. For example, a smaller drive gear means the driven gear is larger in size. If a smaller *drive* gear is used, the unit increases the *driven* force and reduces its speed. If a larger *drive* gear is used, the unit decreases the *driven* force and increases its speed.

The following are examples of gears used by mechanical engineers. An engineer chooses a particular gear type based on a system's required gear speed, size constraints, and the load. See **Figure 20-14**.

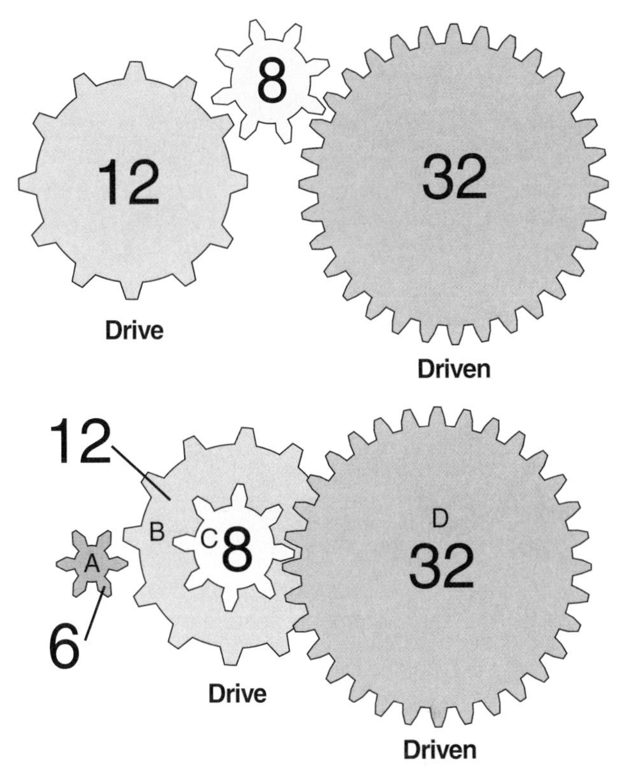

Drive Gear	Driven Gear	Driven Gear
Small	Large	Increased torque Decreased speed
Large	Small	Decreased torque Increased speed

Goodheart-Willcox Publisher

Figure 20-13. Small drive gears are used with large driven gears, and large drive gears are used with small driven gears. Each configuration has a different result.

jarabee123/Shutterstock.com

Figure 20-12. A crank is an arm attached to a rotating shaft.

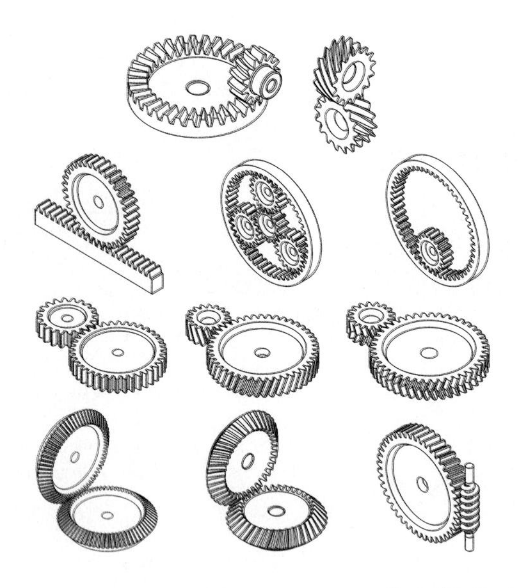

Tewlyx/Shutterstock.com

Figure 20-14. Gears commonly used in mechanical systems.

- **Spur.** This is the most commonly used gear type. A ***spur gear*** contains straight teeth that are parallel to the shaft axis. The gear is used to transmit power from one parallel shaft to another. This type is ideal for transmission gears.
- **Idler.** This gear type transfers motion and direction but does not modify speeds. An idler gear can be used to ensure that two external gears in mesh operate in the same direction.
- **Bevel.** This gear type is used on shafts that are 90° apart. Bevel gears are cone-shaped on the tooth-bearing faces of the gear.
- **Miter.** This is a unique type of bevel gear used at right-angle intersecting shafts to transmit motion and power. This gear operates in unison with other miter gears that contain the same number of teeth and the same pitch. Pitch is related to the distance between the teeth. The gears operate in a 1:1 ratio.
- **Worm.** This type of gear is often used as a speed reducer. The worm gear has at least one complete tooth around the surface, similar to a screw.
- **Helical.** The teeth of helical gears are set at an angle to the axis of rotation. Noise can be reduced as the angle becomes steeper. This type is smoother and less noisy than spur gears.
- **Internal ring gear.** This is a cylindrical internal gear. The gear teeth can mesh inside or outside, often in concert with a spur gear.
- **Gear and rack.** Also called *rack and pinion*, this gear type consists of a bar with linear teeth. It is often used with spur gears. Rotary motion is converted to linear motion or vice versa.

Cams

A ***cam*** is a pear-shaped disk with an off-center pivot point used to change rotating motion into reciprocating motion. A cam with a large lobe (extended portion) creates longer strokes for the reciprocating member. The force is reduced, however. The valves in an internal combustion engine are opened using a cam system.

Pulleys and V-Belts

Pulleys are grooved wheels attached to an axle. They also act as second-class levers. Two pulleys with a belt stretched between them change the speed or power of a motion. This system is called a ***flexible belt drive***. As one pulley turns, the belt moves. This movement, in turn, rotates the second pulley. If the force is applied to the larger-diameter pulley, the smaller-diameter pulley turns faster. The smaller pulley, however, has less power. If the power is applied to the smaller-diameter pulley, the larger-diameter pulley turns slower. The larger-diameter pulley, however, turns over a greater area.

Pulleys and belts drive many machines. Under heavy loads, the belt can slip as it drives the pulleys. The problem is overcome by using gears and a chain drive similar to the one used on bicycles.

As mechanical engineers design systems, they have a number of different belts to choose from, **Figure 20-15**. These belts include the following:

- **V-belts.** V-belts consist of a trapezoidal cross section and are comprised of rubber or molded fabric that allows for a flexible, or bending, action. A steel cord is often used to reinforce a V-belt. These belts are used for quietness and ability to absorb shock. V-belts are classified as high-capacity or standard. A high-capacity V-belt is more durable and designed for less friction.
- **Double V-belts.** Whereas a V-belt transmits power from the wedging action of the belt's tapered sides, double V-belts transmit power from the belt's bottom and top. They are used in situations where pulleys are needed to rotate in opposite directions.

STEM Connection: Mathematics

Calculating Gear Speeds

A mechanical engineer may be required to calculate gear speeds. A gear with a small number of teeth rotates at a fast speed, whereas a gear with a large number of teeth rotates slowly. A driven gear speed depends on the drive gear speed, and the number of teeth on the driven and drive gears.

Driven gear speed is calculated with the following formula:

$$S_2 = \frac{T_1 \times S_1}{T_2}$$

S_2 = speed of driven gear (rpm)

T_1 = number of teeth on drive gear

S_1 = speed of drive gear (rpm)

T_2 = number of teeth on driven gear

For example, a 40-tooth driven gear with an 18-tooth drive gear, rotating at 120 rpm, would rotate at how many rpm? See below for the setup and solving of the equation.

$S_2 = 18 \times 120/40$ $\quad S_2 = 2160/40$ $\quad S_2 = 54$ rpm

You can even work backward for a more informed calculation during the engineering design process. Consider how you would rearrange the equation if you needed to know a different variable.

Figure 20-15. Belts come in a variety of forms, sizes, and shapes. A—Timing belt. B—Flat belt. C—Engine V-belt.

- **Timing belts.** Timing belts consist of teeth that match to a pulley or sprocket. They are used over chain drives to reduce noise and for lubrication needs. Timing belts operate at high speeds.
- **Flat belts.** These belts were commonly used in nineteenth-century power equipment, including motorcycles, sawmills, and conveyers, to transmit power. They are one of the simplest flexible belt drive systems.

Mathematics and science are applied to calculate the speed of a driven pulley. Notice the similarity to the calculations used to determine the speed of a driven gear.

For example, what is the speed of a 16″ driven pulley if the drive pulley is 24″ and drive speed is 250 rpm? The speed of the driven pulley is calculated as follows:

$$S_2 = \frac{P_1 \times S_1}{P_2}$$

S_2 = driven pulley speed (rpm)
P_1 = drive pulley diameter (inches)
S_1 = drive pulley speed (rpm)
P_2 = driven pulley diameter (inches)

$S_2 = 24 \times 250/16 \quad S_2 = 6000/16 \quad S_2 = 375$ rpm

Fluid-Power Systems

Fluid mechanics is the study of how laws of force and motion apply to liquids and gases. This knowledge and understanding is then used to design, improve, and create technological products, such as heating and cooling equipment, pump systems, fans, turbines, pneumatic equipment, and hydraulic equipment.

Fluid-power systems use either liquids or gases to transfer power from one place to another. Systems using air as the transfer medium are called *pneumatic systems.* Liquids (usually oil) are used in *hydraulic systems.* See **Figure 20-16**. Generally, these systems contain two cylinders with movable pistons, a pump, valves to control the flow, and piping to connect the components.

Hydraulic principles and applications

Mechanical engineers must know the principles and applications of ***hydraulics***, a branch of science that considers how water and other liquids function. Specifically, it studies how liquids are conveyed through pipes and pathways, largely as a source of mechanical force. Hydraulics includes two areas. ***Hydrostatics*** is the branch of fluid mechanics that explains how liquids at rest function in regards to the forces exerted on them or by them. ***Hydrodynamics*** explains how forces are exerted on a solid body by a fluid's motion or pressure. There are three basic uses for hydraulic systems. See **Figure 20-17**. Hydraulic systems are often utilized in three main industries: industrial hydraulics, mobile hydraulics, and aircraft hydraulics. For example, a hydraulic press is used in manufacturing to generate a compressive force in order to shape ductile metals such as aluminum, brass, or stainless steel. In the aircraft industry, hydraulics are used throughout the plane, specifically the landing gear during take-off and landing.

Pneumatic principles and applications

As mechanical engineers solve problems, they often consider and apply principles of pneumatics. The force and distance movement calculations in pneumatics are more difficult than those in hydraulics. Liquids do not compress, so nearly all the force is

©iStockphoto.com/dgilder

Home

©iStockphoto.com/donald_gruener

Figure 20-16. Hydraulic cylinders are used to activate the loading boom of a truck. A pneumatic cylinder is used in the nailer.

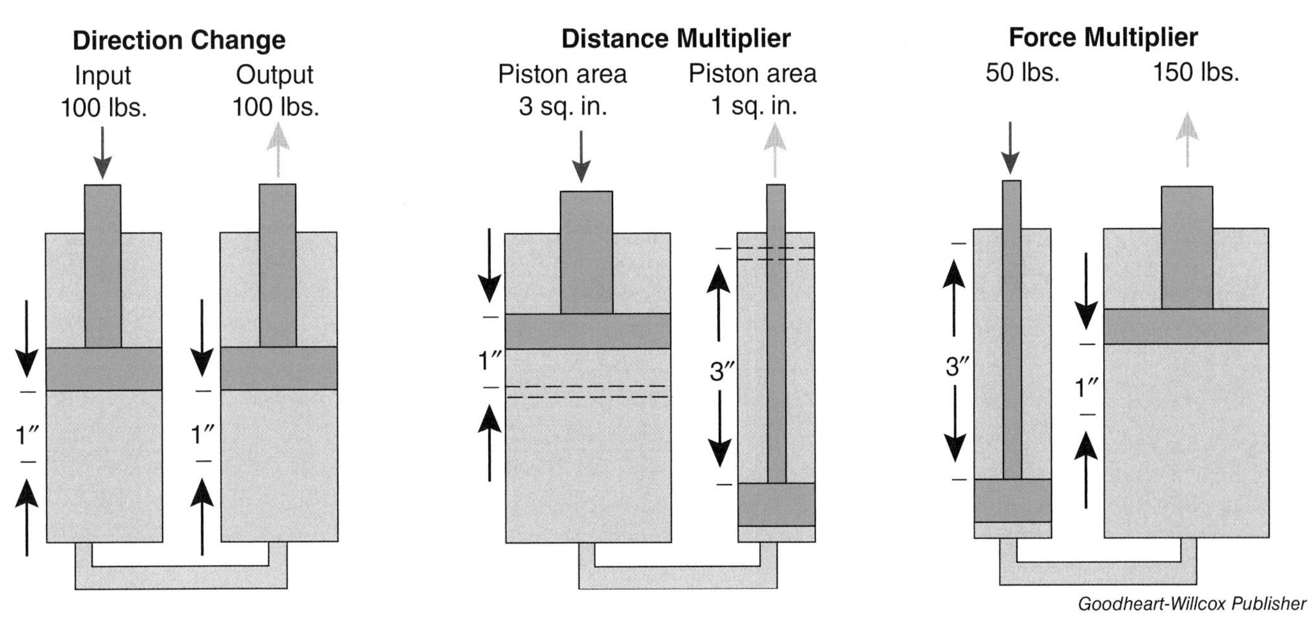

Figure 20-17. Hydraulic systems can change the direction of a force, increase the distance of a force, and increase the strength of a force.

transferred from one cylinder to the other. A slight amount of the force is used to overcome the friction of the piston and the fluid in the pipes.

Air can be compressed, however. Therefore, some of the force in pneumatic systems is used to reduce the volume of the air in the system. The remainder of the force is applied to moving a load.

Thermodynamics

Thermodynamics is the study of heat and temperature and the relation of these factors to work, energy, and power. Mechanical engineers working with thermal systems analyze how forms of energy are converted into other forms. This knowledge is useful as the engineers design, improve, and create technological products and systems such as power plants, air-conditioning or heating units, and automobile engines.

One concept covered in thermodynamics is ***conductive heat transfer***, which states that heat flows from a hot temperature to a cold temperature through a material. This occurs through *conduction*, the movement of heat along a solid material or between two solid materials touching each other. Heat is commonly transferred through a flat surface. If a mechanical engineer can reduce the surface area, heat transfer can be reduced.

The type of material and its thermal conductivity value drastically affect the rate of heat transfer. See **Figure 20-18**. An example of when this is important is in the selection of a material for insulation. The thickness of a material also affects the flow of heat through a material.

Metal	Temperature (°F)	Thermal Conductivity (Btu/hr ft °F)
Aluminum	68	118
Copper	68	223
Iron	68	42
Nickel	68	52
Platinum	68	42
Silver	68	235

Goodheart-Willcox Publisher

Figure 20-18. Thermal conductivity of common materials.

Calculating Heat Transfer

Mechanical engineers can use *Fourier's law* to calculate conductive heat transfer. Fourier's law is expressed by the following equation:

$$Q = \frac{kA(\Delta T)}{L}$$

where:

Q = Heat flow/transfer (in Btu/hour or W, or J/s)

k = Thermal conductivity of the material

A = Area of surface through which heat is conducted (varies)

ΔT = Temperature difference between hot and cold

L = Thickness

Think Green Fuel Cells

The automobiles of the future will probably look much different, both inside and out, from what we see today. Engineers and others have developed a new automobile platform designed for use with a fuel cell (electrochemical-cell) engine and a hydrogen-fuel tank. Many mechanical systems will be eliminated. These controls include the internal combustion engine, exhaust system, and brake and accelerator pedals. Electric motors in the hub of each wheel will propel the chassis, making each vehicle all-wheel drive. All controls will be located in the steering mechanism.

The fuel cell is a system that uses stored hydrogen and oxygen from the air to create electrical energy. Potential residential applications for fuel cells include laptop computers, cell phones, vacuum cleaners, and hearing aids. Used in these ways, a fuel cell provides much longer life than a battery does. A fuel cell can be recharged quickly.

Using fuel cells to power automobiles, however, is more difficult. Recently, researchers have made major breakthroughs in creating a fuel cell with commercial prospects. While researchers and automobile manufacturers work on developing fuel cell vehicles that run on pure hydrogen, gasoline-fed fuel cells will serve as a transitional technology. Some of these, already on the market, cut carbon dioxide emissions by 50% and can get up to 40 miles per gallon of fuel.

Fuel cell engines have many advantages. This engine has no moving parts. The fuel cell engine is a quiet, reliable source of power. This engine will lead to a cleaner environment because the only emission is water vapor. This engine will reduce dependence on foreign oil. Another advantage is that the vehicle, while sitting in the garage, can generate enough electricity to power a home.

However, there are obstacles to overcome. A new fueling infrastructure (replacements for gas stations as we know them) needs to be installed across the country to distribute a usable fuel, such as methanol. However, locations are increasingly being offered for vehicle owners. California is currently leading the United States by specifically funding the effort to make these stations more accessible to the public. Additionally, the current costs to purchase fuel cell–dependent vehicles make it cost prohibitive for many individuals. Hopes are high for the emergence of this new technology.

Fuel cells have many applications beyond powering automobiles. For example, they are used as power sources in remote locations, such as spacecraft and remote weather stations. Fuel cells are particularly useful in this type of application because they are lightweight and compact. They are also reliable because they have no moving parts.

In the following example, a mechanical engineer is designing a system and chooses aluminum as a wall barrier. The thickness of the wall is six inches, and surface length and width is one foot by one foot (remember to convert inches to feet when necessary). The surface temperature on one side is 110°F, and the other side's surface temperature is 60°F. The conductive heat transfer can be calculated as follows, with an answer of 11,800 Btu/hour:

$$Q = (118)(1 \times 1)(110 - 60)/.5$$
$$Q = (118 \times 1 \times 50)/.5$$
$$Q = 5900/.5$$
$$Q = 11{,}800 \text{ Btu/hour}$$

Measuring Work, Power, and Torque

As mechanical engineers design and test mechanical systems, they must also measure work, power, and torque to ensure their product works effectively and efficiently. For example, a mechanical engineer designing a new engine may specifically analyze how the piston rod turns the crankshaft. During this process, a measurement of how much work the rod does on the shaft can be taken to compare and contrast against other engines and goals of the project. Additionally, torque may be measured on the tightening of fasteners for a manufacturing plant to ensure a component is tightened correctly. Lastly, power can be measured to compare products, such as engines, to determine efficiency and output.

Measuring Work

Work is measured by multiplying the weight of an object being moved by the distance the weight was moved. See **Figure 20-19**. The result is the amount of energy needed to move an object from one location to another. In the US Customary system, the result is expressed in ***foot-pounds*** (***ft-lb***). The amount of work completed can be measured with the following formula:

$$\text{work (in ft-lb)} = \text{force, or weight, (in lb)} \times \text{distance (in ft)}$$

If you weigh 140 lb and plan on walking across a 40′ wide room, you would need 5600 ft-lb of energy (140 lb × 40′) to complete the task. Likewise, lifting a 20-lb weight off the floor and placing it on top of a 36″ high table requires 60 ft-lb of energy (20 lb × 3′ [36″]).

In the metric system, work is measured in newtons per meter, or ***joules*** (***J***). The force, or weight, is measured in newtons, and the distance is measured in meters. The metric work formula is the following:

$$\text{work (in J)} = \text{force, or weight (in newtons),} \times \text{distance (in meters)}$$

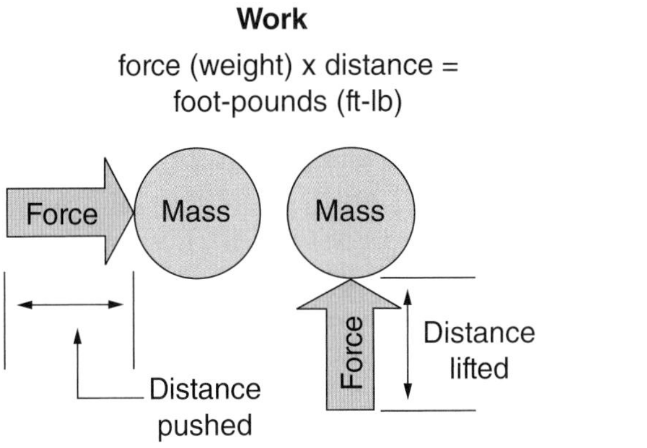

Measuring Power

Work is done in a context of time. Measuring the rate at which work is done gives you the term *power*. See **Figure 20-20**. Power can be calculated by dividing the work done by the time taken:

$$\text{power (in ft-lb/seconds)} = \frac{\text{work done (in ft-lb)}}{\text{time (in seconds)}}$$

The metric version is the following:

$$\text{power (in watts)} = \frac{\text{work done (in J)}}{\text{time (in seconds)}}$$

Two common power measurements are the horsepower and the kilowatt-hour. The term *horsepower* is used to describe the power output of many mechanical systems. The power needed to move 550 lb a distance of 1′ (550 ft-lb) in one second is one ***horsepower***. The factor of time is important to power. A motor that lifts 550 lb in one minute can be smaller than one that lifts 550 lb the same distance in one second. Likewise, the engine that moves a car from 0 to 60 mph in seven seconds must be more powerful than one that does the same job in nine seconds.

The term *horsepower* is used in several different ways. The theoretical, or indicated, horsepower is the rated horsepower of an engine or a motor. This number suggests the maximum power that can be expected from the device under ideal operating conditions. Most often, this amount of power is not available from the device.

The brake horsepower is the power delivered at the rear of an engine operating under normal condi-

Work
force (weight) x distance =
foot-pounds (ft-lb)
Force
Mass
Mass
Force
Distance lifted
Distance pushed

Goodheart-Willcox Publisher

Figure 20-19. Work is done when a force moves a mass over a distance.

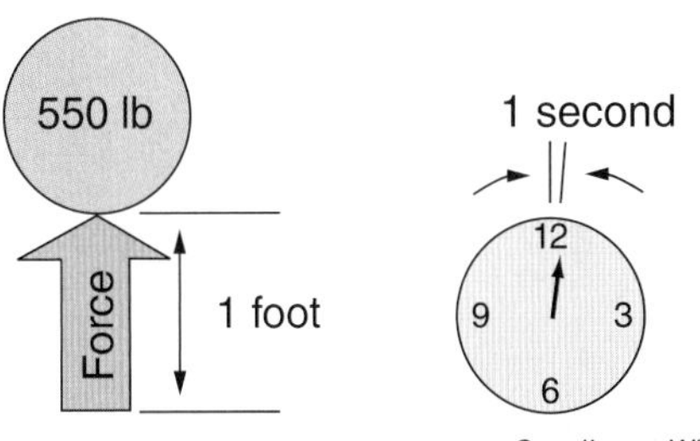

Goodheart-Willcox Publisher

Figure 20-20. Power is work done per unit of time.

tions. Drawbar horsepower is the power delivered to the hitch of tractors. Frictional horsepower is the power needed to overcome the internal friction of the technological device.

A ***watt*** is equal to 1 J of work per second. The work that 1000 watts complete in one hour is one ***kilowatt-hour***. In electrical apparatus, the resistance of the device determines the power consumed. The device's wattage rating is the product of the electrical current flowing through the device and the voltage drop across it. The formula can be expressed as follows:

P (power in watts) = I (current in amperes) × V (electromotive force in volts)

Measuring Torque

Torque is the measure of the twisting force on an object. This force from rotating an object around an axis or fulcrum is not a push or pull, but a twisting force. See **Figure 20-21**. Torque can be calculated as follows:

T (torque in lb-ft) = F (force in lb) × D (distance [inches or feet])

The torque generated by a rotating machine, including a motor, can also be calculated using the machine's horsepower and rpm. The torque (in lb-ft) of a motor, or rotating machine, can be calculated using a constant of 5252 as follows:

$$T = (5252 \times hp)/rpm$$

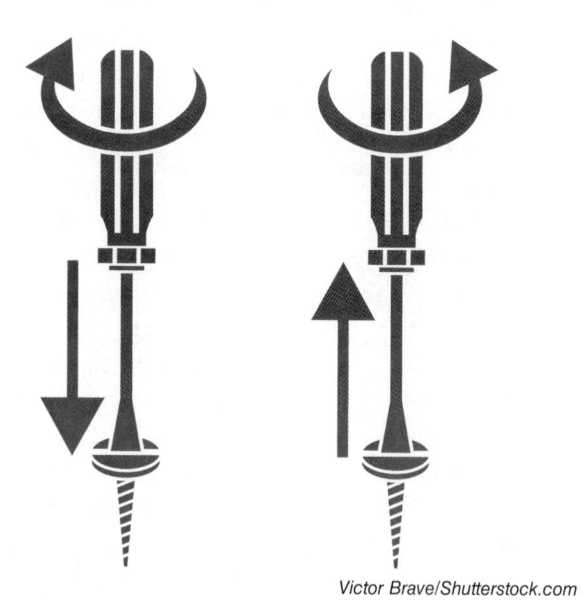

Victor Brave/Shutterstock.com

Figure 20-21. Torque is a twisting force, such as a screwdriver tightening a screw.

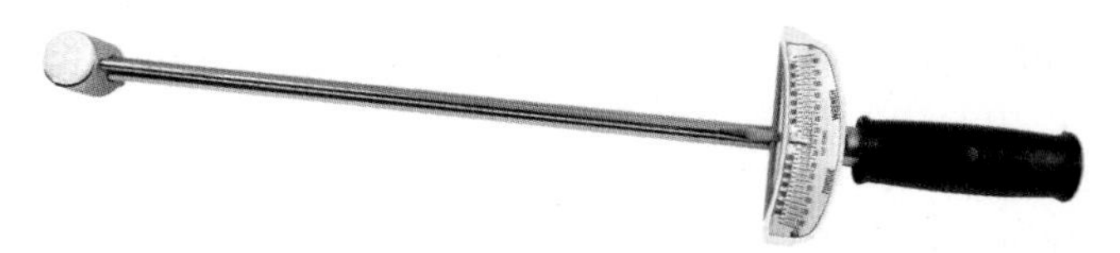

MATRGRIT HIRSCH/Shutterstock.com

Figure 20-22. A torque wrench.

Torque can also be measured with a torque wrench, **Figure 20-22**. A torque wrench allows engineers to accurately apply a specific amount of torque to a fastener. The user can match applied torque to a particular situation based on predetermined specifications for proper tension.

Mechanical Engineering and Materials Science

Materials science involves the study and creation of materials, with an emphasis on solids. Mechanical engineers apply their knowledge and understanding of a material's strength and properties to select materials for products. They consider how a material will perform in a specific environment and under certain conditions, ensuring that personal injury and damage to property does not occur.

Mechanical engineers apply their understanding of materials science when examining why products fail. They are often required to conduct ***solid and structural analysis***, a process specifically devoted to preventing and determining how and why objects fail. ***Static structural failure*** occurs when, loads, or forces, have been applied to an object, and the object either breaks or is deformed due to some criterion. ***Fatigue failure*** occurs when an object fails after being subjected to repeated loading and unloading and progressive brittle cracking occurs until the object fractures. Fatigue failure can be the result of deficiencies in the product. A microscopic flaw may progressively build up to a failure. When conducting structural analysis, either before or after failures, mechanical engineers reference documents that contain standards, or expected values, of material properties. One such source is MatWeb, an online resource that contains a database of material properties.

Mechanical Engineering and Mechatronics

Mechatronics is a field that involves mechanical engineering, electrical engineering, and computer control and information technology. Mechatronics is represented across all engineering fields, but mechanical engineers often receive specific instruction in this specialization. An industrial robot is an example of a mechatronics system, **Figure 20-23.** Mechanical engineers design a mechatronics system by integrating knowledge and principles of electronics, computing, mechanics, and robotics. They are often tasked with creating a simpler, more efficient, economical, safer, or reliable system.

Mechanical Engineering Careers

Students interested in becoming mechanical engineers may specialize in a number of pathways. These pathways include fluid mechanics, machine design and solid mechanics, automation and control, thermal engineering, welding and joining specialization, transportation systems, manufacturing, nanotechnology, robotics, aerospace engineering, energy systems, and biomechanics. Since mechanical engineers are employed in a variety of industries, each pathway offers electives to increase the success in future employment. Even industries that are largely covered by other engineering professions, such as aerospace, civil, electrical, or computer, require a mechanical engineer's specialization to solve technological problems.

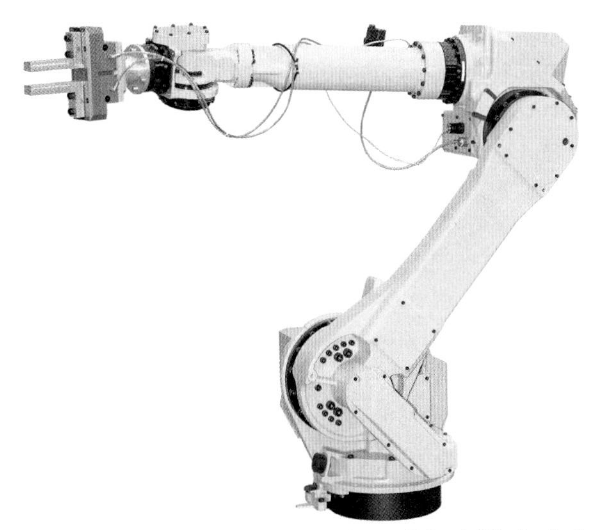

Baloncici/Shutterstock.com

Figure 20-23. Robotics and mechatronics are two fields that overlap.

Mechanical engineers are employed in the following industries:

- **Automotive.** Mechanical engineers work on projects involving multiple vehicle systems. They provide expertise in the design and manufacture of traditional combustion and diesel engines, as well as alternative engine types. They also create and improve drive train components, including transmissions, as well as suspensions and safety systems.
- **Aerospace.** Although the aerospace industry typically employs aerospace engineers to solve technological problems, mechanical engineers assist by applying knowledge of fluid mechanics, designing turbines that effectively transfer heat, and working to create or improve control systems used by aerospace technologies.
- **Computers.** Alongside computer engineers, mechanical engineers contribute to new and improved products by solving problems related to heat transfer of components, packaging of systems and devices, and the design of components.
- **Construction.** Although civil engineers have traditionally been credited with the design of structures, mechanical engineers aid the construction industry. They design heating, ventilation, and air-conditioning (HVAC) systems, design stress analysis tests, and create elevator and escalator systems.
- **Electrical power generation.** Mechanical engineers are employed by electrical power generation industries to aid electrical engineers. They use their knowledge of thermodynamics, work, power, and energy to design mechanical devices, such as turbines, generators, and motors.
- **Petrochemicals.** Mechanical engineers assist chemical and petroleum engineers in the analysis of stress, structures, fluid flow, and efficiency of oil drilling. A primary role of mechanical engineers is the design of oil refineries, including the piping and pressure systems.
- **Robotics.** Mechanical engineers are employed by robotics companies to design mechanical systems and components, such as actuators, motors, and sensors. They are also responsible for the analysis of forces, such as stress and tension, for design products.

Summary

- ✔ Mechanical engineers must have an understanding of energy, work, and power, as well as knowledge of thermodynamics, hydraulics, pneumatics, mechatronics, robotics, drafting, 3-D modeling, forces, torque, and strengths of materials.
- ✔ Isaac Newton's three laws of motion and his development of calculus resulted in Newtonian mechanics, which describes the relationship among force, mass, and motion in the natural world. Newton's methods are still the mathematical foundation for physics and mechanical engineering problems.
- ✔ Forces can be described as a push or pull that results from the interaction between two objects.
- ✔ The three types of motion are linear (movement in a straight line), rotary (spinning around an axis or motion in a circle), and reciprocating (moving back and forth).
- ✔ Power transmission takes the energy generated by a converter and changes it into motion.
- ✔ Mechanical-power and fluid-power systems are used to change the type, direction, or speed of a force.
- ✔ Mechanical methods used to transfer energy include levers, cranks, gears, cams, pulleys, and V-belts.
- ✔ Fluid-power systems use either liquids (in hydraulic systems) or gases (in pneumatic systems) to transfer power from one place to another.
- ✔ Heat flows from a hot temperature to a cold temperature through a material. The rate of heat transfer is affected by the type of material and its thermal conductivity value.
- ✔ Work is measured by multiplying the weight of an object being moved by the distance the weight was moved.
- ✔ Power can be calculated by dividing the work done by the time taken. Two common power measurements are the horsepower and the kilowatt-hour.
- ✔ Torque is the measure of the twisting force on an object.
- ✔ There are many specializations within the mechanical engineering field, and mechanical engineers are employed in a variety of industries.

Check Your Engineering IQ 

Now that you have finished this chapter, see what you learned by taking the chapter posttest.
www.g-wlearning.com/technologyeducation/

Test Your Knowledge 

Answer the following end-of-chapter questions using the information provided in this chapter.

1. *True or False?* Mechanical engineers are required to have knowledge of the strengths of materials.
2. What is described by Newtonian mechanics?
3. What SI unit do mechanical engineers use to describe *force*?
4. What is Newton's first law of motion?
5. _____ motion is moving back and forth.
   A. Linear
   B. Reciprocating
   C. Rotary
   D. Helical
6. Two or more wheels with teeth on their circumferences that work together to change the direction of a rotating force are called _____.
   A. cranks
   B. pulleys
   C. cams
   D. gears
7. *True or False?* A flexible belt drive can change the direction of a motion.
8. Fluid power systems use either _____ or _____ to transfer power from one place to another.
9. The branch of fluid mechanics that explains how forces are exerted on a solid body by a fluid's motion or pressure is _____.
   A. hydraulics
   B. hydrostatics
   C. thermodynamics
   D. hydrodynamics
10. The study of heat and temperature and the relation of these factors to work, energy, and power is called _____.
11. The amount of energy needed to move an object from one location to another is measured in _____.

12. What is *brake horsepower*?
13. Torque is the measure of a(n) _____ force on an object.
14. *True or False?* An industrial robot is an example of a mechatronics system.
15. In the petrochemical industry, mechanical engineers help analyze stress, structures, _____, and efficiency of oil drilling.

Critical Thinking

1. In what situations would a chain drive system be the preferred choice over a flexible drive system? In what situations would a flexible drive system be a better choice?
2. Select two technologies you have encountered today. Illustrate how you would measure their work, energy, and power.

STEM Applications

1. Solve the following equation. A truck is pulling a load a distance of 100 feet. A force of 900 lb is required to keep the truck moving at a constant speed. Calculate the quantity of work done on the truck.
2. Correct the following equation.
   power (in ft-lb) = force, or weight, (in lb) × distance (in ft)
3. Rearrange the following equation to calculate for current.
   P (power in watts) = I (current in amperes) × V (electromotive force in volts)

Engineering Design Challenge

Wind-Powered Electricity Generator

Background

Deposits of fossil fuels (petroleum, natural gas, and coal) may be exhausted in the not-too-distant future. Therefore, finding ways to use inexhaustible energy sources is a major challenge. One important inexhaustible energy source is wind.

Challenge

To understand wind power more fully, build and test a working model of a wind-powered electricity-generation system. Use the drawings and the following procedure to complete this challenge.

Materials and Equipment

- One 3/4″ × 5 1/2″ × 8″ wood base (pine, fir, or spruce).
- One 3/4″ × 18″ strip of thin-gauge sheet metal.
- Two 1/2″ No. 6 sheet metal screws.
- One small electric motor.
- One multimeter.
- 24″ of electrical-hookup wire.
- One three-blade or four-blade propeller (prop).

Procedure

Building the Prototype

1. Obtain a piece of wood with dimensions of 3/4″ × 5 1/2″ × 8″.
2. Cut a strip of sheet metal 3/4″ wide and 18″ long.

Safety

Follow all safety rules your teacher demonstrates!

3. Drill a 1/8″-diameter hole 1/2″ from each end of the strip of sheet metal.
4. Form the sheet metal strip around the motor. See **Figure A**. This forms the motor mount.

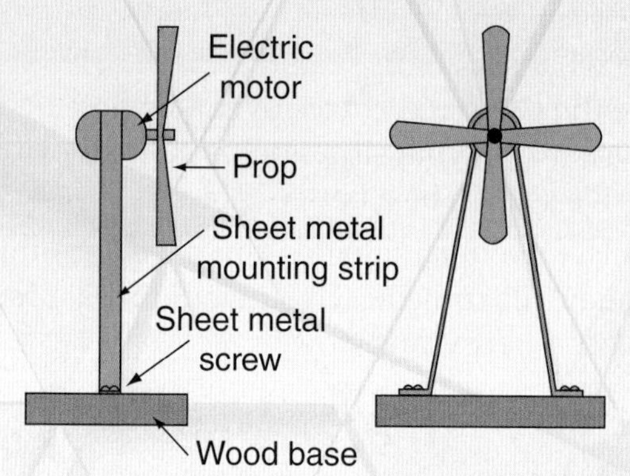

Figure A. A model windmill.

5. Bend the sheet metal strip to form 1″-long tabs that can be mounted to the wood base.
6. Use the two sheet metal screws to attach the motor mount to the base.
7. Attach the prop to the motor.
8. Press the motor into the loop of the motor mount.
9. Attach a piece of hookup wire to each wire coming out of the motor.
10. Solder your joints. Wrap them with electrical tape.
11. Set the multimeter to read current.
12. Attach the wires to the multimeter. See **Figure B**.

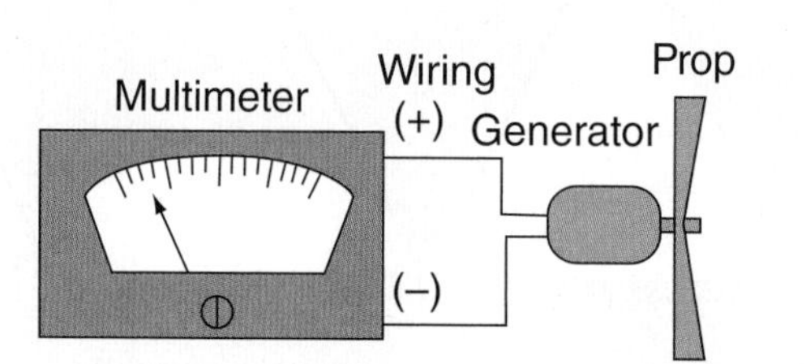

Figure B. A wiring schematic for testing the windmill operation.

Testing the Device

1. Place the wind-powered generator in front of a window fan.
2. Turn the fan on to its low setting.
3. Observe the rotation of the prop.
4. Note and record the meter reading on the multimeter.
5. Repeat steps 3 and 4 on the fan's medium setting.
6. Repeat steps 3 and 4 on the fan's high setting.
7. Turn off the fan.

Optional

1. Change the prop to a model with a different number of blades, or change the angle at which the wind is hitting the propeller to 30° off center.
2. Repeat steps 1 through 4 in the Testing the Device section.
3. Write a report explaining your observations.

CHAPTER 21

Communicating Information and Ideas

Check Your Engineering IQ

Before you read this chapter, assess your current understanding of the chapter content by taking the chapter pretest.
www.g-wlearning.com/technologyeducation/

Learning Objectives

After studying this chapter, you should be able to:

- ✔ Define *information and communication technology*.
- ✔ Explain the purposes and goals of communication.
- ✔ Describe the components of the communication model.
- ✔ Explain the major delivery types of information and communication technology systems.
- ✔ Describe the two main types of telecommunication systems.
- ✔ Explain how computers are connected to share information using the Internet.
- ✔ Identify the features of the World Wide Web.

Technical Terms

amplitude
amplitude modulation (AM)
analog signal
backbones
browser
communication
computer
digital signals
digital technology
domain name
edutainment
electromagnetic force
file-transfer protocol
frequency
frequency modulation (FM)
hyperlinks
hypertext markup language (HTML) tags
infotainment
interference
Internet Protocol (IP) address
modem
network
noise
point of presence (POP)
routers
search engines
server
social networking service
telemedicine
transducers
World Wide Web (WWW)

Close Reading

Before you read this chapter, think about all the ways you use technology to communicate. Then think about how different your relationships would be if such technology did not exist.

Information and communication technologies have greatly impacted our culture, society, and quality of life throughout history. More recently, the rapidly increasing advancement of these technologies has revolutionized the processing of information. See **Figure 21-1**. The ability to access information instantaneously and to immediately communicate it across the world has reshaped the world's economy and workforce into a global one. Technology has enabled many countries to economically compete with each other despite their distance.

Information and communication technologies are intertwined with daily work and leisure activities. In many ways, society has become increasingly dependent on them. Imagine if you did not have access to the Internet or some type of mobile communication device. How would you be able to complete your homework? How would you apply for a job? How would you plan an event? As communication and information technologies evolve, their impacts on society, culture, and the economy will become even more profound.

Information and Communication Technology

The terms *information* and *communication* are often used together because the purpose of communication is to transmit information. Information has little purpose if it cannot be communicated to others.

The term *data* refers to unorganized facts. *Information* is some form of arranged data. Information is the foundation for knowledge. Knowledge enables us to advance as a society and to succeed individually. It enables people to apply information to create ideas or to complete tasks. We are now said to be living in a knowledge-based economy. This means that people who can use their depth of knowledge to create the best ideas will be the most successful in the workforce.

How do you obtain information? Think back over the day. Did you listen to a radio newscast or check the weather on your smartphone? Did you receive text messages or e-mail? See **Figure 21-2**. Did you read a traffic signal or sign? Did you listen to music or watch television? All of these actions allow you to obtain information and illustrate the use of information and communication technologies.

Remember, technology is the application of knowledge, tools, and skills to extend human capabilities. Therefore, information and communication technologies are the systems and products that extend the ability to collect, analyze, store, manipulate, receive, and transmit information or data. A wide range of technologies are used for these purposes. Examples include computers, fiber-optic cables, GPS, televisions, smartphones, social media, printing devices, video cameras, music recording devices, and wireless Internet routers. These technologies and many others communicate or create information. See **Figure 21-3**.

Toria/Shutterstock.com

Figure 21-1. Information and communication technologies enable us to access information at a moment's notice and communicate information around the world in fractions of a second.

George jmclittle/Shutterstock.com

Figure 21-2. Smartphones provide access to the world's information in the palm of your hand.

stoatphoto/Shutterstock.com

Figure 21-3. A wide range of communication processes and technologies are used to collect, analyze, store, manipulate, receive, and transmit information or data.

Interrelationship of Technologies

Information and communication technologies are a central component of our designed world as a whole. Other technologies depend heavily on information and communication technologies to operate. Information and communication technologies are used to control other technological systems. For example, global positioning systems are used to send data to guide and control transportation systems, such as airplanes, drones, and automobiles. Without information and communication systems, the safe and efficient function of these transportation systems would not be possible.

Information and communication technologies also play an important role in biomedical technology. These technologies have improved health-care diagnostics and thus increased life spans and enhanced the quality of life for many. Doctors can receive patient data from medical technicians at a moment's notice, no matter where they are located.

Additionally, these technologies have made telemedicine possible. ***Telemedicine*** is the provision of health-care at a distance through communication technology, **Figure 21-4**. Patient data can be continuously collected using monitoring devices while a patient is at home. A surgeon can provide directions on a procedure to another doctor anywhere in the world. Information and communication technologies eliminate the barriers that distance places on health-care and provides access to medical services that people would not otherwise have.

verbaska/Shutterstock.com

Figure 21-4. This patient is measuring his blood pressure in a virtual physician visit.

Evolution of Information and Communication Technologies

The designed world has evolved to extend human capabilities to send and receive information via technology. As a result, our sensory systems are now surrounded with endless inputs of information. It has been estimated that the information provided in one week's worth of a national newspaper is more than the amount of information a person living in the eighteenth century would have encountered in a lifetime. The impacts of these technological developments have defined the current age of human existence. Just as the Iron Age was classified by the creation of stronger tools and weapons through the use of iron, advancements in information and communication technologies have classified our current era as the information age.

The information age is estimated to have started in the 1950s, after the industrial era, and continues today. This era has resulted in the creation of some of history's greatest inventions, such as the transistor, integrated circuit, computer, communication satellite, and space shuttle. Now, more than ever, the gathering, manipulation, classification, storage, and retrieval of information is central to the workings of society.

Information is quickly sent across computers digitally, represented as 1s and 0s. Advances in

Academic Connection: Communication

The Power of Radio

The author H. G. Wells wrote one of his most popular science fiction books, *The War of the Worlds*, in 1898. The book is about an invasion from Mars. Little did Wells know the uproar the book would cause 40 years later, as a result of the power of the telecommunication system known as the *radio*.

In the 1930s, one of the more popular forms of entertainment was the broadcast of various programs on the radio. Every night, millions of Americans tuned in to hear their favorite comedies, dramas, and news programs. One of the more popular dramas featured plays that Orson Welles directed. For the broadcast of October 30, 1938, he decided to feature an adaptation of *The War of the Worlds*. In adapting the book for the radio, however, he apparently wanted to add more drama. He made the play sound similar to a news broadcast. Welles had fake news bulletins interrupt the play's music. The bulletins, read by actors in dramatic tones, warned that Martians had invaded New Jersey and were intending to destroy the United States. Even though Welles had noted before the program that this was just a play, many had not heard the disclaimer. The program sounded so realistic that some people thought an actual invasion had occurred. They panicked. Some jumped into their cars to run. Others hid in their basements. Still others put wet towels on their heads to avoid the Martians' poison gas.

ChameleonsEye/Shutterstock.com

The disturbance caused a scandal. Some called for more government regulation. The commotion eventually died down. To this day, however, the incident shows the tremendous power involved in using technology to communicate. Can you give another example of this power?

information and communication technologies have increased access to information and knowledge, thus empowering individuals to make decisions that only "experts" made in the past. Through enhancing global connectivity, information and communication technology developments have revolutionized our world just as agriculture and industrialization have done in the past.

Change in information and communication technologies continues to occur more and more rapidly. Forms of communication that you have used as a child may have already become obsolete. Technological advances lead to new products that are quicker or more effective at producing and sharing information. These changes have created opportunities, such as successful multimedia and social networking businesses, telecommuting, telemedicine procedures, and open and free online learning.

However, information and communication technologies can also have negative consequences and raise the following questions. Are people bombarded with too much information? Is it unhealthy to be connected to work 24/7 through e-mail? How much of the information on the Internet is valid? Are we outsourcing too much work? Is our personal and national information protected? It is important for engineers and consumers alike to understand the trade-offs involved in information and communication technologies.

Engineers will continue to expand the range and complexity of information and communication technology. For example, they are tasked with finding new and better ways to enhance the security of cyberspace, improve the teaching of students, minimize cyber terrorism, maintain national security, and ensure personal privacy.

Communication

Communication is the passing of information from one location to another or from one person to another. Communication takes a variety of forms, including phone calls, text messages, billboards, television ads, and face-to-face conversations. Through evolution, humans have become social beings who have a need to share and communicate information. Improved methods for communication have even been linked to increasing human survivability.

When the spoken word is used to communicate, we call it *language,* or *verbal communication.* When a technical means is used to convey information, it is called *information and communication technology.* Humans first communicated with gestures and grunts. Later, they developed language, which increased their ability to communicate. These forms of communication, however, do not involve technology. There is no technical means between the sender of the message and the receiver.

Possibly, the first use of communication technology was cave paintings. The "artists" used sticks, grass, or their fingers to apply paint to cave walls, **Figure 21-5**. The result was a message that could be stored. At some later date, a person could retrieve the message. Over time, communication technologies have become more and more advanced, making it ever easier to transmit, interpret, and store information.

Pichugin Dmitry/Shutterstock.com

Figure 21-5. Famous prehistoric rock paintings of Tassili N'Ajjer, Algeria.

What Is Communicated?

Communication takes place for a variety of purposes. Most often, it is used to convey ideas, exchange information, and express emotions. See **Figure 21-6**.

best4u/Shutterstock.com

Cory A Ulrich/Shutterstock.com

Rob Hainer/Shutterstock.com

Figure 21-6. Communication technology can be used to convey ideas, information, or emotions.

Think Green Forest Stewardship Council

Paper products are used for a variety of reasons, and consequently, a great deal of it is used. Today, most paper is recycled. While recycling is important, another consideration for using paper should be where the paper comes from. The Forest Stewardship Council (FSC) is an organization that has created standards meant to lessen the environmental impact of practicing forestry. The FSC certifies specific forests to be used to create wood products and paper. These forests are managed in a responsible way that protects the environment and provides economic and social benefits.

For companies wishing to use certified materials, the FSC issues a chain-of-custody certificate to verify the materials came from an FSC-certified location. This chain-of-custody certificate allows companies to be more responsible with materials by giving them the ability to track the origin of the materials. The safe and responsible handling of the materials can also be checked with the chain-of-custody certification. As a result of using certified materials, company products may be labeled to indicate that FSC-certified materials were used for the product.

An idea is a mental image of what a person thinks something should or could be. You have ideas about what kind of music is good, how people should behave, what marketing campaign will work the best, or what solution will solve a problem most efficiently. Also, you probably have opinions on how to perform various tasks, such as washing an automobile, riding a bicycle, and mowing a lawn. People also have ideas on such issues as how to protect the environment and whether or not to allow capital punishment. These ideas are shared through communication.

Information is vital to taking an active part in society. It provides a concrete foundation for decision making and action. Information can be as simple as the serving time for lunch or as complex as the moon's effect on the tides.

Ideas and information are important. Feelings, however, are just as vital to many people. Communication media can convey emotions, or feelings. For example, a photograph or image can communicate the excitement of a sporting event. People can communicate affection for each other through social media. Communication media can make us laugh or cry, be excited or calm, or feel good or bad.

Goals of Communication

Each communication message is designed to impact someone. The communication can meet one or more of the following basic goals:

- To *inform/educate* by providing information about people, events, or relationships. We read books, magazines, and newspapers to obtain information. Radio news programs, television news programs, and documentaries are designed to provide information for educational purposes.
- To *persuade* people to act in a certain way. Examples include political campaign advertisements and campaigns urging people to refrain from texting while driving. Print and electronic advertisements, billboards, and signs are typical persuasive communication media.
- To *entertain* people as they participate in or observe events and performances. Television programs, movies, video games, and novels are common entertainment-type communication.
- To *control* a person's actions for a variety of reasons. For example, a traffic light is used to control when a person proceeds at an intersection, thus helping to ensure the safety of motorists and pedestrians at an intersection.
- To *manage* people as they perform certain actions. For example, a speed limit sign is used to help manage the speed of motorists.

Some of the goals of communication can be merged. Two words in our language arise from this merging of goals. The first is ***infotainment***, which means information provided in an entertaining way. You might learn as you watch a quiz show on television or use a computer simulation. Both of these are enjoyable ways of gaining new and useful information. The second term, ***edutainment***, takes communication one step beyond infotainment. Edutainment creates a situation in which people want to gain the information. It is educational content that is designed to have entertainment value. Examples of edutainment include video games with an educational aspect. The television program *Sesame Street* is an example of edutainment.

The Communication Model

Communication can be thought of as a simple process. The communication process through any system is essentially the same. See **Figure 21-7**.

The process of communication starts with the action of encoding a message. Encoding means the message is changed from a form that we can understand into a form that can be easily transmitted safely across a vast distance. The encoding can be a series of bumps on a DVD, an arrangement of letters on a printing plate, or a series of 1s and 0s on a hard drive.

The message is then transmitted to the receiver. Transmission involves a communication channel or carrier. The channel might be electromagnetic waves broadcast through the air, electrical signals carried by a wire, pulses of light on fiber-optic cable, or printed text on paper.

At the other end of the communication channel is the receiver. The receiver gathers and decodes the message, which means the message is converted back to a form that people can understand. Examples of receivers are radios and televisions. Radios change electrical impulses into sound. Televisions change electrical impulses into a series of images. The human mind decodes the written and graphic messages contained in photographs and printed media.

Every information and communication system can be summarized in a basic communication model, **Figure 21-8**. There is a source of information, information in the form of a message to be sent, potential noise that can impede the delivery of the message, the receiver of the message, and the feedback generated from the receiver of the message.

There is a difference between information and noise. ***Noise*** is unwanted sounds or signals that become mixed in with the desired information. Noise is due to interference in the communication channel. ***Interference*** is anything that impairs the accurate communication of a message. Examples of interference are static on a radio, noise in a movie theater, and smudged type in a printed message.

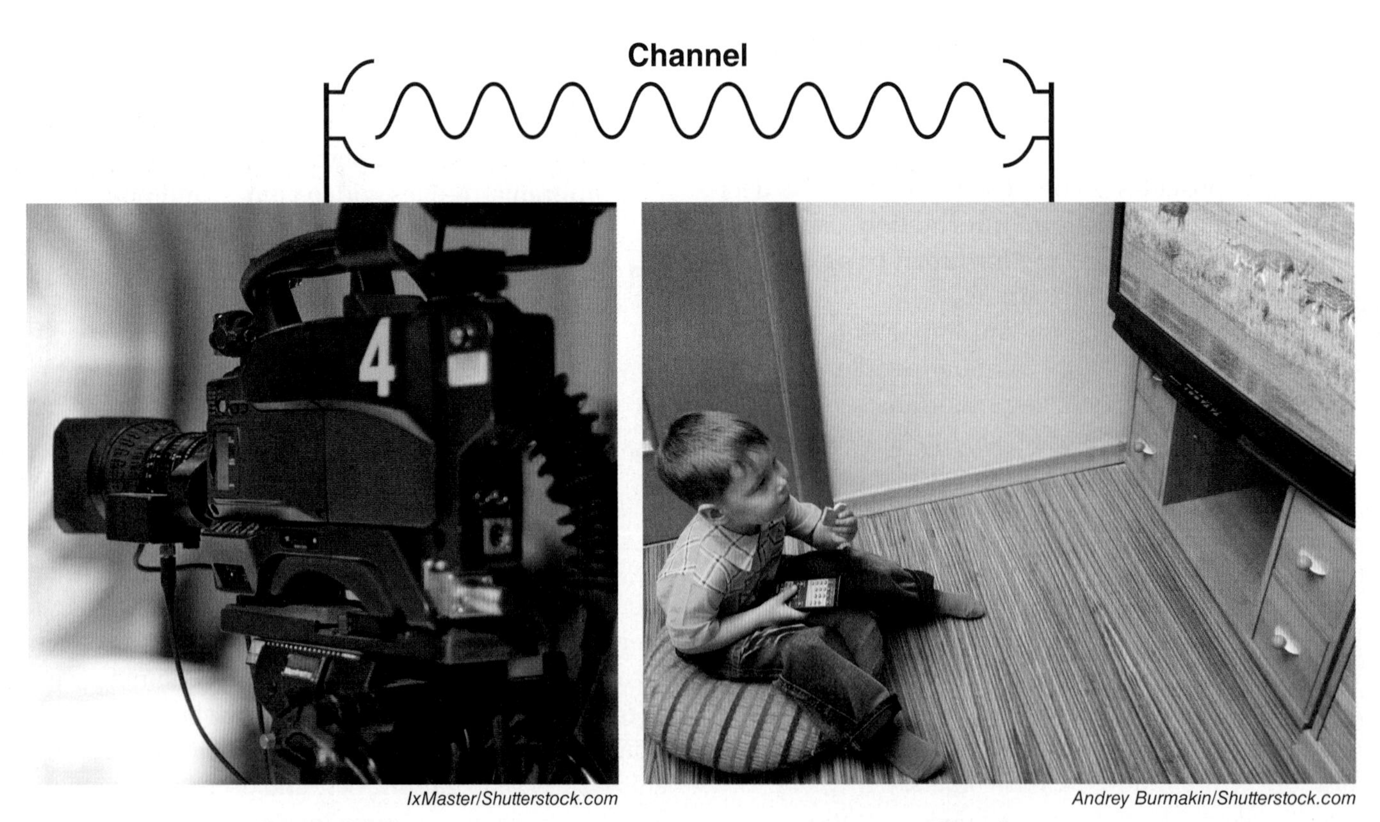

Figure 21-7. A television communication system.

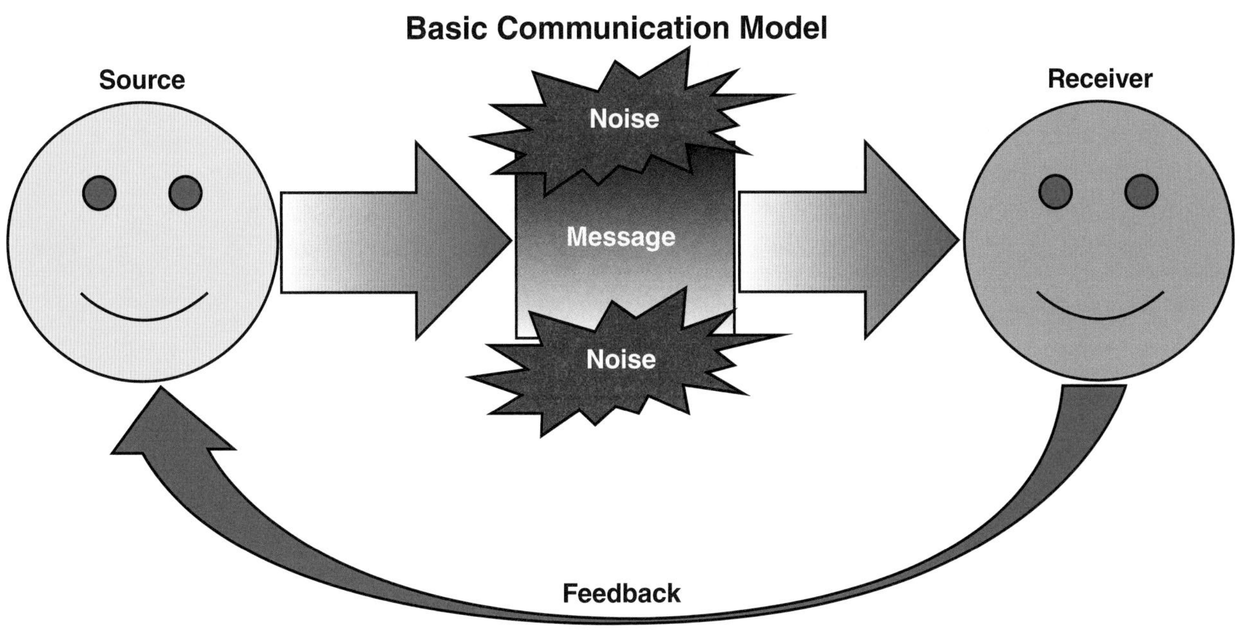

Figure 21-8. The communication model consists of a source of information, information in the form of a message to be sent, potential noise that can impede the delivery of the message, the receiver of the message, and the feedback generated from the receiver of the message.

When you listen to the radio, both information and noise are present. In this example, they are both in the form of sound. The information is the sound you want to hear. The noise is the sound you do not want to hear, such as static. Noise, then, is a type of interference. Also, noise can involve personal taste. Some people call the music you listen to noise. You might feel the same way about the music they prefer. Both of you are correct because any unwanted sound is noise.

The basic communication model shows that the communication process runs in a closed loop system. See **Figure 21-9**. A closed loop system is comprised of inputs, processes, outputs, and feedback. Information

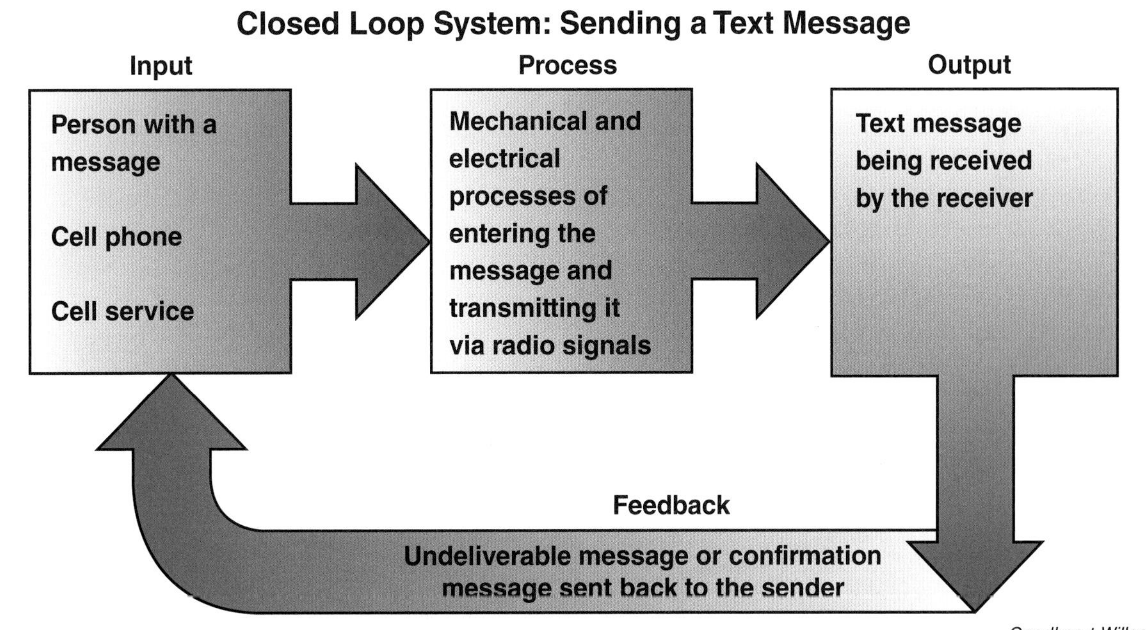

Figure 21-9. An example of sending a text message follows the basic systems model.

and communication systems generally follow this model. For example, sending a text message involves inputs—the phone, the person, and the cell phone service. The process is the mechanical and electrical process of entering the message and transmitting it via radio signals. The output is the end result of the text message being received by the receiver. Finally, the feedback could be either an undeliverable message or confirmation message sent back to the sender.

The basic communication model can be elaborated on to include all of the components of an information and communication technology. These components are a source, encoder, transmitter, receiver, decoder, storage component, and retrieval device. See **Figure 21-10**. In the example of an e-mail, the source of the e-mail is the person typing the message. The encoder is the computer's processor translating the keyboard strokes into the programming language. The transmitter is the wireless connection device linked to the source's computer. The receiver is the router connected to the receiver's computer. The decoder is the receiver's computer translating the binary code to the software being used. The storage component is the receiver's hard drive or e-mail server. The retrieval device is the software application that opens the e-mail.

Information and Communication Systems

Information and communication systems can be classified into four different categories based on their source and receiver. See **Figure 21-11**.

Human-to-human communication works through electronic media and printed products. This type of communication is used to inform, educate, persuade, manage, control, and entertain others.

When you react to the bell indicating the end of a class period, you are participating in machine-to-human communication. Another example of this category is a smoke detector sounding an alarm when there is a fire nearby. This type of communication system is widely used to display machine operating conditions or a diagnostic report.

Keying text into a computer and setting the temperature on a thermostat are examples of human-to-machine communication. This type of communication system starts, changes, or ends a machine's operations.

Computer-controlled operations use machine-to-machine communication. This occurs when a computer directs and controls a device using information it collects through a series of measure-

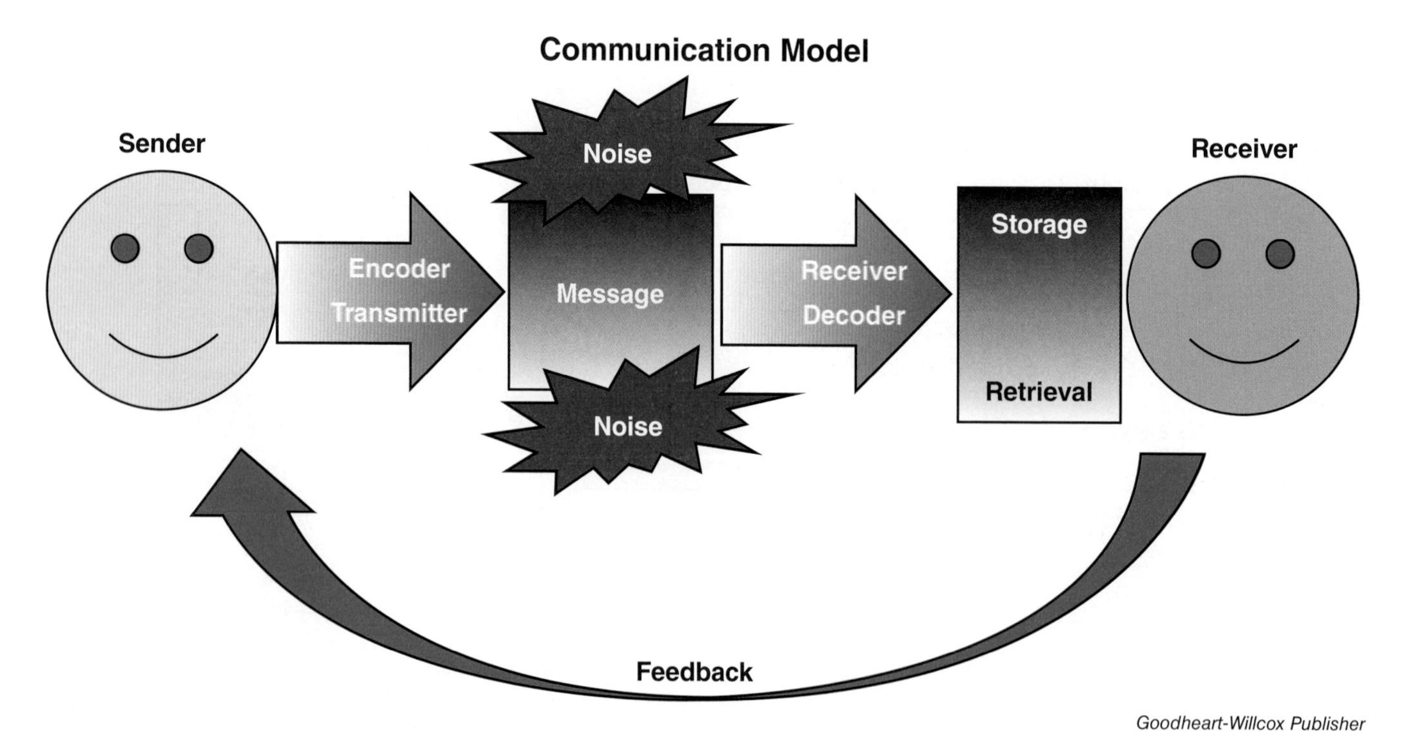

Goodheart-Willcox Publisher

Figure 21-10. The components of an information and communication system.

Types of Communication

Human to Human

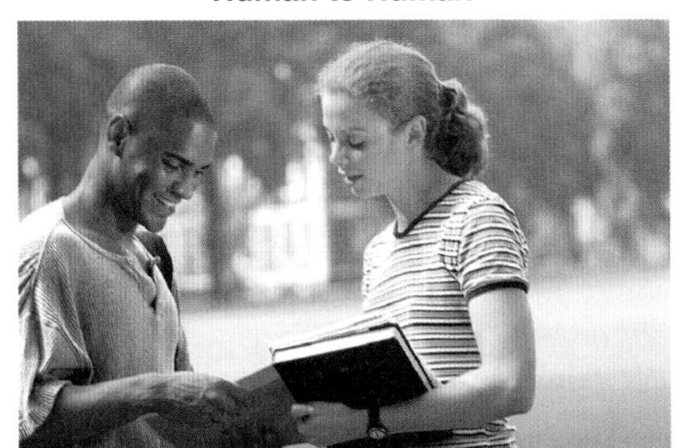

United Parcel Service

Human to Machine

Siemens

Machine to Human

Inland Steel Co.

Machine to Machine

Cincinnati, Inc.

Figure 21-11. The type of communication is based on the sender and the receiver.

ments using sensors. Examples include a computer controlling a printer and a thermostat controlling a furnace. Typical examples of computer-controlled operations are frequently seen in computer-aided manufacturing (CAM), computer-integrated manufacturing (CIM), and robotics.

Information comes in many printed and electronic forms, **Figure 21-12**. The technology used to deliver information can be divided into five main types:

- Printed graphic communication.
- Photographic communication.
- Telecommunication.
- Technical graphic communication.
- Computer and Internet communication.

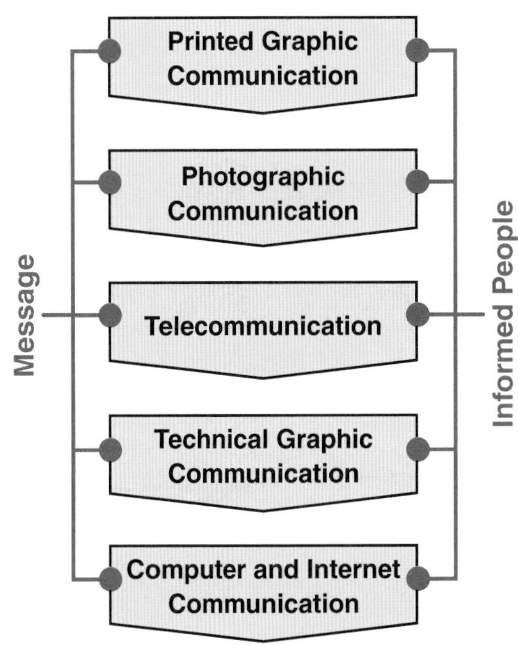

Goodheart-Willcox Publisher

Figure 21-12. We receive information through five basic types of communication-technology systems.

Printed Graphic Communication

Much of information and communication technology was developed to satisfy the need for mass communication. People wanted to tell their messages to large numbers of other people. The first mass-communication system developed was printing, or printed graphic communication. See **Figure 21-13.** The term *printing* originally meant "putting an image on paper with inked type." Most printed communications use an alphabet, or series of symbols developed to represent sounds, to convey the message.

In addition to the alphabet, graphic elements or images, such as drawings, symbols, and icons, are used in printed graphic communication. Combinations of written words and images convey ideas, data, concepts, and emotions. These combinations are designed and laid out to be printed in a manner that visually displays the desired information.

Printing involves putting written words and images on a medium referred to as the *substrate.* Originally, almost all printing was done on paper. Today, a variety of substrates, including paper, glass, plastic, cloth, ceramics, metal, and wood, are used. The result is a broad range of printed products, including newspapers, magazines, books, brochures, pamphlets, labels, stickers, clothing designs, and signs. Each of these is designed and produced by a specific printing process.

Photographic Communication

"A picture is worth a thousand words" is an old saying. This statement suggests that it is often more effective to convey a message visually than to describe it. In some cases, a picture alone is used to communicate an idea or a feeling. In other cases, the picture supplements the written word.

Almost any two-dimensional visual representation is referred to as a *picture.* A photograph is a common type of picture. The act of producing a photograph is called *photography.* See **Figure 21-14.** Using photographs to convey an idea or information is called *photographic communication.*

To distinguish photography from photographic communication, consider the following. Your family takes snapshots at family events and during vacation travels. The main goal of the snapshots is to capture a moment in time. Later, you can look at these pictures and remember those moments. This type of photograph is designed to communicate a historical record.

On the other hand, if you have flown on a commercial jet, you have seen a safety information card. This card contains a series of photographs showing how to fasten the seat belt and exit the plane during an emergency. These photographs are designed to communicate a specific procedure. They are used as photographic communication.

TrotzOlga/Shutterstock.com

Figure 21-13. Books are an example of printed graphic communication media.

Antonio Guillem/Shutterstock.com

Figure 21-14. This photograph communicates specific emotions.

Technology Explained Fiber Optics

Fiber optics: Channeling messages, in the form of light, through glass fibers.

Information has been transmitted with light for many years. One of the first methods used was smoke signals. More recently, flags and flashing lights have been used to convey information. All these techniques, though, are limited to the line of sight between the sender and receiver. Practical use of long-distance light communication came with the invention of optical cables.

Optical cables are channels guiding light waves, through internal reflection, over some distance. Internal reflection means that, when the light waves strike the outer edge of the fiber, they are reflected back toward the center. Optical communication of this type is called *guided optical transmission*. The development of glass fibers for guided optical transmission began in the 1960s.

A typical fiber-optic cable has three layers. The outside layer is a protective plastic coating. The middle layer is called *cladding*. This layer reflects the light waves back into the glass fiber. The inner layer is a strand of glass called the *core*. The individual core strands are as thin as human hairs. Typically, they are about 0.0005″ in diameter. See **Figure A**.

Protective coating
Laser beam
Optical fiber

Goodheart-Willcox Publisher

Figure A. A diagram of a fiber-optic cable.

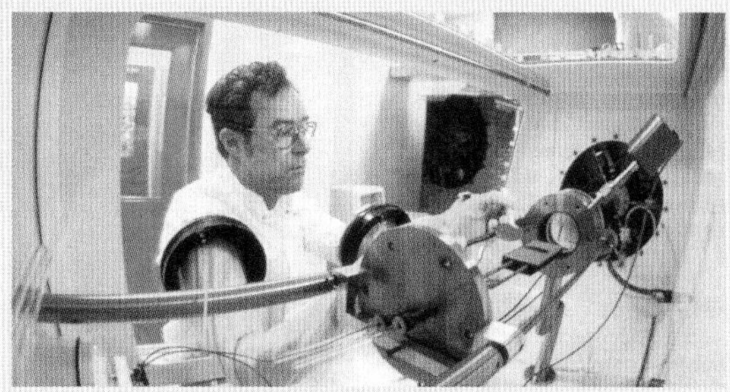

Northern Telephone Co.

Figure B. A worker testing an optical fiber.

Several hundred core strands are bundled into larger cables. Each of these fiber-optic waveguides can carry numerous messages. Billions of bits of information per second can move down an optical fiber. See **Figure B**. This is the same amount of information contained in thousands of independent telephone conversations. Optical fibers are capable of carrying encyclopedias of information per second.

Fiber-optic cables are rapidly replacing copper circuits in telephone systems. They are less expensive to install, smaller in diameter, use less power, and are much more resistant to interference. Phones linked with fiber-optic cable remain unaffected by the flashes of lightning. Fiber-optic systems are especially suited to carrying the digital signals widely used in cable television systems and computer networks.

Fiber optics is finding uses in areas other than communication. Applications include imaging applications, such as bronchoscopes and endoscopes, in medicine, and welding, jet engine, and plumbing-pipe inspections in industry.

Telecommunication

People have always communicated their ideas and feelings. Early communication involved people speaking to one another. Later, writing was developed to record and transmit information. Writing allows us to express and store information, opinions, and concepts. These types of communication, however, do not meet all human needs. People wanted to hear the human voice beyond the limits of face-to-face communication. They wanted to communicate their thoughts and knowledge over great distances. Out of this desire came telecommunication.

Telecommunication is the process of communicating information at a distance. See **Figure 21-15**. Telecommunication includes a number of specific types of communication. Probably the most widely used are radio, television, and the cellular phone. See **Figure 21-16**. This communication implies that a message exists and that hardware (technology) is available to deliver the message.

Andrey_Popov/Shutterstock.com

Figure 21-15. These customer service representatives are using telecommunication to help customers.

VDV/Shutterstock.com

Figure 21-16. Telecommunication means communicating over distance. This satellite-dish receiver is a telecommunication system that aids in capturing messages sent from great distances.

There are scientific principles behind technological devices and systems. The principles of electricity and electromagnetic waves are a part of physics and a key component to understanding telecommunication technologies.

Electrical Principles

In certain situations, electrons travel from one atom to another. (Refer to the discussion of atoms in Chapter 16). This movement, called *electricity*, often takes place in a conductor. Electrons flowing in one direction along a conductor produces direct current (dc). Electrons flowing in both directions along the conductor, reversing at regular intervals, results in alternating current (ac).

Movement of electrons in a conductor creates magnetic lines of force known as ***electromagnetic force***. See **Figure 21-17**. As these lines of force increase and decrease in strength, they can cause electrons to flow in an adjacent wire in a process called induction. Induction is commonly used to change sounds into electrical signals or to change electrical signals into sound. This process is used in microphones and speakers. These technological devices are examples of ***transducers***. Transducers are devices that change energy from one form to another.

Electromagnetic Waves

Understanding electromagnetic waves is vital to telecommunication. Two important characteristics of electromagnetic waves are frequency and amplitude, **Figure 21-18**.

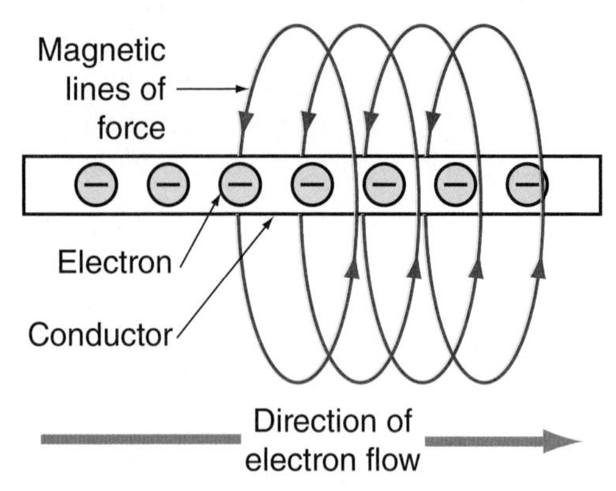

Goodheart-Willcox Publisher

Figure 21-17. Electron movement through a conductor is called *electricity*. This movement creates magnetic lines of force around the conductor.

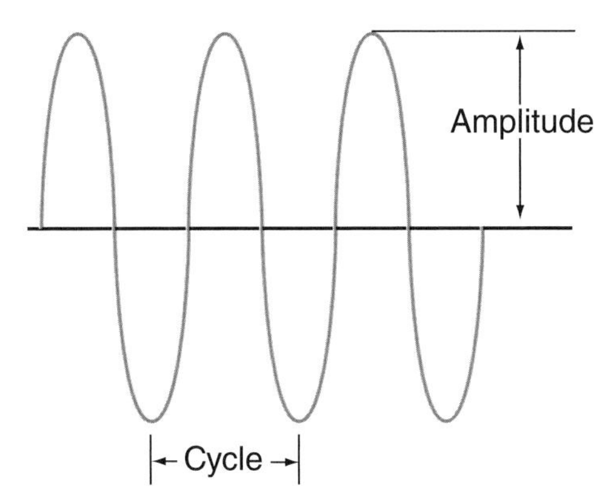

Goodheart-Willcox Publisher

Figure 21-18. The information in radio waves is coded through varying the frequency and amplitude of the wave.

Frequency is the number of cycles (complete wavelengths) passing a point in one second. The number of cycles per second is measured in hertz (Hz). The basic units of measurement in telecommunication are kilohertz (kHz) and megahertz (MHz). One kilohertz (kHz) is 1000 cycles per second. One megahertz (MHz) is one million cycles per second.

Within the electromagnetic spectrum is a series of frequencies called radio waves. These frequencies extend from around 30 hertz to 300 gigahertz (300 billion cycles per second). Below this range of frequencies is a series called extremely low frequency (ELF). This series extends from about 10 Hz to 13.6 Hz. ELF is used for underwater communication.

Radio waves are part of a series of frequencies known as *broadcast frequencies*. These frequencies are used for a wide range of communication systems, including police and fire department radio, broadcast radio, cellular telephone, and television communication.

The Federal Communication Commission (FCC) assigns each type of communication to a range of frequencies. For example, 160.215 to 161.565 kHz is assigned for railroad communication. For 6-meter amateur radio, 50.0 to 54.0 MHz is assigned.

Amplitude measures the strength of the wave. The higher the amplitude is, the stronger the signal is. Telecommunication uses changes in the amplitude of the waves, the frequency of the waves, or both, to carry a communication message.

Types of Signals

Signals are used to carry information from a source to a receiver. All signals contain energy and power. Signals also have a pattern that carries

Academic Connection: History

The Beginning of Photojournalism

The creation of cameras and experiments with photography date back to the 1500s. Not until the mid-1800s, however, did the use of cameras begin to play a major role in how society viewed events. Perhaps the best example of the new power of photographic communication can be seen in the work of Matthew Brady, America's first photojournalist.

When the Civil War broke out in 1861, Brady was a well-known portrait artist. He asked for and received permission to create a photographic record of the war. Brady hired a crew and placed people at sites where soldiers were camped or battles were expected to break out. The resulting photographs were the first visual images ordinary American citizens had ever seen of war as it was occurring.

The images put a human face on the war. People saw pictures of soldiers at their daily chores, as well as pictures of dead combatants lying on the field after a battle. The photographs were dramatic and changed Americans' perceptions. Previous artwork had glamorized and romanticized war. The new images awakened Americans to war's other side, the side that includes both acute boredom and terrible suffering. Can you think of another example of a new technology affecting people's perceptions of reality?

information. The basic types of signals used to communicate information are analog and digital signals.

Thomas Edison developed the first workable device that could record sound in 1877. His machine used a diaphragm attached to a needle to record sound on a foil cylinder. When a person spoke into the diaphragm, it vibrated, causing the needle to scribe a groove into the foil. In the playback mode, the opposite happened. The needle traveled along the groove of the rotating cylinder. The movement of the needle vibrated the diaphragm and reproduced the original sound. Edison's invention was improved over the years and became the common way to record speech and music. This type of recording equipment stores the sound using analog signals.

An ***analog signal*** is a continuous electronic signal carrying information in the form of variable physical values. Information is added to the base signal by amplifying the signal's strength (AM) or varying the signal's frequency (FM). See **Figure 21-19**. AM and FM are discussed later in this chapter.

Formerly, nearly all telephone, television, and radio signals used analog signals because analog signals are fairly easy to create and transmit. However, analog signals are subject to outside forces that can alter the signal pattern, resulting in noise and distortion that make the output different from the input. This major problem led to the development of ***digital signals***, which are finite or discrete, meaning there are a limited amount of values that quantities can have. Digital signals are further discussed in Chapter 22.

Digital technology generates, stores, processes, and transmits data using positive and nonpositive electrical states. The positive state (on) is expressed by a *1*. The nonpositive state (off) is expressed by a *0*. All digital data is, therefore, expressed as a string of *1*s and *0*s.

Electrical signals that are converted into these values of either 1s or 0s are called digital signals. These signals have several advantages over analog signals. Digital signals can be transmitted faster, allowing more information to be moved in a given period of time. They are more accurate and less prone to outside interference (noise and distortion). Digital recordings produce clear sound that is very close to the reproduced signal. They also sound the same no matter how many times they are played. Likewise, digital TV programs have clearer pictures and better sound than analog programs.

Types of Telecommunication Systems

There are many types of telecommunication systems. These systems can be divided into two major types—hardwired systems and broadcast systems. See **Figure 21-20**.

Hardwired Systems

Telecommunication systems have three major parts—a sender, a communication channel, and a receiver. An example of a hardwired system is a landline telephone system, **Figure 21-21**. In this system, the microphone in the mouthpiece changes sound waves into electrical impulses. The frequency

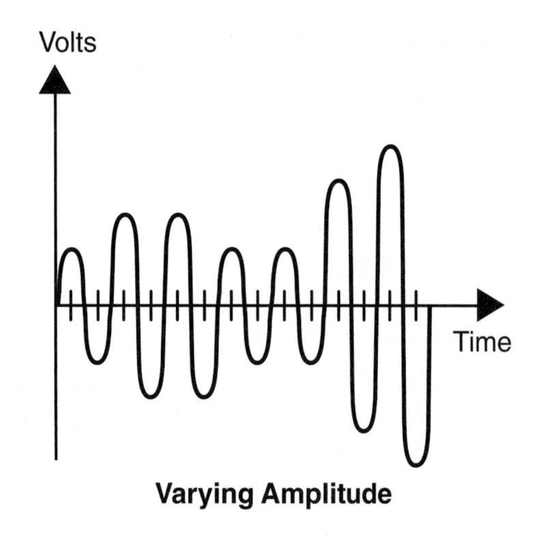

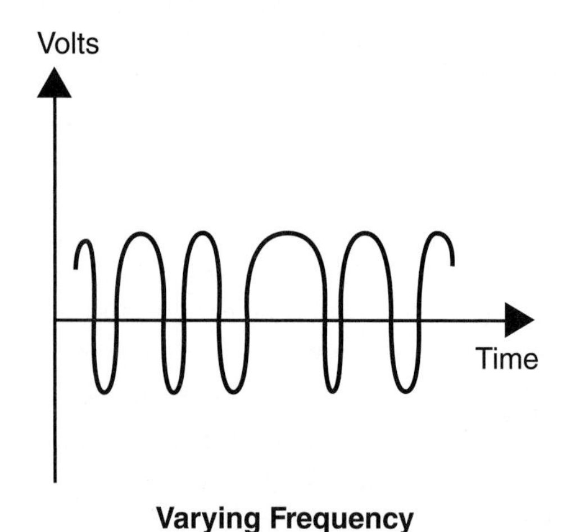

Goodheart-Willcox Publisher

Figure 21-19. An analog signal is a continuous signal varying in strength (amplitude) or in frequency.

Figure 21-20. Telecommunication systems can be grouped as hardwired or broadcast systems. Left—Fiber-optic (hardwired) unit. Right—Communications tower (broadcast).

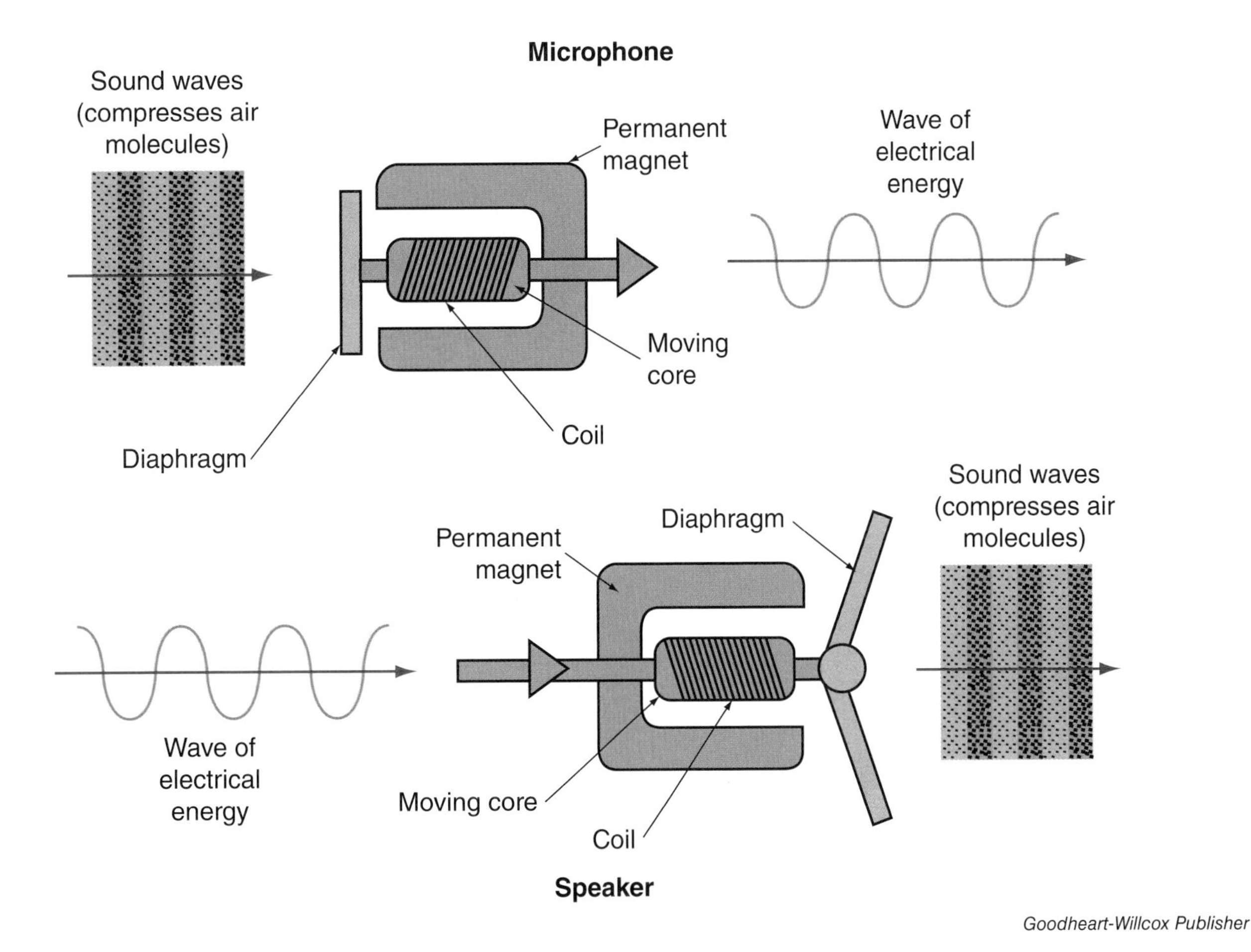

Figure 21-21. Sound waves being transferred into electrical waves by a microphone.

and duration of the electrical impulses are the coded message. These electrical codes are usually conveyed from the sender to the receiver over a permanent waveguide. A waveguide is the physical structure that is used as the channel to carry the message from the sender to the receiver in the form of electromagnetic waves. These guides might be a copper wire or fiber-optic (glass fiber) conductor. In some cases, microwave radio signals take the place of a waveguide for a portion of the circuit.

Think Green Inks

Volatile organic compounds (VOCs) evaporate into the atmosphere, causing environmental and health problems. Conventional inks and ink solvents contain VOCs. The printing industry is taking steps to reduce and even eliminate the amount of VOCs it produces.

As ink dries, it emits chemicals into the atmosphere. One of the first steps taken to reduce VOCs is to change the way inks dry. For instance, UV-curable inks cure instead of dry, so there are no solvents that evaporate into the air. Another step is to change the chemicals used in inks. Vegetable oil-based inks and soy inks are now used as an alternative to conventional inks. These newer inks dry more slowly, but since they do not use petrochemical solvents, they are used in green printing facilities.

Microwaves are often used to send a message between major cities. In these cities, the signals are transferred back onto wire or cable.

Broadcast Systems

Broadcast systems send radio waves carrying the signal through the air from the sender to the receiver. The transmitter (sender) changes sound into a signal containing the message. This signal radiates into the atmosphere from an antenna. Another antenna attached to a receiver gathers the signal. The receiver separates the desired signal from other signals and changes it back into audible sound. See **Figure 21-22**.

Generally speaking, radio signals radiate in all directions from an antenna. Telephone-microwave communication systems use directional antennas to focus the signal to receiving antennas. Two common broadcast systems are radio and television.

Radio communication was the first widespread broadcast medium. Originally, it was called *wireless* because no hardwired connection exists between the sender and the receiver (radio set). In all radio broadcast systems, a carrier frequency radiates from the transmitter. This is the carrier frequency you tune your radio to in order to receive a station. The code for the audible sound is imposed onto this frequency. See **Figure 21-23**.

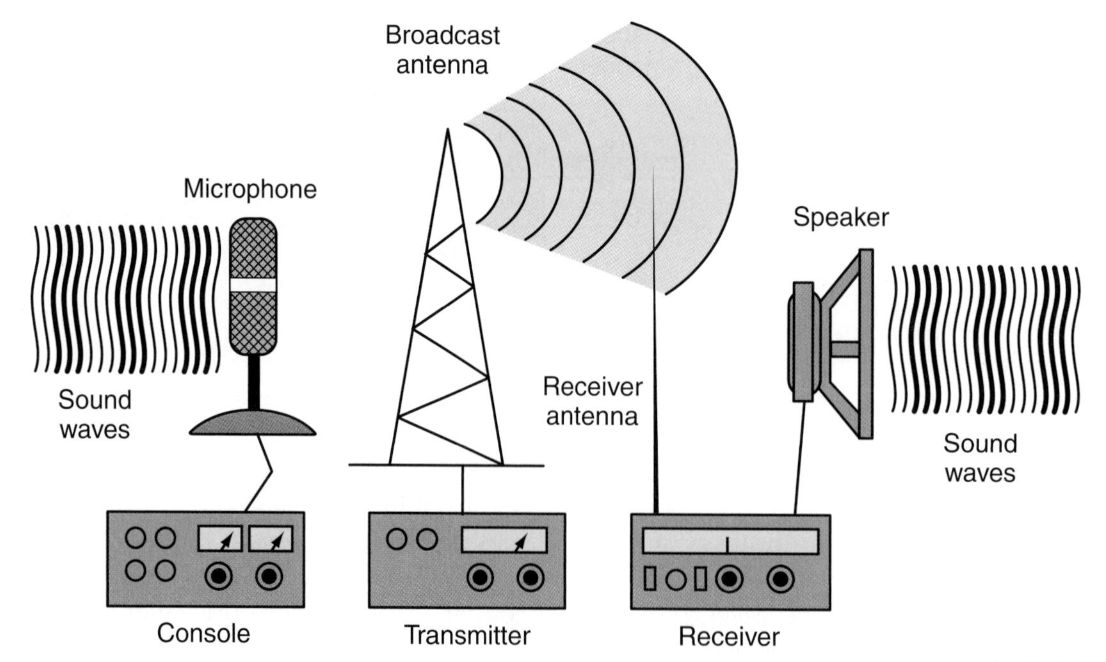

Figure 21-22. A typical radio broadcast system.

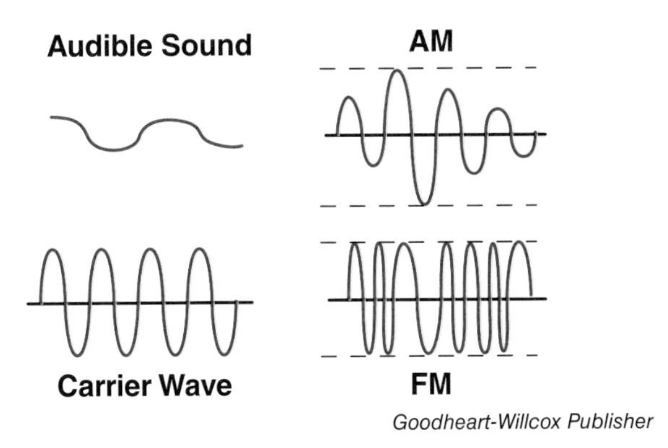

Figure 21-23. Radio waves are changed to carry the message through AM or FM. The amplitudes of the sound wave and the carrier wave are blended in AM. The amplitude of the combined waveform oscillates in a pattern similar to the sound wave. In the frequency modulated waveform, the frequency varies in a pattern with the initial sound wave.

The earliest radios used ***amplitude modulation (AM)*** to code the carrier frequency. These systems merged the message onto the carrier wave by changing the strength of the carrier signal. This type of broadcast radio is assigned the frequencies between 540 and 1600 kHz. Later, radio broadcast systems using ***frequency modulation (FM)*** were developed. These systems encode the message on the carrier wave by changing the wave's frequency. The 200 separate FM radio-broadcast frequencies range from 88.1 MHz to 107.9 MHz.

Television broadcast systems are really two systems in one. Each channel is assigned a bandwidth, which is a set frequency for the signal. See **Figure 21-24**. A radio-like system uses the FM portion of the band to send and receive the audio (sound) portion of the message. The larger portion of the band is assigned to transmit the video part of the program using an AM signal. At the ends of each channel are unused frequencies. These buffer zones keep the signals from one channel from disturbing adjoining channels. Television systems use a microphone to capture the sound and a camera to generate the picture. The television receiver reproduces the program using a speaker for the sound and a screen to display the picture.

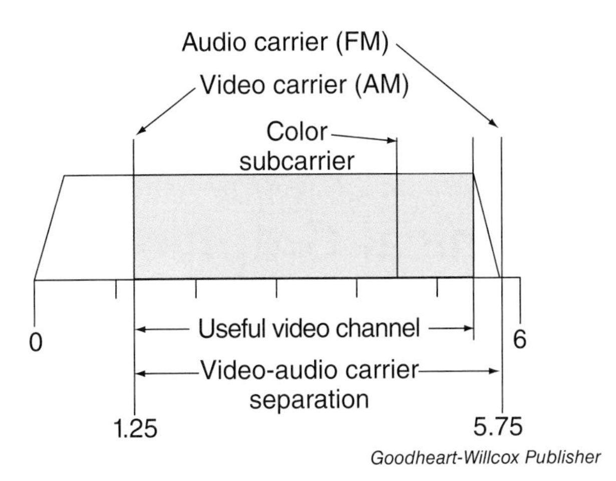

Figure 21-24. This diagram of a traditional (analog) television broadcast channel shows the audio and visual broadcast frequencies.

Cable television systems

For many years, there were three major television networks in the United States. These networks had broadcast stations in major cities that produced signals that could be received only if a home's antenna was in a line of sight from the transmitting antenna. People living in remote and mountainous areas could not receive programs that were becoming a part of everyday American life. In 1948, people living in the valleys of the Pennsylvania mountains could not receive a television signal. They started putting antennas on hilltops and running cables to their houses so they could receive their favorite television programs. This was called a *community antenna television (CATV) system* and was the start of cable television as we know it today. See **Figure 21-25**.

Cable television is a system that transmits signals to televisions through fixed optical fibers or coaxial cables (**Figure 21-26**), as opposed to the through-the-air method used in traditional television broadcasting. Today, cable systems deliver hundreds of channels of television, along with high-speed Internet access.

Satellite television systems

Parts of the country cannot get line of sight reception from television stations and do not have cable television service. This challenge gave rise to satellite television. The common satellite TV systems use an all-digital signal to deliver audio and video. The system's major actions are the following:

1. Programming to be delivered to customers is developed or selected.
2. Broadcast centers receive various programs and beam them to satellites in geosynchronous orbit. This means that the satellite orbits the earth at a speed that matches the rotation of the earth. By doing so, the satellite will remain in the same point of the sky in relationship to

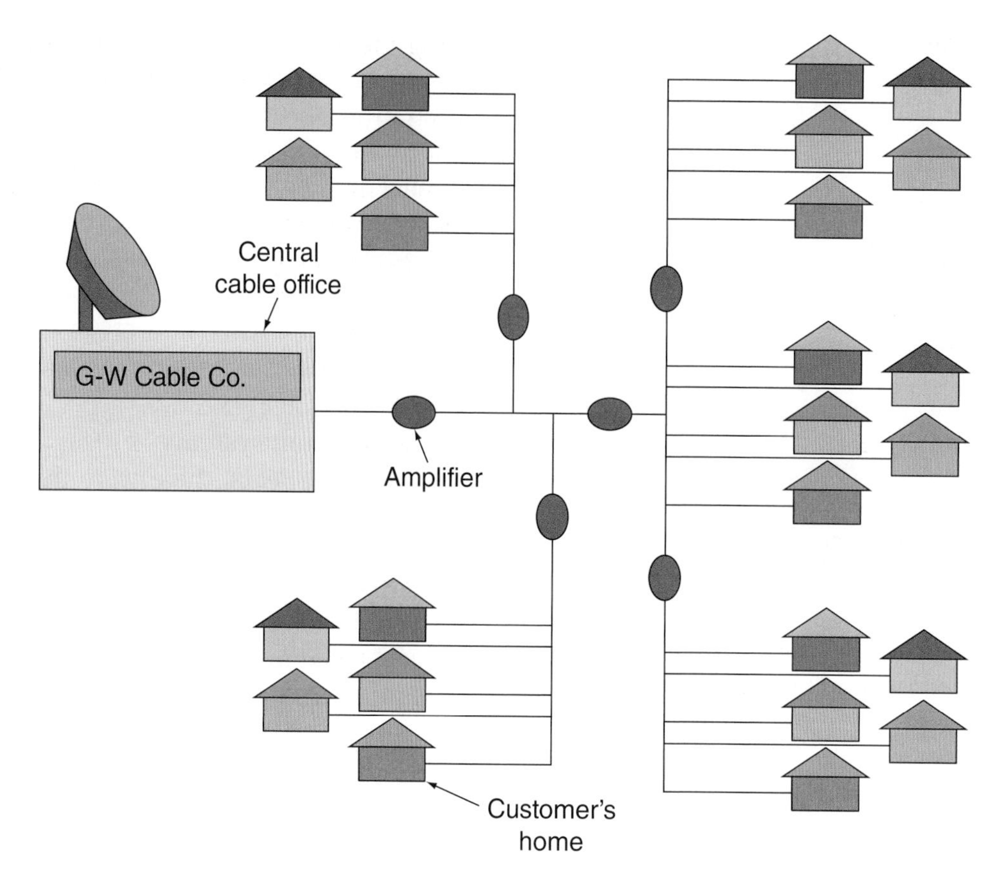

Goodheart-Willcox Publisher

Figure 21-25. This diagram shows a typical cable television system.

zwola fasola/Shutterstock.com

Figure 21-26. Coaxial cables may be used to transmit cable to televisions.

a point on the earth's surface at all times of the day. This enables the broadcast system to continuously send a signal to the same satellite without interruptions due to the earth's rotation and revolution.

3. Satellites receive signals from the broadcast station and rebroadcast them to the customers.
4. Dishes receive the signals from the satellite and relay them to the customer's receiver.
5. Receivers process the signals and relay them to television sets for display.

Technical Graphic Communication

People and companies often want to communicate specific information about a product or the product's parts. This information might convey the size and shape of a part, suggest how parts are assembled to make a product or structure, or indicate how to install and operate a product. Also, it might tell how to adjust and maintain the product. This type of information is often communicated through engineering drawings or technical illustrations. See **Figure 21-27**. Many of the methods for creating these technical forms of communication are discussed in Chapter 5.

Computer and Internet Communication

The world entered the information age with the advent of the widespread use of computers. A ***computer*** is an information-processing machine that performs a number of tasks controlled by a set of instructions called *software*. The software provides instructions for the computer and its applications to gather, process, store, manipulate, and retrieve information for a variety of purposes.

The invention and development of the personal computer rapidly changed how people gather and use information. Computers have enabled the expansion of communication resulting from the networking of computer systems by way of the Internet. The Internet enables computers to share information with one another and allows people to gather information using the World Wide Web and to exchange digital messages. Moreover, the arrival of mobile computing has given people the ability to carry computer systems with them at all times, often in the form of a smartphone, tablet, laptop computer, or even wearable technologies. Mobile computing is the capability of personal computing devices to access, transfer, and process information wirelessly with minimal delay or lag time.

As a result of these developments, computers are now used in all areas of personal and business life, **Figure 21-28**. For example, computers are used to scan the prices of items at the grocery checkout area, maintain financial records, and operate banking machines. They select channels on the television and operate video game systems. Computers maintain temperatures in homes, control microwave ovens, and sequence the cycles in washing machines. They control automobile systems, operate home-security systems, and manage sports scoreboards.

Networks

A ***network*** is a group of computing devices that are connected together and share resources. Through a network connection, a computer can communicate with other computers. Networking allows people to exchange data rapidly and to share output devices and hard disk storage.

Ohio Art Co.

Figure 21-27. This designer is preparing technical illustrations for a new product.

Pressmaster/Shutterstock.com

Figure 21-28. Computers are used in all areas of life.

STEM Connection: Mathematics

Measuring Type

In addition to the customary system and the SI metric system, a third measuring system is used by people employed in the graphic arts. The *American point system* is used to measure such items as the size of type, the lengths of typeset lines, and the space between lines of type.

The two main units of measurement in this system are the point and the pica. A point measures approximately 1/72″. A pica is approximately 1/6″. One pica equals 12 points, and six picas equal approximately 1″.

Generally, points are used to measure the vertical height of type sizes, and picas are used to measure the lengths of lines. The vertical size of a line of type is the measurement from the tops of the letters (ascenders) to the bottoms (descenders). For example, the word *deep* is measured from the top of the *d* to the bottom of the *p*. Line length is measured from the beginning of the first word of the line to the end of the last word of the line.

Those who work with type use a typographer's rule to measure points and picas. Locate one of these rules and use it to measure a line of type in this paragraph. How many picas long is the line? What size type is being used?

Two basic types of networks exist. The first is a local area network (LAN), which is generally used in a single building or site. A LAN connects several personal computers, or workstations, to a server. See **Figure 21-29.** The ***server*** is a device or computer program that stores, processes, and shares data and resources with other devices or clients connected to the network. The second type of network is a wide area network (WAN). These networks cover large geographical areas. They are used to connect computers in distant cities and countries. The largest WAN is the Internet, which enables cloud computing. Cloud computing is a network that uses the Internet to connect to remote servers to store, manage, and process data rather than doing so with the computer's own hardware.

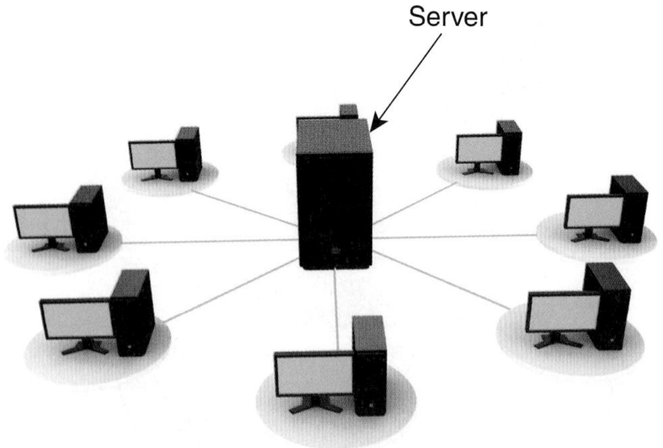

Server

Centurion Studio/Shutterstock.com

Figure 21-29. A LAN connects personal computers to a server.

The Internet

The term *Internet* means "interconnected networks." The Internet is a global communication network that connects individual computer networks. No one really owns it because much of its creation and initial workings emerged from federally funded research. The Internet is a collection of large and small independent networks called *intranets.* Each of these intranets can link hundreds of computers in a company or another institution. The Internet connects all these intranets so they can share selected information. This system provides inexpensive and efficient communication for people all over the world.

Every computer connected to the Internet is part of a network. A home computer is connected into a network formed by the Internet service provider (ISP). The Internet can be described as a network of networks. This complex network is a collection of backbones, access points, and routers.

Large communications companies have built their own communication lines called *backbones*. These ***backbones*** are typically fiber-optic lines that connect regions in the company's information technology (IT) systems. Connection points in each region, called a ***point of presence (POP)***, allow local users to access the company's network. The POP is often a local or toll-free phone number or a dedicated line. The various communication companies connect their systems at network access points (NAPs). NAPs allow a customer of one company to connect with a customer of another company. Many large ISPs interconnect at NAPs in various cities. Therefore, the Internet is a collection of large corporate networks that agree to intercommunicate at the NAPs.

The Internet also contains many ***routers***. These devices determine how to send information from one computer to another.

Internet Access

To use the Internet, an individual must first have access. *Internet access* refers to the way a computer is connected to the Internet. This access is provided in several different ways.

The first and almost obsolete type of access is *dial-up*. In this system, computers are connected to the Internet through a modem. A ***modem*** is a device that can convert data into signals a telephone system can recognize. This device also converts these signals back into data. It uses special software to place a telephone call to a company that provides Internet service (the ISP). See **Figure 21-30**. The ISP modem answers calls from the user's modem. This modem receives and transmits the signals through the telephone lines.

Another type of access is DSL (digital subscriber lines). DSL provides Internet access by using telephone lines to transmit a signal. However, the signal that is transmitted is at a higher frequency than the voice signal for the telephone. Therefore, both voice and data are transmitted at the same time. This enables both computers and telephones to operate using the telephone lines without interfering with one another. DSLs are in decline due to slow data transfer speeds.

Cable is a more popular and faster method of accessing the Internet. Cable modems use cable television coaxial lines to connect to the Internet. Much like DSL, cable modems send data over the wires using unique frequencies or channels. This keeps the Internet communication from interfering with other television signals on the wires or cables.

Fiber-optic cables provide another way to access the Internet. These cables are made of small flexible glass or plastic wires that transmit data in the form of light pulses. The receiver detects the light pulses and generates an electronic signal for the computer to read. High Internet speeds are possible because light travels faster than the electronic signals that the coaxial and telephone lines use.

Wireless Internet access is widely used. The term *wireless* refers to any method in which the Internet service provider delivers Internet access without the use of cables or telephone wires. These methods include the following:

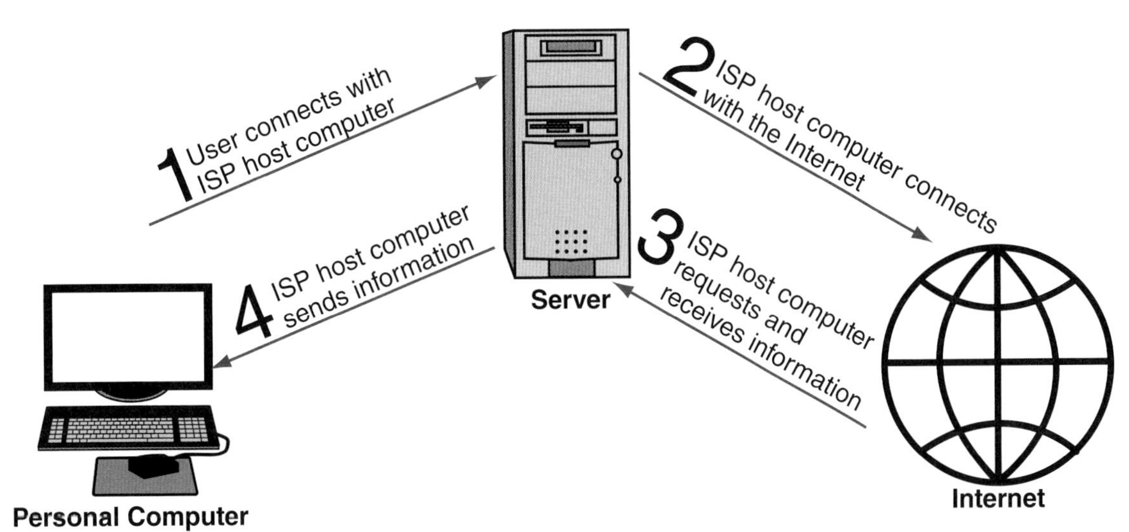

Figure 21-30. Through an ISP, an individual can use a personal computer to access information from Internet sources.

- **Wireless broadband Internet.** An Internet service provider broadcasts Internet connection from a cabled source using radio waves.
- **Mobile Internet.** This type of access works in a similar manner as wireless broadband service. Access for mobile devices is provided through cellular phone service.
- **Satellite Internet.** The Internet signal is sent back and forth between the service provider and computer using a satellite in orbit around the earth as a hub for transmitting the signal. Since the satellite is located above the earth, it can reroute signals across great distances. However, the user's satellite dish must always have a clear line of sight to the actual satellite. This means clouds in the sky can block the signal on occasion. Furthermore, the signal must travel a great distance to the satellite and back, which results in typically slower Internet speeds for homes and businesses. Therefore, satellite is mostly an option for the more rural locations.

Internet Protocol Address

Every computer connected to the Internet has its own identifying number called an ***Internet Protocol (IP) address.*** IP is the computer language used to communicate over the Internet. A typical IP address is a series of numbers, such as 12.225.103.96. The numbers are used to create groups of IP addresses that can be assigned to specific businesses, government agencies, or other entities. These entities include individuals.

An IP address contains two sections. The net section identifies the network to which the computer is connected. The host section identifies the computer on the network.

Internet Domains

When a person uses the Internet, they use a ***domain name*** to locate information. See **Figure 21-31.** A domain name is an address that identifies the location of a web server used to store information for a website or an e-mail account. For example, when you want to access a website, type the domain into your web browser. So if you were to look for ITEEA, you would type in the domain name *http://www.iteea.org/.* Domain names actually mask the true technical address for the web server called an IP address. The domain name is also the root of the more detailed web address for each component of the website that is housed on the web server.

Goodheart-Willcox Publisher

Figure 21-31. Many individuals and organizations have Internet domain names and e-mail addresses.

The more detailed address is called the uniform resource locator (URL). The URL is used to locate more specific web pages, documents, or other items on the World Wide Web. A URL is a series of characters identifying the type and name of a document. This locator also includes the domain name of the computer on which the page is located. For example, if you wanted to find a specific linked page on ITEEA, you would type in the domain name (*http://www.iteea.org/*) with the addition of that link (*http://www.iteea.org/link.page*). For e-mail, you also have an address. For example, an e-mail address on the *https://iteea.org/* domain could be *firstname.lastname@iteea.org*.

The domain name server (DNS) translates domain names. This server changes the name people can read into machine language a computer can read. A domain name has three or four segments. The first series of letters (working backward) is called a top-level domain. In four-segment codes, such as http://www.technology.org.uk, the group of letters on the far right identifies the country. (For example, *uk* means "the United Kingdom," and *au* means "Australia.") The next set of letters from the right identifies the type of site. There were seven original types:

- **com.** Commercial organizations.
- **edu.** Four-year colleges and universities.
- **gov.** Government institutions.
- **int.** International organizations.
- **mil.** Military groups.
- **net.** Organizations directly involved in Internet operations.
- **org.** Organizations other than those directly involved in Internet operations.

A group called the Internet Corporation for Assigned Names and Numbers (ICANN) assigns domain names. ICANN has added more codes because of the need to increase the range of top-level domains available. These codes include aero (for the air-transport industry), biz (for businesses), and info (for unrestricted use). Most likely, even more codes will be developed to allow for new domain names.

In three-segment names, such as http://www.iteea.org, the country code is not used. Instead, the name ends with the three-letter type code. Following from right to left, after the top-level domain code is a second-level domain code. This code identifies the specific organization, agency, or business. In the ITEEA Web address, the "iteea" is the organization identifier.

The left segment of the address contains the host name. In the ITEEA example, the "www" is the host. The host specifies the computer at the site that will receive the message.

The prefix to the domain name (the part of the URL before the colon) indicates the format used to retrieve the document. Hypertext transfer protocol, or the prefix *http*, means the document is on the Internet. If instead, you see the prefix *ftp*, it means the document can be accessed through the ***file-transfer protocol***. This protocol allows the user to retrieve and modify files on another computer connected to the Internet.

Using the Internet

People use the Internet for various activities. Four common uses are the World Wide Web (WWW), e-mail, social media, and electronic commerce (e-commerce).

The World Wide Web (WWW)

The ***World Wide Web (WWW)*** is a computer-based network of information resources. This network is often called the *Web*. The Web was developed in 1993 to allow people to view information and images on the Internet. The Web provides companies, universities, government agencies, and other organizations and individuals a way to present information. Most information on the Web is free; however, some sites charge a subscription fee for user access.

The Web allows individuals to read text, view digital images and video, listen to sounds (music, speeches, and audio books, for example), and access multimedia presentations. The following are some Web features:

- **Web browsers.** Sites on the Web are accessed using a Web browser, **Figure 21-32**. A ***browser*** is a software program that acts as an interface between the user and the Web. Examples include Safari® or Firefox®. Browsers know how to find a Web server on the Internet. They can request a page and deliver it to a personal computer. Finally, browsers format the page so it is correctly displayed on the computer monitor.

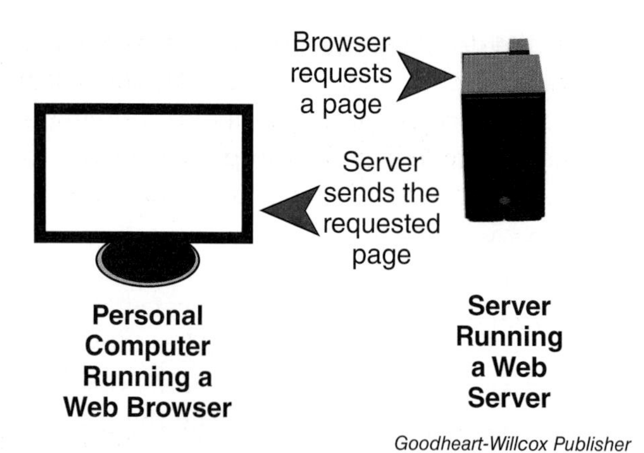

Goodheart-Willcox Publisher

Figure 21-32. People access the Internet through a Web browser.

- **Web servers.** As noted earlier, servers are special computers used to store programs and data for the network. The organizations that develop and share information operate Web servers. The servers hold Web documents and related media. They contain computer software that can respond to a browser's request for a page. These servers deliver the selected page to the Web browser through the Internet. Each document on the server has an address. This address is also a URL. If a site has 100 Web pages, it has 100 URLs, a distinct URL for each Web page.
- **Web pages.** A Web page is a text file created to share information or ideas. This page contains the text of the message and a set of ***hypertext markup language (HTML) tags***. These tags or codes tell the receiving computer how a page should look. See **Figure 21-33**. The tags allow the developer to specify fonts and colors, create headlines, format text, and present graphics on a page.
- **Links.** URLs and HTML tags allow websites to connect a Web page with other pages and websites. These connections, called ***hyperlinks*** or simply links, are underlined phrases, buttons, or other means that can be selected. They allow the operator to point and click on the link to be connected to the selected site or page. Hyperlinks allow users to move between Web pages in no particular order.

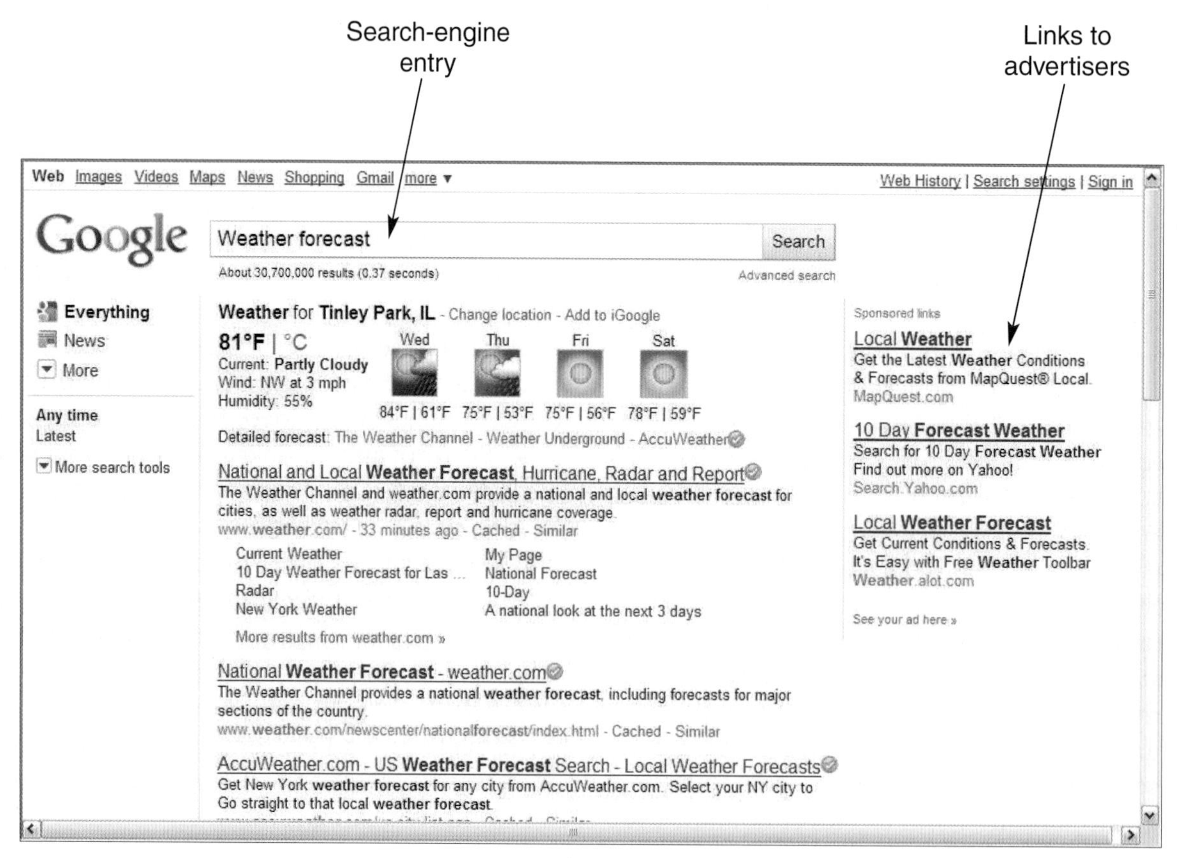

Goodheart-Willcox Publisher

Figure 21-33. The browser page for a search engine.

- **Search engines.** Often, people want to find information about a specific subject but do not know where this information is on the Web. ***Search engines*** are special sites on the Internet that operate on the principle of key words to allow individuals to search the Web by topic. The search-engine operators prepare and maintain an index of the major Internet sites using these keywords. Many different search engines exist.

E-Mail

Every day, billions of personal messages travel over the Internet via e-mail. This computer-based communication system allows an individual to send a message to another computer or a number of different computers. E-mail clients or programs, such as Outlook®, allow a person (client) to interact with an e-mail server computer to compose, send, receive, read, and reply to messages.

The e-mail systems most people use have two servers. The *SMTP e-mail server* handles outgoing mail and operates on the simple mail transfer protocol, which are a set of rules for e-mail transmission. The second server is either an *IMAP* or *POP3 server*, which handles incoming mail. IMAP (Internet Message Access Protocol) or POP3 (Post Office Protocol 3) are standard rules used by e-mail clients or programs to retrieve e-mail from the e-mail server. These protocol allow the computers handling incoming and outgoing mail to talk to one another. They allow the computers to become a communication system.

Social Networking

A ***social networking service*** is an online platform that allows people to share information with one another in a variety of ways on multiple devices. Social networking services include LinkedIn®, Facebook®, Twitter™, Pinterest, Instagram, YouTube, and Wikipedia. Social networking enables people to post and share messages, pictures, and videos, either publicly or privately. Social networking sites enable electronic conversations between two or more people in the form of real-time chat. Users can facilitate ongoing public discussions about particular topics and create discussions, forums, or blogs. Many people subscribe to different social groups, forums, blogs, or discussion boards to stay in constant communication with different groups of people.

Electronic Commerce (E-Commerce)

People have participated in commerce for hundreds of years. Commerce is the exchange of goods and services for money. Since the dawn of the Internet, a new type of commerce called *e-commerce* has developed. E-commerce involves selling products and services over the Internet. This system connects the buyers who want to purchase a good or service and the sellers who have goods and services to sell.

An e-commerce site on the Internet must provide the essential elements of commerce. The first element is merchandising. The product must be displayed so the potential customer can view it. Pictures, videos, and text displays describe the features of the product. Often, a series of links allows the customer to narrow the choices until a single product is selected. Second, the customers must be able to select the items they want to buy. The site generally provides an order form so the customers can purchase the products. Third, the site must have a way to collect money and issue a receipt. Generally, the site provides a way for the customer to use a credit card or an online payment system to pay for the product. An electronic receipt is usually e-mailed to the customer confirming the order.

The e-commerce company must then fulfill the order. The merchandise must be sent to the customer. The company might enable the customer to download their product or send packages through the US Postal Service or by some private package service. Recently, autonomous aircraft called *drones* have been used to test package delivery. A way to return unwanted, damaged, or defective merchandise must also be in place. Finally, the company should provide technical support for complex products.

Summary

- ✔ Information and communication technologies are the systems and products that extend the ability to collect, analyze, store, manipulate, receive, and transmit information.
- ✔ Advancements in information and communication technologies have classified our current era as the information age.
- ✔ Other technologies depend heavily on information and communication technologies to operate. Information and communication technologies are used to control other technological systems.
- ✔ Information and communication technology is the use of a technical means to convey information.
- ✔ Communication is the passing of information from one location to another or from one person to another. Communication is used to convey ideas, exchange information, and express emotions.
- ✔ The goals of communication are to inform/educate, persuade, entertain, and control, and manage the actions of others.
- ✔ The components of an information and communication technology are the source, encoder, transmitter, receiver, decoder, storage component, retrieval device, and receiver.
- ✔ The main types of technology used to deliver information are printed graphic communication, photographic communication, telecommunication, technical graphic communication, and computer and Internet communication.
- ✔ Telecommunication is the process of communicating information at a distance. Telecommunication technologies include radio, television, and the cellular phone.
- ✔ The principles of electricity and electromagnetic waves are a part of physics and a key component to understanding telecommunication technologies.
- ✔ Telecommunication systems can be divided into two major types—hardwired systems and broadcast systems.
- ✔ A network is a group of computing devices that are connected together and share resources. The two basic types of networks are local area networks (LANs) and wide area networks (WANs).
- ✔ The Internet is a collection of large and small independent networks called *intranets*.
- ✔ The Internet is a computer-based network of information resources. Four common uses are the World Wide Web (WWW), e-mail, social media, and electronic commerce (e-commerce).
- ✔ Internet access methods include cable, DSL, fiber-optic, and wireless.
- ✔ The term *wireless* refers to Internet access delivered without the use of cables or telephone wires. Wireless methods include wireless broadband Internet, mobile Internet, and satellite Internet.

Check Your Engineering IQ

Now that you have finished this chapter, see what you learned by taking the chapter posttest.
www.g-wlearning.com/technologyeducation/

Test Your Knowledge

Answer the following end-of-chapter questions using the information provided in this chapter.

1. Information and communication technologies are the systems and/or products that extend the ability to collect, analyze, store, manipulate, receive, and _____ information or data.
2. Communication is often used to convey ideas, exchange information, and express _____.
3. What is meant by the term *infotainment*?
4. *True or False?* All communication systems have transmitters, channels, and receivers.
5. An unwanted sound or signal is called _____.
6. List the four categories of communication based on the source and receiver.
7. _____ is the process of communicating information at a distance.
8. The term _____ is used to describe electrons flowing in one direction along a conductor.
9. *True or False?* A digital signal is a continuous electronic signal carrying information in the form of variable physical values.
10. List the two major types of telecommunication systems.
11. Name the two common types of computer networks.

12. What is a *modem*?
13. What do the two sections of an IP address identify?
14. Define the term *URL*.
15. *True or False?* The prefix *http* in a domain name means the document is on the Web or Internet.

Critical Thinking

1. Select a piece of information or an emotion you want to communicate. Describe four ways you can communicate it and the communication system you would use (printed graphic communication, photographic communication, telecommunication, technical graphic communication, or Internet communication).
2. Develop a communication message persuading a person to your point of view on an issue.

STEM Applications

1. Research a new communication technology and create a technical poster presenting the science behind the system to describe to your class how it actually transfers and receives information.
2. Design a layout for an e-commerce page to sell a new product that you create. Sketch what each page will contain and the links you will use so people can navigate your site. Then use free website creation tools to create your website.

Engineering Design Challenge

Public Service Announcement for Engineering

Background

Information and communication technology is used to deliver information, project ideas, and generate feelings. Often, these goals are incorporated in advertising. Communication can promote a product or an idea by delivering information and persuading people to act in a certain way.

Situation

You are employed as an advertising designer for Breckinridge and Rice Agency. As a public service effort, one of your clients wants to promote high school student interest in engineering and asks you to create an advertisement for a magazine.

Materials and Equipment

- Computer.
- Internet access.
- Graphic design software.
- Digital camera/smartphone.
- Printer.

Procedure

Design a full-page (6 3/4″ × 9 1/2″) magazine advertisement promoting engineering. The ad should encourage students to select at least one class in technology and engineering education to help them understand engineering as it impacts their lives. Develop a theme and prepare a layout. Specify the type size and style, ink colors, and type of and location of the photograph for your design. Then do the following:

1. Print the advertisement as a flyer that can be sent home with students in your school.
2. Convert the layout to a poster. Print copies. Post them on bulletin boards in your school.

TSA Modular Activity

This activity develops the skills used in TSA's Webmaster event.

Promotional Design

Activity Overview

In this activity, you will develop and use the skills necessary to effectively design, build, and launch a website.

Materials

- A computer.
- Web design software.
- Hosting service.
- FTP client.

Background Information

- **General.** Design, build, and launch a website. Consider the functionality of your website as you develop your project.
- **Elements.** The design should feature your school's career and technology/engineering program, the TSA chapter, and the chapter's ability to research and present a given topic pertaining to engineering.

Guidelines

- Your website must be launched on a web server that can be accessed via the Internet 24 hours a day, seven days a week.
- Your website must consist of original web pages:
  - Promote the school's career and technology/engineering program.
  - Promote the school's TSA chapter.
  - Link to the TSA chapter main page.
- Framework systems may be used, but prebuilt templates are not allowed.
- Any copyrighted material, such as text, images, or sound from other sources, must be properly cited with written permission.
- The website must function properly on all browsers and across all devices.

Evaluation Criteria

Your project will be evaluated using the following criteria:

- Design and originality; layout and navigation; graphics and color scheme.
- Career and technology/engineering content and local chapter information.
- Website compatibility with different browsers and devices.
- Appropriate use of Internet and web-based applications.

CHAPTER 22

Meeting Needs through Electrical, Computer, and Software Engineering

Check Your Engineering IQ 

Before you read this chapter, assess your current understanding of the chapter content by taking the chapter pretest.
www.g-wlearning.com/technologyeducation/

Learning Objectives

After studying this chapter, you should be able to:

- ✔ Identify the different engineering disciplines that make electronic devices and communication technologies possible.
- ✔ Identify the components of an electrical circuit.
- ✔ Analyze electrical circuits using Ohm's law.
- ✔ Differentiate between series, parallel, and series-parallel circuits.
- ✔ Explain Kirchhoff's laws of voltage and current.
- ✔ Explain the difference between analog and digital signals.
- ✔ Explain the purpose of logic gates and differentiate between the different types.
- ✔ List the four main units of computer hardware.
- ✔ Understand the function of algorithms and programming languages in the development of software.

Technical Terms 

algorithm
alternating current
anode
apps
batteries
binary code
breadboards
capacitors
cathode
central processing unit (CPU)
circuit diagrams
circuit theory
combinational logic
computation
computer engineering
computer science
current
digital logic
diodes
direct current
electrical circuits
electrical engineers
electrical switch
electrolyte
electromagnetic induction
hardware
integrated circuits (ICs)
Kirchhoff's current law
Kirchhoff's voltage law
light-emitting diode (LED)
memory
Ohm's law
operating systems
parallel circuit
photocells
power source
programming language
rectifier
register
resistors
sequential logic
series circuit
series-parallel circuits
signal
software
software engineering
transistors
voltage

Close Reading

As you read this chapter, identify the ways in which the concepts presented are used in electrical, computer, and software engineering disciplines.

Different types of professionals work together to make electronic devices and communication technologies possible. These include electrical, computer, and software engineers; computer scientists; and other professionals within the fields of information technology and information systems. These professionals work closely together to develop new and faster ways to store, transfer, process, and receive information. This chapter describes the differences between these fields and provides a broad overview of these professions. See **Figure 22-1**.

science photo/Shutterstock.com

nd3000/Shutterstock.com

Figure 22-1. Engineering students repairing an electronic device and programming a robot.

Defining Disciplines

The professions listed in the above paragraph are interrelated but have different purposes. A simplistic way to understand the differences of these fields is to examine computer systems. Computer systems include both physical components (hardware) and operating instructions (software). Electrical and computer engineers develop computer hardware, while software engineers and computer scientists design computer software or programs. Software directs the computer to perform specific tasks, such as writing text, maintaining financial records, or performing calculations. Computer engineers design software too; however, this software is mostly specific to operating the hardware. Furthermore, electrical engineers do more than develop computer hardware. They can be involved in developing any equipment that uses, generates, or transmits electricity. See **Figure 22-2**.

Two other fields are related to electrical, computer, and software engineering and computer science. These fields are information technology and information systems. Professionals in these fields are often involved with the installation, operation, and maintenance of the products and systems developed by engineers and computer scientists. They apply products and systems to best meet the needs of their employer.

Electrical Engineering

Electrical engineers design, test, and oversee the production of devices that use electricity to control, process, store, and receive information, as well as equipment for generating and transmitting power. See **Figure 22-3**. These professionals use mathematics and science combined with design skills to develop large power grids for cities, communication systems, computer systems, automated controls, robotics, motors, generators, health-care equipment, and other electronic or microelectronic devices. Electrical engineering is often considered one of the broadest disciplines of engineering.

Computer Engineering

Computer engineering involves the development, improvement, and supervision of the production and maintenance of computer hardware and

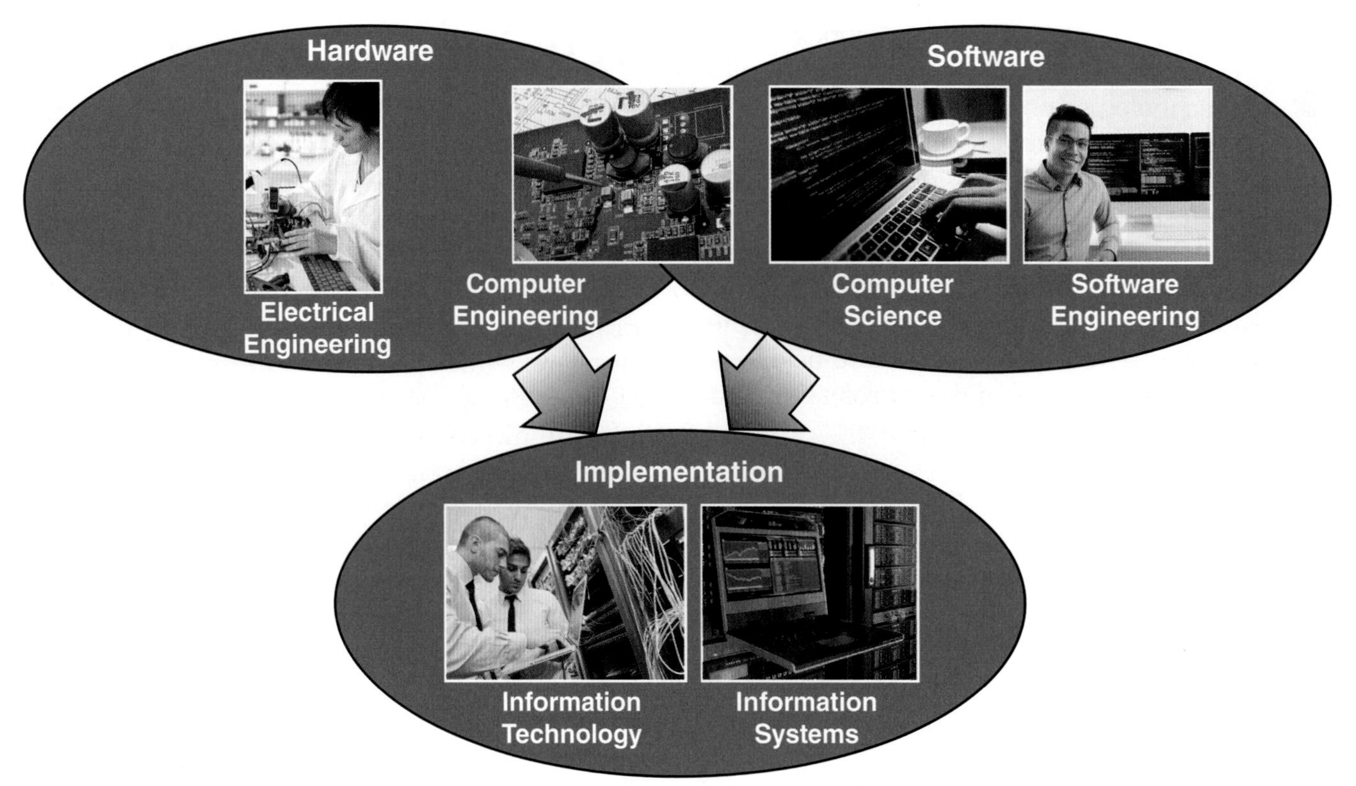

anyaivanova/Shutterstock.com; Vydrin/Shutterstock.com; welcomia/Shutterstock.com; Dragon Images/Shutterstock.com; dotshock/Shutterstock.com; Sashkin/Shutterstock.com

Figure 22-2. Engineers, computer scientists, and IT professionals work together to design, develop, and implement computer hardware, software, systems, and networks.

Photovoir/Shutterstock.com

Figure 22-3. An electrical engineer taking measurements.

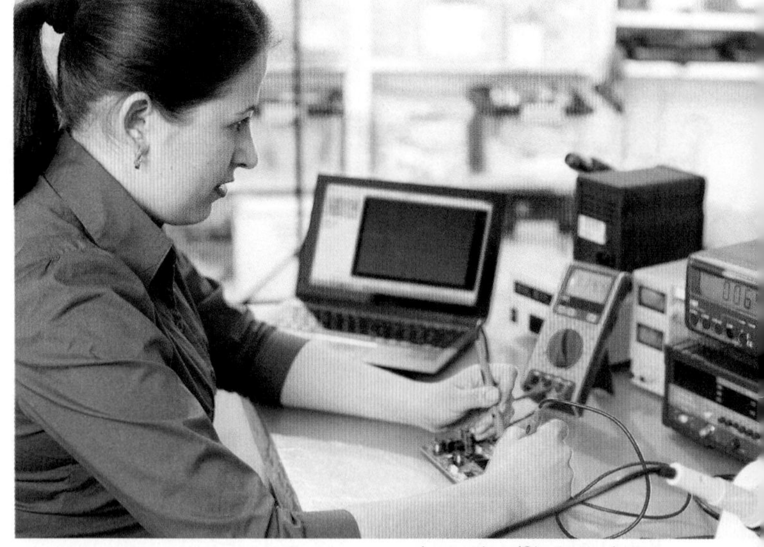

science photo/Shutterstock.com

Figure 22-4. Computer engineering combines electrical engineering and computer science.

software. Therefore, computer engineering is essentially a combination of electrical engineering and computer science. See **Figure 22-4**.

Students preparing to become computer engineers typically take courses from both electrical engineering and computer science. Computer engineers focus on the development of computer hardware that forms the architecture of computer systems and software that is specific to the operation of the hardware, such as device drivers and

Think Green Electronic Waste

Electronics quickly become obsolete, and many are being upgraded on a regular basis. If you invest in an upgrade for an electronic device, what do you do with your older technology? What happens when you throw out DVDs, computers, or smartphones? These devices likely end up in a landfill.

Any type of electronic technology that is thrown out is called *e-waste*. This includes computers and related technology; audio technology; and video technology. E-waste is one of the fastest growing causes of toxic waste in the United States. Each of these electronic devices contains hundreds or thousands of components that are made up of toxic metals such as mercury, beryllium, and lead. As these discarded electronic devices continue to pile up in our landfills, the toxins may cause harm to the environment, as well as to personal health. To reduce e-waste, several electronics manufacturers have initiated recycling programs for their products. Local communities may also collect and recycle e-waste. Additionally, there are valuable metals, such as copper and gold, that can be harvested and reused from these devices. Do you know if your community has an electronics recycling program?

operating systems. The computer hardware they develop is embedded in technological systems such as transportation, manufacturing, construction, health care, and information and communication.

Software Engineering and Computer Science

Computer science focuses on the theoretical study and practical application of computation. ***Computation*** is any type of calculation for processing information. Therefore, computer science is a form of applied mathematics. Computer science became a distinct academic discipline during the 1950s.

Computer scientists focus on designing computational systems and processes to solve problems. They write computer code using different programming languages as a primary means to solve these problems. Computer scientists learn to code, develop algorithms (sets of step-by-step operations to perform calculations, process data, and achieve automatic reasoning), and write software to meet specifications.

Software engineering differs from the computational field of computer science. ***Software engineering*** focuses on using the engineering design process to efficiently produce effective software solutions. Software engineers work in teams and take a holistic approach to developing and improving software solutions for business and industry using the mathematical and computation theories and processes developed by computer scientists. Software engineers design, analyze, test, evaluate, maintain, implement, and update software based on customer specifications.

Evolution of Electrical, Computer, and Software Engineering

Thousands of years ago, ancient Greek philosophers started experimenting with static electricity. But it was not until the seventeenth century that scientists, such as William Gilbert, developed a better understanding of electricity and magnetism. Gilbert developed the word *electricus* to refer to the attraction of objects to one another after being rubbed together to form what we know as static electricity. See **Figure 22-5**. His word eventually evolved into *electric* and *electricity*.

In the eighteenth century, research and development pertaining to electricity expanded. Benjamin Franklin devoted a great amount of time and resources to understanding and controlling electricity. It is said that Franklin attached a metal key to a kite and flew it during a thunderstorm. It is believed that the key was struck by lightning, giving him an electric shock. This event allegedly helped

Alina R/Shutterstock.com

Figure 22-5. Static electricity in action.

him to better understand how electricity occurs and led him to continue to conduct experiments to explain the nature of electricity to the world.

Also late in the eighteenth century, Alessandro Volta created the first battery, called the *voltaic pile*, **Figure 22-6**. The voltaic pile was a major turning point in early electrical engineering because it provided continuous electrical current to be used in electrical circuits. It was made of several cells consisting of a copper and zinc disk separated by an acid-soaked cloth. The voltaic pile would have likely produced over 30 volts.

In the nineteenth century, increased understanding of electrical energy set the foundational knowledge and principles for electrical circuits and devices. For example, George Ohm defined the relationship among voltage, resistance, and current in electrical circuits. Michael Faraday discovered electromagnetic induction, which helped lead to the creation of electric generators and motors. Colleges and universities also began offering degrees in electrical engineering during the nineteenth century.

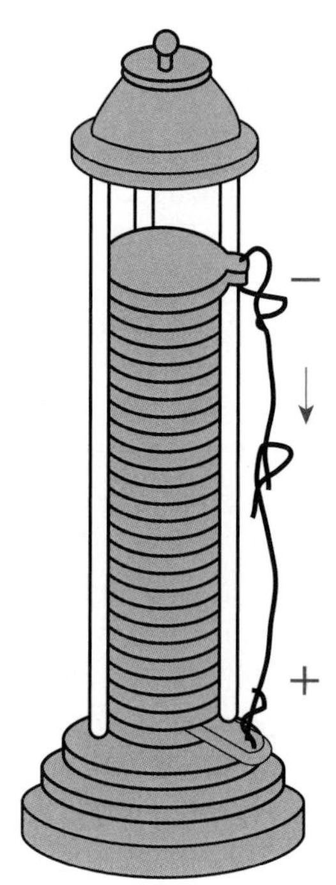

Goodheart-Willcox Publisher

Figure 22-6. Alessandro Volta's voltaic pile was the first electric battery.

Major developments that are responsible for modern-day electrical, computer, and software engineering occurred in the 1900s. The first half of the century saw the invention of many electronic devices, such as radios and televisions. In 1947, William Shockley and his team of researchers invented electrical components that led to the development of integrated circuits, which have revolutionized the world. This development helped launch people into the information age. Just a few decades later, personal computers were being produced. Computer technologies continued to grow increasingly from that point on.

The advancements in electronics, which resulted from electrical engineering, led to the creation of two new engineering disciplines—computer engineering and software engineering. Computers are now integrated into almost every electronic device. As technology advances at an increasing rate, the need for professionals in these engineering disciplines continues to grow.

The Science of Electricity

In order to develop and create new electrical devices, computer systems, and computer software, engineers must understand the science of electricity.

They must understand why electrical charges flow the way that they do and how electrical circuit components interact with each other to perform desired functions. The following sections introduce electrical concepts and principles.

Electrical Circuits

An electrical or computer engineer's work involves developing, testing, analyzing, and improving different types of electrical circuits. Electrical and computer engineering students often study circuit theory, circuit elements, and circuit laws because circuits are involved in every electronic device. They are the components that make computers, smartphones, televisions, and game systems possible.

Electrical circuits are interconnected electrical components that generate and distribute electrical energy for the purpose of controlling, detecting, collecting, storing, receiving, or processing information. They also convert electrical energy into other useful energy forms such as sound, light, heat, or mechanical.

Electrical circuits provide a conducting pathway for electricity to flow from an energy source through a load, which is an electrical component that uses the electricity to produce a desired output.

The circuit in **Figure 22-7** shows a battery connected to an LED (light-emitting diode) using two wires. In this example, the battery provides the electrical energy to power the LED. The LED is the load on the circuit. It uses the electrical energy to produce light. The wires are the conducting paths that allow the electrical energy to flow from one terminal of the battery through the LED and back to the other terminal of the battery, completing the circuit loop.

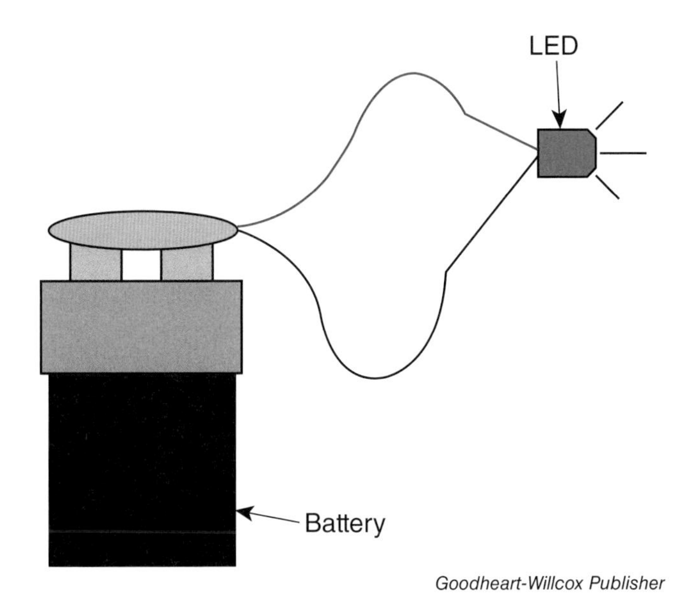

Goodheart-Willcox Publisher

Figure 22-7. A 9-volt battery connected to a blue LED.

All circuits must be a closed loop for electricity to flow from one place to another. Why does electricity flow around the circuit loop? Why does the wire need to go from the power source, to the load, and back to the power source? This movement of electrical energy is explained by circuit theory.

Circuit Theory

Circuit theory is the collection of scientific knowledge used to describe the flow of electrical energy through an electrical circuit. Electrical energy involves the flow of electric charge along a conducting pathway, such as a wire. The flow of electric charge is what you know as *electricity*.

To understand how electricity works in an electric circuit, we must look at the atom. As discussed in Chapter 16, atoms consist of a nucleus, which is made up of one or more subatomic particles called protons as well as a similar number of subatomic particles called neutrons. See **Figure 22-8**. Protons are subatomic particles that have a positive electric charge and neutrons are uncharged subatomic particles. Electric charge is a property of matter that can be measured just as any other property such as volume or mass. Charges can be either positive or negative. However, some particles such as neutrons can have no positive or negative charge, which makes them neutral. It is important to remember that opposite electrical charges attract each other and like electrical charges repel each other. See **Figure 22-9**.

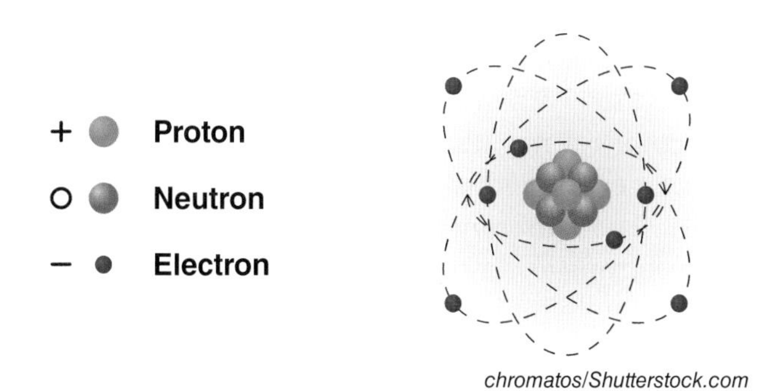

chromatos/Shutterstock.com

Figure 22-8. Planetary model of an atom.

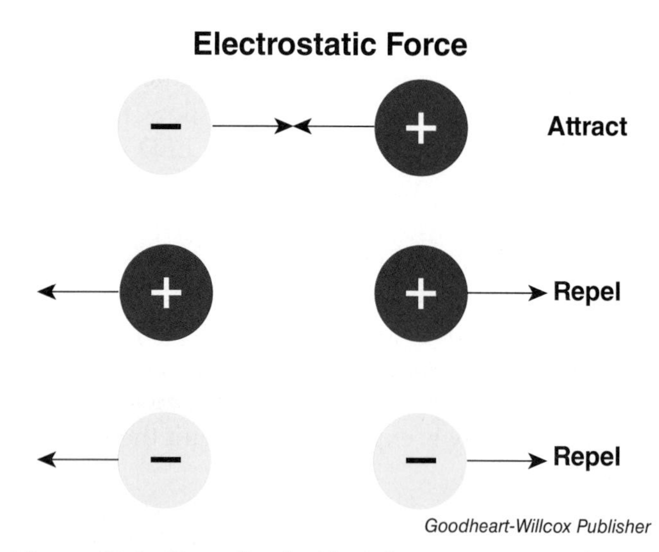

Figure 22-9. Opposite electrical charges attract each other and like electrical charges repel each other. This natural occurrence is called electrostatic force.

In addition to protons and neutrons, the atom is composed of one or more *electrons*. These subatomic particles are bound to an orbit around the nucleus of the atom by an electrical charge. As seen in **Figure 22-8**, the electrons rotate around the nucleus on different orbits. Opposite of protons, electrons have a negative electric charge. Atoms typically have an equal number of electrons and protons. An even number of protons and electrons makes the atom have an overall balanced or neutral electrical charge. The positive charge of the protons cancels out the negative charge of the electrons orbiting around it.

The negative charge of electrons is essential for electricity to happen. This is because electrons can sometimes move from atom to atom. When an electron moves from the orbit of one atom to the orbit of another, *electron flow* occurs. The movement of these electrons with a negative charge along a pathway is electricity. An intentional manmade pathway for these electrons to flow is an electric circuit.

When electrons move between atoms, you are left with unbalanced atoms called *ions*. When an atom loses an electron it becomes a positively charged ion. When an atom gains an extra electron, it becomes a negatively charged ion. Positive ions attract negative ions in hopes to become balanced again. The difference in electrical charges causes the electrons to continue to flow.

For more information about the parts of an atom and their role in creating chemical elements, see Chapter 28.

Wires used in electrical circuits are typically made of copper. The wire is comprised of many copper atoms which each have only one loosely bound valence electron. Free electrons can easily flow between atoms along the copper wire, thus producing electrical current in a circuit.

Current electricity is used to power electronic devices. Current electricity is only possible when electrons are constantly able to flow. In order for this constant flow to occur, a closed electrical circuit must be created using conducting pathways, such as copper wire. This allows the electrons to flow in the same general direction.

In the circuit shown in **Figure 22-7**, one terminal of the battery is connected to the LED by a wire. The LED is connected to the other terminal of the battery by another wire. This provides a closed loop for electrons to flow in the same direction from one terminal of the battery to the other.

Batteries are a common source of energy in electrical circuits. A ***battery*** converts chemical energy to electrical energy. Batteries have three main parts, **Figure 22-10**. The ***anode*** is the negatively charged side of the battery. This terminal is marked with a minus sign (–). The ***cathode*** is the positively charged side of the battery. This terminal is marked with a plus

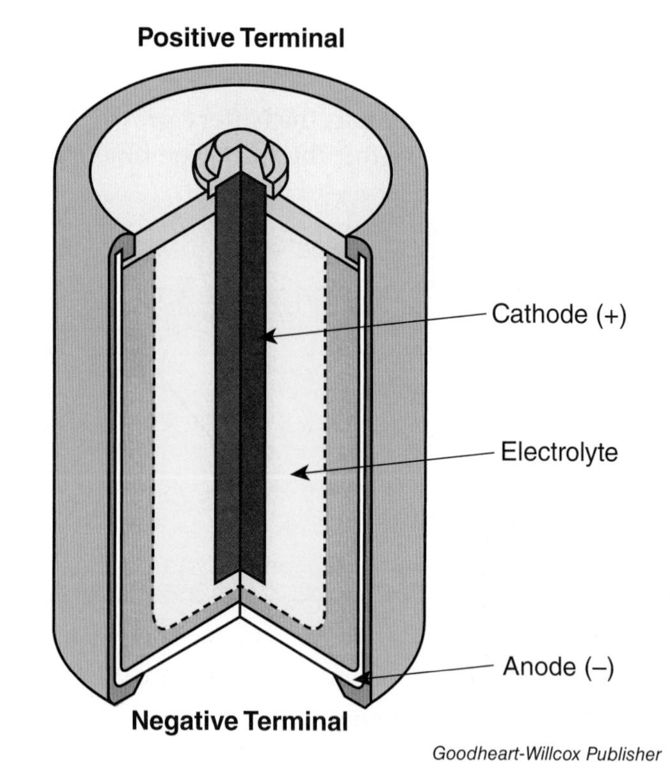

Figure 22-10. The main parts of a battery.

Career Connection Computer Programmer

Computer programmers write, test, and maintain the programs computers follow to perform their functions. These programs are written to specifications established by computer-software engineers and systems analysts. Once a program is designed, a programmer then converts the design into a logical series of instructions the computer can follow. These instructions are coded in a conventional programming language. Programmers update, repair, modify, and expand existing programs. They test a program and make changes until it produces the correct results.

Programmers often are grouped into two broad types—applications programmers and systems programmers. Applications programmers write programs to handle a specific job. Systems programmers write programs to maintain and control computer systems' software.

Rawpixel.com/Shutterstock.com

Students who have strong analytical, mathematical, computational thinking, and problem-solving skills do well in these roles. To prepare, students should complete calculus in high school. Additionally, students should enroll in high school computer science courses as well as technology and engineering courses focused on digital electronics, robotics, or control systems to further their skills in coding and programming.

Bachelor's degree programs in computer science or computer information technology often include the study of digital logic, data structure, algorithms, programming languages, software design, and operating systems.

The number of computer science related jobs is predicted to continue to grow faster than any other STEM-related job over the next couple of decades. Potential projects for computer programmers include:

- Use a variety of programming languages to write software programs.
- Interact with software developers to design new software programs.
- Transform program design ideas into a language that can be processed by a computer.
- Test and debug software programs.

Professional organizations for computer programmers include the Association for Women in Computing; the Computing Research Association; and the Institution of Electrical and Electronics Engineers Computer Society.

sign (+). Chemical reactions in the battery cause electrons to build up on the anode, leaving the cathode with an increased positive charge. Opposite charges are attracted to each other because the atoms want to be balanced. However, the third main part of the battery, the ***electrolyte***, prevents the negative and positive charges on each side of the battery from connecting with each other. The result is an electrical difference between the anode and cathode.

The electrical difference, known as *electrical potential*, means that electrons are waiting to flow toward the positive charge when they are provided a path to do so. Think of the electrons as water being held in a water tower. The water is just waiting to flow out of the tower once a pipe valve is open. While it is waiting, the water is putting pressure on the tank of the water tower. This pressure provides the force for the water to move through a pipe once the valve is open. In an electrical circuit, the pressure of the electrons waiting to move toward the positive charge is called ***voltage***. Voltage provides the force for electrons to move through a circuit. Voltage is measured in volts (V).

In order for the electrons to flow, a conducting pathway must be provided from the negative side of the battery to the positive side, creating a circuit loop. When a copper wire is attached from the negative end of the battery to the positive end, electrons quickly begin to flow from the anode to the cathode in an attempt to balance out the charges. Therefore, after a certain amount of time, the battery no longer has a charge. This is why batteries die. When batteries no longer work, that means there is no longer an electrical difference between the anode and cathode. The charges are balanced so the electrons no longer flow.

If a load is placed in the circuit loop between the anode and cathode of a charged battery, the electrical charge can be used to produce a desired output. For example, an LED will emit light or a motor will turn when the electrons or electricity pass through it. The rate of electron flow through the circuit is called ***current***. Current is measured in amperage (A). Power sources with a higher voltage or greater electrical difference produce a higher current flow.

Ohm's Law

Voltage and current must be properly regulated in electrical circuits. If the voltage and current from a power source are too high, they can burn out components of the circuit. For example, if too many volts and too much current are passed through an electric motor, it will spin so fast and with so much force that parts of the motor will break, bend, or melt. Therefore, it is often necessary to insert resistors into the electrical circuit to slow down the flow of electrons. Resistance, the force that opposes the flow of electrons in a circuit, is measured in ohms (Ω). If the resistance in a circuit is increased, the voltage and current in the circuit decreases. This relationship among voltage, current, and resistance is explained in Ohm's law.

Ohm's law defines the relationship among voltage, current, and resistance in an electrical circuit. The law states that the voltage (also called *electrical potential difference*) between any two points on an electrical circuit is equal to the current between those same two points multiplied by the total resistance between those two points. This is represented as the mathematical equation $V = IR$ where V is the voltage in a circuit, I is the current, and R is the resistance.

Using the Ohm's law equation, you can calculate the current, resistance, or voltage in a circuit when two of the values are known. The current in a circuit is directly proportional to voltage and inversely proportional to the total resistance in a circuit. What this means is that as the voltage in a circuit increases or decreases, the current increases or decreases.

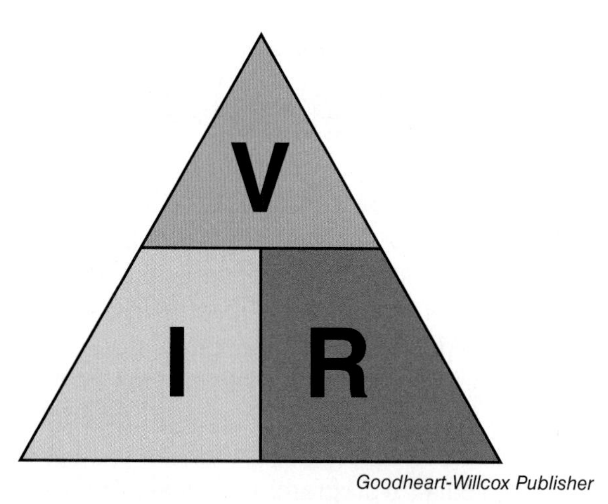

Goodheart-Willcox Publisher

Figure 22-11. Variations of the equation for Ohm's law. Covering up one component of the equation shows you how to determine its value.

The equation for Ohm's law can be rearranged using algebra to solve for the each of the variables as necessary ($V = IR$, $I = V/R$, $R = V/I$). An easy way to remember the different variations of this equation can be seen in **Figure 22-11**. The triangle in this figure shows that voltage equals the product of current and resistance, current equals voltage divided by resistance, or resistance equals voltage divided by current.

An example of the use of Ohm's law is as follows. Calculate the current that will pass through an electric motor when it is connected to a 9-volt battery and a resistor that provides 220 ohms of resistance:

$$I = V/R$$
$$I = (9\ V)/(220\ \Omega)$$
$$I = 0.041\ A$$

Circuit Components and Diagrams

Circuits contain a mix of wires and other components to control the flow of electricity to produce a desired output. See **Figure 22-12**. Since the flow of electricity depends on the properties of different atoms, the materials for electrical components are selected based upon their atomic structure. Three types of materials used for electrical components are conductors, insulators, and semiconductors.

Electrical Components and Their Symbols		
Component		Symbol
Electric generator		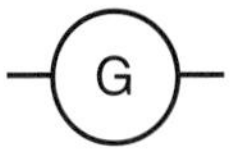 G
Battery		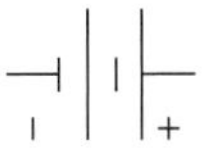 +
Switch		
Resistor		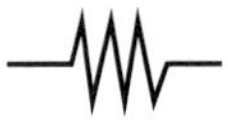
Variable resistor		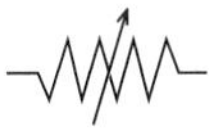
Nonpolarized capacitor		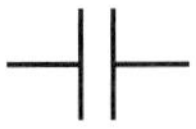
Polarized capacitor		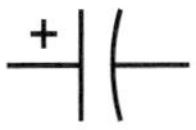 +
Transistor		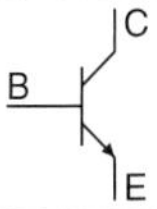 C B E
Diode	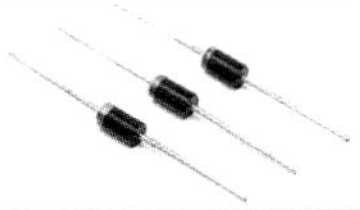	
LED		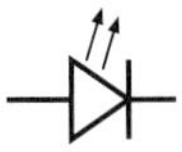
Photoelectric cell		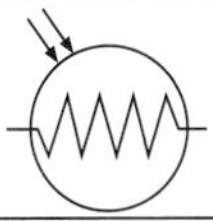
Integrated circuit		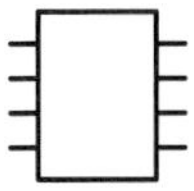
Speaker		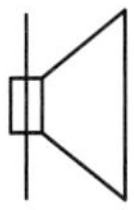
Electric motor	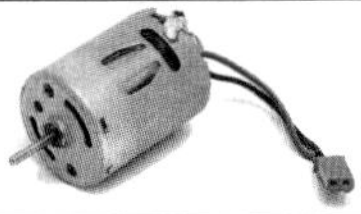	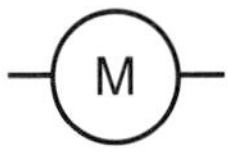 M

Middle column, top to bottom: oleghush/Shutterstock.com; James Hoenstine/Shutterstock.com; mosufoto/Shutterstock.com; Abraksis/Shutterstock.com; optimarc/Shutterstock.com; mosufoto/Shutterstock.com; mosufoto/Shutterstock.com; 3drenderings/Shutterstock.com; NNL_STUDIO/Shutterstock.com; Sergio Delle Vedove/Shutterstock.com; Audrius Merfeidas/Shutterstock.com; Audrius Merfeidas/Shutterstock.com; Fedorov Oleksiy/Shutterstock.com; Bplanet/Shutterstock.com; Right column: Goodheart-Willcox Publisher

Figure 22-12. Basic electrical components and their symbols.

Conductors allow electrons to flow easily because they have three or fewer electrons in their valence orbit. Because these electrons in the outer orbit are held loosely to the atom, they can easily move to another atom or make room for more electrons. Conductors are used in electric circuits to create wires to provide pathways for electricity to move. Copper and gold are good conductors and are often used in electrical circuits. Wires in circuits are often made of copper because copper has only one valence electron.

Insulators have five to eight electrons in their valence orbit. Atoms that have a full or almost full valence orbit prevent the flow of electrons. Insulators are used in circuits to help prevent electricity from flowing to unwanted areas. Typical insulating materials are rubber, plastic, and glass.

Semiconductors have exactly four valence electrons. These materials are neither good at conducting electricity or preventing the flow of electrons. However, their properties can be changed to take on characteristics of either conductors or insulators through a process called *doping*. Doping involves adding impurities into the semiconductor material, which enables engineers to use the materials in components such as diodes and transistors to better control electron flow in a circuit. Therefore, semiconductor materials are used in almost every electronic device. Typical semiconductor materials are carbon and silicon.

Conductors, insulators, and semiconductors are used to create electrical components that serve a variety of purposes. Common electrical components are explained in the following paragraphs.

A ***power source*** is the source of the electron flow in an electrical circuit. A power source can be a battery, generator, solar panel, or other device. Refer to Chapter 19 for an explanation of electricity generation.

An ***electrical switch*** interrupts the flow of electrons from a power source in an electrical circuit. Switches can be open or closed. When a switch is open, the conducting pathway for electrons to flow is severed and the electric current in the circuit is shut off. When a switch is closed, the pathway for electric current is complete and the circuit is on. There are many different types of switches with different purposes. For example, a toggle switch uses a lever to open and close a circuit; a push button switch opens and closes a circuit by pressing and releasing a button; and a selector switch is like a toggle switch with different positions that connects or disconnects different loops in the circuit.

Switches can be used to connect different conducting pathways within a circuit. Instead of interrupting the flow of electrons, these switches provide an alternative pathway for electron flow.

Resistors restrict the flow of electrons in a circuit and reduce the current and voltage by turning it into heat energy. Resistors are available in various shapes, sizes, and resistance values. The resistance produced by a resistor is measured in ohms (Ω).

Every resistor has four color bands that, when interpreted together, indicate the amount of resistance provided by the resistor. The first color band on the resistor signifies the first digit of the amount of ohms the resistor provides. The next band indicates the second digit. The third band indicates the multiplier for the two-digit number to get the final resistance value. The last band, which is either gold or silver, provides the tolerance value of the resistor. Gold indicates that the resistor has a 5% tolerance, meaning that the true resistance value will be within a range of 5% above or below the value indicated by the first three bands. Silver indicates a 10% tolerance. See **Figure 22-13**.

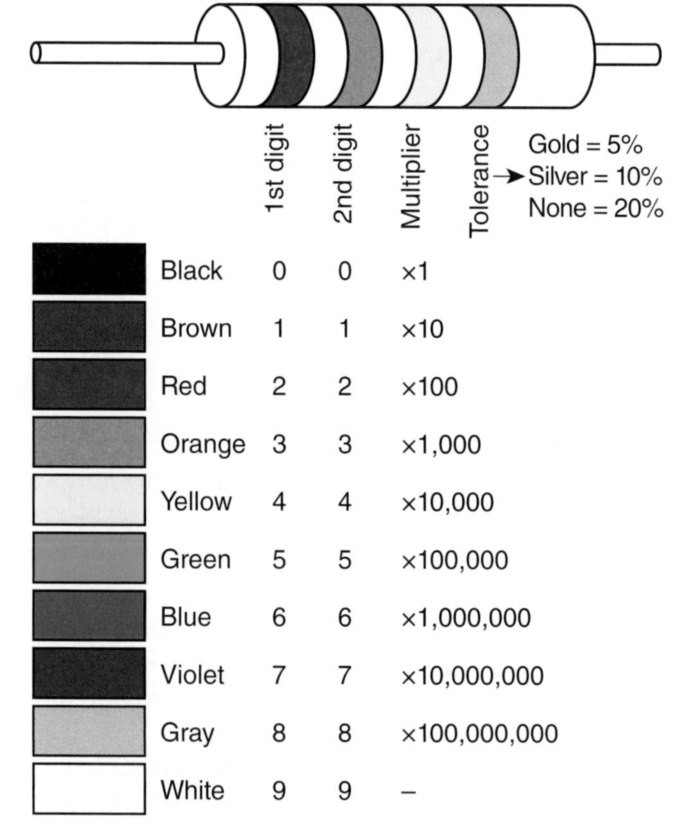

Color	1st digit	2nd digit	Multiplier
Black	0	0	×1
Brown	1	1	×10
Red	2	2	×100
Orange	3	3	×1,000
Yellow	4	4	×10,000
Green	5	5	×100,000
Blue	6	6	×1,000,000
Violet	7	7	×10,000,000
Gray	8	8	×100,000,000
White	9	9	–

Goodheart-Willcox Publisher

Figure 22-13. Resistor color code.

Using the color code, you could determine that a resistor with a red band, brown band, orange band, and gold band is a resistor with a resistance value of 21 kΩ with a tolerance of 5%. There are also resistors that are variable. Variable resistors have a dial that can adjust the amount of resistance they produce.

Capacitors store an electrical charge similar to a battery. However, unlike a battery, a capacitor stores energy in an electric field rather than in a chemical form. This enables a capacitor to discharge its energy more rapidly than a battery that must transform chemical energy to electrical energy. When electrons flow through an electrical circuit with a capacitor, the capacitor will charge by collecting the electrons. When energy is needed, the capacitor can rapidly release all of the stored energy. Because capacitors store charge in a circuit, they can also be used as a timing device, filter, a means to even out voltage, or for a variety of other purposes. Capacitors are available in all shapes and sizes. They vary in capacitance, which is the amount of energy the capacitor can store.

Transistors are a critical component of modern circuits. ***Transistors*** are semiconductor components that can amplify signals for components such as speakers or act as electronic switches.

The use of transistors as electronic switches is the foundation of digital electronics. Transistors provide the foundation for a computer system's memory. When a transistor is switched on, it stores a number value of 1. When it is turned off, it stores a value of 0. Combinations of numerous 1s and 0s are used to represent information or control functions within all digital electronics. This is called *binary code*, which is discussed later in this chapter. Combinations of transistors embedded within integrated circuit chips are used for all types of computing. Millions or more transistors are interconnected to form a computer's memory.

Diodes are electrical components made out of semiconductor materials, such as silicon or selenium. ***Diodes*** allow electrical current to pass through them in only one direction. This function enables a diode to protect the circuit from any voltage surges. Additionally, diodes can be used to convert ac into dc. A ***light-emitting diode (LED)*** is a special diode that releases energy in the form of photons when a suitable voltage is applied. LEDs are used in electronic devices that produce light.

Photoelectric cells, or ***photocells***, are electrical components that produce electric current and voltage when exposed to light. Besides producing electricity, these components can be used to control circuits based on the input of light. The electric current produced depends on the amount of light to which the cells are exposed. The higher the light intensity, the more current is produced. Photocells are used in various types of sensors or sensing equipment.

Integrated circuits (ICs) are electronic circuits formed on a small, thin piece of pure semiconducting material, usually silicon. The tiny electric circuits contain complex combinations of transistors and other electronic components that are shrunk down to a microscopic level. The combinations of electronic components are built to perform particular functions in various types of electronics, such as digital clocks, computers, microwaves, and cellular phones. Because of their small size, ICs have enabled great advancements in the field of microelectronics and computers.

Breadboards are devices that serve as a construction base for prototyping or making experimental models of electrical circuits. See **Figure 22-14**. Breadboards allow electrical components to be connected together in a circuit without being soldered, thus providing a conducting pathway for electric current to flow. A breadboard is useful to test an electrical circuit before soldering

Golubovy/Shutterstock.com

Figure 22-14. Breadboards are used as a temporary construction base for prototyping electrical circuits.

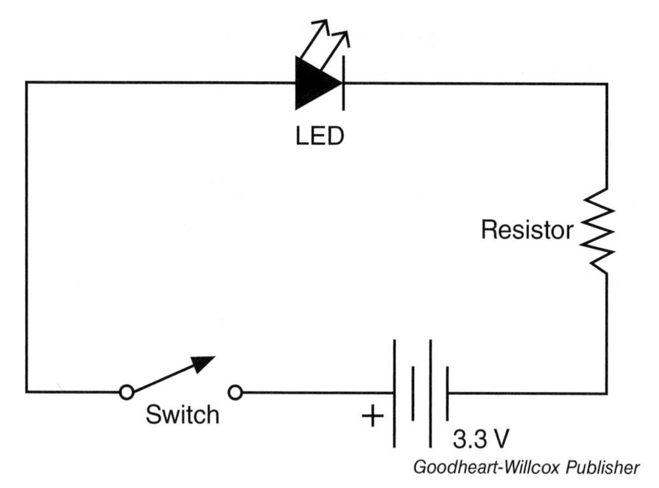

Figure 22-15. A circuit diagram that shows a 3.3-volt battery connected in a circuit to a resistor, LED, and a switch.

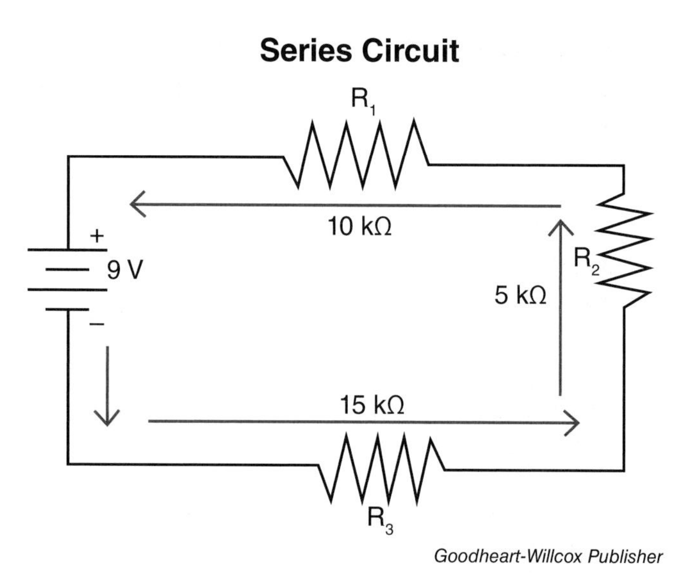

Figure 22-16. An example of a series circuit.

it together by melting a filler metal and flowing it over the connection points of two components.

Circuit diagrams, or electronic schematics, graphically represent an electrical circuit. See **Figure 22-15**. Each type of electrical component can be represented by a symbol, as shown in **Figure 22-12**. Using these symbols is much easier than drawing the actual components.

Circuit Configurations

Components in a circuit can be configured in series or in parallel. In addition, some circuit components may be connected in series while others are connected in parallel.

Series Circuit

In a ***series circuit***, components are connected in a way that provides only a single pathway for electrons to flow. **Figure 22-16** shows a battery connected to three resistors. The resistors in the circuit are connected from end to end, providing only a single pathway for electrons to flow. The electrons flow out of the negative terminal of the battery through resistor 3 and then through resistor 2, followed by resistor 1, and then into the positive terminal of the battery. Because there is only one path for the electrons to flow in a series circuit, the amount of current is the same throughout the entire circuit.

Ohm's law can be used to determine the current in a series circuit. Divide the total voltage by the total resistance. In **Figure 22-16**, there are three resistors, one with 10 kilohms (kΩ) of resistance, one with 5 kΩ, and one with 15 kΩ. The total resistance in a series circuit is equal to the sum of all of the individual resistances:

$$R_{total} = R_1 + R_2 + R_3$$
$$R_{total} = 10\ k\Omega + 5\ k\Omega + 15\ k\Omega$$
$$R_{total} = 30\ k\Omega$$

To find the total current in this example of a series circuit, divide the total voltage of 9 V by the total resistance (30 kΩ):

$$I_{total} = V/R$$
$$I_{total} = 9\ V/30\ k\Omega$$
$$I_{total} = 9\ V/30{,}000\ \Omega\ (1\ k\Omega = 1{,}000\ \Omega)$$
$$I_{total} = 0.0003\ A$$

We now know that the circuit in **Figure 22-16** has a total resistance of 30 kΩ and a total current of 0.0003 A. The total voltage applied to the circuit is 9 V. However, the voltage in a series circuit is not the same at every point. Because electrons are flowing through each component in series, the voltage applied to each component is different as the resistance of each component causes the voltage to drop (or lower). However, for our example, we have found all the values we need in order to use Ohm's law to calculate the voltage drop across each resistor.

As we learned, the current in a series circuit is same throughout the entire circuit. In this example, the current is 0.0003 A. Based on Ohm's law, we multiply the total current and the resistance of each resistor to find the voltage drop for each resistor:

V_{R1} (the voltage drop across R_1) $= I_{total} \times R_1$
$V_{R1} = (0.0003 \text{ A})(10 \text{ k}\Omega)$
$V_{R1} = 3 \text{ V}$

$V_{R2} = I_{total} \times R_2$
$V_{R2} = (0.0003 \text{ A})(5 \text{ k}\Omega)$
$V_{R2} = 1.5 \text{ V}$

$V_{R3} = I_{total} \times R_3$
$V_{R3} = (0.0003 \text{ A})(15 \text{ k}\Omega)$
$V_{R3} = 4.5 \text{ V}$

If you add together the three voltage drops in the circuit (3.0, 1.5, and 4.5), the sum is 9 volts, which is equal to the applied voltage from the battery. This illustrates ***Kirchhoff's voltage law***, which states that the sum of all voltage drops in a circuit is equal to the total applied voltage. This becomes a useful piece of information when you are analyzing any circuit because it provides a known piece of information about a circuit to begin performing calculations. Also, this law is a good way to double-check your calculations in circuit analysis. If the sum of all the voltage drops does not equal the applied voltage, then something was calculated incorrectly. As you are performing circuit analysis calculations, prepare a table of the results. See **Figure 22-17**.

Parallel Circuit

The other configuration for electrical components is parallel. Components connected in a ***parallel circuit*** are connected across common ends or nodes, providing multiple pathways for electrons to flow.

In **Figure 22-18**, the three resistors are connected in parallel, which allows electrons to flow the same across each one. Therefore, the voltage across each component of the circuit is the same. However, because there are multiple pathways for the electrons to flow, the current splits off into each loop across the different components. This means the current is not the same throughout the entire circuit.

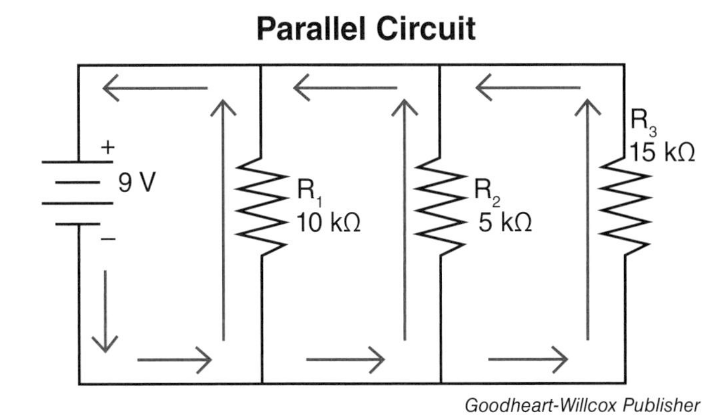

Figure 22-18. An example of a parallel circuit.

Ohm's law can be used to calculate the current across the different resistors shown in **Figure 22-18** because we know that the voltage is the same across each component:

$I = V/R$ (Ohm's law)
$I_{R1} = V/R_1$
$I_{R1} = 9 \text{ V}/10 \text{ k}\Omega$
$I_{R1} = 0.0009 \text{ A}$

$I_{R2} = V/R_2$
$I_{R2} = 9 \text{ V}/5 \text{ k}\Omega$
$I_{R2} = 0.0018 \text{ A}$

$I_{R3} = V/R_3$
$I_{R3} = 9 \text{ V}/15 \text{ k}\Omega$
$I_{R3} = 0.0006 \text{ A}$

Now, you have the voltage, current, and resistance across each resistor in our example of a parallel circuit. It is good practice to prepare a table of the results of the circuit calculations. See **Figure 22-19**.

At this time, the total current in the circuit and the total resistance are unknown. However, another important circuit law, ***Kirchhoff's current law***, states that the sum of the currents across each branch or loop of a parallel circuit is equal to the total current in a circuit. Consequently, the three

	R_1	R_2	R_3	R_{total}
V	3 volts	1.5 volts	4.5 volts	9 volts
I	0.0003 amps	0.0003 amps	0.0003 amps	0.0003 amps
R	10 kΩ	5 kΩ	15 kΩ	30 kΩ

Goodheart-Willcox Publisher

Figure 22-17. Analysis results of the series circuit in Figure 22-16.

	R_1	R_2	R_3	R_{total}
V	9 volts	9 volts	9 volts	9 volts
I	0.0009 amps	0.0018 amps	0.0006 amps	?
R	10 kΩ	5 kΩ	15 kΩ	?

Goodheart-Willcox Publisher

Figure 22-19. Analysis results of the series circuit in Figure 22-18.

currents that were calculated earlier can be added together to find the total current for the circuit:

$$I_{R1} + I_{R2} + I_{R3} = I_{total}$$
$$0.0009\ A + 0.0018\ A + 0.0006\ A = 0.0033\ A$$
$$I_{total} = 0.0033\ A$$

The last unknown value for the parallel circuit in **Figure 22-18** is the total resistance. The total resistance can be found using Ohm's law and the values for the total voltage and current. To find the total resistance, you would perform the following calculation:

$$R = I/V$$
$$R_{total} = V_{total}/I_{total}$$
$$R_{total} = 9\ V/0.0033\ A$$
$$R_{total} = 2.7272\ k\Omega$$

Note that in this parallel circuit, the total circuit resistance is less than any one of the individual resistors in the circuit. In the series circuit described earlier, the total resistance was a sum of all of the resistances in the circuit, which made the total resistance much higher than the resistance of any individual resistor. However, in the parallel circuit, the individual resistance diminishes because of the multiple pathways provided for the electrons to flow. In the example provided, there are three potential loops for the electrons to flow, meaning that a third of the electrons will flow through each loop, causing the resistance to be less. The mathematical equation for calculating the total resistance in a parallel circuit is:

$$R_{total} = 1/[(1/R_1) + (1/R_2) + (1/R_3)\ldots]$$

This equation will work for any number of resistors in a parallel circuit. For the example in **Figure 22-18**, we can perform the following calculation to determine the total resistance. The total resistance is equal to the calculation made using Ohm's law and the total current and total voltage values.

$$R_{total} = 1/[(1/R_1) + (1/R_2) + (1/R_3)]$$
$$R_{total} = 1/[(1/10\ k\Omega) + (1/5\ k\Omega) + (1/15\ k\Omega)]$$
$$R_{total} = 1/[(0.1\ k\Omega) + (0.2\ k\Omega) + (0.06667\ k\Omega)]$$
$$R_{total} = 1/(0.36667\ k\Omega)$$
$$R_{total} = 2.7272\ k\Omega$$

Series-Parallel Circuit

Series-parallel circuits have components that are arranged in series and components that are arranged in parallel. In the circuit shown in **Figure 22-20**, there are two loops for the electrons to flow around. Both loops require the electrons to pass through resistor 1. In this configuration, we consider resistor 2 and resistor 3 to be in parallel with each other and resistor 1 to be in series with the combination of both resistor 2 and resistor 3. With this information, you would conduct the calculations based on the rules for both series and parallel circuits at the points where they are configured in that manner.

Direct and Alternating Current

The electrical circuits that have been discussed up until this point have all used direct current, or dc electrical energy. In ***direct current (dc)***, the electrical charge flows in only one direction. Batteries provide direct current resulting from chemical reactions inside the battery itself. The negative charge (electrons) flows only from the negative terminal of the battery to the positive terminal of the battery. Direct current powers most of our handheld electronic devices.

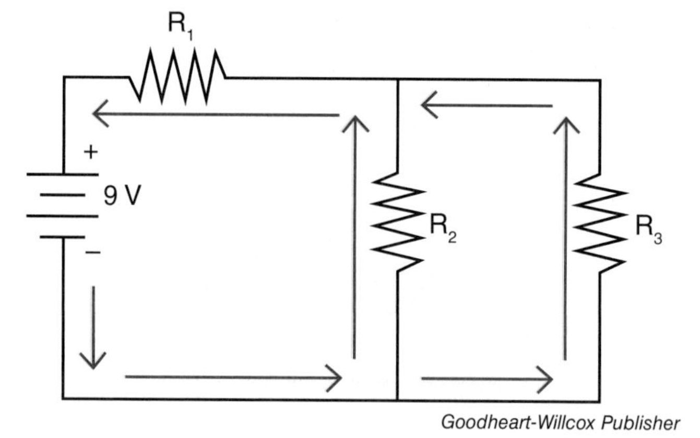

Goodheart-Willcox Publisher

Figure 22-20. An example of a series-parallel circuit.

Conversely, ***alternating current*** (ac) has an electrical charge that changes directions periodically. The voltage also reverses as the direction of current changes. Alternating current is mostly used to transmit high-voltage electricity over long distances. It is the type of electricity that flows through the power lines to power your home. Therefore, the outlets in your house provide alternating current. However, when you plug electronic devices into the outlets in your home to charge them, the alternating current needs to be converted to direct current. A ***rectifier*** is an electrical device used to convert ac into dc by allowing a current to flow through it in one direction only.

Alternating current electricity is produced using an electrical generator, **Figure 22-21**. An electrical generator contains a loop of copper wire and magnets. The loop of wire is rotated or spun within a magnetic field created by the magnets. The rotating motion of the wire loop is provided by a variety of energy sources such as blowing wind, moving water, or steam produced by a power plant.

As the wire spins inside the magnetic field, current is produced along the wire through electromagnetic induction. ***Electromagnetic induction*** produces electromotive force or voltage by continuously moving a conductor, such as copper wire, through varying magnetic fields. The motion causes the electrons to move in the wire based on the attracting and repelling forces exerted by the magnetic field. Because the wire spins past both the negative and positive poles of the magnets, the voltage and current alternate directions as it moves.

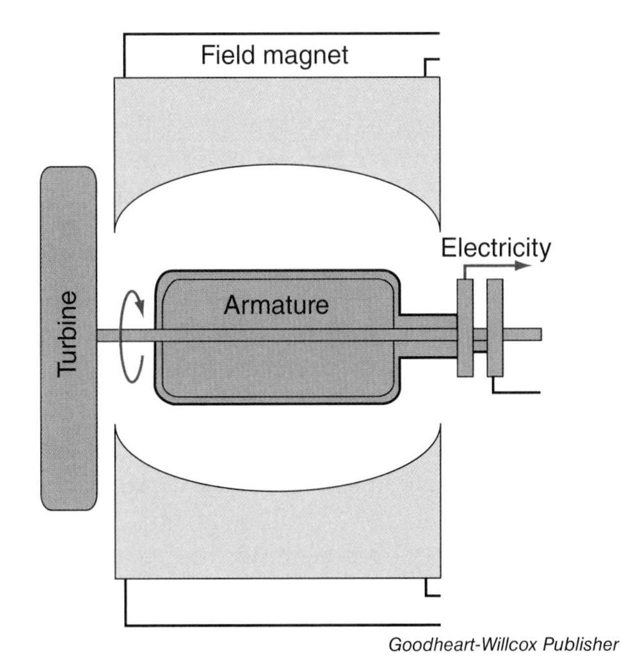

Figure 22-21. An electric generator.

Electrical generators can also produce direct current when they are equipped with a commutator. A commutator reverses the direction of current with each half turn of the generator to produce an output of unidirectional direct current.

Analog and Digital Circuits

We live in a world that is analog. This means that there are infinite possibilities regarding quantities in our world. There are infinite variations in temperature, color, sounds, smells, and so on. Electronics and electrical circuits interact with the analog world to either monitor or control some aspect of it. This is done by receiving, sending, and interpreting signals. A ***signal*** is any measurable quantity used to communicate information. Signals are passed back and forth between devices to share information, such as audio or video. Signals can either be analog or digital.

An analog signal is any continuous quantity that changes over time to express information. Analog signals provide exact values because there is an infinite range of values for different quantities of the physical world. Digital signals are finite or discrete, meaning there are a limited number of values that quantities can have. For example, an analog scale is not limited to how precisely it can display the weight of an object, while a digital scale is limited to the quantity of weight it presents on its segmented display. Therefore, analog signals are more exact than digital signals. See **Figure 22-22**.

Analog signals are difficult to work with in electronics because of the infinite variations. Although there are limited values of a digital signal, many series of these signals can be used to closely represent analog data. As a result, it is much easier to use digital signals in electronics than analog signals.

Digital signals are typically one of two different values used to represent information. In electronic applications, the two possible signal values are 1 and 0. These values can determine whether something is on or off. Using a series of 1s and 0s to represent information is called *binary code*.

In electrical circuits, voltage is typically the varying quantity used to convey information. In analog circuits, the voltage can vary infinitely,

MK photograp55/Shutterstock.com

andrea crisante/Shutterstock.com

Figure 22-22. An analog clock is not limited on how precisely it can display time. A digital clock is limited to the quantity of time it can present on its segmented display.

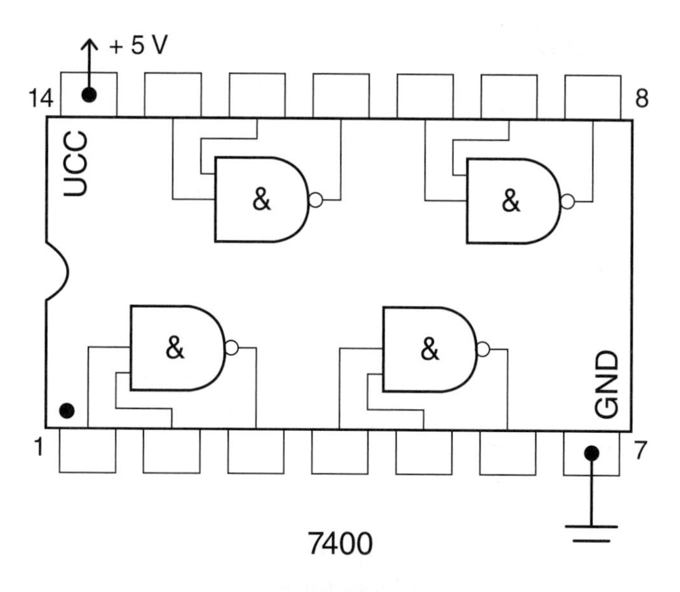

RayPics/Shutterstock.com

Figure 22-23. Logic gates are found within integrated circuit chips.

which makes these circuits difficult to design. Additionally, slight undesired changes in the voltage can produce significant errors in the signal. However, digital circuits operate using discrete voltage values to convey information. Different assigned voltage levels represent just two values (either 1 or 0). A high voltage of typically 5 V represents a 1 and a lower voltage is used to represent a 0. Essentially, these values are the 1s and 0s that are used in binary code. Because the signal is either one value or the other, there is little room for producing errors in the signal transmission. In digital circuits, a combination of transistors, integrated computing chips, and microprocessors are used to process the digital signals.

Digital circuits, the fundamental components of modern computer systems, are based on digital logic. ***Digital logic,*** also known as *Boolean logic,* enables complex decisions to be made within computer systems based on a series of yes or no questions. Understanding digital logic is not only important for electrical and computer engineers but also for computer scientists and software engineers. When programming computer systems, these individuals must know how digital signals are processed to answer the series of yes-or-no questions so the systems can respond as desired.

There are two types of digital logic. The first is ***combinational logic,*** which uses logic gates to make decisions based on present inputs. Logic gates are found within the integrated circuit chips of computer systems, **Figure 22-23**. They are used to perform Boolean algebra based on input signals. Boolean algebra is branch of algebra focused on determining outcomes with input variables that are either true or false or are either 1s or 0s.

Basic logic gates take in two inputs and produce an output that enables the circuit to perform desired functions. Additionally, there is an inverter (NOT gate) that is used to just convert one value to the opposite value. See **Figure 22-24**.

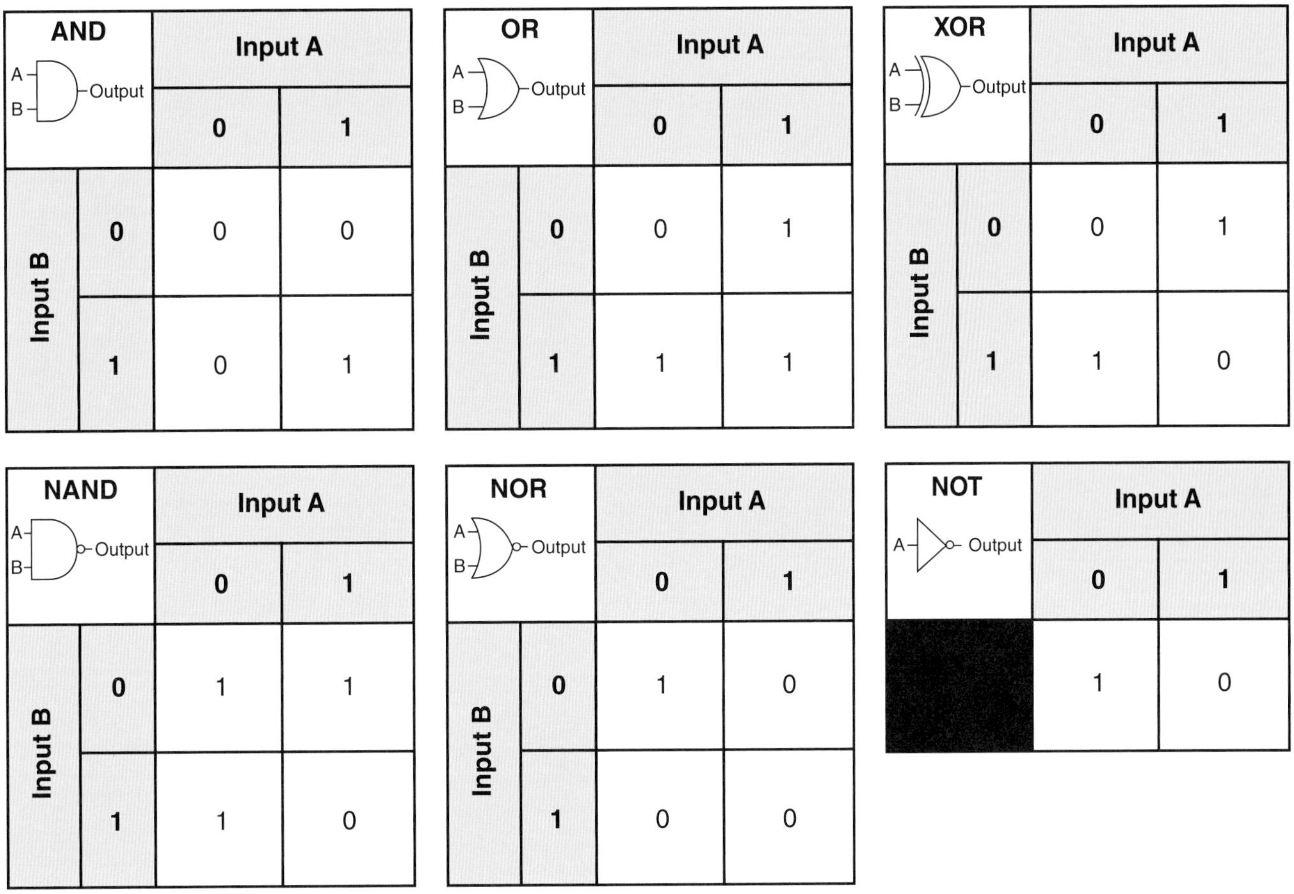

AND		Input A	
		0	1
Input B	0	0	0
	1	0	1

OR		Input A	
		0	1
Input B	0	0	1
	1	1	1

XOR		Input A	
		0	1
Input B	0	0	1
	1	1	0

NAND		Input A	
		0	1
Input B	0	1	1
	1	1	0

NOR		Input A	
		0	1
Input B	0	1	0
	1	0	0

NOT	Input A	
	0	1
	1	0

Goodheart-Willcox Publisher

Figure 22-24. The symbols and truth tables for the five logic gates and inverter.

Each logical gate has a different way of producing an output based on its inputs:

- AND gate. Produces an output signal of 1 if both input signals are 1.
- OR gate. Produces an output signal of 1 if at least one of the input signals is 1.
- XOR gate. Produces an output signal of 1 if only one input signal is 1.
- NAND gate. Produces an output of 1 if at least one input signal is 0.
- NOR gate. Produces an output of 1 if both inputs are 0.
- XNOR gate. Produces an output signal of 1 if the inputs are the same.
- NOT gate. Only has one input and one output. It produces an output that is the opposite of the input. For example, if the input it 1, then the output is 0.

If the conditions of these gates are not met, they will produce an output signal of zero. Truth tables are used to show the different outputs produced by the different combinations of inputs for each logic gate. See **Figure 22-24**.

The different conditions of the logic gates are used in digital circuits to make complex decisions. Logic gates can be used in combination to perform a specific task. As shown in **Figure 22-25**, a series of NAND and OR gates can be used for a basic home alarm system. In this example, the home alarm system monitors the front door, back door, garage, a motion detector, and four windows. Once the alarm is set and the doors, garage, and windows are closed, voltage flows through the circuit because the switches on the different items are closed. Because the switches are closed, the circuits for these items are complete, and voltage is present. When voltage is present, a signal of 1 is produced. One exception is the motion detector. When the motion detector is not sensing any motion, it is off and no voltage is present. This means that the motion detector provides an input of 0. The inverter (NOT gate) is used to switch the 0 to 1 to make all signals in the circuit to be 1 when the alarm is set and not sounding.

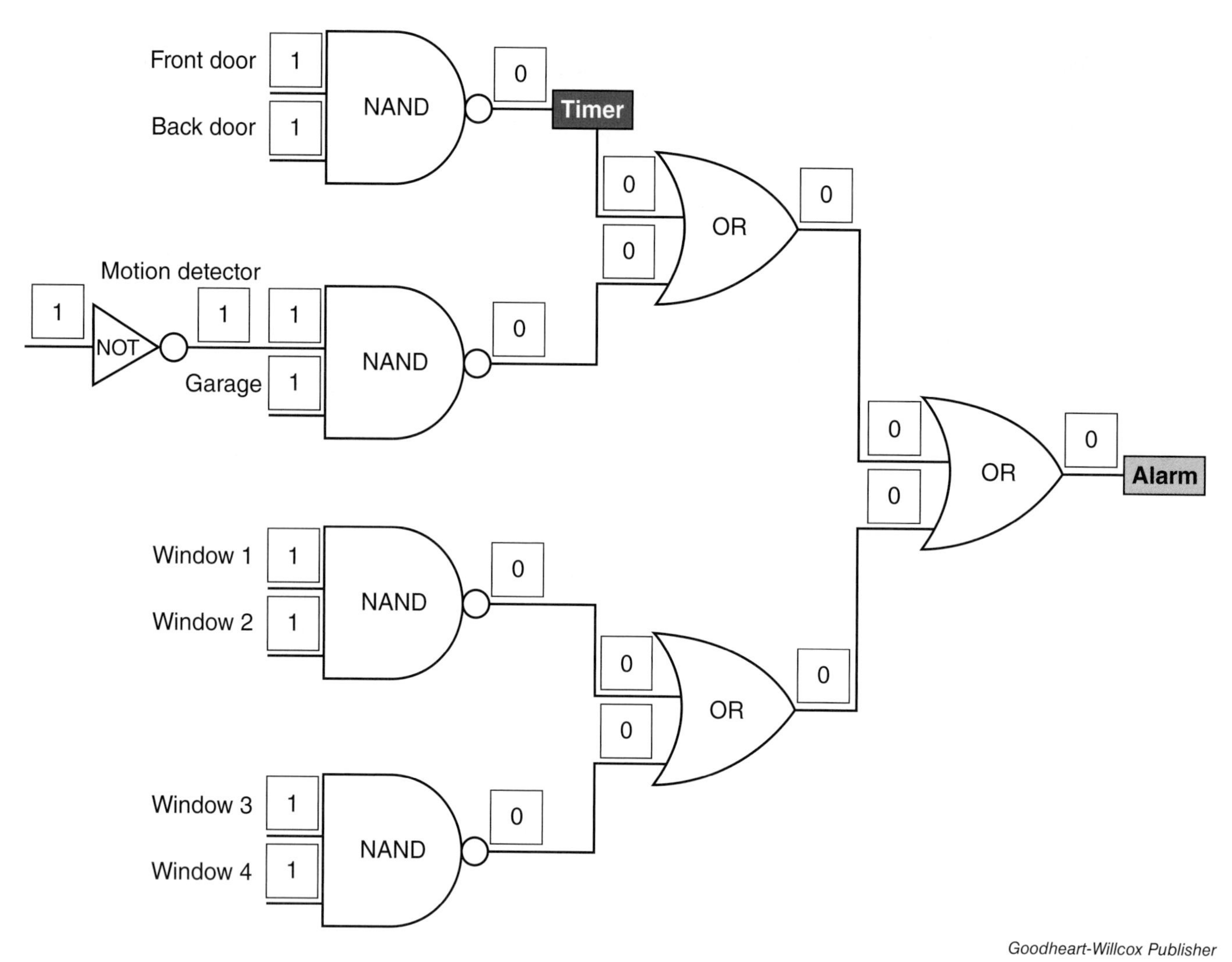

Figure 22-25. A digital circuit diagram for a home alarm system. In this diagram, all the inputs result in the alarm not sounding.

The four NAND gates in the circuit in **Figure 22-25** each convert two inputs of 1 into an output of 0. The resulting four outputs are then connected to two OR gates. Each OR gate converts two 0 inputs into an output of 0. The resulting two outputs from the two OR gates are connected to one last OR gate. This last logic gate converts the two 0 outputs into a 0 output that is then connected to an alarm. In this state, the windows, doors, and garage are closed and the motion detector is not sensing any motion; therefore, the alarm is not sounding. However, if the garage or any windows or doors are opened or motion is detected, an initial input will change and the alarm will sound. For example, if the front door is opened, the circuit loop will be broken, resulting in zero voltage. The lack of voltage will switch the 1 to 0, thus setting off a chain of events to turn on the alarm, **Figure 22-26**.

The other type of digital logic is ***sequential logic,*** in which outputs depend not only on the present input but also on the past inputs. Sequential logic has a memory of the past inputs while combinational logic does not. The memory enables the circuit to store previous outputs to be used in conjunction with new inputs to produce another output. Combinational logic can do many things, but without the memory involved in sequential logic, modern computer systems would not be possible. Therefore, combinational and sequential logic are the essential elements of all digital electronics.

Computer Systems

Computer systems are the heart of all information-processing and communication technologies

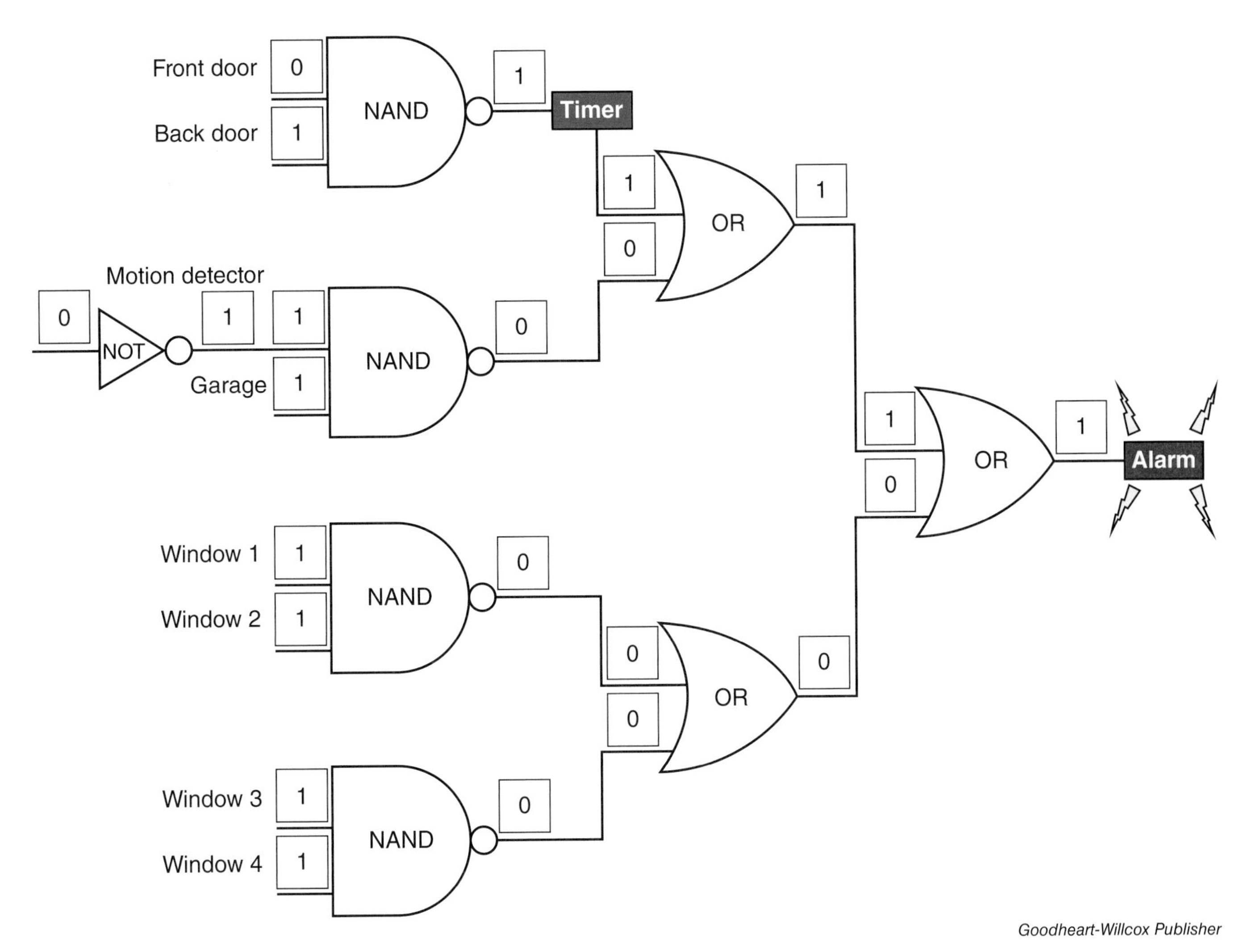

Figure 22-26. A digital circuit diagram for a home alarm system. In this diagram, the front door is open, which results in the alarm sounding.

and perform countless functions. They process financial information and prepare financial reports, maintain schedules and ticket records for airlines, operate point-of-purchase units at supermarket-checkout stands, guide spacecraft to distant planets, maintain the correct fuel-air mixture in automobiles, guide washing machines through their wash-rinse-spin cycles, help prepare layouts for advertising, and control industrial machines. They also provide Internet access and allow for worldwide communication.

Computers include physical equipment (hardware) and operating instructions (software). ***Hardware*** is the computer itself and devices attached to it, **Figure 22-27**. The four main hardware units are the CPU, the memory, the input devices, and the output devices.

Figure 22-27. Computer systems have input devices, a CPU, memory, and output devices.

Academic Connection: History

The Internet

In October 1957, the former Soviet Union stunned the world with its announcement that it had successfully launched the first artificial satellite, called *Sputnik*, into space. No one was more surprised than the Americans. Until then, they had assumed their scientific achievements greatly surpassed those of the Soviet Union. The United States was also concerned because of the military implications behind *Sputnik*. At that time, the United States and the Soviet Union were engaged in what has been called the *Cold War*. Whoever achieved technological superiority might be able to do great damage to the other country's military capabilities. One immediate response by the United States, therefore, was to put more funds into the research and development efforts of the Department of Defense.

One result was the department's creation of the Advanced Research Projects Agency (ARPA). Among other goals, the agency was interested in helping to create technology that would enable computers to communicate with one another and to do so in such a way that some computers could still be in contact, even if others had been shut down because of a military attack. This agency funded the research efforts of scientists at some major corporations and at four universities. These scientists were also working on the idea of computer networking.

After years of testing and retesting various methods, the scientists unveiled the ARPA Network (ARPANET) in 1972. ARPANET was a computer-networking system based on the idea of breaking data down into labeled packets that could be forwarded from computer to computer. This effort led to the development of the network control protocol (NCP) to transfer data. The NCP allows communication between hosts on the same network. Development continued, eventually resulting in the Transmission Control Protocol/Internet Protocol (TCP/IP) technology. Now, computer networks can interconnect and communicate with one another.

Thus, what started as a military concern became a new way of communication for millions of people. ARPANET became the Internet. The Internet continues to expand today.

In 1972, ARPA became the Defense Advanced Research Projects Agency (DARPA). DARPA is still in operation. Can you identify some projects in which it is involved today?

The ***central processing unit (CPU)*** is the working part of the computer that carries out instructions. The CPU is a microprocessor chip, a piece of silicon containing millions of electrical components.

Memory is where the computer stores its data and operating instructions. The computer has two types of memory:

- **Read-only memory (ROM)**. The computer can read ROM, but this type of memory cannot be changed.
- **Random-access memory (RAM)**. Both the computer and the user can read or change RAM.

In addition to the internal memory, data can be stored outside the basic computer circuitry on external storage devices. These devices include external hard drives, flash drives, and DVDs.

Attached to the computer are input and output devices. Input devices allow the operator to enter data into the computer's operating system. Typical input devices are the keyboard and mouse. Other input devices are the following:

- A joystick used in games or simulators.
- A scanner that converts images, such as drawings and photographs.

- A microphone that can be used to gather sound information.
- A touch panel that senses where a person places his finger or a stylus. See **Figure 22-28**.
- A pen with a light that can be used to draw images or select objects from a video display.

The data that computer operations generate is called *output*. Output can be viewed on output devices such as video-display monitors and overhead projectors. See **Figure 22-29**.

Deere & Co.

Figure 22-28. This farmer is using a touch screen to enter data.

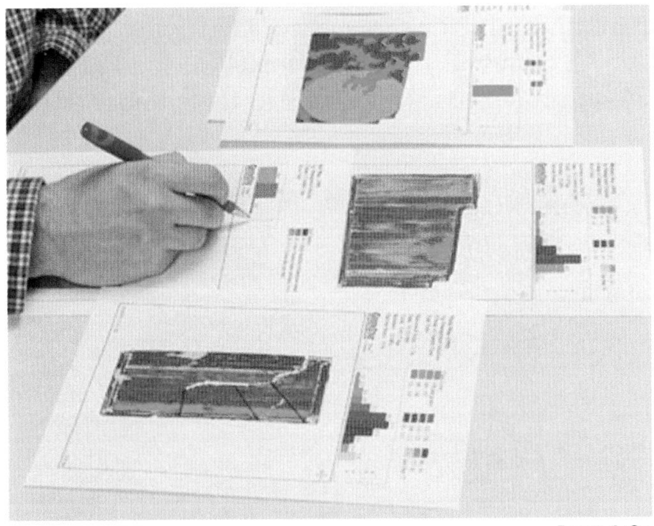

Deere & Co.

Figure 22-29. An inkjet printer produced these pages.

Software is a term used to describe instructions that direct a computer to perform specific tasks. For example, software programs allow people to use computers to write text, prepare drawings, maintain financial records, and perform many other functions. See **Figure 22-30**.

Operating systems are the software that manages the computer's processes, memory, and the operation of all other hardware and software. Operating systems allow you to use the computer without knowing how to communicate in computer code. The three major operating systems that come preloaded on personal computers are Windows®, Linux®, and Apple Mac OS® X. Each of these systems mostly uses a graphical user interface (GUI), meaning that everything you interact with on the screen is a combination of graphics and text. Prior to GUIs, computer commands needed to be typed in line by line on a command-line interface. See **Figure 22-31**.

Application software (***apps***) are computer programs that work with operating systems to help computers perform specific functions or tasks. Apps include business software, such as word processors or spreadsheet programs; graphic software, such as image-editing programs; communication software, such as web browsers; and entertainment software, such as games.

Goodheart-Willcox Publisher

Figure 22-30. Software enables a computer to produce a graphic layout.

```
Microsoft Windows [Version 6.1.7601]
Copyright (c) 2009 Microsoft Corporation.  All rights reserved.

C:\Users\kwarren>help
For more information on a specific command, type HELP command-name
ASSOC          Displays or modifies file extension associations.
ATTRIB         Displays or changes file attributes.
BREAK          Sets or clears extended CTRL+C checking.
BCDEDIT        Sets properties in boot database to control boot loading.
CACLS          Displays or modifies access control lists (ACLs) of files.
CALL           Calls one batch program from another.
CD             Displays the name of or changes the current directory.
CHCP           Displays or sets the active code page number.
CHDIR          Displays the name of or changes the current directory.
CHKDSK         Checks a disk and displays a status report.
CHKNTFS        Displays or modifies the checking of disk at boot time.
CLS            Clears the screen.
CMD            Starts a new instance of the Windows command interpreter.
COLOR          Sets the default console foreground and background colors.
COMP           Compares the contents of two files or sets of files.
COMPACT        Displays or alters the compression of files on NTFS partitions.
CONVERT        Converts FAT volumes to NTFS.  You cannot convert the
               current drive.
COPY           Copies one or more files to another location.
DATE           Displays or sets the date.
DEL            Deletes one or more files.
DIR            Displays a list of files and subdirectories in a directory.
DISKCOMP       Compares the contents of two floppy disks.
DISKCOPY       Copies the contents of one floppy disk to another.
DISKPART       Displays or configures Disk Partition properties.
DOSKEY         Edits command lines, recalls Windows commands, and
               creates macros.
DRIVERQUERY    Displays current device driver status and properties.
ECHO           Displays messages, or turns command echoing on or off.
ENDLOCAL       Ends localization of environment changes in a batch file.
ERASE          Deletes one or more files.
EXIT           Quits the CMD.EXE program (command interpreter).
FC             Compares two files or sets of files, and displays the
               differences between them.
FIND           Searches for a text string in a file or files.
FINDSTR        Searches for strings in files.
FOR            Runs a specified command for each file in a set of files.
FORMAT         Formats a disk for use with Windows.
FSUTIL         Displays or configures the file system properties.
FTYPE          Displays or modifies file types used in file extension
               associations.
GOTO           Directs the Windows command interpreter to a labeled line in
               a batch program.
GPRESULT       Displays Group Policy information for machine or user.
GRAFTABL       Enables Windows to display an extended character set in
               graphics mode.
HELP           Provides Help information for Windows commands.
ICACLS         Display, modify, backup, or restore ACLs for files and
               directories.
IF             Performs conditional processing in batch programs.
LABEL          Creates, changes, or deletes the volume label of a disk.
MD             Creates a directory.
MKDIR          Creates a directory.
```

Goodheart-Willcox Publisher

Figure 22-31. Command-line interface.

Binary, Bits, and Bytes

Computer systems are comprised of digital circuits that operate based on ***binary code***, a language that consists of only 1s and 0s. The 1s and 0s are used to represent numbers, letters, and all other digital information, as well as computer commands. The 1s and 0s are essentially switches. A 1 means *on* and a 0 means *off*.

A series of 1s and 0s can be used to represent different numbers. Much like the decimal system, binary code has different places that represent different values. In the decimal system, there are places for the digits of a number that represent a different power of 10. For example, the number 3152 has a 2 in the 1s place, a 5 in the 10s place, a 1 in the 100s place, and a 3 in the 1000s place. However, the places in binary code increase by powers of 2 instead of 10. Thus, the first place is 1, the second is 2, the third is 4, the fourth is 8, the fifth is 16, the sixth place is 32, and so on. When a 1 is placed in one of these binary places, the value of that place is considered switched on. If a zero is in one of the binary places, the value of that place is considered switched off.

Consequently, if a 1 is placed in a binary place, it indicates that the number represented has the value of that place. If a zero is in one of the places, then the number represented does not include the value of that place. To determine the overall value of the represented number, you add together all of the values of the places that have a 1. For example, "101101" is binary for the number 45. See **Figure 22-32**. The 1, 4, 8, and 32 binary places are switched on because they contain a 1. The 2 and 16 binary places are switched off with a 0. **Figure 22-33** shows the binary code for decimal numbers 0 through 20.

Computers operate using a fixed number of binary digits. These individual digits are called *bits*, which are short for **b**inary dig**its**. A bit can either be a 1 or 0, which is different from the decimal system where a digit can be a value of 0 through 9. Bits are almost always bundled into an 8-bit collection called a *byte*. The eight bits in a byte enable a total of 256 different combinations of 1s and 0s, which can represent numerical values of 0 to 255. A bit is the smallest unit of storage on a computer. However, bytes are typically used to describe information storage. The space that data takes up on a computer, file sizes, and hard disk capacities are measured in bytes.

CPUs are also labeled in bits, such as 32-bit or 64-bit processors. This refers to the number of bits

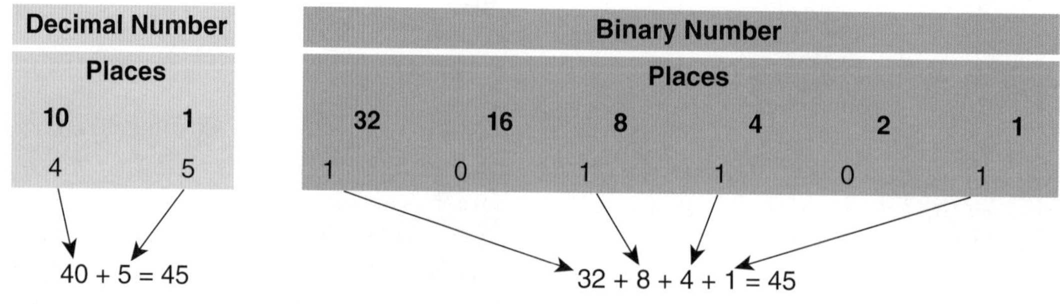

Goodheart-Willcox Publisher

Figure 22-32. 101101 is binary for the number 45.

Decimal Number Places		Binary Number Places					
10	1	32	16	8	4	2	1
	0						0
	1						1
	2					1	0
	3					1	1
	4				1	0	0
	5				1	0	1
	6				1	1	0
	7				1	1	1
	8			1	0	0	0
	9			1	0	0	1
1	0			1	0	1	0
1	1			1	0	1	1
1	2			1	1	0	0
1	3			1	1	0	1
1	4			1	1	1	0
1	5			1	1	1	1
1	6		1	0	0	0	0
1	7		1	0	0	0	1
1	8		1	0	0	1	0
1	9		1	0	0	1	1
2	0		1	0	1	0	0

Goodheart-Willcox Publisher

Figure 22-33. Binary code for decimal numbers 0 through 20.

of data that a CPU can handle for locating information within the computer's memory or the number of instruction sets that it can process. The ***register*** is a component that allows the CPU to access different information or data within the computer's memory by storing the location of the information. These locations are necessary for the CPU to access the correct data within the computer's memory. However, locations are stored in bits, and a register can only store so many bits at a time. Therefore, the number of bits that a register can hold determines how much memory the CPU can access.

The number of locations that a CPU can access is represented by the formula 2^n, where n represents the number of bits the CPU has. For example, a CPU with 32 bits can access 4,294,967,296 locations of information. Additionally, the register holds the instruction sets for the computer, which tell the CPU how to operate. The instruction sets are also stored as bits. The more bits a CPU register can read, the more memory it can access and faster it can process it.

One use of bytes is to represent individual letters of the alphabet and other text characters. To do so, a binary coding system called the *American Standard Code for Information Interchange (ASCII)* is used. ASCII assigns each character, such as upper-case letters, lower-case letters, and punctuation marks, a numerical value so the computer can handle text. The full table of the 128 ASCII characters (characters 0–127) and their decimal values are shown in **Figure 22-34**.

The numerical values for each character are converted into 8-bit code for a computer to read. For example, the ASCII numerical value for "Z" is 90, which means its 8-bit code is 1011010. Recall that 8 bits is equal to a byte. Thus, each character in a text document uses 1 byte of memory. This also includes a byte for each space between words and any commands necessary to control how the text is processed. You can test this yourself by creating and saving a text file on your computer and checking the file size. Type "Hello World" in the text file, save it, and check the file size. You will see that thesize of the file is 13 bytes. If you add a period to the end of "Hello World," the file size increases to 14 bytes.

Large files obviously require a large amount of bytes. Prefixes, such as kilo-, mega-, giga-, and tera-indicate the binary multipliers. See **Figure 22-35**. For

Decimal	Char	Decimal	Char	Decimal	Char	Decimal	Char
0	[NULL]	32	[SPACE]	64	@	96	`
1	[START OF HEADING]	33	!	65	A	97	a
2	[START OF TEXT]	34	“	66	B	98	b
3	[END OF TEXT]	35	#	67	C	99	c
4	[END OF TRANSMISSION]	36	$	68	D	100	d
5	[ENQUIRY]	37	%	69	E	101	e
6	[ACKNOWLEDGE]	38	&	70	F	102	f
7	[BELL]	39	‘	71	G	103	g
8	[BACKSPACE]	40	(	72	H	104	h
9	[HORIZONTAL TAB]	41	)	73	I	105	i
10	[LINE FEED]	42	*	74	J	106	j
11	[VERTICAL TAB]	43	+	75	K	107	k
12	[FORM FEED]	44	,	76	L	108	l
13	[CARRIAGE RETURN]	45	-	77	M	109	m
14	[SHIFT OUT]	46	.	78	N	110	n
15	[SHIFT IN]	47	/	79	O	111	o
16	[DATA LINK ESCAPE]	48	0	80	P	112	p
17	[DEVICE CONTROL 1]	49	1	81	Q	113	q
18	[DEVICE CONTROL 2]	50	2	82	R	114	r
19	[DEVICE CONTROL 3]	51	3	83	S	115	s
20	[DEVICE CONTROL 4]	52	4	84	T	116	t
21	[NEGATIVE ACKNOWLEDGE]	53	5	85	U	117	u
22	[SYNCHRONOUS IDLE]	54	6	86	V	118	v
23	[END OF TRANS. BLOCK]	55	7	87	W	119	w
24	[CANCEL]	56	8	88	X	120	x
25	[END OF MEDIUM]	57	9	89	Y	121	y
26	[SUBSTITUTE]	58	:	90	Z	122	z
27	[ESCAPE]	59	;	91	[	123	{
28	[FILE SEPARATOR]	60	<	92	\	124	|
29	[GROUP SEPARATOR]	61	=	93	]	125	}
30	[RECORD SEPARATOR]	62	>	94	^	126	~
31	[UNIT SEPARATOR]	63	?	95	_	127	[DEL]

Goodheart-Willcox Publisher

Figure 22-34. ASCII table of characters.

Prefix	Unit Symbol	Multiplier / Size
Byte-	B	8 bits
Kilo-	K	2^{10} = 1,024 bytes
Mega-	M	2^{20} = 1,048,576 bytes
Giga-	G	2^{30} = 1,073,741,824 bytes
Tera-	T	2^{40} = 1,099,511,627,776 bytes
Peta-	P	2^{50} = 1,125,899,906,842,624 bytes
Exa-	E	2^{60} = 1,152,921,504,606,846,976 bytes
Zetta-	Z	2^{70} = 1,180,591,620,717,411,303,424 bytes
Yotta-	Y	2^{80} = 1,208,925,819,614,629,174,706,176 bytes

Goodheart-Willcox Publisher

Figure 22-35. Binary multipliers and symbols for each prefix of bytes.

example, *kilo-* indicates a multiplier of 2^{10}, which is equal to 1,024. One kilobyte actually consists of 1024 bytes. If you see a file that is 24 kilobytes, you would multiply 24 by 2^{10} to determine that the file is 24,576 bytes.

Bits and bytes are also used to represent graphical images. An image is made of numerous pixels. A pixel is the smallest unit of an image. Different color pixels are arranged together to form a complete image. The

Technology Explained Virtual Reality

Virtual reality: A computer interface allowing a user to interact with three-dimensional, computer-generated images.

Computer technology has taken one giant step after another over the past 30 years. As computers have become more complex, they have branched off into many different areas. Virtual reality is one of the more spectacular areas of computer technology. See **Figure A**.

Caterpillar, Inc.

Figure A. Employees manipulate a virtual front-end loader.

Virtual reality was born in the 1960s, though in a very crude form. In 1981, the first practical application of a virtual reality system was exhibited. The Visually Coupled Airborne Systems Simulator trained pilots to fly complex, high-speed aircraft. The term *virtual reality* did not, however, appear until the mid-1980s. Jaron Lanier, the founder of VPL Research, coined the term.

This interface allows a user to interact with an environment that a computer creates. In addition to a computer, complex equipment is needed. Computer software, a headset, and a tool relaying the user's movements are required. This tool is often a glove transmitting its relative position and many finger movements to the computer.

Uses for virtual reality are many, with new uses being developed everywhere. The military has used virtual reality to train fighter pilots for some time. Companies use virtual reality to train new employees. Surgeons can have virtual reality training facilities to train on virtual patients with virtual scalpels. Architects allow clients to tour new homes during the design stage. See **Figure B**. Clients can make changes based on what they see before construction begins. Virtual reality has recreational uses as well. Complex systems can be found in expensive arcade games. Simpler systems are available for home video games. As a result of technological advancements, the prospects for virtual reality are endless.

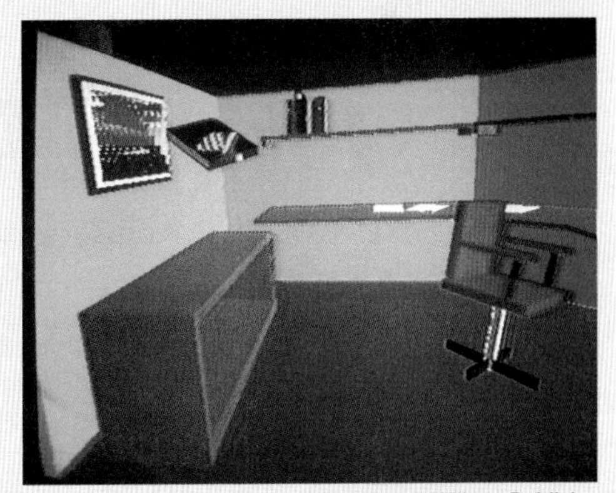

Goodheart-Willcox Publisher

Figure B. Virtual objects can be moved about in this virtual office space.

color of each pixel is made of a combination of red, green, and blue light. The human eye can perceive all colors by mixing just these three colors of light, **Figure 22-36**. The color of each pixel is determined by three bytes of data—one byte represents the value for the intensity of the red light in the pixel, one byte represents green, and one represents blue. Therefore, an image is an arrangement of pixels that each require three bytes of information.

Algorithms

Developing software that instructs computer systems to perform specific tasks requires the formulation of algorithms. An ***algorithm*** is a set of step-by-step instructions used to perform calculations, process data, and automatically answer questions. Algorithms for computer systems are essentially logic written for software to produce a desired output from given inputs, similar to the combinational and sequential logic for digital circuits.

An algorithm can be developed for almost any task. See **Figure 22-37**. In computer systems, algorithms are a set of operations for computer software to accomplish a task or solve a problem. For example, algorithms are used by computer programs to determine the best route from one place to another, to compress data to send large amounts

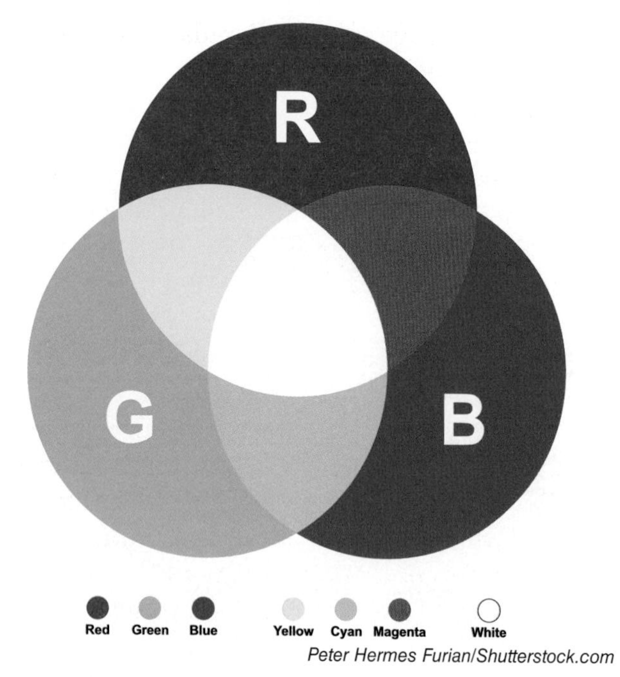

Peter Hermes Furian/Shutterstock.com

Figure 22-36. The human eye perceives all colors by mixing these three colors of light.

of information over the Internet in a short amount of time, and to respond to gamer's input or actions on a video game. The ability to develop and apply efficient algorithms allows engineers and computer scientists to develop significant, functional, and fast computer software. Good algorithms correctly solve a problem in the least amount of time possible.

When computer engineers, software engineers, or computer scientists intend to develop an algorithm for a new computer program or application, they must first clearly define a problem that a written algorithm can solve. They develop a method for translating the procedures into an algorithm. A computer system capable of reading and executing the resulting algorithm is required.

Before a new algorithm is created, research is often done to determine if a specific algorithm already exists to perform the intended task. The existing algorithm is examined to see if it will solve the problem or complete the desired task with all

Peanut Butter and Jelly Sandwich Algorithm
Step 1: Open bread bag
Step 2: If bread is already sliced, remove two slices If not sliced, remove loaf cut two slices
Step 3: Place slices separately on flat surface
Step 4: Open peanut butter jar
Step 5: Pick up butter knife
Step 6: Use knife to remove 1½ tablespoons of peanut butter
Step 7: Spread the peanut butter evenly on the right slice of bread
Step 8: Replace lid on peanut butter jar
Step 9: Repeat steps 4 through 8 using a jar of jelly and left slice of bread
Step 10: Set butter knife down
Step 11: Pick up slice of bread with peanut butter
Step 12: Place the slice of bread with peanut butter, peanut butter side facing down, on top of the slice of bread with jelly
Step 13: If desired, pick up butter knife slice the sandwich in half Otherwise, move on to next step
Step 14: Eat sandwich

Goodheart-Willcox Publisher

Figure 22-37. An algorithm for making a peanut butter and jelly sandwich.

possible inputs for the current situation. If not, the logic of the algorithm can be modified to do so. Algorithm developers also determine if they can minimize the amount of time it takes for a preexisting algorithm to run and the amount of computer memory it requires and modify it as necessary. If a new algorithm needs to be developed, a special language must be used to write it.

Pseudocode, a common language for developing algorithms, is a straightforward, informal, text-based language used to abstractly write a set of procedures for completing a task. Pseudocode is simple and easy-to-read statements that do not have to follow the rules of grammar. It contains well-defined instructions that can easily be used to later write the algorithms in a specific programming language.

Pseudocode consists of basic operations, such as *read input* and *print output*, with conditional operations (instructions to check an input value) and sequential operations (instructions to execute tasks in a specific order) in between, such as the following:

if <condition> then
<statement to be done>
else <statement to be done>

For example, the following pseudocode is a simple algorithm for a program that indicates whether a user passed or failed an exam:

0 *Read Input: Exam Score*
1 *If Score ≥ 60*
2 *Print Output: "Passed"*
3 *Else*
4 *Print Output: "Failed"*

Programming Languages

Once an algorithm for a computer program has been developed using pseudocode, a computer engineer, software engineer, or computer scientist writes the algorithm in a specific programming language. A ***programming language*** is a formal language or set of rules to communicate operations for a computer system to perform. This language expresses the algorithm in a manner that the specific computer system can understand.

Programming languages exist to enable human-to-machine communication for a variety of purposes. For example, programming languages are used to develop mobile apps and model electrical circuits. Each language has a different set of syntax rules. Syntax defines the structure of the language. It is the language's form. Popular programming languages include the following:

- C language was developed in the 1970s and is one of the oldest and most widely used general-purpose programming languages. It is used for a wide range of operating systems and embedded computer applications. This language is the foundation of other popular languages like C++, C#, and JavaScript.
- C++ is used for programming operating systems and hardware platforms. It is an object-oriented language, which means that objects are created rather than sequences of instructions. C++ was designed to enhance C language.
- C# is a general-purpose language that combines the principles of C and C++ for developing software for Microsoft® platforms.
- JavaScript is a programming language that works across all web browsers. It is used in game development and in the creation of interactive web elements.
- Java is an object-oriented language that works across multiple software platforms. It is used for developing games, mobile applications, web content, most enterprise software, and the Android™ operating system.
- Python is a language for developing websites and mobile applications. It is often used by beginning programmers because it does not require as many lines of code to express a program as most others. This language is used for Instagram's web application.
- Ruby is a simplistic programming language used for developing mobile applications and websites. It is often used by beginning programmers.
- PHP (hypertext preprocessor) is a language used for developing web applications and dynamic websites. This language is used for the Facebook® website.
- MATLAB (matrix laboratory) is an interactive software system for performing numerical computations and creating graphical representations of data. However, it also has its own programming language which engineers use to perform complex and expansive computations or calculations to determine information about potential solution designs to real-life problems.
- SQL (structured query language) is a special language used for managing database management systems.

Summary

- ✔ Professionals who work together to make electronic devices and communication technologies include electrical, computer, and software engineers; computer scientists; and other professionals within the fields of information technology and information systems.
- ✔ Electrical engineers design, test, and oversee the production of devices that use electricity to control, process, store, and receive information. They also work with equipment for generating and transmitting power.
- ✔ Computer engineering involves the development, improvement, and supervision of the production and maintenance of computer hardware and software.
- ✔ Computer science focuses on the theoretical study and practical application of computation. Computation is any type of calculation for processing information.
- ✔ Electrical and computer engineering students study circuit theory, circuit elements, and circuit laws because circuits are involved in every electronic device.
- ✔ In an electrical circuit, voltage is the pressure of the electrons waiting to move toward the positive charge. Current is the rate of electron flow through the circuit.
- ✔ Ohm's law defines the relationship among voltage, current, and resistance in an electrical circuit.
- ✔ Electrical circuits may be configured as series circuits, parallel circuits, or series-parallel circuits.
- ✔ Batteries provide direct current resulting from chemical reactions inside the battery. The electrical charge flows only from the negative terminal of the battery to the positive terminal of the battery.
- ✔ Alternating current has an electrical charge that changes directions periodically. The voltage also reverses as the direction of current changes. Alternating current is mostly used to transmit high-voltage electricity over long distances.
- ✔ A signal is a measurable quantity used to communicate information. An analog signal is a continuous quantity that changes over time, providing exact values. Digital signals have discrete values—there are a limited amount of values that quantities can have.
- ✔ Digital circuits operate using discrete voltage values to convey information.
- ✔ Digital logic enables complex decisions to be made within computer systems based on a series of yes or no questions.
- ✔ Computers include physical equipment (hardware) and operating instructions (software). The four main hardware units are the CPU, the memory, the input devices, and the output devices.
- ✔ Operating systems are the software that manages the computer's processes, memory, and the operation of all other hardware and software.
- ✔ Computer systems are comprised of digital circuits that operate based on binary code, a language that consists of only 1s and 0s.
- ✔ Computers operate using a fixed number of binary digits (bits). Bytes are 8-bit collections of bits. Bits and bytes can represent numbers, text characters, and graphical images.
- ✔ Software development requires the creation of algorithms (sets of step-by-step instructions), which are then written in specific programming languages.

Check Your Engineering IQ 

Now that you have finished this chapter, see what you learned by taking the chapter posttest.
www.g-wlearning.com/technologyeducation/

Test Your Knowledge 

Answer the following end-of-chapter questions using the information provided in this chapter.

1. _____ engineers and computer engineers are primarily responsible for developing computer hardware.
2. The theoretical study and practical application of computation is the focus of _____ science.
3. Electrical energy involves the _____ of electric charge along a conducting pathway.
4. *True or False?* Wires used in electrical circuits are typically made of silicon.
5. What are the three main parts of a battery?
6. Ohm's law defines the relationship among voltage, current, and _____ in an electrical circuit.
7. Electrical components that allow electrical current to pass through them in only one direction are called _____.
   A. diodes
   B. transistors
   C. wires
   D. capacitors

8. In photoelectric cells, the electric current produced depends on the amount of _____ to which the cells are exposed.
9. *True or False?* In a series circuit, components are connected across common ends or nodes, providing multiple pathways for electrons to flow.
10. *True or False?* Kirchhoff's voltage law states that the sum of all voltage drops in a circuit is equal to the total applied voltage.
11. Which of the following statements about direct current is *false*?
    A. The electrical charge flows in only one direction.
    B. It powers most handheld electronic devices.
    C. Batteries provide direct current.
    D. It is used to transmit high-voltage electricity over long distances.
12. *True or False?* Digital signals are easier to use in electronics than analog signals.
13. Digital circuits operate using discrete _____ values to convey information.
14. How does combinational logic differ from sequential logic?
15. A(n) _____ gate produces an output signal of 1 if only one input signal is 1.
    A. OR
    B. NAND
    C. XOR
    D. NOR
16. What are the four main hardware units in a computer system?
17. The 1s and 0s in binary code are used to represent _____, letters, and all other digital information, as well as computer commands.
18. What is the name of the binary coding system that assigns numerical values to characters so the computer can handle text?
19. What is an *algorithm*?
20. C++ is an example of a programming _____.

Critical Thinking

1. Analyze the role electronics play in your life today and explain how that role may evolve over your lifetime.
2. Create a diagram to explain how and why electrons flow throughout an electrical circuit.
3. Describe how the periodic table of elements is used to determine the electrical properties of different elements.
4. Explain the relationship among voltage, current, and resistance.

STEM Applications

1. Use pseudocode to develop an algorithm for computing the gas mileage of your vehicle and determining whether the gas mileage is good or not. (Consider good gas mileage to be greater than 24 miles per gallon.)
2. Write the binary numbers for each of the following decimal numbers:
    A. 55
    B. 37
    C. 125
3. Complete the truth table for the following combination of logic gates:

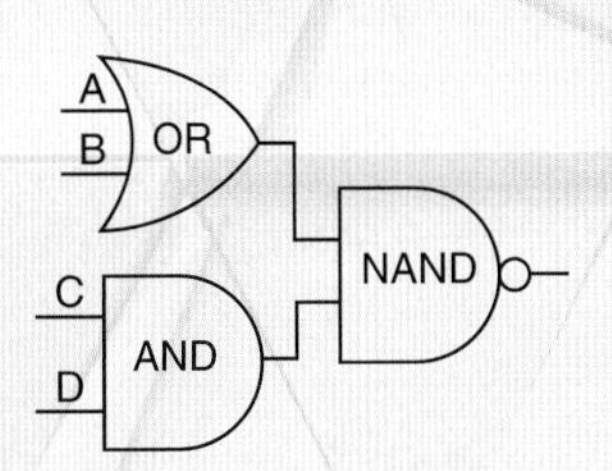

	AB			
	00	01	10	11
CD 00				
CD 01				
CD 10				
CD 11				

4. Complete the table for the circuit provided:

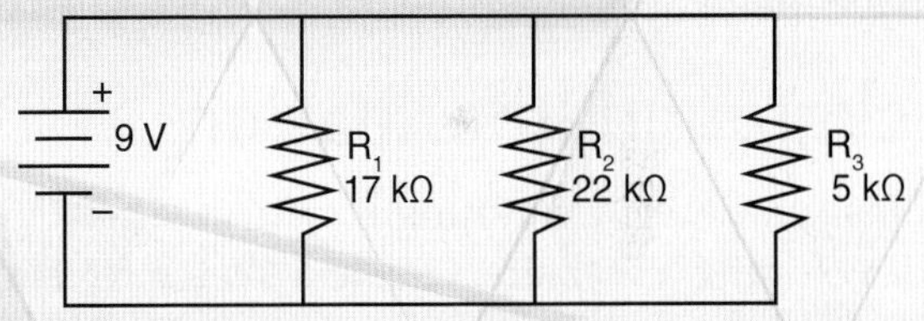

	R_1	R_2	R_3	R_{total}
V				
I				
R				

5. Complete the table for the circuit provided:

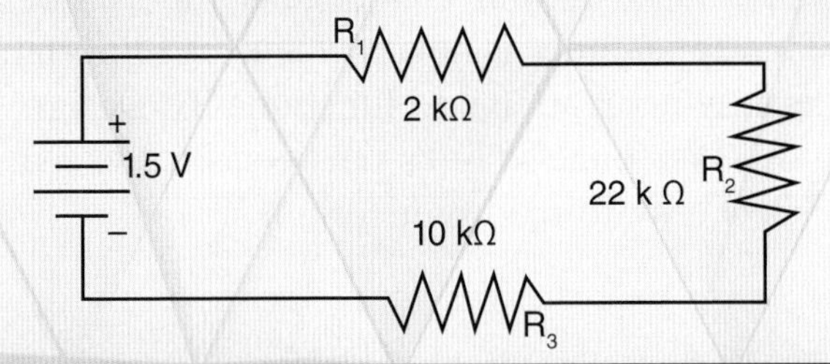

	R_1	R_2	R_3	R_{total}
V				
I				
R				

TSA Modular Activity

This activity develops the skills used in TSA's Webmaster event.

Cyberspace Pursuit

Activity Overview

In this activity, you will create a website composed of four components:

- An overview of your school's technology-education program.
- General information about your school.
- Historical information about your school.
- A page of links to related and interesting websites.

Materials

- Paper.
- A pencil.
- A computer with Internet access and Web page–development software.

Background Information

- **Design.** Careful planning is critical when developing a website. Create a sketch for each Web page. List the elements to be included on each page. Include lines showing the links between Web pages. Consider how a user will navigate within the website to be sure you have the necessary links.
- **Navigation and functionality.** The visual appearance of your Web page is important. This appearance is not, however, as important as easy navigation and functionality. Design your Web page so links can be easily located. Select easy-to-read type for links.
- **Type.** Too many fonts and type sizes can make your Web page unattractive. Vary the size and font based on the function of the text. Titles are meant to draw attention. Blend body type into the overall design and use a typeface, size, spacing, and justification comfortable for reading. If you use an unusual type font, people viewing your Web page might not have that font on their computers. In this situation, another font is substituted. This causes your Web page to have an unintended (and possibly less attractive) appearance. Use common type fonts such as Arial, Times New Roman, Tahoma, and Courier. If you want to use an unusual font, create the text as an image file. Insert the file into the Web page.
- **Images.** Images can add to the visual appeal of your Web pages, but too many images clutter a page and cause it to load slowly. Use a compressed-image file format (such as .jpg) and the lowest acceptable resolution for faster loading.

Guidelines

- Review at least five websites, evaluating the designs of the sites in terms of attractiveness and usability.
- Your website must have a home page containing separate links to each of the four components.
- There is no minimum or maximum number of pages for the individual components.
- The home page and each of the four components must contain both text and graphics.
- All pages must include a link to the home page.

- Use a pencil and paper to prepare a rough sketch for each page and an organizational chart showing how pages are linked.
- After you have developed the rough sketches, create the Web pages.
- Test the completed design to make sure all links work properly.

Evaluation Criteria

Your website will be evaluated using the following criteria:

- Web-page design.
- Originality.
- Content of Web pages.
- Functionality.

CHAPTER 23

Transporting People and Cargo

Check Your Engineering IQ

Before you read this chapter, assess your current understanding of the chapter content by taking the chapter pretest.
www.g-wlearning.com/technologyeducation/

Learning Objectives

After studying this chapter, you should be able to:

- ✔ Explain why transportation is considered a system.
- ✔ Identify the types and components of transportation systems.
- ✔ Summarize the characteristics of transportation pathways.
- ✔ Illustrate the types of transportation vehicles and their support structures.
- ✔ Distinguish among the five systems present in a transportation vehicle.
- ✔ Describe the two major components of a structural system.
- ✔ Identify common propulsion systems and their characteristics used in transportation vehicles.
- ✔ Recognize common guidance and control systems used in transportation vehicles.
- ✔ Explain the major systems of land-, water-, air-, and space-transportation vehicles and their functions.

Technical Terms

airfoil
air-transportation systems
apogee
automatic transmissions
aviation electronics
ballast
blimp
bow thruster
buoyancy
commercial aviation
containerships
control
degrees of freedom
earth-orbit travel
ergonomics
exosphere
fixed-wing aircraft
fuselage
general aviation
geosynchronous orbit
guidance
hovercraft
hydrofoil
inland-waterway transportation
intermodal shipping
internal combustion engine
keel
land-transportation systems
lighter-than-air vehicles
liquid-fuel rockets
manned spaceflight
manual transmission
maritime shipping
merchant ships
mesosphere
outboard motor
ozone layer
perigee
place utility
power-generation system
power-transmission system
propulsion
rotary-wing aircraft
solid-fuel rockets
space-transportation systems
stratosphere
structure
submersible
suspension
tail assembly
tankers
terminals
thermosphere
transaxle
transportation
troposphere
turbofan engine
turbojet engine
turboprop engine
vehicles
water-transportation systems

Close Reading

Before you read this chapter, make a list of all the different types of transportation you're familiar with. As you read the chapter, check your list and add to it if necessary. Additionally, note the different vehicular systems shared by vehicles of all four environments. List their similarities and differences.

The development of transportation and civilization are closely related, **Figure 23-1**. Without transportation, humans would not travel very far. In prehistoric times, when people traveled on foot, a typical journey was from home and back in one day. This distance normally was less than 25 miles (40 km). When people started using animals for transportation, they could venture farther from their homes, sometimes 40–50 miles (64–80 km) in a day. More travel led people to design and build roads. As early as 30,000 BCE, established transportation routes existed.

Sailing ships, followed much later by the railroad, enlarged the area of travel. Still, worldwide travel was available only to wealthy or adventurous people. Travel as we know it today is a recent development. The use of jet-powered aircraft after World War II made worldwide travel available to many people.

Transportation Defined

Transportation is all acts that relocate people or their possessions. Transportation technology provides for this movement, using technical means to extend human ability. This technology extends our ability beyond our own muscle power and our ability to walk and run.

Transportation is also an interaction among physical elements, people, and the environment, **Figure 23-2**. Transportation provides mobility for people and goods. This interaction uses such resources as materials, energy, money, and time. Transportation has a level of risk to cargo and passengers that can result in damage, injury, or death. This interaction has a societal impact, in that it affects such areas as employment and pollution.

Lipskiy/Shutterstock.com

James Steidl/Shutterstock.com

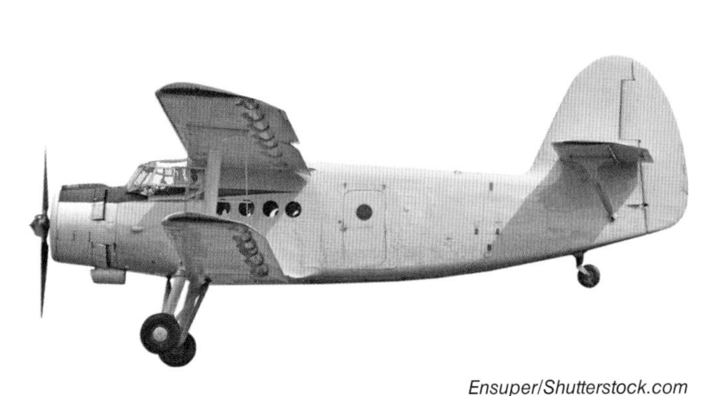

Ensuper/Shutterstock.com

NASA Images/Shutterstock.com

Figure 23-1. Transportation systems have evolved to meet our changing needs. All the transportation vehicles shown here were cutting-edge transportation vehicles at one time.

hxdyl/Shutterstock.com

Figure 23-2. Transportation systems are interactive systems of physical elements (vehicles and pathways, for example), people, and the environment.

Transportation is an important part of human culture. Try to imagine life without well-developed transportation systems. We think of transportation in the same way as food, clothing, and shelter. Transportation has become a basic need. For example, transportation takes us to work, opens up areas for recreation, allows for easy shopping, and helps keep families in touch.

Transportation as a System

Similar to all other technologies, transportation can be viewed as a system. Transportation is a series of interrelated parts. The parts work together to meet a goal. Transportation uses people, artifacts, vehicles, pathways, energy, information, materials, finances, and time to relocate people and goods.

Types of Transportation Systems

People use four environments or modes for transportation. Transportation systems have been developed for land, water, air, and space, **Figure 23-3**.

Land

Humans can move over land with ease. The earliest transportation systems were designed to move people over land. ***Land-transportation systems*** move people and goods on the surface of the earth from place to place. There are three major types of land-transportation systems:

- **Highway systems.** These systems include automobiles, buses, and trucks.
- **Rail systems.** These systems include freight, passenger, and mass-transit systems.
- **Continuous-flow systems.** These systems include pipelines, conveyors, and cables.

Water

Water transportation was the next system developed. Water transportation has grown to be an important mode for moving people and cargo. ***Water-transportation systems*** use water to support a vehicle. Water transportation includes inland waterways (rivers and lakes) and oceangoing systems.

Air

Air transportation became practical in the twentieth century. Orville and Wilbur Wright made the first successful flight in a power-driven airplane in 1903. This flight took place in Kitty Hawk, North Carolina. Since then, air travel has become an enormous industry. ***Air-transportation systems*** use airplanes and helicopters to lift passengers and cargo into the air so they can be moved from place to place. Today, air transportation includes commercial aviation (passengers and freight) and general aviation (private and corporate aircraft).

Christian Mueller/Shutterstock.com

Maximchuck/Shutterstock.com

litabit/Shutterstock.com

Andrey Armyagov/Shutterstock.com

Figure 23-3. Transportation systems operate on land, in water, and through air and space.

Space

Space-transportation systems use manned and unmanned flights to explore the universe. This transportation started in 1957 with the launch of *Sputnik*, a Soviet satellite. Since 1957, space exploration has expanded our knowledge of the universe. Astronauts have traveled to the moon, space shuttles have carried satellites and scientific materials into space, rovers have landed on distant planets, and probes have circled asteroids.

On the horizon is personal space travel. Hypersonic aircraft, or aircraft that is highly supersonic and generally recognized as speeds of Mach 5 and above, will merge air-travel and space-travel technologies. People will then be able to travel anywhere on the globe in a matter of hours.

Transportation-System Components

There are three major components in each transportation system, **Figure 23-4**. They are the following:

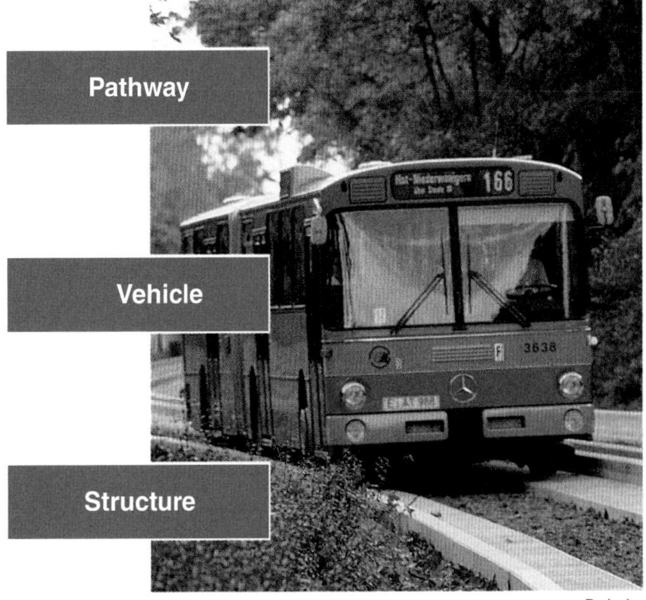

Daimler

Figure 23-4. Transportation systems use pathways, vehicles, and support structures. This photo shows a guided-bus system. The pathway is the concrete guideway. The vehicle is the bus. Support structures are the stations where passengers board the bus.

Career Connection Automotive Technician

Automotive technicians use skill and knowledge to inspect, maintain, and repair automobiles and light trucks. They review the description of a problem, use a diagnostic approach to locate the problem, repair the vehicle, and test the repair. Automotive technicians must be able to use a variety of testing equipment. This equipment includes onboard and stationary diagnostic equipment and computers.

PATIWIT HONGSANG/Shutterstock.com

High school students interested in becoming automotive technicians can take courses such as technology and engineering education, automotive repair, electronics, and computers. Interested students will need to complete some vocational or postsecondary education programs in automotive service for entry-level positions. These programs usually last six months to a year. Many are hands-on programs, centered on topics such as brake maintenance or engine performance.

Individuals who are organized, have strong customer service skills, enjoy troubleshooting, and are hands-on workers will perform well in this field. It is also important for automotive technicians to possess the necessary licenses, certifications, and registrations. Automotive technicians may work as automotive body and glass repairers, diesel service technicians, heavy vehicle and mobile equipment service technicians, and small engine mechanics. One of the largest professional organizations for automotive technicians is the National Institute for Automotive Service Excellence (ASE).

Employment in this area is expected to grow. Automotive technicians may have the following potential projects:

- Identify automotive problems and diagnose using computer software.
- Explain automotive issues to customers.
- Repair and replace worn and damaged parts.
- Perform routine care and maintenance.

- Pathways.
- Vehicles.
- Support structures.

Pathways

Transportation systems are designed to move people and cargo from one place to another. This movement provides ***place utility***, a value provided by being able to move things from one place to another. The things that are moved can be food, products, or people. We are willing to pay someone else to transport us or our goods from place to place.

To deliver place utility, a transportation system links distant locations using a network of pathways. See **Figure 23-5**. Pathways on land are readily visible and include streets, roads, highways, rail lines, and pipelines. These pathways form a network connecting most areas of the United States. Land pathways are all built by people. They fulfill transportation needs.

Water pathways are often less visible than land pathways. Water pathways include all navigable bodies of water, meaning they are deep enough for ships to use. Water pathways can be grouped into two categories: inland waterways and oceans. Inland waterways are rivers, lakes, and bays. They also include human-made canals. These canals allow water transportation in areas lacking navigable, naturally occurring rivers and lakes. Often,

Rail Lines

©iStockphoto.com/buzbuzzer

Canals and Oceans

©iStockphoto.com/photosfromafrica

Air and Space Routes

©iStockphoto.com/Elerium

Figure 23-5. Pathways include roads, rail lines, rivers, canals, oceans, and air and space routes.

the canals use aqueducts to cross streams. Canals use locks to raise or lower watercraft as the terrain changes, **Figure 23-6**.

Oceans provide vast water pathways. Oceans surround the earth's major landmasses. Therefore, an ocean transportation network can easily serve the continents.

Some pathways are hard to see. Air and space routes are totally invisible. They are defined by people and appear only on maps. Airplanes and spacecraft can travel in almost any direction. Only human decisions establish the correct pathway. The pathway is chosen to ensure efficient and safe travel. Airplanes use pathways to move people and cargo easily over land and water barriers.

Vehicles

All transportation systems, except for continuous-flow systems (such as pipelines and conveyors), use vehicles. ***Vehicles*** are technological artifacts

Aqueduct

Aerovista Luchtfotografie/Shutterstock.com

Lock

Dmitry Bruskov/Shutterstock.com

Figure 23-6. Canals are constructed waterways. They include waterways, locks to raise or lower ships, and aqueducts to cross streams and roadways.

designed to carry people and cargo on pathways. They are designed to contain and protect the cargo as it is moved from place to place. The demands on the vehicle change with the type of cargo, **Figure 23-7**.

Passenger vehicles are designed to protect occupants as they travel. Impact-absorbing construction, seat belts, fire protection, and other safety measures should be designed and built into the vehicle. Passenger vehicles also provide such features as proper seating and good lighting. Vehicles that carry people long distances should provide food and restrooms.

Cargo vehicles are designed to protect cargo from damage by motion and the outside environment. Cargo should be cushioned so it is not broken, dented, or scratched. Fumes, gases, or liquids that can affect the cargo must be kept at a distance. Finally, the cargo should be protected from theft and vandalism.

Support Structures

All transportation systems use structures. Many transportation pathways use human-built structures. Rail lines, canals, and pipelines are examples of constructed structures.

Transportation systems need structures other than just pathways. They also need ***terminals***. Terminals are locations where transportation activities begin and end. They are used to gather, load, and then unload passengers and goods, **Figure 23-8**.

Terminals provide passenger comfort and cargo protection before the people and cargo are loaded into transportation vehicles. They also provide connections for various transportation systems. Some allow truck shipping to connect with air, rail, or ocean shipping. Other terminals connect automobile, rail, and bus systems with air-transportation systems. These types of terminals allow for ***intermodal shipping***, **Figure 23-9**. Intermodal shipping is a system that transports cargo on two or more modes of transport before it reaches its destination. An example of intermodal shipping is oceangoing shipping containers being hauled on railcars.

Other structures are used to control transportation systems. These structures are communication

anucha maneechote/Shutterstock.com

Figure 23-8. Transportation terminals are at the start and the end of transportation pathways.

tratong/Shutterstock.com

Matej Kastelic/Shutterstock.com

Figure 23-7. Vehicles, such as aircraft, are designed differently for cargo and people.

Serjio74/Shutterstock.com

Figure 23-9. These oceangoing shipping containers are now traveling on rail. This is an example of intermodal shipping.

towers, radar antennae, traffic signals, and signs. These structures help vehicle operators stay on course and observe rules and regulations.

Vehicular Systems

All vehicles have some common systems. These systems are the structural, propulsion, suspension, guidance, and control systems. See **Figure 23-10**.

Structural System

All vehicles are designed to meet a common goal—to contain and move people and goods. People want to arrive safely and in comfort. Cargo must be protected from the weather, damage, and theft. The ***structure*** of the vehicle helps to do these things. The structural system is composed of the physical frame and covering. The structural system provides spaces for people, cargo, power and control systems, and other devices.

Propulsion System

Transportation vehicles are designed to move along a pathway. This pathway might be a highway, a rail line, a river, an ocean, or an air route. A vehicle must have a force to move it from its starting point to its destination. ***Propulsion*** is this force. The propulsion system uses energy to produce power needed for motion.

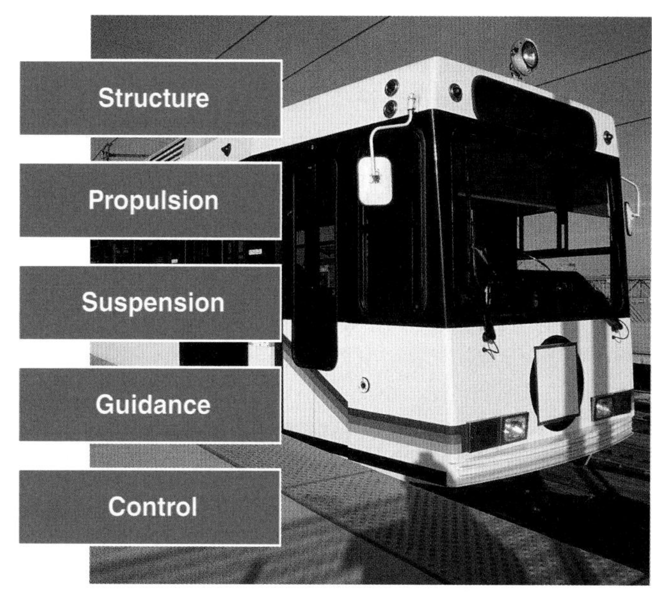

Joseph Sohm/Shutterstock.com

Figure 23-10. Transportation vehicles have five basic systems.

Propulsion systems range from the simple pedal, chain, and wheel system of a bicycle to complex heat engines, such as gasoline, diesel, and rocket engines. See **Figure 23-11**. Several factors

Simple

To

Monkey Business Images/ Shutterstock.com

Airbus

Complex

Figure 23-11. Propulsion systems range from simple to complex.

determine what type of engine will be used in a vehicle. These factors include the following:

- The environment in which the vehicle travels.
- Fuel availability and cost.
- The forces that must be overcome, such as vehicle and cargo weight, rolling friction, and water or air resistance.

The engine must match the job. For example, using a jet engine to propel an automobile is excessive. The capabilities of a jet engine do not match the job of moving a car. Likewise, using a large diesel engine to power an airplane is not a wise choice because the diesel engine is too heavy. Many different sizes and types of engines are used in transportation vehicles.

Suspension System

All vehicles and cargo have weight. This weight must be supported as the vehicle moves along its pathway. ***Suspension*** provides proper support. Suspension systems include the following:

- Wheels, axles, and springs on land vehicles.
- Wings on airplanes.
- Hulls on ships.

Guidance System

Handling any vehicle requires information. Operators must know their locations, speeds, and directions of travel. Information about traffic conditions and rules is also required. The guidance system provides this information.

Guidance information can be as simple as a speedometer reading or a traffic light. Guidance systems can also be quite complex. Instrument landing systems (ILS) and land-based satellite-tracking stations are examples of complex guidance systems.

Control System

A vehicle moves from its origin to its destination to relocate people and cargo. As a vehicle moves, it must be under ***control***, **Figure 23-12**. There are two types of control systems: speed control and direction control. The vehicle can be made to go faster through acceleration or slowed down by braking or coasting. How a vehicle changes its direction depends on its environment of travel. Land vehicles have wheels that turn or they follow a track. Ships have rudders that move the ship. Airplanes have adjustable ailerons, rudders, and flaps. Spacecraft use rocket thrusters.

©iStockphoto.com/grahamheywood

Figure 23-12. Transportation vehicles can have three degrees of freedom: forward and backward; left and right; and up and down. This passenger train has one degree of freedom.

The control system used depends on the ***degrees of freedom*** a vehicle has. For example, rail vehicles have one degree of freedom. They move forward or backward by using speed control. Gravity and the rail eliminate the possibility of up-and-down or left-to-right movement.

Automobiles and ships have more freedom. They can move forward and backward, left or right. These vehicles have two degrees of freedom.

Aircraft, spacecraft, and submarines have three degrees of freedom. These vehicles can move forward and backward and left and right, and can also change altitude by moving up and down.

Land-Transportation Vehicles

Land transportation includes all movement of people and goods on or under the surface of the earth. This transportation includes highway-, material-, rail-, and on-site transportation systems. Highway systems use automobiles, trucks, and buses to move people and cargo. Rail systems move people and cargo from place to place. Material-transportation

Think Green Hybrid Vehicles

Hybrid vehicles use more than one power source. Hybrid cars have smaller gasoline tanks, high-voltage batteries, and convert energy from braking. One reason people choose to drive hybrid vehicles is to save money buying gasoline. The battery provides some power to the vehicle, so the vehicle doesn't use as much gasoline.

Hybrid vehicles are also better for the environment. Because hybrid vehicles consume less gasoline, the use of fossil fuels is reduced. Gasoline does not burn cleanly, and conventional vehicles may emit large quantities of carbon dioxide. Because of the electric power source that works with the gasoline in a hybrid vehicle, the carbon dioxide emissions are reduced.

systems include pipelines and conveyors. On-site transportation systems include systems commonly found in such places as factories, stores, and hospitals. These systems might include forklifts, tractors and carts, chutes, elevators, and conveyors.

Land-Transportation Vehicle Structure

Each land-transportation system requires special vehicles. The structure a vehicle has is based on the vehicle's use. Passenger vehicles are different from cargo-carrying vehicles. All the vehicles, however, have three basic structural units:

- A passenger or operator unit.
- A cargo unit.
- A power unit.

The sizes and locations of these units or compartments vary with the type of vehicle, **Figure 23-13**. Freight and some passenger rail systems place the power and operator units in one vehicle called a locomotive. The locomotive pulls the cargo and passenger units. Mass-transit systems have power units (electric motors) and passenger units in each car. The operator unit is located in the front car.

Standard automobiles and delivery trucks have all three units in one vehicle. Most long-distance trucks, however, place the power and operator units in the tractor. The cargo unit is the trailer attached to the tractor.

Passenger and operator units should be designed for comfort and ease of operation. This requires the use of ergonomic principles. ***Ergonomics*** is the study of how people interact with the things they use. Ergonomic vehicle design requires that seats adjust for people of different sizes. The instruments providing guidance data must be easy to see, **Figure 23-14**. The operating controls, such as for steering and braking, must be easy to reach and operate.

Freightliner Corp.

Figure 23-13. All vehicles have operator or passenger, power, and cargo units.

fztommy/Shutterstock.com

Figure 23-14. This automobile instrument panel was designed to be easy to read.

The vehicle must also have an appropriate structural design. It must provide for safety and for operator, passenger, and cargo protection. Most vehicles have a reinforced frame covered with a skin. The skin protects the vehicle interior from the outside environment. The frame supports the skin and carries the weight of passengers and cargo. This reinforced frame also provides crash protection by absorbing the impact when vehicles are involved in accidents.

Land-Transportation Vehicle Propulsion

Most land-transportation vehicles move along their pathways on rolling wheels. Two systems produce the rotation: power generation and power transmission.

Power-Generation System

The ***power-generation system*** uses an engine as an energy converter. The engine produces the power needed to propel the vehicle. The most common engine in land vehicles is the ***internal combustion engine***. This type of engine burns fuel inside the engine to convert energy from one form to another. The chemical energy in the fuel is first changed to heat energy. The piston and crankshaft then convert the heat energy into mechanical (rotating) energy to move the vehicle.

Land vehicles normally use either four stroke–cycle gasoline engines or diesel engines. Some rail vehicles use electric motors for propulsion. Electrically powered vehicles for highway use are on the horizon.

Most internal combustion gasoline engines make four strokes in a complete cycle. The first stroke is called the *intake stroke*. The intake valve opens, the piston moves down, and a fuel-air mixture is drawn into the cylinder. When the piston is at the bottom of the intake stroke, the valve closes. The piston moves upward, beginning the compression stroke that compresses the fuel-air mixture. When the piston is at the top of its travel, a spark plug ignites the fuel-air mixture. The burning fuel expands rapidly and drives the piston down. This is called the *power stroke*. At the end of the power stroke, the exhaust valve opens. The piston moves upward to force burned gases from the cylinder. The last stroke is called the *exhaust stroke*. The exhaust valve closes, and the engine is ready to repeat the four strokes—intake, compression, power, and exhaust, **Figure 23-15**.

Diesel engines power many large vehicles, such as buses, heavy trucks, and locomotives. The most common diesel engine uses a four-stroke cycle, similar to a gasoline engine, **Figure 23-16**. During the intake stroke, however, only air is drawn in. As the air is compressed, it becomes very hot. When the piston is at the top of the compression stroke,

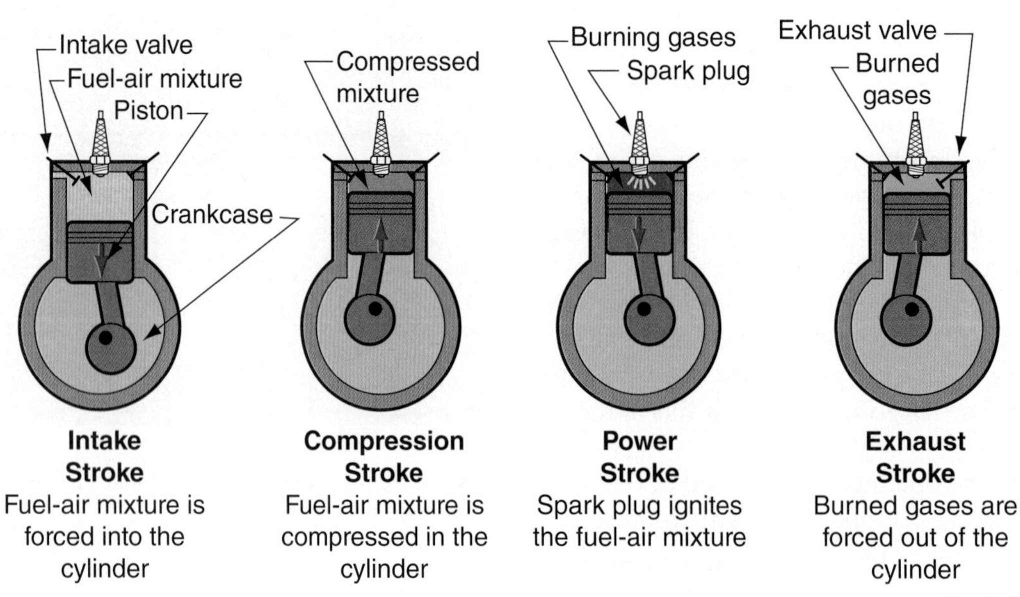

Figure 23-15. The four actions in a four-stroke engine are intake, compression, power, and exhaust.

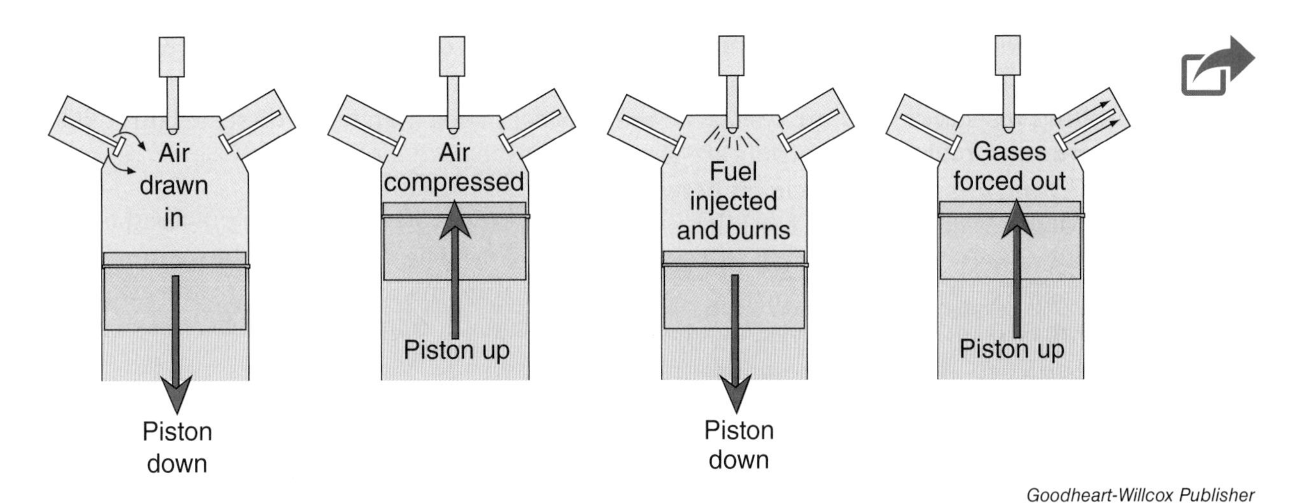

Goodheart-Willcox Publisher

Figure 23-16. How a diesel engine works.

fuel is injected directly into the cylinder. The fuel touches the hot air and burns rapidly. This causes the power stroke, which is followed by the exhaust stroke. These last two strokes are very similar to those of a gasoline engine.

Newer types of propulsion systems include hybrid electric systems. See **Figure 23-17**. This system is at the heart of hybrid electric vehicles (HEVs). The gasoline-electric hybrid propulsion systems contain three major parts:

- A gasoline engine similar to those in standard cars, except it is smaller.
- A high-voltage battery that provides energy to move the vehicle and stores energy for later use.
- A sophisticated electric motor that can be used as a power source to move the vehicle and a generator to charge the vehicle's battery. When the motor is acting as a motor, it draws energy from the battery to power the vehicle. When the vehicle slows, the motor acts as a generator and returns energy to the battery and also provides a braking force (regenerative braking).

Many HEVs reduce fuel use and emissions by shutting down the engine at idle and restarting it when needed.

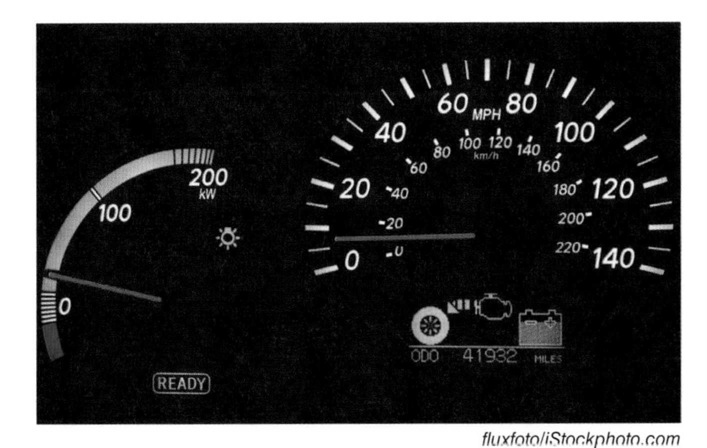

fluxfoto/iStockphoto.com

Figure 23-17. A dashboard readout on a hybrid car showing kilowatts, a speedometer, engine and battery statuses, and an odometer.

Power-Transmission System

A transmission is often an important component in a ***power-transmission system***. This component is located between the engine and the drive wheels. See **Figure 23-18**. Most automobile and truck transmissions are either mechanical or fluid devices. Often, a transmission has several input-to-output ratios, commonly called *speeds*. A five-speed transmission has five ratios. The different ratios allow the engine power to be used efficiently. The

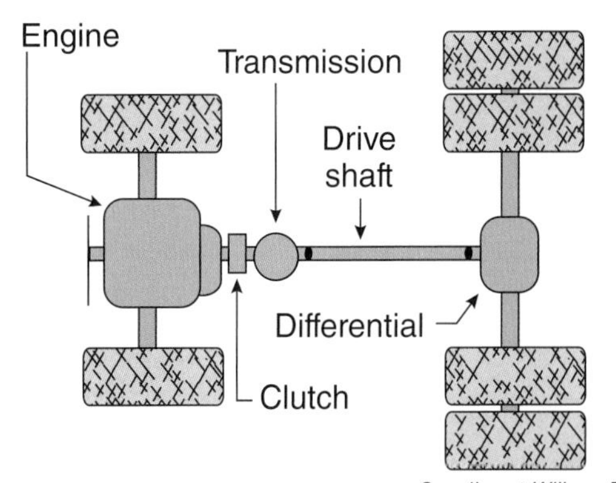

Goodheart-Willcox Publisher

Figure 23-18. The transmission joins the engine to the rest of the drivetrain.

transmission provides high torque (rotating force) at low speeds, as the vehicle starts to move. Later, as gravity, rolling resistance, and air resistance are overcome, low-torque, high-speed outputs are used.

A ***manual transmission*** has a clutch between the engine and the transmission. This allows the operator to disconnect the engine so the transmission ratio can be changed by shifting gears. ***Automatic transmissions*** use valves to change hydraulic pressure so the transmission shifts its input-to-output ratios.

When a vehicle makes a turn, the wheels on the outside must be able to turn faster than those on the inside. The outside wheels turn faster because they have farther to travel. A device called a *differential* allows the wheels to turn at different speeds. The differential is a set of gears that independently drives each axle. The axles are connected to the vehicle's drive wheels. In rear wheel–drive automobiles and trucks, the differential is separate from the transmission. Front wheel–drive cars combine the transmission and differential into a single unit called a ***transaxle***.

Diesel-Electric System

Freight and passenger locomotives use a diesel-electric system to generate and transmit power, **Figure 23-19**. This system uses a very large (8- to 12-cylinder) diesel engine to convert the fuel to mechanical motion. This motion is used to drive an electric generator. The electricity is transmitted to electric motors that are geared to the drive wheels of the locomotive. This motor-generator unit is relatively light. Therefore, extra weight, called ***ballast***, is added to the locomotive to give it better traction. Making the frame of the locomotive out of heavier steel plates than needed creates this extra weight.

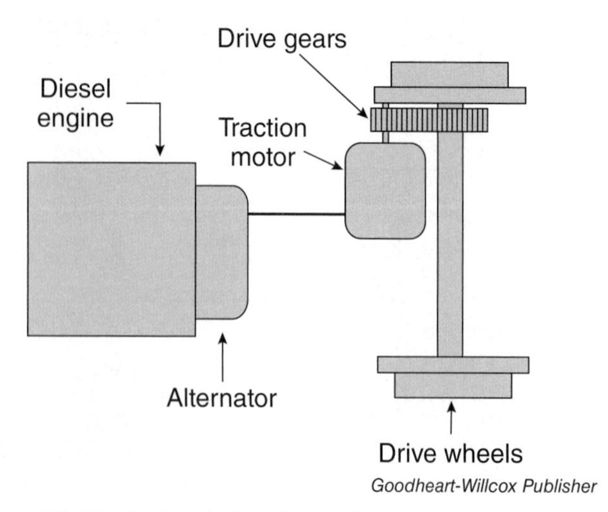

Figure 23-19. A diesel-electric traction system used to power railroad locomotives.

Electric System

Some rail vehicles are powered by only an electrical system. The vehicle has a moving connection that remains in contact with the electric conductor. Overhead wires or a third rail are used as the conductor. The electricity is conducted to motors that turn the vehicle's wheels.

Land-Transportation Vehicle Suspension

Suspension systems keep the vehicle in contact with the road or rail. They also separate the passenger compartment from the drive system to increase passenger and operator comfort. The suspension system of a vehicle has three major parts:

- Wheels.
- Axles.
- Springs and shocks.

The wheels provide traction and roll along the road or rail. They also spread the vehicle's weight onto the road or rail. A wheel can be used to absorb shock from bumps in a road or rail. Often, rubber tires are attached to the wheels to absorb shock and increase traction. Most rail systems use steel wheels and depend on friction between the wheels and rails for traction.

Axles carry the load of the vehicle to the wheels. They support a set of springs that absorb movements of the axles and wheels. Springs, however, create a bouncy ride. Shock absorbers soften this bouncing. They are attached between the axle and the frame to dampen the spring action, **Figure 23-20**. The result is a controlled, smooth ride.

A variation of the suspension system uses torsion bars. Instead of a spring, a steel bar absorbs the movement of the wheels and axles by twisting. The torsion bar untwists to release the absorbed energy. Shock absorbers are used to dampen the motion of the torsion bar.

Land-Transportation Vehicle Guidance and Control

An operator manually controls the speed and direction of most land-transportation vehicles. Throttles and accelerator pedals are used to control

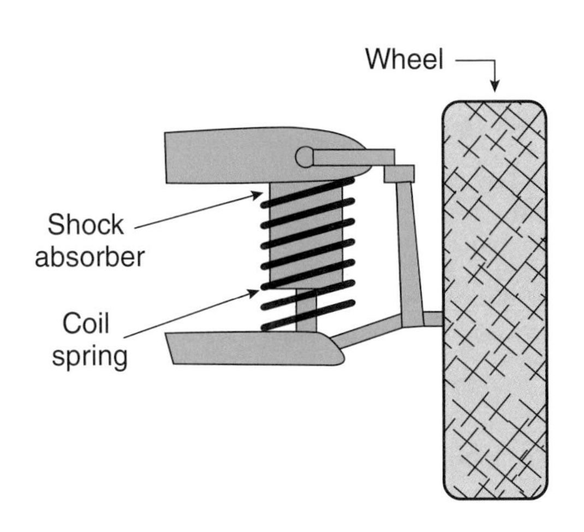

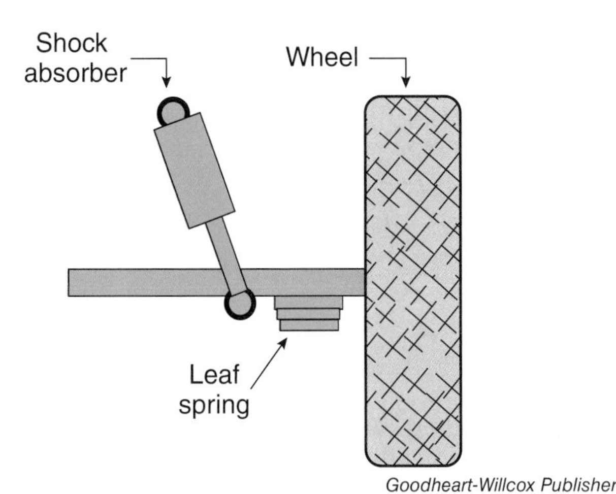

Goodheart-Willcox Publisher

Figure 23-20. Two types of suspension systems used on automobiles.

engine speed. Transmissions are shifted to increase the power being delivered to the drive wheels. Brakes can be applied to slow the vehicle. The wheels can be turned to take the vehicle in a new direction.

These actions require the operator to make decisions. These decisions are made based on visual information and judgment. The operator receives information from signs and signals. See **Figure 23-21**. The operator also observes traffic conditions. Gauges and instruments provide information on vehicle speed and operating conditions. The operator's judgment is important because all this information must be considered and acted on. Inexperienced operators or those influenced by drugs and alcohol might exercise poor judgment. This might result in accidents, human injury, or death.

Water-Transportation Vehicles

Water covers more than 70% of the earth's surface, making water transportation an important form of travel. Water transportation includes all vehicles that carry passengers or cargo over or under water. Water transportation that occurs on rivers, on lakes, and along coastal waterways is called ***inland-waterway transportation***. Water transportation on oceans and large inland lakes, such as the Great Lakes in the United States, is called ***maritime shipping***.

There are three kinds of water-transportation vehicles, **Figure 23-22**. Ships owned by private citizens for their own recreation are called *pleasure*

Joseph Sohm/Shutterstock.com

Grycaj/Shutterstock.com

Figure 23-21. Signs and signals, such as highway signs and rail signals, give operators guidance.

Types of Boats and Ships

Pleasure **Commercial** **Military**

©iStockphoto.com/Terraxplorer ©iStockphoto.com/Toprawman ©iStockphoto.com/DomD

Figure 23-22. The three types of boats and ships are pleasure, commercial, and military.

craft. They allow people to participate in activities such as waterskiing, fishing, sailing, and cruising. Large ships used for transporting people and cargo are called *commercial ships*. Some ships provide for the defense of a country. These vessels are called *military ships* (or *naval ships*) and are owned by the government of a country.

Water-Transportation Vehicle Structure

A ship's intended use influences how it is designed. All ships have two basic parts, **Figure 23-23**. The hull forms the shell that allows the ship to float and contain a load. The superstructure is the part of the ship above the deck. This part contains the bridge and crew or passenger accommodations.

©iStockphoto.com/CaraMaria

Figure 23-23. The two major parts of a ship are the hull and the superstructure.

Similar to land vehicles, ships can be designed to carry either passengers or cargo. Ships that carry people are called *passenger ships*. Cargo-carrying ships are called *merchant ships*.

Passenger ships include oceangoing liners, cruise ships, and for short distances, ferries. Ocean liners and cruise ships are miniature cities. They supply all the needs of the passengers. These ships include such areas as sleeping quarters, kitchens, dining space, recreational areas, and retail shops, such as hair salons and dry cleaners.

Merchant ships are usually designed for one type of cargo. Types of cargo include dry cargo, liquids, gases, and cargo containers, **Figure 23-24**.

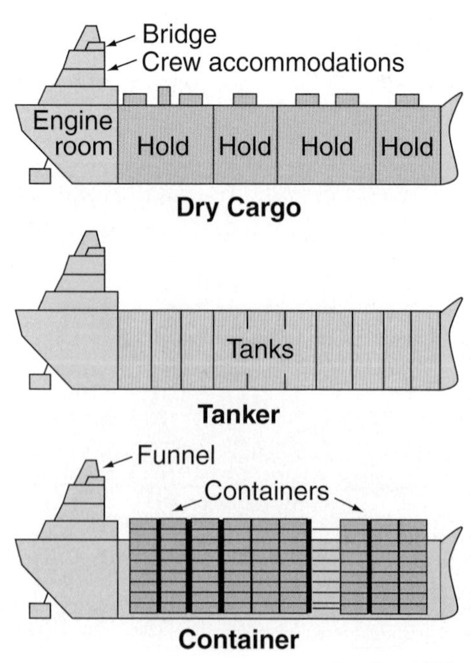

Goodheart-Willcox Publisher

Figure 23-24. Common types of merchant ships.

STEM Connection: Science

Newton's First Law of Motion

Many scientific principles come into play when we use the various kinds of transportation systems now available. Among those principles are Newton's laws of motion. For example, we can use Newton's first law of motion to explain why we need seat belts in our automobiles. Newton's first law of motion states that a moving object continues moving at the same speed, in the same direction, unless some force acts on it. This tendency of an object to resist change is called *inertia*.

Thus, when we are traveling in an automobile at, say, 50 miles per hour, we continue moving at this speed, even when we forcefully apply the brakes of the car. We have applied force (the brakes) to stop the car. Such force works only on the car, however, not on us. Therefore, we need the seat belt because it provides the force needed to stop us from continuing to move and perhaps hitting the windshield. Newton's first law of motion also involves the concept of friction. Using another method of transportation as an example, can you think of a way friction affects this object when it is moving?

©iStockphoto.com/JohnnyLye

Dry-cargo ships haul both crated and bulk cargo. Bulk cargo is loose commodities, such as grain, iron ore, and coal, loaded into the holds (compartments) of ships. Other dry cargo is contained in crates and loaded into various holds of a ship. Land vehicles, such as cars and trucks, are another type of bulk cargo. They are typically transported on special vessels called *roll-on, roll-off (RORO) ships*. The vehicles are driven on and off the ship through large hatches (doors) in the hull.

Tankers are large vessels used to move liquids, such as petroleum, liquefied natural gas, and other chemicals. The liquid cargo is pumped in and out of the tanks. Tankers have also been built to move gases.

Containerships are a newer and faster way to ship large quantities of goods, **Figure 23-25**. The shipments are loaded into large steel containers that resemble semitruck trailers without wheels. The containers are sealed, loaded into the hold, and then stacked on the deck of the ship. The loading process that once took days is now done in hours.

A great deal of shipping on inland waterways is done with tugboats and barges, **Figure 23-26**. Tugboats contain the power-generation and transmission systems and the operation controls. Barges

O M/Shutterstock.com mr.water/Shutterstock.com

tcly/Shutterstock.com

Figure 23-25. Containers are often shipped from inland points and then unloaded at the dock. Later, the containers are loaded onto containerships.

samman58/Shutterstock.com

Figure 23-26. Tugboats pull or push barges to carry large loads on inland waterways.

Folscheid/Shutterstock.com

Figure 23-27. A large marine propeller.

carry cargo. Various types of cargo have specifically designed containers. The tug pushes or pulls a group of barges lashed together.

Water-Transportation Vehicle Propulsion

Most commercial ships are propelled through water by a propeller (prop) driven by a steam turbine or diesel engine. The engine is often located inside the ship. A motor (an internal combustion gasoline engine) attached to the stern (back) of the boat, however, might power small boats. This type of power source is called an ***outboard motor***.

Nuclear reactors power some military ships and submarines. Heat from the nuclear reactor is used to turn water into steam. This steam is used to turn a turbine. The turbine turns the prop using a shaft.

A propeller is a device with a group of blades radiating out from the center, **Figure 23-27**. Props can have from two to six blades. The blades attach to a shaft at the center. Props range in size from 2″ (51 mm) to more than 30′ (9 m) in diameter. Each blade of the prop is shaped to "bite" into the water, very similar to the way a window fan "bites" into the air. The rotation of the prop forces the water past the prop. As we learned earlier, Newton's third law of motion tells us that for every action, there is an equal and opposite reaction. Therefore, the action of forcing the water through the spinning prop causes an opposite reaction that pushes the boat forward.

Engines do not power all boats. Some boats use sails to capture wind. A sail catches the wind and pushes the boat through the water. This method of propulsion has been used for centuries.

Water-Transportation Vehicle Suspension

For water-transportation vehicles, the hull is the primary component of the suspension system. The hull must be carefully designed to ensure that the boat remains stable and afloat. If the hull is not properly designed, a boat can roll onto its side and sink.

The shape of the hull impacts the stability of the boat. Boats with round-bottom hulls are relatively unstable. V-bottom hulls, which are shaped similar to the letter *V*, provide greater stability than round-bottom hulls. Flat-bottom hulls are the most stable, **Figure 23-28**.

The weight distribution in the boat also affects stability. If the boat's load is evenly spread over the hull, the boat should be very stable. If the boat is unevenly loaded, it might lean in the water, and the propulsion system and control system will not work effectively. In addition, an unevenly loaded boat is more likely to sink.

When an object is immersed in a fluid, the fluid tends to push the object up. If you lift your arms away from your body while swimming, you will notice that the water tends to push your arms toward the surface. This tendency is called *buoyancy*. ***Buoyancy*** is the upward force exerted on an object immersed in a fluid.

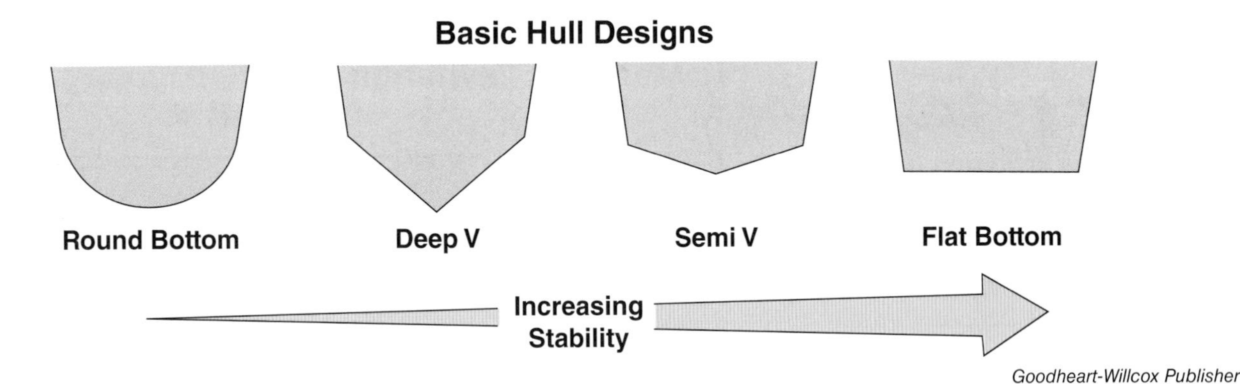

Figure 23-28. Basic hull designs. Flat-bottom hulls are the most stable.

Think of the buoyant force as a force trying to hold an object up in the water. In order for the object to float, the buoyant force must be equal to the weight of the object. If the weight of the object is greater than the buoyant force, the object sinks.

When an object is placed in a fluid, the fluid must move to make room for the object. Displaced fluid is the fluid that had previously occupied the space that the immersed object now occupies. The volume of displaced fluid is identical to the volume of the submerged object. The weight of the displaced fluid can be calculated by multiplying the volume by the density of the fluid. The buoyant force is equal to the weight of the displaced fluid.

To determine whether or not an object will float, you simply compare the weight of the object to the weight of the displaced water (buoyant force). If the weight of the object is greater, the object sinks. If the buoyant force is greater, the object floats.

Three special types of boats use unique suspension principles. The first is the hovercraft, **Figure 23-29**. A ***hovercraft*** is suspended on a cushion of air. Large fans force the air into a cavity under the boat. As the air escapes this pocket, the boat is lifted above the water. Hovercrafts are used over shallow water, swamps, and marshes and where speed is important.

The second type of special boat is the hydrofoil. A ***hydrofoil*** has a normal hull and set of underwater wings called *hydrofoils*. A jet engine provides power for the boat. As the boat's speed increases, the water passing over the hydrofoils produces lift. The lift causes the hull to rise out of the water. The reduced friction between the water and the hull allows the boat to travel faster, while using less fuel.

The third type of special boat is a ***submersible***, or submarine. It can travel on the surface of water and underwater. Allowing water to enter or exit special tanks adjusts the boat's buoyancy. As the tanks fill with water, the submersible becomes heavier, or less buoyant, sinking it into the water. Compressed air can be used to force the water out of the tanks and increase the buoyancy. The submersible then rises to the surface.

Dave Turner/Shutterstock.com

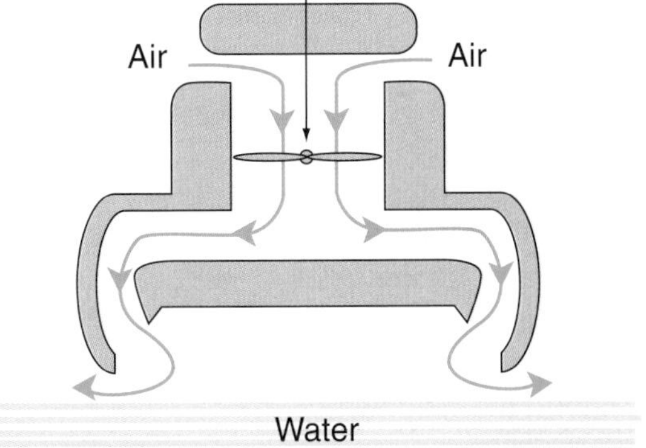

Goodheart-Willcox Publisher

Figure 23-29. A hovercraft is suspended above the water on a cushion of air.

Technology Explained Magnetic Levitation

Magnetic levitation (maglev) train: A transportation vehicle that uses the principles of magnetism to suspend and propel itself along a guideway.

The rough ride and noise produced by traditional rail systems have been a drawback to their use for passenger travel. The steel wheels and rails used in most rail systems cause these issues. Improvements in rails, wheels, and maintenance have made trains quieter. Some trains, such as monorails, use rubber tires or inserts in steel wheels. Some noise is still produced, however, because of the contact between the wheels and rails. In the 1960s, a new land vehicle was developed that did not touch the rails. This type of vehicle is called a *maglev train*. Maglevs use magnetic forces to support and move along special pathways called *guideways*.

Transrapid International

Figure A. Magnetic attraction suspends the vehicle.

There are two types of maglevs: attraction and repulsion. An attraction maglev uses magnets pulling toward each other to support the train, **Figure A**. A repulsion maglev uses magnets pushing away from each other to support the train, **Figure B**. The guideways differ between the two types of maglevs. Attraction maglevs wrap around the guideways. Repulsion maglevs sit in trough-like guideways.

Japan Railways Group

Figure B. Magnet repulsion can also be used to suspend a maglev.

The passenger compartment in maglev trains is quiet and comfortable. Beneath the passenger compartment is the suspension system. Attraction maglevs use electromagnets to support the train. The electromagnets pull toward a rail in the guideway, lifting the train. The amount of electricity is controlled so the electromagnets and rail never touch. Repulsion maglevs use superconducting magnets to induce magnetic fields in coils in the guideway. The maglev floats on these magnetic fields.

A device called a *linear induction motor* (*LIM*) propels maglevs. A LIM works similarly to a standard electric motor, using the principle that like poles of a magnet repel each other, while unlike poles attract. A standard electric motor produces rotary motion. A LIM is laid out flat in a straight line. Thus, a LIM produces linear, or straight-line, motion.

A LIM uses alternating electric current to create magnetic fields. The direction of the electric current changes many times each second, causing the lines of magnetic force to collapse and change direction. This causes the rail to be attracted to the magnet, and the vehicle moves along the rail. The current then switches direction, causing the magnet to repel the rail. This cycle repeats over and over and pushes the maglev along the guideway.

Water-Transportation Vehicle Guidance

The operator of a ship is the captain. The captain is responsible for the ship's course (path) and safety. Captains are guided by a variety of sources. Compasses and charts of rivers, harbors, and oceans are the bases for navigating ships of any kind. In or near a harbor, lighthouses and lighted buoys identify safe channels for navigation or mark hazards. Flags and radio broadcasts communicate weather conditions. Special electronic systems help to pinpoint the ship's location and indicate the depth of the waterway.

Water-Transportation Vehicle Control

The path a ship follows is controlled in various ways, **Figure 23-30**. Ships that have their own power sources are generally guided with rudders, **Figure 23-31**. A rudder is a large flat plate at the stern of the ship. When it is turned away from the ship's course, it deflects the water passing under the hull. This deflection forces the ship into a turn. Unlike most land vehicles, in which the front of the vehicle contains the guidance and control systems,

©iStockphoto.com/aristotoo

Figure 23-31. Note the red rudder under the red-and-white hull. The numbers on the hull are depth (water level) markings.

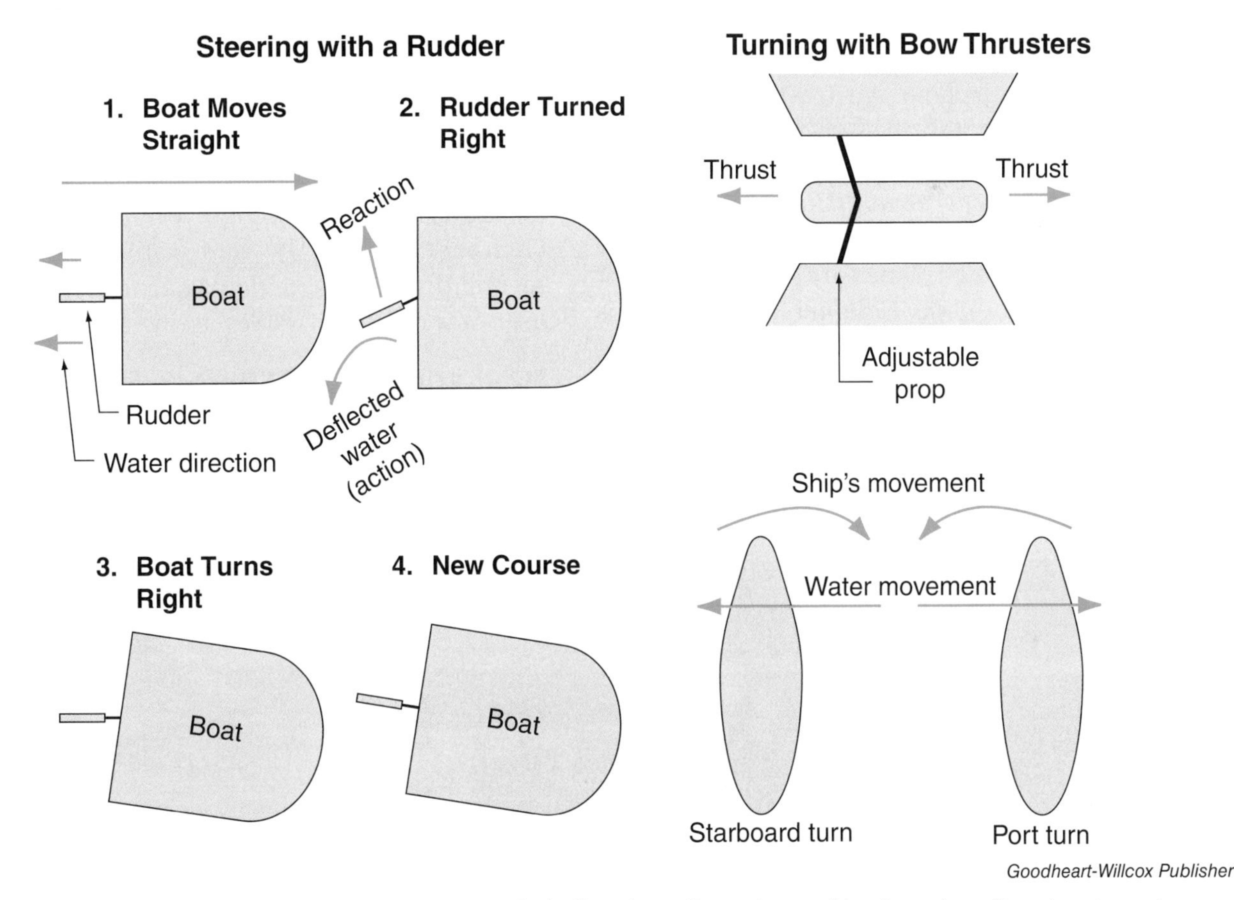

Goodheart-Willcox Publisher

Figure 23-30. Boats and ships use rudders to change their directions. Some large ships have bow thrusters to make small changes while docking.

the back of the ship changes the ship's path and causes the vehicle to turn.

Some large ships have two props or "twin screws." Increasing the speed of one prop creates additional force on that side of the ship. The ship then turns toward the side with the slower prop.

Another way large ships can be turned, especially in docking, is with bow thrusters. A ***bow thruster*** is a prop mounted at a right angle to the keel. A ***keel*** is a flat blade protruding down into the water with the purpose of preventing the boat from being blown sideways by aiding in keeping the boat right-side up. The blades of the prop can be adjusted to provide thrust in either direction. The ship turns opposite the thrust.

Sailboats use both a rudder and sails for control. The rudder uses the force of the water to turn the boat. Sails use forces of the wind to aid the rudder in guiding the boat.

Air-Transportation Vehicles

The newest transportation vehicle in widespread use is the airplane. Air travel has changed from a rare occurrence to an everyday travel option. Air travel can be divided into two groups: general and commercial. ***General aviation*** is pleasure or business travel on an aircraft owned by individuals or businesses. These airplanes are not available for use by the general public. ***Commercial aviation*** is travel on airplanes owned by businesses that make money by transporting people and cargo. These companies include scheduled airlines, commuter airlines, and airfreight carriers.

Air-Transportation Vehicle Structure

Three major types of air-transportation vehicles are lighter-than-air, fixed-wing, and rotary-wing aircraft, **Figure 23-32**.

Lighter-than-Air Vehicle Structure

Lighter-than-air vehicles use either a light gas (such as helium) or hot air to produce lift. These were the first air vehicles to be used by people. The earliest was a hot air balloon developed in France in 1783. This balloon, similar to present-day hot air balloons, was basically a fabric bag with an opening in the bottom. The opening allows warm air from a burner to enter the balloon and displace heavier, cold air. Warm air rises to the top of the bag because it is less dense than cold air. When enough warm air has built up in the balloon, the balloon rises off the ground.

Blimps and dirigibles are more advanced lighter-than-air vehicles. A ***blimp*** is a nonrigid aircraft, meaning it has no frame. The envelope (the bag filled with a light gas, usually helium) determines the blimp's shape. Slung beneath the envelope is an operator-and-passenger compartment. One engine or more is attached to the compartment to give the blimp forward motion.

Dirigibles are rigid airships with metal frames covered with fabric skin. They use hydrogen gas to provide maximum lift. Dirigibles were used for transatlantic passenger service in the 1930s. Hydrogen is highly flammable, however, and several disastrous accidents occurred because of its use. As a result, dirigibles are no longer used.

Fixed-Wing Aircraft Structure

Today, most passenger and cargo aircraft are ***fixed-wing aircraft***. They all use similar structures.

Lighter-Than-Air

Goodyear

Fixed Wing

Grumman Corp.

Rotary Wing

Bell Helicopter

Figure 23-32. The three types of aircraft are the following: lighter-than-air, fixed wing, and rotary wing.

The flight-crew, passenger, and cargo units are contained in a body called the ***fuselage***. One or more wings attached to the fuselage provide the lift necessary to fly. A ***tail assembly*** provides steering capability for the aircraft.

As noted earlier, the Wright brothers developed the first successful fixed-wing aircraft. *Flyer* was first flown in 1903. Later, Charles Lindbergh made a historic transatlantic flight in a fixed-wing aircraft called the *Spirit of St. Louis*.

Rotary-Wing Aircraft Structure

The helicopter is the most common ***rotary-wing aircraft*** in use today. The body of the craft contains the operating and cargo unit and encloses the engine. Above the engine is a set of blades. The blades are adjustable to provide both lift and forward and backward motion. The tail has a second, smaller rotor that keeps the body from spinning in response to the motion of the main rotor.

Helicopters are widely used by the military. In civilian life, they are used for emergency transportation, law enforcement, and in specialized applications. These special jobs include communication-tower erection, high-rise building construction, and aerial logging.

Air-Transportation Vehicle Propulsion

Aircraft use two major types of propulsion systems: props and jet engines. Smaller aircraft use props attached to internal combustion engines. The engine operates on the same principle as the automotive internal combustion engine. Some large aircraft use a variation of the jet engine to turn the prop.

Most business and commercial aircraft are powered by one of three types of jet engines. These types are turbojet, turbofan, and turboprop.

Turbojet Engines

The first type of engine to be used was the ***turbojet engine***, **Figure 23-33.** This engine was developed during World War II. The turbojet engine operates in the following way:

1. Air is drawn into the front of the engine.
2. Air is compressed at the front section of the engine.
3. The compressed air is fed into a combustion chamber.

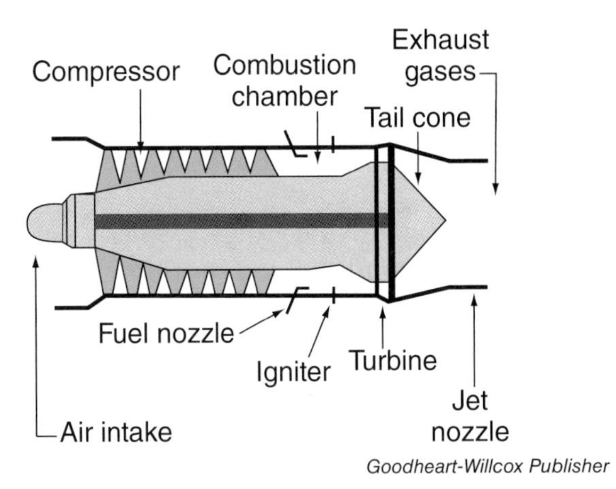

Figure 23-33. Turbojet engines were developed during World War II.

4. The fuel is mixed with the compressed air.
5. The fuel-air mixture is allowed to ignite and burn rapidly.
6. As a result, the rapidly expanding hot gases exit the rear of the engine.

The exiting hot gases serve two functions. The gases turn a turbine operating the engine-compression section. They also produce the thrust to move the aircraft. The turbojet engine operates at high speeds and is used in military aircraft. Early commercial jet airliners also used these engines.

Turbofan Engines

The ***turbofan engine*** is the engine of choice for most commercial aircraft in use today, **Figure 23-34.** The turbofan operates at lower speeds than a turbojet engine does. Also, a turbofan uses less fuel to produce the same power. In this engine, the

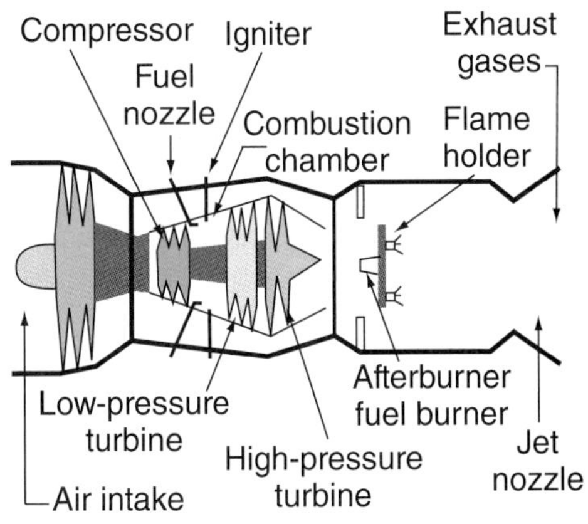

Figure 23-34. Turbofan engines are widely used in modern airliners because they are fuel efficient and deliver high thrust.

turbine drives a fan at the front of the engine. The fan compresses the incoming air. The compressed air is then divided into two streams. One stream of air enters the compressor section. In this section, the air is compressed further, fuel is injected, and the fuel-air mixture is ignited. The other stream of air flows around the combustion chamber. This stream is used to cool the engine and reduce noise. In the rear section of the engine, the exhaust gases and cool air mix. If necessary, additional fuel can be injected and ignited to provide additional thrust. This arrangement is called an afterburner.

Turboprop Engines

Another variation of the jet engine is the ***turboprop engine*****, Figure 23-35**. This engine operates in the same manner as a turbojet engine. The turbine, however, also drives a propeller, providing the thrust to move the aircraft. Turboprop engines operate more efficiently at low speeds than turbojet or turbofan engines do. Therefore, they are widely used on commuter aircraft.

Air-Transportation Vehicle Suspension

Air-transportation vehicles are suspended in the atmosphere. Such a state requires knowledge of the principles of physics. All air-transportation vehicles work based on the fact that an area with less dense air allows heavier air to force the vehicle up. This force must be greater than the force of gravity drawing the craft toward the earth. As with structure, suspension varies according to the type of vehicle.

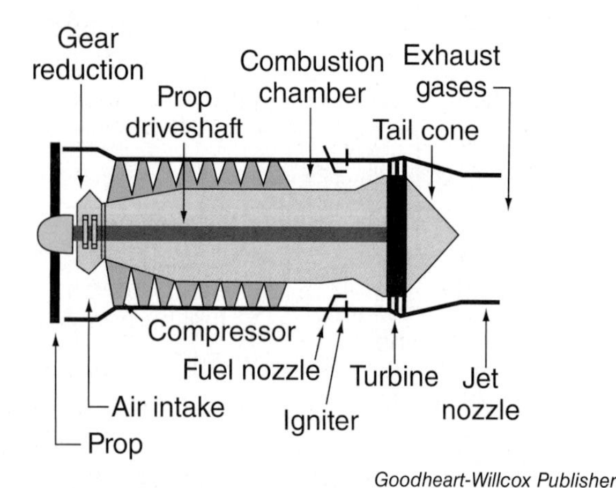

Goodheart-Willcox Publisher

Figure 23-35. Turboprop engines are used on many commuter airline planes.

Lighter-than-Air Vehicle Suspension

Lighter-than-air vehicles use air-weight differences to the make vehicle rise and be suspended in the atmosphere. Two methods are used to cause this difference in weight. The first method uses a closed envelope filled with a light gas. As you learned earlier, blimps use helium, and dirigibles used hydrogen, to generate lift. The vehicle is made to rise and lower in a way similar to a submarine. Inside the vehicle are air tanks or bags. The tanks are filled with outside air, making the aircraft heavier, causing it to descend. Removing the air from the tanks or bags makes the craft lighter, causing it to rise.

In the second method, warm air circulates in an open-ended envelope, causing the craft (the hot air balloon) to ascend and descend, **Figure 23-36**. The balloon has a burner suspended below its opening. To make the balloon rise, the burner is ignited. The flame heats air, which rises into the balloon. The warm, rising air displaces colder, heavier air. The warmer air is less dense and makes the balloon lighter than the air it displaces in the atmosphere. Therefore, it rises until its displaced air is equivalent to its weight. As the balloon rises, it enters air that is less dense (lighter).

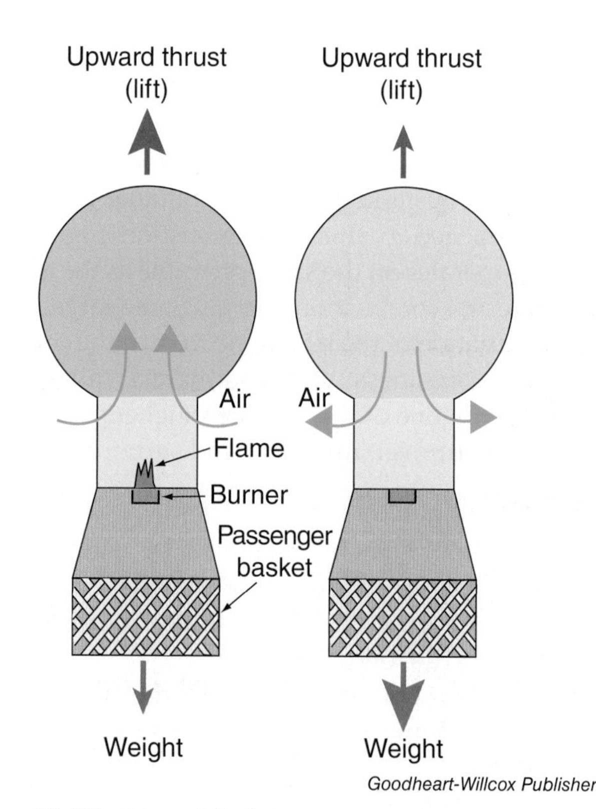

Goodheart-Willcox Publisher

Figure 23-36. A hot air balloon uses warm air to produce lift.

STEM Connection: Mathematics

Calculating Buoyant Force

Buoyant force causes objects to float and is equal to the weight of the displaced water. This force can be calculated by multiplying the volume of displaced water by the density of water. If this force is greater than the weight of the boat, the boat will float.

For example, a fully loaded, flat-bottom boat weighs 50,000 lb. The boat is 20′ long, 8′ wide, and 6′ deep. To determine whether or not the boat will float, we need to calculate the weight of the displaced water and compare it to the boat's weight. First, we calculate the volume of the boat:

Dudarev Mikhail/Shutterstock.com

$$\begin{aligned} \text{volume} &= \text{length} \times \text{width} \times \text{depth} \\ &= 20' \times 8' \times 6' \\ &= 960 \text{ ft}^3 \end{aligned}$$

To determine the weight of the displaced water, multiply the volume of the boat by the density of water. The density of water is 62.4 lb/ft^3 (pounds per cubic foot):

$$\begin{aligned} \text{weight} &= \text{volume} \times \text{density} \\ &= 960 \text{ ft}^3 \times 62.4 \text{ lb/ft}^3 \\ &= 59{,}904 \text{ lb} \end{aligned}$$

The buoyant force (59,904 lb) is greater than the weight of the boat (50,000 lb). Therefore, the boat will float. Will a fully loaded, flat-bottom boat that is 25′ long, 10′ wide, and 6′ deep and weighs 75,000 lb float? Why or why not?

As the air in the balloon cools, it becomes heavier. The burner must continue to operate to keep the balloon aloft. To descend, the burner is turned off. During flight, the operator uses the burner intermittently (turns it on and off) to keep the balloon at a constant height. Running the burner all the time might cause the balloon to rise too high. Not running the burner often enough allows the balloon to descend too low.

Fixed-Wing Aircraft Suspension

As noted earlier, the most common aircraft is the fixed-wing airplane. An airplane has four major forces affecting its ability to fly. See **Figure 23-37**. These forces are the following:

- **Thrust.** This force causes the aircraft to move forward.
- **Lift.** This force holds or lifts the craft in the air.
- **Drag.** This force is the air-resistance force opposing the vehicle's forward motion.
- **Weight.** This force is the pull of gravity causing the craft to descend.

Critical for all flight is lift. Air flowing over the wing of the aircraft generates lift. The wing is shaped to form an ***airfoil***, **Figure 23-38**. The wing separates the air into two streams. The airfoil shape causes the upper stream to move farther than the lower stream. Therefore, the upper stream speeds up. This increased speed causes a decrease in pressure. The high-pressure air below the wing forces the wing up. This gives the plane the required lift.

The greater the slope of the upper surface of the wing, the greater the lift will be. Drag is also increased, however, because of the larger front profile of the wing. An airplane needs greater lift during takeoff and landing. Devices called *flaps* on the leading and trailing edges of the wings are extended to increase lift, **Figure 23-39**. They are also used to slow the aircraft during landings.

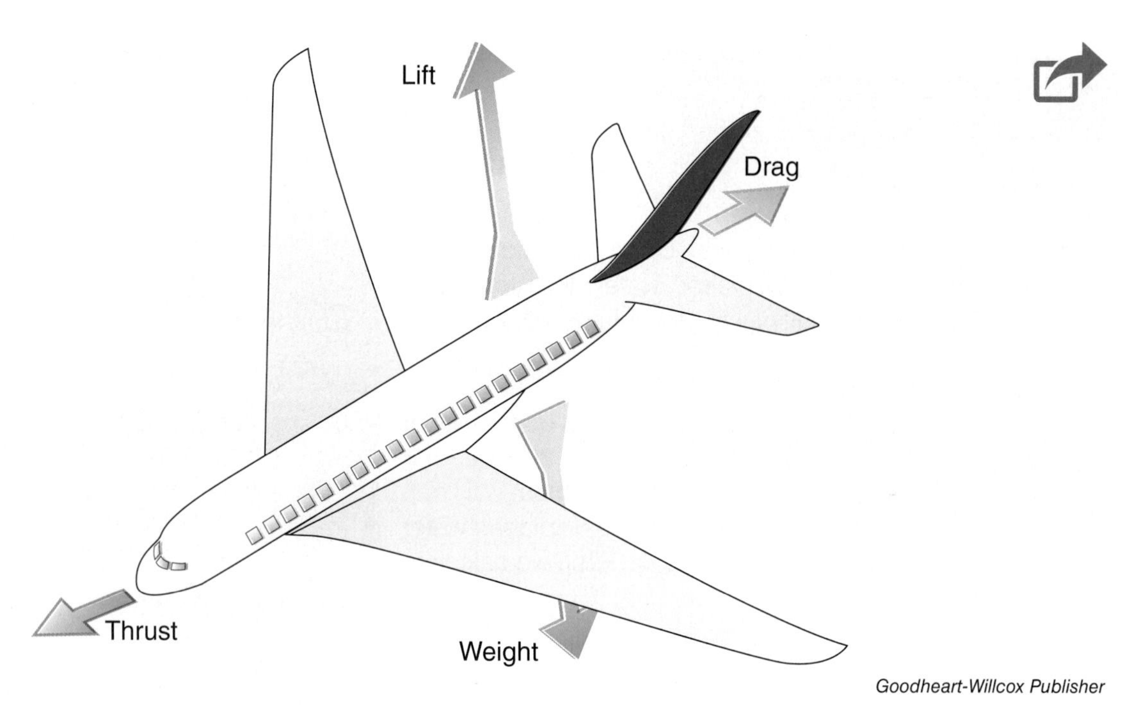

Goodheart-Willcox Publisher

Figure 23-37. Four forces affect flight.

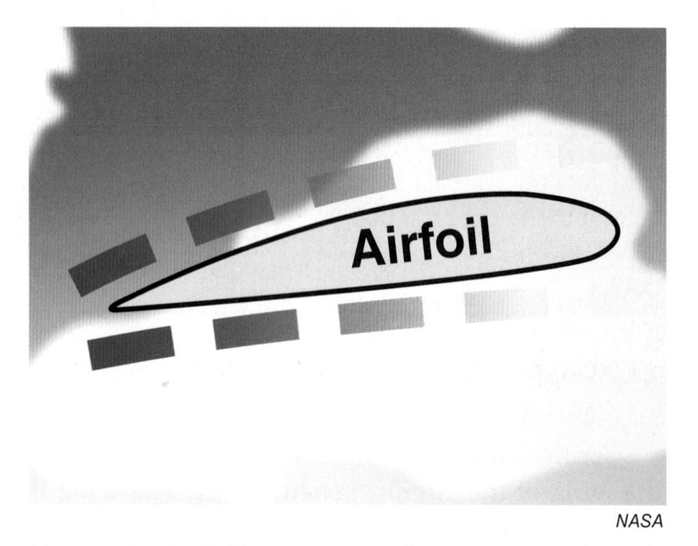

NASA

Figure 23-38. Air flows a greater distance over the top of an airfoil and produces lift.

Rotary-Wing Aircraft Suspension

The blades on a helicopter operate similarly to any other airfoil. As they rotate, they generate lift, and the angle of the blades can be increased to generate additional lift, **Figure 23-40**. This causes the helicopter to rise vertically. When the proper altitude is reached, the pitch (angle) of the blades is changed. The pitch is adjusted so the lift equals the weight. When the weight equals the lift, the helicopter hovers. Reducing the pitch allows the helicopter to descend.

Forward flight is accomplished using a complex mechanism. This mechanism changes the pitch of the blades and tilts the blades as they rotate. In forward flight, the blade is tilted slightly forward. Also, the pitch of the blade is increased in the rear part of the blades' travel. The pitch is reduced as the

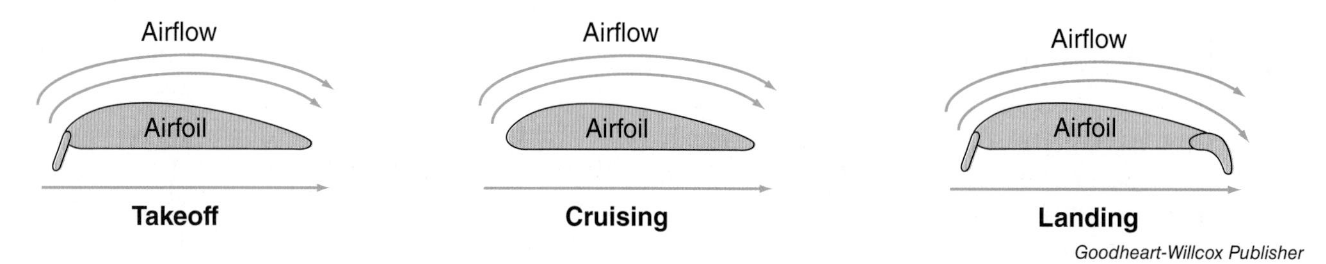

Goodheart-Willcox Publisher

Figure 23-39. Airflow over the wing during takeoff, cruising, and landing. Flaps are extended during takeoff and landing to increase lift.

Blades

©iStockphoto.com/fotoVoyager

Mechanisms

©iStockphoto.com/Macatack

Figure 23-40. Mechanisms adjust the rotating blades or airfoils, allowing the helicopter to adjust its lift and direction of flight.

blades rotate in the front part of their travel. This causes the craft to have more lift in the back than in the front. Therefore, the helicopter leans slightly forward. The combination of the lift and thrust of the rotors causes the helicopter to move forward.

Backward flight is accomplished by reversing the tilt. The rotors are pitched the most in the front part of their travel. This causes the craft to tilt slightly backward and travel in reverse. Helicopters can move left and right in the same way, by tilting the rotor and adjusting the pitch.

Air-Transportation Vehicle Guidance

Guidance for commercial airline pilots comes in many forms. Ground personnel help pilots safely bring their aircraft to the terminal. Control-tower personnel are in radio contact as airplanes taxi and depart. Once the airplane is in the air, regional air-traffic controllers take over. They monitor and direct the aircraft's progress across their region and hand the craft off to controllers in adjoining regional centers as the vehicle leaves their region. See **Figure 23-41**. Operators of general aviation aircraft do not normally use regional air-traffic control centers. They use visual flight rules to govern their movements.

Burben/Shutterstock.com

Figure 23-41. A radar screen helps air-traffic controllers guide and direct aircraft as they travel along their assigned routes.

Air-Transportation Vehicle Control

Various instruments help pilots properly monitor and control aircraft. These instruments are called ***aviation electronics*** (or *avionics*). This term includes all electronic instruments and systems that provide navigation and operating data to the pilot, **Figure 23-42**. The amount of avionics an airplane carries depends on the airplane's size and cost. Large passenger airliners have a wide variety of avionics. A small general aviation airplane has just basic equipment. Avionics include various functions and systems:

- **Communication systems.** Short-distance radio systems allow pilots to communicate with air-traffic controllers and other people on the ground.

Alaska Airlines

Figure 23-42. The avionics in this passenger jet cockpit are extensive.

- **Automatic pilot**. The automatic pilot is an electronic, computer-based system that monitors an aircraft's position and adjusts the control surfaces to keep the airplane on course.
- **Instrument landing system (ILS)**. An instrument landing system (ILS) helps pilots land airplanes in bad weather. ILS uses a series of radio beams to help the pilot guide the airplane to the runway. A wide vertical beam helps the pilot stay on the center of the runway. A wide horizontal radio beam tells the pilot if the plane is above or below the correct approach path. Vertical beams serve as distance markers to tell pilots how far away they are from the runways. Pilots in aircraft without ILS equipment use approach lighting to guide the plane in for landing.
- **Weather radar.** This system provides pilots with up-to-date weather conditions in the path of the aircraft.
- **Navigation systems.** These systems guide the plane along its course and indicate airspeed.
- **Engine and flight instruments.** These instruments give pilots information about how the engine is operating. Flight instruments tell the direction (in three degrees of freedom) the airplane is flying.

Space-Transportation Vehicles

Space travel might be the transportation system of the future. Today, however, its role is limited. Space travel is restricted to conducting scientific experiments and placing communication, weather, and surveillance satellites into orbit.

Types of Space Travel

Space travel has become a part of everyday human experience. Some space travel involves human passengers. Other voyages do not. Some space travel involves spacecraft orbiting the earth, while other missions probe outer space. Space travel can be classified in two different ways:

- Unmanned or manned flight.
- Earth-orbit or outer space travel.

Unmanned and Manned Flights

The first spaceflights were unmanned spaceflights. These flights used rockets to place a payload into orbit. A payload is usually used to describe the items for which a flight is made. Typically, the payload was either equipment for conducting scientific experiments or communication satellites. Today, unmanned space vehicles continue to launch these and other payloads.

A ***manned spaceflight*** carries human beings into space and returns them safely to Earth. Early manned spaceflights used capsules attached to the nose of a rocket to place humans into orbit or into outer space. Later manned flights were then carried out by space shuttles. A shuttle would rocket into orbit, and at the end of the mission, it returned to Earth, gliding in for a landing similarly to an airplane. The next stage of manned flights will be carried out by Orion spacecraft. See **Figure 23-43**.

Earth-Orbit and Outer Space Travel

The two types of space travel are earth-orbit and outer space travel. Communication satellites use ***earth-orbit travel***. They travel at about 18,000 miles per hour (mph) in an elliptical orbit around the earth. This orbit comes closest to the earth at a

NASA

Figure 23-43. An artist's representation of an Orion spacecraft.

point called the ***perigee***. The farthest distance away from the earth is called the ***apogee***.

Satellites can be placed into two types of orbits. One type allows the satellite to circle the earth, viewing a complete path around the globe. This type of orbit can be used for such objectives as weather monitoring, geological and agricultural surveys, and military surveillance.

The second type of orbit is called ***geosynchronous orbit***. In this type of orbit, the satellite travels the same speed the earth is turning. Therefore, the satellite stays over a single point on the globe. This type of orbit is used for communication and weather satellites.

Outer space travel places the spacecraft in earth orbit first. The speed of the spacecraft is then increased to more than 25,000 mph. At this speed, the spacecraft is thrown into a path causing it to move out of the earth's gravitational field. The spacecraft can then travel to distant planets and out of the solar system. *Voyager 1* and *Voyager 2* are examples of this type of spacecraft. They moved out of earth orbit and traveled past a number of planets and out of the solar system.

Space-Transportation Vehicle Structure

A space vehicle and its launch systems have three major parts: a rocket engine to place the vehicle or payload into space; an operating section; and a cargo and passenger compartment. As you have learned, early space programs integrated these components into a single launch vehicle, or rocket. The engine was at the rear. The operating controls were placed above and around the engine. The cargo or passenger capsule formed the nose of the rocket. The space shuttle was a primary mode of space travel for many years and used two distinct systems:

- Two solid-fuel rockets strapped to the external fuel tank.
- The shuttle orbiter.

Rockets

The solid-fuel rockets lift the orbiter off the launchpad and give it initial acceleration. After they exhaust their fuel, they fall away. The solid-fuel rockets parachute into the ocean and are recovered to be used again. At this point, the three engines in the orbiter take over. They use the fuel in the external tank to place the shuttle into orbit. As the shuttle enters orbit, the external fuel tank falls away and burns up as it falls toward the earth.

Shuttle Orbiter

The shuttle is about the size of a small airliner (such as a Boeing 717)—122′ (37 m) long with a 78′ (24 m) wingspan. The shuttle's crew operates the shuttle from the flight deck at the front of the orbiter. The middle of the shuttle is a large cargo bay. This bay can carry nearly 30 tons (27 metric tons) into orbit and almost 15 tons (13.5 metric tons) on reentry.

Reusable Launch Systems

Newer technologies, such as reusable launch systems, are now being used by privately funded ventures. This is done to reduce the time and resources needed for space travel. SpaceX is one company that continues to push the boundaries of space travel. The *Dragon* spacecraft is composed of a capsule and is launched on top of a *Falcon 9* launch vehicle. Many of these technologies can depart, conduct a service mission, and return upright to a launching pad.

Space-Transportation Vehicle Propulsion

Space transportation depends on rocket propulsion systems. The rocket is based on Newton's third law of motion. Thus, rocket engines can be called *reaction engines*. The rocket applies this principle by doing the following:

- Burning fuel inside the engine.
- Allowing pressure from the exhaust gases to build up.
- Directing the pressurized gases out of an opening at one end of the engine.

This action produces motion in the opposite direction from the exiting gases. If you have ever blown up a balloon and allowed it to fly around as the air escapes, you have seen the principle of a rocket engine.

The fuel used in rockets is either liquid or solid. Early rockets were developed in China and India before 1800. The developers used gunpowder as a solid fuel. Robert Goddard designed and built the first modern liquid-fuel rocket in 1926.

Liquid-Fuel Rockets

Today's ***liquid-fuel rockets*** have two tanks. One tank contains the fuel, or propellant. The other contains oxygen. The fuel and oxygen are fed into the combustion chamber. There, they combine and burn to generate the thrust needed to propel the rocket and its payload into orbit.

Liquid-fuel rockets have several distinct advantages. First, the amount of fuel and oxygen fed into the engine can control the amount of thrust. Second, a liquid-fuel rocket engine can be used intermittently. Finally, the rocket engine can be recovered and reused after the spaceflight.

Solid-Fuel Rockets

Solid-fuel rockets use a powder or a sponge-like mixture of fuel and oxygen. See **Figure 23-44**. Once the mixture is ignited, it burns without outside control. Therefore, the thrust cannot be changed. Also, the mixture burns completely. The mixture cannot be stopped from burning once it is started.

©iStockphoto.com/bmcent1

Figure 23-44. This model rocket is launched by a solid-fuel rocket engine.

Space-Transportation Vehicle Suspension

Spacecraft, similar to aircraft, stay aloft because the forces developed by the movement of the craft overcome the forces of gravity. In spacecraft, it is the velocity of the craft that counteracts the pull of gravity from the various solar bodies, including the sun, moon, and other planets.

Space-Transportation Vehicle Guidance

Spacecraft, similar to all other vehicles, require guidance. Guidance is provided by sophisticated instruments that measure three parameters:

- **Velocity.** The speed at which the spacecraft is traveling.
- **Attitude.** The orientation of the spacecraft in space.
- **Location.** The position of the spacecraft in space.

Space-Transportation Vehicle Control

Similar to all other vehicles, spacecraft require control. The major control systems for a spacecraft are velocity control and attitude control. Attitude control keeps the spacecraft correctly oriented in space, while velocity control deals with the speed of the spacecraft.

Velocity control can be used to change the speed or the orbit of a spacecraft. This change can be made in two ways. Bursts from a propulsion system within the spacecraft can be used to increase the speed of the craft. Also, the gravity from other solar bodies can be used to increase the speed of the spacecraft. This action is called *gravity assist*, or the slingshot effect.

Attitude-control systems consist of equipment that measures, reports, and changes the orientation of the spacecraft while the spacecraft is in flight. These systems respond to external forces so the spacecraft does not rotate wildly or move off course. Attitude-control systems contain sensors and actuators. The sensors determine the actual attitude of the vehicle. The actuators change the attitude.

There are many different types of attitude-control devices used in spacecraft. A common device is a thruster, which is a small propulsion device. Another attitude-control device is a momentum wheel. This device is a rotor that is spun in the opposite direction of the natural rotation of the spacecraft. The force of the momentum wheel cancels out the force making the spacecraft rotate. Another device is a solar sail. This device produces thrust as a reaction to reflecting light. Solar sails can be used for both small attitude and velocity adjustments.

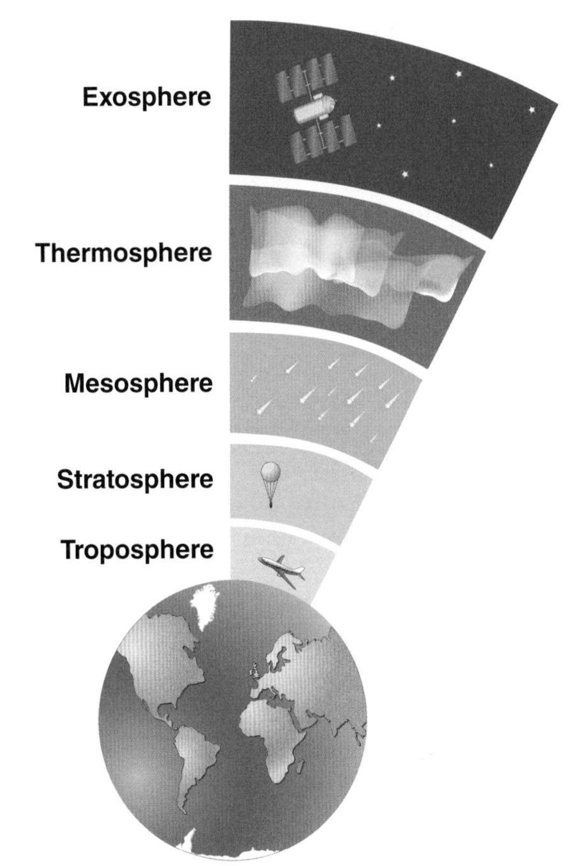

Designua/Shutterstock.com

Figure 23-45. This drawing shows the layers of the earth's atmosphere.

Areas of Operation

Spacecraft and satellites operate in several regions, **Figure 23-45**. The lowest is called the ***troposphere***, which is the first 6 miles (9.7 km) of space above the earth. General aviation and commuter aircraft operate in this region.

Above this region is the ***stratosphere***. This region extends from 7 miles to 22 miles (11 km to 35 km) above the earth. Commercial and military jet aircraft operate in the lower part of this region. The upper part of the stratosphere is called the ***ozone layer***. This layer absorbs much of the sun's ultraviolet radiation. Evidence of damage to this layer has caused great concern about global warming and the health of the planet.

The next layer is the ***mesosphere***. This region extends from 22 miles to 50 miles (35 km to 80 km) above the earth. The ***thermosphere*** lies just above the mesosphere. This region ranges from 50 miles to 62 miles (80 km to 99 km) above the earth. Many satellites operate in this layer of the atmosphere. The last layer is called the ***exosphere*** and blends directly into outer space.

Summary

- ✔ The development of transportation systems allowed people to move in ever increasing distances.
- ✔ Transportation is all acts that relocate humans or their possessions.
- ✔ Transportation is a system because many parts work together to meet a goal.
- ✔ Types of transportation systems include land, water, air, and space.
- ✔ All transportation systems have pathways, vehicles, and support structures.
- ✔ Pathways can be found on land and water, which form a network connecting most areas of the United States.
- ✔ Vehicles are designed to carry people and cargo on a pathway. They are designed to contain and protect the cargo as it is moved from place to place.
- ✔ Support structures include pathways, terminals, communication towers, traffic signals, and signs.
- ✔ Each type of vehicle has structural, propulsion, suspension, guidance, and control systems.
- ✔ Structural systems have two parts: the physical frame and covering. They provide spaces and protection for people, cargo, and other devices.
- ✔ The propulsion system uses energy to create power for motion. The system must match the job.
- ✔ Suspension systems support the weight of the vehicles and cargo.
- ✔ Guidance systems provide information for control.
- ✔ Control systems affect the speed and direction of vehicles.
- ✔ There are two types of control systems: speed control and direction control.
- ✔ Land transportation includes all movement of people and goods on or under the surface of the earth, including highway-, rail-, material-, and on-site transportation systems.
- ✔ Water transportation includes all vehicles that carry passengers or cargo over water or under water.
- ✔ Air transportation moves people and cargo using air travel.
- ✔ Space transportation moves cargo and sometimes people through space.

Check Your Engineering IQ

Now that you have finished this chapter, see what you learned by taking the chapter posttest.
www.g-wlearning.com/technologyeducation/

Test Your Knowledge

Answer the following end-of-chapter questions using the information provided in this chapter.

1. Define the term *transportation*, as used in this chapter.
2. Transportation is a system because it is made up of a series of interrelated ______.
3. List the four types of transportation systems.
4. Define the term *place utility*.
5. Name two groups of water pathways.
6. ______ vehicles are designed to protect occupants as they travel. ______ vehicles are designed to protect cargo from damage by motion and the outside environment.
7. Which of the following statements about terminals is *true*?
   A. Terminals are locations where transportation activities begin and end.
   B. Terminals are used to gather, load, and unload passengers and goods.
   C. Terminals provide connections for various transportation systems.
   D. All of the above.
8. Using more than one mode of transportation to ship cargo is called ______.

Matching questions: Match each system with the correct description. Some letters will be used more than once.

9. Involves degrees of freedom.
10. A speedometer is one example.
11. Includes the physical frame.
12. The wing on an airplane is an example.
13. A bicycle pedal is one example.
14. Provides spaces for cargo.
15. Springs and shocks are examples.

A. Structural
B. Propulsion
C. Suspension
D. Guidance
E. Control

16. What are the three structural units common to all land-transportation vehicles?
17. The most common engine in land-transportation vehicles is the ______ engine.
18. *True or False?* Most land-transportation vehicles are controlled manually by an operator.
19. Ships that carry cargo in sealed steel boxes are called ______.
20. Ships and barges float because of the scientific principle of ______.
21. *True or False?* Friction plays a role in the suspension system of a hydrofoil.
22. What is a fuselage?
23. List the three major types of jet engines.
24. What are the four main forces affecting a fixed-wing airplane's ability to fly?
25. A satellite traveling at the same speed the earth is turning is said to be in ______ orbit.
26. *True or False?* Liquid-fuel rockets have one tank.

Critical Thinking

1. What are the pros and cons to the increasing development of transportation systems that allow people to move in ever increasing distances?
2. What are the ethical considerations of entering space?

STEM Applications

1. Select a transportation system you use. Describe it in terms of the following:
   A. The pathway.
   B. The vehicle.
   C. The structures used.

   Enter your data on a chart similar to the one below.

Type of Transportation Systems		
	Name	**Description**
Pathway		
Vehicles		
Structure		

2. Obtain a variety of bus-, rail-, ship-, or airline-route maps and road maps. Choose a place you would like to visit. Plan a trip that uses intermodal transportation. Be sure your route uses at least two different types of transportation.
3. Using a rubber band for power, design a land vehicle that has all five vehicle systems. Analyze the vehicle. Explain the features of each system.
4. Design a cargo container for a raw egg. The container must be able to withstand the impact of a 15′ (4.6 m) fall.

CHAPTER 24

Meeting Needs through Aerospace Engineering

Check Your Engineering IQ 

Before you read this chapter, assess your current understanding of the chapter content by taking the chapter pretest. www.g-wlearning.com/technologyeducation/

Learning Objectives

After studying this chapter, you should be able to:

- ✔ Define aerospace engineering.
- ✔ Summarize the evolution of flight.
- ✔ Describe the development of aerospace engineering as a discipline.
- ✔ Distinguish between aeronautical and astronautical engineering.
- ✔ Explain the concepts of lift, drag, thrust, and weight and describe how airfoil design is affected by these.
- ✔ Explain aerodynamics, drag, and Bernoulli's principle.

Technical Terms 

aerodynamics
aeronautical engineering
aerospace engineering
angle of attack
angle of incidence
aspect ratio
astronautical engineering
Bernoulli's principle
camber
chord
conservation of energy
drag
fluid dynamics
friction
laminar flow
span
Venturi effect
viscous fluid

Close Reading

Before you read this chapter, consider the types of transportation systems discussed in Chapter 23. As you read, consider how scientific, mathematical, and engineering knowledge is used in the field of aerospace engineering.

Air flight was in its infancy just 100 years ago. But glance into the sky today, in nearly any region of the world, at any moment of the day, and you will likely see at least one aircraft flying above you. Or you may see evidence of an aircraft in condensation trails (or contrails), which are artificial clouds left behind from planes, **Figure 24-1**. Flying is now a part of our culture and daily lives. People travel from one side of the country to another or from one nation to another. Space flights, which were once historic and national events, occur so often now that their launches scarcely make the news.

Advancements made in flight and space travel are the result of scientific and engineering developments made in the field of aerospace engineering. ***Aerospace engineering*** is a highly specialized discipline that applies mathematical, scientific, and engineering knowledge to the development of aircraft and spacecraft, and their related technologies, **Figure 24-2**.

Steve Mann/Shutterstock.com

Figure 24-2. This jet engine was designed by an aerospace engineer.

Aerospace Engineering Evolution

Air transportation became a reality in 1903 when the Wright brothers made the first successful flight in a power-driven airplane. However, there were many centuries of scientific and mathematical breakthroughs and developments that made this historical event possible. These breakthroughs and developments also led to the creation of aerospace engineering.

aleksandr hunta/Shutterstock.com

Figure 24-1. Contrails are often observed as jet engines leave behind hot and humid exhaust that mixes with the atmosphere.

Flight Machines

One of the first recorded attempts at flight was with a kite. It happened over 2000 years ago in China, **Figure 24-3**. Kites and other human-made flight

PaleBlue/Shutterstock.com

Figure 24-3. While kites are now considered recreational, they were once a major technological advancement.

Think Green Reusable Spacecraft

NASA sent spacecraft into space for more than 50 years. The spacecraft returned to Earth by landing safely in the ocean. The systems used to launch the spacecraft were used only once. The goal was to reuse spacecraft. However, the reality was that the engine technologies and materials were not advanced enough to be reused or powerful enough to escape Earth's gravity. Thus, many of the spacecraft were used only once.

In recent years, due to technological advancements in engine technology and materials science, there has been a fundamental shift to use reusable launch systems. Reusable launch systems (RLS) or reusable launch vehicles (RLV) are capable of launching payloads into space multiple times. These designs contrast with launch systems that were previously used only once and then discarded. While the space shuttle was partially reusable, no other fully reusable designs were implemented until recently. Private organizations are currently pursuing and testing RLS. For example, SpaceX has created the *Dragon V2*, a resupply spacecraft. It can land softly, like a helicopter, almost anywhere on Earth and can be used up to 10 times before needing service.

machines were constructed of common materials, such as light wood and paper. The technology slowly and eventually spread to other countries around the world. Historians suggest that initial concepts of flight machines came from people observing the wind's effect on leaves and sails. The process of looking to nature for inspiration is called *biomimicry*.

Other examples of early designs that use the principles of flight include sky lanterns and spinning rotor toys. Sky lanterns, **Figure 24-4**, are basically small hot-air balloons. They are still used today at special events, such as weddings or on holidays. Spinning rotor toys, **Figure 24-5A**, also use the principles of flight. The rotor works in the same way as this maple seed, **Figure 24-5B**, and is another example of biomimicry. As rudimentary as many of these designs were, they served as the foundation of present day flight and aerospace engineering.

Barbora Kristofova/Shutterstock.com

Figure 24-4. Hot air lanterns are still used today at special events or on holidays.

Human Carrying Aircraft

Leonardo da Vinci was an early visionary for the potential of flight. His observations of nature during the fifteenth century, particularly of winged animals such as bats and birds, inspired his designs. **Figure 24-6** depicts a three-dimensional model of one of his most famous designs, the *ornithopter*, or flying machine. This design is modeled on the anatomy of a bird, complete with flapping wings.

In 1783, the Montgolfier brothers of France experimented with lighter-than-air devices. They built and flew a hot-air balloon that rose over 6000 feet and flew more than one mile, **Figure 24-7**. That very same year, two other Frenchmen created a gondola that transported them into the sky, becoming the first people to fly.

Around the same time as the French advancements in flight, hydrogen gas was discovered, resulting in the invention of a hydrogen balloon. Hydrogen, a very light gas, provided greater lift for a given volume,

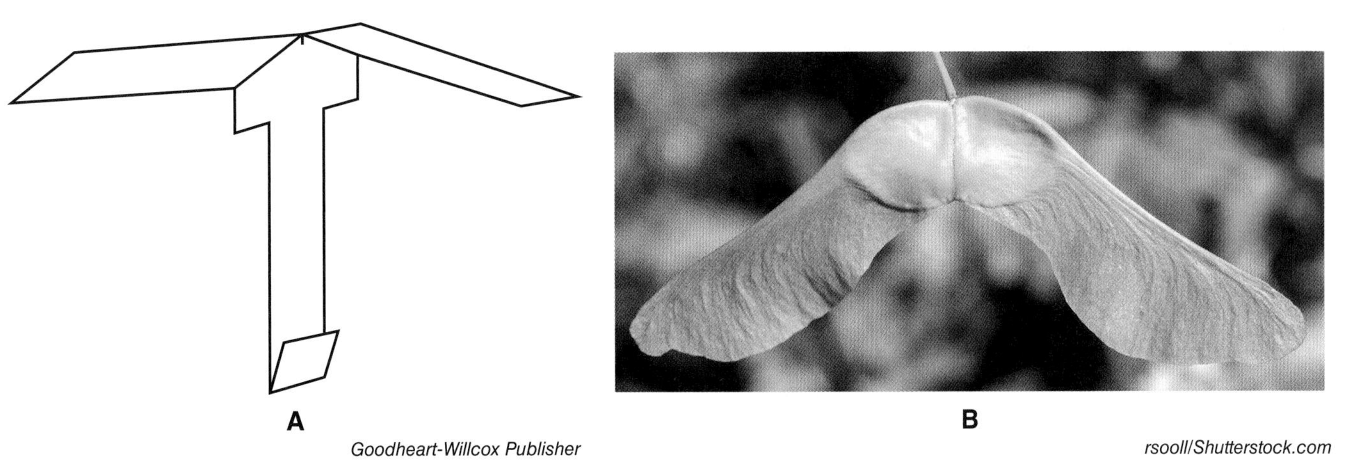

Figure 24-5. A—The rotation direction on this simple rotor can be changed by flipping the tabs. B— Maple seeds spin like helicopter blades once they fall from trees.

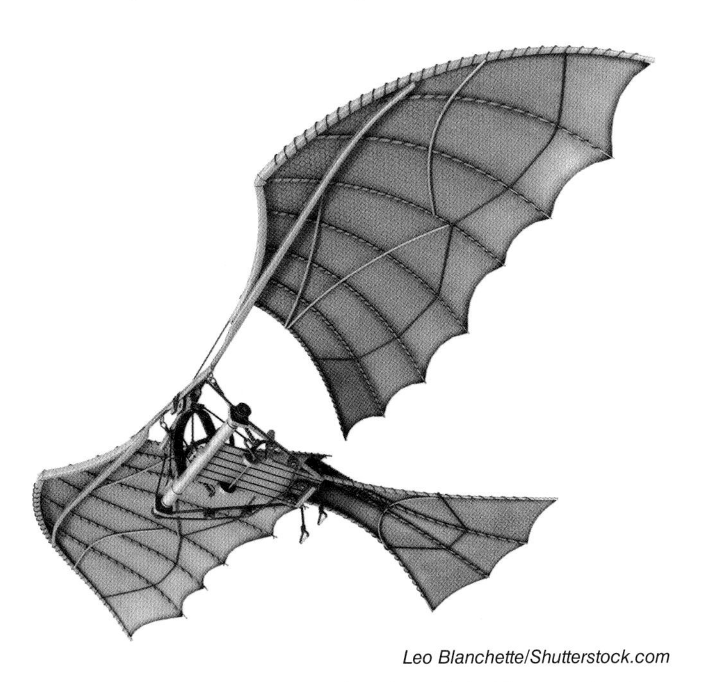

Figure 24-6. A three-dimensional model of Leonardo da Vinci's ornithopter, or flying machine.

so smaller balloons could be used. The balloons could also stay aloft for longer periods of time. Hydrogen, however, is also very flammable and explosive, especially in an atmosphere that contains high traces of oxygen. On May 6, 1937, a large, hydrogen filled German passenger airship, the *Hindenburg*, caught fire and crashed. It is suspected that a spark and leaking hydrogen may have caused the disaster.

These are just a few early advancements in human flight. Yet, alongside the development of theories by physicists in mechanics, particularly fluid dynamics and Newton's laws of motion, they would assist in forming the foundation for present day aerospace engineering.

Figure 24-7. At the time, the Montgolfier brothers' balloon was a major advancement.

Twentieth Century Flight and Space Travel

In 1903, Orville and Wilbur Wright created the first practical machine that could fly. This plane, **Figure 24-8**, was the first controlled, heavier-than-air

Everett Historical/Shutterstock.com

Figure 24-8. The *Flyer* was the first heavier-than-air machine to achieve flight.

flying machine. It had a chain drive with an engine that spun two propellers. Following a traditional design process, the brothers made a series of gliders that they tested repeatedly until they were ready to test the engine-powered *Flyer.* Despite the Wright brothers' innovation in flight, these early machines were not yet powerful enough to meet the demands of travel and transportation.

One event that tremendously improved aircraft was World War I. To meet the needs of World War I, powerful combat planes were required. This resulted in increased research and development. Aircraft structures and engines saw drastic modernizations that may have taken longer to occur in peacetime. Modernization improved flying speeds and ranges. As with other technologies, these military advancements ultimately benefitted the private sector. Larger airplanes soon began carrying freight and passengers on a regular basis. For example, the DC-3 provided a significant portion of all airline traffic throughout the world in 1939, **Figure 24-9.** DC-3s were also used during World War II to transport troops and cargo.

The first jet airliner, the Boeing 707, was built in 1954. By 1967, airlines were carrying over 100 million passengers a year. The volume of traffic became a major concern. New airports had to be built. Commercial airlines demanded larger, but fewer, planes that could move more passengers. Demand for more guidance and traffic control technology impacted, and continues to impact, the electronics industry.

Just as flying toys and gliders were the foundation for the aircraft industry, flight advancements served as the springboard for space travel. The appeal of having satellites controlled from a central location on Earth to communicate, observe weather, analyze crops, and protect against military threats were a few reasons to pursue space travel.

Ivan Cholakov/Shutterstock.com

Figure 24-9. DC-3s had a significant impact on travel.

Despite the desire to travel to space, it wasn't until the late 1950s and early 1960s that it became a reality. This was the result of research, development, and advancements across all technological areas. For example, the 1926 liquid-propellant rocket launch by Robert Goddard provided the solution for propulsion. In October, 1957, *Sputnik I* was launched by the former Soviet Union. It carried a simple radio transmitter and weighed 185 pounds. *Sputnik II,* launched just one month later, weighed over 1000 pounds, and carried a living passenger. Within the next few years, rapid advancements would push the boundaries of space travel, and by 1968, the first human, Neil Armstrong, would step on the moon, **Figure 24-10.**

The Future of Flight and Space Travel

As we continue to push technological boundaries during the twenty-first century, flight and space travel has and will change drastically. Where government agencies, such as NASA and military organizations, were once the leaders and pioneers of space travel, the privatization of space is giving entrepreneurs the opportunity to innovate and contribute. Elon Musk of SpaceX is one such entrepreneur. His company designs, manufactures, and launches increasingly efficient space vehicles. They

Technology Explained Unmanned Aerial Vehicle (UAV)

Unmanned aerial vehicle (UAV): A pilotless aircraft that is controlled by a computer program or remote control.

Lukas Gojda/Shutterstock.com

Unmanned aerial vehicles, often referred to as drones, are becoming increasingly popular for the multiple problems they can solve. Used recreationally and by the military, farming, shipping, and delivery industries, these aircraft do not have a pilot on board. They are controlled by a computer program (autonomously) or by remote.

Although UAVs function differently based on their purpose and environment, they share similarities. For example, most contain many sensors and audio/video technologies, including cameras, radar, defense, infrared, and night-vision detection. Army and military based UAVs may contain multiple aircraft. UAVs are typically controlled by ground stations. A trained crew navigates the craft and constantly analyzes images and video sent back from the drone.

also focus on reusability to reduce costs and development time lines.

Drones are gaining popularity for their wide range of applications, **Figure 24-11**. Also known as unmanned aerial vehicles (UAV), or unmanned aerial systems (UAS), the key distinction is that drones are autonomous, or remotely guided. These technological devices fall into primarily three areas: military, commercial, and recreational. Military drones were first created to keep humans from traveling in dangerous situations. Advancements in digital technologies allowed subsonic (vehicles that fly below the speed of sound) military aviation vehicles to be controlled remotely. In 2015, the longest flight recorded by a UAV, the *AtlantikSolar*, traveled continuously for 81-1/2 hours and 1439 miles (2316 km). There are many potential uses for drones, with both positive and negative effects on society.

Joseph O'Dea/NASA

Figure 24-10. *Apollo 11* landed on the moon in July, 1969. Neil Armstrong was the first person to step on the moon.

Yurchikys/Shutterstock.com

Figure 24-11. Drones are being used in many environments. The camera attached to this drone is used to take aerial photographs.

These developments created, and continue to create, a need for engineers with the knowledge and skills to work as aerospace engineers.

Aerospace Engineering

The field of aerospace engineering has evolved along with the advancements in flight. As these technological developments and the application of mathematics, science, and engineering matured, aerospace engineering has come to focus primarily on two specialties: aeronautical engineering and astronautical engineering.

In ***aeronautical engineering***, engineers are involved with the study and design of aircraft and technologies such as helicopters, gliders, lighter-than-air vehicles, airplanes, and jets. ***Astronautical engineering*** is primarily focused on the study, design, and creation of space technology. An additional interpretation would be that an individual involved with aeronautical engineering considers technologies designed to fly within the earth's atmosphere, while astronautical engineering considers technologies outside the earth's atmosphere.

All aerospace engineers have a strong foundation in science and mathematics. They use this foundation to design space technology and create problem solutions. Aerospace engineers also understand engineering design and development and mechanical engineering concepts. They understand how flight is achieved and how to build aircraft. These concepts are discussed in the following sections.

Fundamentals of Aircraft Design

Chapter 23 introduced the forces of thrust, drag, weight, and lift. These four forces determine an airplane's ability to fly, **Figure 24-12**. Spacecraft are not affected by lift and drag. They experience only weight and thrust. Lift and drag are not present because they are created by movement through the air. Because there is no air in space, lift and drag are absent.

Airfoils

An airfoil is simply a streamlined object that is introduced into airflow. Airfoils are shaped and constructed to take advantage of the response of

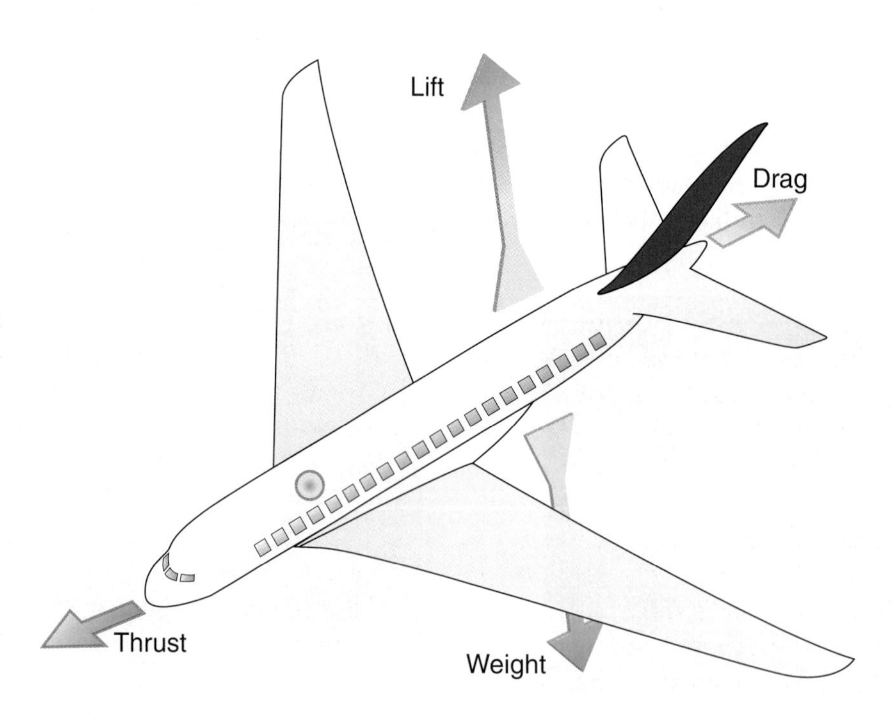

Goodheart-Willcox Publisher

Figure 24-12. The four forces that affect flight.

Career Connection Aerospace Engineer

Aerospace engineers design, test, and analyze aircraft and spacecraft, including satellites, missiles, drones, and unmanned autonomous vehicles (UAV). They may work across a number of industries, examining how products such as automobiles or boats function. They employ their knowledge of mathematics, science, and engineering to solve challenging, real-world problems.

High school students wanting to pursue aerospace engineering should enroll in advanced physics, chemistry, and mathematics courses, along with technology courses, such as computer-aided design.

At the university level, bachelor's degree programs provide students with classroom and laboratory work, as well as field studies and internships. Typical courses include foundational engineering, propulsion, aerodynamics, fluid dynamics, structural engineering, and materials science. In combined bachelor's/master's programs, students gain additional experience and skill application. Students who have strong analytical, business, math, problem-solving, and personal skills should do well in these programs.

Aerospace engineers work in a variety of industries. Traditional career tracks include aerospace product and parts manufacturing, general engineering services, government work, and research and development in the physical, engineering, and life sciences. Less traditional career tracks include navigational, measuring, electromedical, and control instruments manufacturing.

Aerospace engineers held about 73,000 positions as of 2014. Current statistics indicate that employment of aerospace engineers is projected to decline 2% from 2014–2024.

Potential projects an aerospace engineer might expect to work on include designing and constructing aircraft and spacecraft; developing missiles and defense systems; inspecting the manufacture of aircraft and spacecraft; and analyzing and evaluating proposed aerospace designs.

Professional aerospace engineers can join several organizations to network with others in their industry. These include American Institute of Aeronautics and Astronautics, National Society of Professional Engineers, Aerospace Industries Association, and American Society of Engineering Education.

this airflow to physical laws. These features result in two actions being developed from the air mass. The first action forces air downward. The second action forces the airfoil upward. If the airfoil is constructed correctly, it will create a greater lift force than the aircraft weight, and flight will occur.

Air flows a greater distance over the top of an airfoil, producing lift. However, there is more to the story, **Figure 24-13**. ***Camber*** is the characteristic curvature along the upper and lower surfaces of the airfoil. The ***span*** and ***chord*** are measurements from wing tip to wing tip, and leading edge to trailing edge (from the wing center). The relationship between the span and the chord is called the ***aspect ratio***, **Figure 24-14**, and can be found by dividing the span by the chord. ***Angle of attack***, or the angle of the wing as it cuts through the air, also affects lift. Typically, lift increases as this angle increases, **Figure 24–15A**. ***Angle of incidence*** is the angle at which the wings are attached to the fuselage of the plane, **Figure 24–15B**.

Consider the shape of the airfoil, specifically camber. If it resembles a teardrop, **Figure 24-16**, then the velocity and pressure of the airflow would be the same on both sides of the airfoil. However, if the shape resembles half a teardrop and is inclined, then the airflow strikes at an angle. The air flowing over the top surface is forced to move faster, while the bottom surface moves more slowly. This results in increased velocity, reducing the pressure above

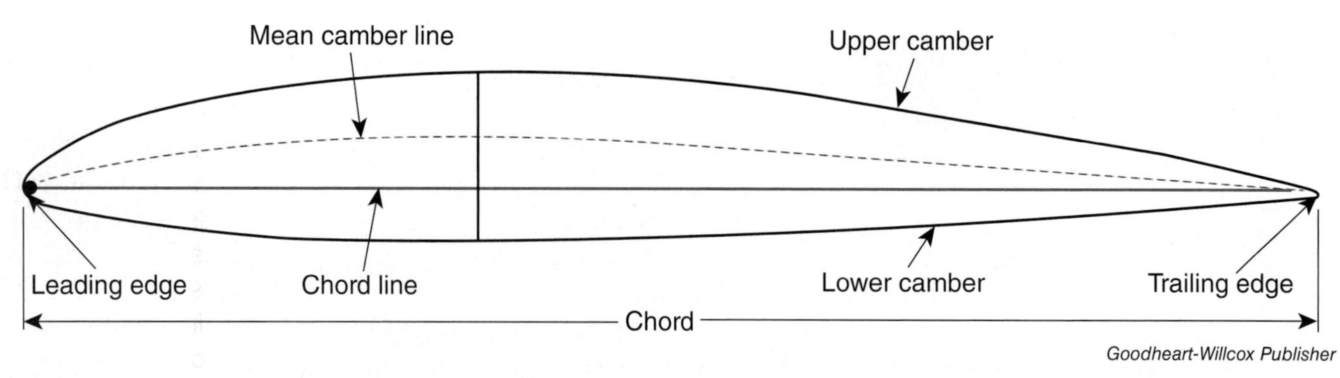

Figure 24-13. Airfoils contain a number of features that all influence lift.

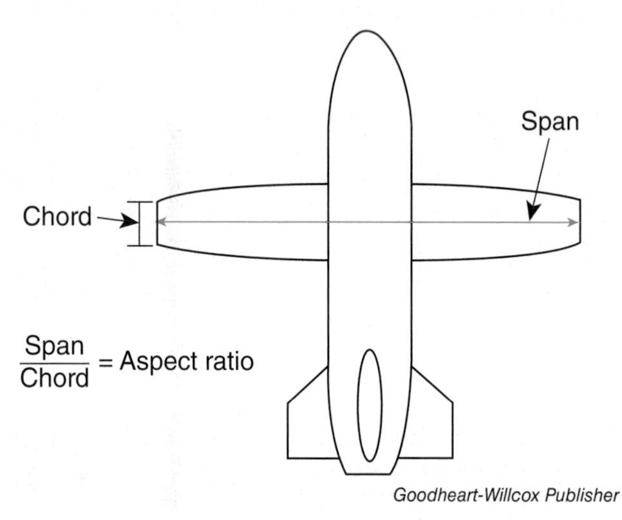

Figure 24-14. Aspect ratio is the relationship between chord and span. For example, a span of 32 and a chord of 4 will have an aspect ratio of 8.

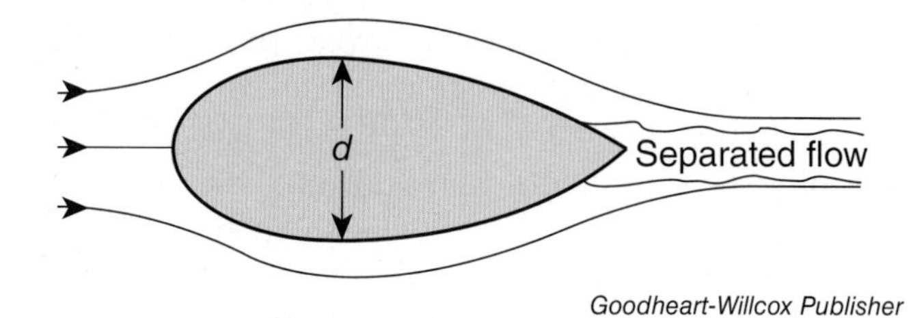

Figure 24-16. This teardrop-shaped airfoil has the same velocity and airflow pressure on both sides.

the airfoil. The air mass below the wing will have positive-pressure lifting action, while the area above the wing will have negative-pressure lifting action.

Many airfoil designs have been tested and not one is able to meet every flight requirement. There is no one-size-fits-all airfoil design. The purpose of the

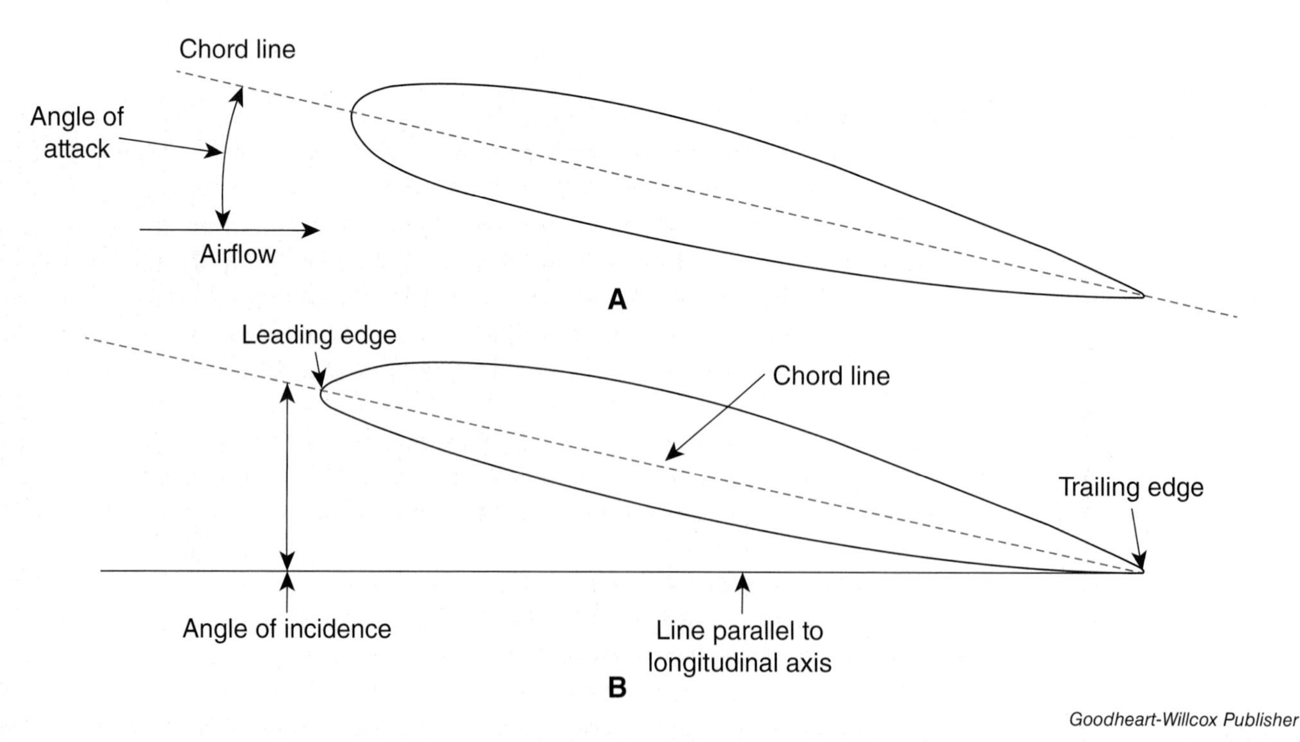

Figure 24-15. A—Angle of attack. B—Angle of incidence.

aircraft, the weight, speed, and flying conditions all dictate the airfoil shape and design. Understanding these core features guides aerospace engineers as they design each airfoil.

Wing Design

There are many shapes, sizes, and forms of wings. Designs are based on differences in pressure and speed and needs of airflow over the surfaces. **Figure 24-17** shows several different wing shapes and forms. What environment and needs may each one address? Would they all function in the same way?

Rotors

Helicopters are a type of rotorcraft, **Figure 24–18**. The thrust and lift required for takeoff is supplied by the rotors. The rotors, or blades, are powered by a motor and generate lift as they revolve around a mast. Spinning rotors provide the lift and thrust needed for the helicopter to leave the ground. In order for the helicopter to lift itself off the ground vertically, the rotors must generate enough lift to compensate for the total weight of the helicopter and the occupants or payload it is carrying. As you read the section on fluid dynamics, consider how the speed and pressure of the spinning rotors cause the helicopter to take off.

Nattapon B/Shutterstock.com

Figure 24-18. Helicopters are able to take off and land vertically, and fly in different directions.

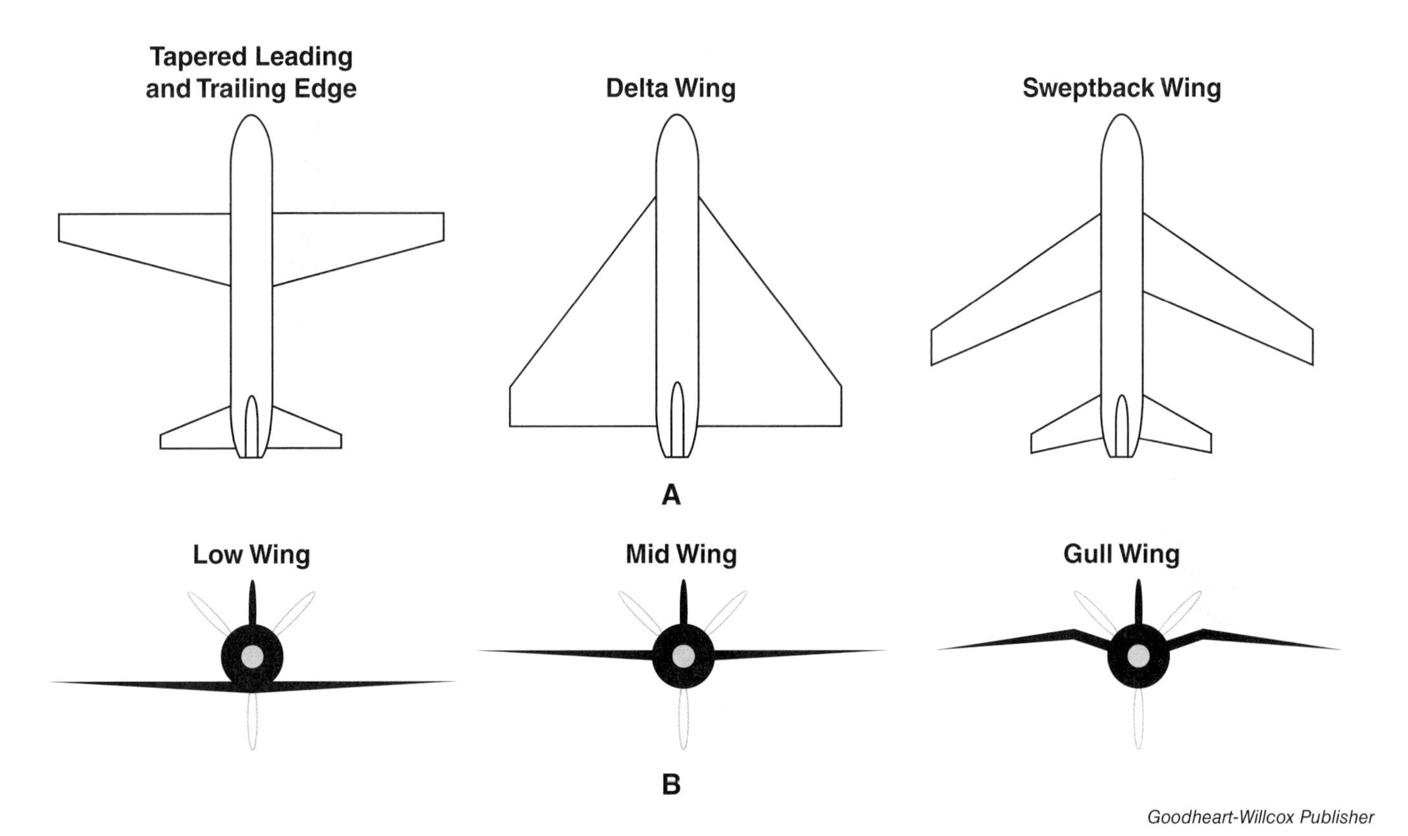

Goodheart-Willcox Publisher

Figure 24-17. A—Wing shapes. B—Wing forms.

STEM Connection: Science

Aerodynamics

Many people use a bicycle simply to get somewhere or to exercise. They do not realize the role science plays in many areas of cycling. For example, the scientific study of the principles of aerodynamics has greatly affected the design of bicycles and bicycle clothing worn by competitive cyclists.

Generally speaking, aerodynamics is the study of the forces acting on an object as it moves through the air or other gaseous fluid. One of the major forces studied in aerodynamics is drag. *Drag* is the term used to describe the aerodynamic force resisting the forward movement of an object. When cycling, we encounter two kinds of drag: frictional drag and form drag. Both forces slow a cyclist.

Scientists counter these forces in two ways. Making the moving object smoother can decrease frictional drag. Streamlining the object can decrease form drag. For example, aero bars lower the rider's torso and help keep the elbows closer to the torso. Drag is decreased because the torso and elbows are not blocking as much airflow.

Designers have also created different kinds of clothing to help with drag. For example, many competitive riders wear such items as skin suits, which are designed to reduce friction.

Bicycles also have other aerodynamic features. Can you think of two of these?

Principles of Flight

Aircraft design is only part of the equation for achieving flight. There are principles and scientific concepts of flight that aerospace engineers must understand in order to design efficient and safe products.

Atmosphere

The atmosphere, **Figure 24-19**, is comprised largely of air and surrounds Earth. Its structural makeup influences flight. Just as land and water affect automotive and boat travel, the atmosphere, a mixture of gases, affects air travel.

The atmosphere has weight and mass that affect flight. The atmosphere (roughly 75% nitrogen, 24% oxygen, and a mixture of other gases) contains gases of varying weights. The lighter the gas is, the higher it will be lifted up in altitude. Oxygen is primarily found between the earth's surface and 35,000 feet.

Air is able to flow and adjust its shape to pressure changes. This is due to the fact that air has a weak molecular cohesion, which means the molecules between are loosely bound and flexible for movement. You have witnessed this if you have ever filled a balloon with air. As you fill it, you can still shape and mold it into any figure you like, until the strength of the balloon is compromised. This ability allows the air to move as the plane and wings move through the air.

Fluid Dynamics

In Chapter 20, you learned that fluid mechanics is the study of how laws of force and motion apply to liquids and gases. ***Fluid dynamics*** is a type of fluid mechanics that studies fluids in motion. It plays a vital role in aerospace engineering because it is the foundation of aerodynamics.

Aerodynamics

Aerodynamics is the study of the way air flows around objects. This can be as the object moves through the air, as air moves over an object, or both.

Wind tunnels are used by aerospace engineers to test and analyze aerodynamics on aircraft and spacecraft, **Figure 24-20**. Wind tunnels can be used to test either models or full-sized objects.

Air is a ***viscous fluid***, meaning that it often sticks to a moving object's surface. This is called ***friction***,

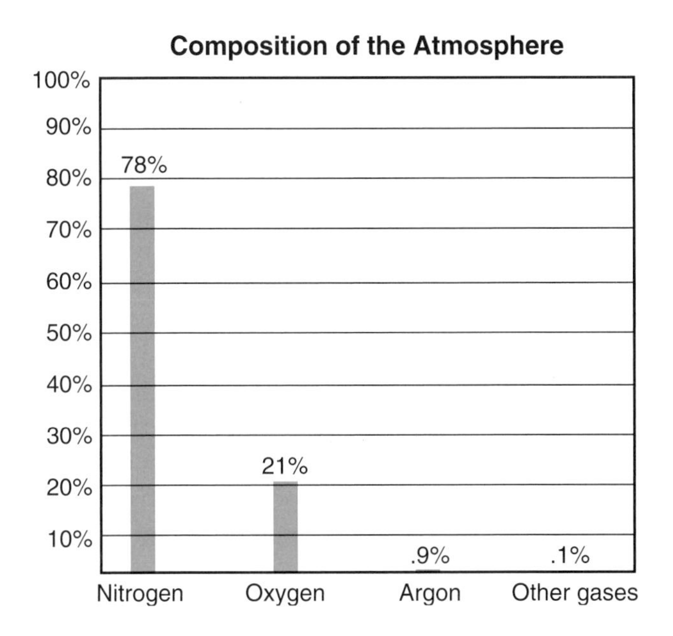

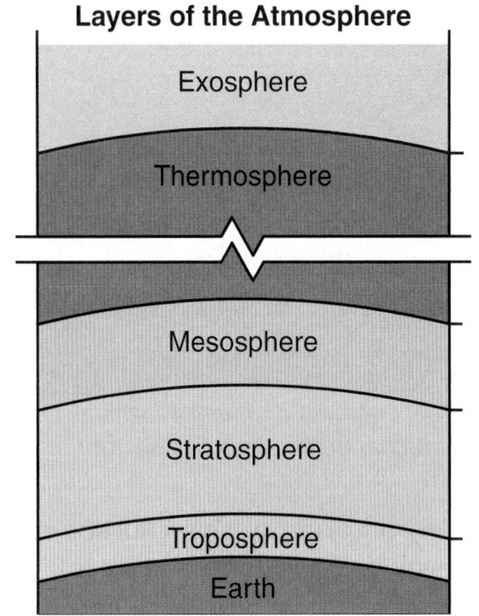

Goodheart-Willcox Publisher

Figure 24-19. The makeup of the atmosphere has considerable implications for flight.

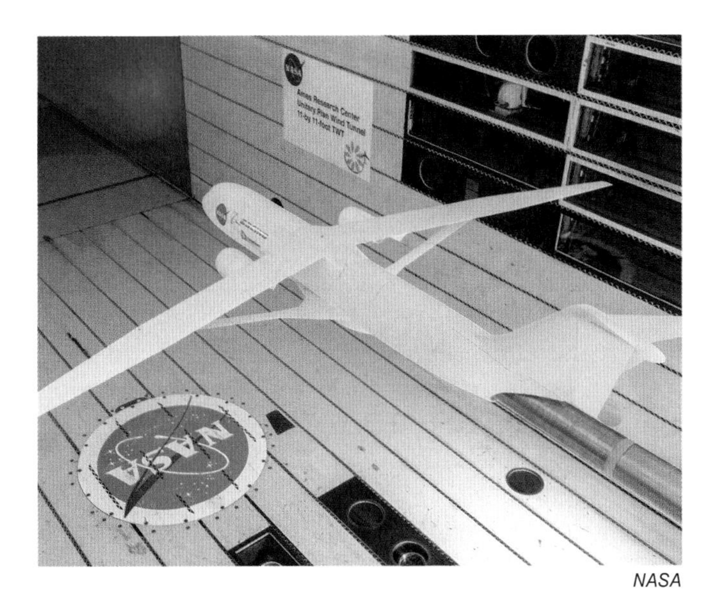

NASA

Figure 24-20. This NASA wind tunnel is used in the development of commercial and military airframes.

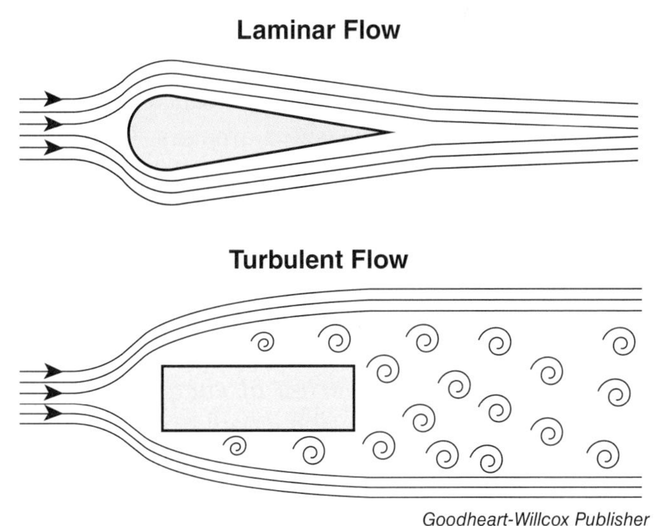

Goodheart-Willcox Publisher

Figure 24-21. Eddy currents are created behind many objects including aircraft, automotive, and marine vehicles.

or a force that acts against motion when two surfaces rub together. Additionally, aircraft is streamlined, which causes eddy currents, **Figure 24-21**. The more streamlined an object is, the more ***laminar flow***, or even and straight flow, it will have.

Drag

Another concept related to aerodynamics is drag. ***Drag*** is an aerodynamic force that acts against the movement of aircraft. Factors that determine the drag on an aircraft include its size and shape, its velocity (speed), the characteristics of the air (viscosity and compressibility), and the angle of the aircraft as it moves relative to the air.

An aircraft has multiple components that contribute to drag. All of these components and the variables mentioned previously must be considered when calculating drag.

Bernoulli's Principle

Bernoulli's principle states that an increase in fluid speed creates a decrease in pressure. Knowl-

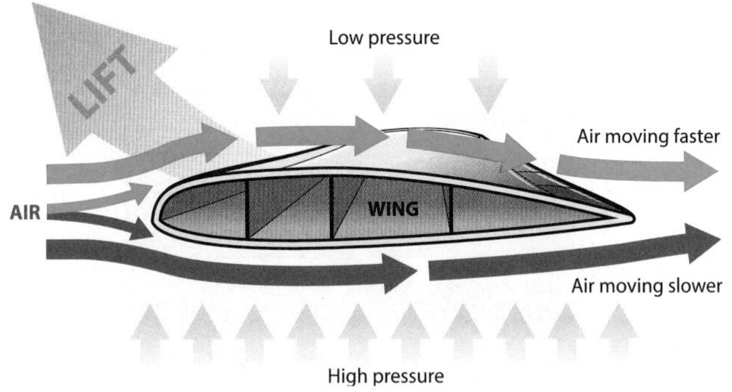

NoPainNoGain/Shutterstock.com

Figure 24-22. Bernoulli's principle explains how velocity and pressure change due to airfoil design.

edge and application of this concept explain why airfoils develop an aerodynamic force, **Figure 24–22.** For example, understanding this concept allows engineers to design wings to achieve lift because knowing how wings are shaped helps ensure that air flows faster over the top of the wing and slower underneath to achieve lift.

Bernoulli's principles can be demonstrated through the ***Venturi effect***. As you can see, the inlet and outlet have the same size diameters. The center is constricted. Air coming in narrows in diameter and then increases in diameter toward the outlet. At the throat, the flow of air increases while the pressure decreases. At the outlet, the pressure increases and flow decreases, **Figure 24–23.**

Another concept that is related to Bernoulli's principle is the ***conservation of energy***. This law states that energy cannot be created or destroyed. The same amount of energy that enters a system must also leave the system.

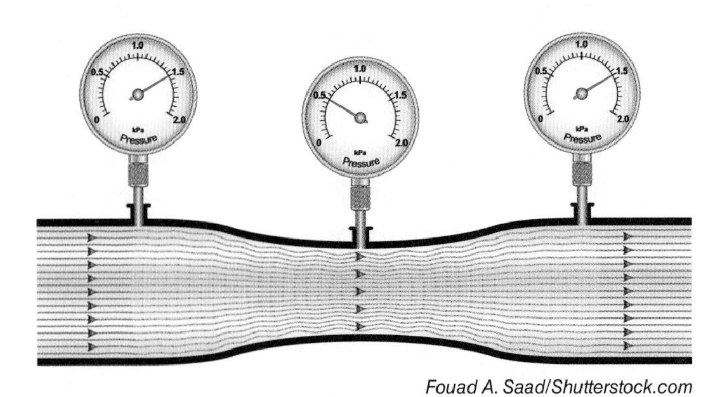
Fouad A. Saad/Shutterstock.com

Figure 24-23. Note the changes in pressure in the tube.

Newton's Laws

Sir Isaac Newton's three laws of motion can be used to explain the movement of an aircraft through air or a spacecraft through space. Newton's first law states that an object at rest will remain at rest until a force is applied to it. Additionally, once the object is set in motion, it will continue in motion, at a constant speed until a force or object strikes it. Once a spacecraft reaches space, there is no gravity. If the spacecraft continues moving at a constant speed in one direction, no additional thrust is required. This is because once the object is set in motion, even without propulsion, space has no friction to slow it down. However, to change direction or speed, additional force, or thrust, is required.

Newton's second law states that as an object's mass increases, so does the force required to make it accelerate. The basic equation for Newton's second law is $f = m \times a$, or force = mass × acceleration.

An airplane's mass is relatively constant except for the fuel that is burned. In this case, Newton's second law would look like:

$$f = m \times (v_1 - v_0) / (t_1 - t_0)$$

m = mass

t = time

v = velocity

In the above equation, the change in velocity divided by the change in time is the explanation of the acceleration. Between two points, the change in velocity and time elapsed allows aerospace engineers to design aircraft and spacecraft to meet specific environmental criteria and constraints.

Newton's third law states that for every action, there is an equal opposite reaction. This law is extremely important in the design of aircraft and spacecraft. It explains the creation of lift from an airfoil or other parts of an aircraft or spacecraft. For example, a jet engine produces thrust through action and reaction. As the engine produces hot exhaust gases, they flow through the back of the engine, producing a thrusting action in the opposite direction, creating propulsion.

Materials Selection

One of the most significant decisions an aerospace engineer has to make is choosing materials to construct models, prototypes, and final designs. These decisions are based on mathematical and scientific knowledge, wind tunnel testing, and the results of those tests. To increase efficiency, engineers choose structural materials that can reduce weight. They also determine the best shape and forms for airfoils and wings, and follow processes and strategies for constructing aircraft, **Figure 24–24**.

Jaromir Chalabala/Shutterstock.com

Figure 24-24. Aircraft structure and construction dramatically affect flight.

Aerospace Engineering in Other Engineering Fields

Aerospace engineers typically work within the aerospace industry. However, their skills are also desirable in many other technological fields.

Aerospace engineers can solve technological problems in the mechanical engineering field. They use their knowledge of fluid mechanics by designing turbines that effectively transfer heat, and by working to create or improve control systems used by aerospace technologies. They may assist in the design and testing of issues surrounding defense contracts, private industries, NASA, and commercial travel companies.

The skill set that aerospace engineers have can be used to assist the automotive industry in the design of fuel-efficient cars. Aerospace engineers are able to design, test, and analyze the aerodynamics of cars to ensure they are as efficient as possible.

The boating and marine industries also have employment opportunities for aerospace engineers. The concept of water flowing around a boat surface is comparable to air flowing over wings. Aerospace engineers can find job opportunities analyzing and testing marine products and providing input for more efficient and effective designs.

Summary

- ✔ Aerospace engineering is a highly specialized discipline that applies mathematical, scientific, and engineering knowledge to the development of aircraft and spacecraft, and their related technologies.
- ✔ Centuries of scientific and mathematical breakthroughs and developments led to power-driven flight and the creation of aerospace engineering.
- ✔ Over 2000 years ago in China, one of the first recorded attempts at flight was made using a kite.
- ✔ Historians suggest that initial concepts of flight machines came from people observing the wind's effect on leaves and sails. This is called biomimicry.
- ✔ In the fifteenth century, Leonardo da Vinci designed a three-dimensional model of a flying machine that mimicked a bird's anatomy.
- ✔ In 1783, two French brothers built and flew a hot-air balloon, while two other Frenchmen created a gondola that transported them into the sky, becoming the first people to fly.
- ✔ The Wright brothers created the first practical airplane in 1903.
- ✔ World War I advanced the development of aircraft.
- ✔ The first jet airplane was built in 1954.
- ✔ Space travel became a reality in the late 1950s and early 1960s.
- ✔ Government agencies were once the leaders and pioneers of space travel, but now private entrepreneurs are contributing and innovating.
- ✔ Aeronautical engineering focuses on the design of aircraft and technologies such as helicopters, gliders, lighter-than-air vehicles, airplanes, and jets.
- ✔ Astronautical engineering focuses on designing and creating space technology.
- ✔ The forces of thrust, drag, weight, and lift determine an airplane's ability to fly.
- ✔ Spacecraft are not affected by lift and drag because there is no air in space.
- ✔ Airfoils are shaped and constructed to create lift.
- ✔ Wing designs are based on differences in pressure and speed, and needs of airflow over the surface.
- ✔ Understanding the principles and scientific concepts of flight allows aerospace engineers to design safe, efficient products.
- ✔ Fluid dynamics is the study of fluids in motion and plays a vital role in aerospace engineering because it is the foundation of aerodynamics.
- ✔ Aerodynamics is the study of the way air flows around objects.
- ✔ Drag is an aerodynamic force that acts against the movement of aircraft.
- ✔ Bernoulli's principle states that an increase in fluid speed creates a decrease in pressure.
- ✔ Sir Isaac Newton's three laws of motion can be used to explain the movement of an aircraft through air or a spacecraft through space.
- ✔ Aerospace engineers must understand the materials used for construction of aerospace models, prototypes, and final designs.
- ✔ Aerospace engineers use their skill set to work in a variety of industries including aerospace, mechanical engineering, automotive, and marine and boating.

Check Your Engineering IQ

Now that you have finished this chapter, see what you learned by taking the chapter posttest.
www.g-wlearning.com/technologyeducation/

Test Your Knowledge

Answer the following end-of-chapter questions using the information provided in this chapter.

1. Define *aerospace engineering.*
2. The process of looking to nature for inspiration is called _____.
3. Name one historical event that contributed to the evolution of flight.
4. How did World War I contribute to the improvement of airplanes?
5. *True or False?* The field of aerospace engineering has evolved along with the advancements in flight.
6. _____ design aircraft and technologies such as helicopters, gliders, lighter-than-air vehicles, airplanes, and jets. _____ focus primarily on designing and creating space technology.

7. What are the four forces that affect flight?
8. What forces are not present in space?

Matching: Match each term with the correct description.

9. Relationship between the span and the chord.
10. Characteristic curvature along the surface of the airfoil.
11. Measurement from leading edge to trailing edge.
12. Angle of a wing as it cuts through the air.
13. Angle at which the wings are attached to the fuselage of a plane.
14. Measurement from wing tip to wing tip.

A. Camber
B. Chord
C. Span
D. Aspect ratio
E. Angle of attack
F. Angle of incidence

15. _____ is a type of fluid mechanics that studies fluids in motion.
16. What is drag?
17. _____ states that an increase in fluid speed creates a decrease in pressure.
    A. Conservation of energy
    B. Venturi effect
    C. Bernoulli's principle
    D. None of the above.

Critical Thinking

1. Bernoulli's principle explains why airfoils develop lift. Provide another everyday context of Bernoulli's principle in action.
2. Explain how you can benefit from the Venturi effect in everyday contexts.

STEM Applications

1. Build and fly either a model airplane or a model rocket.
2. Design various wing shapes and use a hair dryer to test the effect it has on your designs. Use a decision matrix to assess the differences.

CHAPTER 25

Medical and Health Technologies

Check Your Engineering IQ 

Before you read this chapter, assess your current understanding of the chapter content by taking the chapter pretest.
www.g-wlearning.com/technologyeducation/

Learning Objectives

After studying this chapter, you should be able to:

- ✔ Summarize the role of science and technology in wellness.
- ✔ Explain how technology is applied to exercise and sports.
- ✔ Describe different types of health-care professionals.
- ✔ Explain how technological devices are used in diagnosing illnesses and physical conditions.
- ✔ Explain how a new drug is developed.
- ✔ Describe technological devices used to treat illnesses and physical conditions.

Technical Terms 

aerobic exercise
anaerobic exercise
diagnosis
disease
double-blind study
drug
electrocardiograph (EKG) machine
emergency medicine
endoscope
immunization
interventional radiology
medical technology
medicine
over-the-counter drugs
prescription drugs
prosthesis
radiology
therapeutic radiology
ultrasound
wellness
x-rays

Close Reading

How has technology helped you when you were ill or injured? As you read this chapter, list all the ways your health has benefited from technology.

Throughout history, people have been concerned about living longer and better. They have sought ways to cure illnesses, repair damage to their bodies, and improve their health. Each of these challenges has led to the development of new technologies. People have created technological artifacts to care for, improve, and protect their health.

The search for healthy lives can be viewed from two perspectives, **Figure 25-1**. The first is wellness, and the second is illness. ***Wellness*** is the state of being in good health, both mentally and physically. This state is generally achieved through proper diet and regular exercise. Illness can be described as a state of poor health. A disease or sickness might cause illness. In the broadest sense, illness can be extended to include injuries caused by accidents.

Both science and technology play a role in wellness and illness. Science provides knowledge about the human body and health and describes the natural processes the body uses to maintain itself. Technology provides the tools and equipment needed to achieve wellness and treat illness. Technology helps people maintain and restore the body's processes and functions.

Technology and Wellness

Wellness involves actions that keep the body healthy. See **Figure 25-2**. A state of wellness contributes to physical fitness, which is a combination of good health and physical development. The objective of physical fitness is to maximize a person's health, strength, and endurance.

Preventive medicine and treatment help prevent people from becoming ill, but wellness often focuses on what people can do for themselves to maintain their well-being. Four major factors affect wellness: nutrition and diet, environment, stress management, and physical fitness.

Michaelpuche/Shutterstock.com

Figure 25-2. Physical fitness has a positive impact on health.

US Department of Agriculture

Figure 25-1. People are living longer because of wellness programs and modern ways of treating illnesses.

Proper nutrition is important to physical fitness and wellness. See **Figure 25-3**. The energy a person can expend depends on nutrition. If the diet is inadequate, the fitness level drops. Therefore, people should be conscious of the quantity and value of the food they consume.

The environment's effects on the body influence wellness. For example, people should control their exposure to direct sunlight. Technology has been applied to air and water purifiers and has been used to develop sunscreen lotions, sunlight-filtering clothing, and other effective products.

mangostock/Shutterstock.com

Figure 25-3. Wellness programs encourage people to select and eat proper foods.

Stress management is an important factor in wellness. People should be aware of and control their emotional stress. Exercise and recreational activities help to reduce stress. Technologies have been created to help people monitor and reduce their stress levels. These technologies include biofeedback devices, which continuously provide information about a user's physiological functions, such as heart rate; wearable technologies that can help guide a user to maintain a normal breathing rate or stimulate nerves to reduce stress; and smartphone apps that teach stress-reduction strategies.

Becoming and staying fit through physical activity promotes wellness. Technology has been applied to physical fitness through two major areas—exercise and sports.

Technology and Exercise

Exercise can be physical or mental—exerting the body or the mind. See **Figure 25-4**. Exercise is done for training or improving health. Two major types of exercise are anaerobic and aerobic exercise.

Anaerobic exercise involves heavy work by a limited number of muscles. This type of exercise is maintained for short intervals of time. Anaerobic exercises increase strength and muscle mass, but they have limited benefits to cardiovascular health. Examples are weight lifting and sprinting.

Aerobic exercise uses oxygen to keep large muscle groups moving continuously. The exer-

goodmoments/Shutterstock.com

De Visu/Shutterstock.com

Figure 25-4. Exercise prepares people for the demands of work and other life activities.

cise must be maintained for at least 20 minutes. Aerobic exercise uses several major muscle groups throughout the body. This causes greater demands on the cardiovascular and respiratory systems to supply oxygen to the working muscles. Aerobic exercise includes walking, jogging, and swimming. This type of exercise can reduce the risk of heart disease and increase endurance.

Exercise, such as walking and jogging, can be done without special equipment. However, technology has contributed special shoes and clothing to help in these activities. Also, people use exercise equipment to improve their health and well-being. This equipment includes treadmills, stationary bikes, stair-climbers, rowers, and home gyms. See **Figure 25-5**.

Several machines provide lower-body workouts. Treadmills are moving belts that allow people to walk or jog in place. They provide an aerobic fitness workout. Stationary bikes are similar to bicycles, except they do not move. People can obtain the benefits of bicycling without leaving home. In addition, on most of these bikes, the handlebars move and provide resistance, providing an upper-body workout. Stair-climbers allow people to obtain the benefits of climbing when stairs are not available.

Rowing machines simulate a rowboat with oars. They allow people to use their arms and legs without having a boat in the water. Home gyms are multistation exercise machines that allow people to work on many different muscle groups. These machines can provide a complete workout program.

Technology and Sports

Sports are another way to promote wellness and physical fitness. They are games, or competitions, involving skill, physical strength, and endurance. Contests involving physical abilities can be traced back to ancient times. See **Figure 25-6**. Many different types of sports have developed throughout history. Medical proof of the benefits of physical exercise has reinforced the value of sports. This discussion of sports and wellness focuses on recreational sports and not on professional sports.

©iStockphoto.com/Gala98

Figure 25-6. This amphitheater in historic Pompeii was used for sporting events and other activities in ancient times.

nenetus/Shutterstock.com

bikeriderlondon/Shutterstock.com

Syda Productions/Shutterstock.com

Figure 25-5. Among the many types of exercise equipment are stationary bicycles, ellipticals, and treadmills.

All sports require technological products. These products include playing fields or venues, game equipment, and protective equipment. Each game, sport, or contest uses specific technologies. For example, football is played in a very different venue from tennis. Likewise, the game equipment for football (football and goalposts) is different from that for tennis (tennis ball, racket, and net). The clothing and protective equipment are also very different.

Technology and Playing Venues

Each sport requires a venue, such as a playing court, field, or building. These venues are part of our built environment and are the result of construction technology.

Typically, a playing venue has a playing surface, goals, and constraints. See **Figure 25-7**. For example, golf uses grass-covered areas as a playing surface, holes for goals, and water and sand traps as constraints. Basketball uses a wood surface, a concrete surface, or an asphalt playing surface. This sport has a rim and backboard as a goal and painted lines as constraints.

Natural playing surfaces are often improved with technology. For example, special grasses have been developed for golf courses. Fertilizers have been developed and manufactured to encourage the grass to grow. Special lawn-grooming equipment has been developed to maintain the courses. Likewise, snow-grooming equipment prepares and renews the surfaces of ski runs. Snowmaking equipment has been developed to supplement natural snow for skiing. See **Figure 25-8**. This equipment makes snow by breaking water into small particles with compressed air. The water is cooled as it moves through cold air, forms small particles of ice, and is distributed as snow on a surface.

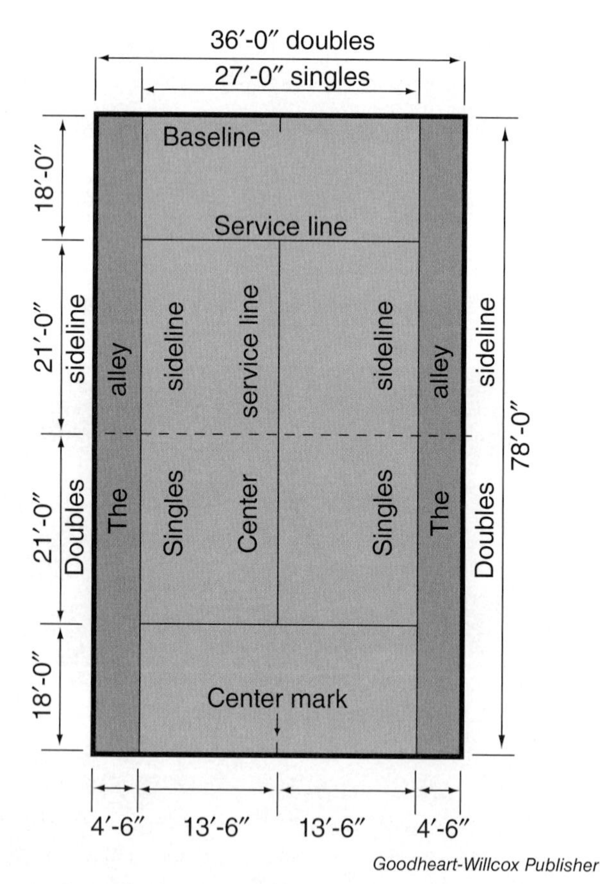

Goodheart-Willcox Publisher

Figure 25-7. A regulation tennis court is 78′ long and 27′ wide for singles matches. For doubles matches, the tennis court should be 36′ wide. These criteria must be met when building an official tennis court.

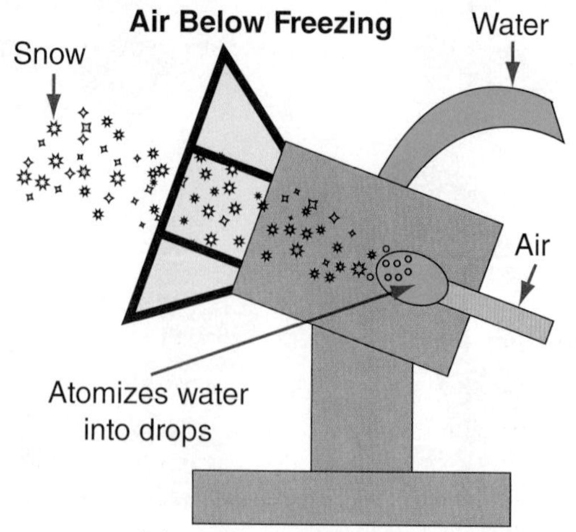

lightpoet/Shutterstock.com; Goodheart-Willcox Publisher

Figure 25-8. A snowmaking machine in action. The drawing shows how the machine works.

Academic Connection: Argumentative Essay

Ethics in Medicine

Argumentative writing is a type of writing that requires people to investigate a topic, collect and evaluate data related to the topic, and establish and present a position on the topic based on evidence. A valid argument also provides facts and data pertaining to any opposing views to the position presented. Scientists and engineers are commonly required to take a particular stance on an issue and defend their position effectively based on logic and fact.

Medical technologies are the tools, systems, and products that are developed to diagnose, treat, and prevent disease and injury. For example, medical technologies can support human physiological and metabolic needs to keep a person alive, and new pharmaceuticals are used to minimize the pain of individuals experiencing health issues. These benefits of technological advancement raise questions. How long should a person with a terminal illness be kept alive? What are the unsafe side effects of a new drug? What happens when the drug is disposed of improperly? In order to make ethical decisions, it is necessary to consider the potential consequences of using medical technologies. The uses of medical technologies are often a major topic for ethical debate and conflict.

Research the topic of a new and controversial medical technology or drug. Also, do research on appropriate and inappropriate uses of medical technology in general. Write an argumentative essay on whether or not the use of the technology or drug you chose should continue. Support your argument with appropriate evidence cited from scientific journals.

Write your essay in the following format:

- I. Introduction
  - a. Introduce the topic
  - b. Present your position
- II. Body
  - a. Topic sentence for Reason 1
    - i. Supporting evidence
  - b. Topic sentence for Reason 2
    - i. Supporting evidence
  - c. Topic sentence for Reason 3
    - i. Supporting evidence
- III. Conclusion
  - a. Restate your position
  - b. Counterarguments
  - c. General conclusions

Technology and Game Equipment

As we noted earlier, each sport has its own playing equipment. Hockey has pucks and sticks. Baseball has bats and balls. Tennis has rackets and balls. These manufactured products are designed and produced to specifications developed by groups who establish rules for playing the sport.

Technology and Protective Equipment

Players in many sports wear special clothing and protective gear. See **Figure 25-9**. Clothing can be made from fabrics that wick perspiration away from the body or shed rain. Special shoes can be developed to provide foot support and absorb the shock that running on hard surfaces causes.

lzf/Shutterstock.com

Alan Bailey/Shutterstock.com

Figure 25-9. Running shoes and catchers' masks are examples of special equipment designed for sports.

Some sports can lead to bodily injury or muscle damage. Participants in these sports wear protective gear. Baseball players wear batting helmets to protect their heads from wild pitches. Football players wear protective devices to protect their heads, necks, bodies, hands, and feet. The game equipment and protective equipment are technological products. They were designed for a specific purpose and have a customer base of amateur and professional athletes.

Protective equipment is extremely important for contact sports, such as hockey or football. Scientific research has led to the understanding of the long term medical impacts that sports injuries can have on the body. For example, research findings indicate that repeated blows to the head or multiple concussions can cause severe brain damage for athletes. These injuries can develop into the disorder now known as *chronic traumatic encephalopathy*. This condition, commonly found in professional athletes, causes the degeneration of brain tissue and the accumulation of proteins that negatively affect the central nervous system. Individuals with this disease experience memory loss, depression, dementia, aggression, and other cognitive impairments.

Technology and Illness

The second focus of health involves the area of ***medicine***, which involves diagnosing, treating, and preventing diseases and injuries. Medicine's goal is to reduce human suffering and physical disability, allowing people to live longer and more active lives. This goal involves health-care professionals working with patients who are ill or injured, as well as an ongoing search for new drugs, treatments, and technologies.

Disease is any change interfering with the appearance, structure, or function of the body. Many different health-care professionals are involved in treating disease and injury. See **Figure 25-10**. These health-care professions include the following:

- Physicians diagnose diseases and injuries. They administer appropriate treatment and advise patients on ways to stay healthy.
- Nurses help physicians diagnose and treat illnesses and injuries. They assist physicians during examinations, treatment, and surgery. Nurses observe and record patient symptoms. They administer medications and provide care in hospitals and nursing homes.
- Nurse practitioners perform some of the basic duties physicians once provided. They diagnose and treat common illnesses and prescribe medication.
- Physician assistants deliver basic health services under the supervision of a physician. They

Physicians
Nurses
Nurse Practitioners
Physician Assistants
Medical Technologists
Dentists
Dental Hygienists
Pharmacists

Syda Productions/Shutterstock.com

Figure 25-10. Health-care professionals work to cure illnesses and to help people stay healthy.

examine patients, order diagnostic tests and x-ray films, and prescribe drugs or other treatments.

- Medical technologists gather and analyze specimens to assist physicians in diagnosis and treatment.
- Dentists diagnose, treat, and help prevent diseases of the teeth and gums.
- Dental hygienists assist dentists in surgery. They also clean teeth and advise patients on proper techniques to prevent tooth and gum disease.
- Pharmacists dispense prescription drugs and advise people on their uses.

This team of health-care professionals depends on technology in their work. They use equipment and techniques to effectively care for patients.

Goals of Medicine

People seek medical care because they are ill or injured. Health-care professionals respond in three major ways. See **Figure 25-11**.

First, the medical personnel diagnose the illness or condition. ***Diagnosis*** is performed by conducting interviews, physical examinations, and medical tests. The diagnostic process tries to determine the nature or cause of the condition.

Medical personnel then treat the illness or condition. Treatment involves applying medical procedures to cure diseases, heal injuries, or ease symptoms.

Health-care professionals also aim to prevent illness or injury. They promote wellness programs that include exercise and proper diet. Another prevention method is immunization, **Figure 25-12**.

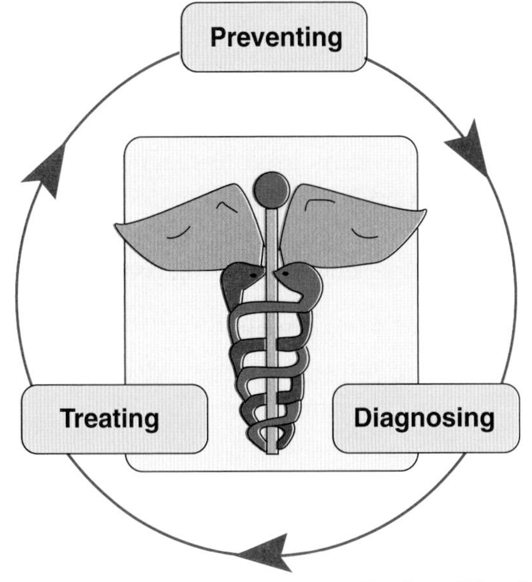

Goodheart-Willcox Publisher

Figure 25-11. Health-care professionals have three major goals.

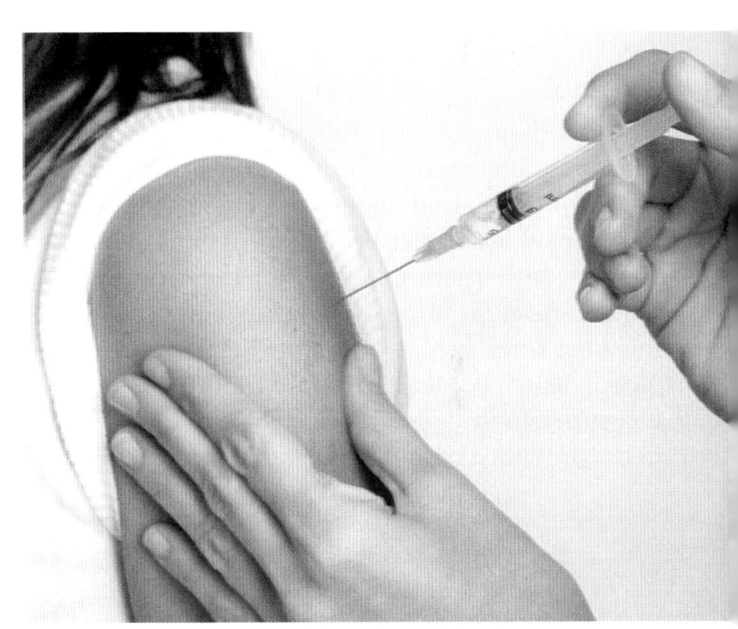

Adam Gregor/Shutterstock.com

Figure 25-12. Immunizations help prevent diseases.

Immunization is a process of exposing the body to small amounts of a bacterium, protein, or virus to cause antibodies to form. These antibodies are cells or proteins that circulate in the blood and attack foreign bacteria, viruses, or fungi that can cause disease.

Technology in Medicine

Physicians and other health-care professionals use tools and equipment to diagnose and treat illnesses and conditions. These tools and equipment are medical technologies as they extend abilities of health-care professionals to deal with medical problems. ***Medical technologies*** are health-care devices, products, processes, and systems used to diagnose, monitor, or treat medical conditions. Medical technologies include pharmaceuticals, vaccines, surgical tools, rehabilitation equipment, and genetic engineering processes.

Technology and Diagnosis

In the past, physicians depended on people to describe their symptoms. From these descriptions, the physicians planned treatment procedures. Physicians found, however, that descriptions are not always accurate and that many are hard to interpret. Over time, physicians and others have developed many different types of diagnostic equipment for more accurate diagnoses. Three major types of diagnostic equipment are as follows:

- Routine diagnostic equipment.
- Noninvasive diagnostic equipment.
- Invasive diagnostic equipment.

Routine diagnostic equipment

Routine diagnostic equipment is used to gather general information about the patient. This equipment includes scales to determine the patient's weight and thermometers to determine body temperature. Routine diagnostic equipment also includes devices that measure the oxygen in the blood, listen to heart rhythm (stethoscopes), and measure blood pressure. See **Figure 25-13**. These items provide a baseline of general information.

Noninvasive diagnostic equipment

Noninvasive diagnostic equipment gathers information about the patient without entering the body. The following paragraphs contain a few examples of the many noninvasive devices that have been designed and built to help diagnose illnesses and physical conditions.

Radiology uses electromagnetic radiation (waves) and ultrasonics (high frequency sounds) to diagnose diseases and injuries. See **Figure 25-14**. In diagnostic radiology, special equipment called *body scanners* or *body-imaging equipment* produce images (pictures) of the body without entering it.

One of the most common diagnostic imaging machines is an x-ray machine. This machine is essentially a camera that uses x-rays instead of visible light to expose the film. ***X-rays*** are electromagnetic waves that are short enough to pass through solid materials, such as paper and human tissue. Denser materials, such as metals and human bones, absorb some or all of the waves. Thus, if you put a piece of film under your hand and then pass x-rays through your hand, the skin and tissue let most of the x-rays pass directly through. The film behind the skin and tissue is completely exposed. The bones, however, absorb most of the x-rays. The film behind them is not exposed completely. When the film is developed, an image of the bones in the hand appears. Any fractures or joint deformities are shown, **Figure 25-15**.

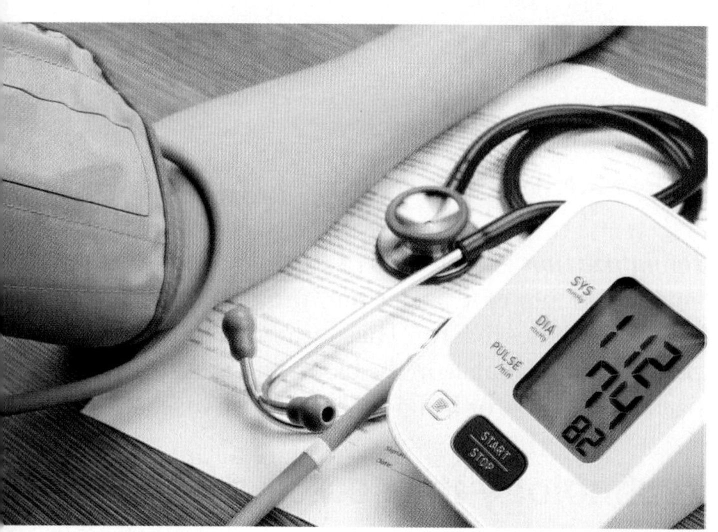

ARTFULLY PHOTOGRAPHER/Shutterstock.com

Figure 25-13. This physician is checking a patient's blood pressure using a routine diagnostic tool.

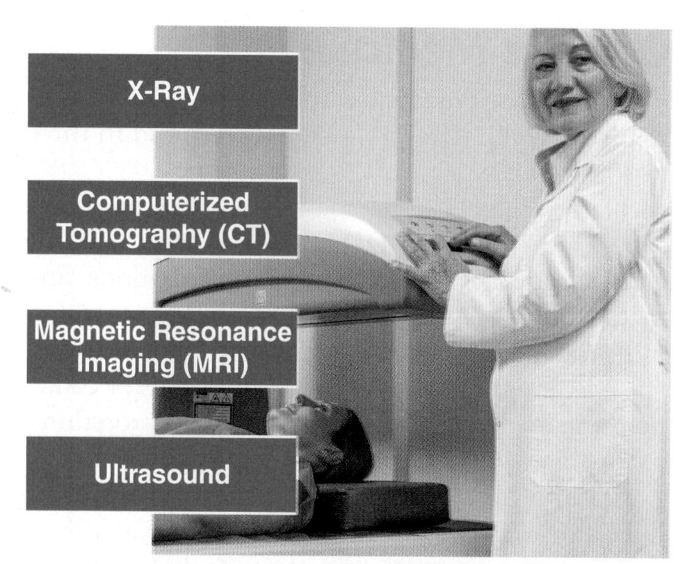

GagliardiImages/Shutterstock.com

Figure 25-14. The four types of radiology are x-ray, CT scan, MRI, and ultrasound.

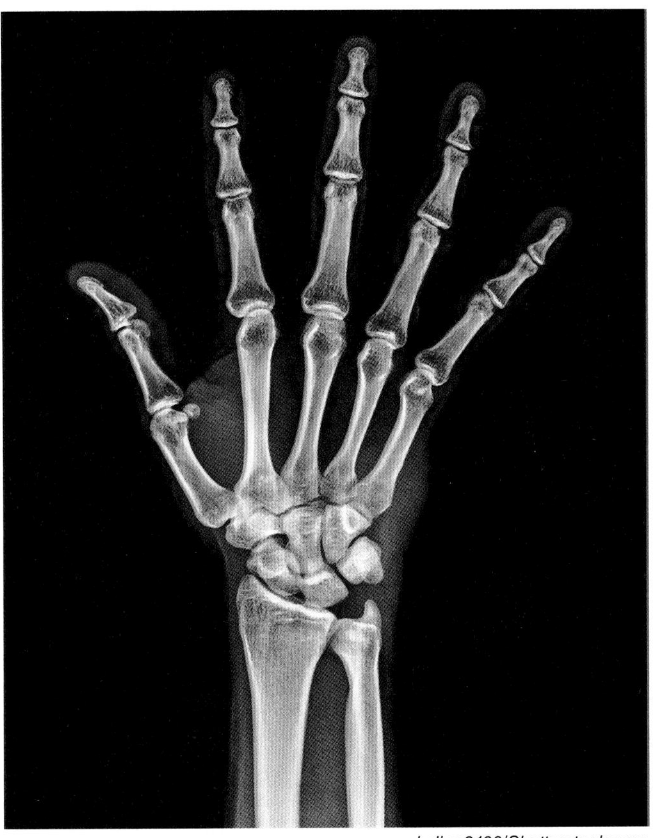

kaling2100/Shutterstock.com

Figure 25-15. An x-ray of a hand.

The first x-ray machines were used to detect fractures in bones and shadows on lungs. Later, substances that absorb x-rays were introduced into the body. Various cavities in the body are filled with a material to aid the x-ray process. This material is more transparent or more opaque when x-rays pass through than the surrounding tissue is. When x-rays are taken, the material brings the particular organ more sharply into view. For example, a patient swallows barium sulfate so gastric ulcers can be located. In other cases, dyes are injected so the heart, kidneys, or gallbladder can be x-rayed. This allows radiographs, or x-ray photographs, of the desired area to be produced.

An x-ray machine has an x-ray tube with a positive electrode (an anode) and a negative electrode (a cathode). See **Figure 25-16**. The cathode is heated and gives off electrons. The electrons travel toward the anode, where they hit and release energy (x-rays). The x-rays are concentrated into a beam that leaves the x-ray tube. The beam is directed at the desired location on the body. There, it passes through the body and exposes light-sensitive film. When the film is developed, an image of the section of the body is revealed. Some newer x-ray machines do not use

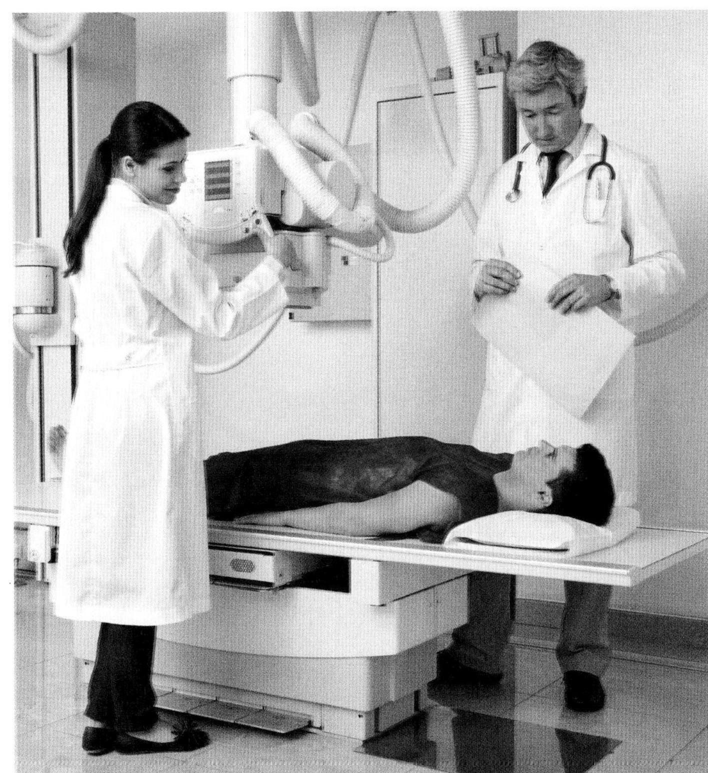

Tyler Olson/Shutterstock.com

Anode absorbs electrons and releases x-rays (energy)

Cathode releases electrons

X-ray tube

X-rays scatter in all directions in the x-ray tube. They are allowed to escape through a narrow window that focuses the beam

X-ray beam passes through the body

X-ray beam exposes film that is located under the body

Goodheart-Willcox Publisher

Figure 25-16. How an x-ray machine works.

Technology Explained Dialysis Machine

Dialysis machine: A device that performs some of the functions of the human kidneys, filtering waste and fluid from the blood.

For any number of reasons, some people's kidneys stop functioning and can no longer remove urea and certain salts from the blood. This condition is fatal unless the person receives treatment. One treatment is called *hemodialysis*, the process of removing waste and excess fluid from the blood. An artificial-kidney machine does this.

The word *dialysis* describes the movement of microscopic particles from one side of a semipermeable membrane to the other side of the membrane. Hemodialysis is a special type of dialysis that involves the blood (hemo-). *Hemodialysis* literally means "cleaning the blood."

In hemodialysis, the blood is pumped out of the patient's body to the artificial-kidney machine. See **Figure A.** In this dialysis machine, blood runs through tubes made of a semiporous membrane. Surrounding the tubes is a sterile liquid called the *dialysate solution*. This liquid is made up of water, sugars, and other components.

As the blood circulates, the red and white cells and other important components are too large to fit through the pores in the membranes. Impurities, such as urea and salt, however, pass through the membrane. See **Figure B.** The dialysate solution, which is discarded, carries them away. Tubes connected to the artificial-kidney machine return the cleaned blood to the bloodstream.

Dialysis is a treatment for people in the late stages of chronic kidney failure. Trained professionals generally perform the procedure. See **Figure C.** Dialysis normally takes three to five hours to complete. Typical patients receive three treatments a week. Dialysis allows these people to maintain many of their normal activities after the treatments.

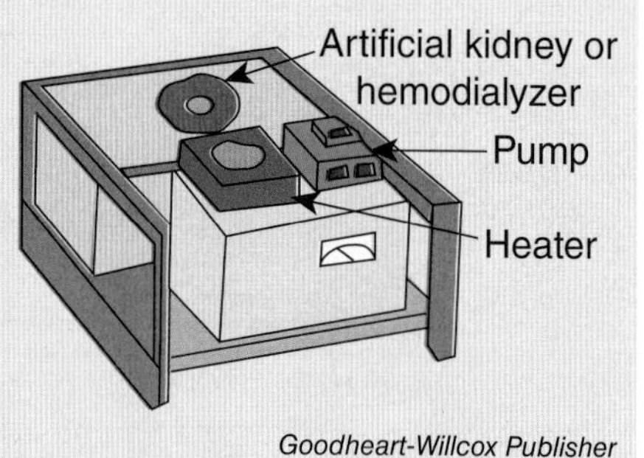

Goodheart-Willcox Publisher

Figure A. The tools used for hemodialysis.

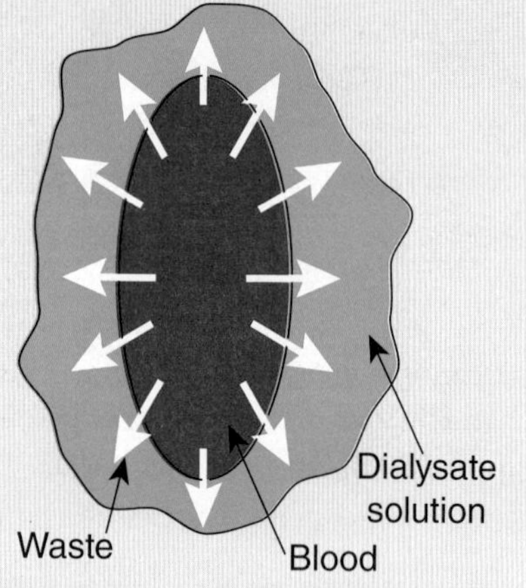

Goodheart-Willcox Publisher

Figure B. Wastes pass from the blood, through the membrane of the tubing, into the dialysate solution.

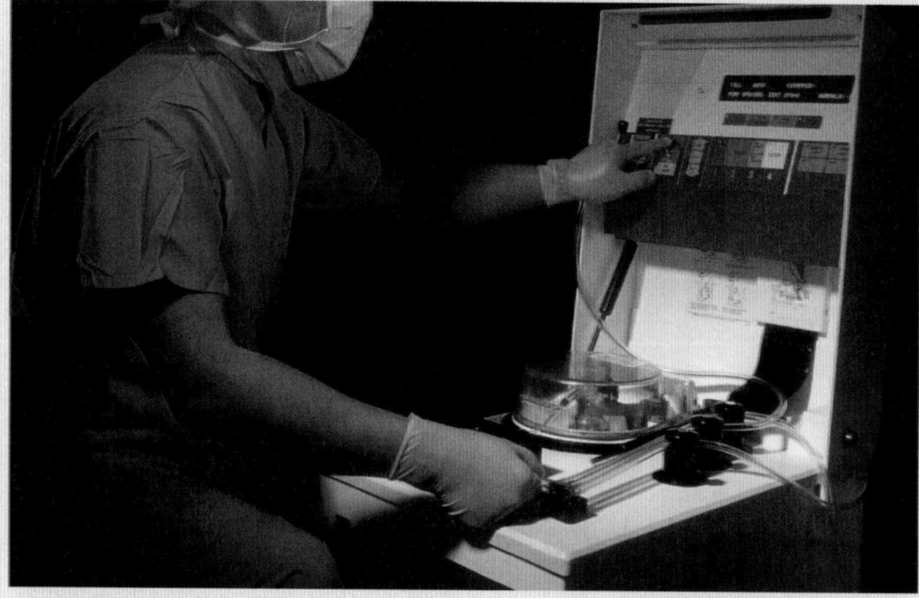

©iStockphoto.com/ElFlacodelNorte

Figure C. A technician prepares a portable dialysis machine for use.

film, but instead produce digital images that can be viewed on computer systems.

The major disadvantage of x-rays is that the images are two-dimensional (flat) images taken of a three-dimensional object (the body). Therefore, depth is not shown. Computerized tomography (CT) scanners (or CAT scanners) were developed to obtain a 3-D image. See **Figure 25-17**. A CT scanner produces images of any part of the body. The scanner rotates around a patient's body and sends a thin x-ray beam at many different points. Crystals opposite the beam pick up and record the absorption rates of the bone and tissue. A computer processes the data into a cross-sectional image of the part of the body being scanned. This digital image can be viewed immediately or stored for later use. Sometimes, contrast dyes are used for CT scans to increase the visibility of the body's tissues. These dyes are either injected intravenously into a patient's circulatory system or ingested orally by a patient.

X-ray radiation can cause damage to the body. Newer imaging techniques, such as magnetic resonance imaging (MRI), eliminate the hazards of radiation. This technique can quickly produce computer-developed, cross-sectional images of any part of the body. The images are developed using magnetic waves rather than x-rays. See **Figure 25-18**.

An MRI unit is a large tube surrounded by a circular magnet. The patient lies on his back on a bed, which is moved into the magnet. The body is positioned with the part to be scanned in the exact center of the magnetic field. An MRI exposes the patient to radio waves in a strong magnetic field. The magnetic field lines up the protons in the tissues. A beam of radio waves then spins these protons. As the protons spin, they produce signals that are picked up by a receiver in the scanner. The computer processes these signals to produce sharp, detailed images of the body.

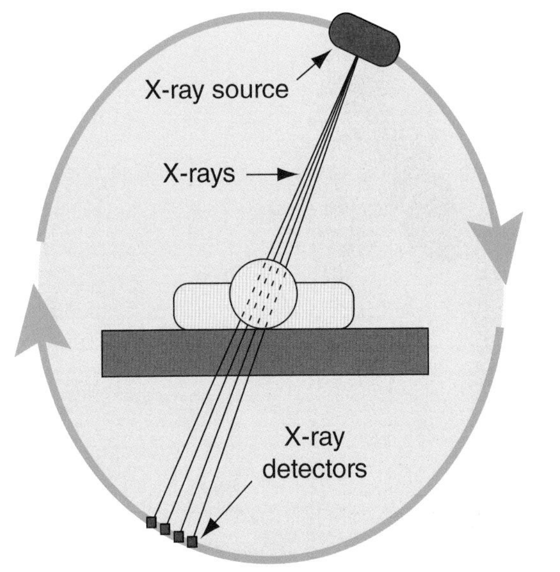

Goodheart-Willcox Publisher

Figure 25-17. A CT scanner rotates around a person's body.

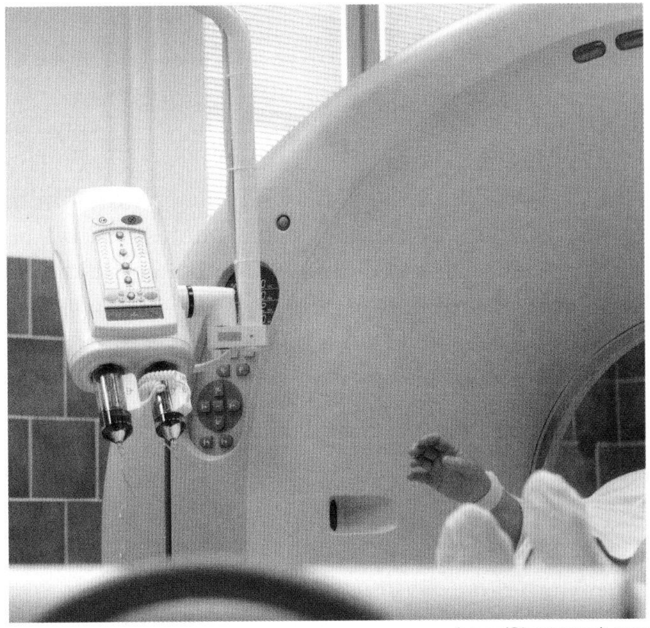

spfotocz/Shutterstock.com

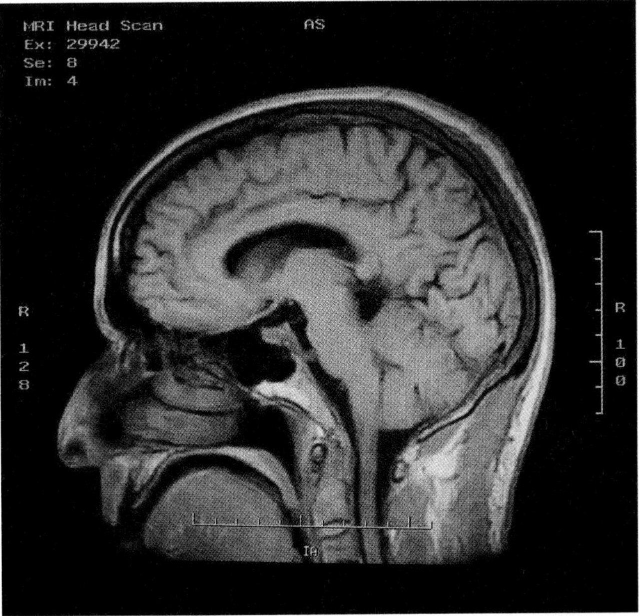

CGinspiration/Shutterstock.com

Figure 25-18. A—An MRI unit uses magnetic waves, rather than x-rays. B—The result of an MRI of a patient's head.

Another imaging technique, ***ultrasound***, uses high-frequency sound waves and their echoes to develop an image of the body. See **Figure 25-19**. The ultrasound machine subjects the body to high-frequency sound pulses using a probe. The sound waves travel into the body. There, they hit a boundary between tissues. This boundary can be between soft tissue and bone or between fluid and soft tissue. Some of the sound waves are reflected back to the probe. Others travel further, until they reach another boundary and get reflected. The probe picks up the reflected sound waves. A computer processes them to produce a still image (photograph) or moving image (television). In a typical ultrasound procedure, millions of pulses are sent each second. The probe is moved along the surface of the body or angled to obtain different images.

Other technological devices are used in diagnosis besides imaging devices. An ***electrocardiograph (EKG) machine***, **Figure 25-20**, produces a visual record of the heart's electrical activity. As the heart works, it sends off very small electrical signals that can be detected on the skin. The EKG machine uses electrodes attached to the skin to capture the signals. The machine amplifies the signals and produces a graph of their values. Health-care professionals read this graph to determine how the heart is functioning.

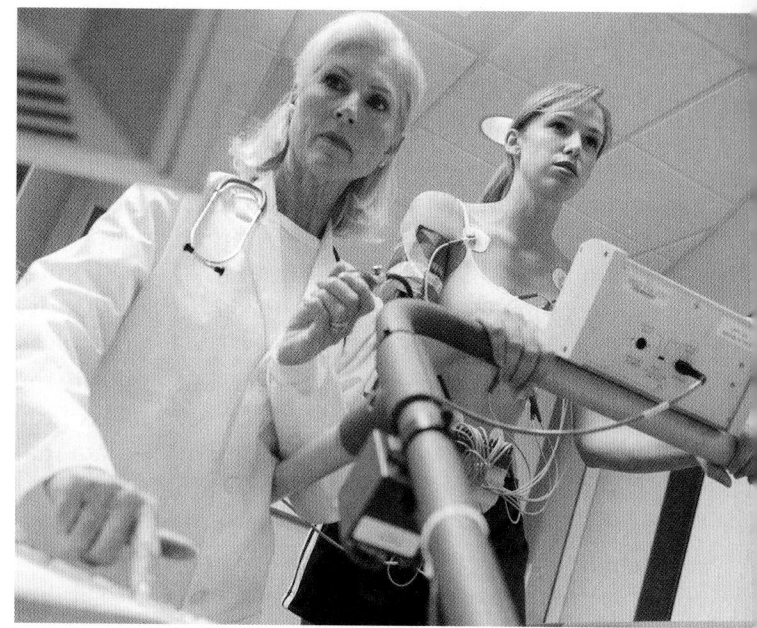

Monkey Business Images/Shutterstock.com

Figure 25-20. This person is being diagnosed with the use of an EKG machine.

Invasive diagnostic equipment

Invasive diagnostic equipment requires puncturing or incising the skin or inserting a foreign object into the body to obtain a diagnosis. For example, invasive diagnosis equipment is used when examining the inside of the nose or drawing and testing a blood sample. A blood test determines the chemical composition of a sample of blood. The sample is tested in a laboratory using a number of different technological procedures. See **Figure 25-21**. Blood tests can detect the presence of specific chemicals associated with a disease. They can also detect an imbalance in the chemical

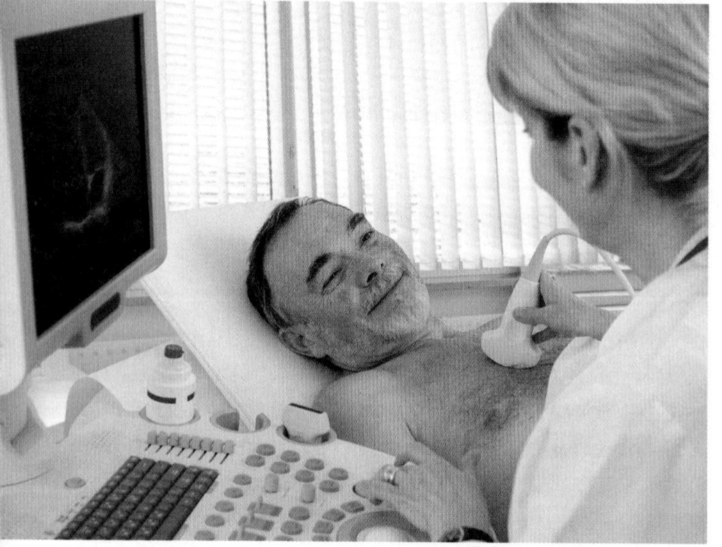

Alexander Raths/Shutterstock.com

Figure 25-19. A technician is using an ultrasound machine to scan the heart of a patient.

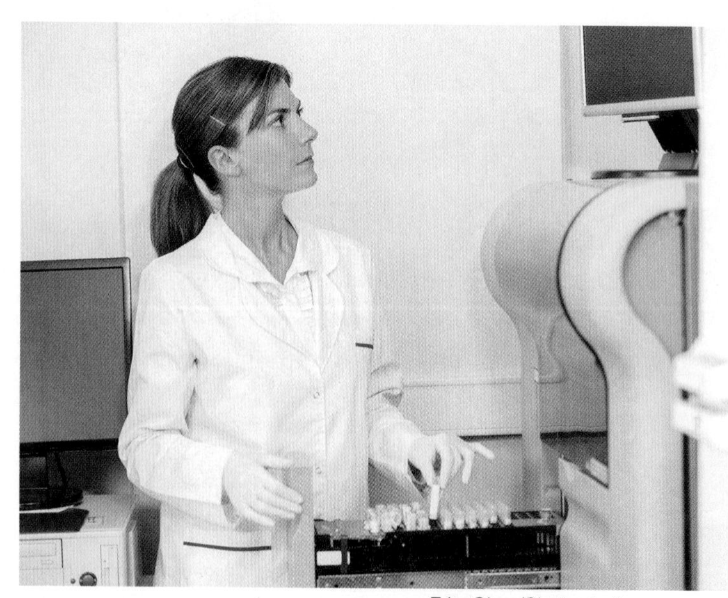

Tyler Olson/Shutterstock.com

Figure 25-21. This technician is testing blood samples.

composition of the blood. This data provides health-care professionals with information needed to treat illnesses or physical conditions.

Invasive diagnostic equipment is used to take a tissue sample (biopsy) for laboratory examination. Many medical conditions, including cancer, are diagnosed by removing a sample of tissue. This tissue is examined by a pathologist, who is a person trained to perform medical diagnoses by examining body tissues and fluids removed from the body. See **Figure 25-22**.

Advances in medical technology have enabled the development of minimally invasive diagnostics. Minimally invasive diagnostic technologies help minimize the size of the incision necessary to obtain the diagnosis. An example of a minimally invasive diagnostic device is the ***endoscope***, which is a device that allows a physician to look inside the body with only a small incision. An endoscope is a narrow, flexible tube containing a number of fiber-optic fibers smaller in diameter than a human hair. The tube can be threaded through a natural opening, such as the throat, or through a small incision. Light sent through the fibers shines on an interior part of the body. This light is reflected back through the fibers to form a series of dots. Each fiber in the tube produces one dot. The dots form a picture of an internal organ or other part of the body.

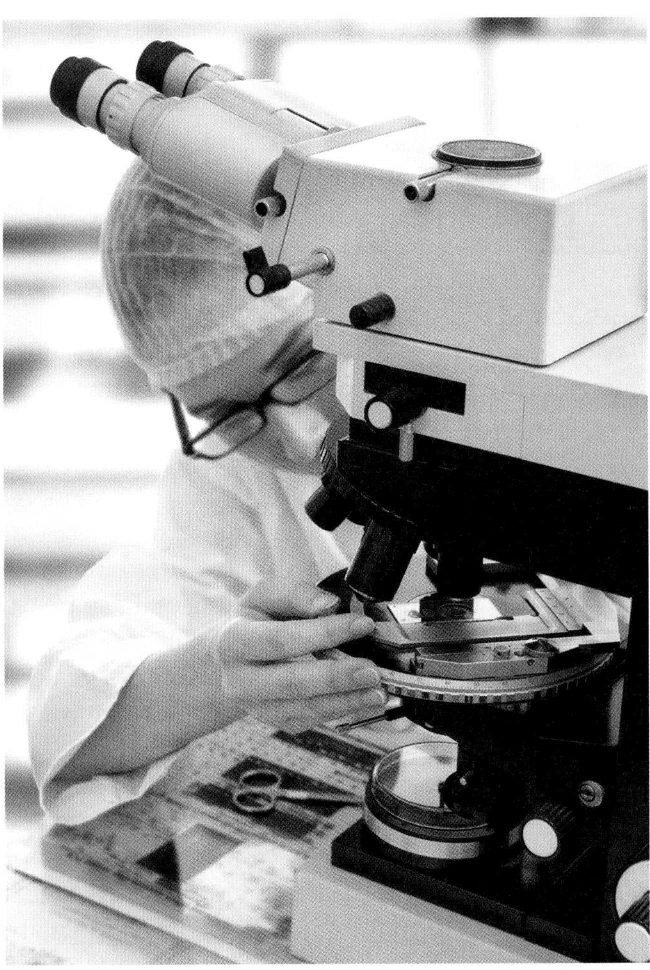

anyaivanova/Shutterstock.com

Figure 25-22. This pathologist is preparing a tissue slide for microscopic examination.

Technology and Treatment

Treatment of illnesses and physical conditions can require drugs, specialized equipment, or both. Both of these approaches are the result of technology. They result from design and production actions.

Drugs and vaccines

Humans have always experimented with substances to treat pain and illness and to restore health. The development of new treatments and prevention techniques begins with the unmet medical needs of people. One common treatment method is drugs.

A ***drug*** is a substance used to prevent, diagnose, or treat a disease. Some drugs can also be used to prolong the lives of patients with incurable conditions.

Throughout history, drugs have improved the quality of life for people. For example, a vaccine, a special type of drug, is a substance administered to stimulate the immune system to produce antibodies against a disease. Vaccines have helped eliminate such diseases as smallpox and poliomyelitis (polio). Drugs can be classified in many ways:

- **Method of dispensing.** Drugs can be classified as prescription or over-the-counter drugs. Those that physicians prescribe and pharmacists dispense are known as ***prescription drugs***. ***Over-the-counter drugs*** are those sold directly to the consumer at retail outlets without a prescription from a health-care professional.
- **Form.** Drugs can be classified as pills, capsules, liquids, or gases. See **Figure 25-23**.
- **Method of administering.** Drugs can be classified as injections (shots), oral drugs, inhalants, or absorbable drugs (patches).
- **Illness or condition treated.** Drugs can be classified as cancer drugs, measles vaccines, or high blood pressure drugs, for example.

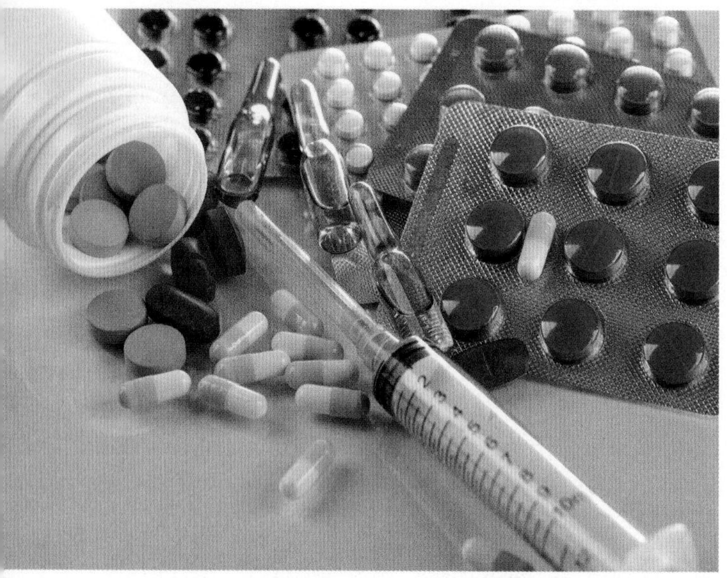

Paul Biryukov/Shutterstock.com

Figure 25-23. Drugs come in many different forms.

Today, most drugs are the products of chemical laboratories, **Figure 25-24**. These synthetic drugs are developed because they can be controlled better than natural drugs can be controlled. Not all synthetic drugs are totally new. Some are developed by altering the structure of existing substances. These new drugs are called *analogs*. Their structure is similar to another compound, but they have a slightly different composition. These new drugs might be more effective than the original drugs or cause fewer side effects.

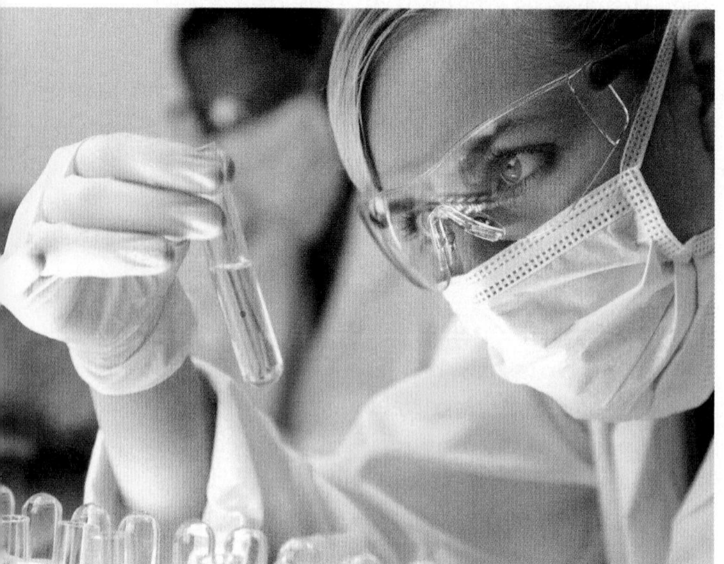

Darren Baker/Shutterstock.com

Figure 25-24. Most drugs are products of chemical laboratories.

A number of new drugs have been developed by using gene splicing or recombinant DNA. This approach joins the DNA of a selected human cell to the DNA of a second organism, such as a harmless bacterium. The new organism can produce the disease-fighting substance. This new substance is extracted from the bacterium and processed into a drug.

The first drug produced using gene splicing was the hormone insulin. This drug was created in 1982 by inserting a human insulin gene into *E. coli* bacteria. The development of any new drug requires great amounts of time and money. Drug development can take ten or more years and several millions of dollars.

The drug-development process generally starts with a need to treat a disease or physical condition. Researchers start with an existing chemical substance that might have medical value. They might work with thousands of different substances before they find one that can serve as a drug.

Once a new substance with potential medical value is discovered, an extensive testing program starts. First, the drug is tested on small animals, such as rats and mice. If the tests show promising results, additional tests are conducted on larger animals, such as dogs and monkeys.

The tests are evaluated in terms of treating the disease and physical condition. Also, the drug must have a low level of toxicity (capability to poison a person). Drugs judged to be effective in animal tests are ready for the next level of testing. At this point, a request is made to the Food and Drug Administration (FDA) to conduct clinical tests.

If the FDA approves the request, the drug can be tested on humans. These tests, called *clinical trials*, are generally conducted in three phases. Each phase can take many months to complete. During the first phase, the drug is given to a small number of healthy individuals. These tests are designed to determine the drug's effect on people. The safety, safe dosage, and side effects of the drug are evaluated.

If the drug passes this test, it moves into the next phase. Here, it is given to a small number of people who have the disease or physical condition that the drug will treat. These individuals are divided into two groups. The first group is given the drug. The second group is given a placebo (an inactive substance, such as sugar). During the test, neither the group nor the researchers know who is receiving the drug or who is receiving the placebo. This research technique is called a ***double-blind study***. A double-blind study prevents participants from influencing the results of the study.

Drugs that pass the second phase move into the third and final phase of testing. Here, the drug is tested with a much larger group of people. The goal of these tests is to determine specific doses, side effects, interactions with other drugs, and other information. The data from these tests is used in drug labeling.

When the third phase is complete, the results of the tests are submitted to the FDA for approval. The agency must decide whether or not the drug is effective and safe. Also, it must weigh the drug's benefits against any risks that might be present. If the FDA determines that the drug meets its criteria, it approves the drug for use.

Medical equipment

Biomedical engineers develop a great deal of diagnostic and treatment equipment. See **Figure 25-25**. These professionals combine engineering with medicine to improve health care. They define and solve problems in biology and medicine. Biomedical engineers design the following devices and systems:

- Life-support equipment, such as cardiac pacemakers, defibrillators, and artificial kidneys.
- Artificial body parts, such as hearts, blood vessels, joints, arms, and legs.
- Computer systems to monitor patients while in surgery and intensive care.
- Sensors for the blood's chemistry.
- Instruments and devices for therapeutic uses, such as a laser system for eye surgery or a catheter to clear out a blocked blood vessel.
- Medical imaging and treatment systems.

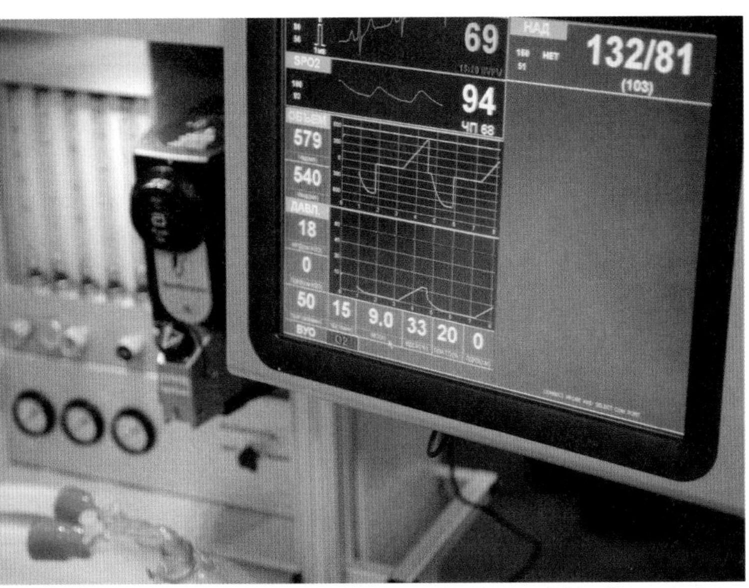

Pavel L Photo and Video/Shutterstock.com

Figure 25-25. This EKG machine is the result of many people's work, including biomedical engineers.

Health-care professionals use technological systems to treat injuries and diseases. Three common examples are radiation therapy, surgery, and emergency medicine.

Therapeutic radiology (radiation therapy) is the treatment of diseases or disorders with radiation. Many types of cancers are treated using radiation therapy. High-energy radiation destroys the cancer cells' ability to reproduce. Normal cells can recover from the effects of radiation better than cancer cells can. Radiation therapy may be only part of a cancer patient's treatment. Treatment may also involve chemotherapy (chemicals or drugs) and surgery.

Interventional radiology uses images that radiology produces for nonsurgical treatment of ailments. These images allow the physician to guide catheters (hollow, flexible tubes), balloons, and other tiny instruments through blood vessels and organs. An example of this approach is balloon angioplasty, which uses a balloon to open blocked arteries.

Surgery is the treatment of diseases and injuries with operations. Most surgery involves manually removing (cutting) diseased tissue and organs. See **Figure 25-26**. However, surgery is also performed to repair broken bones, stop bleeding, and for many other purposes.

Many different technologies are used in surgery. High-frequency sound waves (ultrasound) can be used to break up kidney stones. Lasers use a beam of light to vaporize or destroy tissue. An endoscope can be used with special devices to operate on a particular area of the body. In transplant surgery, organs removed from one person can be implanted into another person. Some devices, such as pacemakers, are implanted in the body.

Biomechanical engineering applies mechanical engineering principles and materials to surgery and prosthetics. For example, biomedical engineers use the knowledge of statics and dynamics to understand

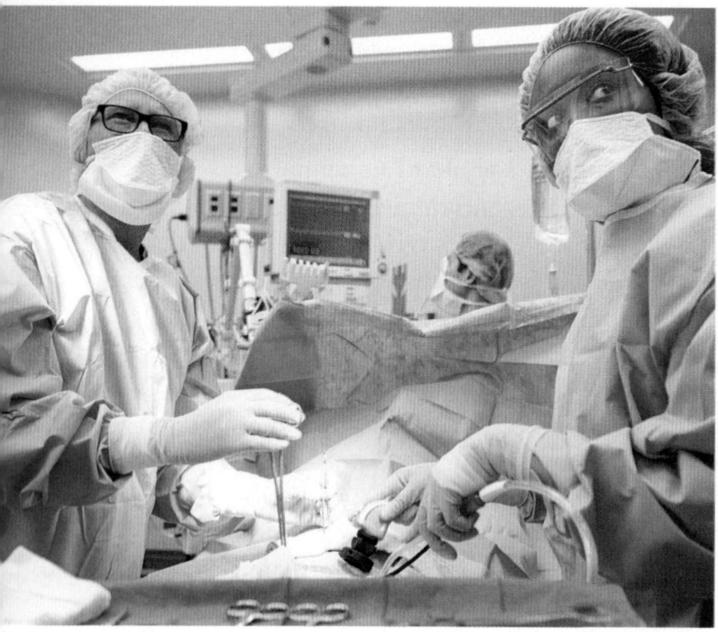

Monkey Business Images/Shutterstock.com

Figure 25-26. Surgery usually involves cutting diseased tissue and organs.

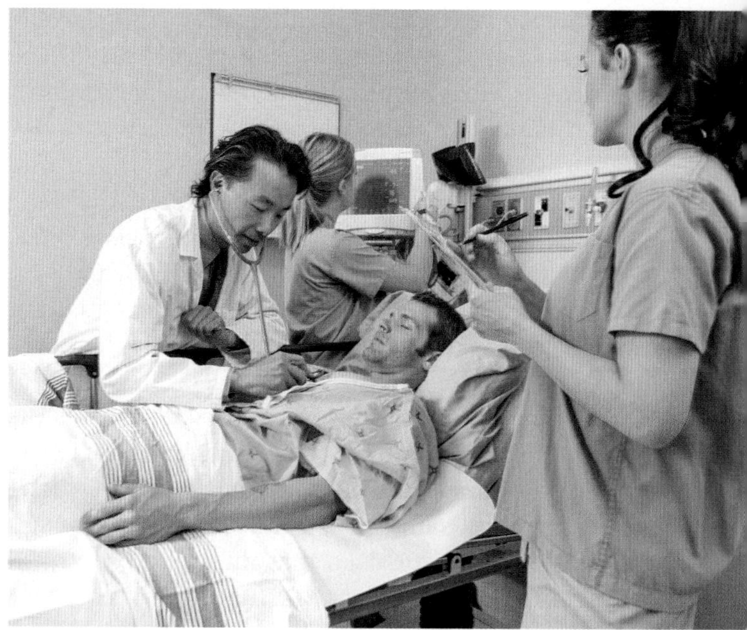

Tyler Olson/Shutterstock.com

Figure 25-27. This patient is receiving treatment in an emergency room.

the rigid components of the body. They use that knowledge to design and attach prosthetics to replace a damaged or missing body part. A practice closely related to surgery is prosthetics. A ***prosthesis*** is an artificial body part, such as an artificial heart or arm. These devices are developed through biomechanical engineering. This branch of engineering applies mechanical engineering principles and materials to surgery and prosthetics.

One of the oldest types of prosthesis is a denture. Dentures are removable, artificial teeth that replace full-mouth (upper or lower) teeth. A similar treatment for a few missing teeth is a tooth-supported bridge. These artificial teeth fill a space between existing teeth. Another type of artificial tooth is a dental implant. Dental implants are replacement teeth set into the gums. They replace one or more missing teeth.

Emergency medicine deals with sudden, serious illnesses and injuries. See **Figure 25-27**. In this area, for example, health-care professionals treat people who are injured in automobile accidents and on jobs. They treat people who have suffered heart attacks and strokes. An emergency room contains many technological devices allowing health-care professionals to diagnose and treat patients. Frequently, the job of these professionals is to determine what is wrong and stabilize the patient's condition. Appropriate treatment can then be administered.

Telemedicine

Telemedicine, the delivery of medical care and diagnosis over long distances, is the result of a convergence of technological advancements in information and communication technologies. These technologies include widespread broadband Internet, video conferencing, virtual presence technologies, artificial intelligence, and robotics. *Virtual presence technologies* are used to make a person feel as if they are actually in a location and situation in which they are not currently present. In addition to giving people easier access to health care, telemedicine also enables medical experts, regardless of their location, to provide input for treatment and diagnosis. See **Figure 25-28**.

Rocketclips, Inc./Shutterstock.com

Figure 25-28. Telemedicine uses technology to deliver medical diagnoses and treatment over a distance.

Summary

- ✔ Science and technology play a role in wellness and illness. Science provides knowledge about the human body and health and describes the natural processes the body uses to maintain itself. Technology provides the tools and equipment needed to achieve wellness and treat illness.
- ✔ Four major factors affect wellness—nutrition and diet, environment, stress management, and physical fitness.
- ✔ Technology has been applied to stress management and physical fitness through two major areas—exercise and sports.
- ✔ Technology has contributed to the development of exercise equipment and to venues, game equipment, and protective equipment used in sports.
- ✔ The field of medicine involves diagnosing, treating, and preventing diseases and injuries.
- ✔ Health-care professionals depend on technology as they diagnose, treat, and seek to prevent diseases and injuries.
- ✔ Three major types of diagnostic medical equipment are routine diagnostic equipment, noninvasive diagnostic equipment, and invasive diagnostic equipment.
- ✔ Noninvasive diagnostic equipment includes imaging equipment such as x-ray machines, CT scanners, and MRI units.
- ✔ Treatment of illnesses and physical conditions can require drugs, specialized equipment, or both.
- ✔ Once a new substance with potential medical value is discovered, an extensive testing program starts. Drugs are first tested on animals. Testing is then done on humans in three phases.
- ✔ Biomedical engineers develop diagnostic and treatment technologies, such as life support systems, artificial body parts, imaging systems, blood chemistry sensors, and patient monitoring systems.
- ✔ Three common technological systems used by health-care professionals are radiation therapy, surgery, and emergency medicine.
- ✔ Radiology may be therapeutic (the use of radiation to treat diseases or disorders) or interventional (the use of radiation for image-guided diagnosis or treatment).
- ✔ Telemedicine, the delivery of medical care and diagnosis over long distances, is the result of a convergence of technological advancements in information and communication technologies.

Check Your Engineering IQ

Now that you have finished this chapter, see what you learned by taking the chapter posttest.
www.g-wlearning.com/technologyeducation/

Test Your Knowledge

Answer the following end-of-chapter questions using the information provided in this chapter.

1. *True or False?* Science provides the tools and equipment needed to achieve wellness and treat illness.
2. List the four major factors involved in wellness.
3. *True or False?* Stair-climbers provide an aerobic fitness workout.
4. Three areas of technological products used in sports are playing venues, game equipment, and _____ equipment.
5. Medicine involves diagnosing, treating, and _____ diseases and injuries.
6. *True or False?* Nurse practitioners can prescribe medication.
7. Blood pressure monitors fall under the category of _____ diagnostic equipment.
8. What is the major difference in the images produced by an x-ray machine and a CT scanner?
9. What is an endoscope?
10. What are two uses of blood tests?
11. *True or False?* An EKG machine is an example of invasive diagnostic equipment.
12. *True or False?* The second clinical phase of an FDA-approved drug involves a double-blind study.

Critical Thinking

1. Think about the types of sports and exercise you participate in. How has technology contributed to those sports and physical activities?
2. Evaluate the procedure in which a new drug is developed and describe potential benefits and risks of this process.
3. What are the challenges and benefits involved in the implementation of telemedicine?

STEM Applications

1. Conduct an experiment to determine the accuracy of various devices for measuring your heart rate (pulse) while at rest and after running. Use different technologies, such as a stethoscope, blood pressure/heart rate monitor, smartphone apps, or other method. Graph your results using a spreadsheet program to compare the results among the different devices. Report your conclusions on how to improve accuracy when measuring heart rates.
2. Choose a professional sport and research the current protective equipment used. Identify an issue with the current equipment and develop an idea for an improvement. Sketch your idea and describe the specific criteria and constraints that you must adhere to when designing the equipment for the human body.

TSA Modular Activity

This activity develops the skills used in TSA's Biotechnology Design event.

Medical Technology

Activity Overview

In this activity, research a contemporary problem or issue related to medical technology and work up a possible solution. Prepare a report that consists of the following items:

- A cover page.
- A definition of the problem or issue.
- A report on the topic (4–10 pages).
- A printout of electronic slides (three slides per page).
- A list of sources and references.

Place your report in a three-ring binder. In addition to the report, prepare an oral presentation that incorporates a PowerPoint® presentation. The presentation should contain both text and graphics. Finally, create a display.

Materials

- A three-ring binder.
- Materials appropriate for your tabletop display. (These will vary greatly.)
- A computer with electronic slide presentation software.

Background Information

- **Selection.** Before selecting the theme for your project, use brainstorming techniques to develop a list of possible themes. Some contemporary topics are as follows:
  - The US healthcare system.
  - Minimally invasive surgery.
  - Home-health testing.
  - Exercise.
  - Immunizations.
  - Diagnostic equipment.
  - Prescription medications.
  - Prevention programs and techniques.
  - Telemedicine.
  - Robotic surgery.
- **Research.** Use a variety of sources to research your theme. Do not rely solely on information you find on the Internet. Use books and periodicals available at your local library. Research the historical developments of the topic. Did an individual or a corporation develop the technology? What were some previous technologies that allowed this technology to become a reality? How did the public receive the technology, and was the response expected?
- **Electronic slide presentation.** When developing your presentation, consider the following design guidelines:
  - Develop a general slide design and use it for all your slides.
  - Keep the design simple. Do not use more than two type fonts. Select easy-to-read type fonts. Be sure the type size is large enough to be seen from the rear of the room in which you will be presenting.
  - Include a title on each slide.
  - Do not attempt to squeeze an abundance of information on a single slide. Create multiple slides instead.

Guidelines

- Focus your research on cultural, social, economic, or political impacts. Address both opportunities and risks.
- The display can be no larger than 18″ deep × 3′ wide × 3′ high.
- If a source of electricity is desired, use only dry cells or photovoltaic cells.
- The oral presentation can be up to 10 minutes in length.

Evaluation Criteria

Your project will be evaluated using the following criteria:

- The content and accuracy of the report.
- The attractiveness and creativeness of the display.
- Communication skills and the presentation design of the oral presentation.

CHAPTER 26

Meeting Needs through Biomedical Engineering

Check Your Engineering IQ 

Before you read this chapter, assess your current understanding of the chapter content by taking the chapter pretest. www.g-wlearning.com/technologyeducation/

Learning Objectives

After studying this chapter, you should be able to:

- ✔ Explain the historical impact of biomedical engineering.
- ✔ Describe the work done by biomedical engineers and biomedical scientists.
- ✔ Identify the different products of biomedical engineering.
- ✔ Describe the main systems of the human body.
- ✔ Understand concepts in mechanics and biomechanics.
- ✔ Explain the role of biomedical engineers in medical imaging.
- ✔ Explain the selection and use of biomaterials in biomedical engineering.

Technical Terms 

appendicular skeleton
arteries
axial skeleton
bioartificial organs
biomaterial
biomechanics
biomedical engineering
biomedical engineers
biomedical imaging
blood
bone marrow
bronchi
capillaries
cardiovascular system
diaphragm
dynamics
genetic engineering
homeostasis
human physiology
kinematics
kinesiology
kinetics
ligaments
lungs
mechanics
medical implants
muscles
musculoskeletal system
myoelectric prosthesis
pectoral girdle
pelvic girdle
plasma
platelets
pulmonary circuit
red blood cells
respiratory system
robotic prosthesis
statics
systemic circuit
tendons
tissue engineering
trachea
transfemoral prosthesis
transhumeral prosthesis
transradial prosthesis
transtibial prosthesis
veins
white blood cells

Close Reading

When reading, it is important to visualize the concepts that are being discussed. It is a good practice to create images or drawings of these visualizations to help develop a deep understanding of the concepts. As you read about the structure and function of the human body, create your own diagrams to depict how the body works and how biomedical devices could aid or replace these processes.

Biomedical engineering involves the design and development of products, systems, and procedures that solve medical and health-related problems. The application of knowledge gained through scientific and technological breakthroughs, combined with engineering design and testing practices, evolved into the formal discipline of biomedical engineering. The practices, processes, and products of biomedical engineering are critical to our health and well-being. They are widely used for diagnosing, treating, and preventing disease and injury.

Evolution of Biomedical Engineering

Solving medical and health problems has been a goal for people throughout history. Biomedical engineering activities can be traced back thousands of years. Archeologists have discovered wooden prostheses in the ancient tombs of Egypt. The ancient Egyptians also used tools made from hollow reeds to listen to a person's internal organs. These primitive tools were the foundation for later medical devices, such as the stethoscope. See **Figure 26-1**. The inventors of simple assistive medical devices, such as crutches, could be considered early biomedical engineers because they produced products to help solve a medical problem.

Medical devices have greatly evolved over time. Early medical devices, **Figure 26-2**, were often developed through the trial and error of medical researchers. However, over time, improved devices have been developed in response to scientific and technological breakthroughs and improved engineering practices.

For example, let's examine the evolution of the hearing aid. An early medical device used to mitigate hearing loss was the ear trumpet. Users placed a horn-like apparatus into their ear to help amplify sound. This device was used until the mid-twentieth century. After the technological breakthroughs made by Alexander Graham Bell and Thomas Edison in regards to the transmission and amplification of sound, electrical tabletop hearing aids were developed. These large devices were eventually improved through engineering practices to be more portable.

The first wearable hearing aid was not made possible until William Shockley and his team of researchers developed the transistor, which is used to amplify or switch electrical power and electronic

Bon Appetit/Shutterstock.com

Figure 26-1. A stethoscope is a medical device used to listen to the sounds of internal organs.

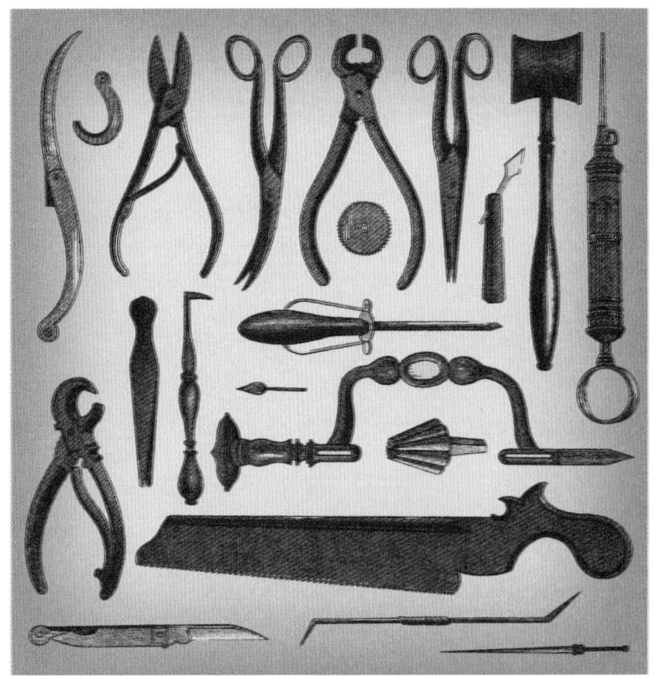

pio3/Shutterstock.com

Figure 26-2. Early medical devices can seem primitive and even scary from today's perspective.

signals. Further technological developments with smaller integrated circuits and more advanced batteries enabled engineers to create nearly invisible hearing aids that fit in the ear canal.

Scientific research resulted in the development of the cochlear implant in the late 1970s. A cochlear implant is an electronic device that is surgically placed into a patient who is deaf or severely hard of hearing. The implant provides a sense of sound to patients who have hearing loss as a result of damaged or lost hair cells in the inner ear. The hair cells in the inner ear are the sensory receptors of the auditory system.

A cochlear implant consists of a microphone, speech processor, amplifier, a signal transmitter, and an electrode that stimulates the nerves in the inner ear. The nerves then send signals to the brain, which enables hearing. The implant replaces the sound inputs to the human body by replicating the different amplitudes and frequencies of sound and delivering that information directly to the brain. This allows people to hear better or even to hear for the first time. See **Figure 26-3**.

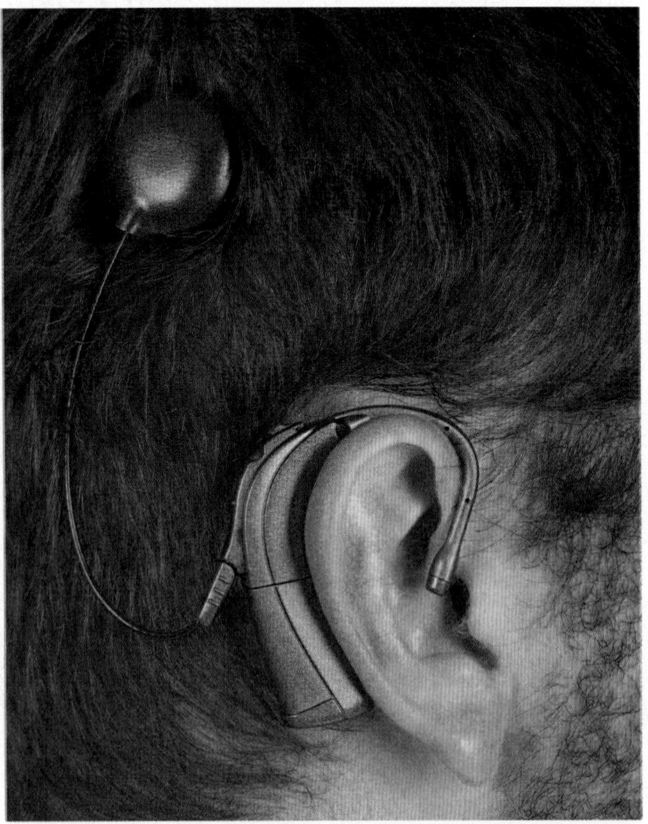

Elsa Hoffmann/Shutterstock.com

Figure 26-3. Cochlear implants are designed to replace the function of a damaged inner ear.

We live in a time when a person's organs can be replaced with artificial ones and damaged body parts can be replaced with robotic ones. Consequently, biomedical engineering is now considered one of the fastest growing engineering disciplines. Due to the work of professionals in this discipline, advancements in medical and health technologies have revolutionized medicine, health care, and biological sciences. These advancements have even made technologies such as the bionic eye, implants that can block pain receptors, augmented reality for surgery, targeted cancer therapy, and the artificially intelligent diagnostic and treatment system.

As a result of enhanced medical and health technologies, the world's aging population continues to increase. People are able to live longer, creating additional health-care and medical needs. Due to increased need and societal demand for better medical care, projected employment growth for biomedical engineers is high.

Biomedical Engineers

Biomedical engineers are responsible for devising new and better medical and health technologies. Biomedical technologies are designed by people with a strong understanding of the science of the human body as well as engineering principles. They combine knowledge of mathematics and science with design skills to plan and create medical equipment in order to meet required criteria and constraints. At the core of any diagnostic, prevention, or treatment practice are scientific and technological breakthroughs made possible through biomedical engineering.

Biomedical engineers design technologies used in various medical procedures. They design the computers used to analyze blood, laser systems used in corrective eye surgery, artificial organs, and medical imaging systems. See **Figure 26-4**. Some biomedical engineers evaluate health-care facilities or review legal matters involved with the improper design or use of medical equipment. In some instances, biomedical engineers work within other biotechnology fields, such as tissue engineering or genetic engineering.

The work of biomedical engineers is used by health-care professionals and biomedical scientists and technicians. For example, when a patient is

science photo/Shutterstock.com

Figure 26-4. Biomedical engineers combine biology and medicine with engineering.

receiving treatment from a doctor, the doctor will first diagnose the patient. To do so, the doctor may collect a blood sample and send it to a lab. In the lab, a biomedical scientist or technician uses certain equipment to analyze the blood. See **Figure 26-5**. That equipment was likely designed and improved by a biomedical engineer. The equipment enables the scientist or technician to analyze the blood and provide the results to the doctor. The doctor uses the results to help properly diagnose the patient and provide appropriate treatment.

The treatment of patients also requires a continual collaboration of medical, engineering, and science professionals. In the case of organ replacement, a scientist determines the organ's functions, which enables a biomedical engineer to design an artificial organ. The doctor must understand all of the scientific information behind the artificial organ design to successfully perform the surgery and monitor the patient.

Another example of this collaboration is the replacement of a patient's damaged hip. Biomedical scientists research how prosthetic hips affect the body. This information helps biomedical engineers develop and improve prosthetic hip designs. Doctors then use the results of this scientific research and engineering development to remove the damaged hip and replace it with a prosthesis. See **Figure 26-6**. Furthermore, biomedical scientists research the effects of the prosthetic hip design after the prosthetics are placed in the body to help engineers determine new ways to improve the designs.

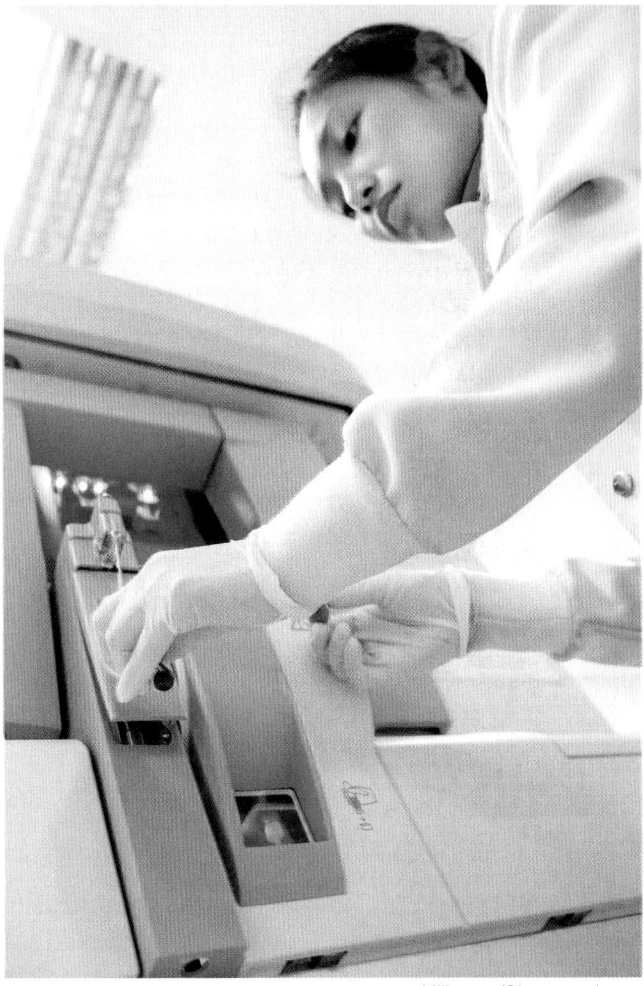

Milkovasa/Shutterstock.com

Figure 26-5. Blood analysis machines are used to perform tests on blood samples to help diagnose potential medical issues.

Biomedical engineers must have a specific four-year degree in biomedical engineering or a background in one of the basic engineering specialties, such as mechanical or electrical. In addition, specialized biomedical training is required. The specialties within biomedical training include biomaterials, biomechanics, medical imaging, rehabilitation engineering, and orthopedic engineering. Most biomedical engineering degree program curriculums give students an overview of these specialties combined with a study of engineering design practices. The study of these specialties enables engineers to design, create, and improve medical and health technologies.

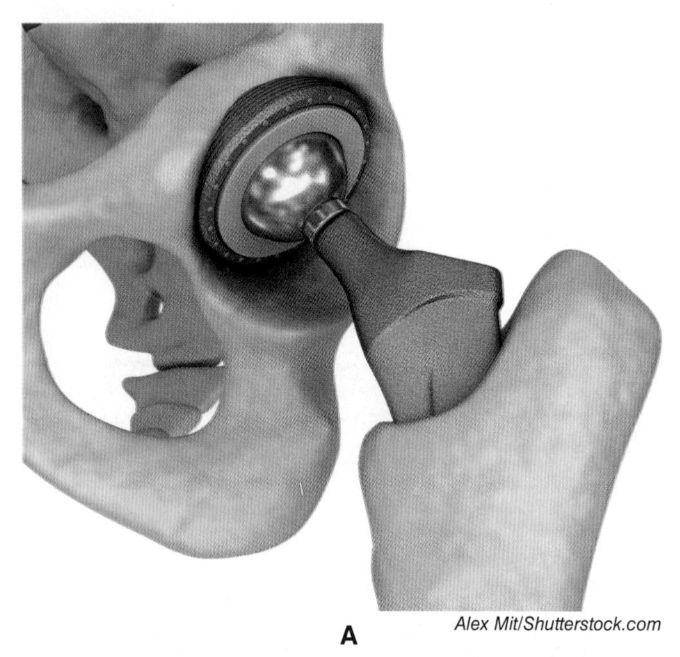

A Alex Mit/Shutterstock.com

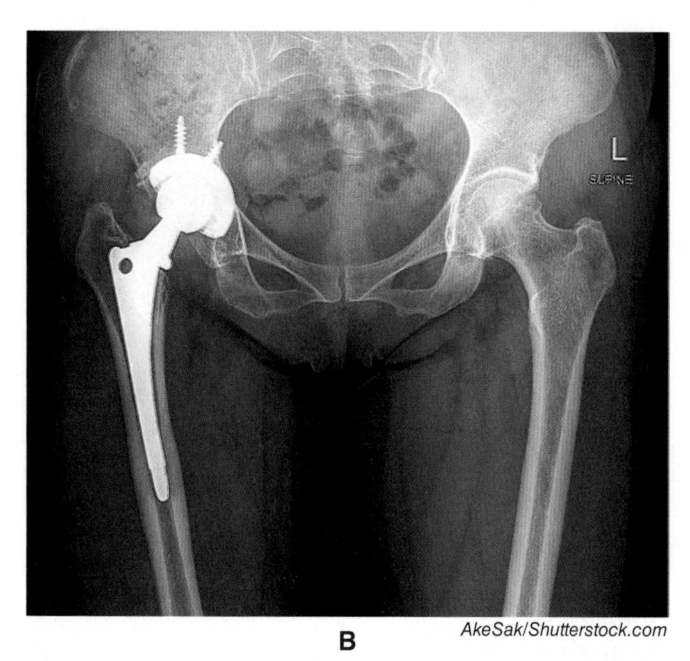

B AkeSak/Shutterstock.com

Figure 26-6. Hip replacements are commonly performed procedures. A—A hip prosthetic. B—An x-ray of a total hip replacement.

Products of Biomedical Engineering

Health-care professionals and biomedical scientists and technicians use products developed by biomedical engineers to help diagnose, prevent, and monitor disease or injury. See **Figure 26-7**. Healthcare professionals also use biomedical products, such as implants, prosthetic devices, and surgical instrumentation, to transplant, replace, or repair body parts as a means of treating disease and injury.

Vasiliy Koval/Shutterstock.com

Figure 26-7. Health-care professionals use devices developed by biomedical engineers to help diagnose, prevent, and monitor disease or injury.

Biomedical engineers design a variety of medical devices to help diagnose disease and injury. Several diagnostic products and systems were discussed in Chapter 25. These products include imaging systems (CTs, MRIs, and ultrasound), laboratory equipment for analyzing blood or tissues, and other diagnostic devices, such as EKG machines and endoscopes.

Medical implants are tissues or devices that are surgically placed inside or on the exterior of the body. Some implants are used to monitor a person's bodily functions. For example, chips can be implanted in a diabetic patient to monitor glucose levels. Other implants, such as chemotherapy ports, deliver medication. Certain implants help support organs or tissues. For example, a pacemaker, **Figure 26-8**, is placed in the body to control abnormal heartbeats using electrical pulses.

Many implants are prosthetic devices designed to replace missing or damaged body parts. These body parts may have been lost due to some type of physical trauma, congenital disorder (birth defect), or disease. Some prostheses are permanent, while others are removed once they are no longer needed.

Prostheses are made from a variety of materials and serve a functional purpose, cosmetic purpose, or both. Cosmetic prostheses help hide injuries

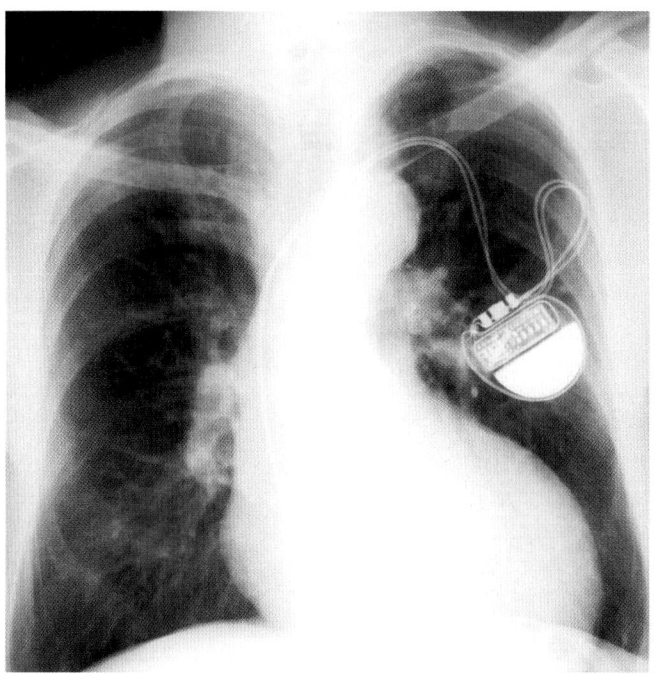

Dario Sabljak/Shutterstock.com

Figure 26-8. A pacemaker is implanted in the body to control abnormal heartbeats by using electrical pulses.

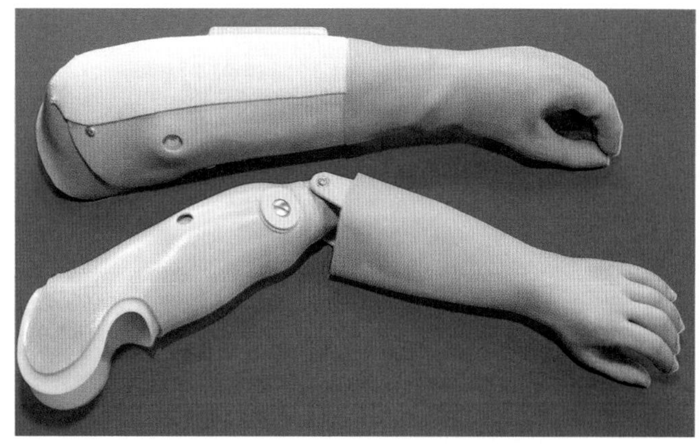

belushi/Shutterstock.com

Figure 26-9. The two types of prostheses for the arm are transradial (top) and transhumeral (bottom).

or disfigurements by mimicking the appearance of a real body part. Functional prostheses replace a missing limb or damaged joint so a patient can perform activities they were unable to do without use of the limb or joint. The most commonly replaced joints are the hip, shoulder, or knee. To replace these joints, surgeons remove portions of the bones that meet at the joints and implant replacement joints made of strong materials that are able to glide over one another.

An arm is replaced by either a transradial or a transhumeral prosthesis. A ***transradial prosthesis*** is attached below the elbow. A ***transhumeral prosthesis*** is used when the elbow joint is missing from the arm. See **Figure 26-9**. When a leg is replaced, a ***transtibial prosthesis*** is used to replace parts missing below the knee, while a ***transfemoral prosthesis*** is used to replace the knee joint as well as the lower part of the leg. See **Figure 26-10**.

The patient can control the prostheses used to replace limbs with harnesses or cables that are operated by other parts of the body or with electric signals from the brain. A prosthesis controlled by electric signals is called a ***myoelectric prosthesis***. Tiny electrodes are attached to muscles, and those muscles are voluntarily contracted by the person to control movements of the connected prosthesis.

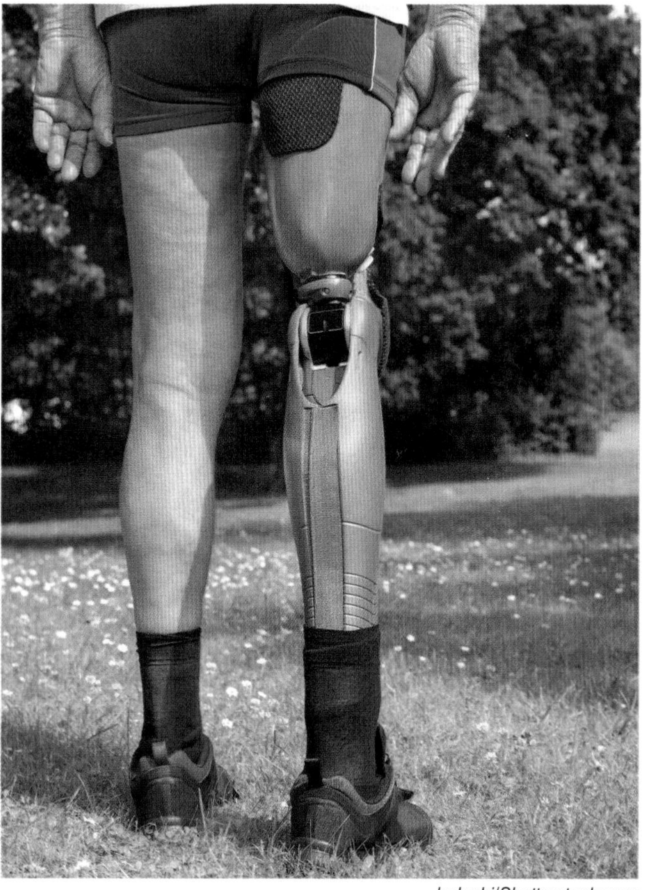

belushi/Shutterstock.com

Figure 26-10. A transfemoral prosthetic.

In addition to limbs, other body parts can be replaced by prostheses. Examples are the ears, nose, breasts, and teeth.

Technological breakthroughs have led to the development of ***robotic prostheses***, which use advanced biosensors, as well as mechanical and electrical components, to make a prosthesis that is

Academic Connection: History

X-Ray Imaging

In 1895, a German physicist, Wilhelm Röntgen, was experimenting with electrical currents in gas-filled vacuum tubes when he accidentally discovered an unknown type of radiation. When passing electricity through the tubes, he noticed that materials outside of the tubes began to glow. He then realized there must be some type of radiation energy traveling through the tube. He called this radiation *x-rays* because he did not quite understand what they were.

Röntgen then realized that different amounts of x-rays passed through different materials, which allowed him to render an image of the different components inside of something, such as the human body. Based on these observations, he created the first x-ray image by passing x-rays through his wife's hand. The x-rays mostly passed through the muscles and flesh of the hand but were blocked by the higher-density bones. This left a perfect shadow of the bones under the skin, thus creating an image of the bones inside the hand. At this point, x-ray imaging was born.

Thomas Edison took interest in Röntgen's work and began developing an x-ray device of his own. One of his assistants, Clarence Dally, worked specifically with this technology and continually tested the device. In the process, he unknowingly exposed himself to harmful radiation. His health deteriorated, and he died only eight years later. However, their work helped to further develop medical x-ray imaging as well as provide the understanding of the dangers of radiation exposure.

capable of natural, agile, and complex functions. Additionally, 3-D printing and scanning technologies have made it possible to create more lifelike and better-fitting prosthetics.

Artificial organs are another product of biomedical engineers. These manmade devices are designed to be implanted into a patient to replace a natural organ and to restore specific biological functions. For example, an artificial heart is used to replace the heart of a patient in need of a transplant. However, this artificial organ is usually limited for use in patients who are awaiting a transplant and whose death is imminent.

Biomedical engineers and scientists are making great strides in developing bioartifical organs, such as livers and kidneys. ***Bioartificial organs*** are produced using living tissues. Biomedical engineers can now design, modify, grow, and maintain living tissues. These living tissues are embedded within synthetic and natural support structures to allow them to perform complex biochemical functions.

Biomedical engineers also develop surgical instrumentation, such as laser cutters, scalpels, retractors, medical drills, injection needles, and surgical scissors. These tools or devices are designed to carry out specific actions during operation or surgery. See **Figure 26-11**. These instruments must be designed to accelerate surgical precision and accuracy, enhance convenience for the surgeon, minimize unnecessary waste, ensure the safety

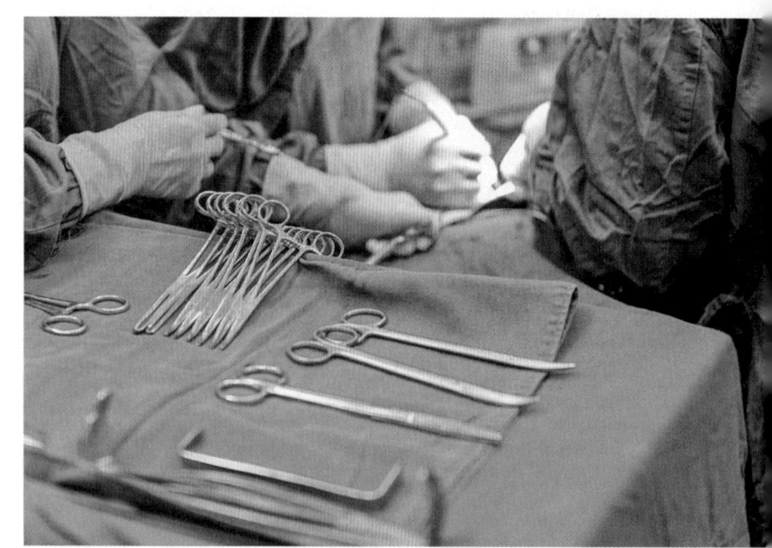

hxdbzxy/Shutterstock.com

Figure 26-11. Instruments used for surgery.

of the patient, maintain sterility, and to increase affordability of medical procedures.

Other biomedical engineering products are designed to support or sustain life. These products are synthetic devices that temporarily replace specific bodily functions. For example, the cardiopulmonary bypass technique takes over the function of the heart and lungs. A heart-lung machine maintains the circulation of blood throughout the body while sustaining the oxygen content of the blood. Other life-support products include mechanical ventilation and feeding tubes.

Human Physiology

In order to develop new and better medical and health technologies, biomedical engineers must understand the science behind the functioning of the human body. ***Human physiology*** is a subdiscipline of biology that specifically studies the structure and function of the human body. It looks at how cells, organs, and muscles work together to perform the necessary chemical and physical functions of a living person. An understanding of the major systems of the body is essential in biomedical engineering in order to design the most effective, safe, and efficient medical products, systems, and procedures.

Cardiovascular System

To design an artificial heart or kidney, a thorough understanding of the cardiovascular system is needed. This system is also referred to as the *circulatory system*. The ***cardiovascular system*** is comprised of the heart, blood, and blood vessels. See **Figure 26-12**. These components work together to carry oxygen, nutrients, hormones, and waste through the body.

The cardiovascular system also distributes heat throughout the body to maintain homeostasis. ***Homeostasis*** is the ability of a biological system to regulate its processes, such as body temperature, to produce conditions that are optimal for survival. If homeostasis does not occur, life does not continue. The components of the cardiovascular system function to balance the optimal conditions of the body to sustain life.

One component of the cardiovascular system is blood. ***Blood*** is a bodily fluid that delivers the

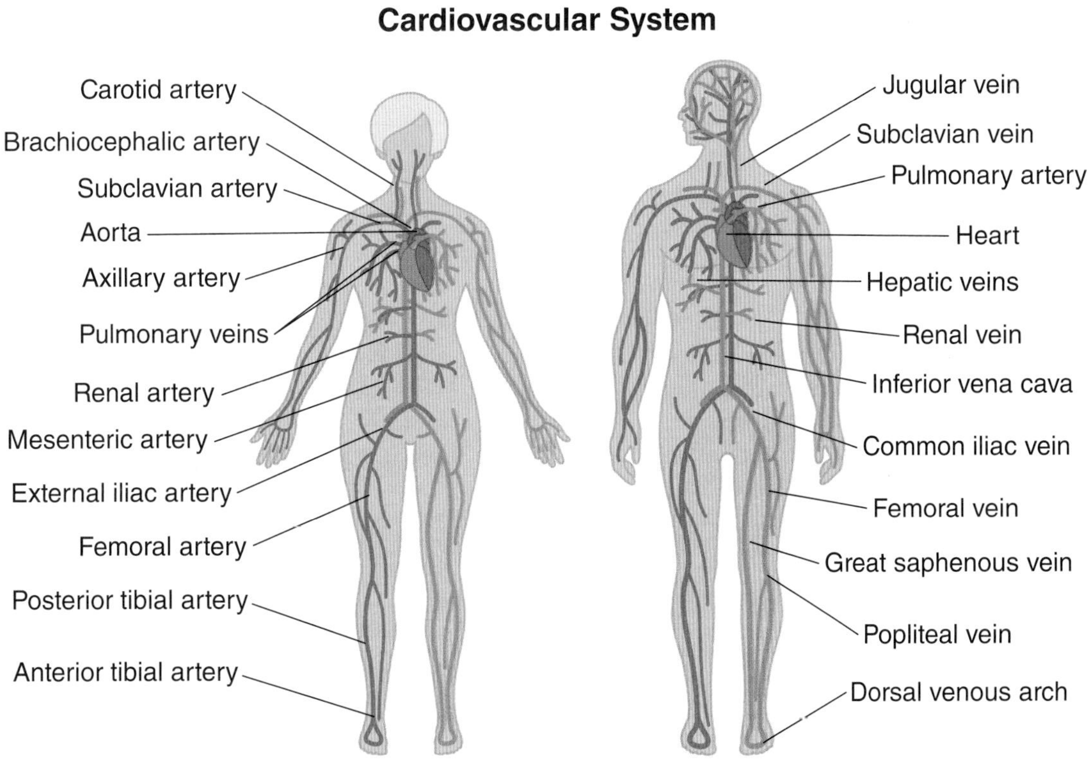

marina_ua/Shutterstock.com

Figure 26-12. The cardiovascular system moves oxygen, nutrients, hormones, heat, and waste throughout the body.

necessary nutrients and oxygen to the cells of the body. As it travels through the body, blood absorbs nutrients from digested food through channels in the small intestines. Oxygen is also diffused into the blood as it passes through the lungs. The oxygen and nutrients are then transferred to the body's cells. Blood also transfers waste products away from the body's cells. The kidneys then remove wastes and extra water from the blood. These wastes become urine.

The four main components of blood are red blood cells, white blood cells, platelets, and plasma. See **Figure 26-13**. ***Red blood cells*** carry oxygen to and remove carbon dioxide from the body's cells, while ***white blood cells*** work to fight off infections. ***Platelets*** are the portion of the blood that helps to clot the blood when a wound occurs. Lastly, ***plasma*** is the liquid component of the blood that helps the other components of the blood flow through the cardiovascular system.

The main purpose of the cardiovascular system is to support life by circulating blood throughout the body. The heart acts like a pump and provides the force needed to distribute blood. The heart is made up of four chambers, **Figure 26-14**. The right and left ventricles are the chambers that push the blood out of the heart when contracted. The right and left atria are the chambers that receive the blood that returns to the heart after it is pumped out to the body. The heart circulates approximately five liters of blood throughout an average adult. To circulate the blood, the heart of an adult beats an average of 72 times per minute.

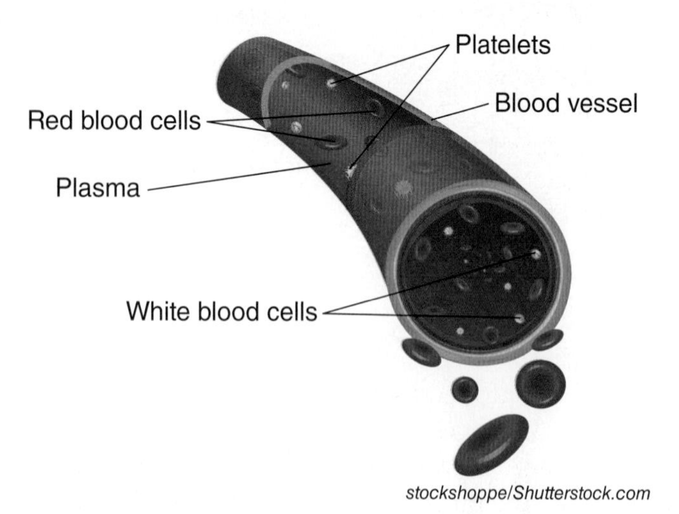

Figure 26-13. Blood vessels transport blood throughout the body. Blood consists of red blood cells, white blood cells, platelets, and plasma.

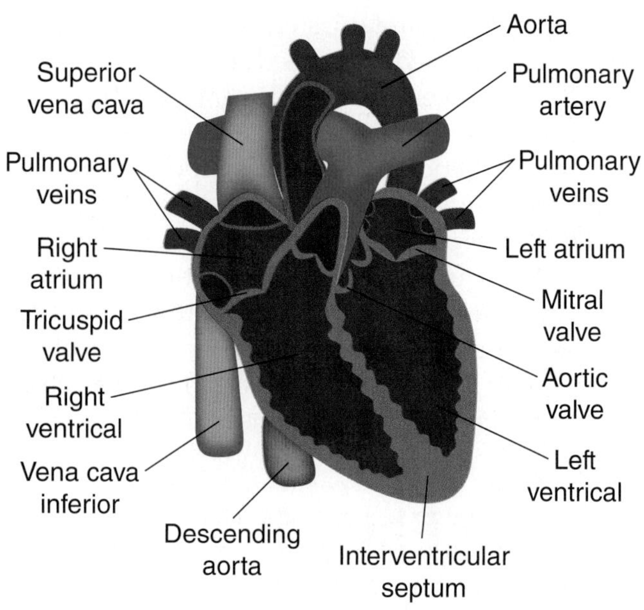

Figure 26-14. A diagram of the human heart.

Blood is pumped in and out of the heart chambers in two different circuits. The ***pulmonary circuit*** carries blood from the heart to the lungs, where oxygen enters the blood through breathing. The blood in this circuit is then circulated back to the heart where it enters the systemic circuit. The ***systemic circuit*** carries the blood from the heart to the cells of the body. The cells then transfer oxygen, nutrients, hormones, and waste with the blood before it is carried back to the heart. See **Figure 26-15**.

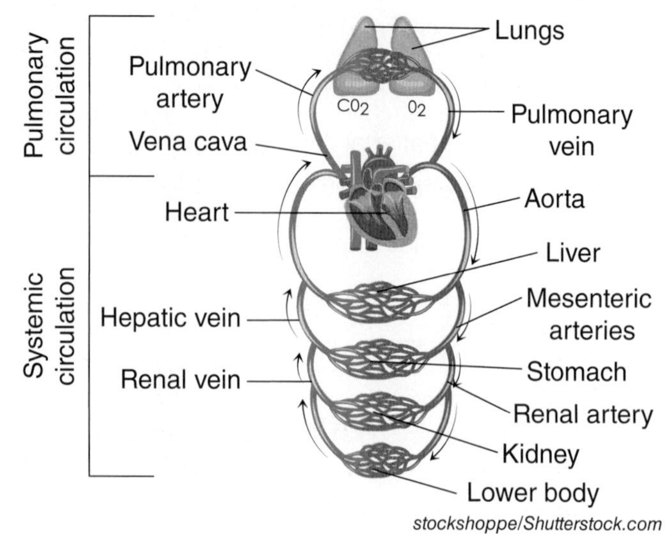

Figure 26-15. Blood flow through the cardiovascular system.

Technology Explained Heart-Lung Machine

Heart-lung machine: A system used to mimic the functions of the heart and lungs of the patient while open-heart surgery is performed.

A heart-lung machine is a medical device used to perform the cardiopulmonary bypass technique, which enables open-heart surgery. See **Figure A**. The heart-lung machine is designed to take over the functions of the heart and lungs while a doctor works to repair a heart. This device diverts blood away from the heart and lungs using a series of tubes. The blood is pumped into a reservoir to hold the blood as it moves through the machine. The reservoir mimics the right atrium of the heart. However, the machine must be designed so it does not hold any more than 1.5 liters of blood at any time. The human body cannot support the functions of its cells if much more than 1.5 liters is out of the body at one time.

Samrith Na Lumpoon/Shutterstock.com

Figure A. A heart-lung machine.

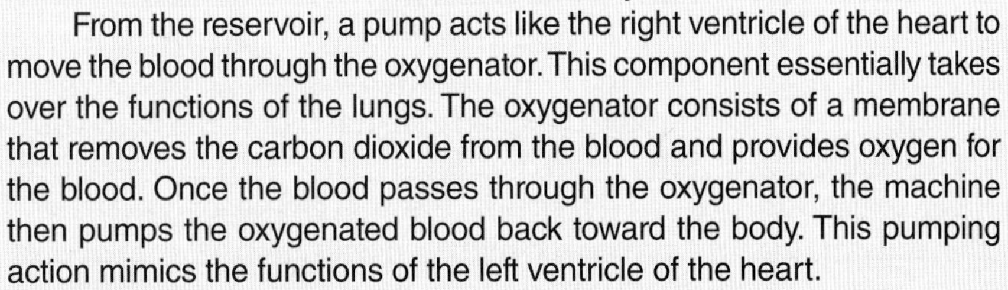

From the reservoir, a pump acts like the right ventricle of the heart to move the blood through the oxygenator. This component essentially takes over the functions of the lungs. The oxygenator consists of a membrane that removes the carbon dioxide from the blood and provides oxygen for the blood. Once the blood passes through the oxygenator, the machine then pumps the oxygenated blood back toward the body. This pumping action mimics the functions of the left ventricle of the heart.

Before the blood reenters the body, it passes through a heat exchanger in order to regulate the temperature of the blood. The blood temperature is typically lowered to slow down cellular metabolism. Cellular metabolism is a series of biochemical reactions that take place within a cell and are necessary to maintain life. These reactions use oxygen as one of the key inputs to either break down molecules to obtain energy for the body or to combine molecules to produce compounds that the body needs. Lower body temperatures cause cellular metabolism to move at a slower rate. Therefore, the oxygen requirements for the blood are reduced, which is helpful during the surgery.

After the blood passes through the heat exchanger, it is pumped back into the body through a series of tubes. The pressure from the pump enables the blood to circulate through the body, delivering oxygen and nutrients to the cells and removing all waste products. As a result, the doctor can perform the open-heart surgery without the heart beating and the interference of blood.

The blood in these circuits travels through three major types of blood vessels. ***Arteries*** are the vessels that carry blood away from the heart. ***Capillaries*** are the vessels that enable the exchange of nutrients, oxygen, and chemicals between the blood and cells of the body. ***Veins*** are the blood vessels that carry the blood back from the capillaries to the heart. Each time the heart beats, blood is pushed in and out of the heart through the different blood vessels. The pressure from the heart causes the vessel leading away from the heart to stretch, which produces the pulse that you can feel on the underside of your wrist. The average pulse rate for adults ranges from 70 to 90 beats per minute.

Respiratory System

The cardiopulmonary bypass technique is an example of why biomedical engineers need to understand the respiratory system. The development of the heart-lung machine required extensive knowledge of how respiration works, since the machine was designed to replace many of the functions of the respiratory system.

The function of the ***respiratory system*** is to bring oxygen into the body and deliver it to every cell, as well as to rid the body of carbon dioxide. Cells need oxygen to break down the sugars from food in order to produce the energy required by the body. This process is called *cellular respiration.* Carbon dioxide is the waste product of this process.

The respiratory system provides oxygen to the blood. The blood moves through the cardiovascular system to deliver the oxygen to the cells. Blood also picks up the carbon dioxide waste from the cells and carries it back to the lungs to be released from the body every time a person breathes out.

The respiratory system consists of several parts with different functions to help deliver oxygen to and remove carbon dioxide from the cells of the body. See **Figure 26-16**. The breathing muscle, called the ***diaphragm,*** is necessary for the body to perform the mechanical process of breathing. The brain and nervous system automatically control the muscles involved in the breathing process. The diaphragm contracts and expands the airtight cavity surrounding the lungs. When the cavity expands, it creates a vacuum, causing air to rush in through the nose or mouth. These openings warm and moisten the air as it enters the body.

After passing the mouth and nose, air flows through the pharynx, which is the body's passageway for air and food. The air then passes into the trachea (windpipe), while a flap called the *epiglottis* stops any food from entering. The ***trachea*** functions as a filter for the air we breathe and channels the air into the two branches of the bronchi. The ***bronchi*** are two air tubes that branch off the trachea into each lung.

The ***lungs*** are the main organs of the respiratory system. The lungs deliver oxygen to and remove carbon dioxide from the blood through diffusion. See **Figure 26-17**.

Diffusion is the movement of molecules from a region of high concentration to a region of low

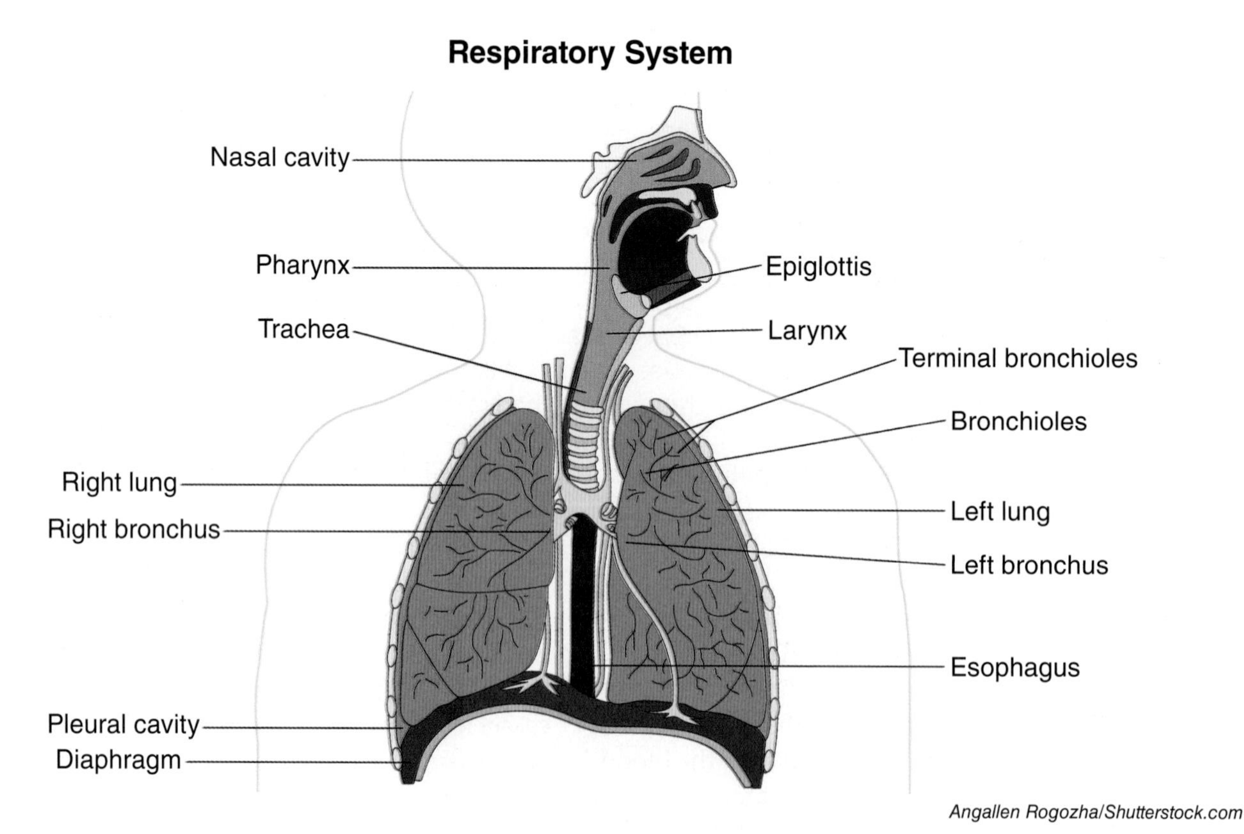

Figure 26-16. Parts of the respiratory system.

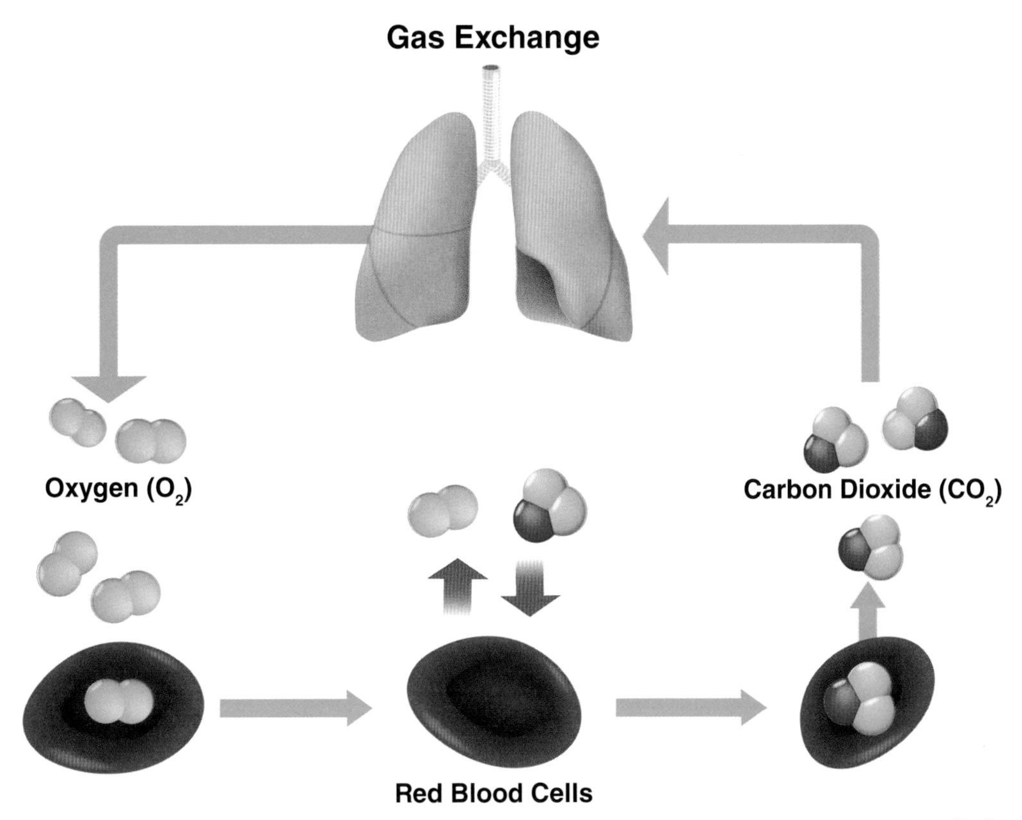

Figure 26-17. The lungs exchange oxygen and carbon dioxide with the blood through diffusion.

concentration. This process occurs in the alveoli of the lungs. The alveoli are tiny hollow cavities that are filled with oxygen during inhalation. The exchange of oxygen and carbon dioxide is possible because tiny capillaries containing blood surround the alveoli. The oxygen diffuses into the blood, while the carbon dioxide diffuses into the alveoli. The alveoli deflate when the diaphragm expands, causing the air cavity surrounding the lungs to decrease in size. The air is pushed out of the body, completing the respiration process.

Musculoskeletal System

Understanding the musculoskeletal system of the body is important for designing or improving medical implants and prostheses because the ***musculoskeletal system*** consists of the body parts that create the framework of the body. This framework supports the body, enables physical motion, and protects vital organs. Medical devices that replace parts of the body are often permanently or temporarily anchored to portions of the musculoskeletal system. The components of the system include bones, connective tissues, and muscles.

The bones of the body form a skeleton, which provides the main support and foundation for the shape of the body. The 206 bones in a human skeleton are made of living cells, which grow and change like the rest of the cells in the body. However, bone cells are embedded within dense layers of minerals and proteins. At the center of many bones is a soft ***bone marrow***, which contains stem cells that produce the majority of the body's red blood cells, white blood cells, and platelets.

The human skeleton can be separated into two main sections. The ***axial skeleton*** is made up of 80 bones that provide support and protection for the brain, spinal cord, and chest cavity. See **Figure 26-18**. This includes the skull bones, ear bones, the hyoid bone, the thorax bones, and the vertebral column. The skull consists of eight plate-shaped cranium bones that create the cranial vault, which houses the brain. Additionally, the skull has 14 facial bones

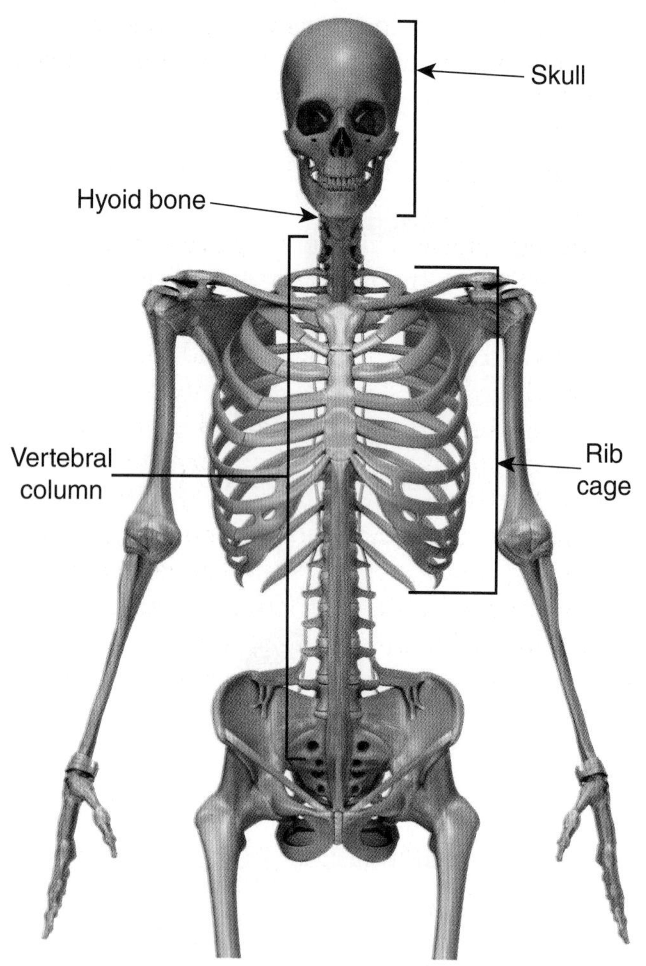

sciencepics/Shutterstock.com

Figure 26-18. Parts of the axial skeleton.

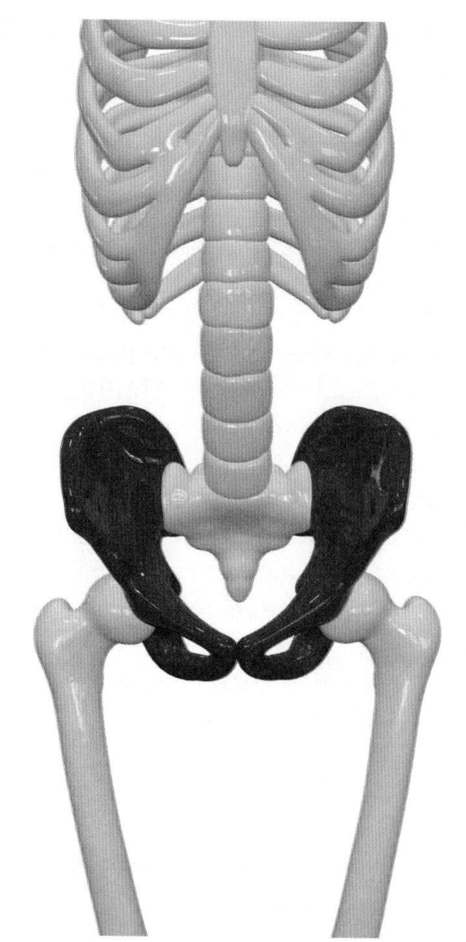

RK Stuff/Shutterstock.com

Figure 26-19. The pelvic girdle of the appendicular skeleton connects the bones of the lower limbs to the axial skeleton.

that form the lower front portion of the skull. Each ear also contains three bones called the *ossicles.* These are the smallest bones of the body that help transmit sounds through parts of the inner ear. The hyoid bone is the irregular shaped bone that forms the bottom jaw. The thorax is comprised of 12 rib bones on each side of the body, as well as the sternum, to form the rib cage. The vertebral column is made up of 24 bones, which are separated into five sections. The uppermost section is made up of 7 cervical vertebrae, followed by the 12 thoracic and 5 lumbar vertebrae. Below the lumbar vertebrae are the sacral and coccygeal bones. These two bones are formed from multiple fused vertebrae that are originally separate at birth.

The ***appendicular skeleton*** is comprised of 126 bones that support the appendages or limbs of the body. The ***pectoral girdle*** connects the bones of the arms to the axial skeleton to provide support for the movement of the upper limbs. The pectoral girdle consists of four bones, the left and right scapula and clavicle. Each arm has three bones—the humerus, ulna, and radius—that connect from the pectoral girdle to the hand. Each hand has 27 bones—8 carpals, 5 metacarpals, 5 proximal phalanges, 4 intermediate phalanges, and 5 distal phalanges.

The pelvic girdle of the appendicular skeleton connects the bones of the lower limbs to the axial skeleton. The ***pelvic girdle*** is comprised of the right and left hipbones. See **Figure 26-19**. Connected to the hip bones are the femur, tibia, and fibula of each leg. The ankles and feet have a total of 52 bones. These consist of the left and right tarsals, metatarsals, proximal phalanges, intermediate phalanges, and distal phalanges.

Where the different bones of the body connect, there are joints. Some of the joints move, while others do not. However, joints are sections of the body that experience wear and tear from physical activities. Therefore, joints are sections that are replaced or repaired using the products of biomedical engineering.

Fixed joints are the joints that do not move. These types of joints, called *sutures*, can be seen in the skull where the cranial bones connect together. See **Figure 26-20**.

Moving joints enable the body to perform physical activity. They allow the parts of the body to twist, bend, and move. See **Figure 26-21**. There are several different kinds of movable joints:

- Ball and socket joints consist of a round end of a bone fitting into an opening of another bone. These joints are found at the hips and shoulders.
- Hinge joints are found at the knees and elbows as well as in fingers and toes. The bones at these joints are molded to each other to allow the bones to glide across each other in only one direction.
- Gliding joints are locations where flat or slightly curved bones meet. These bones can slide across one another in opposite directions. Gliding joints can be found in the vertebrae and in the small bones of the wrists and ankles.
- Pivot joints are locations where bones can rotate about one another. An example of this joint would be the cervical vertebrae of the neck.
- Saddle joints are locations where two bones meet in a manner that is similar to a person riding a horse. This allows movement in all directions without rotational movement. This type of joint is found in the thumb.
- Condyloid joints involve an ovoid end of a bone moving within an elliptical cavity of another bone. These joints can be found in wrists.

Bones are coupled to other bones at joints by connective tissues called *ligaments*. ***Ligaments*** are long, fibrous straps that stabilize and support joints by holding bones in place. These connective tissues cannot usually be regenerated, and thus are often an area in need of medical treatment.

Muscles are a major component of the musculoskeletal system. ***Muscles*** are elastic, fibrous tissues

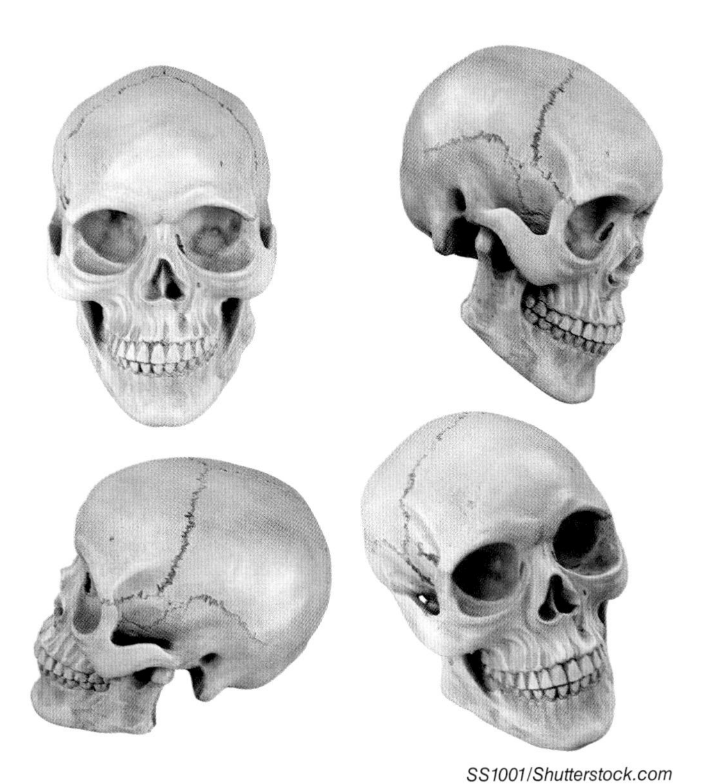

SS1001/Shutterstock.com

Figure 26-20. The skull consists of eight plate-shaped cranium bones.

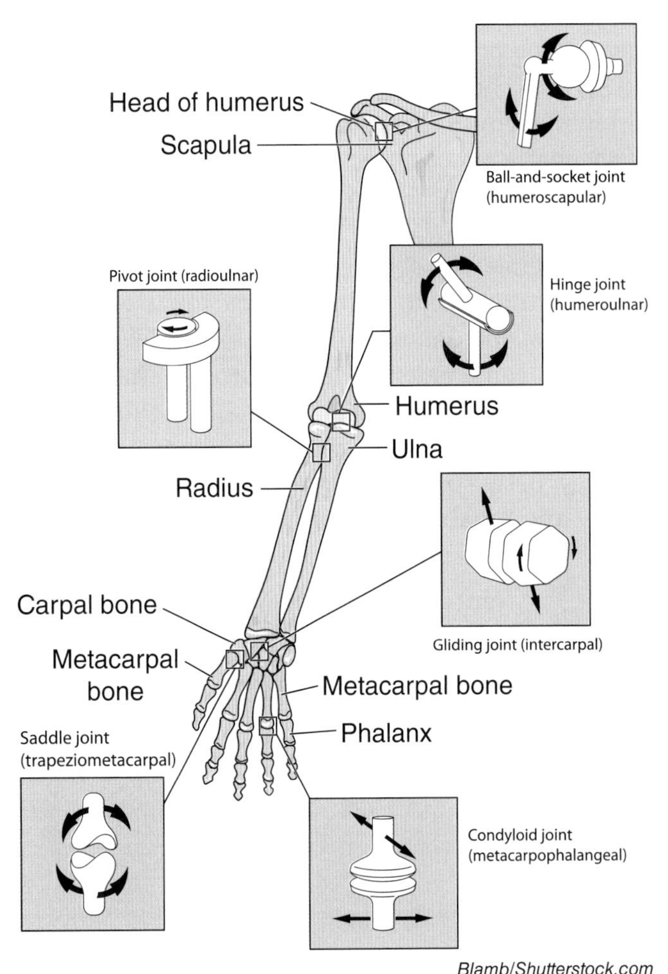

Blamb/Shutterstock.com

Figure 26-21. Types of joints, illustrated by the joints of the upper limb from the shoulder blade (scapula) to the fingers.

that can contract to produce movement. There are three main types of muscles in the human body. The cardiac muscle is the muscle that makes up the heart. Cardiac muscles are not found anywhere else in the body. When the cardiac muscle contracts, it pushes blood through the blood vessels.

Skeletal muscles are voluntary muscles—you control what they do. These muscles often span across joints to help move bones to perform physical activity. Skeletal muscles are connected to bones by connective tissues called ***tendons***. See **Figure 26-22**.

Smooth muscles are involuntary muscles; the body controls them without conscious thought. These muscles typically control the movements needed for different organs to function.

Biomechanics

Think about the movement of the body's limbs and joints and how the muscles and bones are subjected to the forces involved. Then think about developing products to replace these body parts. To do so, a biomedical engineer must have knowledge of biomechanics. ***Biomechanics*** involves the study of mechanical concepts as they relate to the movement or structure of living things. Knowledge of such concepts not only supports the development of replacement body parts but also helps to design safety equipment to reduce injuries in a variety of situations, such as working in hazardous environments, playing sports, or operating vehicles. See **Figure 26-23**. The analysis of the musculoskeletal system during the action of sports is necessary to develop better products and procedures to prevent injury and rehabilitate those who sustain injury.

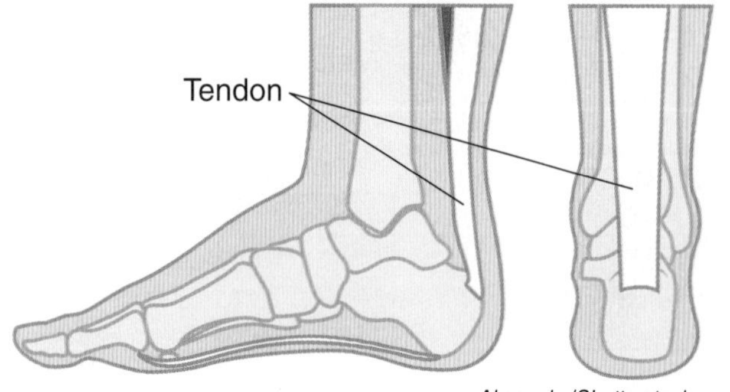

Aksanaku/Shutterstock.com

Figure 26-22. The Achilles tendon connects the calf muscles to the heel bone.

Michelle Donahue Hillison/Shutterstock.com

Figure 26-23. The study of biomechanics is important to help reduce sports injuries.

To effectively design medical implants and devices, biomedical engineers analyze biological systems under the lens of a branch of physics called *mechanics*. Physics is a branch of science that studies energy and matter and how they interact with each other. This study includes the topics of optics, electromagnetism, waves and vibrations, relativity, and thermodynamics, as well as mechanics. See **Figure 26-24**.

Mechanics studies how physical bodies react when forces are applied to them, which is essentially the study of motion. This knowledge is important when designing medical devices or implants like a prosthetic hip, because the artificial hip must move appropriately and withstand the forces involved with physical activities.

Mechanics can be divided into statics and dynamics. ***Statics*** uses Newton's laws of motion to analyze loads placed on objects at rest or at a constant velocity. In engineering, statics is used to analyze forces applied to things like trusses, cables, and chains. Because in statics the object has a constant velocity or is resting, the sum of all of these forces must be zero. With this information, engineers can calculate the magnitudes of the components of forces applied to the object using a series of equations. In biomedical engineering, statics can be applied to the rigid components of the body, as well

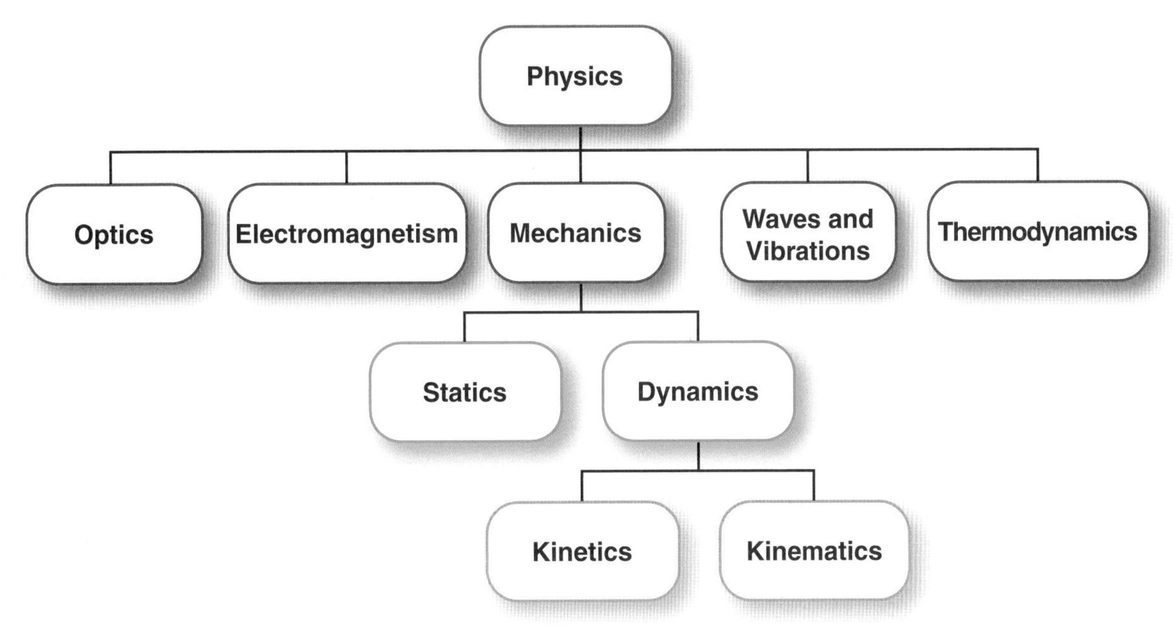

Goodheart-Willcox Publisher

Figure 26-24. Branches of physics.

as the medical device used to replace damaged or missing rigid parts of the body.

Dynamics deals with the analysis of objects that are accelerating as a result of acting forces. This means the sum of all forces on the object are not zero. Dynamics can be divided into two areas: kinetics and kinematics. ***Kinetics*** is the study of forces that cause motion, such as gravity or torque, while ***kinematics*** is the study of describing motion using quantities such as time, velocity, and displacement without the concern of the forces involved.

Kinetics can be divided further into studying linear motion (movement along a line) or angular motion (rotary motion). In biomechanics, the motion of the body is studied. The muscles of the body produce body motion, and motion cannot occur without a force. The muscles provide the force to accelerate parts of the body. The specific study of this human movement is called ***kinesiology***. This study is a combination of kinetics and human physiology.

The design of medical implants or devices requires an understanding of how objects react to forces applied to them and the resulting motion. The human body is constantly in motion and is always experiencing a variety of forces. Therefore, biomedical engineers must design implants and devices that move as intended and can withstand the forces applied to them.

Biomaterials

The capabilities of biomechanical devices rely on the properties and form of the materials they are made of. In addition, the properties of these materials, whether natural or synthetic polymers, metals, ceramics, or composites, must be compatible with the living systems in which they are used. See **Figure 26-25**. This compatibility is called *biocompatibility*. For

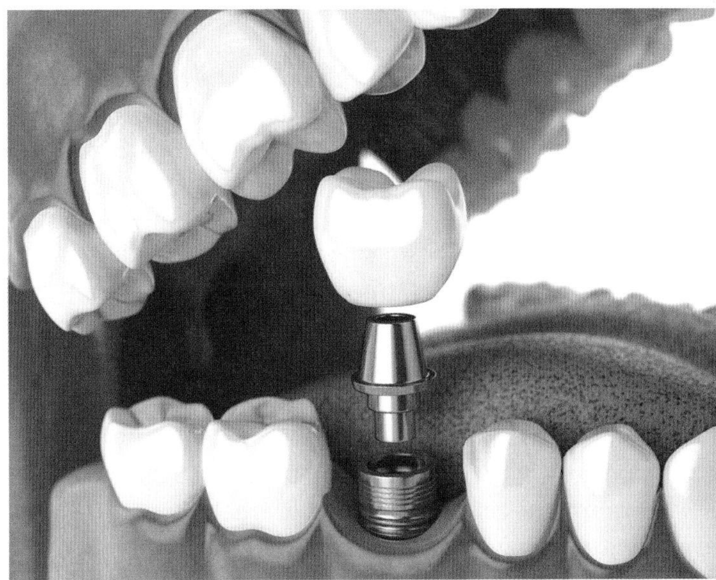

Maxx-Studio/Shutterstock.com

Figure 26-25. The "root" of a dental implant is typically made of titanium.

example, some materials may cause harm to the living cells of the body or may trigger an immune system response, causing the body to reject the device that is implanted.

Choosing the proper materials to make implants or artificial organs requires biomedical engineers to understand the principles of both materials science and biological sciences. A combination of these sciences can be called biomaterials science. A ***biomaterial*** is any material that interacts with living systems. These materials can be natural or synthetic. In biomedical engineering, biomaterials are used for creating medical devices that are designed to replace natural functions of the human body, such as artificial organs or replacement heart valves.

Biomaterials science provides the knowledge to help determine the properties of diverse materials and how they will interact with the human body. If the desired properties are not found, then biomedical engineers and scientists may work together to develop new and advanced materials. Combining the principles of biomaterials science with the concepts of biomechanics, biomedical engineers analyze different materials in order to determine their capabilities for medical applications. Properties such as the material's load bearing capacity or biocompatibility are studied.

Advanced scientific research has made it possible to grow biologically functional tissues using cells and support scaffolds. These tissues are grown to create bioartificial organs that can replace the functions of natural organs. The practice of developing these biologically active materials is known as ***tissue engineering***. See **Figure 26-26**.

Additionally, advances in molecular biology and biochemistry have enabled the manipulation of the genetic material of living organisms. This practice is called ***genetic engineering***. Every living thing carries a genetic code, or blueprint, that precisely determines its traits. Genetic engineers can modify this code for medical purposes, as well as transfer genetic traits between other people and even other species. Altering the genetic materials of living organisms can help prevent or treat disease. Genetic engineering is described further in Chapter 27.

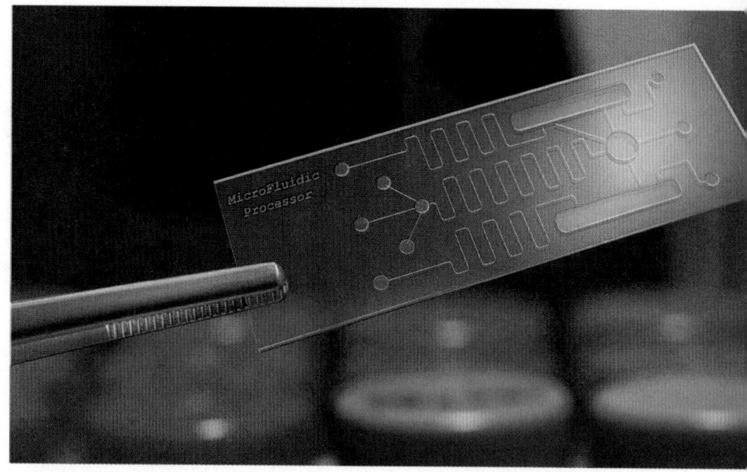

science photo/Shutterstock.com

Figure 26-26. The lab-on-a-chip is an emerging technology used in tissue engineering. Several laboratory operations are integrated on one small-scale chip.

Biomedical Imaging

Biomedical imaging is the practice of creating accurate visual representations of the interior of the human body, **Figure 26-27**. Biomedical imaging can be used as a means to diagnose and monitor disease and injury. This practice is a growing area within the health-care sector and is an important part of the biomedical engineering profession. Biomedical engineers develop the technologies that help diagnose or monitor disease and injury. These

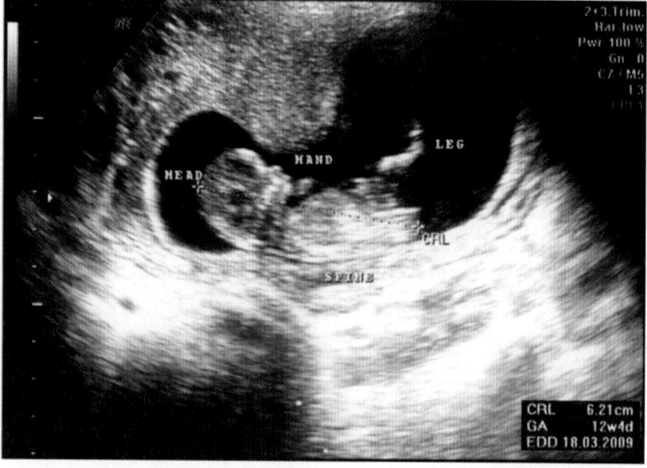

Ngo Thye Aun/Shutterstock.com

Figure 26-27. This ultrasound-imaging device is being used to examine the development of a fetus.

technologies use ways to create visual representations of the body. Examples of biomedical imaging discussed in Chapter 25 are radiography (x-rays), magnetic resonance imaging (MRI), and ultrasound. Another form of imaging is nuclear imaging, which uses radioactive materials and a scanner device to create pictures of areas inside the body. The radioactive materials are administered to a patient and the scanner detects the materials as they move throughout the body.

To develop imaging devices, biomedical engineers must have a solid understanding of electrical engineering principles. They design and experiment with electrical circuits and digital computers to capture visualization data, recognize patterns within the data, and use the information to process accurate images of the body's interior.

Government Regulations and Standards

Biomedical engineering has unlimited potential for improving the health and well-being of living creatures. However, if the practices of this field are not regulated properly, they can pose threats to human health. Therefore, government agencies are responsible for regulating the practices involved with biomedical engineering. Two primary agencies in the United States create guidelines and standards for testing and implementing new biological, medical, and health technologies. These agencies are the Food and Drug Administration (FDA) and the National Institutes of Health (NIH).

Bioethics

The decisions made by engineers affect the well-being of others. This is especially true in biomedical engineering—the products and systems can result in life or death. Therefore, engineers must always adhere to the highest principles of ethical conduct. Advances in medical, health, and related biotechnologies may result in controversial ethical issues.

Ethics surrounding these issues are termed *bioethics* and *medical ethics*. Debates have occurred over issues such as euthanasia, in vitro fertilization, cloning, and gene therapy. Bioethics closely overlaps with medical ethics, which focuses on the morality of different medical treatments. Biomedical engineers must have an understanding of diverse perspectives when developing and implementing new medical and health technologies.

Summary

- ✔ Biomedical engineering involves the design and development of products, systems, and procedures that solve medical and health-related problems.
- ✔ Biomedical technologies are designed by people with a strong understanding of the science of the human body as well as engineering principles.
- ✔ Early medical devices were often developed through trial and error. Over time, improved devices have been developed in response to scientific and technological breakthroughs and improved engineering practices.
- ✔ Health-care professionals and biomedical scientists and technicians use products developed by biomedical engineers to help diagnose, prevent, and monitor disease or injury.
- ✔ Medical implants are tissues or devices that are surgically placed inside or on the exterior of the body. Functions of implants include monitoring bodily functions, delivering medication, and supporting organs or tissues.
- ✔ Many implants are prosthetic devices designed to replace missing or damaged body parts.
- ✔ Understanding of human physiology, including the cardiovascular system, respiratory system, and musculoskeletal system, is important in biomedical engineering.
- ✔ Biomechanics is the study of mechanical concepts as they relate to the movement or structure of living things.
- ✔ Biomaterials science provides the knowledge to help determine the properties of diverse materials and how they will interact with the human body.
- ✔ Biomedical imaging is the practice of creating accurate visual representations of the interior of the human body as a means to diagnose and monitor disease and injury.
- ✔ To develop imaging devices, biomedical engineers must have a solid understanding of electrical engineering principles.
- ✔ Biomedical engineering is subject to government regulations and principles of ethical conduct.

Check Your Engineering IQ 

Now that you have finished this chapter, see what you learned by taking the chapter posttest.
www.g-wlearning.com/technologyeducation/

Test Your Knowledge 

Answer the following end-of-chapter questions using the information provided in this chapter.

1. Products of biomedical engineering are widely used for diagnosing, treating, and _____ disease and injury.
2. *True or False?* Prosthetic limbs were used in ancient Egypt.
3. Biomedical engineers combine knowledge of mathematics and science with _____ skills to plan and create medical equipment to meet required criteria and constraints.
4. What are medical implants?
5. A(n) _____ prosthesis replaces parts of the leg missing below the knee.
   A. transhumeral
   B. transradial
   C. transfemoral
   D. transtibial
6. Bioartificial organs are produced using _____.
7. Human physiology studies the _____ and _____ of the human body.
8. List the three main parts of the cardiovascular system.
9. What is the purpose of the respiratory system?
10. *True or False?* The axial skeleton provides support and protection for the brain, spinal cord, and chest cavity.
11. _____ have the ability to contract and produce movement in the human body.
12. Connective tissues called _____ connect skeletal muscles to bones.
13. The study of human movement is called _____.
    A. physics
    B. kinesiology
    C. kinematics
    D. dynamics
14. *True or False?* Biomaterials can be natural or synthetic.

15. *True or False?* Tissue engineering refers to manipulation of the genetic material of living organisms.
16. A strong understanding of _____ principles is required to develop biomedical imaging devices.

Critical Thinking

1. Describe how a specific technological or scientific breakthrough has led to improved health and wellness.
2. How are mechanics applied in biomedical engineering?

STEM Applications

1. With a partner, perform the following experiment to determine how your pulse rate can indicate that your cardiovascular system is maintaining homeostasis.
   A. Measure each other's pulse rate for 10 seconds.
      i. You can measure the radial pulse by placing two fingers on the inside of your partner's wrist, just under the crease of the wrist at the base of the thumb. Press on the skin and move your fingers around until you feel the blood pushing under the skin. Then count the pulses that occur during a 10-second period. Do not use your thumb to measure someone's pulse. Your thumb has its own pulse and can interfere with your measurement.
      ii. You can also check pulse rates on the side of a person's neck. This is called the *carotid pulse*. It is located below the jaw between the windpipe and the large muscle of the neck.
   B. Using this measurement, calculate the pulse rate per minute.
   C. Have your partner exercise in place for 3 minutes. Then retake his or her pulse and record the measurement.
   D. Continue retaking the pulse measurement every 30 seconds until the pulse rate returns to the same when you first tested it before the exercise.
   E. Reverse roles and have your partner perform these steps on you.
   F. After you have determined the different pulse rates at rest and at different intervals after exercising, answer the following questions:
      i. What happened to the pulse rates immediately after exercising?
      ii. What happened to the pulse at each 30 second interval?
      iii. How do the changes in the pulse rates demonstrate that your cardiovascular system is maintaining homeostasis?
2. Develop a technical report on personalized medicines by reviewing the available scientific literature to answer the following questions:
   A. How can medicines be personalized by leveraging technology, such as rapid diagnostic devices?
   B. How can sensors be used to detect amounts of chemicals or compounds in mixtures or determine the internal condition of the body minute by minute?

Engineering Design Challenge

Prosthesis Challenge

Background

Biomedical engineers develop a great deal of diagnostic and treatment equipment. They also develop artificial parts for human bodies and devices that allow people to do routine tasks.

Situation

You work with a volunteer agency that helps wounded warriors. Many of the veterans that you serve have lost a limb and are awaiting a prosthetic replacement. Therefore, it is difficult for your clients to perform routine tasks. Your challenge is to design a reusable temporary prosthetic device that allows a person who is missing both arms to pick up and move an object. The device should be easy to use and inexpensive to make. It must allow the user to pick up a glass of water and take a drink. The device should only use human power to operate.

Materials and Equipment

- Various prototyping materials (i.e., cardboard, string, wood strips, dowel rods, foam, and fabric); 3-D printed materials are optional.

Desired Outcomes

- A functional prototype that allows a user with no arms to take a drink of water.
- A demonstration of the prototype.
- A final technical drawing of the device.

Procedure

- Remember that the engineering design process is an iterative approach to solving problems. You can and should go back and forth between the different steps.
- Determine your engineering design team based on the class's individual skill sets. As a team, formulate a problem statement based on the background and situation provided above. A good problem statement should address a single problem and answer the questions *who, what, where, when*, and *why*. The problem statement should not imply a single solution to the problem. Your class may create different problem statements based on their own interpretations of the situation and the teams' prior interests and experiences.
- Use your team's defined problem statement as the starting point to creating a valid solution.
- Employ the engineering design process to propose new or improved solutions to the defined problem. Brainstorm and generate ideas.
- Produce prototypes and/or models.
- Evaluate the success of the team's solution and iteratively refine the solution.

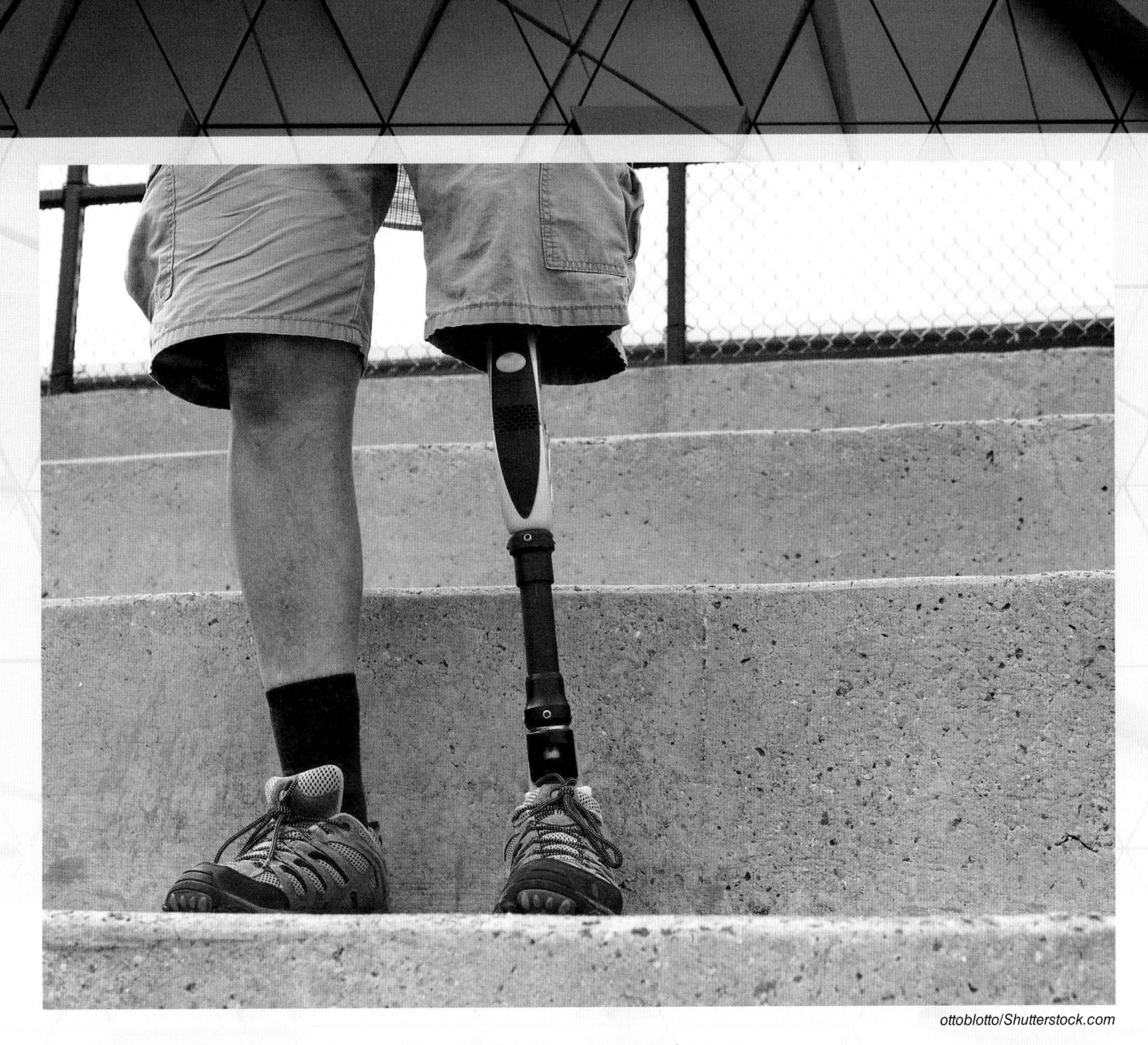

ottoblotto/Shutterstock.com

Prosthetic limbs allow people a degree of freedom they did not have in the past.

CHAPTER 27

Agricultural and Related Biotechnologies

Check Your Engineering IQ 

Before you read this chapter, assess your current understanding of the chapter content by taking the chapter pretest.
www.g-wlearning.com/technologyeducation/

Learning Objectives

After studying this chapter, you should be able to:

- ✔ Define *agriculture*.
- ✔ Compare the roles of science and technology in agriculture.
- ✔ Illustrate the major types of agriculture.
- ✔ Recall major crops grown on farms and equipment used in crop production.
- ✔ Explain how technology is used in hydroponics and aquaculture.
- ✔ Summarize the characteristics of the technology used in raising livestock.
- ✔ Define *biotechnology* and summarize ways it can be used in agriculture.
- ✔ Compare primary and secondary food processing.
- ✔ Describe ways used to process and preserve foods.
- ✔ Summarize the steps that might be used to develop a new food product.
- ✔ Summarize the activities involved in manufacturing a new food product.

Technical Terms 

agricultural technology
agriculture
animal husbandry
aquaculture
aseptic packaging
baler
barbed wire
berries
biotechnology
canning
catalyst
chemical processing
combine
crop production
cultivator
curing
drip irrigation
endosperm
fertilizer
flood irrigation
food-processing technology
forage crops
freezing
fruits
furrow irrigation
gene splicing
genetic engineering
grain
grain drill
hydroponics
irradiation
irrigation
mechanical processing
milling
nonfood crops
nuts
pasteurization
pivot sprinkler
plow
preservatives
primary food-processing technology
secondary food-processing technology
smoking
sprinkler irrigation
swather
thermal processing
tillage
vegetables
windrows

Close Reading

As you read this chapter, make a list of current agricultural trends, and relate them to how they were different before technological advances were made. Also consider the different types of food you ate yesterday, and see if you can determine the types of conversion, preservation, and processing your food underwent.

People have a number of basic needs and wants. Two of the primary needs are food and clothing. Agriculture directly addresses both of these needs. This human activity involves using science and technology to grow crops and raise livestock. ***Agriculture*** is people using materials, information, and machines to produce food and natural fibers. These activities take place on farms and ranches around the world.

Modern farming uses both scientific and technological knowledge. Two important branches of the life sciences support agriculture. These areas of the life sciences are plant science and animal science. They are used in cross-pollinating plants to improve crops and in crossbreeding animals to improve livestock. Other sciences help farms manage their operations. For example, the science of meteorology allows farmers to plan planting and harvesting activities, and to select seasonally appropriate crops and livestock to raise. Knowledge of plant nutrients allows farmers to select appropriate fertilizers. These and other sciences have helped farmers become more efficient.

Technological advancements have created massive changes in farming, **Figure 27-1**. For example, they have caused changes in the size of farms and how farmers go about their work. New and modern machines and equipment allow for greater production. Fewer people grow more food on fewer acres. Technological advances have helped people work with greater ease and efficiency. These advancements can be attributed to ***agricultural technology***. This technology uses technical means (machines and equipment) to help plant, grow, and harvest crops and raise livestock.

Types of Agriculture

Agriculture involves people managing land, buildings and machinery, and crops and livestock. In doing this, people engage in two major types of agriculture: crop production and animal husbandry, **Figure 27-2**.

Crop production involves growing plants for various uses. Crop production provides food for humans, feed for animals, and natural fibers for a variety of uses. This type of agriculture produces trees and plants for ornamental use, such as landscaping. Crop production grows trees for lumber and paper production. This type of agriculture produces basic ingredients for medicines and health-care products. Crop production also provides materials for many industrial processes, such as textile weaving, plywood manufacture, and food processing.

Animal husbandry involves breeding, raising, and training animals. These animals might be used for food and fiber for humans. In some cases, they are used to do physical work. This is especially true

From this...

Everett-Art/Shutterstock.com

To this...

Lurin/Shutterstock.com

Figure 27-1. Technological advancements have greatly changed agriculture.

©iStockphoto.com/nkurtzman

©iStockphoto.com/LUGO

Figure 27-2. Agriculture involves growing crops and raising livestock.

in developing countries. Many animals are also raised as hobbies or are used for riding and racing.

The United States uses more than 40% of its total land for farming activities. A total of about 315,000,000 acres is used for crop production. About half of this land is used for raising soybeans and corn. Another billion acres is used for pastures, ranges, and forests.

Crop Production

Many different crops are raised on farms and ranches in the United States. Some of these crops were originally found in other areas of the world. For example, China and central Asia gave us lettuce, onions, peas, sugarcane, and soybeans. Rice, sugarcane, bananas, and citrus fruits came from other parts of Asia. The Middle East, southern Europe, and North Africa gave us wheat, barley, oats, alfalfa, and sugar beets. Corn, beans, tomatoes, potatoes, peanuts, tobacco, and sunflowers came from North and South America.

The most widely grown major crops in North America are forms of ***grain***, **Figure 27-3**. Grain crops are members of the grass family that have large edible seeds. The commonly grown grains are wheat, rice, corn, barley, oats, rye, and sorghum. All these grains can be used in food products. Corn, barley, oats, and sorghum are widely used in animal feed.

US Department of Agriculture

Figure 27-3. Grain is widely grown as food for people and animals.

Vegetables, another important crop, have edible leaves, stems, roots, and seeds and provide important vitamins and minerals for the daily diet. Vegetables include root crops, such as beets, carrots, radishes, and potatoes. They also include leaf crops, such as lettuce, spinach, and celery. Other vegetables provide food from their fruit and seeds. This group includes sweet corn, peas, beans, melons, squash, and tomatoes.

Fruits and ***berries*** are grown in many parts of the country, **Figure 27-4**, for their edible parts. Major fruit

US Department of Agriculture

Figure 27-4. Berries and fruit are important crops grown on American farms.

Case IH

Figure 27-5. Forage crops provide animal feed.

crops grown in temperate climates include apples, peaches, pears, plums, and cherries. Citrus fruits (oranges, lemons, limes, grapefruits, and tangerines), olives, and figs are grown in warmer climates. Tropical fruits include bananas, dates, and pineapples.

Nuts are grown in specific areas of the country. They are grown for their hard-shelled seeds. Walnuts, pecans, chestnuts, almonds, and filberts (hazelnuts) are grown in temperate climates. Palm oil nuts and coconuts are grown in tropical areas. Peanuts and coconuts are the most important nut crops and are significant sources of food and oil.

Forage crops are grown for animal feed. See **Figure 27-5**. These plants include hay crops, such as alfalfa and clover. Grasses used for pasture and hay are also included in this group.

A number of ***nonfood crops*** are also grown on farms. These crops are not for human consumption. They include tobacco, cotton, and rubber. Nonfood crops also include nursery stock grown for landscape use and Christmas trees.

Technology in Crop Production

Crops are no different from any other living thing. They have set life cycles. Crops are "born" when seeds germinate, continue through growing cycles, and then mature. After a time, they die. To be of benefit to people and animals, crops must be harvested before they spoil or shatter. Farming takes advantage of a plant's life cycle through four major processes: planting, growing, harvesting, and in some cases, storing.

In the past, farming was a very labor-intensive activity. Most of the population was involved in raising crops and animals. Technological advancements during the past 200 years, however, have changed all this in many countries. Now, just a small percentage of the population in developed countries is involved in agriculture. Farming has become equipment intensive. To a large extent, machines and equipment have replaced human and animal labor. Farm equipment is used at all stages of crop production. This equipment can be classified into eight major groups:

- Power or pulling equipment.
- Tillage equipment.
- Planting equipment.
- Pest-control equipment.
- Irrigation equipment.
- Harvesting equipment.
- Transportation equipment.
- Storage equipment.

Power or Pulling Equipment

People have a long history of replacing human power with other power sources. Over most of recorded history, people used animals to pull loads. The invention of the agricultural tractor in 1890 changed this, however. During the 1900s, this new power source replaced animal power on most farms. Today, the farm tractor provides the power to pull all types of farm equipment.

Farm tractors can be either wheel tractors or track machines, **Figure 27-6**. Most agricultural tractors have wheels. Some wheel tractors have rear, power wheels. These tractors usually have smaller front wheels. Other wheel tractors have power to all wheels. These tractors generally have the same size tires on all wheels. For additional traction, both types of wheel tractors might have dual drive wheels.

Track-type tractors are used for special purposes and are generally slower than wheel-type tractors. They are suited for use in muddy fields. These tractors sink less and have less slippage than wheel-type tractors. Also, they cause less compaction of soil and do not create wheel ruts.

Most tractors are designed with the engine, transmission, and gearbox as a single unit. This structure provides a rigid backbone for the machine. The steering and drive wheels are attached to the basic unit. Implements are attached to or pulled behind the tractor.

Tillage Equipment

Soil must be prepared before crops can be planted. Residue from previous crops must be cleared. The seedbed must be conditioned by breaking and pulverizing the soil. This process is called ***tillage***, or tilling the soil.

The most important piece of tilling equipment is the ***plow***. The plow breaks, raises, and turns the soil. This process loosens the ground and brings new soil into contact with the atmosphere.

Three major types of plows are the moldboard plow, the disc plow, and the chisel plow. The moldboard plow is made of a frame and several *plowshares* (part of the plow that actually cuts trenches), **Figure 27-7**. When the plow is pulled through the earth, it cuts and rolls the soil.

Case IH

Figure 27-7. This moldboard plow is cutting and rolling the soil.

Wheel Tractor

Case IH

Track Tractor

Deere and Co.

Figure 27-6. Two types of farm tractors are wheel and track.

Career Connection Agricultural and Food Scientist

Agricultural and food scientists research methods to improve the efficiency and safety of agricultural processes and products. They play a vital role in maintaining a nation's food supply. They may work in a research and development setting, attempting to understand the biological and chemical processes by which crops and livestock grow and use that knowledge to discover ways to improve agricultural products.

Individuals who have strong communication, organizational, and problem-solving skills do well in this field. Students interested in pursuing a degree and career as an agricultural and food scientist should take courses in biology and chemistry, as well as any available agricultural science courses.

Many of the agricultural programs throughout the United States are offered through a land-grant college, which each state has. An undergraduate degree may include courses in chemistry, botany, organic chemistry, and biology. A focus on food science may include food chemistry, food analysis, food microbiology, food engineering, and food-processing operations. Students preparing to be soil and plant scientists take courses in plant pathology, soil chemistry, entomology (the study of insects), plant physiology, and biochemistry. Some states require agricultural and food scientists who specifically deal with soil to be licensed to practice.

Career tracks for this field include food scientist, soil scientist, chemical or biological technician, farmer, rancher, and environmental scientists and technicians. Agricultural and food scientists are expected to have a 5% job growth by the year 2024.

Potential projects for agricultural and food scientists include:

- Conducting research and experiments to improve the productivity and sustainability of field crops and farm animals.
- Creating new food products and developing new and better ways to process, package, and deliver them.
- Studying the composition of soil as it relates to plant growth, and researching ways to improve it.
- Communicating research findings to the scientific community, food producers, and the public.

Disc plows have a frame that has several discs mounted on an axle. See **Figure 27-8**. The discs and axles are set at a steep angle to the direction of travel. When pulled, the discs turn and then cut and loosen the soil.

Chisel plows have a set of shaped chisels attached to a frame. When the plow is pulled through the earth, it breaks up the soil. The plow does not lift and turn the soil, however. Chisel plows are used in grain stubble and where the soil needs little tilling.

Planting Equipment

Once the soil is prepared, the crop can be planted. Planting involves two actions that can be done separately or together: applying fertilizer and planting the crop. ***Fertilizer*** is a liquid, powder, or pellet containing important chemicals that encourage and support plant growth. This liquid, powder, or pellet primarily delivers nitrogen, phosphorus, and potassium. The fertilizer can be applied before, during, or after seeds are planted.

This liquid, powder, or pellet is applied with special equipment or along with a seed planter. In some cases, fertilizer is scattered (broadcast) on the ground before final tilling and planting. A machine with a series of knives can inject liquid and gaseous (anhydrous ammonia) fertilizers into the soil.

Planting the crop might involve putting seeds or starter plants into the fields. Grains, grasses, and many vegetables are started from seeds planted directly in soil. Some vegetables, such as tomatoes,

US Department of Agriculture

Deere and Co.

Figure 27-8. This disc plow is cutting and loosening the soil. Notice the close-up of the discs on the right.

cabbage, cauliflower, and broccoli, are started from seeds in nursery greenhouses.

The most widely used seed planter is the ***grain drill***, **Figure 27-9**. A grain drill is pulled behind a tractor. As it moves across the field, it opens a narrow trench, drops seeds, and closes the trench.

Pest-Control Equipment

Not all plants that germinate grow to maturity. Diseases and insects kill some of them. Weeds crowd others out. (Weeds are any out-of-place plants.) Farmers can reduce these dangers by using chemical and nonchemical pest-control measures. These techniques use special machines and devices.

Deere and Co.

Figure 27-9. A specially designed grain drill is able to plant seeds in soil that has not been tilled.

Chemical control techniques generally use a liquid spray to control weeds and insects. These sprays include herbicides to control weeds and pesticides to control diseases and insects. Ground equipment and aircraft can apply both of these materials.

Nonchemical pest control uses machines and other devices to control pests. A ***cultivator*** has a series of hoe-shaped blades that are pulled through the ground. See **Figure 27-10**. The blades break the crust of the soil, allowing rain and water to enter the soil. The blades also cut off and pull out weeds.

Irrigation Equipment

Some parts of the world receive sufficient rainfall for crops to grow. This is not the case, however, in all parts of the world. Many places receive very little rain,

US Department of Agriculture

Figure 27-10. This field of potatoes is being cultivated to remove weeds and loosen the soil.

or the rain comes at the wrong time, for successful farming. Irrigation systems can be used to support agriculture in these dry or unpredictable climates.

Irrigation is artificial watering to maintain plant growth. Irrigation systems must have the following:

- A constant and dependable source of water.
- A series of canals and ditches to move the water from its source to various crops.
- A way to control the flow of the water in the ditches and canals.

The common sources of irrigation water are lakes, rivers, and underground aquifers. Various methods are used to control the water at its source. A dam can restrict water in a river to form a reservoir, **Figure 27-11**. A dam at the outlet of a natural lake can control the level of the lake and divert water for irrigation. A well and pump can be used to obtain underground water, **Figure 27-12**. Canals and pipes move the water from the source to farm fields. There, one of four irrigation methods can be used to water the crops.

If the land is level, ***flood irrigation*** can be used. These systems use a large quantity of water that advances across the fields. Ditches or pipes bring the water to one end of the field. The water is released from the ditches or pipes through holes along their lengths. As additional water is released, gravity causes the water to flow across the field. In some cases, another set of ditches or pipes carries off excess water.

In row crops, ***furrow irrigation*** can be used. This system uses small ditches, called *furrows*, which are built between the rows of plants. The furrows guide the water as it flows from one end of the field to the other. See **Figure 27-13**. Pipes or siphon tubes are often used to control the flow and direct water to various sections of the field.

Sprinkler irrigation produces artificial rain to water crops, **Figure 27-14**. It is used on uneven

romeo61/Shutterstock.com

Figure 27-12. These pumps supply water for an irrigation system.

Jeff Zehnder/Shutterstock.com

Figure 27-11. A dam can be used to form a reservoir. The water can be used for irrigation, to generate electricity, and for recreation.

US Department of Agriculture

Figure 27-13. The furrows between these rows of potatoes are used to carry irrigated water.

nd3000/Shutterstock.com

Figure 27-14. This sprinkler-irrigation system is being used to water agricultural plants.

Academic Connection: History

The Homestead Act and the Morrill Act

In this chapter and others, we learn how various agricultural and related biotechnologies have changed throughout history. By looking at history from a slightly different perspective, we can also see why these changes occurred. For example, two political events in 1862 forever changed the kinds of crop and livestock production developed in the United States. In May of 1862, Congress, in an effort to promote movement to the West, passed the Homestead Act. This act gave 160 acres of land free of charge to any adult citizen who would live on the land for five years and develop it. In July of 1862, Congress passed the Morrill Act. Under this act, the federal government gave each state 30,000 acres of land for each congressional representative that state had. The land was to be sold, with the resulting monies used to establish colleges of agriculture and the mechanical arts.

Each act influenced agriculture in various ways. For example, because of the Homestead Act, thousands of people moved west. They discovered that the land in the West was ideal for raising livestock. These people also found that the dry conditions were ideal for growing wheat. By the 1900s, the plains area of the United States had become one of the world's major sources of that crop. Moreover, because of the growth of this type of farming in the West and the resulting competition, farmers in the East started turning to dairy and truck farming.

©iStockphoto.com/cascoly

Many colleges and agriculture departments within universities were established because of the Morrill Act. These colleges and departments had the research capabilities to help those in agriculture improve existing methods and products. Their pioneering research efforts continue to improve agricultural and related biotechnologies in the United Sates to this day.

Another event in the 1860s also greatly affected US agriculture. This event was the completion of the transcontinental railroad at Provo, Utah, in 1869. Can you think of some reasons why this event had such an impact on agriculture?

---

ground or where the amount of water applied must be closely controlled. This irrigation method uses less water than flood-irrigation systems do.

In sprinkler irrigation, a pump is used to force water from the source into the main distribution lines. The pressurized water flows through the main lines to a series of lateral pipes. These pipes extend at right angles from the main line. Each lateral line has a series of small-diameter pipes sticking upward at even intervals. At the end of each standpipe is a sprinkler head. The water in the lateral lines enters the sprinkler heads, which spray water onto the land. Valves between the main and lateral lines shut off or control the water flow.

Many sprinkler systems have several straight sprinkler (lateral) lines, which are used to irrigate the fields. Each line applies water to a long, narrow band across the field. To irrigate the strip of land on each side of the sprinkler lines, the water is allowed to run for a set time. The water is then shut off from the lateral lines with a valve, and the lines are disconnected. They are moved by hand or rolled under power to the next position (set) along the main line. Here, they irrigate the next strip across the field. After a series of moves, the entire field is irrigated.

In some cases, many sprinkler lines are used to cover the entire field. When the lines are turned on, the entire field is irrigated at once. This is called a *solid-set sprinkler system*. The solid-set sprinkler system approach eliminates the need to move individual lines. This approach is used where the crop must be watered often or where frost protection must be provided using the sprinkler system.

A ***pivot-sprinkler*** system uses segments of pipes attached to each other, making one long line to deliver water. The line is attached at one end to a water source. This long line pivots around this point on large wheels powered by electric motors. The line moves slowly and constantly in a circle. Sprinkler or mist heads apply the water as the line pivots.

Drip-irrigation systems deliver water slowly to the base of the plants. Drip systems use main lines to bring water near the plants. Individual tubes or emitters extend from the main lines to each plant. These tubes apply water, which soaks into the ground around the roots. This system ensures that each plant is properly watered, and it reduces the amount of water lost to evaporation. Drip systems are used in many orchards and vineyards.

Harvesting Equipment

To be of value, a mature crop must be harvested. Different harvesting machines exist. The most widely used is the grain ***combine***. See **Figure 27-15**. This machine can be used to harvest a wide range of grains and other seed crops.

There are several stages in combine operation, **Figure 27-16**. First, a rotating reel pulls the grain into a cutter bar. There, the tops of the plants containing the grain or seeds are cut off and drawn to the center

Deere and Co.

Figure 27-15. This combine is harvesting a grain crop.

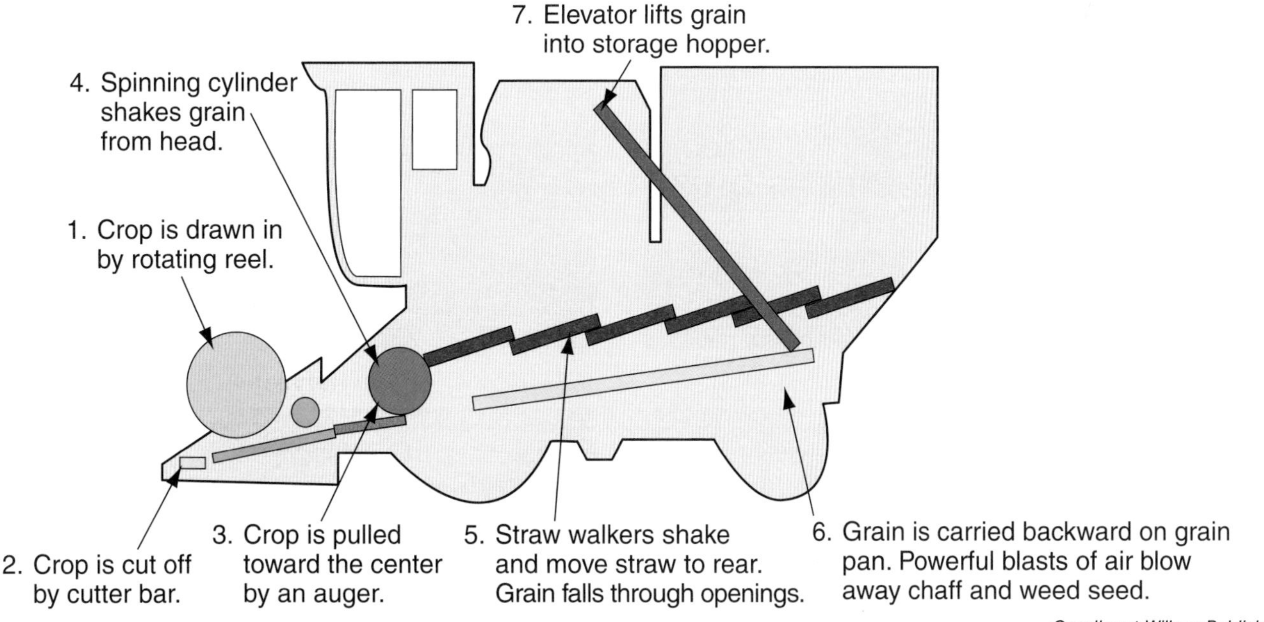

Goodheart-Willcox Publisher

Figure 27-16. Typical combine operation.

of the machine. The heads and straw move into the machine, where a revolving cylinder separates the grain from the heads. The grain and straw move onto straw walkers. These devices move the straw to the back of the machine. As the straw moves, the grain falls through holes in the walkers, onto a grain pan. There, blasts of air blow away chaff and other lightweight materials. An auger lifts the remaining grain from the grain pan into storage hoppers. The unwanted straw and waste materials are conveyed out the back of the machine and drop onto the ground. When the storage hopper is full, another auger unloads the grain onto a truck or wagon.

Specialized harvesting machines have been developed for other crops. See **Figure 27-17**. Cotton is removed from the plants with a cotton picker. The picker strips the cotton bolls from the plant and deposits them in a bin.

Peanut Harvester

Tomato Harvester

Onion Harvester

Potato Harvester

Green Bean Harvester

Sweet Pea Harvester

Cotton Picker

Tea Harvester

Deere and Co.

Figure 27-17. Special-purpose harvesting machines.

Potatoes and onions are dug from the ground with mechanical diggers or harvesters. The crop moves across conveyors, where people remove rocks and dirt clods. The remaining potatoes or onions are conveyed onto trucks. Peanuts are dug using special-purpose harvesters. Corn can be harvested with either a combine or a corn picker that strips the ears from the stalks.

Vegetables can be harvested by special-purpose machines or by hand. Special mechanical pickers are used to harvest green beans, sugar peas, and sweet corn. The vegetables are stripped from the plants in one pass. The crop is conveyed into a bin, which is later dumped into trucks for delivery to canning or freezing plants.

Fruits are often picked by hand and placed in boxes. Special machines might be used for some fruits and nuts, however. These machines shake the tree, causing the fruit or nuts to fall into raised catching frames. The crop is gathered from the frames and hauled from the orchards.

A series of special machines harvests grass and alfalfa hay. The crop must first be cut and laid on the ground to dry. In some cases, a mower might be used to cut the plants and let them fall on the ground. After the hay has dried for at least one day, a rake is used to gather it into ***windrows*** (bands of hay). In large-scale hay operations, a windrower or ***swather*** is used for these processes, **Figure 27-18**. This machine cuts and windrows the hay in one pass over the field.

Deere and Co.

Figure 27-18. This windrower, or swather, is cutting and windrowing hay. Other windrowers are self-propelled, eliminating the need for a tractor.

After the hay in the windrows has dried, it is usually baled. A machine called a ***baler*** is used to gather, compact, and contain the hay. Balers produce either square or round bales. Square bales can weigh up to several hundred pounds. The standard bale, however, is about 4′ long and weighs 75–125 pounds. The bales are bound with either baling twine or wire, making the hay easy to handle and to store.

Transportation Equipment

Harvested crops must be removed from fields and either taken to storage or moved directly to processing plants. Short-distance moves are done using wagons or medium-duty trucks, **Figure 27-19**. Longer hauls are done using semitrucks, railcars, barges, or ships.

Storage Equipment

Many crops are stored on farms or in commercial locations before they are sent to processing plants. Most grain crops are stored in silos or buildings at grain elevators, **Figure 27-20**. Here, the grain is

©iStockphoto.com/ImagineGolf

Figure 27-19. This truck will haul the grain harvest to a storage silo, or elevator.

ArturHenryk/Shutterstock.com

Figure 27-20. A typical rural grain elevator.

Think Green Local Organic Food

You have already learned that organic material comes from a living organism. Food can be organic by using alternative, organic approaches for pesticides and fertilizers. Conventional pesticides and fertilizers may taint soil and water when they break down. They also may release toxins as they evaporate into the air. Organic food eliminates most of the chemicals used to help grow the food and to kill the insects it attracts.

Buying locally grown, organic food also has a more positive environmental impact than buying industry-made food. For example, locally grown, organic food is less likely to be packaged in plastic packaging. As you have already learned, plastic waste does not decompose. Another benefit to the environment from buying locally is that the use of transportation resources is greatly reduced. Trucks transporting food over great distances use a large amount of fossil fuels and have a greater potential for emitting carbon dioxide into the atmosphere.

dumped into a pit with sloping sides. At the bottom of the pit is a grain elevator. The most common type of elevator is a bucket elevator, **Figure 27-21**. This elevator lifts the grain using a series of buckets or pans attached to a moving belt. The moving bucket digs into the pile of grain in the pit and moves the grain upward. When the grain reaches the top of the elevator, it is dumped in a shoot or pipe leading to the top of the silo.

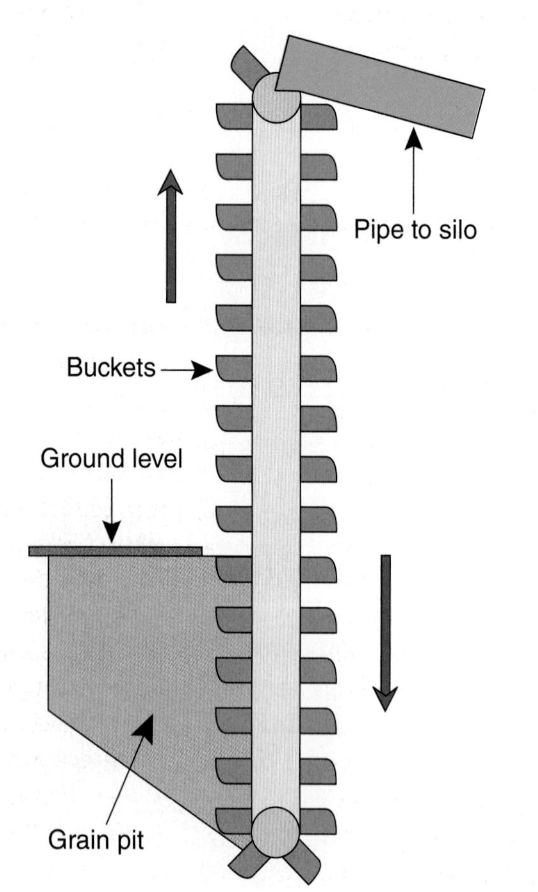

Goodheart-Willcox Publisher

Figure 27-21. How a bucket elevator operates.

Hay is stored in a dry location to protect it from rain and snow. The most common hay-storage structure is a hay barn, **Figure 27-22**. In very wet climates, the sides might be enclosed to provide additional protection.

Many vegetables and fruits are stored in climate-controlled (cold storage) buildings. These insulated buildings use cooling equipment to maintain the crop at an appropriate temperature. The crops are transported from these storage sites to processing plants throughout the country and world, as demand requires.

Andrew Orlemann/Shutterstock.com

Figure 27-22. A hay barn is simply a roof attached to long poles. It does not have enclosed sides or ends.

Hydroponics

Hydroponics is a unique way to grow plants in nutrient solutions without soil, **Figure 27-23**. All hydroponic systems contain:

- Nutrients supplied in liquid solutions.
- Porous materials (peat, sand, gravel, or glass wool) that draw nutrient solution from its source to the roots to support the plants.

Of the various types of hydroponic systems used, the most practical commercial method is a subirrigation system. In this system, the plants are grown in trays filled with a coarse material such as gravel or cinders. At specific times, the materials are flooded with nutrient solution. The solution is allowed to drain off after each flooding. Each flooding feeds the plants as they grow toward maturity.

Hydroponic methods can be used to grow plants in greenhouses. They can also be used in areas where the soil or climate is not suitable for the crop. For example, hydroponics is used to grow tomatoes during winter months.

Animal Husbandry

A major activity of agriculture is raising livestock, or animal husbandry, **Figure 27-24**. This includes the breeding and care of cattle, horses, swine (pigs), sheep, goats, horses, and poultry (chickens and turkeys). These animals are raised to provide meat, milk, or materials for clothing and for recreational purposes.

©iStockphoto.com/dstephens

Figure 27-23. Hydroponic lettuce being grown in a greenhouse.

zlikovec/Shutterstock.com

Figure 27-24. Raising livestock is a major agricultural activity.

Animal Husbandry Technology

Most livestock are raised on single-purpose farms. These farms include cattle ranches; dairies; and swine, horse, and poultry farms. Raising livestock involves various technologies. These include the following:

- Constructing and maintaining livestock buildings.
- Constructing and maintaining fences and fencing to establish feedlots and pastures.
- Operating and maintaining buildings and machines used to feed animals.
- Constructing, operating, and maintaining animal-waste facilities.

Livestock Buildings

Most livestock operations require specialized buildings. These buildings are used to house animals and feed-processing equipment. The types of buildings vary with the type of livestock being raised. Barns and sheds are built to protect animals from the weather. Some facilities house animals while they give birth. Hog-factory buildings contain animals in pens, as the animals are raised for meat. Poultry houses contain chickens and turkeys raised for food. Laying houses or henhouses hold chickens raised for egg production. Stables house horses raised for pleasure riding and racing. The processes for constructing agricultural buildings are the same as are used for other buildings. These processes are discussed in Chapter 17.

STEM Connection: Science

Genetic Engineering

The technology behind today's genetically engineered food products can be traced to scientific experiments conducted 150 years ago. In the mid-1800s, Gregor Mendel, an Austrian botanist and monk, began experimenting with the breeding of garden peas. He studied the traits of various pea plants. Mendel then started breeding and crossbreeding these plants to see what characteristics would be reproduced in succeeding generations. His findings on dominant and recessive genes led to the development of the science of genetics.

We now use the principles Mendel developed to produce better and more cost-effective foods. For example, because of genetic engineering, dairy cows injected with a growth hormone developed from genetically engineered bacteria produce greater amounts of milk. Beef cattle injected with the hormone have leaner meat. Genetically altered bacteria have also been used to make crops more insect resistant.

Genetic engineering has been used to great benefit. Some people have expressed concerns, however, about the practice. Can you think of a problem that might result from genetic engineering? If one does not immediately come to mind, you might want to check a local news source (newspaper, television, magazine). The pros and cons of various types of genetic engineering are frequently in the news.

Fences and Fencing

Farmers and ranchers construct barriers to separate their land from their neighbors' land, to separate fields, and to contain livestock. Fences are the most commonly used barriers.

Early barriers were made of rock, tree limbs, mud, and other natural materials. Later, smooth wire was adopted as a fencing material. It was made by drawing hot iron through dies (steel blocks with holes in them). The smooth wire did not always keep animals in enclosures, however. This led to the development of ***barbed wire***, a smooth wire with pointed wire added along its length, **Figure 27-25**. Barbed wire was such a popular innovation that more than 570 patents were issued for different designs. This relatively inexpensive fencing method opened the western United States to ranching, farming, and settlement.

Farmers and ranchers currently use several types of fences, including rail, barbed wire, woven wire, and electric. The type of fence selected depends on the livestock and crops being raised.

Rail fences are often used as border fences around farm buildings or homes. They are also

eltoro69/Shutterstock.com

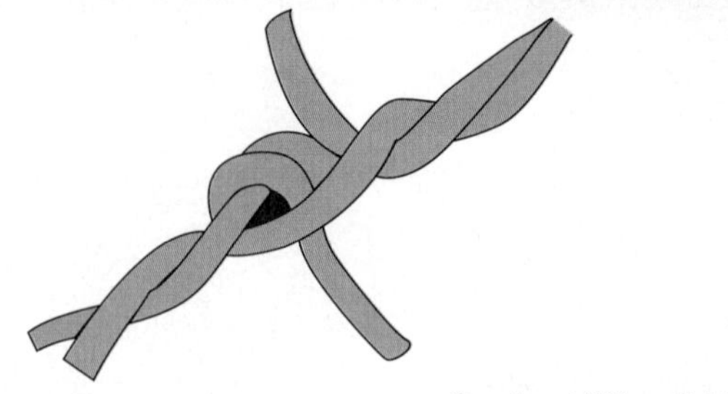

Goodheart-Willcox Publisher

Figure 27-25. Barbed wire was first used in the United States in the 1860s. Top—Barbed wire coils. Bottom—Reproduction of the 1874 patent drawing for barbed wire.

popular on horse farms. Rail fences are made from posts set in the ground with boards or rails attached between the posts. The common materials used are treated wood, painted wood, vinyl-coated wooden boards, and polyvinyl chloride (PVC) plastic.

Barbed wire fences have wood or steel posts with three to five strands of barbed wire attached to them. Posts are normally spaced 10′–12′ apart. Woven-wire fences also use posts, but these fences have woven wire attached to them. The woven wire consists of a number of horizontal smooth wires that vertical (stay) wires hold apart. The horizontal wires are arranged with narrow spacing at the bottom and wider spacing at the top. The height of the fence depends on the size and jumping ability of the contained animals. Typically, the fences are 26″–48″ tall.

Cable-wire fences consist of 3/8″ steel-wire cables stretched from one anchor post to another. Heavy springs are attached to one end of each cable to absorb shock. The other end is rigidly attached to another anchor post. Each cable passes through holes in a number of posts set between the two anchor (end or corner) posts. See **Figure 27-26**.

Electric fences are temporary or permanent fences that use electrical charges to contain animals in a field. This type of fence can be a separate fence or a strand of electric wire added to another type of fence.

Buildings and Machines for Feeding

Most livestock farms have livestock-feeding equipment and buildings. The buildings contain machines that grind and mix feed for the animals. As noted earlier, hay barns might be built for hay storage. Silos are used to contain and protect grain. Animals eat from feed troughs or bunkers. Water is provided using manufactured pumps and tanks.

US Department of Agriculture

Figure 27-26. The pipe at the top of this cable-wire fence adds strength to the entire fence.

Animal Waste–Disposal Facilities

A major challenge for large-scale livestock production is animal-waste disposal. The waste must be controlled so it does not pollute streams, lakes, and underground water. Livestock farmers must plan to properly collect, store, treat, and apply animal waste to land. They must identify sites for waste disposal and use appropriate land-application procedures.

This process requires technological actions, using a combination of structures and practices serving the animal-feeding operation. Typically, this requires collection both of animal wastes and of other kinds of wastes, including feed and litter. In many cases, a lagoon is built to contain these wastes. This lagoon is a confined body of wastewater that holds animal and other waste. This waste is periodically removed from the lagoon and applied to land.

The type of equipment used to apply livestock waste to land depends on the type and consistency of the waste. Dry litter can be applied with a box (manure) spreader. Lagoon waste is often handled in two forms: wastewater and slurry. Wastewater contains less than 2% solids. A slurry mixture is agitated sludge and wastewater. Many farmers apply wastewater using their regular irrigation systems. Slurries require special pumping equipment and sprinklers that have large nozzles.

Another technique is to inject the waste into the soil using special application equipment. Waste is hauled to the field and inserted into the soil using knife applicators, **Figure 27-27**. Animal manure can also be dried and sold to homeowners for lawn and garden fertilizer.

Aquaculture

Aquaculture is the process of growing and harvesting water (aquatic) organisms in controlled conditions. This growing and harvesting involves raising fish, shellfish, and aquatic plants. Aquaculture is considered an agricultural activity, but many differences exist between aquaculture and traditional agriculture. Aquaculture mainly produces protein crops. Traditional farming focuses on starchy staple crops.

Tonko Oosterink/Shutterstock.com

fabiodevilla/Shutterstock.com

Figure 27-27. Lagoon waste is driven out to the field and then applied using a special-purpose applicator. The applicator applies the waste by knifing it into the soil.

Aquaculture has waste-disposal problems not encountered in traditional livestock operations. In traditional livestock operations, animal waste can be disposed of off-site. This waste can be spread on fields as fertilizer. In contrast, animal waste in aquaculture stays in the ponds. Careful management is required to maintain the water quality in the ponds.

This growing and harvesting is also different from traditional fishing. Traditional fishing is called *capture fishing*. This fishing involves locating and capturing fish in their natural habitat (oceans, lakes, and rivers). In contrast, aquaculture requires human action in raising fish.

Aquaculture requires machines and equipment to perform a number of tasks. The controlled environment must be constructed. Most aquaculture is done in ponds dug in the earth. These ponds usually have water inlets and outlets. They are stocked with young fish. Fertilizer is often added to the ponds. This promotes the growth of food supplies for the fish. The fish are fed a diet promoting their growth. When they have reached market size, they are harvested. In a complete harvest, the pond is drained. All the animals are removed from the pond. In a partial harvest, some of the animals are removed using a net, **Figure 27-28**.

Beneda Miroslav/Shutterstock.com

Figure 27-28. This worker is harvesting carp at a fish farm.

Agriculture and Biotechnology

People have used scientific activities to improve plants and animals for hundreds of years. They have used selective breeding of livestock and cross-pollination of plants to create new or improved animals and plants. In recent history, technology has also been used for this goal. This technology, called ***biotechnology***, is the use of biological agents in processes to create goods or services. Biological agents are generally microorganisms (very small living things), enzymes (a special group of proteins), or animal and plant cells. They are used as ***catalysts*** (substance used to cause a reaction) in the selected process. The catalyst, however, does not enter into the reaction itself.

Biotechnology is a fairly new term, but is an ancient practice. In approximately 4000 BCE, the Egyptians used biotechnology to produce bread.

In more recent times, during World War I, scientists used an additive to change the output of a yeast-fermentation process. The result was glycerol, instead of ethanol. The glycerol was a basic ingredient in explosives manufacturing. During World War II, scientists used the fermentation process to produce antibiotics.

Biotechnology has had a major impact on modern agriculture. Agricultural biotechnology is being used to create, improve, or modify plants, animals, and microorganisms, **Figure 27-29**. Agricultural biotechnology is being used to produce new pest-resistant and chemical-tolerant crops that resist diseases through ***genetic engineering***.

Biotechnology is a major factor in increasing crop yields. This technology has helped produce more food per acre. Through development of cotton plants that are resistant to major pests, more plants grow to maturity, increasing yields.

Biotechnology can also be used to promote human health. The nutritional value of foods can be improved using genetic engineering. Genetic engineering is based on the fact that every living thing carries a genetic code (blueprint) precisely determining its traits. This code was linked to a major discovery called *recombinant DNA*. The structure of DNA is a double helix (spiral) structure. This spiral structure consists of a jigsaw-like fit of biochemicals. The two strands have biochemical bonds between them.

The DNA molecule can be considered a set of plans for living organisms. This molecule carries the genetic code determining the traits of living organisms. Scientists can use enzymes to cut the DNA chain cleanly at any point. The enzyme selected determines where the chain is cut. Two desirable parts can then be spliced back together. This produces an organism with a new set of traits. The process is called ***gene splicing***, **Figure 27-30**.

©iStockphoto.com/chemicalbilly

Figure 27-29. This scientist is collecting tissue from a genetically modified plant.

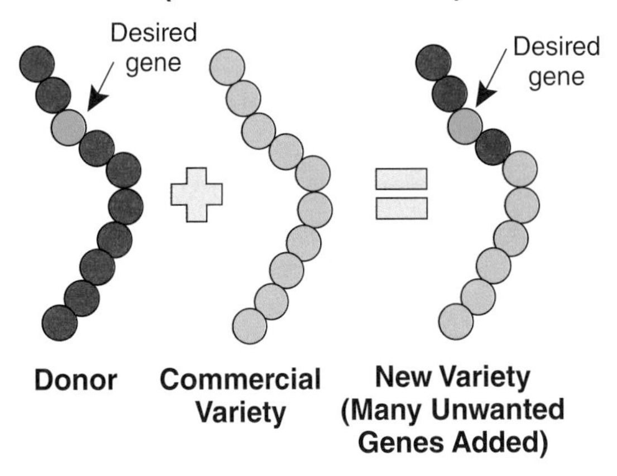

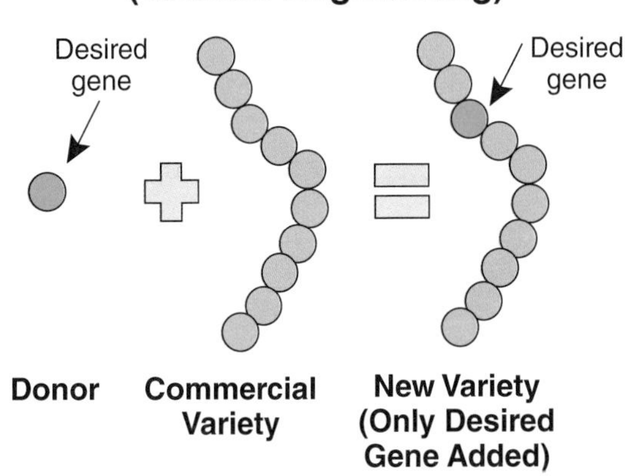

Goodheart-Willcox Publisher

Figure 27-30. Genetic engineering can introduce a desired gene into an existing plant.

This process allows scientists to engineer plants that have specific characteristics. For example, resistance to specific diseases can be engineered into the plant. This can reduce the need for pesticides to control insect damage to crops.

Gene splicing has received many headlines in newspapers and magazines. This activity is controversial. Some people think it makes life better. Others think we should not change the genetic structures of living things.

Food-Processing Technologies

Humans have a number of basic needs and wants. Many of these needs are met by agricultural technologies and related biotechnologies. A fundamental human need is having an adequate food supply. Food is needed to sustain life and promote growth. Without a proper supply of food, human life is not possible, **Figure 27-31**. Food sustains life by providing seven basic nutrients:

- **Carbohydrates.** Carbohydrates provide the body with its basic fuel.
- **Proteins.** Proteins are amino acids that provide the building material cells need to maintain their structure and growth.
- **Fats.** Most of the fats we eat are triglycerides, which are stored in muscle cells and fat cells and are burned as fuel.
- **Vitamins.** Vitamins are organic substances essential in the life of most animals and some plants. The human body needs many different vitamins including vitamin A, vitamin B, vitamin B_1 (thiamin), vitamin B_2 (riboflavin), vitamin B_3 (niacin), vitamin B_6, vitamin B_{12}, folic acid, vitamin C, vitamin D, vitamin E, vitamin K, pantothenic acid, and biotin.
- **Minerals.** Minerals are elements that the body needs to create specific molecules. Common minerals the body needs are calcium, chlorine, chromium, copper, fluorine, iodine, iron, magnesium, manganese, molybdenum, phosphorus, potassium, selenium, sodium, and zinc.
- **Fiber.** Fiber is a substance people eat that their bodies cannot digest. The three major fibers in food are cellulose, hemicellulose, and pectin.
- **Water.** The human body is about 60% water. About 40 ounces of water are lost each day and must be replaced. We replace this water through our consumption of moist foods and drinks.

PosiNote/Shutterstock.com

Figure 27-31. The nutrients in food keep your body working properly.

To supply a reliable source of food, people have developed food-processing technology. ***Food-processing technology*** uses knowledge, machines, and techniques to convert agricultural products into foods that have specific textures, appearances, and nutritional properties. These technological actions transform animal and vegetable materials into safe and edible food for humans. Also, food-processing technology includes processes used to make food more tasty and convenient to prepare. In addition, it involves actions used to extend the shelf lives of perishable foods.

Food-processing technology includes two basic types of processes:

- **Primary food processing.** These processes are technological processes that change raw agricultural materials into food commodities or ingredients.
- **Secondary food processing.** These processes are technological actions that convert food commodities and ingredients into edible products.

Primary food-processing technology produces the basic ingredients for food. For example, this type of technology changes wheat into flour. Primary food-processing technology changes animal carcasses into hamburger and transforms raw milk into pasteurized milk.

Secondary food-processing technology is used to make finished food products. For example, it converts flour and other ingredients into bread and changes hamburger and other ingredients into lasagna. Both of these types of processes are critical to our production of food.

Primary Food Processing

Most agricultural products are not eaten directly from the fields in which they are grown. They are processed using machines to change their form, appearance, or usefulness. Primary food processes include both material conversion (processing) and food preservation.

Material Conversion

A great deal of food we eat is processed in one way or another. This processing might be as simple as washing, grading, and packing fresh fruits and vegetables, **Figure 27-32**. In other cases, the produce might be preserved using one of several techniques. The fruits and vegetables might be canned or frozen for later use. They might be dried or cured. These and other preservation techniques are discussed later. All these actions constitute technology. They involve using machines to change the condition of the food.

Many food-processing techniques convert raw agricultural products into a different form. They might change the product so it has a different nature and appearance. For example, milk might be changed into cheese or butter. Corn and soybeans might be processed to produce cooking oils. Sugarcane and sugar beets might be refined into granulated sugar. These processes convert the original farm product into an entirely new ingredient.

Food can be processed in many ways. These ways can be grouped under the three headings of mechanical processing, thermal (heat) processing, and chemical processing. ***Mechanical processing*** uses machines to change the form of the food product physically. This processing might crush, slice, grind, or scrape the material to form a new ingredient. ***Thermal processing*** uses heat as the primary energy to convert a food. This processing might use the energy to melt, cook, blanch, or roast the material. ***Chemical processing*** uses energy to cause a basic chemical change in the food. Chemical processing includes pickling, fermenting, coagulating, or other similar actions.

Mechanical Processing: Flour Milling

Flour is finely ground grain, such as wheat, rye, corn, or rice. Wheat flour is the most commonly produced flour in the Western world. The composition of this flour depends on the type of wheat used and the milling processes employed. One commonly used milling process has four major steps, **Figure 27-33**:

1. **Grain receiving and cleaning.** Wheat is transported from farms to the mill, where it is unloaded and stored in bins. From the bins, the wheat is moved to cleaners, which remove all impurities. Magnetic separators remove iron and steel particles. Screens remove larger stones, sticks, and other materials. Air blasts remove lighter impurities. Special separators remove other grains (such as oats) and weed seeds. The clean wheat is stored in hopper bins to await milling.

Northwest Cherry Growers

Figure 27-32. Food processing can be as simple as receiving, washing, and grading produce.

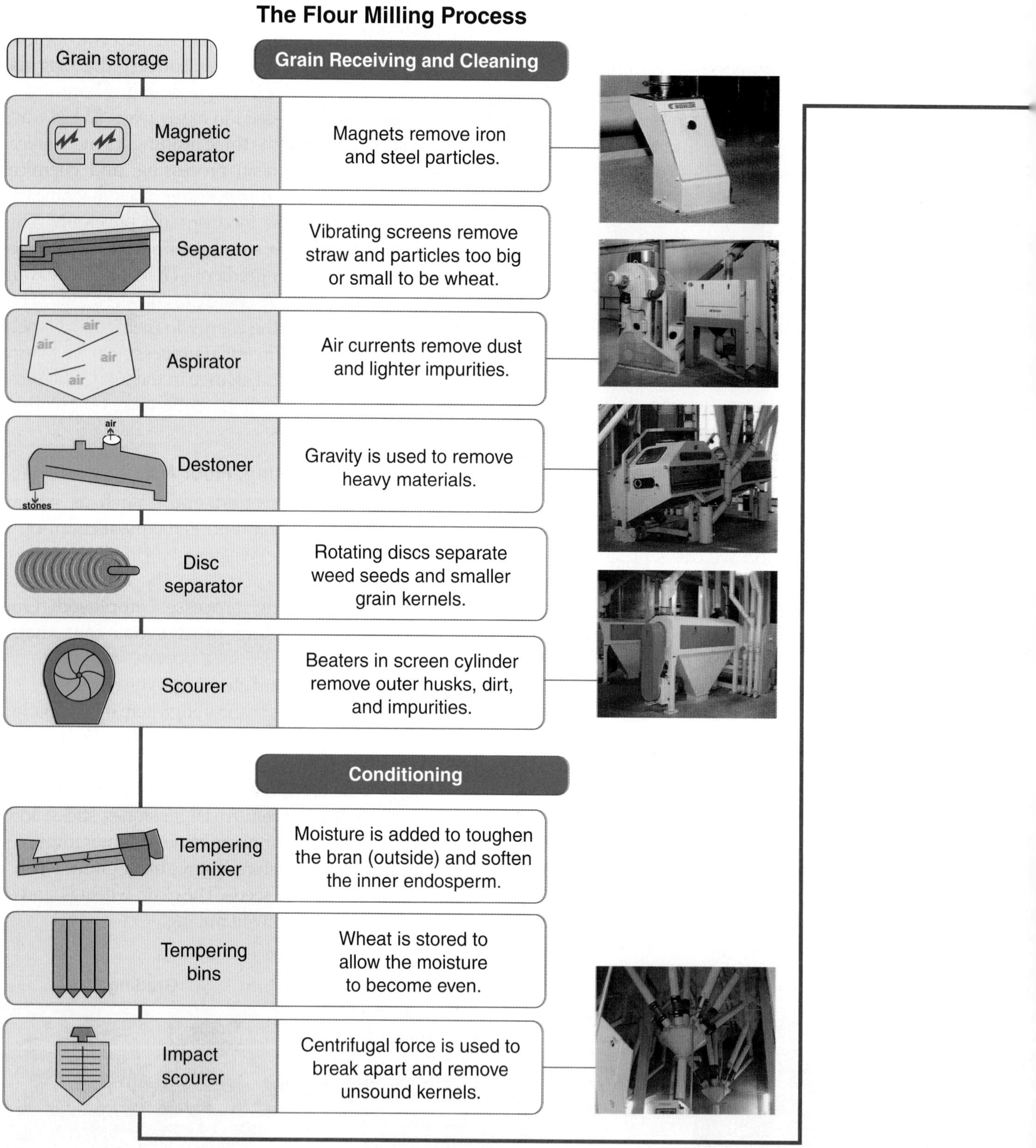

The Flour Milling Process

Grain storage

Grain Receiving and Cleaning

Magnetic separator	Magnets remove iron and steel particles.
Separator	Vibrating screens remove straw and particles too big or small to be wheat.
Aspirator	Air currents remove dust and lighter impurities.
Destoner	Gravity is used to remove heavy materials.
Disc separator	Rotating discs separate weed seeds and smaller grain kernels.
Scourer	Beaters in screen cylinder remove outer husks, dirt, and impurities.

air

stones

Conditioning

Tempering mixer	Moisture is added to toughen the bran (outside) and soften the inner endosperm.
Tempering bins	Wheat is stored to allow the moisture to become even.
Impact scourer	Centrifugal force is used to break apart and remove unsound kernels.

Buhler Group; North Dakota Mill

Figure 27-33. The flour-milling process involves four steps.

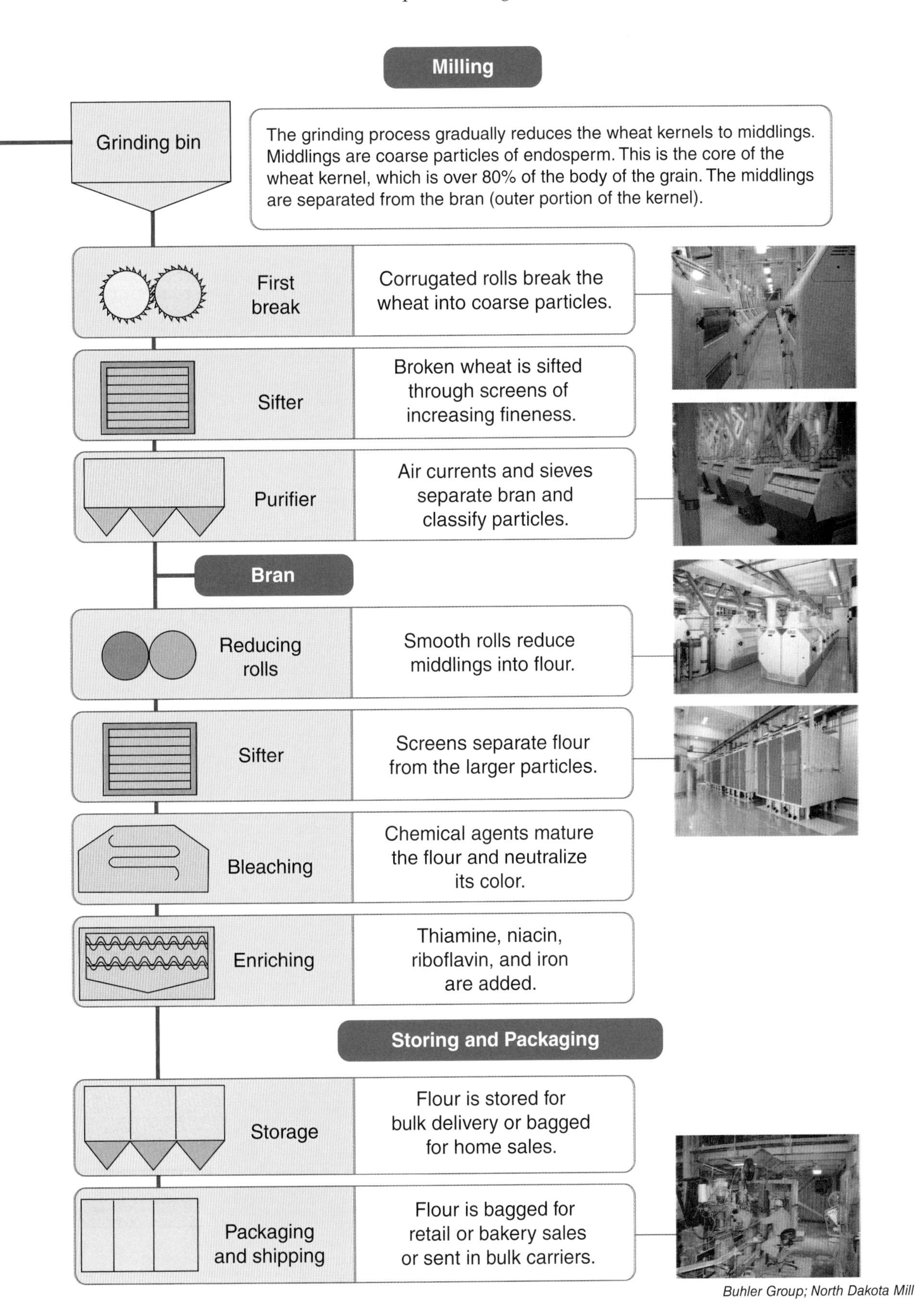

Figure 27-33. *(Continued)*

2. **Conditioning.** The clean wheat is prepared for milling by a conditioning process. Water is added to make it easier to separate the bran (outside hull of the grain) from the endosperm (flour portion) of the wheat kernel. The wet wheat is stored in tempering bins. From the tempering bins, wheat is moved to the flour mill for ***milling*** (grinding or processing).
3. **Milling.** The conditioned wheat is milled using roller mills, sifters, and auxiliary equipment. The first mill is called the *first break*. This mill uses corrugated rolls to break wheat into coarse particles. A sifter and purifier then separate the bran and classify the remaining particles (called *middlings*) by size. Middlings are the coarse particles of endosperm. ***Endosperm*** is the core of the wheat kernel. The middlings move to reducing rolls, which change them into flour. The flour moves through additional sifters and reducing rolls until a uniform product emerges. This product is bleached to produce even-colored flour. Vitamins and minerals might be added to produce enriched flour.
4. **Storage and packaging.** The milled, bleached, and enriched flour is conveyed to steel tanks. From there, it might be moved in tank cars to bakeries or other commercial customers. Flour for retail customers is packaged in bags and sent to grocery stores.

Thermal Processing: Coffee Roasting

Coffee is the fruit of a tree. Originally, coffee was a food, not a drink. People mixed coffee beans with animal fat to produce a high-energy food. Later, the beans were roasted and ground to be used to make a drink.

Coffee beans are found inside the red cherry-shaped fruit of a coffee tree. Inside the fruit are two coffee beans. They are referred to as the *greens*.

The coffee beans are picked, dried, and shipped to the roasting factory, **Figure 27-34**. On arrival, the coffee is inspected for freshness and color. The inspected coffee beans are sent to a coffee pulper. This machine contains a metal cylinder with stripping knobs that remove the husks from the coffee beans. The beans are placed in a fermentation tank, where the film covering them is removed. They are then washed and dried. In the next stage, the parchment cover is removed from the green beans. The coffee is then graded and classified according to size, weight, and quality. The graded green coffee is then roasted, **Figure 27-35**. Basically, two types of coffee-roasting techniques exist: the traditional

patpitchaya/Shutterstock.com

Figure 27-34. These coffee beans are being sun dried before processing.

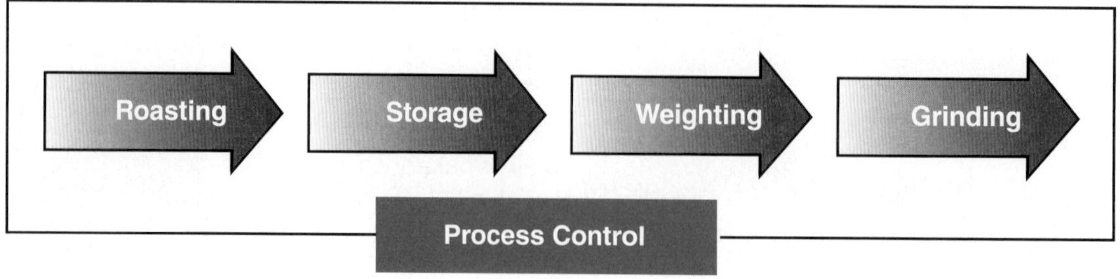

Goodheart-Willcox Publisher

Figure 27-35. The four steps in the coffee roasting-and-grinding process.

drum-roasting, or barrel-roasting, technique and air roasting.

In the drum-roasting process, the green coffee beans are heated to a temperature of about 550°F in large rotating drums. The beans tumble in the cylinder and are roasted.

The air-roasting technique roasts the coffee beans at the desired temperature using hot air. They are constantly moving in the air blast, which allows them to roast evenly.

In both cases, the first stage of roasting turns the green beans to a yellowish color as they dry. They begin to smell similar to toast or popcorn. In the second step, called the *first crack*, the beans double in size. They pop very similarly to popcorn and become a light-brown color.

In the third step, the beans begin to brown as their oils start to emerge. A chemical reaction caused by the heat and oils produces the flavor and aroma of coffee. A second pop occurs several minutes later. At this point, the beans are fully roasted and can be moved to storage.

The roasted coffee moves out of storage and is weighed. This coffee is then ground using mechanical grinding machines. Grinding produces the most flavor from the bean when the coffee is prepared. The ground coffee is packaged in airtight bags or cans to help maintain its freshness.

Chemical Processing: Cheddar Cheese Manufacturing

Cheese is a dairy product made from milk protein. This product is a coagulated, compressed, and ripened curd of milk that has been separated from the whey (moisture in milk). Cheese is made in hundreds of types and forms, **Figure 27-36**. The following describes how cheddar cheese is made using a chemical conversion process:

1. **Receiving milk.** The cheese-making process starts with milk being delivered from local dairies. The milk moves from the farm to the cheese factory in refrigerated milk trucks. On arrival, it is tested for quality, flavor, and odor.
2. **Cooking.** The fresh milk is pasteurized to destroy any harmful bacteria, and is then pumped into cooking vats. Special starter cultures (bacteria) are added to the warm milk and change a small amount of the milk sugar into lactic acid. Coloring is added to produce a consistent cheese color. Finally, a substance to promote coagulation and firm the curd is added.
3. **Cutting the curd.** After about 30 minutes, the vat of milk sets up. In this action, a soft curd, consisting of casein and milk fat, is formed. Stainless steel knives cut the soft curd into 1/4″ pieces. The temperature in the cooking

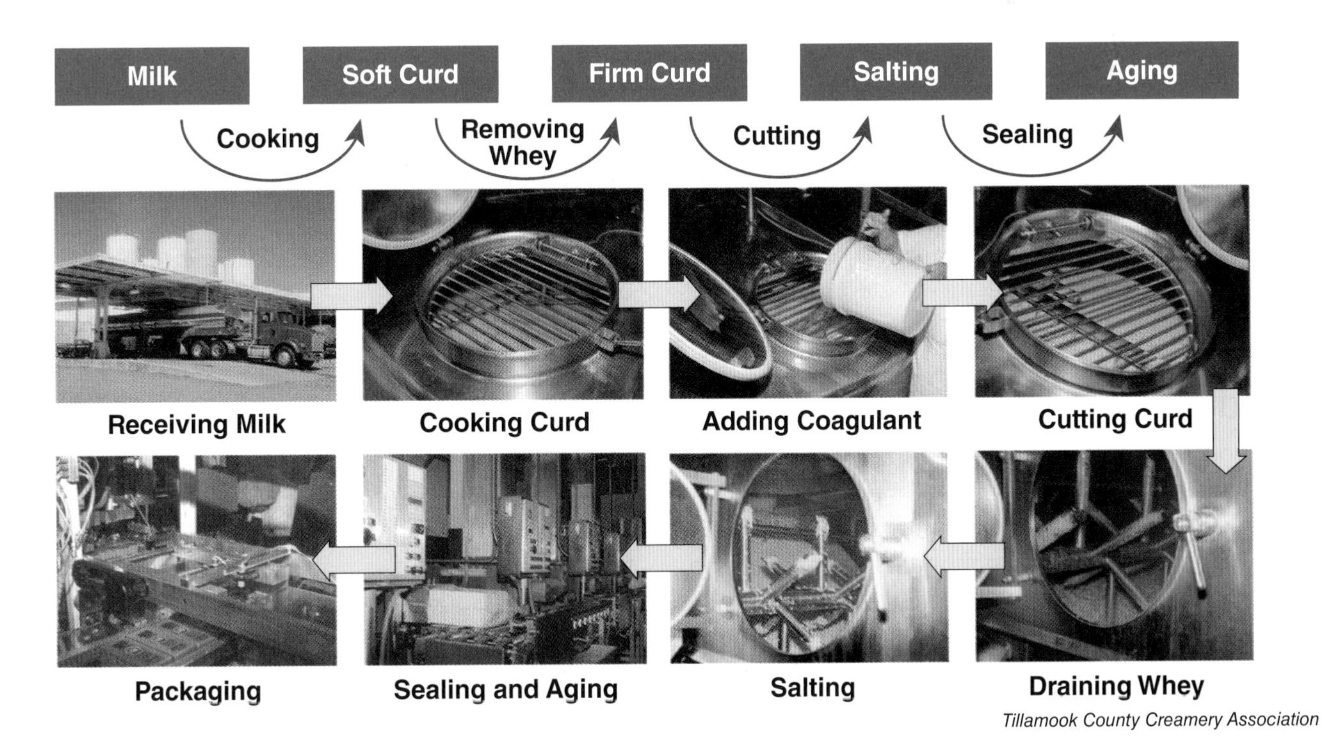

Tillamook County Creamery Association

Figure 27-36. In the cheese-making process, milk curds are cooked, separated from the whey, cut, and sealed.

vat is then raised to about 100°F to drive out moisture and firm the curd. The liquid part in the vat is called *whey*.

4. **Making cheddar.** The curds and whey are pumped to the cheddar-making machine. The whey is removed, and the curd is matted on a wide belt inside the machine. During the cheddaring process, a chemical change occurs. The curd particles adhere to each other, giving them a stringy consistency. When the proper acidity is reached in the curd, the cheddar mat is forced through the curd mill, which chops the large slabs into small, 3″ pieces.
5. **Salting.** The curd chunks are passed through a salting chamber. In this machine, a thin layer of salt is applied to the surface of each curd. The salted curd is stirred, and it absorbs the salt. When the salt is completely absorbed, the curds are transferred to pressing towers, where a vacuum draws off the excess moisture in the cheese.
6. **Sealing and aging.** After 30 minutes in the pressing towers, large blocks of cheese are cut from the bases of the towers. These blocks are placed in plastic bags. A vacuum draws out air, and the package of cheese is sealed. The block is now contained in an airtight, moisture-proof bag. The sealed blocks are transported to a cooling room and held for about 24 hours at temperatures near 38°F. From there, they are placed on pallets and stored for aging and curing at about 42°F.
7. **Packaging.** After aging for a specific number of days, the cheese blocks are removed from their aging bags. Each batch of cheese is tested and graded for quality. See **Figure 27-37**. The blocks are cut into specified sizes and packaged for various markets.

Wisconsin Milk Marketing Board

Figure 27-37. This inspector is checking cheese for unwanted odors.

Food Preservation

Food-preservation technologies are used to keep food safe for consumption. They can reduce or eliminate the effects of bruising and insects. Preservation controls microorganisms such as bacteria, yeast, and molds. These techniques can destroy unwanted enzymes (proteins) present in raw foods. The enzymes cause chemical and physical changes that naturally occur after harvesting and contribute to food spoilage.

Food-preservation technologies also help eliminate moisture or temperature conditions that encourage the growth of microorganisms. Microorganisms can produce unwanted changes in the taste and appearance of the food. They can also cause food-borne illness.

Food processing can improve the nutritional value of foods. For example, high heat can destroy unwanted factors present in many foods. Prolonged boiling destroys the harmful lectins present in such foods as red kidney beans. However, some preservation techniques can reduce the quality of food. They can produce food that has lower nutritional value and poorer texture and flavor. For example, even freezing foods to preserve them has trade-offs, depending on the time between being picked and being frozen.

A number of different technologies are used for food preservation. These technologies include drying, curing and smoking, canning, aseptic packaging, refrigeration and freezing, controlled-atmosphere storage, fermentation, pasteurization, irradiation, and preservatives.

Drying

Drying is the oldest method used to preserve food. Drying removes water that microorganisms need to grow. These microorganisms can cause food spoilage. There are three commonly used drying methods. Sun drying allows foods to dry naturally

©iStockphoto.com/ivanstar

Figure 27-38. These chili peppers have been sun dried in New Mexico.

in the sunlight. See **Figure 27-38**. Hot-air drying exposes food to a blast of hot air. Freeze-drying uses a vacuum chamber to draw water out of frozen foods. In freeze-drying, water is removed from the food by a process in which ice changes from a solid directly to a vapor, without first becoming a liquid.

Dried foods keep well because the moisture content is so low that spoilage organisms cannot grow. Drying is used on a limited range of foods, including milk, eggs, instant coffee, and fruits. This method will never replace canning or freezing as the primary way to preserve foods. Canning and freezing retain the taste, appearance, and nutritive value of fresh food better than drying does.

Curing and Smoking

Curing and smoking are techniques used to preserve meat and fish, **Figure 27-39**. ***Curing*** involves adding a combination of natural ingredients to the meat, such as salt, sugar, spices, or vinegar. This process is used to produce products such as bacon, ham, and corned beef. ***Smoking*** is a process that adds flavor to meat and fish, while preserving them. This process involves slowly cooking the meat or fish over a low-heat fire.

©iStockphoto.com/davidf

Figure 27-39. This salami is being cured in warm, humid conditions, encouraging growth of a bacteria needed in the fermentation process.

These techniques preserve food by binding or removing water. This causes the water to be unavailable for the growth of microorganisms. Both curing and smoking produce a distinctive flavor and color in food. In many cases, these techniques eliminate the need for refrigeration.

Canning

Canning is a preservation method that seals food into glass jars or metal cans. This method preserves food by heating it in a vacuum-sealed container. The process removes oxygen from the container, kills microorganisms in the food, and destroys enzymes that can spoil the food.

During the process, the can is filled with food. See **Figure 27-40**. Air is removed to form a vacuum, and the container is sealed. The food is heated and then cooled to prevent the food from becoming overcooked. Low-acid foods, such as meats and most vegetables, are heated to 240°F–265°F. Foods with higher acid levels, such as fruits, are heated to about 212°F.

In many cases, the food is blanched before canning. In this process, the food is heated with steam or water for a short period of time, and then quickly cooled. Heating inactivates enzymes that can change the food's color, flavor, or texture.

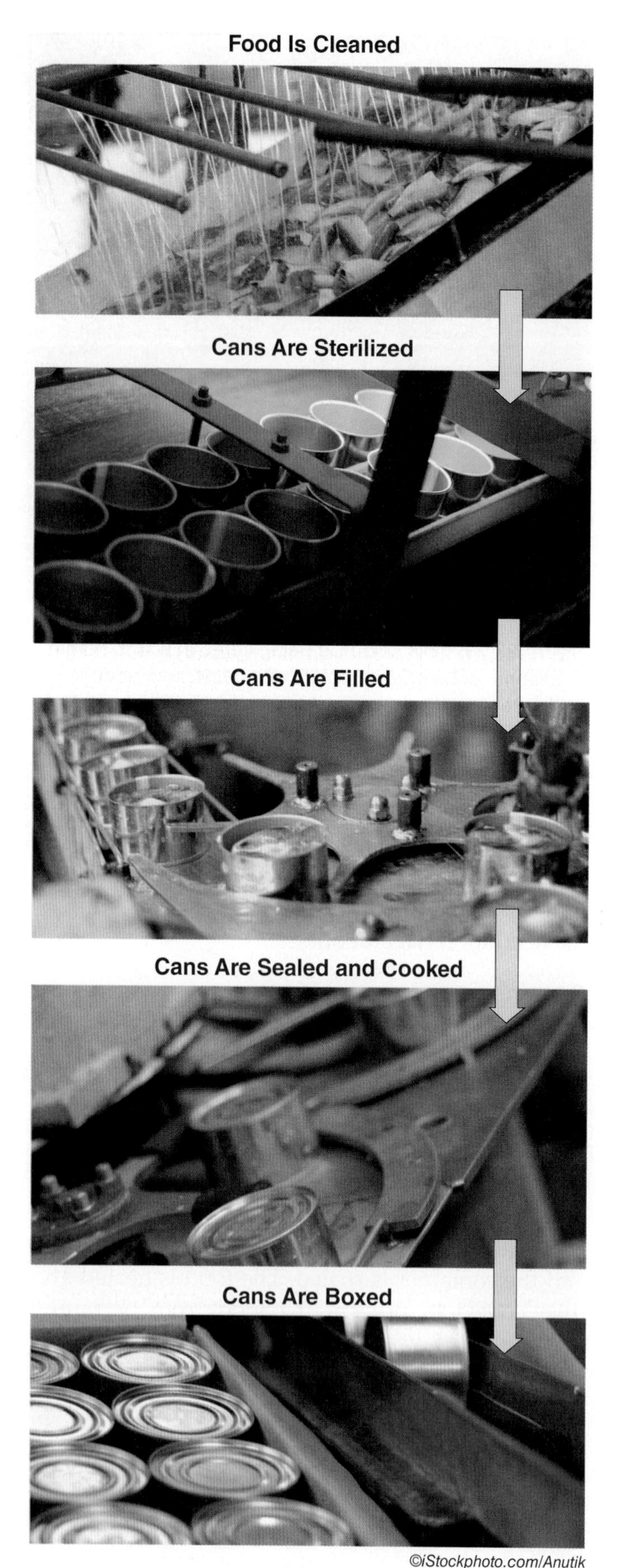

©iStockphoto.com/Anutik

Figure 27-40. The major steps in the canning process.

Blanching reduces the volume of the vegetables by driving out gases.

Canning is used to preserve a wide variety of foods, including fruits, vegetables, jams and jellies, pickles, milk, meats, soups, sauces, and juices. Canned foods can be stored without refrigeration for a long time. Canning affects the color, texture, flavor, and nutrient content of foods because of the high temperatures used.

Aseptic Packaging

Aseptic packaging is commonly used for packaged milk and juice because they keep for long periods of time without refrigeration. This process, similar to canning, uses heat to sterilize food. Unlike canning, however, the package and food are sterilized separately.

The containers are sterilized with hydrogen peroxide, rather than with heat. This allows the use of lower-cost containers that heat sterilization would destroy. The most commonly used containers are plastic bags and foil-lined cartons. The food is sterilized more rapidly and at lower temperatures than in canning. This allows the food to have better flavor and to retain more nutrients.

Refrigeration and Freezing

Storage at low temperature slows many of the reactions that cause foods to spoil. Two low-temperature preservation techniques are refrigeration and freezing. Refrigeration maintains foods at temperatures from 32°F–40°F. ***Freezing*** keeps the foods at or below 32°F.

Refrigeration does not cause chemical or physical changes to food. Still, it preserves foods for a short period of time. Foods that should be refrigerated include most types of milk, eggs, meats, fish, and some fruits and vegetables.

Freezing is used on a wide range of foods, **Figure 27-41**. This process allows foods to be stored for longer periods because the process reduces enzyme activity and the growth of microbes. The ice crystals that form in the food disrupt the food's structure, however. When thawed, the food has a softer texture. Also, foods deteriorate rapidly after thawing because organisms in the food attack the cells injured by the ice crystals.

Making and Freezing Commercial French Fries

J.R. Simplot

Step 1 Getting the potatoes—Potatoes come directly from the producers or from large warehouses. They are checked for solids content, grade, and sugar content.
Step 2 Peeling the potatoes—Batches of potatoes are put into a large, hot, pressurized tank. After a set time, the pressure is quickly released, and the potato skin is said to "fly off." Potatoes are removed from the tank and sprayed with high-power water jets to remove any still clinging peels.
Step 3 Inspecting the potatoes—Peeled potatoes pass an inspection line where sorters remove any defective potatoes.
Step 4 Cutting the potatoes—Potatoes go through a pump that propels them (at about 50 mph) at stationary blades that chop the potatoes into strips.
Step 5 Inspecting the strips—Small parts left from the outer edges of the potatoes are removed. Remaining strips are automatically inspected, and strips with black bits are removed. This is done at the rate of about 1000 strips and chips each second!
Step 6 Blanching—Inspected strips are blanched on a moving conveyor chain that carries them through a large vat of hot water. This process removes excess sugars and gives the strips a consistent, uniform color.
Step 7 Drying—Blanched strips are partially dried as they are conveyed past blasts of hot air from both the top and bottom.
Step 8 Partial frying—Strips are cooked for about 90 seconds in hot oil. The process is called *par fry*, or *partial fry*.

J.R. Simplot

Step 9 Deep freeze—Strips enter blast freezing, where the French fries travel down a conveyor surrounded by air cooled to about –40°F.
Step 10 Packaging—After freezing, the product is bagged or boxed and shipped to customers.

Figure 27-41. Freezing French fries allows them to be stored for a longer period of time.

Controlled-Atmosphere Storage

Fruits and vegetables can be stored in sealed environments where temperatures and humidity are controlled. In some cases, the composition of the air around the fruit is changed. Oxygen levels are often decreased, while carbon dioxide levels are increased. This change reduces the chance of food spoilage. This controlled environment slows the reactions leading to decomposition and decay. Controlled atmospheres can extend the storage lives of fruits and vegetables by several months.

Fermentation

Fermentation uses microorganisms to break down complex organic compounds into simpler substances. Such compounds as alcohol and acids are produced in fermentation. These compounds act as preservatives that reduce further microbial growth.

In some cases, fermentation spoils the product. In other cases, it is desirable. In these cases, microorganisms are added to foods. For example, in the manufacture of yogurt and cheese, bacteria convert a sugar found in milk (lactose) into lactic acid. Fermentation is also used to produce cheese, yeast bread, soy sauce, and cucumber pickles.

Pasteurization

Pasteurization uses heat to kill harmful microorganisms. Pasteurization is commonly used for milk and fruit juices. Milk is usually heated to 145°F for 30 minutes to kill microorganisms, **Figure 27-42**.

In newer processes, this technique is called *ultra-high-temperature (UHT) pasteurization*. UHT pasteurization uses higher temperatures and shorter heat times for foods in sterile packaging. The foods are heated to 280°F for two to four seconds. This rapid sterilization allows the food to have better flavor and retain more nutrients.

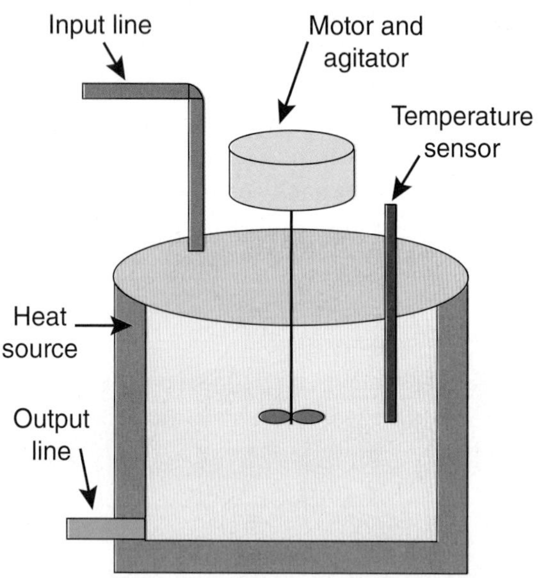

Goodheart-Willcox Publisher

Wisconsin Milk Marketing Board

Figure 27-42. Batch pasteurizer operation.

Irradiation

Irradiation uses gamma rays or x-rays to kill most molds and bacteria that might be in the food. This technology is also used to delay the ripening of fruits and the sprouting of vegetables. The delay allows the produce to be stored for longer periods of time. In this process, the food passes through a chamber where it is exposed to high-energy rays. Irradiation involves little heating and does not change the taste, texture, or nutritional value of food.

Preservation

In preservation, ***preservatives*** (chemicals) are added to food in small amounts. These preservatives act in two ways: they delay food spoilage, and they ensure that the food retains its quality. The first method uses sugar (jams and jellies), vinegar (pickles and meats), and salt (hams and bacon). The second method uses acids, sulfur dioxide, and other agents to slow the growth of microorganisms in food.

Secondary Food Processing

Secondary food-processing technology converts the ingredients produced in primary food-processing activities into edible products. This technology involves combining and processing ingredients and food products to change their properties. There are thousands of products created through secondary food-processing activities.

These products can be grouped in a variety of ways, **Figure 27-43**. One common grouping is:

- Starch products
  - Bread
  - Cakes and cookies
  - Crackers
  - Pasta
- Dairy products
  - Milk, buttermilk, and cream
  - Butter
  - Cheese, cottage cheese, and yogurt

Hurst Photo/Shutterstock.com

Figure 27-43. Food products can be grouped into seven major categories.

- Meat, poultry, and fish products
  - Fresh meat, fish, and poultry
  - Processed products (hot dogs and lunch meats, for example)
- Fruit and vegetable products
  - Fruit preserves and fillings
  - Vegetable dishes
- Oil products
  - Cooking oils
  - Margarine and low-fat spreads
  - Salad dressings
- Sweets and candies
- Drinks

These products are created in two steps: food-product development and food-product manufacture.

Food-Product Development

Many products can be made using similar basic ingredients. For example, flour and water are the basic ingredients for bread, biscuits, cookies, and pasta. Other ingredients are added to create the uniqueness of each product.

Companies use the food-product development process to create an array of new or modified food products. Food-product development is done in a series of steps that moves products from an original idea to the consumer, **Figure 27-44**:

1. **New food product–idea identification.** A company identifies consumer trends and eating patterns. They study new products on the market. The company investigates advancements in food-preparation techniques and uses that information to brainstorm possible new products and line extensions (products building on existing products).
2. **Idea development.** The company develops a number of new recipes. Ingredients are specified and costs are established.
3. **Small-scale testing.** The company makes several versions of the product, usually using slightly different ingredients or processes. The products might be prototypes in the company's test kitchens. Company employees or small focus groups are asked to test the products and provide feedback.

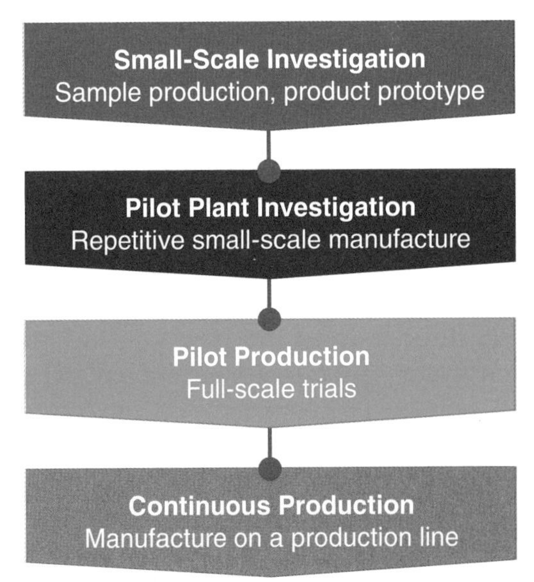

Goodheart-Willcox Publisher

Figure 27-44. Companies follow a series of production steps when creating a new food product.

4. **Sensory evaluation.** The company asks trained personnel to comment on the appearance, taste, smell, and texture of the products.
5. **Product modification.** The company alters the composition and recipe of the product to incorporate the results of the initial test and sensory evaluations.
6. **Pilot plant production.** The company produces the item using small versions of the equipment that can be used in full-scale manufacture.
7. **Consumer and sensory evaluation.** The company tests the output of the pilot production to determine its potential for profitability.
8. **Product specification.** The company develops the final product specifications, including final ingredients and methods of production. These specifications are used in producing each batch of the product.
9. **Pilot production.** The company readies the production plant for the product. Production tests are conducted to determine the effectiveness of the production processes.
10. **Continuous or large-scale product production.** The company releases the product for large-scale production. The actual manufacture is often done in major processes, such as measuring, mixing, and cooking. The company carefully controls each of these processes to maintain product quality, promote food purity and safety, and reduce waste.
11. **Product packaging.** The company places the product in appropriate packages. Individual packages are placed in shipping containers that are ready to be sent to the wholesaler or retailer.
12. **Product promotion.** The company announces the availability of the product through advertising and product promotion.

Not all companies develop products using this procedure. For example, some companies move products directly from the "test kitchen" stage to trials on factory production equipment. The decision on how to proceed depends on competitive pressure, economic factors, the size of the production facility, and the type of product being developed.

Food-Product Manufacture

Making a consumer food product involves a number of manufacturing processes. The procedure most foods go through includes some of the following:

1. **Storing.** This involves containing and protecting the condition of the raw materials.
2. **Cleaning.** This involves removing foreign matter from ingredients used in the product.
3. **Sorting and grading.** This involves assessing the quality of the ingredients to be used in making the product, **Figure 27-45.**
4. **Measuring.** This involves weighing and measuring the wet and dry ingredients to be used in the product.
5. **Size reduction.** This involves trimming, slicing, and crushing the basic ingredients.
6. **Compounding.** This involves mixing and combining the ingredients for the product.
7. **Heat processing.** This involves cooking and cooling the mixture to form the product.

These technological processes produce a nutritional and easy-to-use consumer product. Initially, food-processing technology produced a food unit, such as bread, pasta, crackers, or stewed tomatoes. These products were combined at home to prepare dishes and complete meals. Now, however, people are demanding food that is easier and quicker to prepare. Today, food-processing technology provides both single unit–type foods and finished dishes. These finished dishes might be a meal entrée, such as lasagna, or a complete meal, such as a frozen dinner.

Secondary Food Processing Example: Pasta Making

The goal of pasta production is to convert flour into an edible product. Pasta production includes

Vladimir Nenezic/Shutterstock.com

Figure 27-45. These raspberries are being sorted and graded before entering the manufacturing line.

blending the ingredients to form dough, kneading and mixing the dough, forming the final product, and then drying it, **Figure 27-46**.

1. **Blending the ingredients.** The basic ingredients of pasta are durum wheat semolina and water. These ingredients are measured and added to a batch mixer. Additional ingredients, such as eggs, spinach, tomato powder, and flavorings, can be added at this time. They are mixed under vacuum to distribute moisture evenly throughout the flour. The semolina starts to absorb water and forms a crumb structure.
2. **Kneading and mixing the dough.** The lumps of semolina are kneaded (pressed together) to form a basic dough. This action forces the crumb structures together, which fuses the particles. The result is consistent dough.
3. **Forming the product.** An extruder is used to form the pasta shape. This machine uses the screw principle to push dough through a shaped die. These molds have round or oval holes through them to produce rods for products, such as spaghetti. A steel rod can be placed in the center of each hole to form a tube-shaped product, such as macaroni. Grooves in the screw scoop up the dough. The rotating screw pushes the dough toward the die opening. As the material passes through the die, it is shaped and formed. Cutters cut the developing tube of product to length. Other pasta products, such as noodles, are formed in flat sheets. The shapes are stamped out of the sheets using shaped cutting dies.
4. **Drying the product.** The cut products pass through long tunnel dryers to remove excess moisture. Inside the dryers, very hot, moist air removes the moisture. The product is cured.

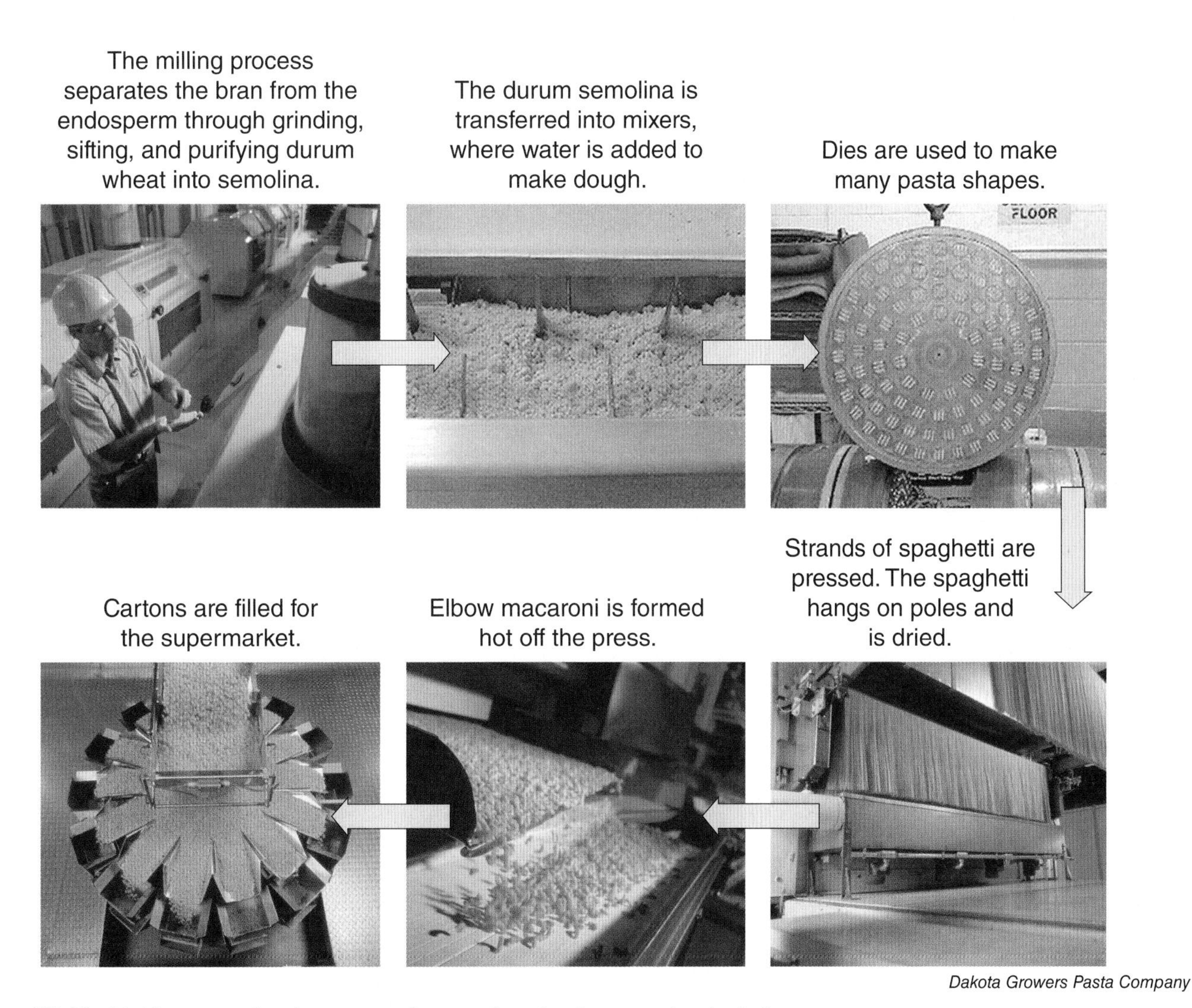

Figure 27-46. Making pasta involves several secondary food-processing techniques.

Summary

- ✔ Agriculture includes two major activities taking place on farms and ranches: crop production and animal husbandry.
- ✔ Crops produced in the United States include grains, vegetables, fruits, berries, nuts, forage crops, and nonfood crops.
- ✔ Crop production involves the use of various types of machinery and equipment: power or pulling equipment (tractors), tilling equipment, planting equipment, pest-control equipment, irrigation equipment, harvesting equipment, transportation equipment, and storage equipment.
- ✔ Hydroponics is a way to grow plants in nutrient solutions without soil.
- ✔ Raising livestock involves technologies such as constructing and maintaining livestock buildings and fences; operating and maintaining buildings and machines used to feed animals; and constructing, operating, and maintaining animal-waste facilities.
- ✔ Agriculture also employs biotechnology, which applies biological organisms to production processes.
- ✔ Food provides seven basic nutrients: carbohydrates, proteins, fats, vitamins, minerals, fiber, and water.
- ✔ Food-processing technology involves converting agricultural products into foods that have specific textures, appearances, and nutritional properties.
- ✔ There are two basic types of food-processing technology: primary-food processing technology and secondary-food processing technology.
- ✔ Primary food processing involves two steps: material conversion and food preservation.
- ✔ Secondary food processing involves the development of the food products and their manufacture.
- ✔ Food-product development is done in a series of steps that moves products from an original idea to the consumer. The steps vary among companies and deciding which steps to follow is based on a variety of factors.
- ✔ Food-product manufacture usually involves storing, cleaning, sorting and grading, measuring, size reductions, compounding, and heat processing.

Check Your Engineering IQ

Now that you have finished this chapter, see what you learned by taking the chapter posttest.
www.g-wlearning.com/technologyeducation/

Test Your Knowledge

Answer the following end-of-chapter questions using the information provided in this chapter.

1. Define *agriculture*.
2. Name one example each of how science and technology are used in agriculture.
3. What are the two types of agriculture?
4. Members of the grass family that have edible seeds are called _____.
5. Grass and hay crops grown for animal feed are called _____.
6. The four agricultural processes that use technology are planting, growing, harvesting, and _____.
7. List five types of farm equipment.
8. _____ is a liquid, powder, or pellet containing important chemicals that encourage and support plant growth.
9. Identify each type of irrigation system listed here.
   A. Uses small ditches built between rows of plants to guide water as it moves from one end of a field to the other.
   B. Produces artificial rain to water crops.
   C. Uses ditches or pipes to bring water to one end of a field, where it is released through holes along the lengths of the ditches or pipes.
   D. Delivers water slowly to base of plants.
10. Growing plants in nutrient solutions without soil is called _____.
11. Name one kind of livestock farm.
12. Name one kind of fencing used to contain livestock.
13. What is the purpose of a silo?
14. The practice of growing and harvesting fish in controlled conditions is called _____.
15. Define *biotechnology*.

16. The process of producing new pest-resistant and chemical-tolerant crops that resist diseases is known as _____.
17. Define *food-processing technology.*
18. Food processes that make basic food ingredients are called _____. Food processes that make finished food products are called _____.
19. What are the major steps in milling flour?
20. Coffee roasting is a type of _____ processing.
    A. mechanical
    B. thermal
    C. chemical
    D. engineered

Matching: Match each description with the correct food-preservation technology. (Answers may be used more than once.)

21. Using x-rays to kill mold and bacteria.
22. Heating food in a vacuum-sealed container.
23. Removing water by placing food in sunlight.
24. Storing food in sealed areas with controlled environments.
25. Using chemicals to preserve food.
26. Treating meat with salt and adding flavor.
27. Storing food at low temperatures.
28. Using heat to kill microorganisms in milk and juices.
29. Sterilizing a package and then placing sterilized food in it.
30. Using microorganisms to break down food's structure.
31. Using blanching to drive out gases.
32. Removing water by exposing food to hot air.

A. Drying
B. Curing and smoking
C. Canning
D. Aseptic packaging
E. Refrigeration and freezing
F. Controlled-atmosphere storage
G. Fermentation
H. Pasteurization
I. Irradiation
J. Preservation

33. The step of altering the composition of a food product to incorporate the results of the initial test is called _____.
34. Producing a scaled-up food product using small versions of the equipment is called _____.
35. List the major manufacturing processes used in making a food product.

STEM Applications

1. Farmers use many types of equipment during the growing season. Some equipment is used throughout the growing season. Other equipment is used only part of the season. Using a table similar to the one shown, list the names of various pieces of farming equipment (plow or combine, for example) in the season or seasons it is used.

Spring	Summer	Fall

2. Genetic engineering has been controversial when used to modify food crops. Research the topic of genetically modified products (such as corn) and make two lists: one listing the benefits and one listing the risks. Compare the lists and try to form an opinion on the subject.

Engineering Design Challenge

Dehydrating Food

Background

Drying is one of the oldest food-preservation methods. Originally, it was done by salting food and then drying it in the sunshine, in open rooms, or on stoves. In 1795, the first dehydrator was developed to dry (dehydrate) fruits and vegetables. Today, dried foods have become a multimillion-dollar industry.

Foods can be dehydrated using sunshine, a conventional oven, or an electric dehydrator.

- Solar drying is a type of sunshine drying in which solar energy is used to heat a specially designed unit that has adequate ventilation for removing moist air. This type of unit can develop temperatures up to 30° warmer than the outside temperature. The higher temperatures reduce the drying time.
- Oven drying is the easiest way to experiment with dehydration. This drying requires little initial investment. Oven drying produces darker, more brittle, and less flavorful foods, however, than foods dried in a dehydrator.
- Electric dehydrators use trays to hold the food being processed. The units have a heat source and ventilation system. They produce a better product than other drying methods produce.

Drying times in conventional ovens and dehydrators vary considerably, depending on the amount of food dried, the food's moisture content, the room temperature, and the humidity. Some foods require several hours, and others might take more than a day. It is important to control air temperature and circulation during the drying process.

Temperatures between 120°F and 140°F are recommended for drying fruits and vegetables. If the temperature is too low or the humidity is too high, the food dries too slowly to prevent microbial growth. If the temperature is too high initially, a hard shell might develop on the outside, trapping moisture inside. If the temperature is too high at the end of the drying period, the food might scorch.

Challenge

You are a researcher for a food-processing company, and you want to develop a dried-fruit snack hikers can take with them on outings. Select a mixture of several fruits and vegetables. Experiment with drying them to develop a trail mix snack.

Materials and Equipment

- An electric dehydrator.
- One pound of fruit and vegetables.

Procedure

1. Carefully read the operating instructions for the electric dehydrator.
2. Select fresh, good-quality fruits and vegetables.
3. Trim away inedible and damaged portions.
4. Cut fruits and vegetables into halves, strips, or slices that will dry readily. The strips should be 1/8″–1/4″ thick.
5. If you are drying vegetables, they need to be blanched. Heat them in water to a temperature high enough to neutralize the natural enzymes.

6. Treat most fruits by dipping them in antioxidants, such as citric and ascorbic acid mixtures.
7. Preheat the dehydrator to 125°F (52°C).
8. Place a single layer of food on each tray.
9. Stack the trays in the dehydrator. Close the unit.
10. Gradually increase the temperature to 140°F (60°C). It takes 4–12 hours to dry fruits or vegetables in a dehydrator.
11. Examine the food often and turn trays frequently. Turn larger pieces, such as apricot halves, halfway through the drying time. Move pieces at the sides of the trays to the center.
12. After the drying process is complete, condition the food. This is done to equalize (evenly distribute) the moisture left in the food after drying. To condition a food, do the following:
    A. Allow the food to cool on trays.
    B. Pour the food into a large nonporous container until the container is about two-thirds full.
    C. Cover the container. Place it in a warm and dry place. Stir the contents at least once a day for 10 to 14 days.

Dried fruits can be eaten as snacks or soaked for one to two hours to rehydrate them. Most vegetables are refreshed with water before use. This can be done by soaking them in water for one to two hours, adding two cups of boiling water for each cup of food, or adding dried vegetables directly to soups or stews.

TSA Modular Activity

This activity develops the skills used in TSA's Biotechnology Design event.

Agriculture and Biotechnology Design

Activity Overview

In this activity, you will research a contemporary problem, issue, or technology related to agriculture or biotechnology; prepare a report; and create a display. You will prepare an oral presentation incorporating presentation software. Your report must be contained in a three-ring binder and consist of the following items:

- A cover page.
- The definition of the problem, issue, or technology.
- A report on the topic (4–10 pages).
- A printout of presentation slides (three slides per page).
- A list of sources and references.

Materials

- A three-ring binder.
- Materials appropriate for a tabletop display. (These will vary greatly.)
- A computer with PowerPoint® presentation software (or similar).

Background Information

- **Selection.** Before selecting the theme for your project, use brainstorming techniques to develop a list of possible themes. Some contemporary topics include the following:
  - Waste management.
  - DNA testing.
  - Soil-conservation techniques.
  - The Human Genome Project.
  - Genetically modified food.
  - Cloning.
  - Aquaculture.
  - Food-production techniques.
  - Irradiation.
- **Research.** Use a variety of sources to research your theme. Do not rely solely on information you find on the Internet. Use books and periodicals available at your local library. Research the historical developments of the topic. Did an individual or a corporation develop the technology? What were some previous technologies that allowed this technology to become a reality? How did the public receive the technology, and was the response expected?
- **Digital presentation.** When developing your presentation, consider the following design guidelines:
  - Develop a general slide design. Use it for all your slides.
  - Keep the design simple. Do not use more than two type fonts. Select easy-to-read type fonts. Be sure that the type size is large enough to be seen from the rear of the room in which you will be presenting.

- Include a title on each slide.
- Do not attempt to squeeze an abundance of information on a single slide. Create multiple slides instead.

Guidelines

- Research should focus on any cultural, social, economic, or political impacts. Both opportunities and risks should be addressed.
- The display can be no larger than 18′ deep × 3′ wide × 3′ high.
- If a source of electricity is needed, use only dry cells or photovoltaic cells.
- The oral presentation can be up to 10 minutes long.

Evaluation Criteria

Your project will be evaluated using the following criteria:

- The content and accuracy of the report.
- The attractiveness and creativeness of the display.
- Communication skills and the presentation design of the oral presentation.

CHAPTER 28

Meeting Needs through Chemical Engineering

Check Your Engineering IQ

Before you read this chapter, assess your current understanding of the chapter content by taking the chapter pretest.
www.g-wlearning.com/technologyeducation/

Learning Objectives

After studying this chapter, you should be able to:

- ✔ Define *chemical engineering*.
- ✔ Describe how the history of chemical engineering was influenced by scientific advancements and industrial development.
- ✔ Define chemistry in relation to other physical sciences.
- ✔ Describe how atomic structure provides elements with their chemical and physical properties.
- ✔ Use the periodic table of elements to determine the atomic structure of different elements.
- ✔ Explain how matter interacts with other matter through chemical reactions, and how matter forms ionic, covalent, and metallic bonds.
- ✔ Explain why chemical equations must be balanced.
- ✔ Compare the different states of matter.
- ✔ Contrast organic chemistry and biochemistry.

Technical Terms

alkali metals
alkaline earth metals
allotropes
anion
atomic number
atoms
Aufbau principle
biochemical engineering
biochemistry
Bohr atomic model
cation
chemical
chemical compounds
chemical elements
chemical engineering
chemical engineers
chemical equations
chemical reactions
chemistry
coefficients
covalent bonds
deoxyribonucleic acid (DNA)
electric charge
electron configuration
electron orbitals
electrons
energy
fermentation
gas
groups
halogens
ion
ionic bond
law of conservation of mass
liquid
matter
neutrons
noble gases
organic chemistry
organic compounds
oxidation
periodic table of elements
periods
petrochemicals
physical science
plasma
products
protons
reactants
solid
stoichiometry
transition elements
valence electrons
valence orbital

Close Reading

As you read this chapter, make an outline of the various chemistry concepts and note how each plays a role in chemical engineering.

Chemicals are essentially any type of matter. However, people generally use the term ***chemical*** to refer to a compound or substance that has a specific molecular composition that is produced or used for a specific purpose. Developing chemicals, using chemicals, and changing the chemical structure of materials and living things help us extend our human capabilities. See **Figure 28-1**. These actions allow us to produce stronger materials and better fuels, create more food, enhance food preservation, and improve medicines.

It is hard to imagine what life would be like without the knowledge and skills to use chemical processes to produce products that meet our needs and desires. The professionals who specialize in this area are chemical engineers.

Chemical Engineering

Chemical engineering is the application of chemistry, biology, physics, and mathematics to solve problems that involve the conversion of raw materials and chemicals into more useful forms of food, medicine, fuels, and many other products. Not only do chemical engineers develop these products, they also design the processes necessary to produce them. Chemical engineering is one of the more broad engineering disciplines, as are mechanical and electrical engineering.

History of Chemical Engineering

Chemical engineering activities can be traced back for centuries. For example, the process of fermentation has been used for thousands of years to produce desired food and beverages, such as yogurt, bread, and cheese. ***Fermentation*** is a process in which an organism, such as yeast or bacteria, metabolizes a carbohydrate, such as a starch or sugar, and converts it into an acid or alcohol. When yeast metabolizes sugar, chemical reactions occur through digestion to provide energy for the yeast while producing ethanol and carbon dioxide. These by-products of fermentation are used to give food products their desired qualities, **Figure 28-2**.

Although this chemical process has been used throughout history, chemical engineering was not an actual profession until the beginning of the twentieth century. This is when chemistry was applied to industrial applications. For example, the fermentation process was then used at the industrial scale to produce products other than food or beverages, such as acetone, **Figure 28-3**, and butanol. Both are often used as solvents.

Comaniciu Dan/Shutterstock.com

Figure 28-1. This student is analyzing a chemical structure model of a material.

MilosR/Shutterstock.com

Figure 28-2. Fermentation is used to produce food and beverages, including bread, yogurt, and cheese.

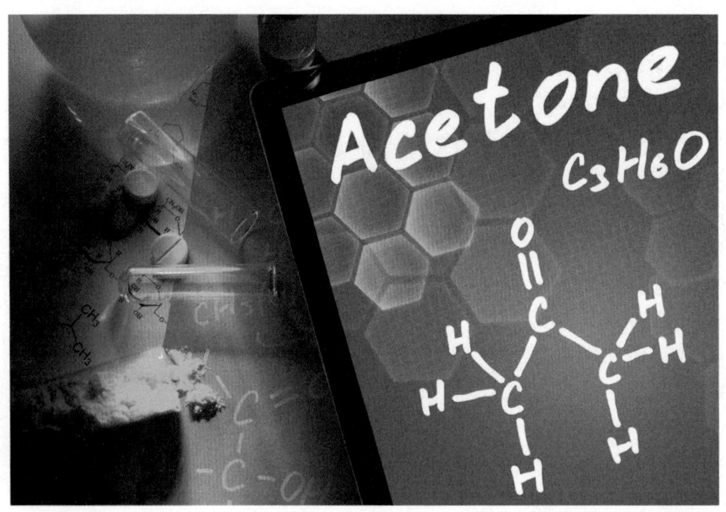

designer491/Shutterstock.com travis manley/Shutterstock.com

Figure 28-3. The fermentation process is used at the industrial scale to produce products other than food or beverages. Acetone is one such product. It is used as a solvent.

As the twentieth century progressed, more scientific advancements were made. Chemical engineers developed many new products and processes by converting raw materials into more useful forms for specific needs. These new products and processes were often created to solve issues encountered during the various wars of the twentieth century. For example, during World War II, chemical engineers developed processes for producing high-octane fuels necessary for aircraft and synthetic rubber needed for airplane tires.

Scientists also advanced their understanding of ***deoxyribonucleic acid (DNA)***, the genetic material of living organisms. This advancement helped to launch biochemical and genetic engineering. See **Figure 28-4**. All of these achievements helped shape our world today and solidify the profession of chemical engineering.

What Do Chemical Engineers Do?

Each engineering discipline is responsible for the design and development of different types of products and processes, which often overlap. People often associate specific items with each discipline. For example, chemical engineers are thought to produce chemicals, and more specifically chemicals made from crude oil, or ***petrochemicals***. However, this only covers a small percentage of what professionals in chemical engineering actually do.

Chemical engineers develop new materials and chemicals, by using energy to create a basic chemical change in raw materials or chemicals. They work with the earth's chemical elements to design a variety of chemical, fuel, food, material, and pharmaceutical products. For example, in the pharmaceutical industry they may create new, affordable drugs. In the agricultural industry, they

bikeriderlondon/Shutterstock.com

Figure 28-4. Understanding DNA helped to launch biochemical and genetic engineering.

Career Connection Chemical Engineers

Chemical engineers apply their knowledge of mathematics and science, specifically chemistry, to develop new materials, chemicals, foods, and pharmaceuticals by altering and combining natural materials or chemicals. The demand for chemical engineers is projected to continuously grow.

To become a chemical engineer, one must obtain a bachelor's degree in chemical engineering, **Figure A**. Most universities or colleges that offer engineering programs offer degree programs in chemical engineering. Graduates who hold this degree are prepared to oversee, design, monitor, and enact procedures to subject raw materials to chemical and physical changes in order to produce higher-value material products. This can involve producing products, such as adhesives, detergents, and plastics, from a raw material such as ethylene.

anyaivanova/Shutterstock.com

Figure A. Chemical engineers hold bachelor's degrees in chemical engineering. Many universities or colleges have engineering programs.

Chemical engineers typically work in offices and laboratories at industrial plants, refineries, and various other locations. They direct operations, develop new products, improve processes, and solve a variety of onsite problems.

High school students interested in pursuing a chemical engineering career should enroll in advanced chemistry and mathematics courses, along with technology and engineering coursework. They may even explore options for early college opportunities in chemistry and calculus.

Professional chemical engineers can join several organizations to network with others in their industry. These include American Institute of Chemical Engineers, National Society of Professional Engineers, and the Institution of Chemical Engineers.

may develop new herbicides or pesticides. Chemical engineers also design the processes and equipment used to process chemicals in a variety of industries, such as materials production, medicine, manufacturing, construction, energy, and even waste and water treatment facilities, **Figure 28-5**. For example, chemical engineers may design new procedures for extracting and refining oil and natural gas to be used for electrical energy production.

Chemical engineers work on a wide range of products, processes, and systems. They apply their knowledge of mathematics and science, and specifically chemistry, to solve technological problems in the most effective and efficient manner possible, using engineering practices.

RGtimeline/Shutterstock.com

Figure 28-5. Chemical engineers develop products and design the processes and equipment used for chemical processing in a variety of industries.

Chemical Engineering Fundamentals

In order to produce, develop, or use specific chemicals, fuels, pharmaceuticals, foods, and other products, chemical engineers must understand matter and energy and how they interact with other matter and energy. They must also understand the principles of the physical sciences, specifically chemistry, to develop different products and processes that meet our needs and wants. The following sections will introduce the basic concepts and principles of chemistry.

Chemistry

Chemical engineering became a distinct field through the industrial application of chemistry in the late nineteenth century. Chemistry is one of the physical sciences. ***Physical sciences*** study forces, motion, energy, matter, and their interactions. Branches of physical sciences include physics, chemistry, and astronomy. See **Figure 28-6**. All of these branches are intertwined and interconnected with one another. ***Chemistry*** is the study of matter and its interactions with other matter and energy.

Matter is the term used for anything that has mass or takes up space. It is the physical world that surrounds you. ***Energy*** is an attribute of matter. Energy comes in various forms and affects the matter that possesses it, acquires it, or loses it. Energy is typically used to do some form of work.

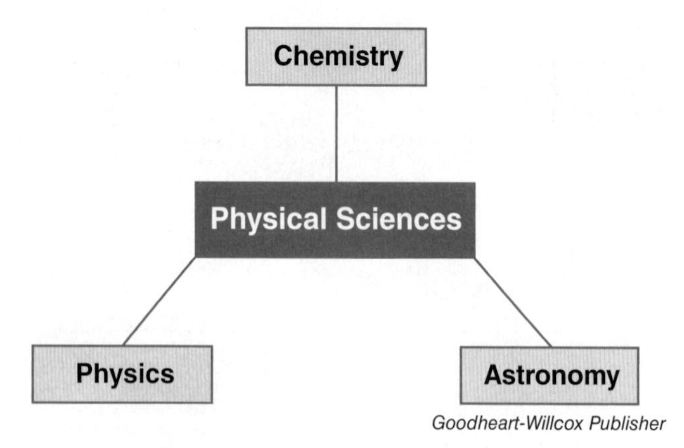

Figure 28-6. Physical sciences study forces, motion, energy, matter, and their interactions.

Structure and Properties of Matter

Every material is made up of matter. Chemical engineers use their knowledge of chemistry to convert materials into more usable forms. They must also select the best materials to use for specific applications. To do so, chemical engineers must understand the different properties of the materials available to them. Properties, such as hardness, plasticity, and conductivity, are characteristics that describe materials. These material properties are based on the structure of matter.

All matter comprises different types of atoms. The atoms of each material interact with the other atoms in that material, along with the atoms of other matter. The atomic structure of matter determines how materials react with other matter because the atomic structure is what gives matter its properties and governs its functions.

Basic Atomic Structure

Atoms are the smallest distinct unit of matter. They are the basis for everything in our universe. As you learned in previous chapters, atoms consist of three basic parts, or subatomic particles: electrons, protons, and neutrons, **Figure 28-7**.

Protons are subatomic particles that have a positive electric charge. ***Neutrons*** are uncharged subatomic particles. ***Electric charge*** is a measurable property of matter. Charges can be either positive or negative. Some particles, such as neutrons, can have no positive or negative charge, which makes them neutral. In addition to protons and neutrons, the atom is composed of one or more ***electrons***, which are subatomic particles with a negative electric charge. These subatomic particles are bound by an electrical charge to an orbit around the nucleus of the atom.

Refer again to **Figure 28-7**. The electrons rotate around the nucleus on different orbits called ***electron orbitals***, rings, or shells. The ***Bohr atomic model*** depicts the electron orbitals of an atom in two dimensions. One bromine atom (Br), **Figure 28-8**, has 35 protons in the nucleus and 35 electrons orbiting the nucleus. Also note that the electrons move in four different orbitals around the nucleus. Orbitals

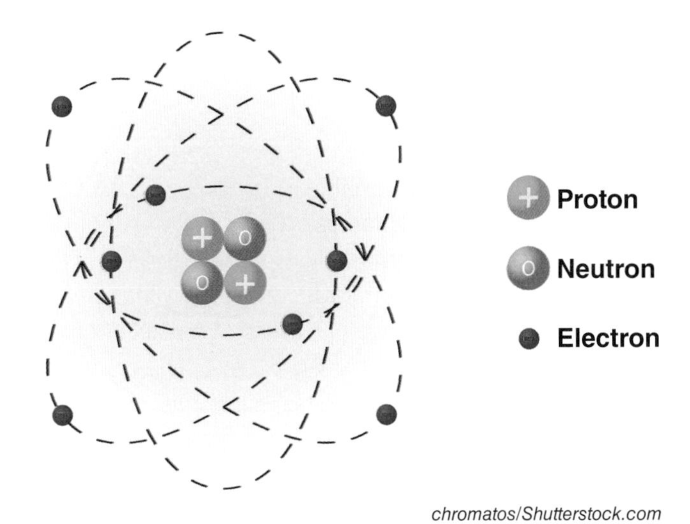

chromatos/Shutterstock.com

Figure 28-7. A planetary model of an atom.

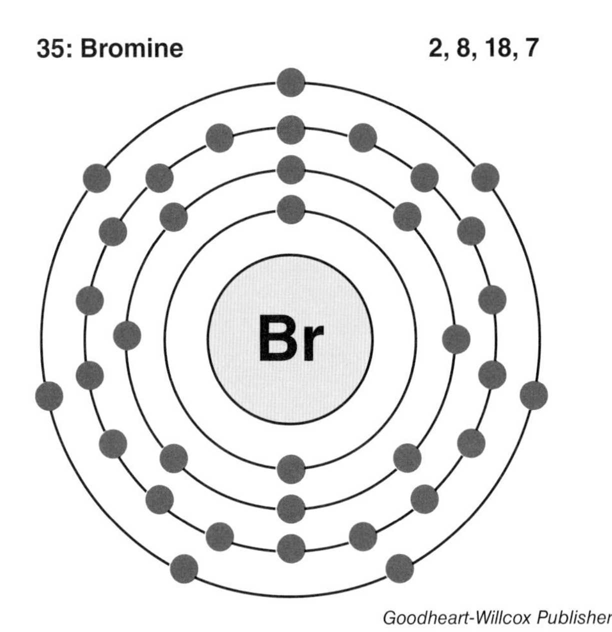

Goodheart-Willcox Publisher

Figure 28-8. A Bohr atomic model of a bromine (Br) atom. The atomic number indicates that this atom has 35 protons in the nucleus and 35 electrons orbiting the nucleus.

are located at various distances around the nucleus. The electrons in the orbits closest to the nucleus have a stronger attraction to the protons and are more bound to the atom. Electrons in more distant orbitals have a weaker attraction to the protons, making it possible for them to move from one atom to another.

The number of electrons in each orbital varies. The ***Aufbau principle*** determines how the electrons are configured in these orbitals. The principle states that electrons fill the orbital closest to the nucleus and then fill successive orbitals farther away from the nucleus. The closer electron orbitals hold a smaller number of electrons, while the more distant orbits can hold a larger number of electrons. The Aufbau principle indicates that electrons are typically distributed across the various potential orbitals as follows: the closest orbit of an atom can typically hold two electrons; the next orbit can hold eight, the next 18, the next can hold 32, then 50, and then 72. See **Figure 28-9**. However, the outermost orbital of any atom can typically only hold up to eight electrons. This is called the ***valence orbital***, shell, or ring. The number of electrons in the valence orbital determines how atoms interact with each other. Atoms act to achieve a stable electron configuration, meaning that they want to have a full valence orbital of eight electrons. To do so, atoms give electrons to, take electrons from, or share electrons with other atoms. When atoms have a full outer orbital, the atoms are stable.

Look again at the bromine (Br) atom. Notice that the orbits are filled with electrons. This atom has 35 electrons. The innermost orbits are filled first. Therefore, the first orbit is filled with two electrons. The next orbit is filled with eight electrons. The third orbit is filled with 18 electrons. The first three filled orbits account for 28 of the 35 electrons of the bromine (Br) atom. This leaves seven electrons that orbit the nucleus in the valence orbital. Electrons located in the valence orbit are called the ***valence electrons***. The valence electrons are the most important because they are responsible for chemical bonds, which is how atoms combine to form chemical compounds. This will be discussed in more detail later in this chapter.

Electron Orbitals	
Orbital Number	**Maximum Electrons**
1	2
2	8
3	18
4	32
5	50
6	72
Valence orbital	8

Goodheart-Willcox Publisher

Figure 28-9. Electron orbital configuration.

Think Green Green Household Cleaners

One reason to clean a house, office, or other indoor space is to disinfect it, which can help its occupants stay healthy. Conventional household cleaners, however, combine several types of substances that may have adverse effects on the environment. Ammonia, which is often used as a cleaning agent, can taint the water supply. Bleach, commonly found in cleaners, can also contaminate water. In both of these cases, while the chemicals are not directly harmful to humans, they will kill fish and possibly other animals.

There are effective household cleaners, however, that do not use these toxic chemicals. Alternatives include carbonate- and plant-based materials (such as vegetable oils). These materials are not harmful to the environment or to animals, and they biodegrade more easily. There are no volatile organic compounds (VOCs) emitted with green household cleaners. Some types of green cleaners can also be made at home.

Chemical Elements

The earth's 118 chemical elements are made of atoms. ***Chemical elements*** are the building blocks of matter. They are pure substances that cannot be broken down into simpler substances. These elements are shown in the periodic table of elements, **Figure 28-10**.

Each chemical element is unique and is made of atoms with a specific number of protons in its nucleus. The number listed with each element on the periodic table of elements is its ***atomic number***.

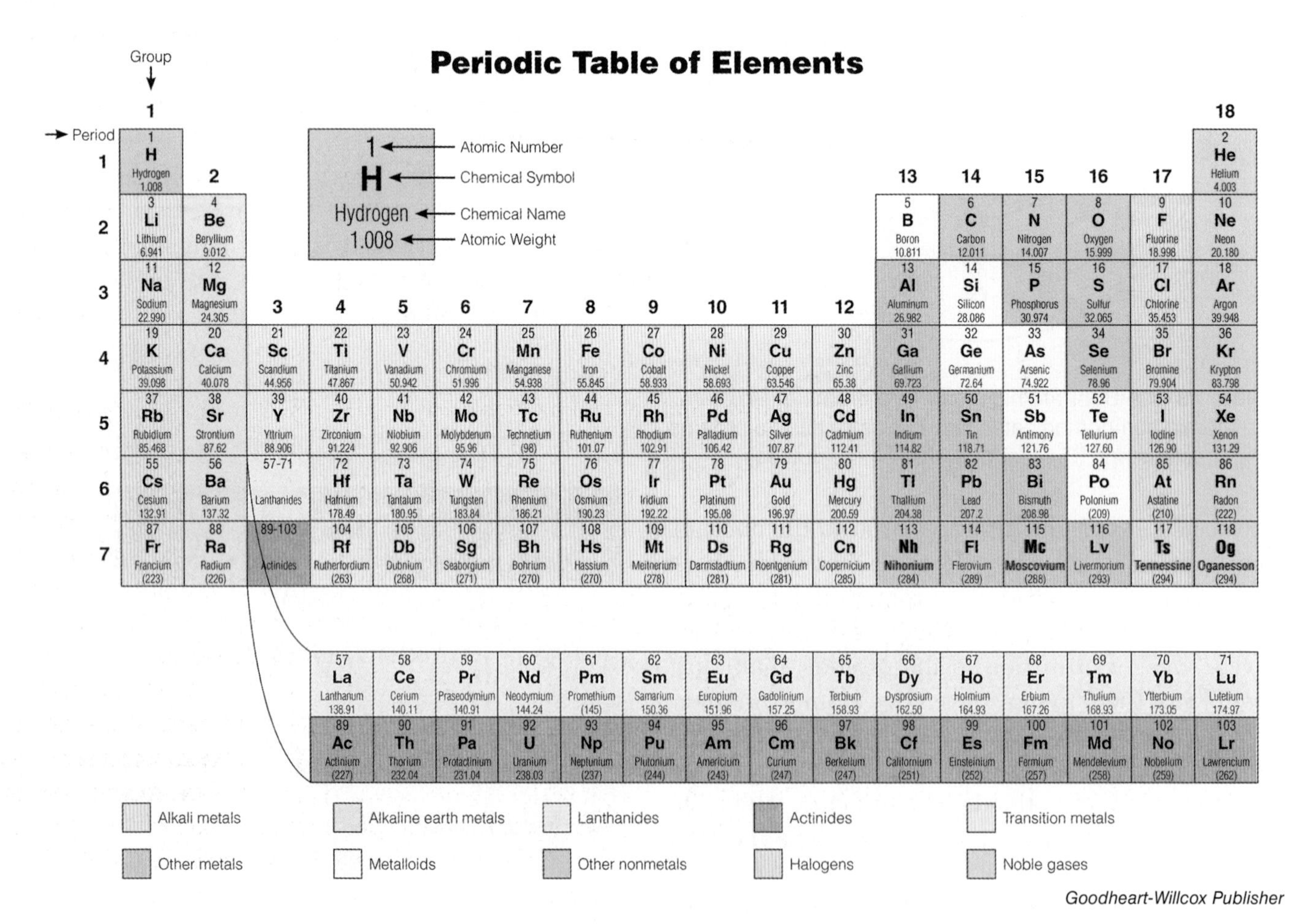

Figure 28-10. The periodic table of elements.

This number indicates the number of protons in the nucleus of its atom. The number of protons determines the characteristics of an element because it governs how it reacts with atoms of all other chemical elements. This has to do with the atom's electrical charge. As you read earlier, the subatomic particles of an atom each have a different electrical charge. Electrons have a negative charge, protons have a positive charge, and neutrons have no charge or a neutral charge. Typically, the atoms of each element have a neutral charge. This means that it has the same number of protons as it does electrons. Therefore, the atomic number can also tell us how many electrons are in an atom of the element as well. For example, a potassium (K) atom, **Figure 28-11**, has 19 protons and 19 electrons.

The number of protons in an atom of an element does not change. Therefore, the atomic number of an element will never change. If it did, it would no longer be that element. However, the number of electrons of an atom can change. When it does, this produces an ion. An ***ion*** is an atom that has more electrons than protons or an atom with fewer electrons than protons. This results in an atom that no longer has a neutral charge. An atom with more electrons than protons is a negatively charged ion called an ***anion***. An atom with fewer electrons than protons is a positively charged ion called a ***cation***. See **Figure 28-12**.

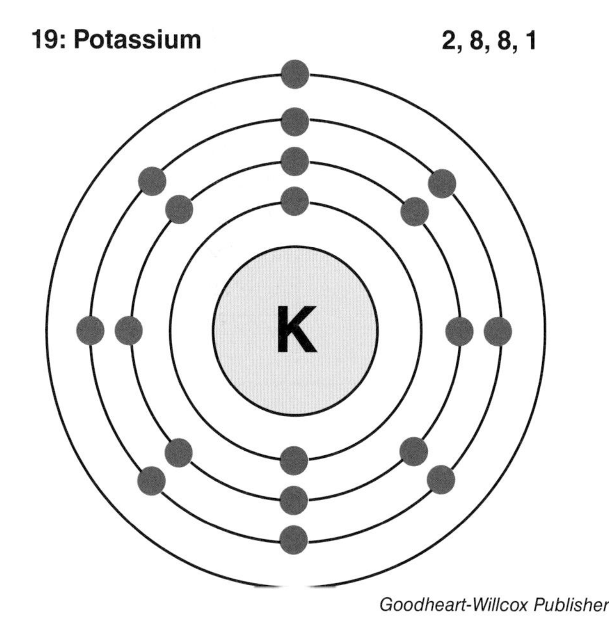

Goodheart-Willcox Publisher

Figure 28-11. A potassium (K) atom has 19 protons and 19 electrons.

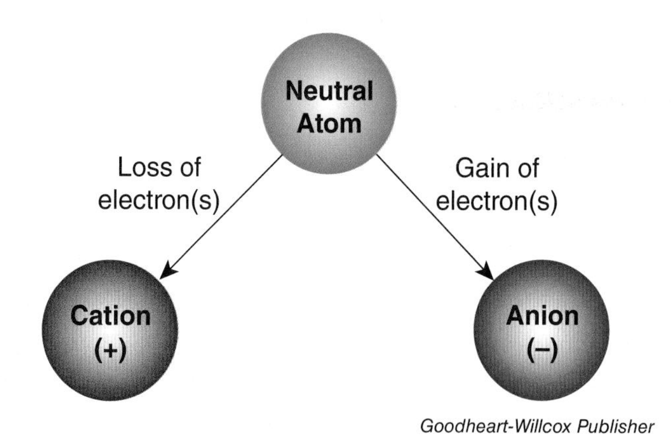

Goodheart-Willcox Publisher

Figure 28-12. When an atom gains or loses electrons, it is no longer electrically neutral and is called an ion. A positively charged ion is called a cation. A negatively charged ion is called an anion.

Periodic Table of Elements

The ***periodic table of elements*** is a grid used to organize the different chemical elements. Refer again to **Figure 28-10**. In this grid, the columns are called ***groups*** and the rows are called ***periods***. The elements within each group and period have specific properties or characteristics. For example, lithium (Li) and sodium (Na) are in group (column) 1 and share similar characteristics. Lithium (Li) is also in the same period (row) as beryllium (Be) and shares similar characteristics. However, beryllium (Be) does not share the same group or same period as sodium (Na) and, therefore, does not share all of the same characteristics as it does with lithium.

All elements within the same period have the same number of atomic orbitals. The top period only has one orbital for its electrons. As you move down the table, each period adds another orbital. The maximum number of orbitals for an atom of an element is seven.

All elements within the same group have the same number of electrons in their outermost orbital. These are the electrons that are responsible for chemical bonding with other elements to form molecules and chemical compounds. *Molecules* are any combination of atoms that are bonded or stuck together. ***Chemical compounds*** are any substances that consist of two or more different atoms that are bonded or stuck together, **Figure 28-13**. Atoms bond together when they gain, lose, or share electrons. These molecules and compounds can have very different properties from the individual elements themselves.

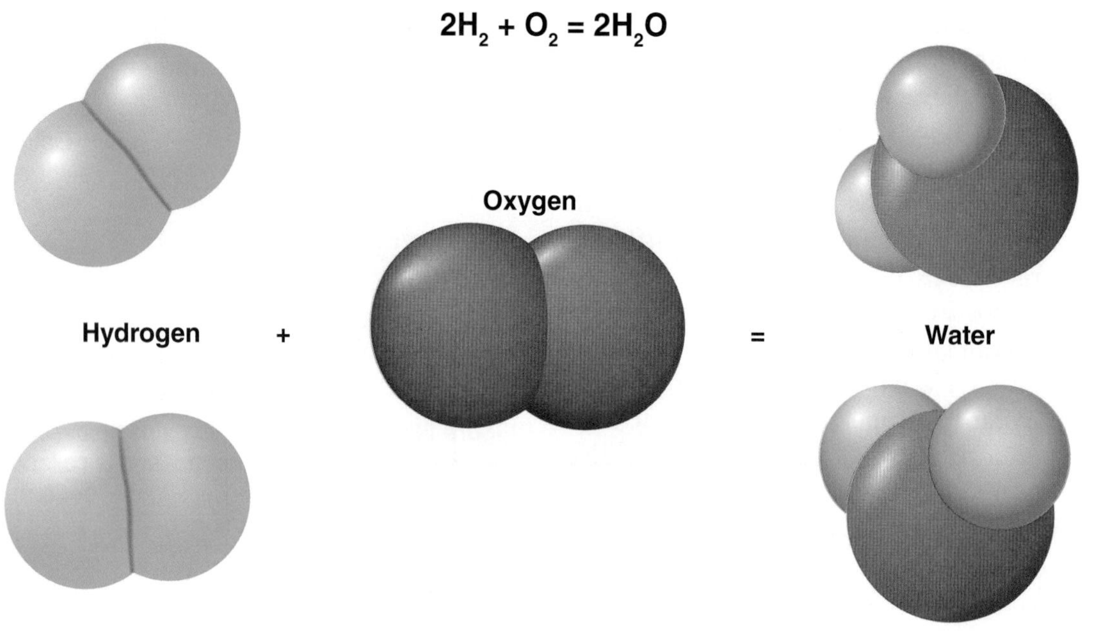

Designua/Shutterstock.com

Figure 28-13. Water molecules are a compound of hydrogen and oxygen atoms.

Read the periodic table from left to right. Elements in group 1 have one valence electron, **Figure 28-14**, and are known as ***alkali metals***, except for hydrogen, which is a nonmetal. Elements in group 2 have two valence electrons, **Figure 28-15**, and are known as ***alkaline earth metals***. Because the atomic structure of groups 1 and 2 hold only one or two electrons in their valence orbitals, they are highly reactive elements that readily give up their valence electrons to form chemical bonds.

Look now at groups 13 through 18. Group 13 has three valence electrons. As you move across toward group 18, add one more valence electron for each group. Group 18 has a full valence orbital with eight valence electrons. Group 18 are the ***noble gases***,

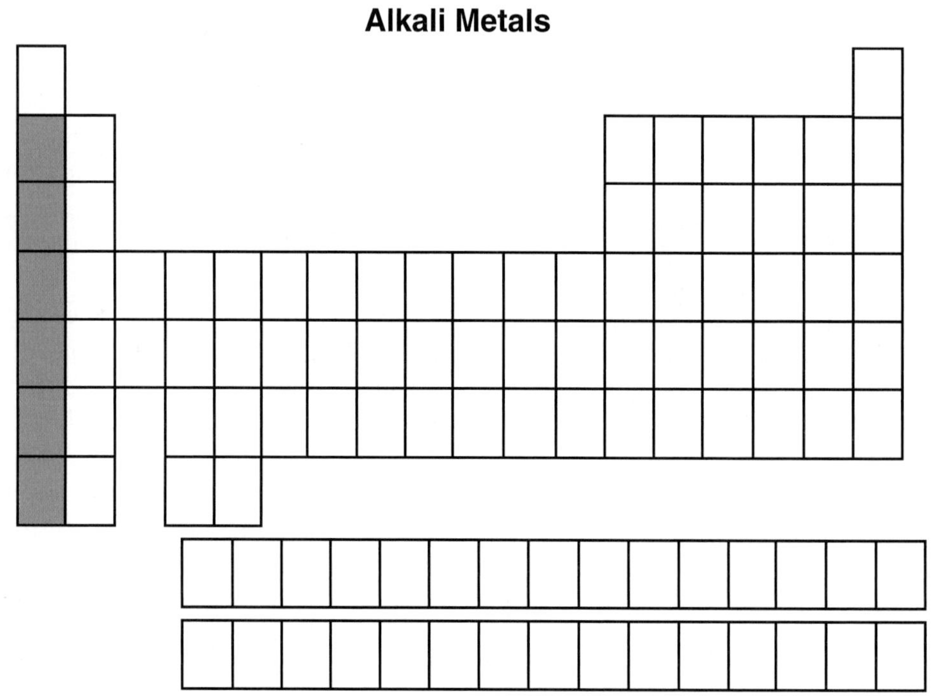

Goodheart-Willcox Publisher

Figure 28-14. The alkali metals are lightweight, shiny, soft, and malleable.

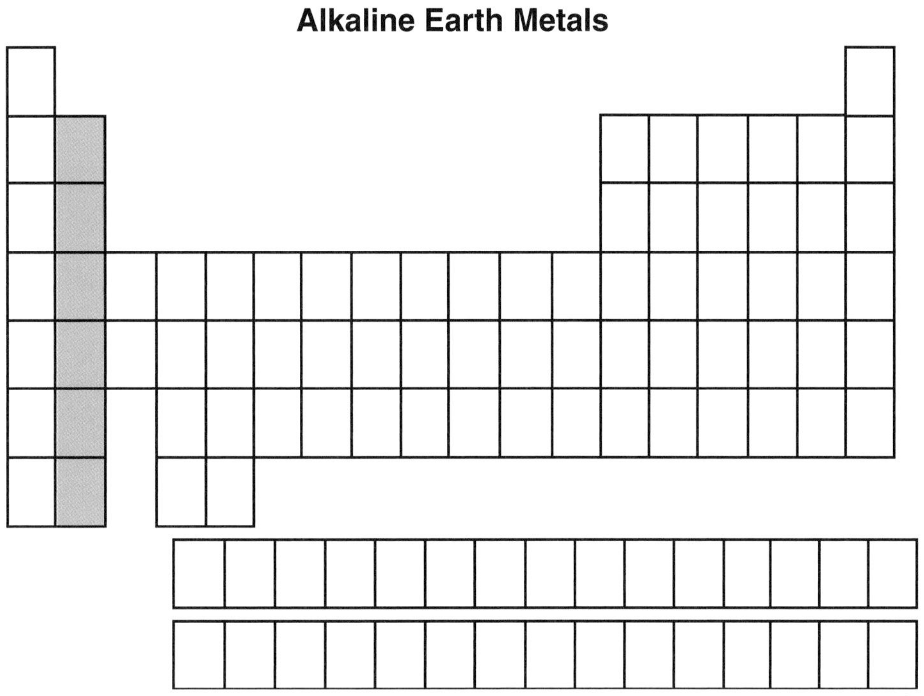

Goodheart-Willcox Publisher

Figure 28-15. The alkaline earth metals have two electrons in their outermost orbital.

which all share similar properties, **Figure 28-16**. Because of their atomic structure, which has a full valence orbital, they typically do not react with other elements. These atoms are stable and do not need to share, take, or give away any electrons. Helium (He) has only two electrons, so it cannot have eight electrons in its outer orbital. While it does not fit the criteria of this group, it is very stable with only two electrons, so it has the same properties of the noble gases, and is therefore included in group 18.

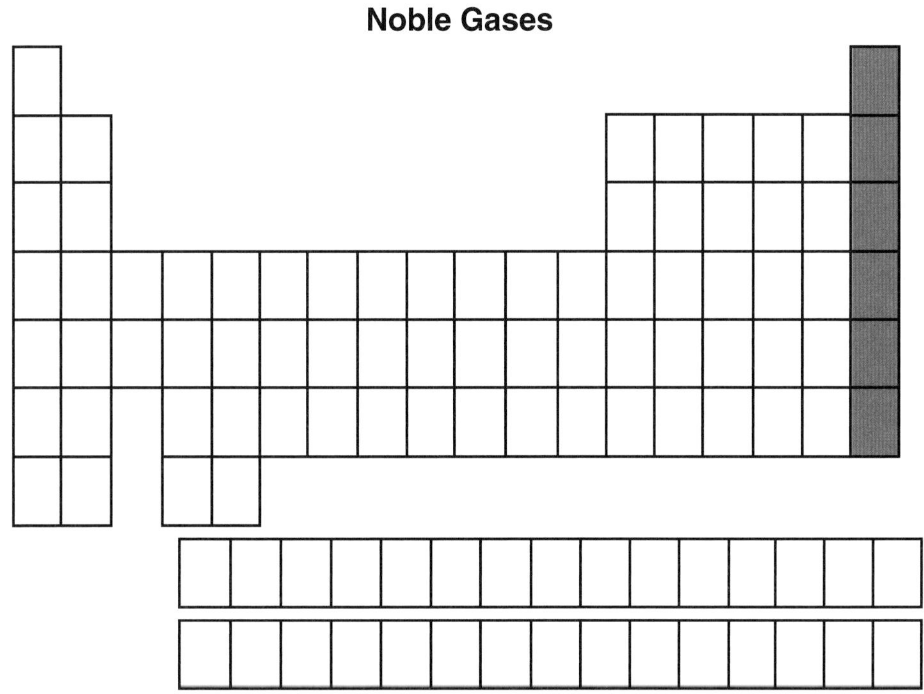

Goodheart-Willcox Publisher

Figure 28-16. Group 18 is the family of elements known as the noble gases. The atoms of these elements have eight electrons in their outermost orbital.

STEM Connection: Science

Irradiation

The question of whether or not we should consume irradiated food is hotly debated. Some groups fear that the process produces harmful chemicals. Others argue that irradiated foods might cause cancer. Many scientists, however, believe that irradiation is both safe and beneficial.

In order to make informed choices regarding this subject, we should know what irradiation of food involves. Irradiation itself is a form of radiant energy, which, as we learned in earlier chapters, is energy in the form of electromagnetic waves. The electromagnetic waves used in irradiation are primarily gamma rays, UV rays, and x-rays. Scientists have found that, when these rays penetrate food, chemical bonds are broken and the chemical composition of the food changes.

The change in chemical bonds can affect food in many ways. For example, when a potato is exposed to a low level of these rays, the change in chemical bonds causes sprouting to stop. However, at higher levels of exposure, the changes in chemical bonds within the potato can eliminate insects and destroy harmful bacteria.

Food irradiation is sometimes chosen over cooking food using thermal energy because, even though chemical changes occur in both processes, fewer bonds are broken in irradiation. Therefore, the food is much fresher. Irradiation is also sometimes preferred to the addition of chemicals to destroy pests or extend storage life because of the possible hazards involved with certain chemicals.

With irradiation, food stays fresh longer and in better condition. We still use other forms of food-preservation technologies, however, with particular foods. For example, salt is used to preserve bacon, and heat (pasteurization) is used to keep milk fresh. Can you think of some reasons for using these methods, as opposed to irradiation?

Group 17 then has seven electrons in valence orbital. These are called ***halogens***. See **Figure 28-17**. Because halogens have only one empty space for an electron in their valence orbital, they readily gain electrons and react with metals to form chemical compounds. They are typically harmful or lethal to biological organisms. When they react with metals, they form salts and when they react with hydrogen, they form acids. Groups 13, 14, 15, and 16 contain nonmetallic elements, such as nitrogen (N) and oxygen (O), metalloid elements, such as silicon (Si), and other metals, such as tin (Sn) and aluminum (Al).

Groups 3 through 12 and elements 57 through 102 represent ***transition elements***, **Figure 28-18**. These elements essentially break the rules with the number of electrons that can be in each orbital. The atoms of the transition elements can add additional electrons to the second to last orbital, as well as their valence orbital. Unlike the typical maximum number of electrons for each orbital (2-8-18-32-50-72 and 8 for the valence orbital), transition elements can have up to 32 electrons in their second-to-last orbital. This can alter the typical arrangement of electrons in the two outermost orbitals for these elements. This is important to remember because this also determines the properties of these elements and how they interact with other elements. Based on this atomic structure, the electrons in the two outer shells can chemically bond with other elements to form chemical compounds, unlike the other groups of elements, which only bond with other elements using only their valence orbital.

Chemical Reactions and Bonds

As you can see, the periodic table of elements is organized to help us understand the atomic structure of each element and, more importantly, the configuration of its electrons. ***Electron***

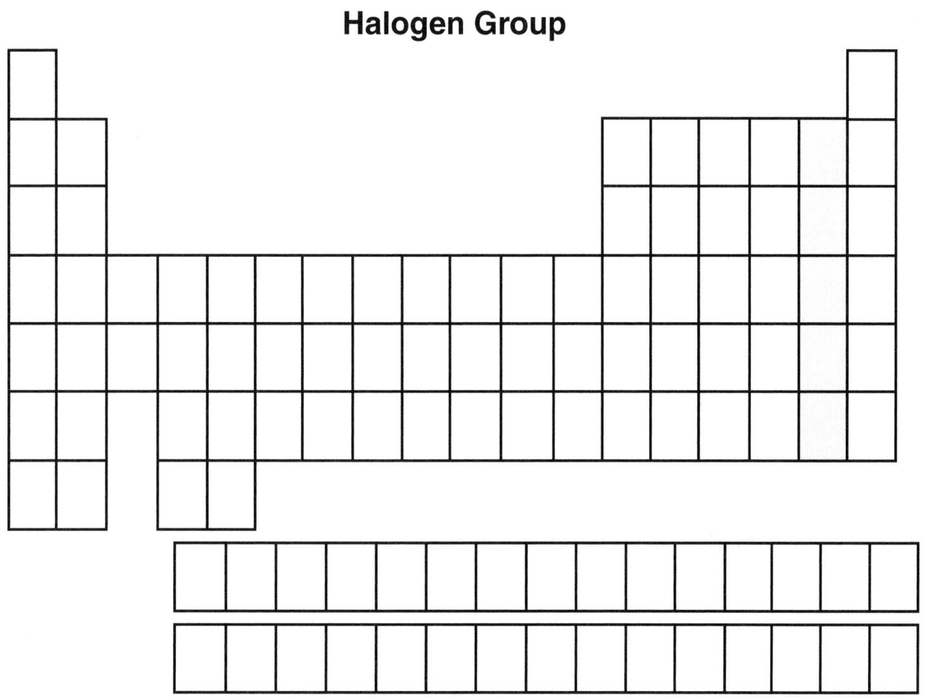

Figure 28-17. Group 17 contains the family of elements known as halogens. With seven electrons in their outermost orbital, they are highly reactive elements.

configuration, which is the arrangement of electrons within an atom's orbitals, dictates the element's properties in a chemical reaction. ***Chemical reactions*** are the processes that either bond atoms together or break bonded atoms apart. These reactions create or transform matter that exhibits different properties than it did before the reactions, **Figure 28-19**.

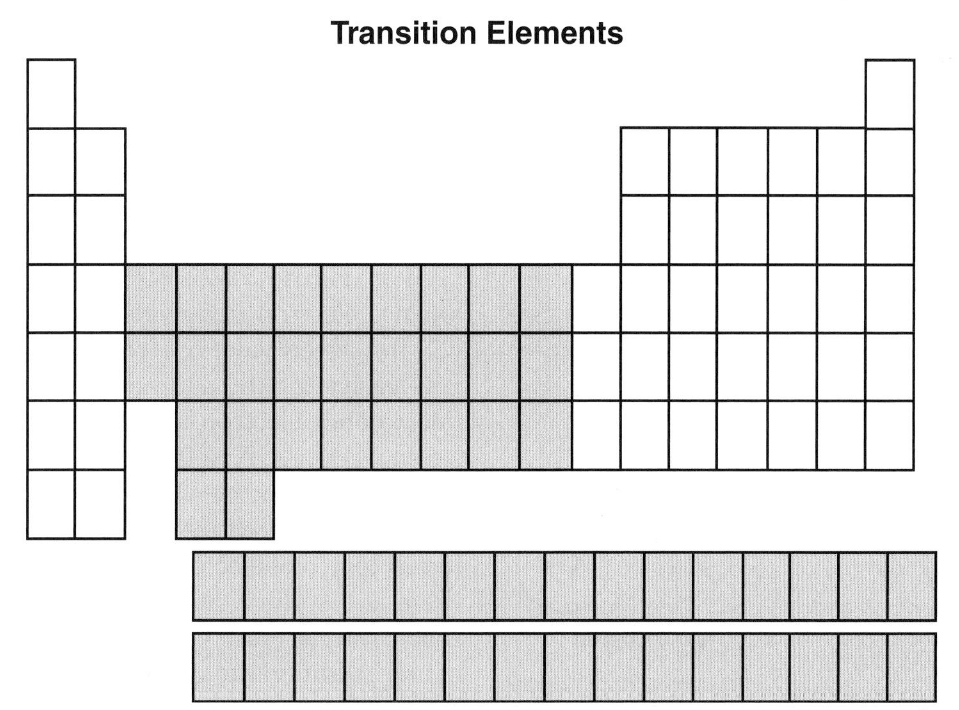

Figure 28-18. Transition elements are represented in groups 3–12 on the periodic table and are unique because they can hold valence electrons (electrons they use to bond with other elements) in more than just their outermost or valence orbital.

mikeledray/Shutterstock.com

Figure 28-19. A chemical reaction between CO_2 (carbon dioxide), H_2O (water), and NH_3 (ammonia).

When a chemical reaction occurs that bonds atoms, they form molecules. Molecules are a group of two or more atoms bonded through atoms gaining, losing, or sharing electrons in order to have a full valence orbital and to become electrically neutral. Molecules can consist of atoms of the same element, such as two oxygen atoms (O_2), or ozone (O_3), **Figure 28-20,** or of different elements, forming chemical compounds such as water (H_2O—two hydrogen atoms, one oxygen atom). See

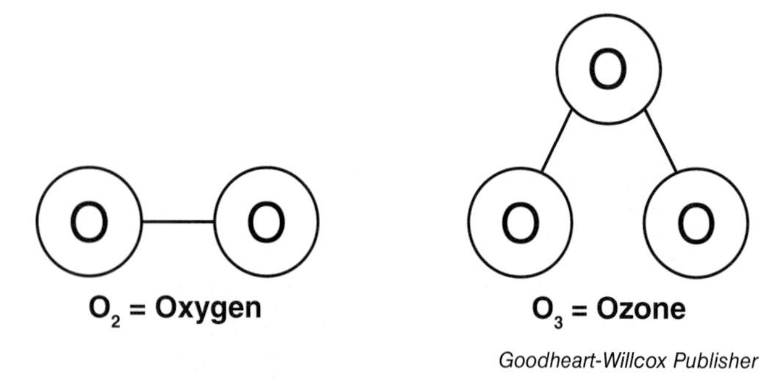

Goodheart-Willcox Publisher

Figure 28-20. Atoms of the same element can bond in different ways to form different types of molecules with different properties.

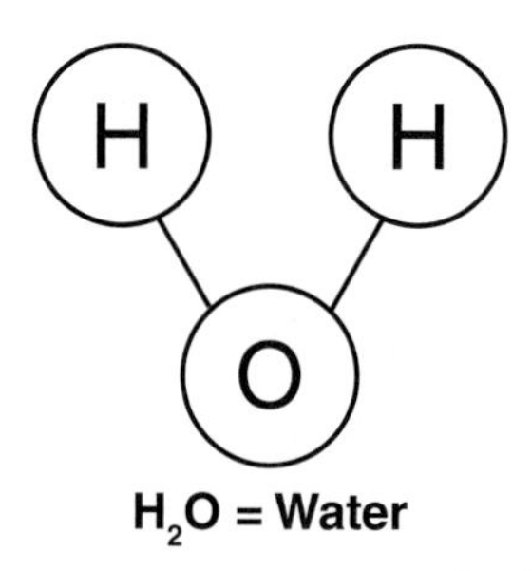

Goodheart-Willcox Publisher

Figure 28-21. Atoms of different elements bond to form many different molecules, such as water, that have different physical and chemical properties.

Figure 28-21. Chemical compounds are molecules that consist of two or more different, bonded atoms. As a result, all compounds are molecules but not all molecules are compounds.

The important thing to understand is that different molecules can have very different properties from the individual elements themselves. Understanding how different elements react with one another enables chemists and chemical engineers to develop new and better materials and products. They can also use this knowledge to determine how materials or products will react with the elements of the environment in which they are used.

Why and how do atoms bond? Atoms bond together because they are looking for a stable electron configuration and a neutral charge. They need a full outer orbital to be stable. This involves losing, gaining, or sharing electrons. Therefore, if there is an atom that wants to give away an electron and an atom that wants to take an electron, then they can benefit from interacting with each other. As seen in **Figure 28-22**, sodium (Na) has only one valence electron in its valence orbital and chlorine (Cl) has seven valence electrons. To reach stability, it is easier for the sodium atom to dispose of its one valence electron rather than take in chlorine's seven electrons. And it is also easier for the chlorine atom to gain one electron than to give away seven. Therefore, if a sodium atom is near a chlorine atom, it will give one electron to the chlorine atom. Now, there is a sodium atom that has one less electron than protons, making it a positively charged ion, or cation. Additionally, there is a chlorine atom with one more electron than protons, making it a negatively charged ion, or anion. Opposite electrical charges attract each other. So now, the sodium cation and

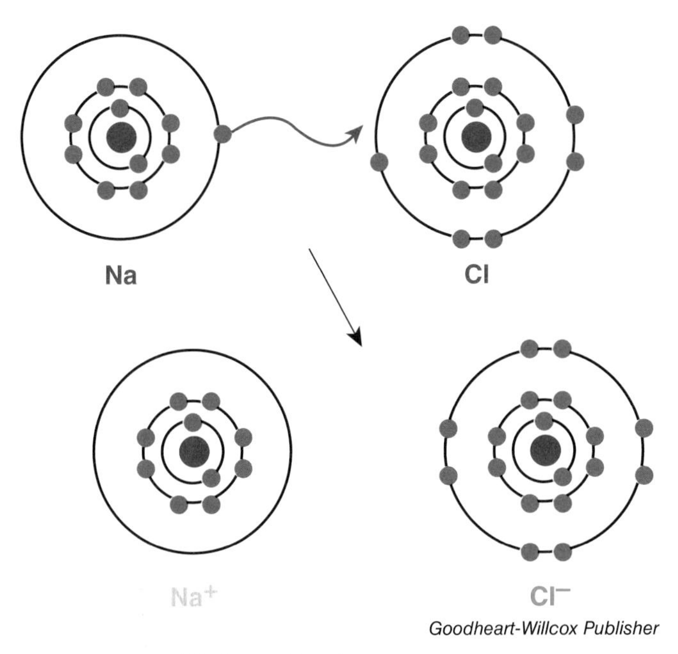

Figure 28-22. While stable, these atoms are now unbalanced electrically, making the sodium atom a cation and the chlorine atom a negative ion (anion).

the chlorine anion stick together or bond because of their electrical charges. This bonding forms the chemical compound sodium chloride (NaCl), or table salt. This is called an ***ionic bond*** because it was the result of the formation of ions, **Figure 28-23**. The opposite charges of different ions cause them to attract each other, forming ionic bonds and thus creating chemical compounds.

Chemical engineers need to understand this type of bond with regard to material properties. Ionic bonds are very strong. They are hard to break. However, the sodium chloride bond can be broken by placing it in water. It dissolves into sodium ions and chlorine ions because the water interrupts the electrical attraction between the different ions. Also note that materials that consist of ionically bonded atoms are not the best choice for use in water applications.

Another type of bond is a covalent bond. ***Covalent bonds*** occur when atoms actually share pairs of electrons to have a full valence orbital. This is different from ionic bonds, which are a result of ions attracting each other based on their electrical charges. This can occur when there are atoms surrounding each other that have similar electron configurations. For example, if two oxygen (O) atoms are next to each other, they can share electrons to essentially fill both of their outer orbitals and bond. The oxygen (O) atom has six valence electrons in its valence orbital. So two oxygen (O) atoms can bond together to share two electrons, which will make them both seemingly have a full valence orbital of eight electrons even though the total electrons between the two only equals 12. However, two electrons will be shared, **Figure 28-24**, to make each atom act as if each valence orbital is full. In this case, the covalent bond forms an oxygen molecule, O_2. The same type of covalent bonding can be seen in the molecule for water (H_2O), **Figure 28-25**.

Metallic bonds are another type of chemical bond. A common characteristic of all metals is that they have many electrons in their outer orbitals and willingly share them. So when atoms of a metal such as iron (Fe) are grouped together they create a sea of electrons between them, which are not within any particular atom's orbit. See **Figure 28-26**. The

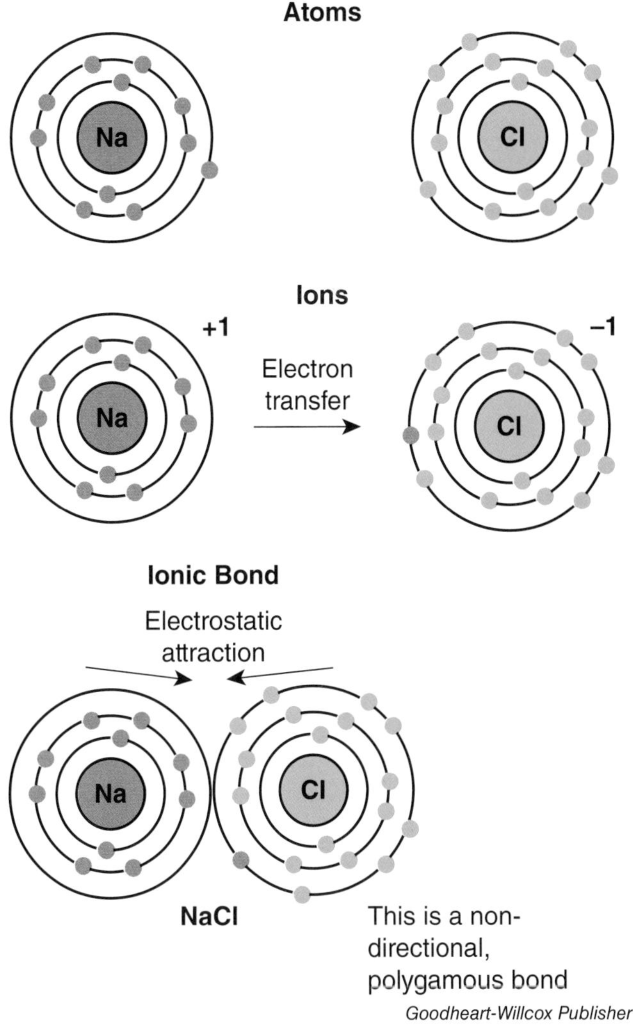

Figure 28-23. An example of an ionic bond.

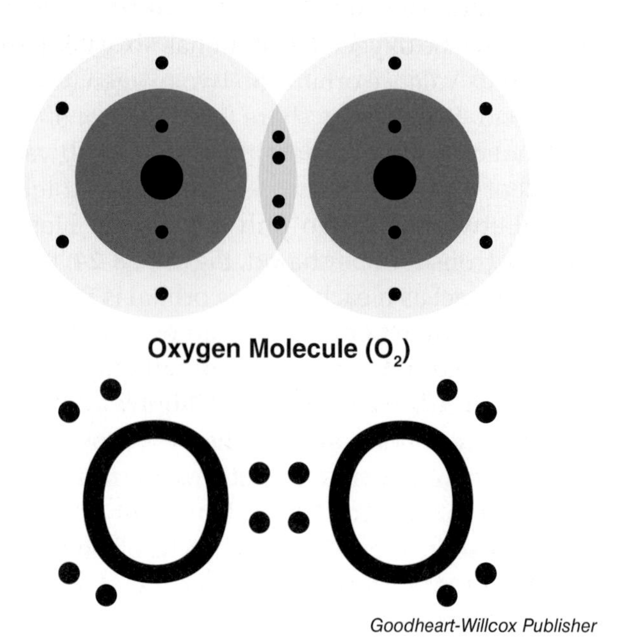

Figure 28-24. Two oxygen atoms share two electrons to form an oxygen molecule, making them each seem to have a full valence orbital. This is called covalent bonding.

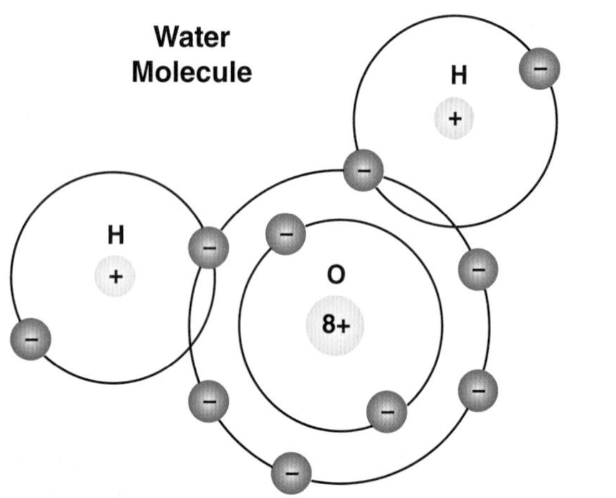

Figure 28-25. One water molecule is formed by covalent bonds between two hydrogen (H) atoms and one oxygen (O) atom.

electrons are free to move about the metal molecule that is held together by the bond. The bond is created by the positive nucleus of the atom being electrically attracted to the negative charge of the electrons. This type of bond gives metals many of their characteristics, such as good electrical conductivity, because the electrons are free to move throughout the metal molecules. This also makes metal bendable, or malleable. Because the electrons can move, the bonds are flexible. This is unlike rigid, inflexible ionic bonds that create brittle material. How would a chemical engineer use this knowledge to design products?

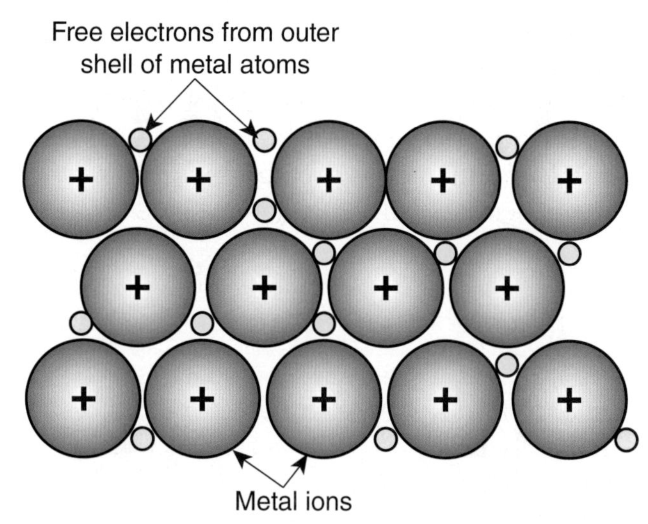

Figure 28-26. Metallic bonds occur when metal atoms share many free electrons, which attract the positively charged nuclei toward the negatively charged free electrons.

Chemical reactions can also break the bonds of molecules to form different molecules. A common example of this type of chemical reaction is seen when steel rusts, which is called ***oxidation***. This is a reaction between an oxygen molecule (O_2) found in the atmosphere and the iron (Fe) in the steel. The reaction breaks down the different bonds of atoms in the steel and in oxygen and combines the atoms to form a new molecule called iron oxide, or Fe_2O_3.

Why is rusting a concern? One reason is because it changes the appearance of the steel and makes it look rough and orange, **Figure 28-27**. A more important reason rusting is a concern, however, is that steel oxidation makes metal weaker. Oxidation replaces strong bonds that form the steel with molecules of

Figure 28-27. Rust is a result of oxidation.

iron oxide, a flaky, powdery material. The chemical reaction essentially eats away at the steel, making it weak. This would be a safety concern if that oxidizing steel was used to support a bridge. Understanding what chemical reactions may occur with different materials in different environments is important to any engineer, especially chemical engineers.

Chemical Equations

In order to better understand chemical reactions in order to control them, chemical engineers represent these reactions as equations. ***Chemical equations*** are a standard method to represent what occurs in a chemical reaction. The general format for a chemical equation is:

$$\text{Reactants} \rightarrow \text{Products}$$

The ***reactants*** are the substances that begin a chemical reaction, and the ***products*** are the substances that are produced as a result of the reaction. The reactants are always on the left side of the equation, followed by an arrow pointing to the products of the reaction. The arrow indicates that the reactants have transformed into the products due to the reaction. Recall that during a chemical reaction, the chemical bonds between the reactants break and new bonds are formed, creating different products of the reaction. The chemical reaction of oxidation or rusting can be represented as:

$$\text{Iron} + \text{Oxygen} \rightarrow \text{Iron oxide}$$

Although this equation represents the reactants and products of the equation, it does not provide much information about the reaction. Therefore, chemical engineers use symbols and coefficients in the equation to represent the reaction. Using chemical symbols and coefficients helps these professionals better analyze chemical reactions.

Let us look at the reaction that produces carbon dioxide (CO_2). Carbon reacts with the oxygen molecule (O_2). The bonds between the two oxygen atoms of the oxygen molecule are broken and new bonds are formed between the carbon atom and the two oxygen atoms. Using symbols, the chemical equation for this reaction is:

$$C + O_2 \rightarrow CO_2$$

Looking at this like a mathematical equation, we know the chemical equation must be balanced. So, just like the equation $2x + 3x = 5x$, the reactants must equal the products. Therefore, each side of the chemical equation must have the same number of each type of atom. The ***law of conservation of mass*** states that matter cannot be created or destroyed. During a chemical reaction, the matter is transformed or changed but the total mass of the matter remains the same. Look back to the chemical equation for carbon dioxide; is the equation balanced? Yes, it is. There are the same number of carbon atoms and oxygen atoms on each side of the equation. The atoms are just arranged differently on each side.

Chemical reactions also include ***coefficients***, which represent the number of atoms and molecules needed for a balanced representation of a chemical reaction. Looking again at the example of rusting steel, you can see that iron + oxygen → iron oxide. The chemical symbols for the reactants are Fe for iron and O_2 for the oxygen molecule. The chemical symbol for the iron oxide molecule is Fe_2O_3. Put these symbols into the equation:

$$Fe + O_2 \rightarrow Fe_2O_3$$

Is the equation balanced? The answer is no. One iron (Fe) atom and two oxygen (O) atoms do not equal two iron (Fe) atoms and three oxygen (O) atoms. In order for this reaction to take place, multiple iron (Fe) atoms and oxygen molecules (O_2) must interact to form iron oxide (Fe_2O_3). This is where coefficients are necessary to represent the chemical reaction. To balance this equation, at least four iron (Fe) atoms must interact with three oxygen molecules (O_2) to form two iron oxide molecules (Fe_2O_3).

Placing coefficients in this equation results in:

$$4Fe + 3O_2 \rightarrow 2Fe_2O_3$$

The 4, 3, and 2 in the equation are the coefficients. Four iron (Fe) atoms and three oxygen molecules (O_2) are the fewest number of atoms and molecules that can react and produce a whole amount of iron oxide molecules. The reactions produce only whole molecules or atoms because a fraction of a molecule would not actually be that molecule, and a fraction of an atom is not possible.

Look again at the chemical equation, and count the number of each atom on each side of the equation. Are they equal to each other? Yes, they are. There are four iron (Fe) atoms on the reactant side (4Fe) and four iron (Fe) atoms on the product side ($2Fe_2$, or 2×2 iron atoms). The reactant side also has six oxygen (O) atoms ($3O_2$ or 3×2 oxygen atoms) and the product side has six oxygen atoms ($2O_3$ or 2×3 oxygen atoms). The equation is balanced and we now know that oxidation requires at least four iron (Fe) atoms and six oxygen (O) atoms.

Balancing equations to determine the amount of substances needed to fulfill the requirements of the chemical reaction is called ***stoichiometry***. Stoichiometry is a part of chemistry that studies the amount of matter that is involved in a chemical reaction. It is a critical tool used by chemical engineers. As you learned, chemical reactions depend on how much matter makes up the reactants. This amount must equal the amount of the products produced through the reaction.

States of Matter

In the previous sections, you read about the structure of matter and how matter can experience chemical changes through chemical reactions. Matter also experiences physical changes. These physical changes do not typically change the chemical composition of different compounds. They can, however, change the state of the matter. Matter can exist in different states or phases, such as a liquid, solid, gas, or plasma. See **Figure 28-28**. These different states all behave in different ways and have different physical properties. However, the molecules that change state do not change chemically. For example, you can freeze water and make it a solid or boil water and turn it into a gas, but the water molecule is still the same. Ice and water vapor are still H_2O.

The state of matter can be changed by adding energy to it or taking it away, typically through freezing or heating. For example, frozen water is a ***solid***, which has a fixed shape and volume. Solids are made of atoms that are densely packed together. This is because when an element is cooled, the movement of the atoms slows down. Melted ice becomes a ***liquid***. Liquids have a fixed volume but can assume any shape. The atoms of a liquid are more spread out than those in a solid. This is because when energy is added to the ice through heat, the atoms move more rapidly. When the water is heated and transformed into water vapor, it becomes a ***gas***. Gases have no fixed volume or shape; they expand to fill whatever container they are in. Gas atoms are even more spaced apart than in a liquid. This is because when heat is applied to water, the atoms become more excited and move around even more. This movement causes the atoms to spread out more. If the water vapor contacts a cool surface, such as a window, the vapor cools and then condenses, changing back to a liquid.

A fourth state of matter is plasma. ***Plasma*** is similar to gases, as it does not have a definite shape or definite volume. However, it is different because it can form structures when subjected to a magnetic field. Plasmas are made up of groups of unbound positively and negatively charged particles that are highly reactive to electrical and magnetic forces. This means the electrons are free to move around the matter in the state of plasma. A gas can only become a plasma if large amounts of energy are applied to remove the electrons. Conversely, a plasma can become a gas if the energy is removed. Therefore, plasmas are not commonly found on Earth. However, plasmas are found throughout the universe, specifically within our sun and other stars.

Within the same physical state, chemical elements also have the capability of existing in different forms. These different forms of the same element in the same physical state are known as

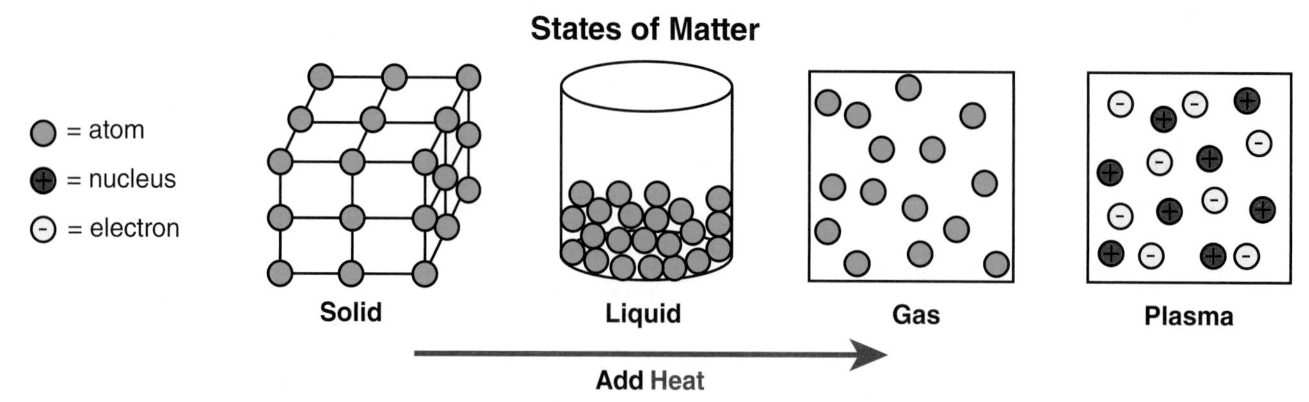

Goodheart-Willcox Publisher

Figure 28-28. As heat is applied to molecules, the atoms become more active and spread apart from one another. This changes the physical properties of the material. When extreme heat is applied, electrons become unbound and are able to move freely around the matter.

allotropes. Allotropes of a chemical element can exhibit very different physical properties from the typical form of the element. This is because allotropes are different structural modifications of the molecules of different chemical elements. Essentially, the atoms of molecules bond together in a different manner.

A good example of element allotropes can be seen by examining carbon (C). Carbon atoms can bond together in one arrangement that forms diamonds and in another arrangement that forms graphite, **Figure 28-29**. A diamond is formed from a tetrahedral arrangement of carbon atoms. Graphite is formed from a hexagonal arrangement. These different arrangements give these different materials very different physical properties. Diamond is very hard, transparent, and bright while graphite is soft, opaque, and dull. Also, diamond is rare and graphite is common. Understanding the allotropes of various elements is important to being able to select, use, and even develop the best materials for specific applications.

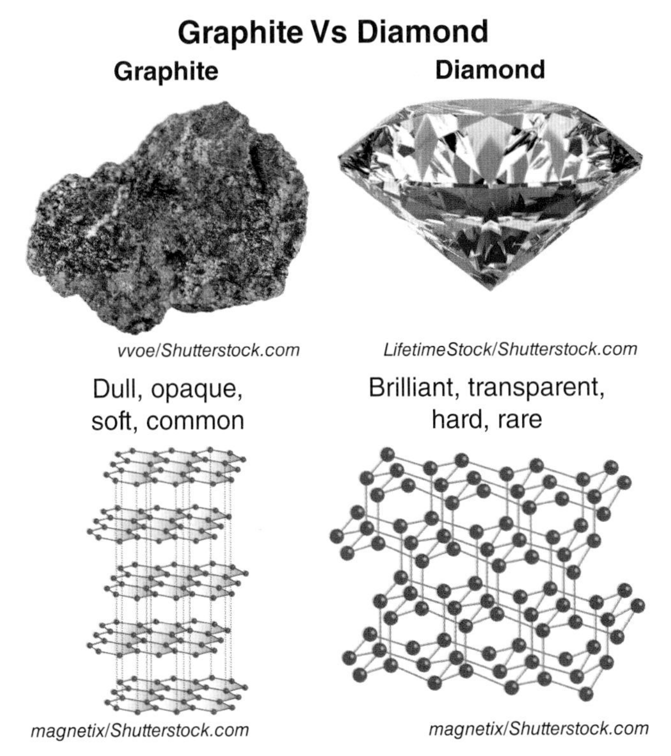

Figure 28-29. Molecules of elements can take on different structural configurations known as allotropes. The different structures give the molecules different properties.

Organic Chemistry and Biochemistry

Chemistry also involves the study of biological phenomena. Just as all matter is made up of atoms bonded together, so are all living organisms. Chemistry principles can be used to determine how the matter of living things interacts with all other matter. The knowledge of these interactions is a huge benefit to all technological fields, especially medicine and agriculture. There is even a branch of chemical engineering focused specifically on the design and development of products and processes involving biological organisms or molecules. It is called ***biochemical engineering***. Biochemical engineers may develop new fertilizers to increase agricultural output or develop ways to grow animal cells to create bioartifical organs. To do these things, engineers require a thorough understanding of organic chemistry and biochemistry.

Organic chemistry is a subdiscipline of chemistry that studies structures, properties, and reactions of organic compounds. ***Organic compounds*** are molecules that contain carbon. Carbon-based molecules can be found inside and outside of living things. These carbon-based molecules can include proteins, carbohydrates, and nucleic acids such as deoxyribonucleic acid (DNA). They can also include hydrocarbons, such as methane and crude oil, which are by-products of decomposing organic matter.

Biochemistry is the study of the chemical compounds and processes that take place in living organisms. It specifically studies the atoms and molecules that make up cells and make the functions of life possible. All living things use the same chemical compounds to sustain life. These compounds include sugars, proteins, and enzymes. Biochemical processes occur in ongoing cycles in living organisms to produce the compounds that they need to survive.

Both of these areas of chemistry examine organic matter. However, organic chemistry is a much broader field, as it studies any compound containing carbon. Biochemistry focuses specifically on the chemical processes within living creatures. That being said, knowledge of both areas is required for chemical engineers to work in specific industries, especially in medicine, biotechnology, and agriculture.

Summary

- ✔ The term *chemical* refers to a compound or substance that has a specific molecular composition that is produced or used for a specific purpose.
- ✔ Chemical engineering is the application of chemistry, biology, physics, and mathematics to solve problems that involve the conversion of raw materials and chemicals into more useful forms of food, medicine, fuels, and many other products.
- ✔ Chemical engineering was not an actual profession until the beginning of the twentieth century. This is when chemistry was applied to industrial applications.
- ✔ The advancement of understanding about DNA helped to launch the fields of biochemical and genetic engineering.
- ✔ Chemical engineers develop new materials and chemicals by using energy to create a basic chemical change in raw materials or chemicals.
- ✔ Chemistry is the study of matter and its interactions with other matter and energy.
- ✔ All matter comprises different types of atoms.
- ✔ The atoms of each material interact with the other atoms in that material, along with the atoms of other matter.
- ✔ Atoms are made up of electrons, protons, and neutrons.
- ✔ Electrons rotate around the nucleus on different orbits called *electron orbitals*.
- ✔ The Aufbau principle determines how the electrons are configured in these orbitals.
- ✔ Every element has an atomic number that indicates the number of protons in the nucleus of its atom.
- ✔ The number of protons in an atom of an element does not change.
- ✔ The periodic table of elements is a grid used to organize the different chemical elements.
- ✔ Chemical reactions are the processes that either bond atoms together or break bonded atoms apart.
- ✔ Molecules are a group of two or more atoms bonded through atoms gaining, losing, or sharing electrons in order to have a full valence orbital and to become electrically neutral.
- ✔ Ionic bonds are the result of the formation of ions. Ionic bonds are strong and hard to break.
- ✔ Covalent bonds occur when atoms share electrons in order to have a full valence orbital and to be stable.
- ✔ Metallic bonds have many electrons in their outer orbitals and willingly share them. Because the electrons can move, the bonds are flexible.
- ✔ Chemical equations are a standard method to represent what occurs in a chemical reaction.
- ✔ Balancing equations to determine the amount of substances needed to fulfill the requirements of the chemical reaction is called *stoichiometry*.
- ✔ Matter can exist in different states or phases, such as a liquid, solid, gas, or plasma. Each state behaves in different ways and has different physical properties.
- ✔ Organic chemistry is the study of the structures, properties, and reactions of organic, or carbon-based, compounds.
- ✔ Biochemistry is the study of chemical compounds and processes that take place in living organisms.

Check Your Engineering IQ

Now that you have finished this chapter, see what you learned by taking the chapter posttest.
www.g-wlearning.com/technologyeducation/

Test Your Knowledge

Answer the following end-of-chapter questions using the information provided in this chapter.

1. Define *chemical engineering*.
2. Chemical engineering became an actual profession when chemistry was applied to _____ applications.
3. List three branches of physical science.
4. *True or False?* Chemistry is the study of matter and its interactions with other matter and energy.
5. What are the three subatomic particles of an atom and what is the electrical charge of each?
6. What information does the atomic number of an element provide?
7. When a(n) _____ loses or gains an electron, it becomes a(n) _____. A(n) _____ is an atom that has gained an electron and a(n) _____ is an atom that has lost an electron.

8. Based on its location on the periodic table of elements, how many electron orbitals does a sulfur (S) atom have?
9. Based on its location on the periodic table of elements, how many valence electrons does tin (Sn) have?
10. _____ are any combination of atoms that are bonded together, while _____ are combinations of two or more different atoms bonded together.
11. Which family of elements are the least reactive, based on the number of electrons in their valence orbital?
12. Define *chemical reaction*.
13. What are the three main types of chemical bonds?
14. What is the general format for a chemical equation?
15. Name the four states of matter.
16. _____ is the study of the structures, properties, and reactions of carbon-based compounds.

Critical Thinking

1. Explain how chemical engineering is used in various industries.
2. Describe how atomic structure provides elements with their chemical and physical properties.
3. Describe the difference between organic chemistry and biochemistry. Explain how each is applied in chemical engineering.

STEM Applications

1. Use the Aufbau principle to create a Bohr atomic model of a krypton (Kr) atom.
2. Balance the following chemical equations by finding the coefficients:
   A. C_3H_8 + ____ O_2 → ____ CO_2 + ____ H_2O
   B. ____ S_8 + ____ O_2 → ____ SO_3
   C. ____ H_2O + ____ O_2 → ____ H_2O_2
3. Set up a scientific experiment to determine how steel wool rusts under different environmental conditions, such as dry air, saltwater, and vinegar. Then answer the following questions:
   A. Does the wool rust differently in each environment?
   B. If so, why does it rust differently?
   C. Why are metal objects often painted?

Engineering Design Challenge

Exothermic Reactions

Background

Prematurely born infants often do not have enough body fat to regulate their body temperature. External heat sources, such as incubators, are used to help maintain their body temperature. In some areas of the world, however, incubators and the electricity they need to operate are not easily accessible. To solve this problem, a team of students designed a portable incubator called the *Embrace Warmer*. This product uses a phase-change material (a substance that can store and release large amounts of heat) within a baby-sized sleeping bag to keep babies warm, helping them to survive. Designing the product and ensuring that it was safe to use required the knowledge and application of chemistry and materials science.

Chemical reactions that produce heat are called exothermic reactions. These reactions give off heat because the energy needed to initiate the reaction is less than the energy that is released. Therefore, the energy is released as heat into the environment. Remember, energy cannot be created or destroyed; it only changes forms. The energy in the reaction is not destroyed; it is just changed to heat. The heat released from chemical reactions can be used for specific purposes. For example, hand warmers use an exothermic chemical reaction to keep your hands warm. The reactants within the hand warmer react with the air, resulting in the production of heat.

Situation

You have learned about a problem that required the use of heat to help babies maintain their internal body temperature. You also learned that some chemical reactions produce heat. How can you use this knowledge to solve another problem?

Challenge

Identify a problem where heat is needed. Research different types of exothermic reactions to determine what types of reactants can be used to produce heat. Then, propose a solution to the problem that uses chemical reactions to produce the necessary heat.

Beginning the Process

- Remember that the engineering design process is an iterative approach to solving problems, meaning you can and should go back and forth between the different steps.
- Identify a problem that requires the production of heat and write a problem statement. A good problem statement should address a single problem and answer the questions who, what, where, when, and why.
- Research exothermic reactions and identify different materials that can be used as reactants to produce heat.
- Create a detailed sketch of your product that uses an exothermic reaction to solve the problem that was identified.
- Propose your solution to your class and instructor.

TTstudio/Shutterstock.com

Many chemical engineers are employed in the oil and gas industries.

SECTION 5
Technology and Society

Tomorrow's Technology Today

Artificial Ecological Systems

Your neighborhood greenhouse contains hundreds of species of plants. However, those plants are likely only the species that can grow and thrive in your local environment. In order to see an array of plants not native to your surroundings, you would have to find an artificial ecological system or closed ecological system; that is, a type of greenhouse that does not rely on anything from the outside to help its plants grow. Closed ecological systems are controlled environments.

Scientists have been attempting to study plants in closed ecological systems for decades. The BIOS-3 facilities in Russia were used to conduct various experiments until 1984. The Biosphere 2 structure in Arizona has had various owners since 1991 and has been conducting research since it was constructed. It is the largest closed ecological system in the world. Biosphere 2 has been used to research agriculture, recycling, and human health.

Another artificial ecological system is the Eden Project, which is located in Cornwall, England, and is open to the public. The Eden Project is not considered a closed ecological system. It consists of two separate biomes and an open-air garden, which house over one million plants from around the world.

The multidomed greenhouses of the Eden Project are the Rainforest Biome and the Mediterranean Biome. The Rainforest Biome consists of several domes representing the environment of a tropical rain forest. It contains hundreds of trees and other plants from rain forests in Africa, Asia, South America, and Australia. Plants include rubber, cocoa, vanilla orchids, and bamboo. The second biome is another multidomed greenhouse. The Mediterranean Biome has plants from temperate rain forests in California, southern Africa, and the Mediterranean. Here, you can find grapevines, olives, orange groves, and hundreds of colorful flowers. The Eden Project's main goal is to educate people about the natural world, especially about using natural resources efficiently so they will continue to be available in the future.

While the Eden Project is the only one of its kind here on Earth, other projects have begun in order to study plant life and to find ways of using resources more efficiently. MELiSSA, or Micro-Ecological Life Support System Alternative, uses similar research methods to other closed ecological systems. However, its purpose is to use the results for long-term space missions rather than to apply the results on Earth. The European Space Agency has been heading this project, which has been in work since 2000. The project is divided into various phases, which start with research on Earth and go through applying that research in structures in space.

These artificial ecological systems have been used for research for decades. The Eden Project and MELiSSA have taken the research some steps further. With these projects making a difference, you can expect to see similar projects around the world in the coming years.

While studying this chapter, look for the activity icon to:

- **Assess** your knowledge with self-check pretest and posttests.
- **Practice** technical terms with e-flash cards, matching activities, and vocabulary games.
- **Reinforce** what you learn by submitting end-of-chapter questions.
  www.g-wlearning.com/technologyeducation/

CHAPTER 29

Technology and Engineering: A Societal View

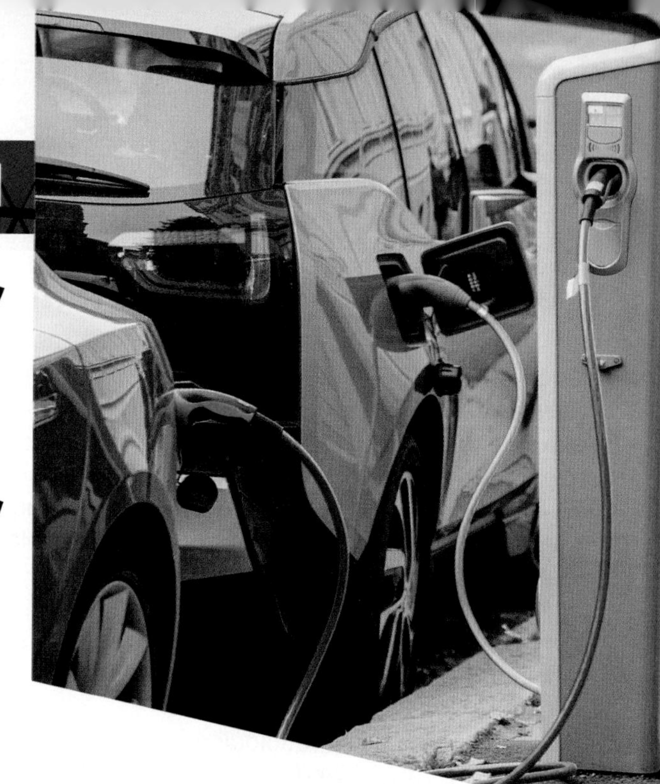

Check Your Engineering IQ

Before you read this chapter, assess your current understanding of the chapter content by taking the chapter pretest.
www.g-wlearning.com/technologyeducation/

Learning Objectives

After studying this chapter, you should be able to:

- ✔ Describe how technology is used to help to control natural forces.
- ✔ Compare and contrast futuring and traditional planning.
- ✔ Differentiate between divergent thinking and convergent thinking.
- ✔ Identify the four major types of futures considered in futures research.
- ✔ Understand the role of nonrenewable, renewable, and inexhaustible energy sources in technology and engineering.
- ✔ Describe the factors that contribute to environmental crisis.
- ✔ Recall actions that companies can take to remain competitive in a world economy.

Technical Terms

aquifers
biological future
environmentalists
futuring
greenhouse effect
human-psyche future
pollution
social future
technological future

Close Reading

What do you think the term *futuring* means? As you read this chapter, create a list of the various ways engineers use this term to consider technology's changing impacts on society.

Technology is the product of human knowledge and ability. Technology is designed through engineering practices to help people modify and control the natural world. Also, technology is the sum total of all human-built systems and products. Technology is the human-built world. See **Figure 29-1**.

Technology is neutral. By itself, technology does not affect people or the environment. How people use technology determines if it is good or bad. How we use technology can help or harm the world around us.

Technology and Natural Forces

Natural forces affect, and sometimes disrupt, human life. Hurricanes wreck ships and destroy

Figure 29-1. The application of technology has greatly changed our world and our way of life.

coastal settlements. Tornadoes can level entire sections of towns and cities. Floods wash away homes and carry away vital topsoil, reducing the productivity of croplands. Fires burn buildings, crops, and forests. Earthquakes shake structures until the structures collapse. In all these natural events, people can be killed. Their possessions can be destroyed.

One of the earliest uses of technology was to harness natural forces. People started to design a human-built world that could reduce and harness destructive natural forces. Dams hold back floodwater and produce energy in hydroelectric plants. Fire is tamed to process industrial materials, heat homes, and cook. See **Figure 29-2**. Geothermal power plants capture Earth's inner heat. Wind is used to generate electricity. Solar energy provides heat and electricity for businesses and homes.

Dmytro Gilitukha/Shutterstock.com

Photografeus/Shutterstock.com

Figure 29-2. People use technology to harness and control the forces of nature. Fire can destroy a forest or can be harnessed for useful tasks, such as cooking a pizza.

Controlling and using natural forces, however, is not enough. Humans are starting to understand that having control over Earth's future carries with it serious responsibilities. Many people realize that humans must protect the environment and the plants and animals that live with us on this planet. We must live in harmony with the natural world.

Technology and the Future

Since technology is a product of human activity, humans can control it. To do this, they must have an idea of the kind of future that is desired. A research technique called ***futuring***, or *futures research*, helps people select the best of many possible courses of action for the future. See **Figure 29-3**. Futuring emphasizes five distinct features:

- **Alternate avenues.** The futurist looks for many possible answers, rather than one answer.
- **Different futures.** A futurist looks for an entirely new future, while traditional planning sees the future as a refinement of the present.
- **Rational decision making.** Traditional planners rely heavily on statistical projections. They use mathematical formulas and scientific knowledge to help them predict events and

Peshkova/Shutterstock.com

Figure 29-3. People can use futuring to determine the type of future they want.

Career Connection Environmental Engineer

Environmental engineers are concerned with problems related to the environment. They address issues that surround humans, animals, and nature, such as pollution and contaminated drinking water. Environmental engineers predict and identify possible hazards and create solutions for containing and reducing the impact. They spend a significant amount of time in the field, rather than inside a lab. This allows them to collect the samples they use for testing.

High school students interested in environmental engineering should take calculus, chemistry, and biology courses to prepare them for a highly respected college program. Additionally, courses in technology and engineering can provide the foundation for the engineering program they choose in college.

A bachelor's degree in environmental engineering or related field is required. Licensure requirements and exams include the Fundamentals of Engineer exam and Principles of Practice of Engineering. Environmental engineers with a master's degree can expect greater job prospects.

Those interested in an environmental engineering career should possess critical-thinking and problem-solving skills. They should also have strong communication abilities and be able to work with others.

Environmental engineers work in a variety of fields, including civil engineering and manufacturing engineering. They also work for government agencies, such as the Environmental Protection Agency (EPA). However, they are able to work in a diverse selection of positions including data collection, technician, and administrator.

Environmental engineering is expected to grow through the year 2024 as a result of increasing regulations from the federal government and professional organizations.

An environmental engineer may work on the following types of projects:

- Creating plans that recommend solutions to local environmental concerns, such as storm water programs.
- Providing expertise in formulating management plans and regulations related to the environment.
- Investigating, documenting, and reporting on environmental incidents.
- Assessing potential environmental incidents specific to local municipalities.

Professional organizations include AAEES and National Society of Professional Engineers.

impacts. A futurist uses logical thinking and considers the consequences when making decisions.

- **Designing the future.** Futurists are not concerned with improving present or past practices. They do not see the future as a variation of the present. Futurists focus on predicting a possible future that can be created.
- **Interrelationships.** Traditional planners use linear models that suggest that one step leads to the next step. Futurists see alternatives, cross-impacts, and leaps forward.

Using a futures research approach, exciting new technologies can be developed. Futurists have a dual view. One view is a present challenge or problem. The other view is the world of the future. Looking at both of these views requires a combination of short-term and long-term goals. The most important tool to develop futures is the human mind, which is the only tool capable of reasoning and making value judgments.

Types of Thinking in Futuring

Futuring requires two types of thinking. The first is divergent thinking, which allows people to let their minds be free to explore all possible and, in many cases, impossible solutions. Divergent thinkers look for interrelationships and connections. A number of futures emerge from this activity.

One possible future, however, must be selected as the one that will be created. This requires convergent thinking. The final solution must receive focus and attention. This solution's positive and negative impacts must be carefully analyzed.

Types of Futures

The analysis of a solution must be done with four types of futures in mind. The first is the ***social future***, which suggests the types of relationships people want with each other. The second is the ***technological future***. This looks at the type of human-built world we desire, such as structures, transportation systems, and electronics. The third is the ***biological future***. This future deals with the types of plant and animal life we want. See **Figure 29-4**. The biological future might hold new developments in the use of plant and animal products, especially those from the sea. The fourth is the ***human-psyche future***. This future deals with the mental condition of people. The human-psyche future stresses the spirit, rather than the mind—attitude, instead of physical condition. This future is concerned about how people will feel about life and themselves.

Christopher Wood/Shutterstock.com

Figure 29-4. New technologies impact the environment. The survival of polar bears is threatened by a rapid loss of sea ice due to greenhouse gases.

The different types of futures are interconnected. For example, new technologies directly impact how people relate to one another. Television dramatically changed family life, recreation, and the entertainment industry. Technology also changes the natural environment. Acid rain caused by automobile emissions and coal-burning electric plants has destroyed forests in Canada and the eastern United States. See **Figure 29-5**. Likewise, technology has changed how we view ourselves. Some people feel threatened by technology. Other people feel empowered by it. Along with futurists, engineers must consider the effect of the technology they create on all the types of futures. Often, they cannot predict what the outcome will be.

Technology's Challenges and Promises

This chapter contains examples of how new technologies have impacted our lives and how they will impact the future. Later, you can use these examples as you evaluate other technologies. Three widely discussed issues are energy use, environmental protection, and global economic competition. Let's look at these issues in terms of their challenges and promises.

Mary Terriberry/Shutterstock.com

Figure 29-5. Damage to trees caused by acid rain.

Technology Explained Organic Light-Emitting Diodes (OLED)

Organic light-emitting diode (OLED): Energy-efficient, solid-state devices that use thin layers of organic materials to create light.

Television sets (TVs) and electronic displays are continuously being improved as a result of advancements in science and engineering. One of the newest and most advanced TVs available for consumers uses organic light-emitting diodes (OLEDs). See **Figure A**.

While previous TVs, such as plasma displays, had many advantages over traditional CRTs, there were some disadvantages that OLED technology overcame. For example, plasma TV panels were often heavy and not economically manufactured in larger sizes. Conversely, OLEDs can be produced in large sizes and are very energy efficient.

Alternatives to plasma TVs include LCD (liquid crystal display), DLP (digital light processing), and LED. However, OLED (organic light-emitting diode) displays are said to provide a brighter, higher quality display while consuming less electricity than traditional LEDs or LCDs, making them better for the environment.

cigdem/Shutterstock.com

Figure A. Gas-plasma displays are made of a phosphorescent material contained between two thin surfaces.

OLEDs are solid-state devices, which means they are built solely of solid materials that allow electrons to flow while containing no moving parts. OLED displays are primarily composed of thin layers of organic molecules that create light. Components commonly include a substrate, an anode, organic layers, a cathode, a conducting layer, and an emissive layer.

An OLED works by infusing voltage across the cathode and the anode. As a result, the cathode layer receives electrons while the anode loses electrons. When this happens, the conductive layer ends up positively charged, while the emissive layer is negatively charged. The positive holes, or where electrons should be, jump between layers. This movement releases an energy burst in the form of light. The energy bursts happen so fast that the result appears to be continuous light.

Energy Use

The supply and use of energy resources are important because almost everything in the human-built environment depends on energy. Engineers are increasingly pressed to consider energy use in the technologies they create. For example, new products such as refrigerators, automobiles, and air-conditioning units need to be created to use less energy.

Nonrenewable Energy Resources

Many of the energy resources we use are exhaustible, or nonrenewable, resources. The supply is finite—there is a limited quantity of the resource available. Examples of exhaustible energy resources are coal and natural gas.

Another nonrenewable resource is petroleum, which is the fuel that powers most transportation vehicles. Our society's focus on the automobile as its primary transportation vehicle has global impacts. The citizens of North America represent a small percentage of the world's population; however, they use a large percentage of the world's petroleum. Also, automobiles discharge great quantities of pollutants into the atmosphere.

Nuclear power provides a long-lasting source of energy. These power plants do not leak the toxic gases into the atmosphere that chemical-consuming power plants produce. Yet, the waste from nuclear power is more dangerous and longer lasting.

The shrinking supply of nonrenewable resources is a major concern, particularly in the case of petroleum. We are challenged to reduce our dependence on these resources by finding alternative energy resources. See **Figure 29-6**. Some alternatives to petroleum are hydrogen, solar, wind, and electric vehicles.

Renewable Energy Resources

One alternative is to shift to renewable resources—the products of farming, forestry, and fishing. However, these resources have life cycles and, therefore, are in limited supply at any one time. For example, in many less developed countries, wood is the primary fuel for cooking. As the population grows, people must roam a greater distance to find the firewood they need. Thus, large regions are being stripped bare of trees. Also, using wood for fuel eliminates it as a source of building material.

Likewise, corn can be used to make a fuel called *ethanol*. However, if we shift large quantities of corn from food production to fuel production, world hunger might be worsened.

Inexhaustible Energy Resources

Another solution is to make greater use of inexhaustible resources. The most common of these are solar energy, wind energy, and water energy. Electricity can be generated using any or all of these three energy sources, but a large expenditure of money and human energy is needed. Generating electricity using these energy sources also covers large tracts of land with solar and wind generators.

It takes time to develop the technology to fully use inexhaustible energy resources. For example, solar-powered automobiles have been developed, although consumer use may be years away. In addition, solar planes have been developed. A solar plane has completed a 17-leg journey around the world.

ArtisticPhoto/Shutterstock.com

Shift from Exhaustible Resources

Malzasgphoto/Shutterstock.com

to Renewable Resources

Kajano/Shutterstock.com

and Inexhaustible Resources

Figure 29-6. The future requires a shift from exhaustible energy sources to renewable and inexhaustible sources.

Using Energy Efficiently

Another solution to our energy problems is to use energy more efficiently. Engineers create energy-efficient technologies, but this is not the total answer to energy conservation. Members of society should carefully consider the way energy is used. Should people drive to work alone in a personal car? Should people heat their homes to 75°F (24°C) in the winter and cool them to 65°F (18°C) in the summer? Should we make buildings more costly by using more insulation and installing double- and triple-glazed windows? Should people be strongly encouraged, through taxes or fees, to use public transportation instead of their cars? Should gas-guzzling cars be removed from the market?

Environmental Protection

Open space, clean air, land for a home or farm, and safe drinking water were once viewed as birthrights for people in North America. People simply expected them to be available, but today, we know better. The understanding that unwise use of technology can threaten our quality of life was slow to come.

However, environmental issues have been brought into sharper focus in recent decades. People have participated in what can be called an *environmental revolution*. It has been realized that the natural environment has a direct effect on the safety and health of people. See **Figure 29-7**. The long-term survival of any civilization is based on wisely managing natural resources.

Protecting the environment involves studying the relationship between the human population and the use of technology. This relationship directly impacts a number of environmental conditions. These include the climate and the supplies of food, water, energy, and material resources.

Many scientists say we are creating an environmental crisis. This means we must take action or the environment will be permanently damaged. Three important forces contributing to this crisis are overpopulation, resource depletion, and pollution.

Overpopulation

From the dawn of civilization to the end of the 1700s, the world population grew to about 1 billion people. During the 1800s, the population increased at a higher rate. By 1900, it reached a total of about 1.7 billion people. According to the US Census Bureau, the world population increased from 3 billion in 1959 to 6 billion by 1999, a doubling that occurred over 40 years. The current world population is approximately 7.5 billion people, and is predicted to increase to approximately 9 billion by the year 2044.

The world's population is not evenly distributed. See **Figure 29-8**. The number of people per square mile of cultivated land varies greatly. In Europe, population density is three times that of North America. In China, it is eight times that of North America.

David Tadevosian/Shutterstock.com

Figure 29-7. A healthy environment, with unpolluted air and water, is essential for a good quality of life.

lazyllama/Shutterstock.com

Figure 29-8. A crowded favela (shantytown) in Rio de Janeiro, Brazil.

Think Green Shopping Bags

The way consumers transport products has an environmental impact. It is a small but meaningful change that can make a big difference.

When you make your purchases at a store, you may have the option of putting your merchandise in paper bags or plastic bags. Paper bags are not always reusable, but you can recycle them when you have finished using them. If you throw them out, they can decompose naturally. If you choose plastic bags, you can typically save and reuse them. If they cannot be reused, however, they should be recycled. Recycling plastic bags uses less fossil fuels than producing new plastic bags. Plastic bags should not be thrown out.

Another option you have is to invest in reusable bags and to carry at least one with you whenever you shop. Because there is no recycling or waste associated with reusable bags, using these bags has a lesser effect on the environment.

This is a "good news and bad news" situation. The good news is people are living longer. Technology has given us more food, better health care, better disease control, and better sanitation. The bad news is that this is true only in developed countries. Third World countries are experiencing most of the population growth. The economies of these countries cannot support rapidly growing populations, leading to tremendous hardships. Many people in Third World countries go to sleep hungry each night. Millions of people starve to death each year.

As we continue to control diseases, improve health care, advance agricultural abilities, and use diplomacy instead of war to solve international conflicts, the population grows even faster. The increased population contributes to resource depletion and pollution. A concerted effort is needed to respond to growth of the world's population. For example, technologies to improve and create desalination plants can assist in making seawater potable and resolve issues concerning access to water.

Resource Depletion

Each living person places demands on the resources of the planet. Many material resources have finite supplies. Once used, they are gone forever. These resources include metal ores, petroleum, natural gas, sulfur, and gypsum. As with energy resources, we can shift from using nonrenewable resources to renewable resources. These renewable resources are primarily the food and fiber produced by farming, forestry, and fishing. See **Figure 29-9**. The fertility and availability of land and the productivity of the oceans, however, limit the supply of these resources. If we take too much from the water and land, the water and land will be damaged.

For example, intensive forestry practices are used in many parts of the world. The logs, limbs, and bark of the tree are taken away to be processed into products. Nothing is left behind. However, leaving nothing behind is unwise. The forest floor is left without the limbs and the bark. These parts of the tree normally break down and provide nutrients for other plants and trees. As a result of the bare forest floor, the soil becomes less fertile. Coupled with the acid rain falling on many forests,

digitalreflections/Shutterstock.com

Figure 29-9. Farming is one source of renewable material resources and food.

this produces trees that are not as healthy as before. The production of wood fiber is reduced.

Likewise, intensive farming practices deplete the soil. Soil depletion makes heavy applications of commercial fertilizers necessary. Also, using prime farmland for housing forces us to cultivate less productive areas. We strain precious water resources to irrigate crops growing in the desert and draw the water from rivers and wells. This, in turn, reduces the amount of water downstream and depletes aquifers. ***Aquifers*** are underground water-bearing layers of rock, sand, or gravel.

A problem related to these issues is land use. Throughout the country, conflicts are raging between ***environmentalists*** (people who are concerned about the protection of the environment and leaving it unspoiled) and commercial interests. See **Figure 29-10**. An example of one controversy is the spotted owl habitat in the forests of Oregon and Washington. Commercial interests want to cut virgin timber and manage the forest for maximum fiber production. Environmentalists want most of these forests left untouched as a habitat for the owls.

Debate continues over how much wilderness is enough. Some people want large tracts of rangeland and forestland to remain untouched. They want only hiking trails in these areas. Few people dispute the need to save some unspoiled areas. The debate is over the size and number of these areas. There are no clear-cut answers. There are only opinions of what is right or wrong, good or bad.

As with energy, short-term solutions to resource depletion lie in better use of material, land, and water resources. For example, with the aid of technology, various aspects of the environment can be monitored to provide information for decision making. One such technology is the use of drones to conduct crop analysis or monitor the effect of runoff from streams. Additionally, the alignment of technological processes with natural processes, such as the use of hydroelectric dams, solar panels, and wind turbines, maximizes performance and reduces negative impacts on the environment. Decisions regarding the implementation of technologies involve the weighing of trade-offs between predicted positive and negative effects on the environment. Consider the effect hydroelectric dams and wind turbines have on animals, such as fish and birds.

We must use our land, but we must not abuse it. Some of it should be set aside as nature preserves. Other areas should be considered multiuse land, combining recreational and commercial uses. Hunters, campers, ranchers, farmers, and loggers could all use this land. Finally, some tracts of land should be devoted to commercial, residential, and transportation uses.

We must reconsider how we use Earth's resources and learn to use all materials more efficiently. This includes buying fewer items we really do not need and buying products that will last. Maintaining and repairing products whenever possible, instead of discarding them, is also important. Finally, when a product can no longer be used, the materials must be recycled.

Pollution

People living in some major cities rarely see clear skies. Haze reduces the visibility. Smog has a negative

Environmentalism

haveseen/Shutterstock.com

Commercialism

Anna Moskvina/Shutterstock.com

Figure 29-10. There is a continuing debate over land use in the United States. On one side are those who want land kept in a wilderness state. On the other side are people who want more land for lumber or agricultural production.

effect on health. Water in many parts of the world is unsafe to drink. In some areas, the land is contaminated with hazardous waste. People cannot live on this land or travel over it. All these things are called pollution. ***Pollution*** is the introduction of a harmful material into the environment. See **Figure 29-11**. Pollution is most often a product of human activity.

This product of human activity sometimes has been brought to our attention in dramatic ways. In 1969, the Cuyahoga River in northern Ohio literally caught fire. This river runs through the city of Cleveland. The Cuyahoga River had become severely polluted with debris and flammable liquids. The incident made many people more aware of the pollution issue. This fire led to the passage of new laws to protect the waterways and efforts to clean up streams, rivers, and lakes.

Pollution affects more than the air we breathe, the land we walk on, and the water we drink. Many scientists think it is changing the climate of our planet. In the spring of 1983, scientists observed a brown layer of air pollution over the Arctic region. Later, other scientists discovered a hole in the ozone layer over the Antarctic. This hole has grown larger over the years. Since that time, a controversy has raged. Some scientists say the hole lets more UV light into the atmosphere than before. They believe these rays, combined with increased levels of carbon dioxide and other gases, cause Earth to retain more heat. Scientists call this problem the ***greenhouse effect***. This causes global warming, with higher land and water temperatures. Scientists are concerned that a warmer climate will melt the polar ice caps and cause the oceans to rise. In turn, large coastal areas, including some major cities, might be flooded.

Other scientists feel that global warming is part of a natural cycle. They say warming and cooling have occurred a number of times over history. Global warming, however, is an area of great concern. Changes have been called for in technological actions. For example, certain chemicals in air conditioners, refrigerators, foam containers, and aerosol sprays are blamed for part of the problem. These chemicals are being removed from the market as replacements are developed.

A plan is needed to protect the atmosphere and the ozone layer. Failure to act might result in a significant change in the world's climate. The greenhouse effect could turn lush farmland into desert, cause the extinction of some species of plants and animals, and create widespread human suffering.

Global Economic Competition

Another issue directly related to society and technology is distribution of wealth and industrial power. Some historians suggest that technical knowledge and power have moved steadily around the world from east to west. At one time, China was a global power and the source of many innovations. This center of power moved across what is now India into the area around the Mediterranean. That area was the dominant economic center about 2000 years ago. Later, northern Europe became the economic leader, with the Industrial Revolution of the 1800s. The dominance of the United States followed in the 1900s. Now, the area called the *Pacific Rim* has become more important. Japan, Korea, China, and other Far East countries are challenging the industries of North America and Europe. See **Figure 29-12**. This represents nearly a full cycle of industrial development around the globe.

Today, we understand economics better than at any time in history. This allows countries on

Figure 29-11. Pollution affects the land, water, and air we depend on for life.

Sakarin Sawasdinaka/Shutterstock.com

Figure 29-12. Pacific Rim countries challenge the industries of North America and Europe. Japan is a major exporter of automobiles and consumer electronics.

the back side of the economic wave to resist losing economic power. These countries can take a number of actions to overcome a loss of economic power, including the following:

- **Change management styles.** Companies are redefining the roles of workers and managers. They are creating teams that design and produce products. These teams include designers, engineers, production workers, quality control specialists, marketing people, and managers. Managers are not seen as bosses. They help others do their work better and with greater ease.
- **Increase use of computers.** Competitive companies are replacing manual labor with computer-aided or computer-controlled work. See **Figure 29-13**. Examples include the following:
    - CAD systems make drawings easy to produce, correct, and store.
    - Computer-integrated manufacturing (CIM) ties many manufacturing actions to the computer. Computer systems monitor machine control, quality control, parts movement, and an array of other operations.
    - 3-D printing allows quicker modeling and prototyping of new products.
    - JIT inventory control uses a computer system to monitor material orders so supplies arrive at the plant just before they are needed. Also, finished products are made only when they are needed.
    - Computer-controlled purchasing and warehousing operations, such as warehousing systems in which robot vehicles store and retrieve parts.
- **Produce world-class products.** Successful companies make products to meet the needs of customers around the world. The products must function well, be fairly priced, and deliver excellent value to the customer. No longer can a local or national area provide a safe, protected market for a company. Political forces are causing the world to become one large market for all countries.
- **Use flexible, automated manufacturing systems.** Traditional manufacturing is based on long production runs. Semiskilled workers on a manufacturing line make a single product. Automated production lines called *flexible manufacturing systems* are replacing this type of production. See **Figure 29-14**. Such computer-controlled systems can produce a number of different products with simple tooling changes. Flexible manufacturing is cost-effective for small quantities of products.

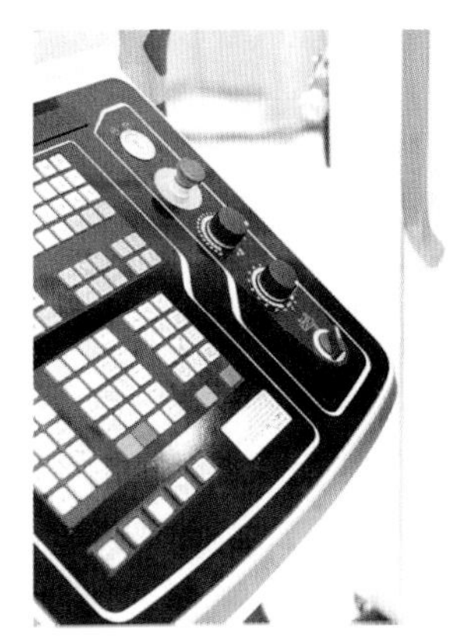

Pixel B/Shutterstock.com

Figure 29-13. In many business and industrial operations, computer applications have replaced manual methods.

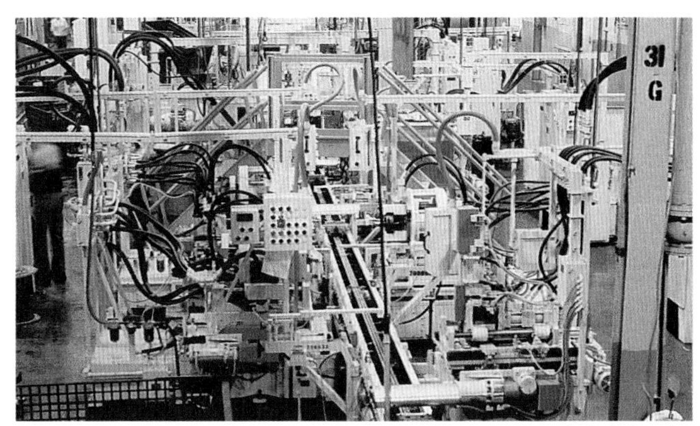

Arvin Industries

Figure 29-14. Today, products are often produced on flexible, automated production lines. Computer control allows fast, simple changeover to produce different types of products.

Summary

- ✔ Technology is neutral. How people use technology determines if it is helpful or harmful.
- ✔ One of the earliest uses of technology was to harness natural forces.
- ✔ Control over Earth's future carries with it serious responsibilities. The environment, plants, and animals must be protected.
- ✔ Futuring requires two types of thinking. Divergent thinking involves exploring all possible and even impossible solutions, resulting in a number of futures. Convergent thinking focuses on the final solution, and the solution's positive and negative impacts are carefully analyzed.
- ✔ The four types of futures are the social future, the technological future, the biological future, and the human-psyche future.
- ✔ Along with futurists, engineers must consider the effect of the technology they create on all the types of futures.
- ✔ Energy concerns include shifting from use of nonrenewable energy sources to renewable and inexhaustible energy sources and using energy more efficiently.
- ✔ The natural environment has a direct effect on the safety and health of people. The long-term survival of any civilization is based on wisely managing natural resources.
- ✔ Overpopulation, resource depletion, and pollution are three major challenges that must be addressed in order to protect the environment.
- ✔ Companies use various methods to compete in the global economy. These include changing management styles, increasing computer-aided or computer controlled work, producing world-class products, and employing flexible, automated manufacturing systems.

Check Your Engineering IQ

Now that you have finished this chapter, see what you learned by taking the chapter posttest.
www.g-wlearning.com/technologyeducation/

Test Your Knowledge

Answer the following end-of-chapter questions using the information provided in this chapter.

1. *True or False?* Technology, by itself, does not affect people or the environment.
2. An early use of technology was to control _____ forces.
3. *True or False?* A futurist sees the future as a refinement of the present.
4. Describe the two types of thinking used in the futuring process.
5. The type of future that examines the type of human-built world desired is the _____ future.
   A. social
   B. biological
   C. technological
   D. human-psyche
6. Which of the following is *not* an inexhaustible energy resource?
   A. wood
   B. the sun
   C. wind
   D. water
7. Due to their _____, the products of farming, forestry, and fishing are in limited supply at any one time.
8. *True or False?* Rapid population increase contributes to resource depletion and pollution.
9. What is the greenhouse effect?
10. *True or False?* Flexible manufacturing systems are based on long production runs in which semiskilled workers make a single product on a manufacturing line.

Critical Thinking

1. What might happen if humans continue to consume nonrenewable energy resources at an increasing rate?
2. Do you believe that global warming is a natural occurrence or a result of human-created technology? List evidence to support your point of view.
3. Compare and contrast two alternative transportation fuels. List the pros and cons of each.
4. What steps has your local area taken to use energy and material resources efficiently? What steps that have not been taken do you think would help to conserve resources?

STEM Applications

1. Consider a technology you encounter every day. Explain the effect of this technology and the positive and negative consequences it may have on the various types of futures.
2. Explain challenges and possible solutions for how resource depletion and pollution can be overcome.

Engineering Design Challenge

Resource Depletion

Background

Natural resources such as oil, forestry products, and coal are used to harvest food, construct shelters, and generate energy. Yet, as the global population continues to rise, the increasing consumption of these natural resources is neither sustainable nor good for the environment. The United Nations Human Development Report indicates that the United States contains around 5% of the global population, yet consumes nearly 40% of the world's resources. This places a significant amount of responsibility on the citizens of the United States.

Situation

As an environmental engineer and concerned citizen of your community, you have been tasked with designing alternative homes that are more environmentally sound. This requires you to research and investigate current home construction (the process and final design) and identify ways in which homes can be designed to minimize their negative impact on the environment.

You will create two final products. The first is a model house design, which can be either a 2-D blueprint, 3-D computer model, or an actual physical model. The second is a new set of standards and regulations for your township's planning and zoning department.

Desired Outcomes

- A detailed 2-D design, 3-D design, or model of an environmentally sound home or alternative construction process that meets the following criteria:
  - Accommodates a family of four.
  - Reduces consumption of at least two natural resources.
  - Includes the use of two environmentally friendly resources.
  - Improves the efficiency of the home.
  - Is reasonably priced and relatively inexpensive.
- An explanation of how the engineering design process was used to develop the design. This can be done using a poster, presentation, or video.

Materials and Equipment

- Graph paper.
- Construction supplies: 1/4 × 1/4 × 36 balsa wood, 1/4 × 1/2 × 36 balsa wood, 1/8 × 3 × 36 balsa wood, styrofoam, cardboard, PVC pipe, foam insulation, plywood, garbage bags, dowel rods, etc.
- Computer and CAD software.
- Glue gun.
- Fasteners.

Procedure

- Use the engineering design process to guide you through the process of defining your problem and creating a viable solution to it.
- Remember the engineering design process is an iterative approach to solving problems. You can and should go back and forth between the different steps.
- Define the problem based on the background and situation provided above.
- Gather information related to the problem that has been defined.
- Apply the knowledge you obtain to propose a new or improved solution to the problem. This will be a natural progression from brainstorming, to generating ideas, to developing a design, to evaluating the design's potential success.
- Document all of the work that you do to create a solution design in your engineering notebook.

esbobeldijk/Shutterstock.com

This community uses solar panels for some of their electrical power needs.

CHAPTER 30

Technology and Engineering: A Personal View

Check Your Engineering IQ 

Before you read this chapter, assess your current understanding of the chapter content by taking the chapter pretest.
www.g-wlearning.com/technologyeducation/

Learning Objectives

After studying this chapter, you should be able to:

- ✔ Illustrate how jobs have evolved from the colonial period, through the Industrial Revolution, to the information age.
- ✔ Describe the typical levels of technology-based jobs.
- ✔ Identify the factors that should be considered when selecting a job.
- ✔ Describe the three factors affecting job satisfaction.
- ✔ Explain how individuals exercise control over technology.
- ✔ Describe the three major concerns people have to which technology can contribute to a solution.
- ✔ Provide examples of some technological activities introduced as fiction that have become actual processes.

Technical Terms 

activists
flexible manufacturing
information skills
language and communication skills
lifestyle
people skills
personal skills
society
socioethical skills
thinking skills

Close Reading

As you read this chapter, think of different jobs in technology that appeal to you. What do you think the job requirements are for these jobs? What kinds of skills are necessary?

Your personal life is highly dependent on technologies developed by engineers and others. In the last century, life has changed dramatically for citizens of developed countries. We live in different housing, travel on different systems, have different products to purchase, and communicate in ways far different from in the past. In most instances, these new technologies are more efficient, effective, and less demanding on the environment. To help investigate the changes affecting everyday life in the twenty-first century, this chapter examines five areas:

- Technology and lifestyle.
- Technology and employment.
- Technology and individual control.
- Technology and major concerns.
- Technology and new horizons.

Technology and Lifestyle

Each person lives in a specific way, or has a lifestyle. A ***lifestyle*** is what a person does with business and family life. See **Figure 30-1**. People pursue different lifestyles, but overall, lifestyles are changed with the advancement of technology.

Blend Images/Shutterstock.com

Figure 30-1. What kind of lifestyle do you think this father and son enjoy?

Colonial Life and Technology

The lifestyle during America's colonial period was a harsh contrast to that of today. Housing was simple and modest. Most products were designed and produced to meet basic human needs. Few decorative items were available, and those were mostly owned by the wealthy. Transportation systems included horses, animal-drawn wagons, and simple boats. Communication systems were crude and often face-to-face.

Most people lived on farms. A few people practiced basic crafts. These people were the carpenters, blacksmiths, and other tradespeople needed to produce basic products required by the community. The men and boys did most of the fieldwork on the farm and practiced the trades. Women and girls tended gardens and did household work. See **Figure 30-2**.

Everyone had to work long hours, six days a week, to raise small amounts of food. The results were often dramatically affected by weather changes and lack of scientific knowledge to alter their processes for producing food. Most colonial families set aside the seventh day for church activities. People did not take vacations and celebrated few holidays.

With the exception of slaves, however, most people were their own bosses. They owned their farms and stores. These people practiced their crafts as independent workers. Each person was an owner, a manager, and a worker rolled into one.

Brian McEntire/Shutterstock.com

Figure 30-2. This woman is trimming the end of a hand-dipped candle, a common task in colonial life.

The Industrial Revolution and Technology

The Industrial Revolution of the mid-1800s changed this lifestyle. A series of events took place. First, advanced technology was developed for the farm. This technology included the moldboard plow, the reaper, and the steam tractor. These and other devices made farmers more efficient. Fewer people could farm more land and produce more food to sustain the society. The term ***society*** refers to a collection of people who generally live in a community, region, or nation and share common customs and laws.

Second, the movement to the land west of the Appalachians in the United States allowed for larger and more efficient farms. During this period, the percentage of the workforce engaged in farming began dropping rapidly. At the start of the Industrial Revolution, more than 90% of the workforce were engaged in farming. Today, fewer than 3% of workers are employed on farms.

Third, a large number of people from Europe immigrated to the United States during this time. These immigrants, plus the farmers who were no longer needed to till the soil, provided a vast labor supply. This labor supply was a basic resource for the factory system that was then being developed. The local tradespeople working in their shops could no longer meet the demand for goods. Centralized manufacturing operations were replacing the functions of the tradespeople. See **Figure 30-3**. Features of these operations included the following:

- **Professional management.** This management established procedures, employed resources, and supervised work.
- **Division of labor.** This division assigned portions of the total job to individual workers. Each worker did only part of a total job, allowing the workers to quickly develop the specialized skills needed to do the assigned tasks.
- **Continuous-manufacturing techniques.** These techniques increased production speed. Raw materials generally entered a production line at one end. Finished products left the line at the other end.
- **Material-handling devices.** These devices were used to move the products from workstation to workstation. Workers remained at their stations. The products moved to them.

Photographee.eu/Shutterstock.com

Figure 30-3. Many current manufacturing plants are based on the principles of the Industrial Revolution.

- **Interchangeable parts.** These parts allowed the production of large quantities of uniform products.

Low wages and poor working conditions in the factories caused widespread worker unrest. Labor unions were formed to give the workers a voice in determining working conditions and pay rates. Bloody battles erupted between the workers and management. The government usually supported managerial positions, resisting the unions' attempts to deal with the issues.

These conflicts were finally settled with changes in governmental attitudes, new laws, and different management stances. This led to a strong industrial period for the country. Broad employment opportunities characterized the period. The workers enjoyed a high standard of living. The 40-hour, five-day workweek with a number of holidays and paid vacation time became fairly common.

The Information Age and Technology

The development of the computer changed the Industrial Revolution. During the industrial period, the company that could efficiently process the greatest amount of material was the most successful. This required major investment in large continuous-manufacturing plants. The huge automobile- and steel-manufacturing operations characterized these plants. These plants employed thousands of people and used millions of tons of materials.

The computer allowed the development of a new type of manufacturing called ***flexible manufacturing*** that can quickly and inexpensively respond to change. People with few skills are replaced with computer-controlled machines. See **Figure 30-4**. The workers who remain have more training and motivation to work. They accept change and responsibility more readily than the workers of the Industrial Revolution. Another technological development related to the information age that is affecting manufacturing is the Internet of Things (IOT). The IOT is the Internet of physical devices and other technological products through smart devices. This requires the use of software, sensors, and network connectivity to enable objects to communicate and achieve a task.

Also, management styles have changed greatly. Management is less distant from the workers. The entire workforce is seen as a team, with each person having an area of responsibility. See **Figure 30-5**. Managers might be responsible for setting goals and controlling money. Workers are responsible for producing products. Everyone is responsible, however, for work procedures and product quality.

These management style changes have given people a new lifestyle. Those people who are able to change and adjust to the demands of the new age are better informed, work more with their brains than their muscles, and have more control over their work. See **Figure 30-6**.

wavebreakmedia/Shutterstock.com

Figure 30-5. In today's industry, managers and workers work as a team. They cooperate to reach company goals.

Cincinnati Milicron

Figure 30-4. This robot is placing cartons on a pallet. Semiskilled workers formerly did this task.

Billion Photos/Shutterstock.com

Figure 30-6. Workers in modern industry more often work with their brains than with their muscles.

Technology and Employment

Lifestyle and employment are closely connected. Most people need the money they earn through working to afford the type of life they want.

Some general requirements for most jobs in the future can be identified. Almost all technical jobs will require some type of postsecondary education, which includes technical training in career centers and community colleges, a college or university education, or an advanced degree. See **Figure 30-7**. In addition, workers in the information age must be willing to do the following:

- Continuously pursue additional education and training throughout their work lives.
- Accept job and career changes several times during their work lives.
- Work in teams and place team goals above personal ambitions.
- Exercise leadership, be self-driven, and accept responsibility for their work.

Types of Technical Jobs

A wide variety of jobs require technical knowledge. They include five levels, **Figure 30-8**. These levels are as follows:

Phovoir/Shutterstock.com

Figure 30-7. Specialized training beyond high school is required for many technical jobs.

- **Technically trained production workers.** These workers include the people who process materials and make products in manufacturing companies, erect structures, and operate transportation vehicles.
- **Technicians.** Technicians work closely with production workers, but do more specialized jobs. They typically set up and repair equipment, service machinery, conduct product tests and laboratory experiments, work in dental and medical laboratories, and do quality control testing.
- **Engineering technologists.** These highly trained technical employees form the bridge between the engineers who design systems and the workers and technicians who must implement the systems.
- **Engineers.** Engineers have advanced technical knowledge. They design products and structures, conduct research, and develop production processes and systems.
- **Managers.** Technically trained managers can ensure successful implementation by setting goals, plot courses of action, and motivate people to work together. They are people-oriented leaders who also have technical knowledge.

Selecting a Job

In selecting employment, people should consider whether they are compatible with the job requirements. Consider whether your level of education meets the level required by the job. Consider your freedom to organize the tasks assigned to the job and the level of accountability that goes with the job. Consider the balance of working with data, machines, and people. See **Figure 30-9**. Whenever possible, a job should match your interests and abilities.

When deciding on employment opportunities, consider lifestyle, job requirements, and job satisfaction. Each of us wants to live comfortably. A person's job has a direct impact on both life at work and life away from it. For example, some people like to travel and meet new people. An industrial-sales job can provide these aspects. A factory job that pays well and has good holiday and vacation benefits can also meet this need.

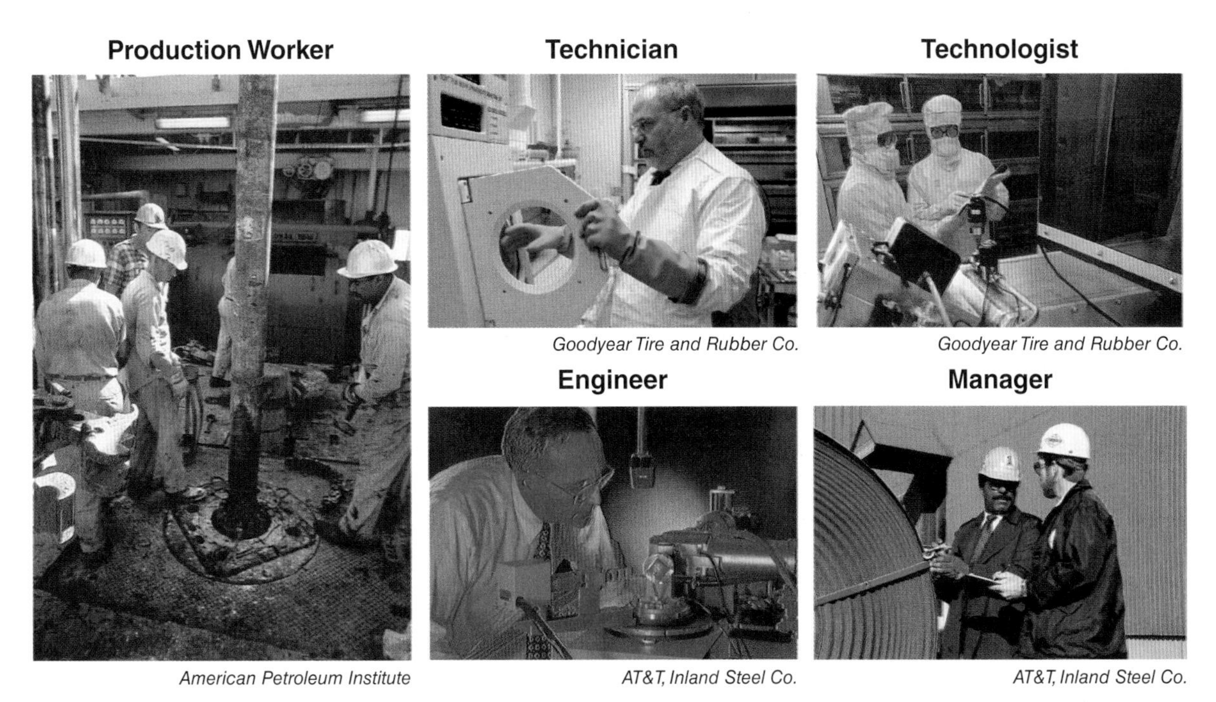

American Petroleum Institute; Goodyear Tire and Rubber Co.; Goodyear Tire and Rubber Co.; AT&T, Inland Steel Co.; AT&T, Inland Steel Co.

Figure 30-8. People with technology backgrounds or an aptitude for technology can become production workers, technicians, technologists, engineers, or managers. Many people work in more than one of these areas during their careers.

Job satisfaction is a description of a worker's happiness with the job. Three factors that strongly affect job satisfaction are values, recognition, and pay. A job should match the values of the person doing it. Some jobs allow for more visible recognition than others.

Job Skills

Each person seeking a job has a set of job skills. These are the activities a person does well. Each job has a set of skill requirements. For example, an engineer must be able to use mathematics, science, and technology to solve problems. An accountant possesses mathematics and accounting skills. A taxi driver needs to know the city's streets, major hotels, and tourist sites and how to drive a car. The challenge is for people to find jobs matching their own sets of skills.

In addition to specific job skills, all applicants need general job skills. General job skills are transferable skills, meaning that they can be used in many different jobs and taken from one job to another. General job skills should be developed during grade school and high school. Most specific job skills are gained through additional education and training activities.

General job skills can be divided into six major groups. These groups are language and commu-

Rocketclips, Inc./Shutterstock.com; SpeedKingz/Shutterstock.com; Alexander Raths/Shutterstock.com

Figure 30-9. Every job involves working with machines, people, and data in varying degrees. Which areas do you prefer?

Think Green Energy-Efficient Lighting

Conventional incandescent bulbs use a great deal of energy to give light, and they also waste energy by giving off heat. As a result, the United States, along with other countries, have enacted laws to phase out incandescent lightbulbs. This has affected consumers who were required to buy alternatives after 2014. Compact fluorescent lamps (CFLs) and light-emitting diode (LED) lamps may be used to replace most incandescent bulbs. When used and disposed of properly, the use of alternative bulbs makes a great deal of difference to the environment.

Compact fluorescent lamps use gas and ballasts to produce light. LEDs use semiconductors that help produce photons, creating light. CFLs and LEDs do not use as much energy to produce the same amount of light as incandescent bulbs. The wattage used is much lower with the same results. Because of the low energy usage, they also last much longer than incandescent bulbs. CFLs and LEDs help the environment because they use less energy, and less electricity needs to be generated. LEDs can be disposed of or recycled. However, you cannot just throw out CFLs because of the mercury inside. This can contaminate the ground and water supply. Be sure to take your used CFLs to a recycling center to be disposed of properly.

nication skills, thinking skills, information skills, socioethical skills, people skills, and personal skills.

Language and communication skills are the abilities to read, write, and speak a language. These skills include the following:

- **Reading.** Reading skills involve the ability to locate text information and then identify and comprehend relevant details.
- **Writing.** Written-communication skills include the ability to write clearly and for a variety of audiences and purposes.
- **Speaking.** Oral-communication skills involve the ability to explain and present ideas in clear, concise ways to a variety of audiences.

Thinking skills involve the ability to use mental processes to address problems and issues. These skills include the following:

- **Problem solving.** Problem-solving skills include the ability to recognize a problem, create possible solutions, implement a selected solution, and evaluate the solution.
- **Creative thinking.** Creative-thinking skills involve the ability to use imagination to develop and combine ideas or information in new ways.
- **Decision making.** Decision-making skills include the ability to identify a goal and alternate courses of action, evaluate the advantages and disadvantages of each alternative, and then select the best action.
- **Visualizing.** Visualization skills involve the ability to envision a three-dimensional object that an engineering drawing or a schematic drawing presents.

Information skills are the abilities to locate, select, and use information. These skills include the following:

- **Information literacy.** Information-literacy skills include the ability to locate, evaluate, use, and cite information properly.
- **IT.** IT skills involve the ability to use computer systems to acquire, organize, and present information. See **Figure 30-10**.

Socioethical skills involve understanding the implications of actions on people, society, and the environment. These skills include the following:

- **Ethics.** Ethical skills include the ability to see the implications of actions on people, society, and the environment and then to act in legally and morally responsible ways.
- **Sociability.** Social skills involve the ability to show interest in, an understanding of, and respect for the feelings and actions of others.

People skills are the abilities needed to work with people in a cooperative way. These skills include the following:

ESB Professional/Shutterstock.com

Figure 30-10. IT skills are required for most jobs.

- **Leadership.** Leadership skills include the ability to encourage and persuade people to accept and act on a common goal.
- **Teamwork.** Team skills involve the ability to work cooperatively with others to complete a project. See **Figure 30-11**.
- **Diversity.** Diversity skills include the ability to work with people from different ethnic, social, or educational backgrounds.

Personal skills involve the ability to grow and manage personal actions on a job. These skills include the following:

- **Self-management.** Self-management skills involve the ability to manage time, adapt to change, be aware of personal responsibilities, and maintain physical and mental health.
- **Self-learning.** Self-learning skills include the ability to recognize the need for new knowledge and seek ways to meet the demand.
- **Goal.** Goal-oriented skills involve the ability to set goals and work toward meeting them.

In addition to general and specific job skills, employers look for proper work attitudes. Employers look for cooperation, dependability, a good work ethic, and respect. This means the employees are expected to do the following:

- Cooperate with supervisors, other employees, and customers.
- Arrive at work on time and complete tasks in a timely manner.
- Put in an honest day's work for a day's pay.
- Show respect for others, the company, and themselves.

Today's workplace emphasizes equality—that is, the idea that all employees are to be treated alike. Harassment (an offensive and unwelcome action against another person) and discrimination (treating someone differently due to a personal

Rawpixel.com/Shutterstock.com

Figure 30-11. Teamwork is an essential general job skill.

Career Connection Technology and Engineering Education Teachers

Technology and engineering teachers primarily teach in high school classrooms but can find elementary positions as well. They can also work as county supervisors, state directors, or professors.

Teachers must have the ability to effectively organize and deliver material to students. Effective teachers are good communicators. Communication skills are necessary to facilitate comprehension and understanding, whether the subject is technology, mathematics, science, reading, or English. In addition, technology and engineering education teachers must know and understand how people have developed, produced, and used tools, materials, and machines to control and improve their lives. These teachers are developed in technology teacher education programs and are certified to teach by the state in which they work.

DGLimages/Shutterstock.com

High school students interested in this career can prepare themselves by taking a number of courses including computer-aided drafting, robotics, foundations of technology, introduction to engineering design, and fundamentals of engineering. They should also take STEM courses, including chemistry, biology, algebra, and trigonometry.

A bachelor's degree in education with a concentration in technology and engineering education is required for this career. Courses related to manufacturing, construction, robotics, drafting and design, biotechnology, transportation, and communication can be taken. In addition, pedagogy (the method and practice of teaching) courses and student teaching allows teachers to practice what they have learned in a classroom with students. Additionally, courses in STEM, such as college algebra and chemistry 101, are also highly beneficial.

Successful technology and engineering educators enjoy working with students, collaborate well with other teachers, are effective planners, and are good problem solvers. These individuals enjoy working with tools and technological systems.

Technology and engineering teachers may work on the following types of projects:

- Creating and planning lessons and instructional activities.
- Assessing and evaluating student learning.
- Writing curriculum and deliver professional development to other teachers.
- Organizing and supervising technology student associations.

Technology and engineering teachers are in constant demand and find employment throughout the United States. Professional organizations that pertain to this field include the International Technology and Engineering Educators Association, the American Society of Engineering Education, and the American Educational Research Association.

characteristic, such as age, sex, or race) are not tolerated. These negative behaviors often result in termination of employment.

Technology and Individual Control

Technology holds great promise and hidden dangers. People make the difference. We often say, "They should control this," "They should do that," or "They should stop doing something else." The harsh reality is that *they* will never accomplish anything. Only when someone says *I*, instead of *they*, does anything meaningful get done.

The future lies in the hands of people who believe they can make a difference. Examples of such people are Thomas Edison, the Wright brothers, and George Washington Carver. These individuals did not wait for a group to do something. They pursued their own visions of what was important and needed by society. See **Figure 30-12**. This type of action requires people to understand technology. People must also comprehend the political and economic systems directing technology's development and implementation.

Everett Historical/Shutterstock.com

Figure 30-12. Thomas Edison developed many devices that had a great impact on everyday life.

The Role of the Consumer

Individuals control technology in a number of ways. The first is through the role of the consumer. This involves the following:

- Selecting proper products, structures, and services.
- Using products, structures, and services properly.
- Maintaining and servicing the products and structures they own.
- Properly disposing of worn or obsolete items.

Consumer action causes appropriate technology to be developed and helps inappropriate technology disappear. See **Figure 30-13**. When a product or service does not sell, production of that product or service stops.

Diyana Dimitrova/Shutterstock.com

Figure 30-13. The types of products and structures demanded by consumers make a difference. Solar heating, for example, is becoming increasingly popular. This solar-heated home saves energy without harm to the environment.

Political Power

An individual also has political power. All companies must operate under governmental regulations. These regulations include the Occupational Safety and Health Act, the Pure Food and Drug Act, the Environmental Protection Act, and the Clean Water Act. Few laws are passed solely because an elected official thinks they are important. Most of them come from people asking their elected representatives to address a problem or concern.

Activism

Finally, an individual can make a difference as an activist. ***Activists*** use public opinion to shape practices and societal values. Activists are people who are practicing their constitutional rights to freedom of speech and assembly. As a citizen, you have the right to meet with others to discuss and promote your point of view on a subject.

Technology and Major Concerns

People are worried about many aspects of the human-built world, including pollution, energy use, and unemployment. Engineers act to resolve these issues. Some areas of concern are at the forefront of our attention, and are resolved through direct action. Others are resolved through indirect action. Indirect action is the creation of new, unrelated products or processes that addressed the concerns. Major concerns involving technology include the following:

- **Nuclear power and nuclear-waste disposal.** Is nuclear-power generation an appropriate activity? How do we design and build safe and reliable nuclear power plants? Can the radioactive waste that nuclear power plants produce be disposed of safely?
- **Technological unemployment.** Should technology that causes unemployment be applied? Do companies have a responsibility to provide training and other benefits to workers who become unemployed through the adoption of new technologies? Should foreign products that cause technological unemployment be barred from domestic markets? What are the individual worker's responsibilities in seeking training to deal with changing job requirements?
- **Genetic engineering.** Is it right to change the genetic structure of living organisms? Who decides what genetic engineering activities are appropriate? Is it acceptable to alter the genetic structure of humans? How are religious and technological conflicts going to be dealt with?
- **Energy use.** How can we reduce our dependence on petroleum? Can society's reliance on the private automobile be changed? See **Figure 30-14**. What alternate energy sources and converters should be developed? Should mass transit be financed with tax money? How can we control the environmental damage caused by burning fossil fuels? Should we place high taxes on gasoline?
- **Land use.** What are the rights of landowners? Should the desires of the majority overrule individual landowner rights? What responsibilities do governmental officials have for public lands? How do you balance environmental protection issues with economic issues? See **Figure 30-15**.

kentoh/Shutterstock.com

Figure 30-14. Greater use of mass transit will be needed in the future to minimize environmental damage and relieve traffic congestion in urban areas.

- **Pollution.** Should strict pollution controls apply equally to individuals and companies? Should products polluting the environment be banned from manufacture and use? What type of evidence is needed before a product can be banned? Should there be a pollution tax on fuels that damage the environment? How should the economic and social impacts of banning products be handled? Should we limit the use of wood as a fuel? Should the solid waste (garbage) from one state be allowed to enter another state?

This book does not attempt to provide answers to these or other global problems. The right answer according to one person or group is often rejected by other people and other groups. The best we can hope for is an answer most people support.

Enrique Alaez Perez/Shutterstock.com

Figure 30-15. Economic considerations must be balanced with concern for the environment as the population continues to grow.

Technology and New Horizons

In the 1800s, Jules Verne wrote a fictional book called *20,000 Leagues Under the Sea*. This story deals with traveling under the ocean in a submarine, a feat that was impossible at the time. In the 1940s and 1950s, people read a fictional comic strip in which the characters used the seemingly ridiculous means of rocket ships to travel in space. What is fiction today might become an everyday part of life for future generations.

The rapid evolution of technology means that many ideas that seem impossible now will be commonplace in the near future. Many futuristic ideas have been proposed, such as mining the resources of outer space and our oceans. Will we have colonies on the moon or other planets that extract and process precious mineral resources? Our oceans are the last of the vast resource beds on Earth. Will they be mined, or will concerns regarding pollution keep them off-limits?

Manufacturing in space has been tried on a small scale, **Figure 30-16**. In the future, we might build complex manufacturing systems in space.

Experiments have shown that protein crystals can be grown in the microgravity (very low gravity) of space. Earth-grown crystals are often small and flawed. Crystals grown in space can be larger and more complex. Some experiments involve the growth of zeolite crystals. These crystals might be used in portable kidney-dialysis machines and in cleanup efforts with radioactive waste. Earth's atmosphere and gravitational pull adversely affect some manufacturing processes. Will these processes be moved into space, and will we have space stations where manufacturing is routinely done?

Commercial space travel is another future technological activity. To date, space travel has mostly been government financed. Most of it is restricted to military and scientific missions or communication satellite launching. Modern space exploration combines government organizations and private companies. Private citizens have paid to have a ride into space. Funding for NASA has been reduced dramatically since the 1960s. This has opened the door for companies such as SpaceX to develop space technology. Additionally, the Commercial Orbital Transport Services (COTS) program has

NASA

Figure 30-16. NASA astronaut Barry Wilmore is holding a 3-D printed ratchet wrench. The wrench was created on the 3-D printer located on the International Space Station.

encouraged the development, testing, and creating of replacements for the aging Space Shuttles. Will we routinely be traveling into space and back some day, and will future generations take vacation trips to the moon or nearby planets?

Additionally, other emerging technologies include artificial intelligence. For example, biometrics are increasingly enabling more advanced interactions between humans and machines. These interactions include image, touch, speech, body language, and facial recognition. Furthermore, these technologies are rapidly becoming more and more sophisticated.

These are only a few examples of possible technological advancements. Society has moved from the horse-drawn wagon to space travel in a single lifetime. What might be next?

Technology Explained Warm-Up Jackets

Warm-Up Jacket: A jacket often worn by athletes that uses a special material to give off heat.

If you have ever taken part in sports such as track or basketball, you probably have worn a warm-up jacket over your uniform. A warm-up jacket does not actually warm your body—you generate the heat. The jacket merely traps it to keep you warm. Now, however, there is a jacket that actually does provide the heat to warm up the person wearing it. This makes it ideal for people who must be outdoors in frigid conditions, such as military personnel, mountain climbers, hunters, and construction workers.

The principle involved is similar to that used in an electric blanket. An electric current causes wires in the fabric to give off heat. Whereas an electric blanket uses fairly large wires and has to be plugged into a wall outlet to operate, the jacket has very thin wires heated by two small batteries. The wearer's movements are not restricted.

The breakthrough that made the new jacket possible was the development of microthin carbon fibers—electrical conductors thinner than human hairs. The microfibers can be woven right into the cloth of the jacket and are as soft and flexible as the fabric itself. They can even make a trip through the washing machine without harm. Two rechargeable batteries supply the power. Together, the batteries weigh less than one-half of a pound.

Heating wires are concentrated in the chest and back areas. Research has shown that heating the chest is the most effective method. Heating the chest warms the body's core, where the heart and lungs are located. The blood circulates through the core, carrying heat to all parts of the body. Heating the back provides an additional level of comfort. The batteries for one manufacturer's jacket can presently provide five hours of heating at the low setting, two-and-one-half hours at the medium setting, or one-and-one-half hours at the high setting. Improvements in battery technology will eventually make much longer periods possible.

To test the effectiveness of the system, a company that makes clothing and equipment for mountain climbers conducted a dramatic demonstration. The company testers outfitted a climber with a jacket, placed him in a deep crevasse on a glacier, and then buried him in snow. He was told to wait until the extreme cold caused his body to start shaking violently and then switch on the jacket. Within minutes, the jacket warmed his body enough to stop the shaking.

Several companies are marketing these jackets to consumers. The technology extends to other articles of clothing, such as gloves, socks, and vests. In addition, a plastic-based fabric has been developed that may be used in the future to make clothing that will cool the wearer. Researchers are working to create a smart fabric that can adjust its temperature based on the wearer's needs.

Summary

- ✔ Advancements in technology affect how people work, their social life, and their recreational activities.
- ✔ During the Industrial Revolution, centralized manufacturing operations replaced the functions of the tradespeople who made goods in their shops.
- ✔ In the information age, computers allowed the development of flexible manufacturing that quickly respond to change. Computer-controlled machines replaced some workers and management styles changed.
- ✔ Educational requirements for most jobs in the future include technical training in career centers and community colleges, a college or university education, or an advanced degree.
- ✔ Five levels of technical jobs are technically trained production workers, technicians, engineering technologists, engineers, and managers.
- ✔ Considerations in selecting a job include matching level of education with the education level required by the job; the freedom to organize the tasks assigned to the job; the level of accountability on the job; and the balance of working with data, machines, and people.
- ✔ General job skills include language and communication skills, thinking skills, information skills, socioethical skills, people skills, and personal skills.
- ✔ The future lies in the hands of people who believe they can make a difference and pursue their own vision of what is needed by society.
- ✔ An understanding of the political and economic systems directing technology's development and implementation is important.
- ✔ Consumer action causes appropriate technology to be developed and helps inappropriate technology disappear.
- ✔ Concerns about technology can be addressed through political power and activism.
- ✔ Major concerns about technology include nuclear power and nuclear-waste disposal, technological unemployment, genetic engineering, energy use, land use, and pollution.
- ✔ The rapid evolution of technology means that many ideas that seem impossible now will be commonplace in the near future.

Check Your Engineering IQ

Now that you have finished this chapter, see what you learned by taking the chapter posttest.
www.g-wlearning.com/technologyeducation/

Test Your Knowledge

Answer the following end-of-chapter questions using the information provided in this chapter.

1. Before the Industrial Revolution of the mid-1800s, more than _____ percent of the workforce was engaged in farming.
2. List five features of the centralized manufacturing operations used during the Industrial Revolution.
3. In the information age, _____ manufacturing was developed.
4. What are the five categories of technical jobs described in this chapter?
5. When considering a job, you should look at whether it places emphasis on data, machines, or _____.
6. Three factors that strongly affect job satisfaction are values, _____, and pay.
7. What are four types of thinking skills?
8. The ability to recognize the need for new knowledge and seek ways to meet the demand is a(n) _____ skill.
   A. information literacy
   B. self-learning
   C. socioethical
   D. leadership
9. *True or False?* Consumer action can result in the development of appropriate technology.
10. People who use public opinion to shape practices and societal values are called _____.

Critical Thinking

1. What are the possible positive and negative effects of commercialization of space travel?
2. How have new technologies affected your lifestyle? Have the effects been good or bad?
3. Select a technological idea that is now considered fiction. How could this technological idea be used as an everyday item or process in the future?

STEM Applications

1. Read a book about people who developed new technology. Write a report on their attitudes toward change and criticism, as well as their inventions or innovations.
2. Examine the world around you. Consider how the Internet of Things (IOT), could be used to improve your lifestyle, making it more efficient and productive.

Technological Impacts Commercial

Background

Humans can control the use of technology. How a technological object is used can create benefits or drawbacks. Each person should strive to use technology wisely to make the future better, protect the environment, and help people live in harmony with nature. Many people, however, are unaware of the impacts technology has on individuals and society.

Situation

You are the communications director of a citizens group. Your group is concerned about public-policy issues. This group has determined that people in your community are not participating in the local recycling program. Design a 60-second public-service commercial for television. The commercial needs to explain the importance of recycling to your community. Identify the materials that can be recycled and the benefits of recycling.

Materials and Equipment

- Storyboard forms.
- Pencils.
- Felt-tip pens.
- A video camera.
- A video recorder.
- A video monitor.

Procedure

- Design the commercial
  1. Select the theme of the commercial.
  2. List the major points to be emphasized in the commercial.
  3. Develop a storyboard for the commercial. Use photocopies of **Figure A**.
  4. Write a script for the actors.
  5. Develop a shot chart for the director and cameraperson to follow.
- Produce the commercial
  1. Recruit and select actors for the commercial.
  2. Present the script to the actors. Have them rehearse their parts.
  3. Walk through the commercial with the director. Have the actors and camera operators block (plan) their movements.
  4. Record the commercial.
  5. Edit in any titles you need.
  6. Present the commercial to an audience. Ask for their reactions.

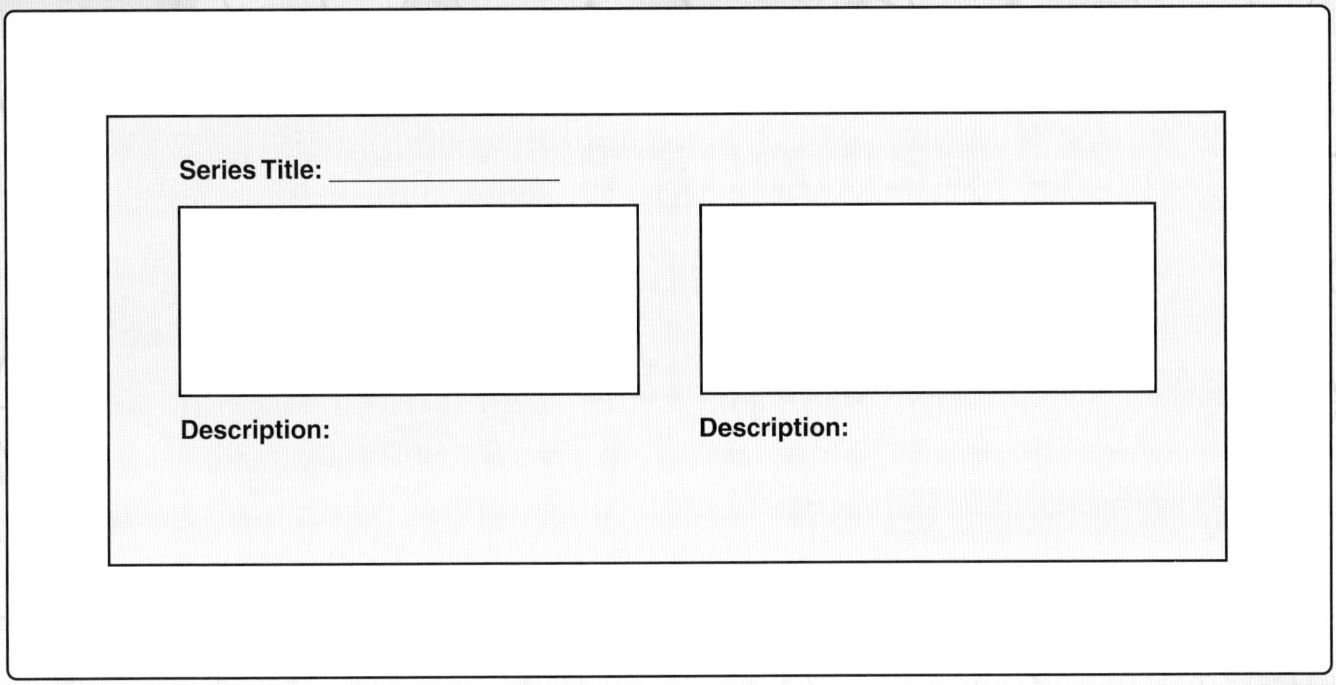

Series Title: ______________

Description:

Description:

Goodheart-Willcox Publisher

Figure A. The storyboard layout.

CHAPTER 31

Managing and Organizing a Technological Enterprise

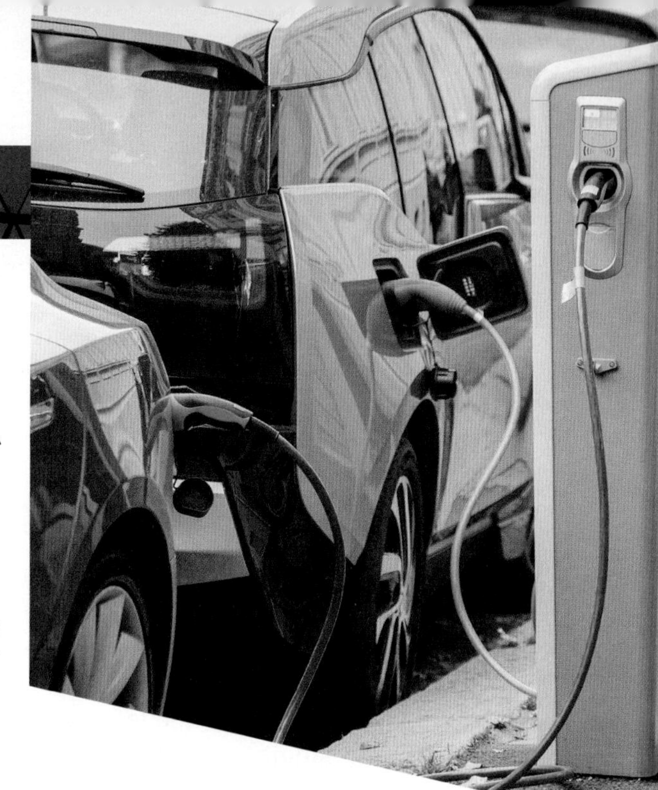

Check Your Engineering IQ

Before you read this chapter, assess your current understanding of the chapter content by taking the chapter pretest. www.g-wlearning.com/technologyeducation/

Learning Objectives

After studying this chapter, you should be able to:

- ✔ Compare and contrast entrepreneurship and intrapreneurship.
- ✔ Describe the four functions of management.
- ✔ Identify the different levels of management.
- ✔ Explain the risks and rewards associated with being involved with a company.
- ✔ Differentiate between the three main forms of business ownership.
- ✔ Describe the role of a board of directors in a corporation.
- ✔ Explain equity financing and debt financing.

Technical Terms

actuating
board of directors
bonds
bylaws
charter
chief executive officer (CEO)
corporation
debt financing
dividend
entrepreneur
equity financing
inside directors
intrapreneurship
limited liability
management
middle management
operating management
outside directors
partnerships
private enterprises
proprietorship
public enterprises
supervisor
top management
unlimited liability
vice presidents

Close Reading

Make a list of various businesses in your area. As you read this chapter, determine the type of ownership of each business.

Have you ever thought about where technology comes from? Technology is a product of the human mind, developed by people to serve people. At one time, most technology was developed, produced, and used by one person. In the modern world, however, organization is necessary. People use complex systems to develop and produce technology. See **Figure 31-1**.

Technology and the Entrepreneur

At the base of many of the systems used to develop and produce technology are entrepreneurs. An ***entrepreneur*** is someone who takes the financial risks of starting a small business. See **Figure 31-2**. Entrepreneurs look beyond present practices and products. They see new ways to meet human needs and wants. By focusing on what the customers value, entrepreneurs develop systems and products to meet desires and expectations. They might change the entire way something is being done.

An example of entrepreneurship is the McDonald's® restaurant chain. The first McDonald's restaurant was a small hamburger stand in southern California. The original owners had developed some innovative ways to make and sell their product. They did not, however, look beyond their local market. An outsider, Ray Kroc, saw greater possibilities. Under his leadership, the fast-food business was born. Kroc and his managers carefully studied the various jobs and developed special management techniques. They standardized the product, created effective training programs, and developed the chain into a worldwide organization.

mavo/Shutterstock.com

Figure 31-2. Entrepreneurs have the vision to recognize consumer wants and to devise ways to meet them with new products.

NASA

Figure 31-1. Complex systems and a large organization are needed to produce a product, such as this intricate space robot.

The previous example illustrates how entrepreneurs improve the use of resources and create new products or markets. The entrepreneurial spirit is the spirit of innovation. People starting another beauty shop, delicatessen, or bakery are taking financial risks and might become successful business operators. They are not, however, innovators. These people do not deal with change as an opportunity to produce a new product or service.

Entrepreneurship involves an attitude and an approach. It entails searching for opportunities for change and responding to them. Large companies often encourage entrepreneurship within their organizations. The term ***intrapreneurship*** has evolved to describe the application of entrepreneurial spirit and action within an existing company structure. For example, a company may have time dedicated specifically to research and development, where staff are encouraged to explore new ideas that allow the company to remain competitive.

Technology, Engineering, and Management

Engineers are purposeful and intentional when creating technology. Therefore, a system is developed to solve a problem or meet an opportunity. Identifying and responding to the need for change is only one part of developing technology. The production and use of technology must be managed. Therefore, technology is a product of managed human activity. See **Figure 31-3**. ***Management*** is the act of planning, directing, and evaluating any activity. See **Figure 31-4**. This act can be as simple as managing personal expenses or as complicated as managing an industrial complex. Management involves authority (the privilege to direct actions) and responsibility (accountability for actions).

Managers have the responsibility to make decisions to ensure that the business is successful. Their authority might include hiring personnel, purchasing materials, developing products, and setting pay rates. Managers are responsible for protecting the rights of a company's owners, workers, and customers. This might include securing product patents, investing company funds wisely, providing a safe work environment, and producing a quality product.

The Functions of Management

To carry out their duties, company managers perform four important functions. See **Figure 31-5**. These functions are the following:

bikeriderlondon/Shutterstock.com

Figure 31-3. The actions of these workers must be managed to produce boats efficiently.

stockphoto mania/Shutterstock.com

Figure 31-4. Planning is an important part of managing a business. These managers are using a model of a boiler system as they plan a factory expansion.

Figure 31-5. Managers plan, organize, actuate, and control to ensure company activities are successful. This manager is performing the checking function by gathering information with her employees.

- **Planning.** Planning includes establishing and communicating a vision and mission. Long-term and short-term S.M.A.R.T (specific, measureable, attainable, realistic, timely) goals for the company or parts of it are set. A

STEM Connection: Mathematics

Calculating Interest Rates

When starting a business, it is important to understand how to calculate interest rates for a business loan. Without knowledge of the effect of average percentage rate (APR), a borrower could end up in serious debt and out of business. The APR is a rate that is charged to an individual for borrowing, or earning, money, and is based on a yearly percentage, distributed over the loan.

The following examples illustrate the effect of different interest rates on the payment and expenses of a loan.

Example 1: $20,000 loan, four years to repay it, 6% yearly interest rate. This would result in interest of $2,545.63.

Example 2: $20,000 loan, four years to repay it, 4% yearly interest rate. This would result in interest of $1,675.89.

As you can see, having a lower interest rate can significantly reduce the overall costs of a loan.

OSABEE/Shutterstock.com

course of action to meet those goals is selected. Planning activities often result in an action plan, strategic plan, or plan of work. Such a plan lists what needs to be done, who will do it, and when it is to be done.

- **Organizing.** Organizing involves structuring the company or workforce to address company goals. Typical activities include developing organizational charts, establishing chains of command, and determining the company's operating procedures. Organizing ensures that people, materials, and equipment are in place to meet the action plan.
- **Actuating.** Initiating the work related to the action plan is known as ***actuating***. This function can include training employees, issuing work orders, providing a motivational work environment, and solving production problems. Actuating causes plans to take form. Products and structures are built or services are provided.
- **Controlling.** Comparing results against the plan is the controlling function. Control actions ensure that resources are used properly and outputs meet stated standards. Typical terms applied to this area are *inventory control*, *production control*, *quality control*, and *process control*.

The Authority and Responsibility of Management

Managerial functions are carried out through an organizational structure, **Figure 31-6**. This structure typically begins with the owners of the business. The owners have ultimate control, or final authority, over company activities. The business is their company. They can hire or fire personnel, set policies, or close the business.

Top Management

In most large companies, however, the owners are not the managers. They frequently have other jobs or interests. Also, they often do not have the skills needed to manage a large, complex business. Such owners delegate responsibility to full-time managers. In many companies, the top manager is the president or the ***chief executive officer (CEO)***. ***Top management*** is responsible for the entire company's operation. In very large companies, two different people usually hold these titles. In smaller companies, the same person might hold the two titles. People at this level have day-to-day control of the company.

Few people can manage a company by themselves. Therefore, the top managers employ other

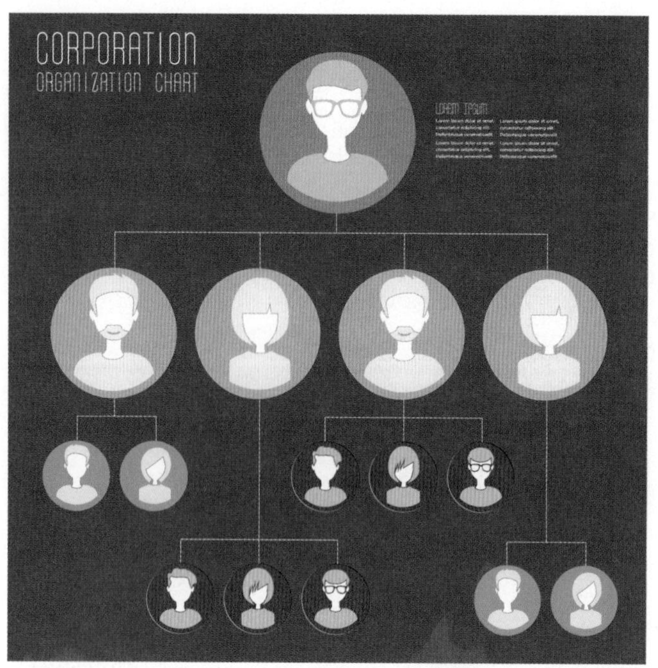

Danielala/Shutterstock.com

Figure 31-6. Managers have specific levels of authority and responsibility within a company.

bikeriderlondon/Shutterstock.com

Figure 31-7. Managers that directly oversee specific operations in the company are often called *supervisors*.

managers to assist them. The number of managers and levels of management vary with the size and type of the company. Larger companies generally have ***vice presidents*** who report to the president or CEO. Each vice president is responsible for some segment of the company. The segment might be a functional area, such as marketing, finance, production, or engineering. In other cases, the vice president has a regional responsibility, such as foreign sales or West-Coast operations.

Middle Management

Most vice presidents have a scope of responsibilities. A number of managers may report to a vice president. Regional sales managers might report to a vice president for sales. Plant managers often report to vice presidents in charge of production. This level of management is often called ***middle management***. Middle management is below ***top management*** (the president and vice presidents) but above operating management.

Operating Management

The lowest levels of management directly oversee specific operations in the company. Managers at this level might be supervisors on the production floor, district sales managers, or human resources directors. They are the managers who are closest to the people who produce the company's products and services. These managers are often called ***supervisors*** or ***operating management***. See **Figure 31-7**.

Risks and Rewards

As noted earlier, most technology is developed and produced through the planning of industrial companies. The company designs, engineers, and produces the products, structures, transportation services, and communication media we depend on daily.

Everyone involved with a company is subject to risks and rewards. Owners risk their money to finance the company. For example, Peter Theil, the former PayPal CEO and investor, provided an initial loan to assist Mark Zuckerberg with creating Facebook®. Although Peter surely provided advice and guidance, his contribution was largely taking a financial risk in hope for long-term gains. Banks also accept a level of risk. They and other lending institutions make loans to finance company growth. Employees risk missing other employment opportunities by working for the company. Consumers risk their money when they buy a product.

In return, the risk takers expect a reward. The owners want their investments to grow. They also expect periodic financial returns for the use of

Burlingham/Shutterstock.com

Figure 31-8. Banks earn interest on business loans.

their money. Banks expect interest to be paid on their loans. See **Figure 31-8**. Employees expect job promotions, pay raises, safe working conditions, and job security. Consumers expect performance and value in return for the money they spend on products and services.

Forming a Company

Companies are organized and operated under the laws and mores (accepted traditions and practices) of our society. The following are important features involved in the formation of a company:

- Selecting a type of ownership.
- Establishing the enterprise.
- Securing financing.

Selecting a Type of Ownership

Business enterprises can be divided into two different sectors—public and private. ***Public enterprises*** are controlled by the government or a special form of corporation. These enterprises are operated for the general welfare of society and cannot, or should not, make a profit.

A police department is an example of a public enterprise. A police department is run similarly to a business. Police are commissioned to protect people and their property. They have managers (police chiefs and captains) and workers (patrol officers, traffic officers, and narcotics officers). If a police department had to show a profit and attract private investment, some aspects of law enforcement might get cut back. This might limit the department's market to the people who could pay for the service. The police would solve the crimes showing the most potential for profit. See **Figure 31-9**.

Another example of public ownership is road construction. We pay for it through taxes. Each segment of road, however, does not have to show a profit. If it did, we would not have many of our rural roads.

Individuals or groups of people own ***private enterprises***. This ownership can be through a direct means of investment or through an indirect means, such as a pension or an investment fund. Owners of private enterprises invest their money, take risks, and hope to reap a profit. Within legal limits, the owners are free to select business activities, produce the products and services they choose, and divide the profits as they see fit.

Private enterprises can be publicly held or privately held. A publicly held enterprise is one in which the public can purchase a portion of ownership in the form of shares of stock, or fractional ownership of a company. Individuals or a group of people own privately held enterprises, and stock is not offered for sale to members of the public. There are three main types of private-business ownership. These are discussed in the following sections.

©iStockphoto.com/slobo

Figure 31-9. Police and fire departments are examples of public enterprises.

Proprietorships

A ***proprietorship*** is a business with a single owner who has complete control of the company. See **Figure 31-10**. The owner sets goals, manages activities, and has the right to all business profits. A proprietorship is a fairly easy type of ownership to form.

Proprietors might have difficulty raising money, especially to start a new business. The company's finances are typically limited to the owner's personal wealth or borrowing ability. Banks might be hesitant to loan a large quantity of money to unproven businesses or individuals.

An additional problem of this type of business is limited knowledge. Many individuals do not have the skills and knowledge needed to run all aspects of the business. This causes inefficiency in operating the company. Finally, the proprietor is responsible for all the debts the business incurs. Business income and liabilities are not separate from the proprietor's personal finances. This is a major disadvantage called ***unlimited liability***.

Proprietorships are the most common type of ownership in the United States. They are generally small retail, service, and farming businesses. Thus, the dollar impact of this form of ownership on the economy is considerably less than the impact of large corporations.

©iStockphoto.com/jwilkinson

Figure 31-10. Many small retail and service businesses are sole proprietorships.

Partnerships

A second form of private ownership is the *partnership*. See **Figure 31-11**. ***Partnerships*** are businesses that two or more people own and operate and thus have more sources of money to finance the company. Also, the interests and abilities of the partners can complement each other. One partner might be strong in production. Another might have sales skills.

More than one active owner in a business can cause confusion, however. Employees might receive conflicting directions. The partners can disagree when making important business decisions. Also, a partnership, similar to a proprietorship, has unlimited liability. This is a particularly touchy problem, since one partner can commit the entire partnership to financial risk.

Corporations

The third form of ownership is the corporation, **Figure 31-12**. A ***corporation*** is a business in which investors have purchased partial ownership in the form of shares of stock. Investors can be individuals, other companies, or groups (such as in a pension plan or an investment club). Legally, the corporation is similar to a person. This business can

©iStockphoto.com/LivingImages

Figure 31-11. Many small businesses, such as farms, are partnerships.

Career Connection Top Executives

Top executives set specific goals and objectives for companies and direct the overall operations of businesses. They formulate policies and strategies so these objectives are met. The range of titles for top executives includes chief executive officer (CEO), chief operating officer (COO), chair of the board, and president. Executives typically have spacious offices and large support staffs. They might travel to national and international sites to monitor operations and meet with customers and staff members.

High school students who are interested in becoming a top executive should take business courses and participate in programs that develop soft skills. These skills include communication, decision making, leadership, management, problem solving, and time management.

Many top executives have a bachelor's degree or an advanced degree. The degree is typically in business administration or a subject related directly to their field. Many top executives have a master's of business administration (MBA) or a doctoral degree. Executives can expect to move up within an organization as they gain experience or complete advanced degree programs.

A top executive must have excellent personal skills and an analytical mind. Executives must be able to communicate clearly, provide leadership, and make difficult decisions.

Employment growth for top executive positions varies by industry. The job outlook for top executive positions is largely related to growth and expansion of companies and industries. There will be tough competition for many of these openings.

A top executive may work on the following types of projects:

- Negotiate contracts between organizations.
- Plan and oversee financial budgets of a company.
- Manage multiple employees and direct the work of a corporation.
- Establish and implement a company's vision, mission, and strategic plan.

Organizations for top executives include the American Management Association, Financial Executives International, and National Management Association.

©iStockphoto.com/Freezingtime

Figure 31-12. Most large businesses are corporations.

own property, sue, be sued, enter into contracts, and contribute to worthy causes.

Generally, the investor-owners of a corporation do not manage the business. They employ professional managers for this task. The owners invest their money and expect to receive a ***dividend*** (periodic payment from the company's profits) in return.

Since a corporation is a legal person, the company is responsible for the debts. (The company's owners are not responsible.) This feature is called *limited liability*. ***Limited liability*** means if the company fails, an owner's loss is limited to the amount of money he or she has invested.

Establishing the Enterprise

Once the type of ownership is selected, the company must be established. There are few legal requirements for proprietorships and partnerships. In many cases, all that is needed is a license from the city or other local government where the business will operate.

Corporations are a different story. They are often large businesses that can have a serious financial impact on people and communities. For this reason, their formation is placed under state control. Each state establishes its own rules for forming a corporation. Most states require the completion of the steps shown in **Figure 31-13** when a corporation is formed.

Articles of Incorporation

A corporation, similar to a person, must be born. This process is begun by filing articles of incorporation with one of the states. The articles of incorporation serve as an application for a corporate ***charter*** (a "birth certificate" for the corporation), detailing the fundamental components of a company. This may include the objectives, its structure and its strategic plan. The state usually asks for the company name, type of business the company plans to enter, location of the company offices, and type and value of any stock that will be issued.

Corporate Charter

The articles of incorporation are filed with the appropriate state office. State officials then review the articles to determine if the business will operate legal activities and provide customers with appropriate products or services. If they believe the business meets all state laws, a corporate charter is issued. This allows the company to conduct the specified business in the state. All the other states will recognize the corporate charter and allow the company to conduct business within their borders.

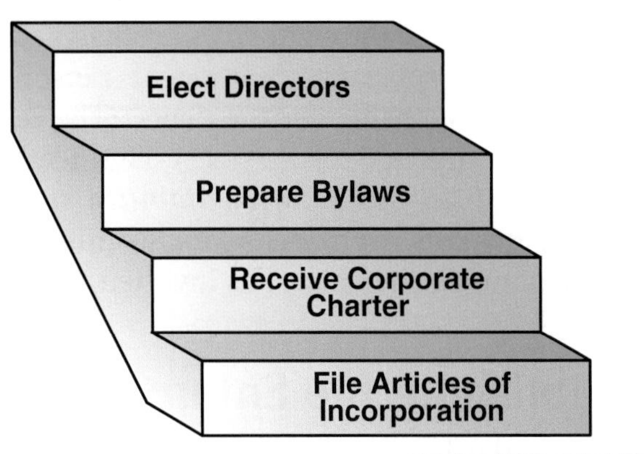

Goodheart-Willcox Publisher

Figure 31-13. The steps in forming a corporation.

Bylaws

An incorporated business must have a set of ***bylaws***. These are the general rules under which the company operates. A set of bylaws includes the information contained in the charter—the name of the company, purpose of the business, and location of the corporate offices. In addition, the bylaws list the following:

- The corporate officers.
- The duties of, terms of office of, and method of selecting corporate officers.
- The number of directors, as well their duties and terms of office.
- The date, location, and frequency of the board of directors' meetings.
- The date and location of the annual stockholders' meeting.
- The types of proposals that can be presented at the annual stockholders' meeting.
- The procedure for changing the bylaws.

Board of Directors

The charter and bylaws allow the company to operate. The stockholders, however, want their investments to be wisely managed. This requires oversight and supervision. Many companies have hundreds or thousands of stockholders. Few stockholders can be, or wish to be, involved in managing the company. Therefore, a ***board of directors*** is elected to represent the interests of the stockholders. The directors are responsible for forming company policy and providing overall direction for the company.

A typical board of directors includes two groups of people. ***Inside directors*** are the top managers of the company. ***Outside directors*** are outside the managerial structure and are not involved in the day-to-day operation of the company. They are selected to provide a different view of the company's operation.

Directors are elected using a voting system similar to our political system. The main difference is that companies do not use a one-person and one-vote rule. Instead, they use a one-share and one-vote procedure. Each share of stock (equal portion of the total company) has a vote assigned to it. Stockholders each have as many votes as the shares of stock they own. Therefore, those who own a larger portion of the company and accept a larger risk have a greater say in forming company policy.

Securing Financing

There is more to starting a company than completing paperwork. The company needs money to operate. The two basic methods of raising operating funds are equity financing and debt financing, **Figure 31-14**.

Equity Financing

Equity financing involves selling portions of ownership in the company. This is an important way in which corporations are financed. The company's charter authorizes the company to sell a specific number of shares of stock. Investors can buy these and, as a result, become owners of part of the company. Owners receive certain rights with the shares they own. These rights include the following:

- The right to attend and to vote at the annual stockholders' meeting.
- The right to sell their stock to another individual.
- The right to receive the same dividend per share as other stockholders.
- The right to a portion of the company's assets (property and money) if the company is liquidated.

Financing Company Operations

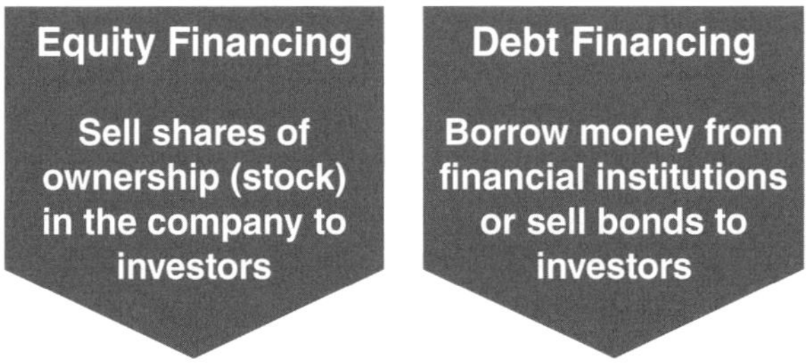

Equity Financing	Debt Financing
Sell shares of ownership (stock) in the company to investors	Borrow money from financial institutions or sell bonds to investors

Goodheart-Willcox Publisher

Figure 31-14. Two major methods of financing are available to build and expand a company.

Debt Financing

Debt financing involves borrowing money from a financial institution or private investors. Banks and insurance companies loan corporations money to finance new buildings, equipment, and the company's daily operations. They charge interest for the use of their money. These banks and insurance companies charge their best (safest) customers a lower interest rate. This rate is called the *prime interest rate* (*prime*). Other borrowers pay a rate higher than the prime rate. This rate is often quoted in terms of the prime, such as "prime plus 2%."

Corporations can also sell debt securities called ***bonds***, which are a type of loan in which the issuer owes the holders a debt. Based on the terms of the bond, the issuers are expected to pay the holders interest and/or repay in its entirety at a later date. These securities are often sold in fairly large denominations, such as $5,000. The bonds are usually long-term securities. This means they will be in force for 10 to 20 years. The company pays quarterly or yearly interest on the face value (original value) of the bonds. At maturity (the end of the bond's term), the company pays back the original investment.

Summary

- ✔ An entrepreneur is someone who takes the financial risks of starting a small business. The entrepreneurial spirit is the spirit of innovation.
- ✔ The application of entrepreneurial spirit and action within an existing company structure is called intrapreneurship.
- ✔ Management is the act of planning, directing, and evaluating any activity. Management involves authority (the privilege to direct actions) and responsibility (accountability for actions).
- ✔ Management functions include planning, organizing, actuating, and controlling.
- ✔ Three levels of management in large companies are top management, middle management, and operating management.
- ✔ Everyone involved with a company is subject to risks and rewards.
- ✔ Formation of a company involves selecting a type of ownership, establishing the enterprise, and securing financing.
- ✔ Public enterprises are operated by the government or a special form of corporation for the general welfare of society and cannot, or should not, make a profit. Owners of private enterprises invest their money, take risks, and hope to reap a profit.
- ✔ Three main types of private-business enterprises are proprietorships, partnerships, and corporations.
- ✔ A proprietorship has a single owner who has complete control of the company. The owner has unlimited liability for business debts.
- ✔ A partnership is a business owned and operated by two or more individuals.
- ✔ To establish a corporation, articles of incorporation are filed with a state. The articles of incorporation serve as an application for a corporate charter, which is required for the corporation to operate in that state and other states.
- ✔ The bylaws of an incorporated business are the general rules under which the company operates.
- ✔ A corporation's board of directors is elected by stockholders and represents their interests. The directors form company policy and provide overall direction for the company.
- ✔ Equity financing involves selling portions of ownership in the company. Debt financing involves borrowing money from a financial institution or private investors.

Check Your Engineering IQ

Now that you have finished this chapter, see what you learned by taking the chapter posttest.
www.g-wlearning.com/technologyeducation/

Test Your Knowledge

Answer the following end-of-chapter questions using the information provided in this chapter.

1. Entrepreneurship within an existing company is referred to as _____.
2. Management involves _____ and responsibility.
   A. credibility
   B. authority
   C. discipline
   D. personality
3. Describe the management function called *actuating.*
4. *True or False?* In most larger companies, the owners are the managers.
5. *True or False?* Everyone involved with a company takes risks with expectation of receiving a reward.
6. List the three forms of business ownership.
7. *True or False?* Private enterprises can be publicly held or privately held.
8. Which of the following statements about a proprietorship is *false*?
   A. It is the most common type of ownership in the United States.
   B. The proprietor is responsible for all the debts the business incurs.
   C. A single owner has complete control of the company.
   D. A charter is required to start the business.
9. A(n) _____ can own property, enter into contracts, sue, and be sued.
10. Which form of business ownership provides the owners with limited liability?
11. The general rules under which a corporation operates are included in the _____.
    A. bylaws
    B. charter
    C. articles of incorporation
    D. state license

12. The board of directors of a corporation represents the interests of the _____.
13. Equity financing involves the sale of _____.

Critical Thinking

1. What is the current prime interest rate? How might this rate affect the economy and decisions to start a business?
2. When forming a company, what are the advantages and disadvantages of having many levels of management?
3. When would an educational television station be considered a public enterprise?

STEM Applications

1. Assume your class is going to produce and sell popcorn at school basketball games. Develop an organization chart for the enterprise. Indicate who has the most authority and the chain of command from that person to all other members of the enterprise.
2. Contact or visit the office of a local stockbroker. Find out if you can obtain brochures or other simple explanations of stocks and bonds and how they are traded. Use the information as the basis for a written report. You might want to invite the person to speak to your class.

Engineering Design Challenge

Forming a Company

Background

The companies that produce technological devices and products are the products of human actions. They are formed and structured to efficiently use resources to produce artifacts meeting wants and needs.

Situation

Students have mentioned they need a way to be informed about important sporting and social events at school. They also want a way to publicize happenings they feel are important. From these comments, you have concluded that an inexpensive, personalized calendar will meet their needs and earn you a profit. This type of calendar can be produced with limited finances by using new computer software. See **Figure A**.

Challenge

Organize a company to produce a nine-month calendar. The calendar should span the school year and have selected days personalized. Consider the tasks to be completed and the managerial structure needed to complete them. Be sure to recognize that there are production and marketing tasks. Also, there are two distinct phases of the company's operations. These phases might require two different organizations. One operation can finance the company, design the calendar, and sell calendar entries. The other operation can maintain financial records, produce the calendars, and sell the finished products. See **Figure B**.

April

Sunday	Monday	Tuesday	Wednesday	Thursday	Friday	Saturday
					Honor Society Dance -7:30 1	2
3	Jim Brown's Birthday 4	5	6	7	Baseball at Southside 8	9
10	11	12	13	14	15	16
17	Spring Break Starts 18	19	20	21	End of Spring Break 22	23
24	25	26	Senior Class Pictures 27	28	Baseball vs Westfield - Home 29	30

Goodheart-Willcox Publisher

Figure A. Sample calendar.

Prepare an organization chart and write a one-page job description for each job on the chart. Also, develop a set of goals for each major department in the company.

Materials and Equipment

- Poster board.
- Pens.
- Paper.
- Note cards.
- Crayons.
- Colored pencils.
- Rulers.
- Card stock.

Beginning the Process

- Utilize the engineering design process to guide you through the process of defining your problem and creating a viable solution to it.
- Remember, the engineering design process is an iterative approach to solving problems, which means people can and should go back and forth between the different steps.
- Define the problem based on the background and situation provided above.
- Gather information related to the problem that has been defined.
- Apply the knowledge you obtain and employ the engineering design process in order to propose new or improved solution to the problem defined from the situation provided. This will be a natural progression from brainstorming to generating ideas, developing a design, and evaluating the design's potential success.
- Document all the work you do to create a design solution.

Figure B. Furnace operation tasks.

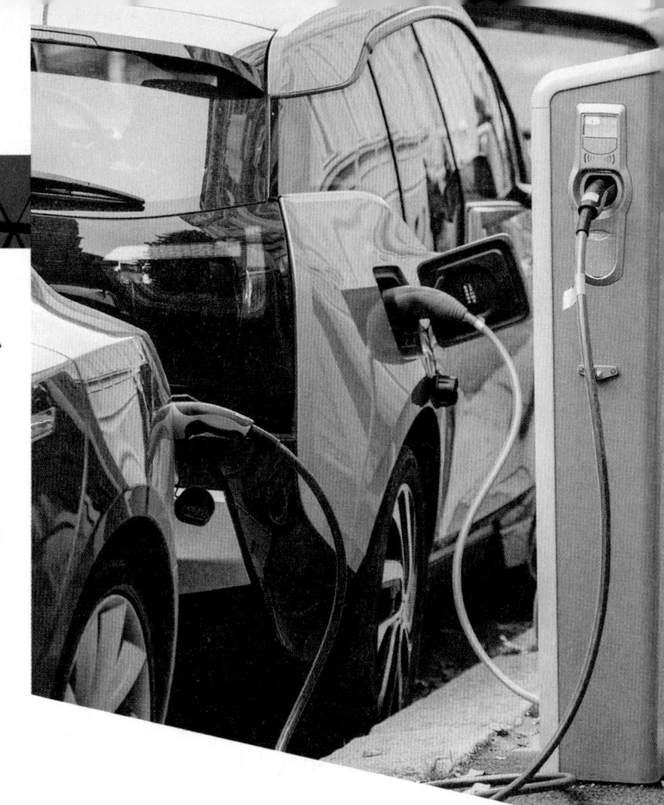

CHAPTER 32

Operating a Technological Enterprise

Check Your Engineering IQ 

Before you read this chapter, assess your current understanding of the chapter content by taking the chapter pretest.
www.g-wlearning.com/technologyeducation/

Learning Objectives

After studying this chapter, you should be able to:

- ✔ Identify the five major societal institutions.
- ✔ Describe the functions of research and development.
- ✔ Identify four common manufacturing systems.
- ✔ Describe the three important activities involved in production of products and structures.
- ✔ Describe the four important activities involved in marketing.
- ✔ Describe the three main industrial relations programs.
- ✔ Explain how money is managed in a company.

Technical Terms 

accounting
advertising
applied research
apprenticeship training
architectural drawings
assembly drawings
basic research
benefit
bill of materials
commission
detail drawings
development
distribution
economic enterprises
employee relations
expenses
flexible manufacturing
free enterprise
grievances
income
industry
inspection
labor agreements
labor relations
marketing
market research
packaging
public relations
research
retained earnings
salary
sales
systems drawings
value
wage

Close Reading

There are different areas of activity in technological enterprises. To better understand the areas of activity discussed in this chapter, make an outline of the main points as you read.

Society is made up of major parts called *institutions*. One major societal institution deals with the economic (goods and services) activities of society. Within this economic domain are organizations called *economic enterprises*, of which industry is one type. In this chapter, you will explore a number of topics dealing with technological enterprises and, more specifically, industry.

Societal Institutions

Over time, humans have developed a complex society to meet their wants and needs. Within this society are five basic institutions, **Figure 32-1**. They are the following:

- **Family.** This institution provides the foundation for social and economic actions. Family is the basic unit within society.
- **Religion.** This institution develops and communicates values and beliefs about life and appropriate ways of living.
- **Education.** This institution communicates information, ideas, and skills from one person to another and from one generation to another.
- **Politics and law.** This institution establishes and enforces society's rules of behavior and conduct.
- **Economy.** This institution designs, produces, and delivers the basic goods and services the society requires.

All of these institutions use technology because they are concerned with efficient and appropriate action. They apply resources to meet human wants and needs. People in each institution use technical means to make their jobs more efficient.

Almost all technology, however, originates in the economic institution. For example, teachers working in the educational institution use computers to make their teaching more efficient. The computer, however, is not a product of the educational institution. Computers are products of the economic institution. In the same way, politicians, who are part of the politics and law institution, use television and printed material to help win an election. However, television, printing presses, and all associated communication devices are not developed in the political system. They are outputs of the economic institution.

Economic Enterprises

Technology is most often directly associated with economic enterprises. ***Economic enterprises*** are organizations engaging in business efforts directed toward making a profit. See **Figure 32-2**.

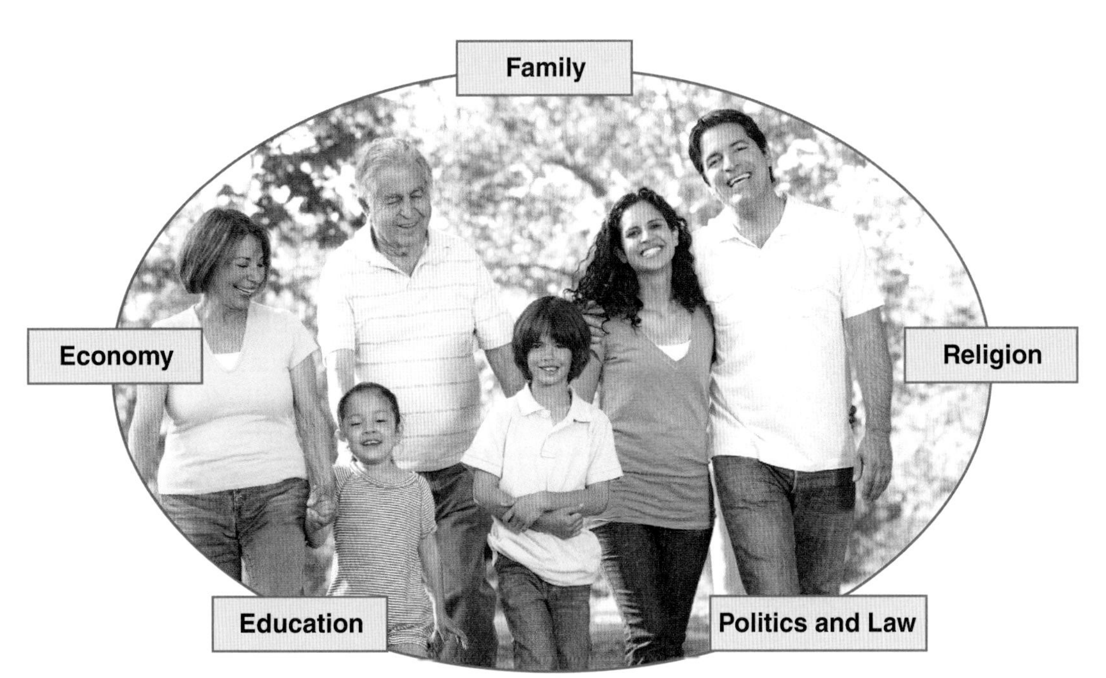

Monkey Business Images/Shutterstock.com

Figure 32-1. Five basic institutions are important to our society.

zhu difeng/Shutterstock.com

Figure 32-2. There are four major types of economic enterprises. Only industrial enterprises actually develop and produce technology.

This economic activity includes all trade in goods and services paid for with money.

In a ***free enterprise*** economic system, the principles of supply and demand determine production and prices. The government has little direct impact on markets. Businesses are free to produce any products. As consumer demands change over time, some businesses are created and some businesses close. To meet current demands, some businesses modify their production. In this system, business plays the important roles of owning production equipment and processes, determining product supply, and reacting to market demands.

Not all economic activity develops technology. Commercial trade includes the wholesale and retail merchants who form the link between producers and consumers. These traders do not change the products they distribute. Their purpose is to make these products easily available for people to buy.

Banks, insurance companies, and stockbrokerage firms provide financial services. They protect our wealth, buy and sell stocks and bonds, or insure our lives and possessions. Again, these companies use technology but do not develop it.

Many service businesses repair products and structures. They service and maintain technological devices. These businesses extend the useful lives of such devices. They do not, however, develop the devices.

Industry...
• Is part of the economic system.
• Uses resources.
• Produces products structures, and services.
• Intends to make a profit.

Usmanov Ramil/Shutterstock.com

Figure 32-3. Industrial enterprises carry out the design and production of goods, such as these products used by students.

Almost all of the design, development, and production of technology takes place within the type of economic enterprise called *industry*. See **Figure 32-3**. The term *industry* can have several meanings. One definition groups together all businesses making similar products. Thus, we read about the steel industry or the electronics industry. In this book, however, a more restricted definition is used. ***Industry*** is "the area of economic activity that uses resources and systems to produce products, structures, and services with the intent to make a profit."

Areas of Industrial Activity

In each industry, a number of actions take place. These actions are designed to capture, develop, produce, and market creative ideas and solutions. These technological activities form the link from the inventor or innovator to the customer. The thousands of individual actions that cause a new product or service to take shape can be gathered into five different areas of managed activity. See **Figure 32-4**. These areas are the following:

- **Research and development.** These activities are performed to find and create new or improved products and processes.

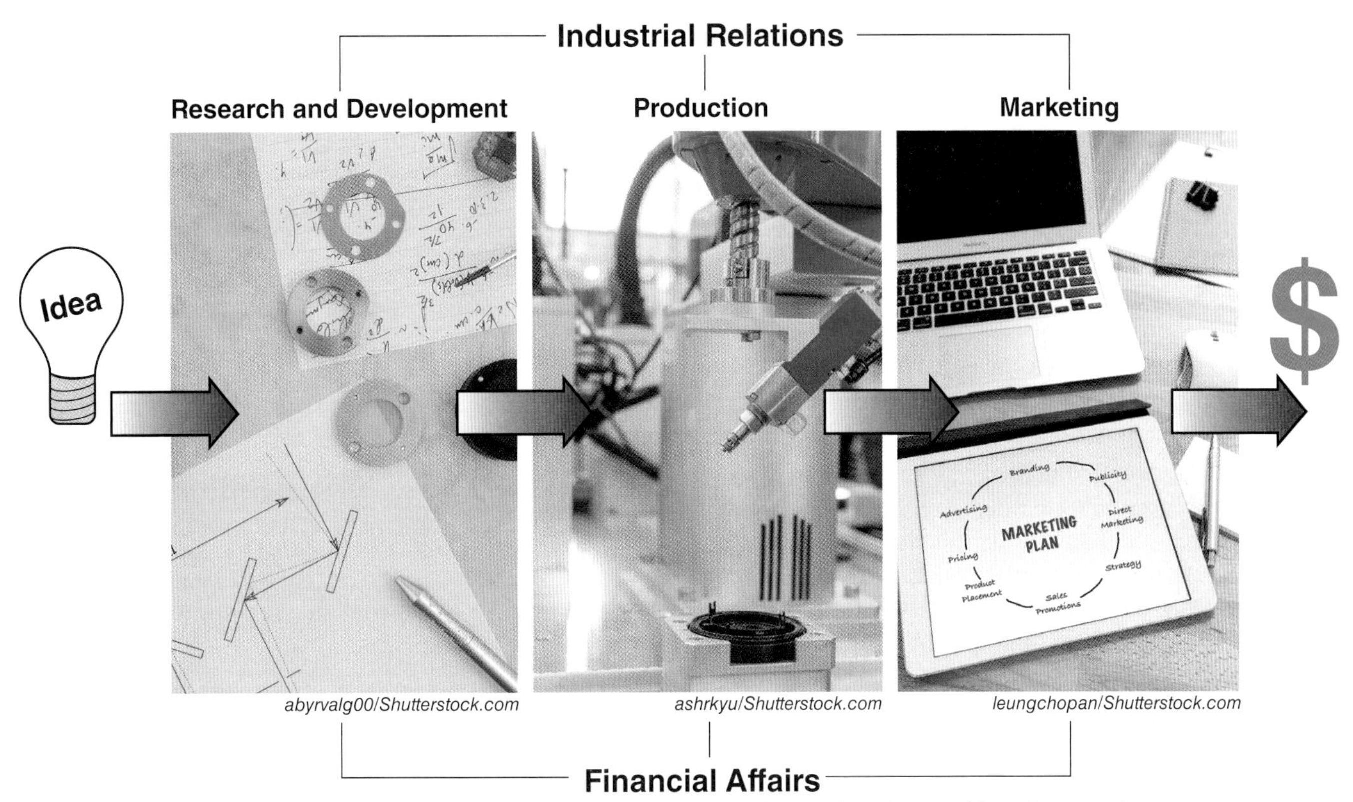

abyrvalg00/Shutterstock.com ashrkyu/Shutterstock.com leungchopan/Shutterstock.com

Figure 32-4. There are five major areas of managed activities in industry that change ideas into products, structures, or services. Notice how each area relates to the others.

- **Production.** These activities develop methods for producing products or services and produce the desired outputs.
- **Marketing.** These activities encourage the flow of goods and services from the producer to the consumer.
- **Industrial relations.** These activities develop an efficient workforce and maintain positive relations with the workers and the public.
- **Financial affairs.** These activities obtain, account for, and pay out funds.

The first three activity areas are product- or service-centered. They directly contribute to the design, production, and delivery of the planned outputs. The other two areas are support areas. Industrial relations provides human or personnel support. Financial affairs provides monetary support.

Research and Development

Research and development can be viewed as the "idea mill" of the enterprise. In this area, employees work with the true raw material of technology, human ideas. They convert what the mind envisions into physical products and services. These actions can be divided into three steps—research, development, and engineering.

Research

Research is the process of seeking and discovering knowledge. See **Figure 32-5**. Research explores

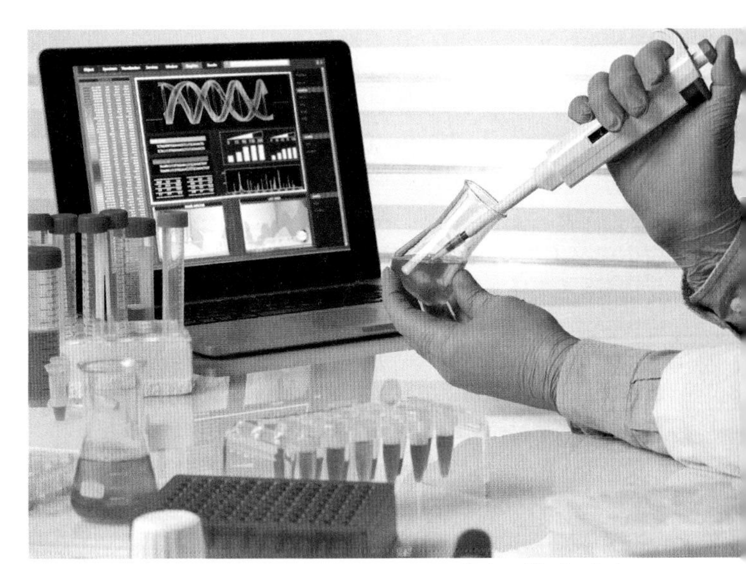

angellodeco/Shutterstock.com

Figure 32-5. Research seeks to discover knowledge.

Career Connection Technical Illustrators

Technical illustrators create art to communicate ideas, thoughts, or feelings. They typically create pictures for publications, such as books, magazines, sales brochures, and advertisements, and for commercial products, such as textiles, wrapping paper, stationery, greeting cards, and calendars. Increasingly, illustrators prepare work directly on computers in a digital format. Industrial enterprises, advertising agencies, publishing companies, and design firms employ illustrators.

Chuck Rausin/Shutterstock.com

High school students interested in pursuing a degree and career as a technical illustrator should take classes in AutoCAD®, graphics arts, Adobe® Creative Suite® (Photoshop®, Illustrator®, InDesign®), Microsoft® Office, art, and drafting.

Many companies expect applicants to have at least an associate's degree in technical illustration or a related field. Knowledge of computer graphics and skill in using visual-display software are important. Applicants should have strong communication skills, creativity, and knowledge of an employer's product, vision, and market.

Technical illustrators can work in a variety of fields. Examples include marketing, construction, food and beverage, clothing, and engineering.

Technical illustrators may work on the following types of projects:

- Working with clients to determine criteria and constraints for new products.
- Collaborating with managers, designers, and engineers to gain greater understanding of a product.
- Creating new ideas that can solve a company's problem.
- Consulting with multiple clients, providing input for various stages of new product development.

Professional organizations for technical illustrators include the Society of Technical Illustrators and the American Design Drafting Association.

the universe systematically and with purpose. This process determines, to a large extent, what technology we will have in the future. Research determines the type of human-built world in which we will live.

Two types of research are basic research and applied research. ***Basic research*** seeks knowledge for its own sake. We conduct basic research to enlarge the scope and depth of human understanding. People working in basic research are not concerned about creating new products. Their focus is on generating knowledge. ***Applied research*** seeks to reach a commercial goal by selecting, applying, and adapting knowledge gathered during basic research. The focus of applied research is on tangible results, such as products, structures, and technological systems.

The two types of research complement one another. Basic research finds knowledge, and applied research finds a use for this knowledge. For example, basic research might develop knowledge about the reactions of different materials to high temperatures. Applied research might then determine which material is appropriate for the reentry shield of a spacecraft.

Development

Development uses knowledge gained from research to derive specific answers to problems. Knowledge is converted into a physical form. The inputs for development are two-dimensional types of information, such as sketches, drawings, or reports. The outputs are models of three-dimensional artifacts, such as products or structures.

Two types of development are product or structure development and process development, **Figure 32-6**.

Product or structure development involves the application of knowledge to design new or improved products, structures, and services. Development might result in a totally new product or structure or improve one that already exists. For example, the bicycle was originally a product from the 1800s. The 10-speed bicycle was later developed from the standard bicycle. Likewise, the laser printer uses many processes originally developed for photocopiers.

Process development devises new or improved ways of completing tasks in manufacturing, construction, communication, agriculture, energy and power, medicine, or transportation. Process development might result in something totally new, such as fiber-optic communication. Process development might also improve existing processes. An example is the development of metal inert gas (MIG) welding, which was derived from standard arc welding.

Boeing

Figure 32-6. The two types of development, product and process, were used in designing this aircraft.

Engineering

Developed products and structures must be built. Developed processes must be implemented. To do these things, people need information. Product and structure engineering is responsible for this activity. Engineering develops the specifications for products, structures, processes, and services. This is done through two basic activities—engineering design and engineering testing.

Engineering design conveys the information needed to produce a product or structure. The three main types of documents used are engineering and architectural drawings, bills of materials, and specification sheets.

Engineering drawings convey the characteristics of manufactured products. See **Figure 32-7**. A set of engineering drawings includes the following:

- ***Detail drawings*** conveying the sizes, shapes, and surface finishes of individual parts.
- ***Assembly drawings*** showing how parts go together to produce assemblies and finished products.
- ***Systems drawings*** showing the relationship of components in mechanical, fluidic (hydraulic and pneumatic), and electrical and electronic systems.

Architectural drawings are used to specify characteristics of buildings and other structures, **Figure 32-8**. They include floor, plumbing, and electrical plans for a structure. The drawings also include elevations showing interior and exterior walls.

Drawings, however, do not convey all the information needed to build products and structures. The people who implement designs need to know

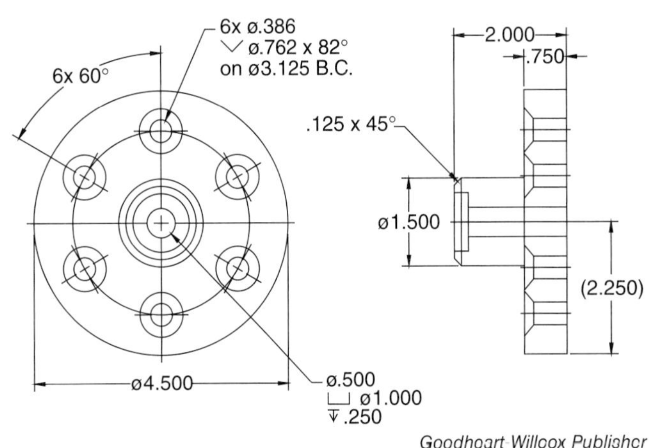

Goodheart-Willcox Publisher

Figure 32-7. Engineering drawings provide detailed information about products or parts to be manufactured.

Menna/Shutterstock.com

Figure 32-8. This engineer is reviewing architectural drawings for a warehouse under construction.

Click and Photo/Shutterstock.com

Figure 32-9. Making a guitar to a customer's specifications is an example of custom manufacturing.

the quantities, types, and sizes of the materials and hardware needed. This information is included on a ***bill of materials***. Finally, data about material characteristics are contained on specification sheets.

Engineering testing is done during the design stage of a new or redesigned product or system. The goal of engineering testing is to ensure that the product or system meets design criteria and customer expectations.

Production

Research and development conceptualize and specify ideas for products and structures. Production then manufactures or constructs the physical item. A number of different manufacturing systems are used to produce products and structures, including the following four common systems:

- **Custom manufacturing.** This system produces a limited quantity of a product to a customer's specifications. Generally, the product is produced only once. The custom-manufacturing system requires highly skilled workers and has a low production rate. This makes it an expensive system to operate. Examples of custom-manufactured products are tailor-made clothing, some items of furniture, and some musical instruments. See **Figure 32-9**.
- **Intermittent manufacturing.** Intermittent (job lot) manufacturing involves the manufacture of a group of products to the company's or a customer's specification. The parts move through the manufacturing sequence in a single batch. All parts are processed at each workstation before the batch moves to the next station. Often, repeat orders for the product are expected. This manufacturing activity is relatively inexpensive. However, considerable setup time is required between batches of new products.
- **Continuous manufacturing.** In continuous manufacturing, a production line manufactures or assembles products continuously. The materials flow down a manufacturing line specifically designed to produce those products. See **Figure 32-10**. The parts flow from station to station in a steady stream. This type of manufacturing handles a high volume and has relatively low production costs. Continuous-manufacturing lines are fairly inflexible, however, and can be used for only a few different products. Many are dedicated to a single product.
- **Flexible manufacturing.** ***Flexible manufacturing*** is a computer-based manufacturing system combining the advantages of intermittent manufacturing with the advantages of continuous manufacturing. Thus, it makes possible short runs with low unit-production cost. Machine setup and adjustment are computer controlled, permitting quick and relatively inexpensive product changeovers.

©iStockphoto.com/Kativ

Figure 32-10. This cutting-and-packaging line for cheese is part of a continuous-manufacturing process.

Similar production systems are used in the other technologies, such as construction. Some homes are built to an owner's specifications (custom-built). Other dwellings are built in tracts with a common plan used to make a large number of buildings. See **Figure 32-11**.

An analogy can be made in transportation. Driving an automobile is similar to custom manufacturing. Automobiles are flexible. They are, however, relatively expensive. Rapid transit buses and trains are more similar to continuous manufacturing. They move people on set lines at lower costs.

The actual production of products and structures can be divided into three important tasks—planning to produce, producing, and maintaining quality of the product or structure.

Planning

Planning is the backbone of most production systems. Planning determines the sequence of operations needed to complete a particular task and the needs for human, machine, and material resources. It involves assigning people and tasks to various workstations.

FloridaStock/Shutterstock.com

Figure 32-11. Homes built in tracts with a common plan used to make a large number of dwellings.

Closely associated with planning and scheduling is production engineering. Production engineers design and install the system used to build the product or structure. They are concerned with the physical arrangement of the machines and workstations needed to produce the product.

Producing

Producing is the actual fabrication of the product. See **Figure 32-12**. In manufacturing, it involves changing the forms of materials to add

Dmitry Kalinovsky/Shutterstock.com

Figure 32-12. Pharmaceutical mixing tanks on a production line.

Think Green Sustainability Plan

Throughout this text, you have learned about the different types of environmental impacts technology can have and the ways in which individuals can change these impacts. In much the same way, companies can use greener alternatives to make a difference. Several businesses now include a sustainability plan in their business plans. A sustainability plan contains guidelines and procedures that help companies use, develop, and protect resources in a way that meets needs without harming the environment.

A company's sustainability plan may come from individual employee ideas, but the whole company puts these ideas into action. There are no set guidelines for all businesses to use in creating sustainability plans. The general goals typically focus on reducing the use of materials, energy, and water. Another common goal is to reduce the amount of pollution output by the company. The ways companies achieve these goals are similar to how individuals work to lessen their environmental impacts.

to the worth, or value, of the materials. Producing activities include locating and securing material resources, producing standard stock, and manufacturing the products.

Product manufacturing processes are used to change the sizes, shapes, combinations, and compositions of materials. These processes include casting and molding, forming, and separating processes that size or shape materials. Conditioning processes change the internal properties of the material. Assembling processes put products together. Finishing processes protect or beautify the products' surfaces.

Construction processes are used to produce buildings and heavy engineering structures. Typical construction processes include preparing the building site, setting foundations, erecting superstructures, enclosing and finishing structures, and installing utility systems.

Communication processes are used to produce graphic and electronic media. Generally, communication messages are designed, prepared for production, produced, and delivered. This is done through the processes of encoding, storing, transmitting, receiving, and decoding operations.

Transportation processes are used to move people and cargo. They are used in land, water, air, and space systems. Typical transportation production processes include loading, moving, and unloading vehicles.

Maintaining Quality

Throughout these processes, a standard of perfection is maintained. Quality control includes all systems and programs that ensure the outputs of technological systems meet engineering standards and customer expectations.

Often, people think "quality" means smooth, shiny, and exactly sized. This is not always true. A smooth, shiny road makes a poor driving surface. Cars would have difficulty controlling and braking on such a surface. Likewise, holding the length of a nail to a tolerance of ±.001" is inappropriate. The cost of manufacturing to that tolerance is too high for the product. The important quality consideration in a nail is holding power, not exact length. Quality can be measured only when a person knows how the product or part is to be used. The product's function dictates quality standards.

An important part of a quality control program is inspection. The ***inspection*** process compares materials and products with set quality standards. There are three phases of an inspection program, **Figure 32-13**. The first phase inspects materials and purchased parts as they enter production operations. The second phase inspects work during production. The final phase inspects the end product or structure.

Inspection can be done on every product or on a representative sample of the products. Expensive, complex, or critical components and products are subjected to 100% inspection. This means every part is inspected at least once. Products such as aircraft components and some medical devices are examples of outputs receiving 100% inspection.

Less expensive and less critical parts receive random inspections. A sample of the product is

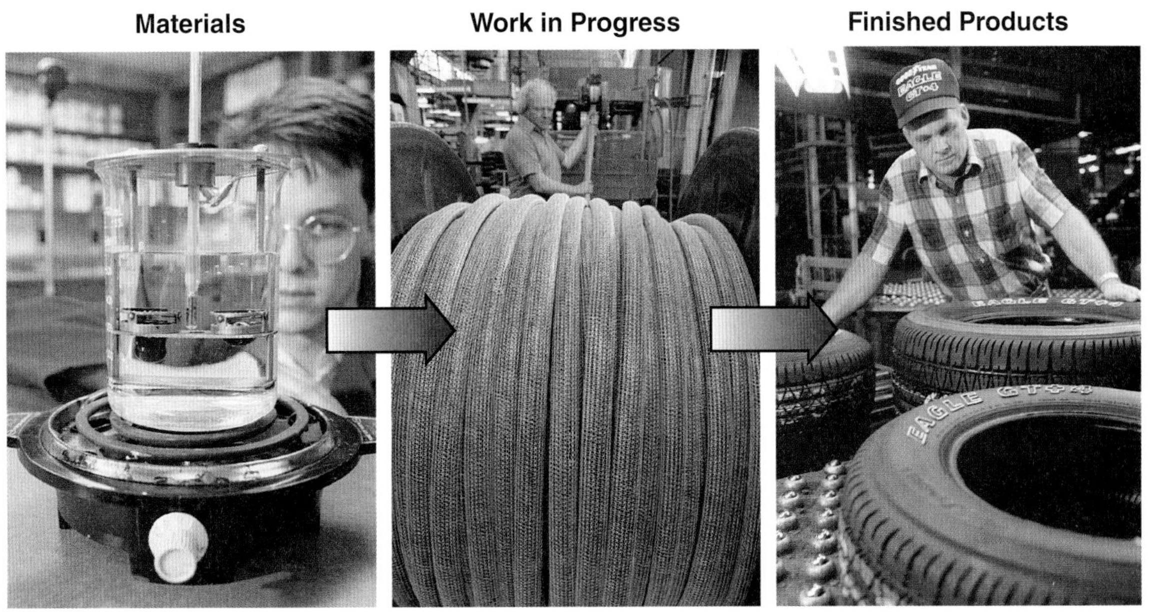

Goodyear Tire and Rubber Co.

Figure 32-13. Quality control inspects materials entering the plant, work in progress, and finished products.

selected representing a typical production run. The sample size and the frequency of inspections are determined using statistics (mathematically based predictions). Inspection of representative samples is part of a program called *statistical quality control*.

The selected sample is inspected. If it passes, the entire run is accepted. If the sample fails to meet the quality standards, the entire run is rejected. Rejected production lots can be dealt with in various ways. The run can be sorted to remove rejects (parts failing to meet the standards). The whole run can be discarded. The run can also be reworked.

Random inspection is used whenever it is cost-effective. Often, the cost of 100% inspection outweighs the value of this type of inspection. For example, it would be expensive to use 100% inspection on roofing nails. The user of the nails can discard the occasional defective nail that slips past random inspection without endangering the product or people.

Production and Value

In addition to the aspects of planning, producing, and maintaining quality, production activities must deliver products at a reasonable cost. This challenge requires attention to both price and value. Delivering products at a reasonable cost also requires attention to cutting unnecessary costs out of production activities.

In addition to quality, customers expect value. Price and value are two different things. Price is what someone must pay to buy or use the product or service. Initial prices are established by businesses and reflect market conditions. The customer determines value. ***Value*** is a measure of the functional worth the customer sees in the product. The customer expects the product or structure to deliver service and satisfaction equal to or greater than its cost. Answering the question, "Was the product worth what I paid for it?" can establish the product's value.

A number of new production systems reduce product cost and, in turn, increase product value. See **Figure 32-14**. Computer-aided design (CAD) reduces product design and engineering costs. Just-in-time (JIT) inventory control systems schedule materials to arrive at manufacturing when the materials are needed, reducing warehousing costs. Flexible manufacturing reduces machine-setup time.

Similar systems are used in the other technologies. Computer scheduling is used in construction to ensure that human and material resources are effectively used. Computer ticketing reduces transportation costs. Computer systems make layout and preparation of color illustrations for printed products more economical than traditional paper or human input–based systems.

Marketing

Products, structures, and services are of little value to companies unless the companies can sell them to customers. The products and structures

©iStockphoto.com/seraficus

Figure 32-14. This automatic control center in a newspaper-printing plant reduces costs by maintaining quality and reducing labor.

must be exchanged for money. ***Marketing*** efforts promote, sell, and distribute products, structures, and services. Specifically, marketing involves four important activities:

- ***Market research*** gathers information about the product's market. This can include data about who will buy the product and where these people are located, in addition to their ages, genders, and marital status. Also, market research can measure the effectiveness of advertising campaigns, sales channels, or other marketing activities.
- ***Advertising*** includes the print and electronic messages promoting a company or its products. This activity can also present ideas to promote safety or public health. Advertising is designed to cause people to take action (buy a certain product) or think differently (buckle your seat belt while riding in an automobile). Closely related to advertising is ***packaging***, which involves designing, producing, and filling containers. See **Figure 32-15**. The packages are designed to promote the product through colorful or interesting designs. Packaging also protects the product during shipment and display. Finally, the packages must include information that helps the customer select and use the product wisely.
- ***Sales*** is the activity involving the physical exchange of products for money. This activity includes sales planning, which involves developing selling methods and selecting and training sales personnel. Sales also includes

Goodheart-Willcox Publisher

Senol Yaman/Shutterstock.com

Nutrition Facts

Serving Size 3 oz. (85g)

Amount Per Serving	As Served
Calories 38	**Calories from Fat** 0
	% Daily Value
Total Fat 0g	0%
Saturated Fat 0g	0%
Cholesterol 0g	0%
Sodium 0g	2%
Total Carbohydrate 0g	3%
Dietary Fiber 0g	8%
Sugars 0g	
Protein 0g	

Vitamin A 270% • **Vitamin C** 10%

Calcium 2% • **Iron** 0%

Percent Daily Values are based on a 2,000 calorie diet. Your daily values may be higher or lower depending on your calorie needs:

	Calories	2,000	2,500
Total Fat	Less than	65g	80g
Sat Fat	Less than	20g	80g
Cholesterol	Less than	300mg	300mg
Sodium	Less than	2,400mg	2,400mg
Total Carbohydrate		300g	375g
Dietary Fiber		25g	30g

Goodheart-Willcox Publisher

Figure 32-15. Packages are designed and produced to perform multiple functions. A package promotes the product, protects the product, and contains information about nutritional content and other topics.

Technology Explained Wind Tunnel

Wind tunnel: A device used to test the aerodynamics of vehicles and structures under controlled conditions.

Wind resistance affects vehicles as they move people and cargo. This resistance has an impact on the operating efficiencies of the vehicles. In designing these vehicles, engineers subject models to tests to maximize the efficiencies of the designs. One important test instrument in this quest is the wind tunnel.

Originally, wind tunnels were used solely to test airfoils during aircraft-design activities. The uses of these devices have since been expanded. With current concern for fuel efficiency, wind tunnels are now used extensively to ensure that vehicles offer the least amount of wind resistance possible. Wind tunnels are also used to test the wind patterns over and around buildings and various other structures. See **Figure A**.

©iStockphoto.com/ktsimage

Figure A. A wind tunnel is often used to test the design of an automobile.

A wind tunnel is designed to pass high-speed air over a full-size or scale model of a vehicle or structure. An important design requirement is that a smooth and uniform flow of air be produced in the tunnel. To accomplish this, a number of wind tunnel designs have been produced. See **Figure B**.

In most wind tunnels, fans or turbines develop airflow in large ducts. The diameter of the ducts increases as the air travels away from the fans. This reduces the airspeed, as well as frictional losses. The tunnels have mitered corners, which reduces the wind loss as the airflow changes directions. Also, in high-speed tunnel models, the air passes through cooling tubes to remove the heat it gains while passing through the fans or turbines.

The air produced by the fans has a swirling pattern. This airflow creates unreliable test results. Therefore, the airflow passes through an air-smoothing unit. This unit is a series of tubes removing the swirls and directing the air in a straight line.

As the air reaches the test chamber, the diameter of the tunnel shrinks rapidly. This causes the speed of the airflow to increase. The amount of the decrease in the tunnel's diameter controls the final airspeed in the test chamber.

Models in the chamber are carefully tested. Anemometers test the airspeed. Smoke might be introduced to visually observe the flow patterns of the air as it passes structures or vehicles. The vehicle itself might be attached to instruments to measure the lift and drag it develops as air passes.

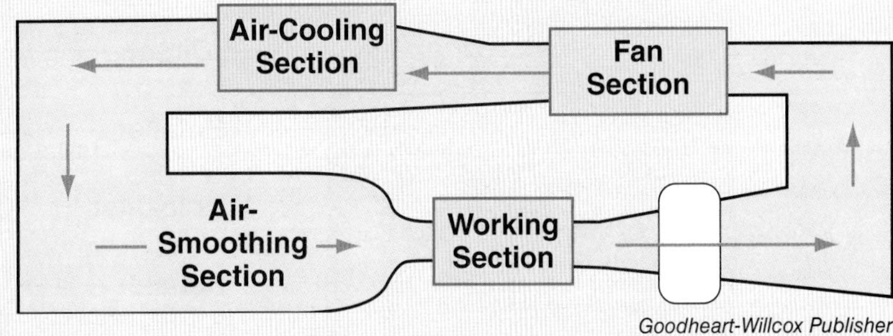

Goodheart-Willcox Publisher

Figure B. The path the airflow follows in a closed loop wind tunnel design.

the act of selling—approaching customers, presenting the product, and closing the sale. This series of steps is all part of sales operations. The end results of sales operations are an order from the customer and income for the company.

- *Distribution* is physically moving the product from the producer to the consumer. The consumer might be another company or a retail customer. Consumer products follow at least three common channels. See **Figure 32-16**.
  - **Direct sales.** The product moves from the producer directly to the consumer. Sales of homes, cosmetics, and kitchenware may use this channel. Methods of direct sales methods include in-home shopping parties and presentations of goods.
  - **Retail.** The producer sells the products or service directly to a retailer. The retailer then makes the items available to the customer. Franchised businesses, such as new-automobile dealers and some restaurants, are examples of this distribution channel. The producer regulates the number of sales outlets and the quality of service those outlets provide.
  - **Wholesale.** The producer selling products to wholesalers who buy and take possession of the products. The wholesalers, in turn, sell their commodities to retailers. The retailers then sell the commodities to the customers. In this channel, producers have little control over the retailers who are selling their products.

Research and development, production, and marketing activities require money and people to make everything work. These requirements are the responsibility of two managed support areas—industrial relations and financial affairs.

Industrial Relations

People are fundamental to all company operations. They work for the company, buy the company's products, and pass laws regulating company operations. Companies, therefore, are very concerned about their relationships with people. They nurture positive relationships with people by engaging in industrial relations (human resources) activities. These activities can be grouped under three main programs. See **Figure 32-17**. These programs are the following:

- *Employee relations* programs recruit, select, develop, and reward the company's employees.
- *Labor relations* programs deal with the employees' labor unions.
- *Public relations* programs communicate the company's policies and practices to governmental officials, community leaders, and the general population.

Employee Relations

All companies have employees. These employees do not magically appear, fully qualified to work. They are the result of managed employee-relations activities. These activities select, train, and reward

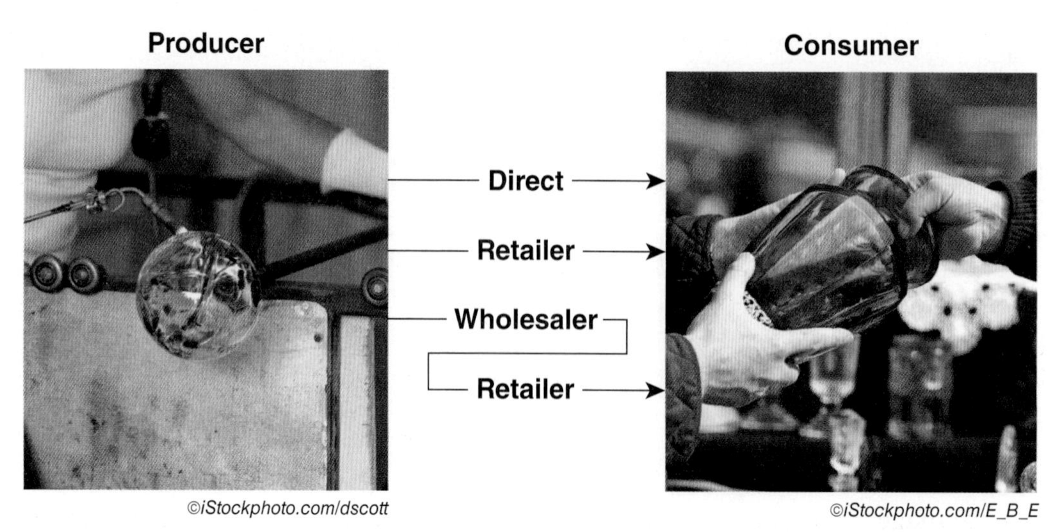

Figure 32-16. There are three different paths moving products from producers to consumers.

Monkey Business Images/Shutterstock.com

Figure 32-17. Industrial relations personnel develop and administer three types of people-centered programs.

the people producing products and managing operations.

Selecting employees

The first action in the employee-relations process is employment, which involves determining the company's need for qualified workers. Applicants are then acquired through a searching process called *recruiting*. Recruiting can be done through newspaper advertisements, school recruiting visits, employment agencies, or even through online recruiting methods such as social networking. Other applicants might come to the company seeking employment on their own.

Next, job applicants go through a screening process. Screening allows qualified people to be selected from the applicant pool. Generally, this selection process starts with an application form to gather personal and work-experience data. The promising applicants are interviewed to gather additional information. See **Figure 32-18**. Some jobs require special abilities and knowledge. In these cases, applicants are given a test to find those who qualify. Successful applicants gain employment.

Training employees

Few new employees are ready to begin work without training. Most need at least some basic training. Some employees might need special instruction. Basic information about the company and its rules and policies is provided to all workers. This is called *induction training* or *orientation*.

Special job skills can be provided through one of three programs. Simple skills are generally taught through on-the-job training. In this method, experienced workers or managers train new workers at the new workers' workstations. More specialized skills

Antonio Guillem/Shutterstock.com

Figure 32-18. Applicants with promising qualifications are called in for an interview.

might be developed in classroom training sessions. In classroom training, qualified instructors provide information and demonstrate practices that each employee must learn.

Highly skilled workers are developed through ***apprenticeship training***. Apprentices receive a combination of on-the-job and classroom training over an extended period of time. Apprenticeships usually last from two to four years, depending on the skills to be learned.

In all three types of training, workplace safety is stressed. New employees are informed about company safety rules and shown how to work safely. In many cases, they are tested on safe work practices. Safety training is vital to providing a safe workplace. Federal and state agencies provide companies with safety rules and regulations and are responsible for enforcing them. A principal source of these regulations is the Occupational Safety and Health Administration (OSHA).

Executives and professional employees also receive training. This training might be called *executive development, sales training,* or *managerial training*. Since the work these employees do does not involve making the products, most training sessions are given in classroom settings. See **Figure 32-19**. In many companies, the total training program is called *human resource development (HRD)*.

Rewarding employees

People want to be recognized and rewarded for their work. Companies recognize and reward

Iakov Filiminov/Shutterstock.com

Figure 32-19. These managers are receiving classroom training to improve communication skills.

people in two ways. First, they pay employees a wage, salary, or commission as a direct reward for work accomplished. A ***wage*** is a set rate paid for each hour worked. A ***salary*** is payment based on a longer period of time, such as a week, month, or year. Wage earners are often called *hourly workers.* Hourly workers are usually the production workers who build products, erect buildings, print products, or provide transportation services.

Salaried employees are usually technical and managerial workers. They develop products, engineer facilities, maintain financial records, and direct the work of other people. These employees often have more formal education than hourly workers. Salaried employees are held accountable more for the amount of work they do than the hours they work.

Some salespeople are paid in a different way. Instead of wages or a salary, they receive a ***commission*** for each sale they make. The commission is usually a percentage of the total dollar value of the goods sold. See **Figure 32-20**.

The second type of reward is called a ***benefit***. Benefits are the insurance plans, paid time off, holidays, and other programs the company provides. These items cost the company money and, therefore, are a part of the total pay package for an employee. Some companies also make use of special rewards called *bonuses* or *incentives*. These rewards are typically awarded to employees for performance exceeding what is expected or for suggestions leading to improvements in efficiency, productivity, or workplace safety.

STEM Connection: Mathematics

Calculating Bids

Members of the Rodriguez family want to add a 10′ × 10′ screened-in porch to their house. They ask Mr. Murphy, the sole proprietor of the A-Able Construction Company, to provide a bid for the work. Mr. Murphy decides he needs to charge enough to cover his costs and earn a 20% profit to finance new ventures. He calculates his bid as follows:

Rent equipment to clear the land and pour the foundation and footings.		$ 400.00
Purchase materials (cement, lumber, screening, shingles, felt paper, fasteners, and a door).		$2900.00
Hire two laborers for approximately one week (eight hours per day) at $25.00 per hour each.		$2000.00
	Subtotal	$5300.00
Earn a 20% profit ($5300.00 × 20%).		$1060.00
	Total	$6360.00

What would Mr. Murphy bid for the job if he found out that equipment rental is $550, instead of $400, and if he decided he needed to make a 25% profit?

bikeriderlondon/Shutterstock.com

Figure 32-20. Real estate agents sell houses and earn a commission.

Labor Relations

In many larger companies, labor unions represent the employees. These companies require a labor-relations program. This program works on two levels. First, ***labor agreements***, called *contracts*, are negotiated between the company and the union. Labor agreements establish pay rates, hours, and working conditions for all employees covered by the contract. The agreements cover a specific period of time, generally ranging from one to three years.

During the contract period, disputes often arise over the contract's interpretation. These disputes are called ***grievances***. This is the second level of work for labor relations. Labor-relations officials work with union representatives to settle the grievances.

Public Relations

Companies hire people, pay taxes, and have direct impacts on communities. Company managers form policies and have practices that they feel benefit the company. These practices often are subject to government regulation and might be affected by community pressures.

A company's public relations program is designed to gain acceptance for company operations and policies. The program informs governmental officials about the need for, and impact of, laws and regulations. Public relations also communicates with community leaders so local actions do not hamper the company's legitimate interests. Finally, public relations communicates with the general public. This communication presents the company as a positive force in the community. A positive image improves the company's ability to sell its products and to hire qualified workers.

Financial Affairs

Just as a company needs people, it also needs money. Companies buy materials and equipment, pay wages and salaries, and rent or buy buildings. Taxes and insurance premiums must be paid. These actions can be shown in a flowchart, **Figure 32-21**. The chart shows that management employs people to use machines to change the form of materials. Employees, machines, and materials are paid for with money. The products produced are converted in the marketplace into money.

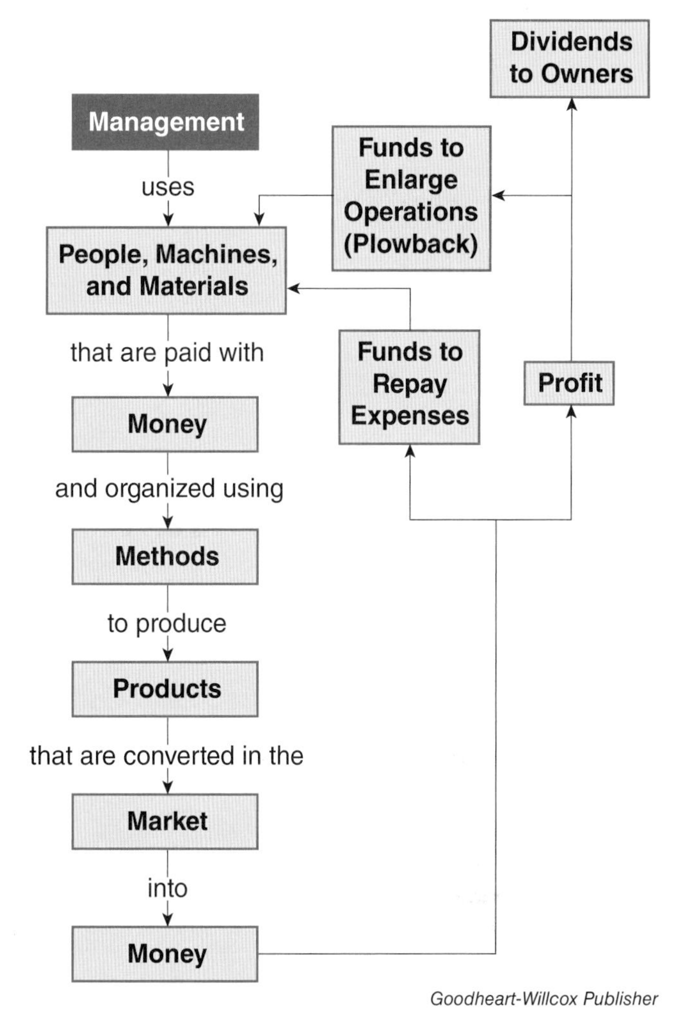

Goodheart-Willcox Publisher

Figure 32-21. This flowchart shows how money cycles within company activities. Start at the upper-left corner and read down.

The money that pays for resources (people's time, materials, and machines) is called ***expenses***. Money is also the end result of sales; this money is called ***income***. The difference between income and expenses is profit or loss. The goal of a company is to have more income than expenses. This makes a company profitable.

Profits must be managed because they serve two important purposes. First, they can become ***retained earnings***, which are profits the company holds and uses to enlarge its operations. Profits are an important source of money for financing new products, plant expansions, and mergers. Many companies pay out another portion of the profits as dividends. These dividends are quarterly or annual payments to the stockholders. They are the reward for investing in the company and sharing the risks of owning a business.

Managing the use of the money a company processes is the responsibility of financial-affairs employees. They raise money, pay for insurance, collect from customers, and pay taxes. These employees also keep records of the financial transactions of the company. This area is called ***accounting***. Each financial action is recorded as either an income item or an expense item.

Finally, employees in financial affairs purchase the materials, machines, and other items needed to operate the company. Purchasing officers seek items that best meet company needs with respect to price, quality, and delivery date.

The Industry-Consumer Product Cycle

We have been looking at company activities as a linear action. These activities are presented as starting with research and development, followed by production, and finally marketing. This path is correct if you look at a single model or version of one product or structure. Our system, however, is much more complex. The economy is described as dynamic, or always changing. Products are developed, produced, and sold. Consumers select, use, maintain, and discard the products. See **Figure 32-22**. They communicate their satisfaction or dissatisfaction with current products. This might cause companies to redesign existing products or, in some cases, to develop new ones. These new products are sold. They in turn are selected, used, maintained, and discarded. This cycle continues with a constant array of new products being developed and obsolete products disappearing.

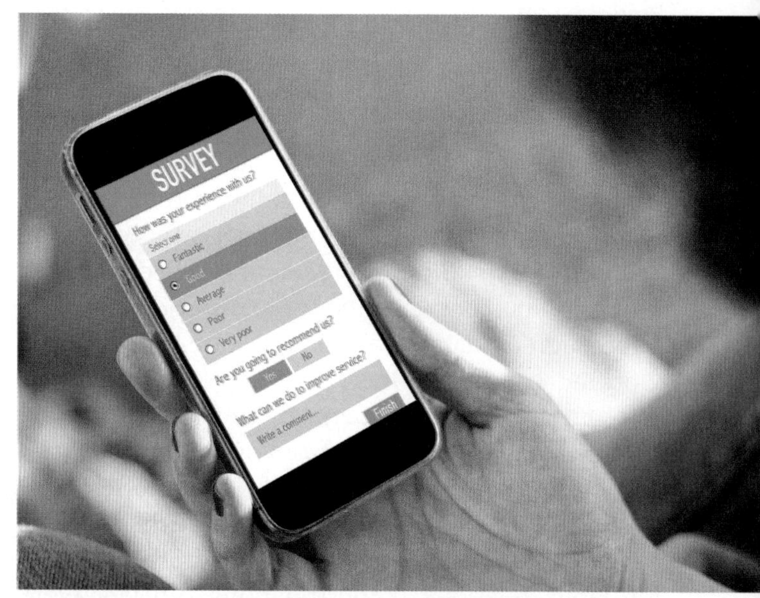

Georgejmclittle/Shutterstock.com

Figure 32-22. Customers' reactions to products cause companies to continually design, produce, and sell new items.

Diego Cervo/Shutterstock.com

Running a successful business requires attention to detail, including accounting tasks.

Summary

- ✔ The five basic societal institutions are family, religion, education, politics and law, and economy.
- ✔ Economic enterprises are organizations engaging in business efforts directed toward making a profit.
- ✔ Industry is the area of economic activity that uses resources and systems to produce products, structures, and services with the intent to make a profit.
- ✔ The five main areas of managed activity in industry are research and development, production, marketing, industrial relations, and financial affairs.
- ✔ Basic research seeks knowledge for its own sake. Applied research seeks to reach a commercial goal by selecting, applying, and adapting knowledge gathered during basic research.
- ✔ The production of products and structures involves planning to produce, producing, and maintaining the quality of the product or structure.
- ✔ Marketing activities include market research, advertising, sales, and distribution of products, structures, and services.
- ✔ The three main industrial relations activities are employee relations, labor relations, and public relations.
- ✔ Financial affairs activities obtain, account for, and pay out funds.
- ✔ Profits can become retained earnings, which the company holds and uses to enlarge its operations, or dividends, which are paid to stockholders.
- ✔ Industry redesigns existing products and develops new ones to satisfy the needs and wants of customers.

Check Your Engineering IQ

Now that you have finished this chapter, see what you learned by taking the chapter posttest.
www.g-wlearning.com/technologyeducation/

Test Your Knowledge

Answer the following end-of-chapter questions using the information provided in this chapter.

1. Which of the following is *not* a societal institution?
   A. Religion.
   B. Sports.
   C. Family.
   D. Education.
2. Businesses directed toward making a profit are called ______.
3. How are production and prices determined in a free enterprise system?
4. Industry is the area of economic activity that uses ______ and ______ to produce products, structures, and services with the intent to make a profit.
5. In terms of technology, why is industry different from the other types of economic enterprises?
6. What is the difference between basic research and applied research?
7. *True or False?* Development can result in a totally new product or improve one that already exists.
8. What three main types of drawings are used by engineers to convey the information needed to produce a product?
9. List the four types of manufacturing systems.
10. What are the three phases of an inspection program?
11. List three functions of packaging.
12. *True or False?* Franchised businesses typically purchase products and necessary services through a wholesaler.
13. What are the three main programs in industrial relations?
14. ______ training involves a combination of on-the-job and classroom training over an extended period of time.

15. An employee who is paid a percentage of the total dollar value of the goods sold is receiving a ______.
    A. benefit
    B. salary
    C. commission
    D. wage
16. To show a profit, a company must have more ______ than expenses.
17. List two basic ways that profits are used by a company.
18. List the four steps on the consumer side of the industry-consumer product cycle.

Critical Thinking

1. Summarize the relationships among the five main managed areas of activity within a technological enterprise.
2. Describe how expenses, income, and profit are related in a modern technology enterprise.

STEM Applications

1. Select a simple product, such as a kite. Apply the principles of research and development, production, and marketing to design, produce, and advertise it.
2. Set up a production line for cake pops, cookies, or another food product. Describe how you will plan for the product, produce the product, and maintain quality.

Engineering Design Challenge

Operating a Company

Background

Companies involve a series of independent tasks that have been integrated into functioning enterprises to achieve goals. Each task must be planned for and carried out with efficiency.

Materials and Equipment:

- Computers.
- Software (illustrator software, image editing software, presentation software, etc.).
- Poster board.
- Pens.
- Pencils.

Situation

You have formed a company to produce a nine-month, personalized calendar for your school. In this activity, you will use the organization chart and department goals you prepared to operate a company that produces and markets the calendars. Schedule and complete the tasks required to produce and market the calendars.

Procedure

Schedule and complete the several tasks required to produce and market the calendars.

- Design and development department
  - Obtain software that can be used to produce a personalized calendar.
  - Follow the instructions to produce a calendar for a single month. This will acquaint the department members with the operation of the software.
  - Establish the layout for the calendar.
  - Produce a common layout sheet for the marketing group to use in selling calendar entries.
- Calendar-entry marketing department
  - Determine the selling price for a calendar entry.
  - Develop a calendar-entry order form.
  - Make posters to promote the sale of calendar entries.
  - Sell calendar entries.
- Production department
  - Receive calendar-entry forms from the marketing department.
  - Enter data on the calendar layouts.
  - Print a proof of the calendar.
  - Submit the proof to marketing for approval.
  - Correct the calendar entries.
  - Print the master calendar.
  - Reproduce the calendar.
- Calendar marketing department
  - Produce and distribute advertisements for the sale of calendars.
  - Select and train calendar salespeople.
  - Sell calendars.
  - Maintain sales records.
- Finance department
  - Set budgets for company operations.
  - Sell stock. Maintain stockholder records.
  - Purchase materials and supplies.
  - Maintain all financial records.
- Executive committee (president and vice presidents)
  - Set deadlines for important activities.
  - Monitor progress in completing tasks.
  - Set budgets.
  - Establish selling prices for calendar entries and finished calendars.
  - After the calendar copies have been sold, close the company. Liquidate, or dissolve, the assets.

Rawpixel/Shutterstock.com

Managing a technological enterprise requires careful planning and organizational skills.

CHAPTER 33

Understanding and Assessing the Impact of Technology

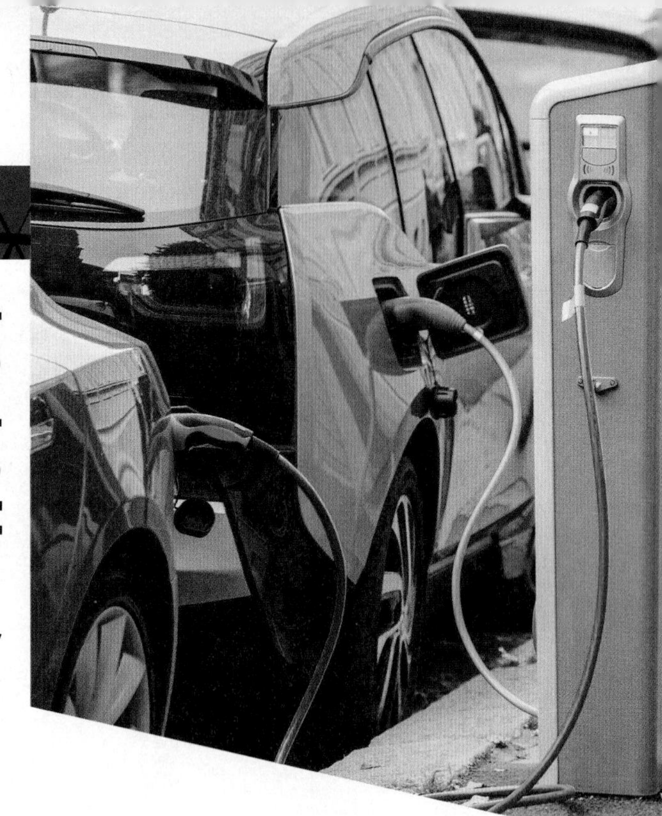

Check Your Engineering IQ

Before you read this chapter, assess your current understanding of the chapter content by taking the chapter pretest.
www.g-wlearning.com/technologyeducation/

Learning Objectives

After studying this chapter, you should be able to:

- ✔ Explain how using a technological product is different from using a technological service.
- ✔ Describe the five steps involved in the proper use of technological products.
- ✔ Summarize how people can effectively select an appropriate technological service.
- ✔ Describe the ways in which societies assess technological advancements.

Technical Terms

repairing
servicing
technological products
technological services
technology assessment

Close Reading

As you read this chapter, think of how the technological products you use meet your needs. Do you spend enough time assessing your products' usefulness and effects?

People working in various industries design and make technological products and systems. This group of people is a part of a larger community or nation, whose citizens hold a variety of different jobs. One job that all people share, however, is using technological outputs. We all use the products of technological effort every day to meet our needs. In doing so, our lives, as well as the world around us, are impacted. The use and impacts of technology involve two major technological actions:

- **Using technology.** Selecting and operating tools, machines, and systems to modify or control the environment.
- **Assessing technology.** Measuring and reporting the impacts of technological use on people, society, and the environment.

Using Technology

People are constantly generating needs and wants. They might want to travel to a vacation spot, cook food faster, or improve their health or energy. These and thousands of other needs and wants can be met with technological devices and systems.

Technological outputs can be grouped into two categories—technological products and technological services. See **Figure 33-1**. ***Technological products*** are the artifacts people build. They can be a manufactured product, a constructed structure, or a communication medium (such as a book or CD recording). ***Technological services*** are outputs that we use, but do not own. These include transportation and communication services, such as airline travel and television programming.

Using products and systems make life easier, more enjoyable, and more productive. Using products and systems can, however, also create frustration and negative impacts on people or the environment. This tension between good and bad, positive and negative, means that people should use technology wisely.

How do people know that they are using technology properly? Proper use of technology involves understanding and applying five steps. See **Figure 33-2**.

Using Technological Products

When we use technological products, we complete a series of tasks. We might do these tasks ourselves or have another person do them for us. The tasks are the "steps of life" of the product. These steps start when we obtain the product and end when the product is no longer of use. Simply put, the steps in using technological products are the following:

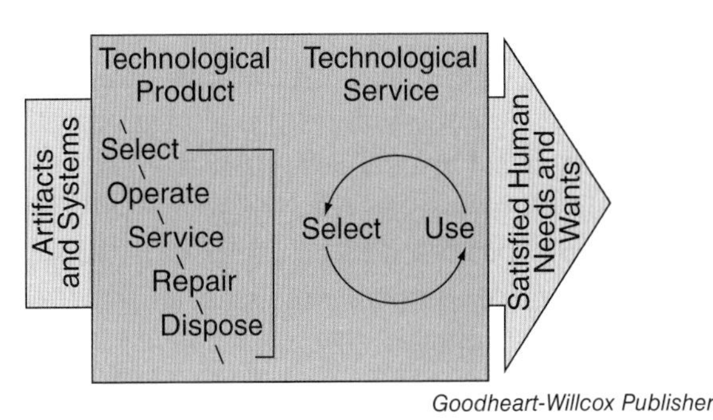

Goodheart-Willcox Publisher

Figure 33-2. The five steps involved in using technological products.

Service

Matej Kastelic/Shutterstock.com

Product

Ollyy/Shutterstock.com

Figure 33-1. Technological outputs can be products or services. These passengers in flight are using a technological service. The headphones and tablet being used by the woman are examples of technological products.

1. Selecting appropriate devices and systems.
2. Operating devices and systems properly.
3. Servicing devices and systems.
4. Repairing broken parts and systems.
5. Disposing of worn-out or obsolete devices and systems.

When we use technological services, we do not use all these steps.

Selecting Appropriate Devices and Systems

Look around you. You are surrounded with technological products. Someone selected each of these. They were not the only products that could have served their purposes. Consider electric lighting. People can choose from incandescent, halogen, and fluorescent lights. These lights can be of different lighting levels (brightnesses), such as 60 watts, 75 watts, or 100 watts. People can select ceiling, floor, or table lights.

Some people use trial and error to make product selections. They buy a product and try it. If it does not work well, they replace it with another product with different features. This is an inefficient and expensive approach to selecting an appropriate product.

A better selection method is to approach the challenge logically, **Figure 33-3**. First, the person should determine the exact need being addressed. The need should then be described in terms of the product's operation, price, quality, and similar features. From these descriptions, a list of alternative products meeting the need can be made. The features of each product should be identified. A list of advantages and disadvantages (pros and cons) of each product should be developed from these features. For example, purchase price, operational features, and maintenance requirements for each product can be identified. The ease of use can be explored. The product's safety can be examined. Likewise, the appearance and styling can be considered. From this list of features and pros and cons, the most appropriate product can be selected. The most appropriate product might not be the cheapest or best-operating product. This product should, however, be the one most closely meeting the overall need.

Iakov Filimonov/Shutterstock.com

Define need

Describe need (establish criteria)

List advantages and disadvantages

Compare products

Select best product

Figure 33-3. Product selection follows a logical sequence.

Operating Devices and Systems

Once a product has been selected and purchased, the new owner must learn how to operate the device. Operation of some products requires little or no training. See **Figure 33-4**. The owners already know all they need to know to operate the product. Few of us need to learn how to use a new pen or pencil.

Other products are replacements for an older model. Again, the owners know a lot about the product's operation. They need only to review the new product's operation and new features. For example, maybe you can effectively use a version of a word processing program. If you buy a new version, you need to learn how to use only the program's new features.

Some products are new to the owners. These owners have never used the product or one similar to it. They then have a lot to learn. This includes learning how to do the following:

- Unpack and set up the product.
- Adjust the product for different operations.
- Correctly and safely operate the product.
- Care for and maintain the product.
- Obtain service and repair.

There are several ways to learn this information. One way is to carefully read the owner's manual. See **Figure 33-5**. The new owner should read and study the manual to obtain information about the five elements listed earlier. Only after learning this information should the owner operate the device.

Another way to obtain this information is through training. This training can be from a person skilled in using the device. For example, a salesperson in a computer store can provide basic information about setting up and using a simple

Little Training Is Needed

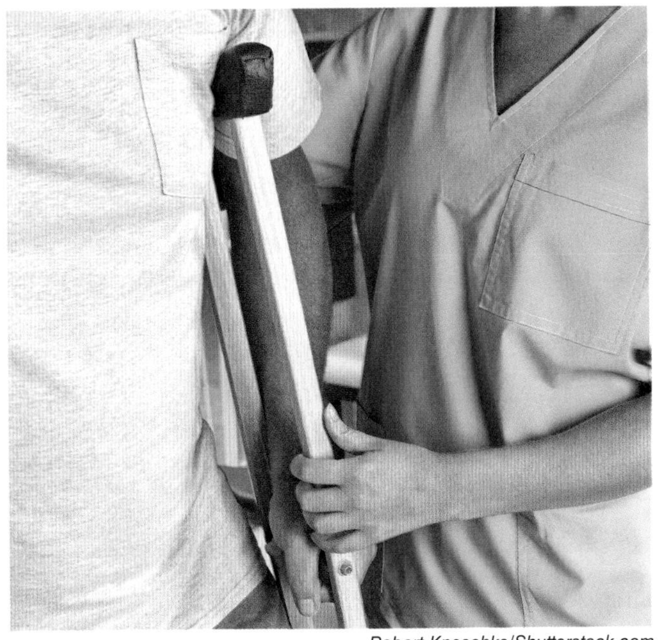

Robert Kneschke/Shutterstock.com

Significant Training Is Needed

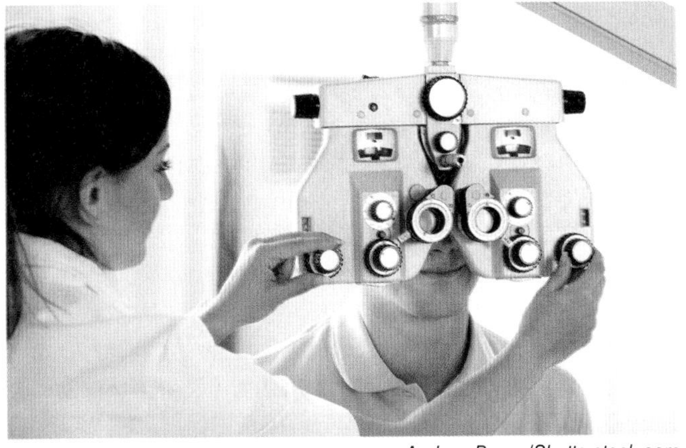

Andrey_Popov/Shutterstock.com

Figure 33-4. Most people need little instruction on how to use crutches. The doctor, however, needed training to use the eye-testing equipment.

computer system. New owners can also seek more formal instruction. They can attend a seminar or class that provides instruction for using the device.

Servicing Devices and Systems

Many products need periodic maintenance or service during their lifetimes. ***Servicing*** is doing routine tasks that keep the product operating. This includes cleaning the product, oiling moving parts, making simple operating adjustments, replacing filters, and other similar tasks. See **Figure 33-6.** Most owners' manuals contain information needed to complete periodic servicing.

Sergey Ryzhov/Shutterstock.com

Figure 33-5. An owner's manual contains valuable information.

Checking Oil

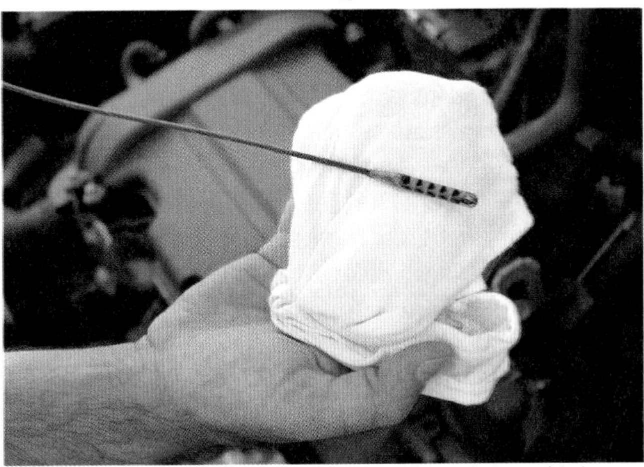

©iStockphoto.com/vladacanon

Adding Oil

©iStockphoto.com/skodonnell

Figure 33-6. Servicing involves simple tasks to keep equipment working, such as checking the oil level and adding oil as needed in an automobile.

Career Connection Landscape Architects

Landscape architects design outside areas, such as residential yards, public parks, college campuses, shopping centers, and golf courses. They help plan the locations of buildings, roads, and walkways. They also determine the arrangement of plant materials, such as lawns, flowers, shrubs, and trees. Landscape architects also work with environmental scientists and foresters to conserve or restore natural resources. They create detailed plans for sites. These plans include soil slopes, vegetation, walkways, and landscaping details, such as fountains and statues. Often, these plans are prepared and presented using CAD systems and video simulation.

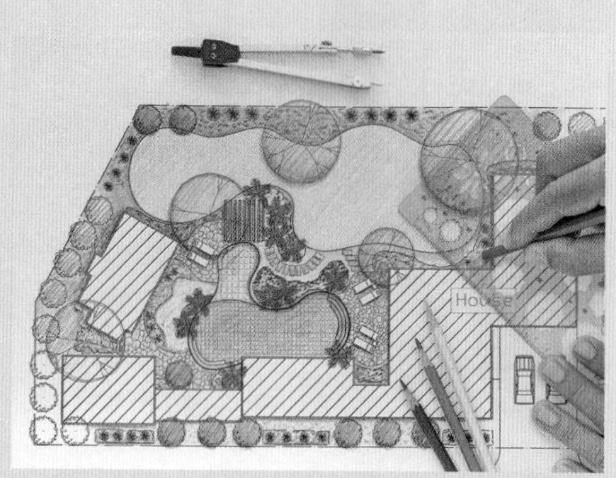

Toa55/Shutterstock.com

High school students interested in becoming a landscape architect should take technology and education courses in computer-aided drafting and engineering design.

Entry into the field requires a bachelor's or master's degree in landscape architecture. A successful landscape architect has creative vision, artistic talent, and good oral and written communication skills. Individuals who are creative, good problem solvers, and have the appropriate licenses and certifications do well in this field. Career tracks include landscape architect, landscaper, surveyor, and mapping technician.

Landscape architects may work on the following types of projects:

- Create site plans and cost estimates for local municipalities.
- Design and prepare graphic illustrations using CAD programs.
- Investigate environmental and land issues related to drainage and landscape.

Professional organizations for people employed in this field include the American Society of Landscape Architects and Council of Landscape Architectural Registration Board.

Repairing Devices and Systems

Occasionally, products stop functioning properly and require repair. ***Repairing*** devices involves replacing worn or broken parts. See **Figure 33-7**. Repairing also includes making major operating adjustments. The owner can sometimes do this type of work; however, it most often is done by trained mechanics and technicians. For example, repairing an automobile transmission, a television set, or a furnace is beyond most people's ability or interest. Also, as products have fewer moving parts and as solid drive devices continue to enter the market, repairs increasingly need to be made by experts.

Thomas Andreas/Shutterstock.com

Figure 33-7. Repairing involves replacing broken and worn parts. This bicycle shop employee is performing a repair.

Disposing of Devices and Systems

Few products last forever. Most products have useful lifetimes, after which they wear out or become obsolete. When this happens, the owner

Nick Starichenko/Shutterstock.com

Figure 33-8. These bins allow people to appropriately recycle various types of discarded items and materials.

should properly dispose of the product or material. See **Figure 33-8**.

A product can be disposed of by donating it to a group that has use for it. For example, many charities and schools have use for computer systems that are not state-of-the-art. Many businesses that replace operating systems with newer, faster models donate the older systems to such groups.

Worn or obsolete products can be sold or given to a recycling operation. A group or company dismantles the product to retrieve good parts or materials. For example, automobile wrecking yards recycle good parts by removing and selling them to people who are repairing similar models. They also sell scrap metals to steel mills and aluminum processors.

Some products cannot be reused or recycled. The product might not have any useful parts. Recycling the materials might be impossible or might not be cost-effective. For example, old dinnerware has little value. Most ceramic materials cannot be recycled. Likewise, broken concrete slabs have little use or value. In this case, the materials should be sent to a landfill.

Using Technological Services

Selecting and using a technological service requires the customer to complete fewer tasks than using technological products. We can only select, use, and evaluate the service. Most of us do not operate the airline or the television network. We do not have a product to service, repair, or discard.

If you plan to travel across the country, you need to establish your need (goal). This might be using the least amount of time to travel from point A to point B or seeing as many historical sights as possible.

Rawpixel.com/Shutterstock.com

Figure 33-9. Once a goal is defined, various ways of arriving at that goal should be examined.

Once the need is defined, the possible ways to make the trip should be established. You can look at a road map and decide if you can make the trip in a car. See **Figure 33-9**. Also, you can explore flight, bus, and train schedules, routes, and prices. Armed with this information, you can choose a method of travel.

After the trip is completed, you can evaluate the trip. Was it pleasant? Did you get where you were going on time? In short, you can decide if the service met your needs. If not, the next time you travel, you can select a different mode of travel or choose a different company to take you to your destination.

You can use the same *select-use-evaluate* model for any technological service. This model can be applied to a television program. If you want an entertainment program appropriate for all your family to watch at 7:00 p.m. on Tuesday, you can search the program listings, read the summaries, and select a program. You can then watch it. After seeing the program, you can decide if it was entertaining and appropriate for your family. If not, you will select a different program the next Tuesday evening.

Assessing Technology

Technological systems have existed for several centuries. One of the most complex technological systems is a city. This system arose during the Middle Ages and gathered people in one spot.

Technology Explained Earth-Sheltered Building

Earth-sheltered building: A structure built into the earth to take advantage of the insulation value and thermal properties of soil.

A number of different building techniques can be used to make buildings more energy-efficient. One way is to use earth-sheltered construction, in which soil partially or completely covers the dwelling. The soil's thermal qualities keep the house at an even temperature. The temperature of soil a few feet below the surface does not vary much throughout the year.

Using the earth to shelter a dwelling is not a new idea. Native Americans in Arizona used earth-sheltered buildings in 1000 CE. This was 400 years before Columbus discovered America. Using protective cliff faces to shelter communities was common throughout the southwestern United States. Today, a number of these sites have been uncovered and restored. See **Figure A**.

MBoe/Shutterstock.com

Figure A. This cliff dwelling at the Walnut Creek National Monument in Arizona uses earth sheltering.

Earth-sheltered construction requires careful planning. Special problems exist, such as moisture, loads from soil, and orientation for solar heat gain. Buildings can be partially earth-sheltered or totally covered. See **Figure B**. In the Northern hemisphere, the building is normally sited so it faces south. This is so the house can absorb solar heat through large windows on the south face. The wall of windows helps make the front rooms light and airy. These are the daytime living rooms. Bathrooms and bedrooms can be placed toward the back of the structure. These rooms are normally used at night. At this time, artificial light is needed.

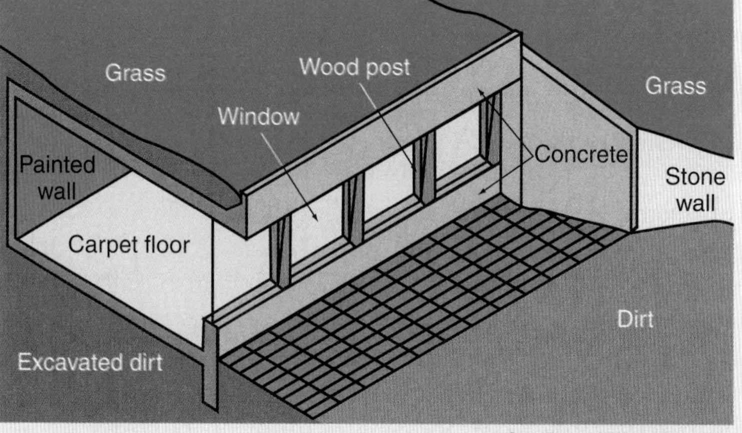

Goodheart-Willcox Publisher

Figure B. The design for an earth-sheltered house that has the roof covered with earth. This house would be built using reinforced concrete.

Earth-sheltered construction can be used for more than housing. It is also appropriate for theaters, shopping malls, convention centers, and warehouses. The functions of these structures do not require windows or exterior views.

Cities provided the workers for the Industrial Revolution. This provision launched a new way of working called *division of labor.* The task of making something was divided into a large number of jobs. Each job was assigned to a single worker. This model replaced the skilled worker who made a product from start to finish.

During the Industrial Revolution, agricultural productivity increased. Medical science was developed. Most people saw technological advancements as a positive force. Life became better as the years passed. People were making more money. Many could work fewer hours and have more time for family activities. New and improved products appeared almost daily. Homes became larger and more comfortable. Better health care was increasing life expectancies.

However, World War I and the Great Depression caused some people in the United States to question technology's effects on people. A growing number of critics suggested that many technological products had harmful aspects. The era of smog, caused largely by pollutants from automobile exhausts and industrial plants, created concerns, **Figure 33-10**. Industrial wastes polluted lakes, rivers, and groundwater. Pesticides, such as DDT, were entering the food chain.

Several methods have been developed to address technological impacts. Individuals evaluate devices and systems according to their personal needs. This is, however, a personal view. A different type of evaluation is called ***technology assessment***. In this type of evaluation, groups of people evaluate the impacts of technology on people, society, and the environment.

Technology assessment places the task of controlling technology on the government and the courts. These agencies provide a way of assessing the effects of technological innovations on human life. They conduct studies and review the studies done by other groups. In particular, they evaluate the economic, ethical, social, health, environmental, and political effects of various technological products and systems.

One model of technology assessment suggests what groups need to do. See **Figure 33-11**. First, the underlying goal for the technology needs to be identified. This should clearly describe why the technology was developed or used. The goal should list the desired outcomes for any new or improved technology. For example, the goal might be a fuel-efficient way to transport people quickly and safely from city-center hotels to the local airport.

Second, a way to measure the success in meeting the goal needs to be developed. This requires criteria to clearly describe the proposed outcomes. Using the airport-transport example, several criteria can be developed. The system should be convenient to the major hotels in the downtown area. This transport should be fast and clean. The system should have a schedule that meshes with the major flights at the airport. This transport should produce at least 50% less emissions than auto travel produces. The system should be cost-effective. This transport should run on public right-of-ways.

VLADJ55/Shutterstock.com

Figure 33-10. Air pollution being produced by an oil refinery.

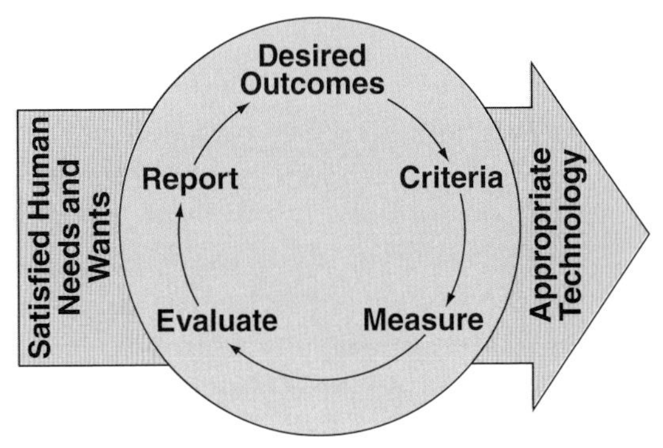

Goodheart-Willcox Publisher

Figure 33-11. Technological assessment can be conducted in five steps.

Third, the impact of the technological innovation should be monitored and measured. A method to gather data needs to be developed. The system should gather information about each criterion of success. Again, using the transport example, interviews with riders can determine the convenience, cleanliness, and speed of the service. Monitoring equipment can measure vehicle emissions. Financial records contain data about cost-effectiveness. The system's schedules and airline-flight schedules contain data about the interface between the two modes of travel.

Fourth, people must review the data and draw conclusions. They need to evaluate the results of the measuring and monitoring action. The results must be measured against the criteria.

Finally, the assessment group must prepare a report. This report should present the problem and the criteria used to explore it, summarize the findings, and present the positive and negative aspects of the technology being assessed. Most importantly, the report should recommend action. For example, it might recommend that the technology be abandoned, altered, or maintained.

The findings of technological assessment can be applied to political and legal actions. Local, state, and national laws and regulations can be developed. Lawsuits can be filed when the actions are found to be against existing laws and regulations. People can support or boycott the product or service.

Summary

- ✔ Using technology involves selecting and operating tools, machines, and systems to modify or control the environment.
- ✔ Assessing technology involves measuring and reporting the impacts of technological use on people, society, and the environment.
- ✔ Technological products are the artifacts people build. Technological services are outputs that people use but do not own.
- ✔ Using technological products involves selecting appropriate devices and systems, operating devices and systems properly, servicing devices and systems, repairing broken parts and systems, and disposing of worn-out or obsolete devices and systems.
- ✔ Technological services are selected, used, and then evaluated.
- ✔ In a technology assessment, groups of people evaluate the impacts of technology on people, society, and the environment.
- ✔ The steps in technology assessment are determining the goal of the technology, measuring its success in meeting the goal, monitoring and measuring its impact, reviewing data and drawing conclusions, and preparing a report.

Check Your Engineering IQ

Now that you have finished this chapter, see what you learned by taking the chapter posttest.
www.g-wlearning.com/technologyeducation/

Test Your Knowledge

Answer the following end-of-chapter questions using the information provided in this chapter.

1. Assessing technology means _____ and _____ the impacts of technological use on people, society, and the environment.
2. What are the two major categories of technological outputs?
3. Technological _____ are the technological outputs that people use, but do not own.
4. The five steps involved in using technological products are selecting, operating, servicing, _____, and disposing.
5. Why are fewer steps needed to properly use a technological service than to use a technological product?
6. The first step in selecting an appropriate product is to determine the exact _____.
7. List five types of information that can be obtained from an owner's manual.
8. List two ways of disposing of unwanted products other than putting them in a landfill.
9. Evaluating the impact of technology on people, society, and the environment is called _____.
10. What are the five steps in assessing a technological product or system?

Critical Thinking

1. Why is it beneficial to understand products before you purchase them? Provide a personal example of one case in which you did not fully understand a product you purchased.
2. Is it becoming more difficult to operate technology? Why or why not?
3. Identify the pros and cons of recycling old technologies.

STEM Applications

1. Solar-energy technology has failed to make the impact many people expected. For many years, this technology has been called a promising technology. What are the barriers to solar technology?
2. Hybrid cars use a combination of a gasoline engine and an electric motor and generator for power. These cars have high fuel efficiency and low emissions. Many people believe we should shift quickly to hybrid vehicles. What government and industry action is needed to make this happen?

Rawpixel.com/Shutterstock.com

Assessing technology's impact is an important task for society.

TECHNICAL TERMS

A

abutment: An element of an arch that stops the thrust of the arch by tying the ends of the arch to the ground. (18)

accounting: The area of a company that keeps records of financial transactions. (32)

acoustical properties: How a material reacts to sound waves. (16)

active collectors: The use of mechanical systems to circulate fluids through solar collectors. (19)

active voice: Writing style that focuses on the subject that performs an action. (8)

activists: People who use public opinion to shape practices and societal values. (30)

actuating: Initiating the work related to the action plan, causing plans to take form. (31)

additive manufacturing: The use of data from 3-D modeling software to direct the building of an object through the placement of successive layers of material. (14)

adjusting device: Type of controller that modifies a system to produce better outputs. (12)

advanced materials: Innovative materials that are created to meet the needs of high-tech applications. (16)

advertising: The print and electronic messages promoting a company or its products, as well as those promoting safety or public health. (32)

aerobic exercise: Exercise that uses oxygen to keep large muscle groups moving continuously. The exercise must be maintained for at least 20 minutes. (25)

aerodynamics: The study of the way air flows around objects. (24)

aeronautical engineering: The study and design of aircrafts and technologies such as helicopters, gliders, lighter-than-air vehicles, airplanes, and jets. (24)

aerospace engineering: A highly specialized discipline that requires the application of mathematical, scientific, and engineering knowledge to develop air and spacecraft related technologies. (24)

aesthetics: The beauty of a structure. (18)

agricultural technology: Technology that uses technical means (machines and equipment) to help plant, grow, and harvest crops and raise livestock. (27)

agriculture: 1—Growing plants and raising animals to be used for food or other technological products. (13); 2—People using materials, information, and machines to produce food and natural fibers. (27)

airfoil: an object designed to produce some directional motion when in movement relative to the air. (23)

air-transportation systems: A system that uses airplanes and helicopters to lift passengers and cargo into the air so they can be moved from one location to another. (23)

algorithm: A set of step-by-step instructions used to perform calculations, process data, and automatically answer questions. (22)

alkali metals: Group 1 elements that have one valence electron (excluding hydrogen) and are highly reactive. (28)

alkaline earth metals: Group 2 elements that have two valence electrons and are highly reactive. (28)

allotrope: Different forms of the same element in the same physical state. (28)

Note: The number in parentheses following each definition indicates the chapter in which the term can be found.

alloy: A metal comprised of two or more metallic elements. (16)
alternating current: An electrical charge that changes directions periodically. (22)
amplitude: The measurement of the strength of a wave. (21)
amplitude modulation (AM): Radio broadcast system that codes the carrier frequency by merging the message onto the carrier wave by changing the strength of the carrier signal. (21)
anaerobic exercise: Exercise that involves heavy work by a limited number of muscles. The exercise is maintained for short intervals of time. (25)
analog signal: A continuous signal varying in strength (amplitude) or in frequency. (21)
analytical system: Data-comparing device that makes comparisons mathematically or scientifically. (12)
angle of attack: The angle of the wing as it cuts through the air; affects lift. (24)
angle of incidence: The angle at which the wings are attached to the fuselage of the plane. (24)
animal husbandry: The process of breeding, raising, and training animals. (27)
anion: A negatively charged ion. (28)
anode: The negatively charged side of the battery. (22)
apogee: The point in orbit at which an object is farthest from the earth. (23)
appendicular skeleton: The section of the skeleton that supports the appendages or limbs of the body. (26)
applied research: Research that seeks to reach a commercial goal by selecting, applying, and adapting knowledge gathered during basic research. (32)
apprenticeship training: Training consisting of a combination of on-the-job and classroom training over an extended period of time. (32)
apps: Computer programs that work with operating systems to help computers perform specific functions or tasks. (22)
aquaculture: Process of growing and harvesting water (aquatic) organisms in controlled conditions. (27)
aquifers: Underground, water-bearing layers of rock, sand, or gravel. (29)
arbors: Spindles, or shafts, used to hold table-saw blades and milling cutters. (6)
arch: A curved structural component used to support a load while spanning a space. (18)
arch bridge: A type of bridge that uses curved members to support the deck. The arch can be above or below the deck. (17)
arch dam: A type of dam that spreads the pressure from the dammed water onto the walls of the canyon where the dam is built. (17)
architects: Professionals who design structures. (18)
architectural drawings: Drawings used to specify characteristics of buildings and other structures. They include floor, plumbing, and electrical plans for a structure. (32)
architecture: The art and science of designing and building structures, including aesthetics and its spatial and functional layout. (18)
arteries: The vessels that carry blood away from the heart. (26)
artifacts: Human-made object. (1)
artificial intelligence (AI): The intelligence exhibited by a manufactured device or system. (14)
aseptic packaging: Preservation method that uses heat to sterilize food. The package and food are sterilized separately. (27)
aspect ratio: The relationship between the span and the chord; found by dividing span by the chord. (24)
assembling: Bringing materials and parts together to make a finished product. (11)
assembly drawings: Engineering drawings that show how parts fit together to make assemblies and finished products. (5, 32)
astronautical engineering: The study, design, and creation of space technology. (24)
atom: The smallest distinct unit of matter; consists of three basic parts: electrons, protons, and neutrons. (28)
atomic number: The number of protons that an element contains. (16) The number listed with each element in the periodic table of elements. (28)
Aufbau principle: Principle that determines how the electrons are configured in these orbitals. (28)
automated guided vehicles: Computer-controlled devices that use sensors to navigate from one place to another and transport materials, parts, and final products throughout a facility or warehouse. (14)
automatic control systems: Technological control systems that can monitor, compare, and adjust systems without human interference. (12)
automatic storage and retrieval systems: A series of computer-controlled conveyors and vehicles that transport and store materials, parts, or final products. (14)
automatic transmissions: A transmission system that uses valves to change hydraulic pressure so the transmission shifts its input-to-output ratios. (23)
auxiliary view: Additional sketched views of an object needed for complex parts. (5)
aviation electronics: The instruments that help pilots monitor and properly control aircraft. (23)
axial skeleton: The section of the skeleton that provides support and protection for the brain, spinal cord, and chest cavity. (26)

B

backbones: Lines, typically fiber-optic, that connect regions in a company's information technology (IT) systems. (21)

baler: Farming machine used to gather, compact, and contain hay. (27)

ballast: Extra weight added to a locomotive to give the locomotive better traction. (23)

band saws: Saws that use a blade made of a continuous band of metal. Most bands have teeth on one edge. The band travels around two wheels, which gives a continuous linear cutting action. (6)

barbed wire: Smooth wire with pointed spikes added along its length. Used as fencing. (27)

basic research: Research that seeks knowledge for its own sake. (32)

batch production: Production systems that manufacture products in groups or batches instead of individually or continuously. This system can produce many similar products with different variants. (15)

battery: A source of energy in electrical circuits that converts chemical energy to electrical energy. (22)

beam bridge: A type of bridge that uses concrete or steel beams to support the deck. (17)

bearing surface: The ground on which the foundation and building will rest. (17)

bending moment: Bending that occurs in a beam because of moment. (18)

bending stress: Materials that experience both tension and compression. (18)

benefits: The insurance plans, paid time off, holidays, and other programs that a company provides to its employees. (32)

Bernoulli's principle: Concept that states that an increase in fluid speed creates a decrease in pressure. (24)

berry: Crop grown in many parts of the country and cultivated for its edible parts. (27)

bill of materials: A document that contains the quantities, types, and sizes of materials and hardware needed to build a product or structure. (32)

binary code: Language that consists of only 1s and 0s. The 1s and 0s are used to represent numbers, letters, and all other digital information, as well as computer commands. (22)

bioartificial organs: Organs produced using living tissues. (26)

biochemical conversion: The use of enzymes to break down carbohydrates, such as cellulose found in plants, into sugars. (19)

biochemical engineering: A branch of chemical engineering that focuses on the design and development of products and processes involving biological organisms or molecules. (28)

biochemistry: The study of the chemical compounds and processes that take place in living organisms. (28)

biofuels: Fuels created from the burning of organic materials. (19)

biogas: Gas formed by the breaking down of organic wastes by bacteria in the absence of oxygen. The gas is predominantly highly flammable methane. (19)

biological future: The type of future that is concerned with the types of plant and animal life that will exist. (29)

biomaterial: Any material that interacts with living systems. (26)

biomechanics: The study of mechanical concepts as they relate to the movement or structure of living things. (26)

biomedical engineering: The design and development of products, systems, and procedures that solve medical and health-related problems. (26)

biomedical engineers: Engineers who combine knowledge of mathematics and science with design skills to plan and create medical equipment in order to meet required criteria and constraints. (26)

biomedical imaging: The practice of creating accurate visual representations of the interior of the human body as a means to diagnose and monitor disease and injury. (26)

biomimicry: The study of processes and designs that are in nature. The characteristics of nature are then imitated in new materials. (16)

biotechnology: 1—The technology associated with producing genetic materials. (10); 2—The application of living organisms, in whole or in part, to create technology. (13); 3—Using biological agents in processes to create goods or services. (27)

blimp: A lighter-than-air, nonrigid aircraft. (23)

blood: A bodily fluid that delivers the necessary nutrients and oxygen to the cells of the body and transfers waste products away from the body's cells. (26)

board of directors: Group of people elected to represent the interests of stockholders in a corporation. The directors are responsible for forming company policy and providing overall direction for the company. (31)

Bohr atomic model: A two-dimensional depiction of electron orbitals. (28)

bonding: Assembly process that holds plastic, metal, ceramic, and composite parts to each other by cohesive or adhesive forces. (15)

bonds: A type of loan in which the issuer owes the holder a debt. (31)

bone marrow: Soft center of a bone that contains stem cells, which produce the majority of the body's red blood cells, white blood cells, and platelets. (26)

bottom-up manufacturing: The manufacture of products by building them up from the atomic and molecular level. (20)

bow thruster: A maneuvering prop mounted at a right angle to the keel on a large ship. (23)

brainstorming: A process in which two or more people work together to generate ideas in order to solve an identified problem. (5)

breadboards: Devices that serve as a construction base for prototyping or making experimental models of electrical circuits. They allow electrical components to be connected together in a circuit without being soldered. (22)

broach: A machine that uses a tool with many teeth. Each tooth sticks out slightly more than the previous tooth. As the tool is passed over a surface, each tooth cuts a small chip. (6)

bronchi: The two air tubes that branch off of the trachea into each lung. (26)

Bronze Age: Period of civilization that began around 3000 BCE in which humans used copper and copper-based metals as the primary materials for tools. (1)

browser: A software program that acts as an interface between the user and the Web. (21)

bubble plans: Rough drawings used when brainstorming a structure's layout. (18)

buckling: A structural failure that occurs when a structural element bends under compression. (18)

buildings: Structures erected to protect people and machines from the outside environment. (11)

buoyancy: The upward force exerted on an object immersed in a fluid. (23)

buttress dam: A type of dam that has a solid upstream side and is supported on the downstream side with a series of supports, or buttresses. (17)

bylaws: The general rules under which a corporation operates. (31)

C

cam: A pear-shaped disk with an off-center pivot point used to change rotating motion into reciprocating motion. (20)

camber: The characteristic curvature along the upper and lower surface of an airfoil. (24)

canning: Preservation method in which food is sealed into glass jars or metal cans. (27)

cantilever: A long, projecting beam that is supported on only one end. (18)

cantilever bridge: A type of bridge that uses trusses extending out in both directions from the support beams. The arms transmit the load to the center. (17)

capacitors: Electrical components that store an electrical charge similar to a battery, but discharge energy more rapidly. (22)

capillaries: The vessels that enable the exchange of nutrients, oxygen, and chemicals between the blood and cells of the body. (26)

cardiovascular system: System of the body in which the heart, blood, and blood vessels work together to carry oxygen, nutrients, hormones, and waste through the body. (26)

casting and molding: Introduction of a liquid material into a mold cavity where it solidifies into the proper size and shape. (11)

casting and molding processes: Secondary manufacturing processes that give materials shape by introducing a liquid material into a mold. (15)

catalysts: Substance used to cause a reaction. (27)

cathode: The positively charged side of the battery. (22)

cation: A positively charged ion. (28)

ceiling joists: Joists that rest on the outside walls and some interior walls to support the ceiling. (17)

ceilings: Top, inside surface of a room. (17)

central processing unit (CPU): The working part of the computer that carries out instructions. (22)

ceramics: A class of materials characterized by high brittleness, poor electrical and thermal conductivity, very high melting points, and nonflammability. (16)

charter: A "birth certificate" for a corporation, detailing the fundamental components of a company. This may include the objectives, its structure, and its strategic plan. (31)

charts: Models that show the relationship among people, actions, or operations. (5)

chemical: A compound or substance that has a specific molecular composition and is produced or used for a specific purpose. (28)

chemical bonding: The valence electrons of atoms are attracted to one another and are transferred or shared between the atoms. (16)

chemical compounds: Any substances that consist of two or more different atoms that are bonded or stuck together. (28)

chemical conditioning: The use of chemical actions to change the internal properties of a material. (15)
chemical elements: The building blocks of matter; pure substances that cannot be broken down into simpler substances. (28)
chemical energy: Energy stored within a chemical substance. (19)
chemical engineering: 1—Engineering discipline that involves the conversion of raw materials into useful products. (3); 2—The application of chemistry, biology, physics, and mathematics to solve problems that involve the conversion of raw materials and chemicals into more useful forms of food, medicine, fuels, and many other products. (28)
chemical engineers: Professionals who develop new materials and chemicals by using energy to create a basic chemical change in raw materials or chemicals. (28)
chemical equations: A standard method to represent what occurs in a chemical reaction. (28)
chemical processing: Food processing method that uses energy to cause a basic chemical change in food. (27)
chemical properties: A material's characteristics as a result of chemical reaction or change. (16)
chemical reactions: The processes that either bond atoms together or break bonded atoms apart. (28)
chemistry: The study of matter and its interactions with other matter and energy. (28)
chief executive officer (CEO): The top manager in a company. (31)
chord: The measurement from leading edge to trailing edge (from the wing center). (24)
chucks: Attachments used to hold and rotate drills and router bits. (6)
circuit diagrams: Graphical representations of electrical circuits. Also called *electronic schematics*. (22)
circuit theory: The collection of scientific knowledge used to describe the flow of electrical energy through an electrical circuit. (22)
circular saws: Saws that use a blade in the shape of a disk with teeth arranged around the edge. (6)
civil engineering: Engineering discipline that focuses on design and improvement of structures such as bridges, skyscrapers, roadways, and dams. (3)
civil engineering structures: Structures primarily designed by civil engineers. (11)
civilized conditions: Circumstances that allow people to exert their will on the natural scene. (1)
classification: Method for developing preliminary design solutions that involves dividing the problem into major segments, and then reducing each segment into smaller parts. (5)
clear-cutting: Logging method in which all trees, regardless of species or size, are removed from a plot of land that is generally less than 1000 acres. (14)
closed-loop control system: A system that uses feedback to adjust the process. (12)
coal: A combustible solid composed mostly of carbon. (14)
coefficients: Figures that represent the number of atoms and molecules needed for a balanced representation of a chemical reaction. (28)
combinational logic: Digital logic that uses logic gates to make decisions based on present inputs. (22)
combine: Farming machine used to harvest a wide range of grains and seed crops. (27)
commercial aviation: An industry that makes money by transporting people and cargo in airplanes. (23)
commercial buildings: Structures used for business purposes; can be publicly or privately owned. (17)
commercial structures: Stores and offices used to conduct business. (11)
commission: Form of payment to an employee that is usually a percentage of the total dollar value of the goods sold. (32)
communication: The passing of information from one location to another or from one person to another. (21)
composite materials: Combinations of natural and synthetic materials that are mixed to create items with other desirable properties. (10)
composites: Materials comprised of multiple materials with different chemical or physical properties. (16)
compression: Force that pushes on material and squeezes it. (18)
computation: Any type of calculation for processing information. (22)
computer: An information-processing machine that can store information and use programs that can be changed. (6, 21)
computer engineering: Engineering field that focuses on the development, improvement, and supervision of the production and maintenance of computer hardware and software. (22)
computer-integrated manufacturing (CIM): The use of computers to control the entire production process. The integration relies on closed-loop control systems. (14)
computer science: The theoretical study and practical application of computation. (22)
concept map: A tool used to graphically organize information. (4)
conceptual models: Model that illustrates ideas for specific structures and products, showing general views of components and their relationships. (5)

conditioning: Using heat, chemicals, or mechanical forces to change the internal structure of material. (11)
conditioning processes: Secondary manufacturing processes that change the internal properties of a material. (15)
conduction: The movement of heat along a solid material or between two solid materials touching each other. (19)
conductive heat transfer: The flow of heat from a hot temperature to a cold temperature through a material. (20)
conductor: Material composed of atoms with only one valence electron that is loosely bound and can become easily free to conduct electrical current. (16)
conservation of energy: Law that states that energy cannot be created or destroyed. (24)
constraints: 1—Limitations. (3); 2— The limits (such as capital) on a design. (5)
construction technologies: The design, assembly, and maintenance of structures. (13)
consumers: People who financially support technological systems by spending money on products or services. (10)
containerships: Ships that carry large quantities of goods sealed in steel containers. (23)
continuous production: Production systems that produce a large quantity of similar products in the least amount of time possible without interruption. (15)
control: 1—The feedback loop that causes management and production activities to change through evaluation, feedback, and corrective action. (9); 2—A vehicular system that makes it possible to control the speed and direction of a vehicle. (11)
control theory: An engineering and mathematical way of thinking that focuses on monitoring and correcting systems behaviors using output performance measures as feedback to control the system. (12)
convection: The transfer of heat between or within fluids. This heat transfer involves the actual movement of the substance. (19)
convergent thinking: Method used during the development of preliminary design solutions that narrows and focuses (converge) ideas until the most feasible solution is found. (5)
conversion and processing: The final step in the agricultural process that involves changing food products into foodstuffs. (11)
conveyor systems: Systems that move materials, parts, or products in a fixed path along a production line. (14)
corporation: 1—A legal entity formed by people to own a company. Investors purchase ownership in the form of shares of stock. (10); 2—A business in which investors purchase partial ownership in the form of shares of stock. Legally, a corporation is similar to a person—it can own property, sue, be sued, enter into contracts, and contribute to worthy causes. (31)
covalent bonds: Chemical bonds that occur when atoms share electrons to have a full valence orbital and be stable. (28)
creativity: 1—Thinking in a way that is different from the norm. (3); 2—The ability to see a need or way to make life easier, and design systems and products to meet a need or desire. (10)
creed: A set of beliefs that is established to guide a person's actions. (3)
criteria: A listing of the features (such as safety) to be met by a design. (5)
critical thinking: A type of thinking that uses mental abilities to identify, analyze, and evaluate a situation or problem. (4)
critical value: Amount of pressure that causes buckling. (18)
crop production: The process of growing plants for various uses, including food for humans, feed for animals, and natural fibers for a variety of uses. (27)
cultivator: A machine used to control weeds. It uses a series of hoe-shaped blades pulled through the ground. (27)
curing: Preservation method that adds a combination of natural ingredients to meat. (27)
current: The rate of electron flow through the circuit. (22)
customized production: Production systems that individually produce unique products that are tailored to meet the specific needs of the customer. (15)
cutting motion: Action that causes excess material to be cut away from the work. (6)
cylindrical grinder: A type of grinder in which the workpiece is held in a chuck or between centers and rotated, and a grinding wheel is rotated in the opposite direction. The opposing rotating forces produce the cutting motion. The grinding wheel is fed into the work to produce the feed motion. (6)

D

data: Raw facts and figures people and machines collect. (10)
data-comparing device: Device used to compare the data gathered by monitoring devices against expectations. (12)

dead load: Any permanent component of a building such as the floors, walls, roof, or beams. (18)
debt financing: A method for raising money by borrowing it from financial institutions or private investors. The debt must be repaid. (10, 31)
decision matrix: A chart used to record a rating for how well a design meets a desired criterion. (4)
decoding: Process in which coded information is changed back into a recognizable form. (11)
degrees of freedom: The limited number of ways or directions a vehicle can move. (23)
delayed outputs: Outputs that occur at a later date than expected or planned. (12)
deoxyribonucleic acid (DNA): The genetic material of living organisms. (28)
descriptive methods: Procedure used to record observations of present conditions. (5)
design: A thinking and reasoning process that involves devising a new solution, one that has not previously been found or defined. (4)
design analysis: Analysis that helps engineers, designers, and decision makers assess the design's purpose and likely acceptance in the marketplace in order to choose the best design. (7)
desirable outputs: Outputs that benefit people. (12)
detail drawings: Engineering drawings that convey the sizes, shapes, and surface finishes of individual parts. (32)
detailed sketches: Sketches that communicate the information needed to build a model of a product or structure. They can also be used as a guide to prepare engineering drawings for manufactured products and architectural drawings for constructed structures. (5)
development: 1—Products and services designed through technological innovation. (1); 2—The use of knowledge gained from research to derive specific answers to problems; the conversion of knowledge into a physical form. (32)
diagnosis: 1—Using knowledge, technological devices, and other means to determine the causes of abnormal body conditions. (11); 2—The process of determining the nature or cause of an illness or injury by conducting interviews, physical examinations, and medical tests. (25)
diagrams: Models that show the relationship of components in a system. (5)
diaphragm: The breathing muscle that contracts and expands the airtight cavity surrounding the lungs. (26)
digital logic: Type of logic that enables complex decisions to be made within computer systems based on a series of yes or no questions. Also called *Boolean logic*. (22)
digital signals: Signals that are finite or discrete, meaning there are a limited amount of values that quantities can have. (21)
digital technology: The generation, storage, processing, and transmittal of data using positive (on) and nonpositive (off) electrical states. Digital data is expressed as a string of 1s and 0s. (21)
dimension: A measureable feature of an object such as length, width, or height. (3)
diodes: Diodes are electrical components made out of semiconductor materials, such as silicon or selenium. Diodes allow electrical current to pass through them in only one direction. (22)
direct active solar system: An active solar system that does not have a heat exchanger. The water circulated in the system is used as domestic hot water, flowing directly to such areas as household faucets, washing machines, and showers. (19)
direct current: An electrical charge that flows in only one direction. (22)
direct-gain solar system: A solar system that allows the radiant energy to enter the home through windows, heating inside surfaces. (19)
direct-reading measurement tools: Measurement tools that an operator manipulates and then reads. (6)
disease: Any change interfering with the appearance, structure, or function of the body. (25)
distance multipliers: Simple machines that increase the amount of movement applied to the work at hand. (10)
distribution: Physically moving a product from the producer to the consumer. (32)
divergent thinking: Method used during the development of preliminary design solutions that imagines as many different (divergent) solutions as possible. (5)
dividend: Periodic payment from the company's profits to investor-owners. (31)
domain name: An address that identifies the location of a web server used to store information for a website or an e-mail account. (21)
double-blind study: A research technique in which neither the group being tested nor the researchers know who is receiving the active substance or who is receiving the placebo. (25)
drag: An aerodynamic force that acts against the movement of aircraft. (24)
drilling: The process of obtaining materials from the earth by pumping them through holes drilled into the earth. (11)
drilling machines: Machines that produce or enlarge holes using a rotating cutter. (6)

drip-irrigation: Irrigation system that delivers water slowly to the base of plants. (27)
drug: A substance used to prevent, diagnose, or treat a disease. (25)
drywall: Gypsum wallboard used to cover interior walls. (17)
ductility: The ability of a material to be stretched or lengthened into wire before failure occurs. (16)
durable products: Products designed to operate for a long period of time. (15)
dynamic loads: Rapidly or abruptly changing, impermanent forces acting on a structure. (18)
dynamic process: Process that changes and improves. (1)
dynamics: Branch of mechanics that deals with the analysis of objects that are accelerating as a result of acting forces. (26)

E

earth-orbit travel: A type of space travel used to launch communication satellites. (23)
economic enterprises: Organizations engaging in business efforts directed toward making a profit. (32)
edutainment: Educational content that is designed to have entertainment value. (21)
elastic range: The range between rest (no force applied) and the yield point of a material. (15)
elastomers: Elastic polymers that can stretch, but return to their original shape when released. (16)
electrical and electronic controller: Type of adjusting device that uses electrical devices (switches, relays, and motors) and electronic devices (diodes, transistors, and ICs) to adjust machines or other devices. (12)
electrical and magnetic properties: Characteristics that describe a material's ability to conduct electrical current by allowing electrical charge to transfer from atom to atom, as well as its magnetic permeability (ability to retain magnetic forces). (16)
electrical circuits: Interconnected electrical components that generate and distribute electrical energy for the purpose of controlling, detecting, collecting, storing, receiving, or processing information. (22)
electrical energy: Energy created by electrons moving along a conductor. (19)
electrical engineering: Engineering discipline that focuses on electricity and electromagnetism. (3)
electrical engineers: Engineers who design, test, and oversee the production of devices that use electricity to control, process, store, and receive information, as well as equipment for generating and transmitting power. (22)
electrical or electronic sensors: Sensors used to determine the frequency of or changes in electric current or electromagnetic waves. (12)
electrical switch: An electrical component that interrupts the flow of electrons from a power source in an electrical circuit. (22)
electric charge: A measurable property of matter; charges can be positive or negative. (28)
electrocardiograph (EKG) machine: A machine that produces a visual record of the heart's electrical activity. (25)
electrolyte: The component of a battery that prevents the negative and positive charges on each side of the battery from connecting with each other. (22)
electromagnetic force: Magnetic lines of force created by movement of electrons in a conductor. (21)
electromagnetic induction: The production of electromotive force or voltage by continuously moving a conductor, such as copper wire, through varying magnetic fields. (22)
electromechanical controllers: Type of adjusting device that uses electromagnetic coils and forces to move control linkages and operate switches to adjust machines or other devices. (12)
electron configuration: The arrangement of electrons within an atom's orbitals that dictates the element's properties in a chemical reaction. (28)
electron orbitals: 1—Areas within an atom where electrons have a high potential of being located. (16); 2—The rings on which electrons orbit a nucleus. (28)
electrons: Subatomic particles with a negative charge. (28)
elements: Substances that cannot be broken down any further through chemical reactions. (16)
elevations: Front, sides, and back views of a building shown on an architectural drawing. (18)
emergency medicine: Medical treatment that deals with sudden, serious illnesses and injuries. (25)
empirical research: Research, based on the scientific method, which generates new knowledge about particular phenomena. (8)
employee relations: Company programs that recruit, select, develop, and reward the company's employees. (32)
encoding: The process of changing a message into a format that can be transmitted. (11)
endoscope: A device that allows a physician to look inside the body. The device consists of a narrow, flexible tube containing a number of fiber-optic fibers smaller in diameter than a human hair. (25)
endosperm: The core of a wheat kernel. (27)

energy: 1—The ability to do work. (9, 13); 2—An attribute of matter that comes in various forms and affects matter. It is typically used to do some kind of work. (28)

energy converter: A device that has energy as its input and its output. (19)

engineered wood products: Composite materials made from lumber production process waste materials and adhesives. (14)

engineering: The practical application of science, mathematics, and technological know-how to solve problems in the most efficient way possible. (2)

engineering design: 1—Problem-solving process that follows a specific set of practices including defining problems, developing and using models, planning and conducting investigations, analyzing and interpreting data, using computational thinking, designing solutions, arguing based on evidence, and collecting, assessing, and communicating information. (3); 2—The design of either a technological product or system as a solution. (4)

engineering design process: Method used to solve problems by designing a product or system that meets a desired goal while adhering to established constraints and taking into consideration factors such as potential impacts, risks, and benefits. (4)

engineering notation: A modification of scientific notation in which all of the powers of 10 are multiples of three. This method allows numbers to more easily conform to International System of Units prefixes. (3)

engineering notebooks: Notebooks in which all notes, research, brainstorming work, sketches, test procedures, test data, and any other thoughts or ideas related to a project are recorded. (8)

engineers: People who conduct research and apply scientific and technological knowledge to the design and development of products, structure, and systems. (2)

entrepreneur: An individual or group of people who accept the financial risks of creating a small business and focus on what customers value in order to develop systems and products to meet desires and expectations. (10, 31)

environmental engineering: Managing the use of natural resources to minimize the negative impacts that human activity can have on the environment. (18)

environmentalists: People who are concerned about the protection of the environment and leaving it unspoiled. (29)

equity financing: Raising money by selling partial ownership in a company. (10, 31)

ergonomics: 1—The science that considers the size and movement of the human body; mental attitudes and abilities; and senses. (7); 2—The study of how people interact with the things they use. (23)

estimations: Rough calculations. (3)

ethics: Principles of conduct that govern the actions of an individual or group. (3)

executive summary: A brief document written to inform business administrators, executives, funders, and other interested parties of a project that is being proposed or is underway. (8)

exhaustible materials: Materials that occur naturally on the earth in finite amounts. Once they are depleted, they cannot be replaced. (10)

exosphere: The last layer of space above the earth that blends directly into outer space. (23)

expenses: The money a company spends to pay for resources (people's time, materials, and machines). (32)

experimental methods: Procedure used to compare different conditions. One condition is held constant, while the other is varied. (5)

experiments: Established research procedures carried out in a controlled environment to prove or disprove a hypothesis. (2)

extrusion: A process in which metals are heated and pushed through a die to produce long continuous shapes for specific purposes. (14)

F

fascia board: A board used to finish the ends of the rafters and the overhang. (17)

fatigue failure: The failure of an object after being subjected to repeated loading and unloading. Progressive brittle cracking occurs until the object fractures. (20)

feedback: The process of using information about the outputs of a system to regulate the inputs. (9)

feed motion: The action bringing new material into the cutter, allowing the cutting action to be continuous. (6)

fermentation: Process in which an organism, such as yeast or bacterial, metabolizes a carbohydrate, such as a starch or sugar, and converts it into an acid or alcohol. (28)

fertilizer: Liquid, powder, or pellet that contains important chemicals that encourage and support plant growth. (27)

figure: Any type of graph, chart, drawing, photograph or graphic material that depicts some type of information necessary to a project. (8)

file-transfer protocol: The protocol that allows a user to retrieve and modify files on another computer connected to the Internet. (21)

finances: The money and credit necessary for the economic system to operate. (9)

finishing: Coating or modifying the surfaces of parts or products to protect them, make them more appealing to consumers, or both. (11)

finishing processes: Techniques that protect products and enhance their appearance. Some change the surface of the product, while others involve the application of a coating. (15)

first-class lever: A lever in which the fulcrum is between the load and the effort. (10)

fixed-wing aircraft: Passenger and cargo aircraft. (23)

fixture: A device that secures a workpiece while machining operations are performed. (15)

flammability: The ease with which a material will ignite or burn. (16)

flexible belt drive: Two pulleys with a belt stretched between them change the speed or power of a motion. (20)

flexible manufacturing: 1—Type of manufacturing that can quickly and inexpensively respond to change. (30); 2—A computer-based manufacturing system combining the advantages of intermittent manufacturing with the advantages of continuous manufacturing. (32)

flood irrigation: Irrigation system that sends large volumes of water across a field. (27)

floor joists: Horizontal support pieces that extend across the structure and carry the weight of the floor. (17)

flowchart: A diagram that uses symbols to represent a sequence of actions or operations of a complex system such as a manufacturing facility. (15)

fluid dynamics: A type of fluid mechanics that studies fluids in motion; the foundation of aerodynamics. (24)

fluidic controller: Type of adjusting device that uses fluids to adjust machines or other devices. (12)

fluid mechanics: The study of how laws of force and motion apply to liquids and gases. (20)

fluid mining: Mining method in which two wells extend into a mineral deposit. Hot water is pumped down one of the wells. The water dissolves the mineral and is forced up the other well. (14)

food-processing technology: Using knowledge, machines, and techniques to convert agricultural products into foods that have specific textures, appearances, and nutritional properties. (27)

footing: Low, flat pad that a foundation sits on. (17)

foot-pound (ft-lb): The amount of energy needed to move an object from one location to another. (20)

forage crops: Plants grown for animal feed. (27)

force: Pushing or pulling applied to an object as a result of an interaction with another object. (18)

forced-air heating system: A type of indirect heating method that heat air as a conduction medium. (17)

force multipliers: Simple machines that increase the force applied to the work at hand. (10)

forming: Using force applied by a die or roll to reshape materials. (11)

forming processes: Secondary manufacturing processes that apply force through a forming device to cause the material to change shape. (15)

form utility: The change in the form of materials to make them more valuable. (14)

Forstner bits: Two-lipped woodcutters that produce a flat-bottomed hole. (6)

fossil fuels: Material resources formed from the remains of prehistoric living organisms. These resources are called hydrocarbons because they are mixtures of carbon and hydrogen. (14)

foundation: The base of a structure. (11)

fracture point: The point at which a material breaks due to the amount of force applied to it. (15)

free body diagram: Diagram showing where forces are applied. (18)

free enterprise: An economic system in which the principles of supply and demand determine production and prices. (32)

freezing: Preservation method that keeps foods at or below 32°F. (27)

frequency: The number of cycles (complete wavelengths) passing a point in one second. (21)

frequency modulation (FM): Radio broadcast system that encodes the message on the carrier wave by changing the wave's frequency. (21)

friction: A force that acts against motion when two surfaces rub together. (24)

fruits: Crop grown in many parts of the country and cultivated for its edible parts. (27)

fulcrum: The support on a lever on which a lever arm rests and turns. (10)

functional analysis: Type of analysis that evaluates the degree to which a product, structure, or system operates effectively under the conditions for which it was designed. (7)

furrow irrigation: Irrigation system that uses small ditches built between rows of plants (furrows) to guide water from one field to the other. (27)

fuselage: The body of an aircraft that carries people or cargo. (23)

futuring: A research technique that helps people select the best of many possible courses of action for the future. (29)

G

Gantt chart: Tool that illustrates a project schedule using horizontal bars that represent each phase and activity that makes up the project. (3)

gas: 1—Material that easily disperses and expands to fill any space. It has no physical shape, but does occupy space and has volume. (10); 2—State of matter with no fixed volume or shape. (28)

gears: Two or more wheels with teeth on their circumferences that work together to change the direction of a rotating force. (20)

general aviation: Travel for pleasure or business in an aircraft owned by an individual or business. (23)

gene splicing: Process of producing an organism with a new set of traits. (27)

genetic engineering: 1—Manipulation of the genetic material of living organisms. (26) 2—Process producing new pest-resistant and chemical-tolerant crops that help combat diseases. (27)

genetic materials: 1—Materials produced by living things. They have life cycles and can be regenerated. (10); 2—Material resources obtained during the normal life cycles of plants or animals through farming, fishing, and forestry. (14)

geosynchronous orbit: A type of orbit in which a satellite travels at the same speed that the earth is turning. (23)

geotechnical engineering: The study of the way soil behaves under the stresses and strains created by soil or rock. (18)

geothermal system: A type of climate-control system that uses coils buried a few feet down in the soil to heat or cool a building. (17)

grain: Crops that have large edible seeds; member of the grass family. (27)

grain drill: A type of seed planter pulled behind a tractor. (27)

grant: A sum of money that is provided by the government, a private foundation, or a public corporation for a specific purpose. (8)

graphic communications: A communications process in which messages are visual and have two dimensions. (11)

graphic models: Conceptual models, graphs, charts, and diagrams that each serve a specific purpose. (5)

graphs: Models that organize and plot data. (5)

gravity dam: A type of dam with an upstream vertical wall and a sloping downstream wall. (17)

green buildings: Structures that are environmentally responsible, sustainable, and resource efficient. (13)

greenhouse effect: The combining of UV light rays with increased levels of carbon dioxide and other gases, causing Earth to retain more heat. (29)

grievances: Labor disputes between a union and an employer over a contract's interpretation. (32)

grinding machines: Machines that use bonded abrasives (grinding wheels) to cut the material. (6)

groups: Columns in the periodic table of elements. (28)

growth: A step in the agricultural process that involves providing feed and water for animals or cultivating and watering (irrigating) crops. (11)

guidance: A vehicular system that gathers and displays information so the vehicle can be kept on course. (11, 23)

H

halogens: Chemical elements that have seven electrons in their valence orbital. They readily gain electrons and react with metals to form chemical compounds. (28)

hand tools: Simple, handheld artifacts requiring human-muscle power, air, or electric power to make them work. They can be classified by their purpose. (10)

hardness: A measurement describing the resistance of a material through indentation, scratch, and rebound tests. (16)

hardware: Physical equipment that includes a computer and the devices attached to it. (22)

hardwood lumber: Lumber produced from deciduous (leaf-bearing) trees. It is sold in standard thicknesses and is available in random widths and lengths. (14)

harvesting: A step in the agricultural process that involves removing edible parts of plants from trees and stock and butchering animals to produce meat and other consumable products. The process of gathering genetic (living) materials from the earth or bodies of water at the proper stage of their life cycles. (11)

header: Part of framework that carries the weight from the roof and ceiling across door and window openings. (17)

heat of combustion: The quantity of energy released when a material is burned. (16)

heat pump: A cooling and heating system that works by capturing heat in the atmosphere. (17)

heat-treating: A thermal conditioning process used on metals in which the metal is heated and allowed to cool slowly. (15)

heavy engineering structures: Structures primarily designed by civil engineers. (11)

historical methods: Procedure used to gather information from existing records. (5)

homeostasis: The ability of a biological system to regulate its processes, such as body temperature, to produce conditions that are optimal for survival. (26)

horizontal drilling: Drilling a well along a curve to reach petroleum or natural gas deposits that cannot otherwise be tapped. (14)

horsepower: A term used to describe the power output of many mechanical systems. One horsepower is the power needed to move 550 lb a distance of 1′ (550 ft-lb) in one second. (20)

hot water heating: A type of indirect heating method that uses water to carry heat. (17)

hovercraft: A special type of boat suspended on a cushion of air. (23)

human physiology: A subdiscipline of biology that studies the structure and function of the human body. It looks at how cells, organs, and muscles work together to perform necessary chemical and physical functions. (26)

human-psyche future: The type of future that is concerned with the mental condition of people—how they will feel about life and themselves. (29)

hydraulic fracturing: Pumping a high volume of water underground to create pressure that fractures shale rock formations to release trapped pockets of natural gas. (14)

hydraulics: A branch of science that studies how liquids are conveyed through pipes and pathways, largely as a source of mechanical force. (20)

hydrodynamics: The branch of fluid mechanics that explains how forces are exerted on a solid body by a fluid's motion or pressure. (20)

hydrofoil: A special type of boat that has a normal hull and a set of underwater wings. (23)

hydroponics: Growing plants in nutrient solutions without soil. (27)

hydrostatics: The branch of fluid mechanics that explains how liquids at rest function in regards to the forces exerted on them or by them. (20)

hyperlinks: Underlined phrases, buttons, or other means that can be selected that allow an operator to point and click on a link to be connected to the selected site or page. (21)

hypertext markup language (HTML) tags: Tags or codes tell the receiving computer how a page should look. (21)

hypothesis: A proposed explanation of a situation. (2)

I

ideation: A process in which designers create many possible answers by letting their minds create solutions. (11)

immediate outputs: Products or services designed for use now. (12)

immunization: A disease prevention technique in which the body is exposed to small amounts of a bacterium, protein, or virus to cause antibodies to form. (25)

inclined planes: Simple machines that have sloped surfaces used to make a job easier to do. (10)

income: The money a company earns as the end result of sales. (32)

indirect active solar system: A solar system that has a series of collectors, each with a black surface to absorb solar energy. Above or below this surface is a network of tubes, or pipes. Water is circulated through these channels. As the fluid passes a warm black surface, it absorbs heat. The warm water is then pumped to a heat exchanger. (19)

indirect-gain solar system: A solar system in which sunlight heats the surface of a Trombe wall, and the air between the wall and the glass panels becomes heated by energy radiating from the wall. Warm air rises and flows into the building through openings at the top of the wall, creating convection currents that draw cooler, heavier air into the openings at the bottom. This new air, in turn, is heated and rises. (19)

indirect-reading measurement tools: Measurement tools in which sensors gather the measurement data, the data is processed by computers or other automatic devices, and the final measurement is displayed on a digital readout, computer screen, or printout. (6)

industrial buildings: Structures that house manufacturing processes. Protect the machines, materials, and workers that make products. (17)

Industrial Revolution: Time period that began in England in the late 18th century and is characterized by the introduction of and improvements to power-driven machinery. (1)

industrial structures: Buildings that house machinery used to make products. (11)

industry: The area of economic activity that uses resources and systems to produce products, structures, and services with the intent to make a profit. (32)
inexhaustible energy: The type of energy that cannot be entirely used up or consumed. (10)
information: Data that has been sorted and categorized according to its type, for human use. (9, 10)
information age: Current period in time characterized by information-based industries and the transmission of information. (1)
information and communication technologies: Technologies that provide the ability to record, store, manipulate, analyze, and transmit data across various modes. (13)
information processing: 1—The gathering, storing, manipulating, and retrieving of information. (6) 2—A communication process that manipulates data to produce useful information. (11)
information skills: The abilities to locate, select, and use information. (30)
infotainment: Information provided in an entertaining way. (21)
inland-waterway transportation: Transportation on rivers and lakes, and along coastal waterways. (23)
innovations: Refinements or improvements made to preexisting products or processes that better solve a problem. (2)
inorganic materials: Materials that do not come from living organisms. (10)
inputs: Resources that go into a technological system and are used by the system. (9)
inside directors: The top managers of a corporation. (31)
insolation: A term used to describe the solar energy available in a specific location at any given time. (19)
inspection: The process of comparing materials and products with set quality standards. (32)
insulator: A material that does not conduct electrical current under typical conditions, due to valence electrons that are tightly bound to an atom. (16)
integrated circuits (ICs): Electronic circuits formed on a small, thin piece of pure semiconducting material, usually silicon. (22)
intellectual property: Any creation of the mind that can be legally protected. (8)
intended outputs: Products or services designed and produced with specific goals in mind. (12)
interference: Anything that impairs the accurate communication of a message. (21)
intermodal shipping: A system that transports cargo on two or more modes of transport before it reaches its destination. (23)
internal combustion engine: A common power source in land vehicles, in which fuel is burned inside the engine to convert energy from one form to another. (23)
Internet: A global communication network that connects individual computer networks. (6)
Internet Protocol (IP) address: An identifying number assigned to each computer connected to the Internet. (21)
interventional radiology: Radiology techniques that use images for nonsurgical treatment of ailments. (25)
intrapreneurship: The application of entrepreneurial spirit and action within an existing company structure. (31)
inventions: New, useful product or process that solve some type of problem and did not previously exist. (2)
inventory control: Process that compares finished goods in warehouses and products in process with sales projections. (12)
ion: 1—An atom that gains or loses a valence electron and becomes unbalanced. (16); 2—An atom that has more electrons than protons or an atom with fewer electrons than protons. (28)
ionic bond: A chemical bond that is the result of the formation of ions. (28)
Iron Age: Period of civilization that began around 1200 BCE in which humans used iron and steel as the primary materials for tools. (1)
irradiation: Preservation method that uses gamma rays or x-rays to kill molds and bacteria that might be in food. (27)
irrigation: Artificial watering to maintain plant growth. (27)
isometric sketches: Sketches in which the angles formed by the lines in the upper-right corner are equal. Each angle is 120°. The object is shown as if it is viewed from one corner. (5)
iterative: Repetitive manner. (3)

J

jig: A device that guides the machines or tools performing manufacturing operations. (15)
joules (J): A measurement that is one newton per meter. (20)
judgmental system: Data-comparing device that allows human opinions and values to enter into the control process. (12)
just-in-time (JIT) manufacturing: A production strategy in which companies purchase the parts for their goods only as they are needed. (14)

K

Kaizen: A continuous improvement philosophy that promotes constant change in manufacturing operations based on employee input. (14)
keel: A flat blade protruding from the bottom of a boat into the water with the purpose of preventing the boat from being blown sideways by aiding in keeping the boat right-side up. (23)
kilowatt-hour: The work that 1000 watts complete in one hour. (20)
kinematics: The study of describing motion using quantities, such as time, velocity, and displacement, without the concern of the forces involved. (26)
kinesiology: A combination of kinetics and human physiology that studies human movement. (26)
kinetic energy: Energy in motion; energy that is involved in moving something. (19)
kinetics: The study of forces that cause motion, such as gravity or torque. (26)
Kirchhoff's current law: The sum of the currents across each branch or loop of a parallel circuit is equal to the total current in a circuit. (22)
Kirchhoff's voltage law: The sum of all voltage drops in a circuit is equal to the total applied voltage. (22)
knowledge: Information that people learn and apply. (10)

L

labor agreements: A contract that establishes pay rates, hours, and working conditions for all employees covered by the contract. (32)
labor relations: A company's programs that deal with the employees' labor unions. (32)
laminar flow: Even, straight flow of fluid in layers. (24)
landscaping: Trees, shrubs, and grass that are planted to help prevent erosion and improve the appearance of the site. (17)
land-transportation systems: A transportation system that moves people and goods on the surface of the earth from place to place. (23)
language and communication skills: The abilities to read, write, and speak a language. (30)
lathes: Turning machines that produce cutting motion by rotating the workpiece. Linear movement of the tool generates the feed motion. (6)
law of conservation of mass: Law that states that matter cannot be created or destroyed. (28)
lean manufacturing: A systematic strategy to reduce operation costs and improve the overall consumer value of products by eliminating wastes and inefficiencies throughout the manufacturing process. (14)
lever: A simple machine that multiplies the force applied to it and changes the direction of a linear force. (10)
lever arm: A rod or bar on a lever that rests and turns on a fulcrum. (10)
lifestyle: What people do with business and family life—their work, social, and recreational activities. (30)
ligaments: Long fibrous straps that stabilize and support joints by holding bones in place. (26)
light-emitting diode (LED): A special diode that releases energy in the form of photons when a suitable voltage is applied. (22)
lighter-than-air vehicles: Air vehicles that use either light gas or hot air to produce lift. (23)
limited liability: A company in which an owner's loss is limited to the amount of money he or she has invested. (31)
linear motion: 1—Cutting or feed motion in which the cutter or work is moved in one direction along a straight line. (6); 2—Movement in a straight line. (20)
liquid: State of matter that has a fixed volume but can assume any shape. (28)
liquid-fuel rockets: A rocket with two tanks: one with fuel, the other with oxygen. (23)
liquids: Visible, fluid materials that will not normally hold their sizes and shapes. They cannot be easily compressed. (10)
literature review: A scholarly paper that presents an extensive explanation of the current knowledge related to a topic based on a thorough examination of related literature, or written works. (8)
live load: The weight of items such as furniture, appliances, people, and equipment. (18)
loads: Weights that act on a structure, including wind, settling ground, earthquakes, gravitational field pull, changing temperatures, and heavy snowfalls. (18)
lumber: A flat strip, or slab, of wood. (14)
lungs: The main organs of the respiratory system. They deliver oxygen to and remove carbon dioxide from the blood through diffusion. (26)

M

machines: 1—Tools that amplify the speed, amount, or direction of a force. (9); 2—Artifacts that transmit or change the application of power, force, or motion. They can be simple or complex. (10)
machine tools: Machines used to make other machines. (6)

machining: Separating process in which excess material is removed in small pieces. The motion of a tool against a workpiece removes the material. (15)

magnetic (electromagnetic) sensors: Sensors used to determine whether or not changes are occurring in the amount of current flowing in a circuit. (12)

management: The act of planning, directing, and evaluating any activity. (31)

management processes: All the actions people use to ensure that the production processes operate efficiently and appropriately. (9)

managers: People who organize and direct the work of others in businesses. (10)

manned spaceflight: A spaceflight carrying people into space and returning them safely to earth. (23)

manual control systems: Technological control systems that require humans to adjust the processes. (12)

manual transmission: A transmission that has a clutch between the engine and the transmission. (23)

manufactured home: Home that is built in a factory, usually in two halves. Once completed, the home is transported to the building site. (17)

manufacturing enterprise: An organization that contains all of the components needed to operate and maintain a facility to convert materials into finished products with the goal of generating a profit. (15)

manufacturing technologies: Processes, systems, and procedures to create the products people need or want. (13)

maritime shipping: Water transportation on oceans and large inland lakes. (23)

marketing: A company's efforts to promote, sell, and distribute products, structures, and services. (32)

market research: Activity that gathers information about a product's market. A measure of the effectiveness of advertising campaigns, sales channels, or other marketing activities. (32)

mass customization: The ability of manufacturing systems to produce a large output of customized products at a low cost. (14)

mass production: The process of creating a large number of standardized products efficiently by separating the assembling process into small and easily managed steps. (14)

material processing: Changing the form of materials. (6)

materials: Substances from which artifacts are made and categorized as natural, synthetic, or composite. (9)

materials science and engineering: The study of solid materials at the atomic level. (16)

mathematical models: Models that show relationships through formulas. (5)

mathematics: The study of measurements, patterns, and the relationships between quantities, using numbers and symbols. (2)

matter: Any substance that has mass or takes up space. (28)

measurement: The practice of comparing the qualities of an object to a standard. (6)

mechanical advantage: A measure of the ability to amplify the amount of effort exerted using some type of mechanical device. (2)

mechanical conditioning: The use of mechanical forces to change the internal structure of a material. (15)

mechanical controller: Type of adjusting device that uses cams, levers, and other types of linkages to adjust machines or other devices. (12)

mechanical energy: Mechanical energy is energy that an object possesses in relation to its motion or due to the position it is in. (19)

mechanical engineering: An engineering discipline that involves designing, improving, and working with mechanical systems, or machines that cause or involve movement. (3, 20)

mechanical fastening: Assembly process that uses mechanical forces, such as friction, to hold parts together. (15)

mechanical processes: The use of mechanical forces, such as compression or shearing, to change the form of natural resources. (14)

mechanical processing: Food processing method that uses machines to change the physical form of a food product. (27)

mechanical properties: A material's reaction to physical forces. (16)

mechanical sensor: Sensor used to determine the position of components, force applied, or parts movement. (12)

mechanics: 1—Skilled workers in service operations. (10); 2—Branch of physics that studies how physical bodies react when forces are applied to them. (26)

mechatronics: A multidisciplinary field that is largely a combination of mechanics and electronics. (20)

medical implants: Tissues or devices that are surgically placed inside or on the exterior of the body. (26)

medical technologies: Technologies that create the tools used in the prevention, diagnosis, monitoring, and treatment of illness, as well as repair of injury. (13)

medical technology: The narrow field of gathering and analyzing specimens to assist physicians in diagnosing and treating illnesses. (25)

medicine: Diagnosing, treating, and preventing diseases and injuries. (25)

memory: A computer's storage area for data and operating instructions. (22)

merchant ships: Cargo-carrying ships. (23)

mesosphere: A layer of the earth's atmosphere that extends from 22–50 miles (24–80 km) above the earth. (23)

metals: Materials with these characteristics: strong electrical and thermal conductivity; solid at normal temperatures; shiny, hard, malleable, and ductile; difficult to burn; high density. (16)

metamaterials: Materials created by scientists to exhibit characteristics that cannot be found in nature. (20)

metric system: A measurement system based on decimals (multiples of the number 10). (6)

micrometer: A precision measuring tool used to measure diameters. (6)

Middle Ages: Period of civilization that began in Europe about 400 CE and continued until about 1500 CE. (1)

middle management: Level of management in a company that is below top management (the president and vice presidents) but above operating management. (31)

milling: Grinding or processing. (27)

mind map: A diagram that helps engineers and designers generate ideas by visually organizing information or ideas around a central topic or theme. (4)

minerals: Substances with a specific chemical composition that occurs naturally. (14)

mining: The process of obtaining materials from the earth through shafts or pits. (11)

mobile devices: Small computing devices, typically small enough to be handheld, that have a display screen. These devices access the Internet through wireless services. (6)

mock-up: A physical model that represents only how a product or structure will look. It is used to evaluate the styling, balance, color, and other aesthetic features of a technological artifact. (6)

modeling: Simulation of actual events, structures, or conditions. (5, 6)

modem: A device that can convert data into signals a telephone system can recognize. This device also converts these signals back into data. (21)

molecules: The grouping, or bonding, of two or more atoms. (16)

moment: The rotation of a beam about an axis as a result of an applied force. (18)

monitoring device: Sensor that gathers information about an action being controlled. (12)

multiview drawings: One or more views of an object in one drawing in order to fully communicate the shape and size of that object. (5)

muscles: Elastic, fibrous tissues that can contract to produce movement. (26)

musculoskeletal system: Bones, connective tissues, and muscles that create the framework of the body. (26)

myoelectric prosthesis: A prosthesis that is controlled by electric signals. Tiny electrodes are attached to muscles, and those muscles are voluntarily contracted by the person to control movements of the connected prosthesis. (26)

N

nanotechnology: The manipulation of individual atoms and molecules of materials. (16)

natural gas: A fossil fuel resource that occurs in porous rock. It is a combustible gas composed of light hydrocarbons. (14)

natural materials: Materials that can be refined and combined to make products. (10)

natural phenomena: Any observable events that happen in nature and are not human made. (2)

natural resources: Materials that occur naturally on the earth. (10)

natural systems: Groups of elements that work together in nature to perform some type of function not under the direct control of humans. (9)

network: A group of computing devices that are connected together and share resources. (21)

neutrons: Subatomic particles with a neutral charge. (28)

noble gases: Chemical elements that have a full valence orbital. They are stable and typically do not react with other elements. (28)

noise: Unwanted sounds or signals that become mixed in with the desired information. (21)

nondurable products: Products designed to be used only for a short period of time. (15)

nonfood crops: Plants that are not grown for human consumption but for other uses, such as tobacco and nursery plants. (27)

nuclear energy: Energy produced from the splitting of atoms (fission) or the combination of two atoms into a new, larger atom (fusion). (19)

nuts: Crop grown in specific areas of the country and cultivated for its hard-shelled seeds. (27)

O

oblique sketch: Sketch that shows the front view of an object as if a person is looking directly at it. (5)

Ohm's law: A law that defines the relationship between voltage, current, and resistance in an electrical circuit. It is represented by the equation $V = IR$, where V is the voltage in a circuit, I is the current, and R is the resistance. (22)

open-loop control system: A system that does not use output information to adjust the process. (12)

operating management: The lowest levels of management. These managers directly oversee specific operations in the company and are closest to the people who produce the company's products and services. (31)

operating systems: The software that manages the computer's processes, memory, and the operation of all other hardware and software. (22)

optical properties: How a material reacts to visible light waves. (16)

optical sensors: Sensor used to determine light level or changes in light intensity. (12)

optimism: A tendency to look at the more favorable side of events or to expect the best outcomes in various situations. (3)

optimization: Engineering practice of making something as fully effective or as perfect as possible using mathematical procedures and based on the specifications and constraints of the design. (2)

optimize: Improve. (3)

organic chemistry: A subdiscipline of chemistry that studies structures, properties, and reactions of organic (carbon-based) compounds. (28)

organic compounds: Molecules that contain carbon. (28)

organic materials: Materials that come from living organisms. (10)

orthographic assembly drawings: Drawings that use standard orthographic views to show parts in their assembled positions. (5)

orthographic projection: Projection of a single view of an object onto a drawing surface in which the lines of projection are perpendicular to the drawing surface. (5)

outboard motor: A type of power source attached to the stern of a boat. (23)

outputs: Results, good or bad, of the operation of any type of system. (9)

outside directors: Directors who work outside the managerial structure and are not involved in the day-to-day operation of the company. (31)

over-the-counter drugs: Drugs sold directly to the consumer at retail outlets without a prescription from a healthcare professional. (25)

oxidation: Chemical reaction that causes steel to rust. (28)

ozone layer: The upper part of the stratosphere. (23)

P

packaging: Designing, producing, and filling containers that hold a company's products. (32)

parallel circuit: A circuit in which components are connected across common ends or nodes, providing multiple pathways for electrons to flow. (22)

partnership: A form of private ownership in which two or more people own and operate the company. (10, 31)

passive collectors: Solar collectors that directly collect, store, and distribute the heat they convert from solar energy. (19)

passive voice: Writing style that focuses on the subject that experiences an action rather than the subject that performs the action. (8)

pasteurization: Preservation method that uses heat to kill harmful microorganisms. (27)

patent: A property right granted by the government to inventors or innovators, allowing them exclusive use to make and sell their inventions or innovations. (2, 8)

pathways: Vehicular support system that contains the structures that vehicles travel. (11)

pectoral girdle: Set of bones that connects the bones of the arms to the axial skeleton to provide support for the movement of the upper limbs. (26)

pelvic girdle: The right and left hipbones. (26)

people skills: The abilities needed to work with people in a cooperative way. (30)

performance tests: A type of test that focuses the final measurable performance characteristics. (7)

perigee: The point at which an orbit comes closest to the earth. (23)

periodic table of elements: A grid used to organize the different chemical elements. (28)

periods: Rows in the periodic table of elements. (28)

personal skills: The ability to grow and manage personal actions on a job. (30)

perspective sketch: A sketch that shows the object as the human eye or a camera sees it. Lines are drawn to meet at a distant vanishing point on the horizon. (5)

petrochemicals: Chemicals made from crude oil. (28)

petroleum: A fossil fuel resource that is an oily, flammable mixture of different solid and liquid hydrocarbons. (14)

photoelectric cells: Electrical components that produce electric current and voltage when exposed to light. They can also be used to control circuits based on the input of light. Also called *photocells*. (22)
physical properties: The characteristics due to the structure of a material, including size, shape, density, moisture content, and porosity. (16)
physical sciences: The study of forces, motion, energy, matter, and their interactions. (28)
physical tests: Tests that assess whether features of a new technological product can withstand physical demands such as tensile stress, compression, and torsion. (7)
pictorial assembly drawings: Drawings that show the assembly using oblique, isometric, or perspective views. (5)
pile foundation: Type of foundation that uses piles driven into the ground until they reach solid soil or rock. Used on wet, marshy, or sandy soils. (17)
pivot-sprinkler: Irrigation system that rotates around a central point that is attached to a water source. (27)
place utility: A value provided by being able to move things from one place to another. (23)
planing machines: Metalworking machine tools that produce flat surfaces. The workpiece is reciprocated under the tool to generate the cutting motion. The tool is moved one step across the work for each cutting stroke. (6)
planning: Management process that involves developing goals and objectives. (11)
plan view: The top view of a building that includes cross-sectional views of different levels. (18)
plasma: 1—The liquid component of the blood that helps the other components of the blood flow through the cardiovascular system. (26); 2—State of matter than has no definite shape or volume, but can form structures when subjected to a magnetic field. (28)
plastic range: The range to which a material can be stretched, compressed, or bent without being destroyed. (15)
platelets: The portion of the blood that helps to clot the blood when a wound occurs. (26)
plow: Piece of tilling equipment that breaks, raises, and turns the soil. (27)
point load: A load that is applied in one spot on a beam. (18)
point of presence (POP): Connection points let local users access a company's network, usually through a local or toll-free phone number or a dedicated line. (21)
pollution: The introduction of a harmful material into the environment. (29)
polymers: Materials containing many repeating molecules. (16)
post and beam construction: A horizontal element, or beam that spans a gap and rests on two columns or posts. (18)
potable water: Water that is safe for drinking. (17)
potential energy: Energy stored for later use. (19)
power: 1—The rate at which work is accomplished or energy is changed from one form to another. (11); 2—The amount of energy consumed over time by a technological system. (13)
power-generation system: A system that uses an engine as an energy converter. (23)
power source: The source of the electron flow in an electrical circuit, such as a battery or generator. (22)
power transmission: Taking the energy generated by a converter and changing it into motion. (20)
power-transmission system: A system that controls and directs the power of an engine to do work. (23)
precision measurement: Measurement made to 1/1000″ to 1/10,000″ in the US customary system or to 0.01 mm in the metric system. (6)
predictive analysis: 1—Determining how well a proposed solution solves a problem, before the solution is actually produced. (2); 2—A method used to solve problems and test initial ideas through the application of mathematical or scientific concepts before building models, prototypes, or final designs. (5)
prescription drugs: Drugs that physicians prescribe and pharmacists dispense. (25)
preservatives: Chemicals added to food in small amounts to delay food spoilage and ensure food quality. (27)
prevention: Using knowledge, technological devices, and other means to help people maintain their health. (11)
primary food-processing technology: Technology that produces the basic ingredients for food. (27)
primary manufacturing: The resource processing systems of manufacturing. (14)
primary processing: The conversion of material resources into industrial materials. (11)
prime mover: A device that changes a natural source of energy into mechanical power. (19)
primitive conditions: Situation in which primitive people existed with nature, without attempting to control nature or improve the natural condition. (1)
private enterprises: Enterprises owned by individuals or groups of people who invest their money, take risks, and hope to reap a profit. (31)

problem-solving process: The use of thinking and reasoning skills to consider and find solutions for a situation that has an identified goal state, readily available solutions, and one or more reasonable pathways for solving. (4)

processes: The steps needed to complete a series of tasks in a technological system. (9)

product design: The development of a product or structure to the needs of customers. (4)

production processes: Actions completed to perform the function of the technological system. (9)

production workers: People who process materials, build structures, operate transportation vehicles, service products, or produce and deliver communication products. Often categorized as unskilled, semiskilled, or skilled workers, based on training and work experience. (10)

products: Substances that are produced as the result of a chemical reaction. (28)

profit: Money remaining after all expenses are paid. (9)

programmable logic controller (PLC): A device that uses microprocessors to control machines or processes. (14)

programming language: A formal language or set of rules to communicate operations for a computer system to perform. (22)

propagation: A step in the agricultural process that allows biological organisms to reproduce. (11)

proprietorship: A business in which a single owner has complete control of the company. (31)

propulsion: A vehicular system that generates motion through energy conversion and transmission, enabling a vehicle to move from one location to another. (11, 23)

prosthesis: An artificial body part. (25)

protons: Subatomic particles that have a positive charge. (28)

prototype: A full-size and fully functional model of a system, assembly, or product that is built to test the operation, maintenance, or safety of the design. (6)

public enterprises: Enterprises controlled by the government or a special form of corporation. These enterprises are operated for the general welfare of society and cannot, or should not, make a profit. (31)

public relations: A company's programs that communicate the company's policies and practices to governmental officials, community leaders, and the general population. (32)

pulleys: Grooved wheels attached to an axle. (10)

pulmonary circuit: The circuit that carries blood from the heart to the lungs, where oxygen enters the blood through breathing. The blood in this circuit is then circulated back to the heart. (26)

Q

qualitative: Type of data collection method that involves numbers and objective data. (7)

quality control: The process of setting standards, measuring features, comparing them to the standards, and making corrective actions. (6)

quantify: Expressing situations as numbers, equations, and inequalities. (2)

quantitative: Type of data collection method that involves words or subjective data. (7)

quantities: Amounts, numbers. (2)

R

radiant energy: Energy in the form of electromagnetic waves. (19)

radiation: Heat transfer using electromagnetic waves. (19)

radiology: The use of electromagnetic radiation (waves) and ultrasonics (high frequency sounds) to diagnose diseases and injuries. (25)

rafters: Angled boards resting on the top plate of the exterior walls. (17)

rapid prototyping: The fabrication of a scaled part or model of a product using three-dimensional computer-aided design (CAD) data. (14)

reactants: Substances that begin a chemical reaction. (28)

reaction forces: Forces that resist the applied load at the beam supports. (18)

receive: To recognize and accept information. (11)

reciprocating motion: Cutting or feed motion in which the tool or the work is moved back and forth. (6, 20)

rectifier: An electrical device used to convert ac into dc by allowing a current to flow through it in one direction only. (22)

red blood cells: Blood cells that carry oxygen to and remove carbon dioxide from the body's cells. (26)

refined sketch: Sketch that merges ideas from two or more rough sketches, blending the different ideas into a unified whole. (5)

register: A computer component that allows the CPU to access different information or data within the computer's memory by storing the location of the information. (22)

reinforced concrete: Concrete that has wire mesh or steel bars embedded into it to increase its tensile strength. (17)

Renaissance: Period of civilization that began in the early 1300s in Italy and lasted until 1600; a time of great cultural advancement. (1)

renewable energy: Energy source that can be used up, but can also be replaced within the normal life cycle of the energy source. (10)

repairing: Replacing worn or broken parts in a product or system. (33)

research: The process of seeking and discovering knowledge. (32)

residential buildings: Structure in which people live; can be single-family or multiple-unit dwellings. (17)

residential structures: Buildings in which people live, such as homes, townhouses, condominiums, and apartment buildings. (11)

resistors: Electrical components that restrict the flow of electrons in a circuit and reduce the current and voltage by turning it into heat energy. (22)

respiratory system: The system that brings oxygen into the body, delivers it to cells, and rids the body of carbon dioxide. (26)

retained earnings: The profits a company holds and uses to enlarge its operations. (32)

retrieve: To bring back information that has been stored. (11)

robotic prosthesis: A prosthesis that uses advanced biosensors, as well as mechanical and electrical components. It is capable of natural, agile, and complex functions. (26)

robots: Mechanical devices that can perform tasks automatically or with varying degrees of direct human control. (14)

rotary motion: 1—Cutting or feed motion in which round cutters are used or the work is spun around an axis. (6); 2—Spinning around an axis, or motion in a circle. (20)

rotary-wing aircraft: An aircraft that develops lift by spinning an airfoil. (23)

rough sketches: Incomplete, unrefined sketches used to communicate design solutions. (5)

routers: Devices that determine how to send information from one computer to another. (21)

rule: A rigid or flexible strip of metal, wood, or plastic with measuring marks on its face that is used for linear measurement. (6)

S

salary: Payment based on a period of time longer than an hour, such as a week, month, or year. (32)

sales: The activity involving the physical exchange of products for money. (32)

sawing machines: Machines that use teeth on a blade to cut material to a desired size and shape. (6)

scale model: A three-dimensional miniature physical representation of a design solution that maintains accurate relationships between all aspects and features of the design. (6)

schematic drawings: Systems drawings used to show the way parts in a system relate to each other and work together. (5)

science: The study of the natural world. (2)

scientific method: Method to structures scientific research in a way to ensure valid experimental results can be obtained. (2)

scientific models: Simulations based on existing or accepted findings related to the natural world. (5)

scientific notation: A way to express numbers that includes just the digits of a number with the decimal point placed after the first digits, followed by a multiplication of a power of ten that will put the decimal back in the correct place. (3)

screw: An inclined plane wrapped around a shaft. (10)

scroll saws: Saws that use a straight blade that is a strip of metal with teeth on one edge. The blade is clamped into the machine. The machine then moves the blade up and down to produce a reciprocating cutting motion. (6)

search engines: Special Internet sites that operate on the principle of key words to allow individuals to search the Web by topic. (21)

secondary food-processing technology: Technology that produces finished food products. (27)

secondary manufacturing: Product-specific manufacturing systems that turn industrial materials into final products. (14)

secondary processes: The process of changing industrial materials into industrial equipment and consumer products. (11)

second-class lever: A lever in which the load is between the effort and the fulcrum. (10)

seed-tree cutting: Logging method in which all trees, regardless of species, are removed from a large area, except for three or four per acre that are used to reseed the area. (14)

seismographic study: The creation of and measurement of shock waves bouncing off various rock layers in order to construct a map of rock formations to aid in searching for petroleum and natural gas. (14)

selective cutting: Logging method in which mature trees of a desired species are selected and cut from a plot of land. (14)

semiconductors: Material that has electrical conductivity between an insulator and conductor. (16)

sensor: Device that detects or measures physical quantity and responds to it by recording information and initiating some other function. (11)

separating: Using tools to shear or machine away unwanted material. (11)

separating processes: Secondary manufacturing processes that remove excess material to make an object of the correct size and shape. (15)

sequential logic: Digital logic in which outputs depend not only on the present input but also on the past inputs. (22)

series circuit: An electrical circuit in which components are connected in a way that provides only a single pathway for electrons to flow. (22)

series-parallel circuits: Circuits that have components that are arranged in series and components that are arranged in parallel. (22)

server: A device or computer program that stores, processes, and shares data and resources with other devices or clients connected to the network. (21)

servicing: 1—The maintenance, repair, and reconditioning that structures require to maintain their integrity. (11); 2—Doing routine tasks that keep a product operating. (33)

shaping machines: Metalworking machine tools that produce flat surfaces. The tool moves back and forth over the workpiece to produce the cutting motion. The work is moved over after each forward-cutting stroke to produce the feed motion. (6)

shear: Unaligned forces that push two segments of an object in opposite directions, causing them to slide over each other. (18)

shear force diagram: Diagram that illustrates how shear forces occur at different points on a beam, in order to determine where shearing may be greatest. (18)

shearing: Separating process in which opposing edges of blades, knives, or dies are used to fracture the unwanted material away from the work. (15)

sheathing: Plywood, fiberboard, or rigid foam sheets used to cover exterior walls. (17)

signal: Any measurable quantity used to communicate information. (22)

significant figures: The digits of a number that are considered to be reliable as a result of a calculation or measurement. (3)

sill: A wood piece attached to the top of a foundation. (17)

site preparation: Step in the construction process that involves removing existing buildings, structures, or vegetation that interferes with locating the new structure. (11)

slab foundation: Type of foundation that becomes the floor of a building, allowing the weight of the building to be spread over a wide area. Used on soft soils. (17)

smart materials: Materials that have properties that can be modified in a controlled manner by external forces. (16)

smoking: Preservation method that adds flavor to meat and fish, while preserving them. (27)

social future: The type of future that is concerned with the types of relationships people want with each other. (29)

social networking service: An online platform that allows people to post and share messages, pictures, and videos either publicly or privately. (21)

social systems: Organized groups of individuals that have a variety of goals, functions, or characteristics. (9)

society: A collection of people who generally live in a community, region, or nation, and share common customs and laws. (30)

socioethical skills: The skills that involve understanding the implications of actions on people, society, and the environment. (30)

soffit: Aluminum, vinyl, or plywood material that is installed to enclose the underside of an overhang. (17)

software: A term used to describe instructions that direct a computer to perform specific tasks. (22)

software engineering: An engineering field that focuses on using the engineering design process to efficiently produce effective software solutions. (22)

softwood lumber: Lumber produced from needle-bearing trees and produced in nominal sizes. (14)

sole plate: The strip at the bottom of a framed wall. (17)

sole proprietorship: A company owned by one person. (10)

solid: 1—Material that holds its size and shape and can support loads without losing its shapes. (10); 2—State of matter that has a fixed shape and volume. (28)

solid and structural analysis: A process specifically devoted to preventing and determining how and why objects fail. (20)

solid-fuel rockets: A rocket that uses a powder or spongelike mixture of fuel and oxidizer. (23)

solid mechanics: 1—The study of how stresses are distributed as they are applied to certain materials and the resulting strains on the materials. (18); 2—The analysis of the behavior of solid materials or systems when they are subjected to stresses, loads, and other external forces. (20)

solid model: Complex computer model that takes into account both the surface and interior substance of an object. (5)

space-transportation systems: A transportation method that uses manned and unmanned flights to explore the universe. (23)

spade bits: A drilling tool with flat cutters on the end of a shaft. The bottoms of the cutters are shaped to produce the cut. (6)

span: The measurements from wing tip to wing tip. (24)

specifications: Criteria or standards, such as quality of craftsmanship, materials, or precision. (7)

spread foundation: Type of foundation that sits on a low, flat pad called a footing. Used on rock and hard soils. (17)

sprinkler irrigation: Irrigation system that produces artificial rain to water crops. (27)

spur gear: A gear containing straight teeth that are parallel to the shaft axis. The gear is used to transmit power from one parallel shaft to another. (20)

square: A tool with a blade that is at a right angle to the head. Squares are commonly used to take measurements at 90° angles. (6)

stakeholder: Any person or group that has a vested interest in the project, such as managers, government agencies, creditors, owners, or funders. (8)

standard measurement: Measurement made to the foot, inch, or fraction of an inch in the customary system or to the nearest whole millimeter in the metric system. (6)

standard stock: Industrial materials made from converted material resources. (14)

static loads: The weight of a structure itself (dead load) and weight added to the structure under normal use (live load). (18)

statics: Branch of mechanics that uses Newton's laws of motion to analyze loads placed on objects at rest or at a constant velocity. (26)

static structural failure: The failure of an object that occurs when loads or forces have been applied to an object and the object either breaks or is deformed due to some criterion. (20)

stoichiometry: A part of chemistry that studies the amount of matter that is involved in a chemical reaction. (28)

Stone Age: The earliest period of known civilization that began about two million years ago and in which humans used stone as tools. (1)

store: To retain information for later use. (11)

strain: The relative change in size or shape of an object due to stress caused by forces. (18)

stratosphere: The region in the earth's atmosphere located above the troposphere. (23)

stress: The force per unit of area or unit of length of an object. (18)

stroke: In an internal combustion engine, the movement of a piston from one end of a cylinder to another. (19)

structural engineering: The study of the framework of structures: designing, analyzing, and constructing structural components or assemblies to resist the stresses and strains of loads and forces that affect them. (18)

structure: Vehicular systems that provides spaces for devices in vehicles and protects people and cargo. (11, 23)

studs: Upright framing members nailed to the sole plate. (17)

subfloor: A base, usually made from plywood or particleboard, that is installed on top of the joists and acts as a base for finished flooring material, such as carpet, tile, or wood. (17)

submersible: A special type of boat that can travel on the surface of or under water. (23)

subtractive manufacturing: The removal of material from raw stock until a desired shape and size is reached. (14)

superstructure: The framework of a building or tower that is constructed on a foundation, or pipes used in pipeline, road surfaces, airport runways, and railroad tracks. (11)

supervisor: The lowest levels of management. These managers directly oversee specific operations in the company and are closest to the people who produce the company's products and services. (31)

support staff: People who carry out tasks such as keeping financial records, maintaining sales documents, and developing personnel systems. (10)

support system: External operations and facilities that maintain transportation systems. (11)

surface grinder: Type of grinder in which a rotating grinding wheel is suspended above the workpiece. The work is moved back and forth under the wheel to produce the cutting motion and moved slightly, or indexed, after each grinding pass to produce the feed motion. (6)

surface mining: Mining method used when a coal vein is not very deep underground. After surface layers of soil and rock are stripped from above the coal, the coal is dug up with giant shovels. Also called *open-pit mining*. (14)

surface models: Computer model that shows how a project will appear to an observer. Used mostly to develop sheet metal products. (5)

suspension: A vehicular system that keeps a vehicle held in or on the medium it is traveling by producing the proper support for the weight of the vehicle and cargo (land, water, air, space). (11, 23)

suspension bridge: A type of bridge that uses cables to carry loads. (17)
swather: Farming machine that cuts and windrows hay in one pass over a field. (27)
synergism: The working together of two or more individuals that produces an effect greater than the sum of individuals. (5)
synthetic materials: Materials that are manufactured. (10)
system: Sets of interacting elements that work interdependently to form a complete entity that has an overall function or purpose. (3); 2—Any entity or object that consists of parts, and each of those parts has a relationship with all other parts and the entity as a whole. (9)
system design: The arrangement of components to produce a desired result. (4)
systemic circuit: The circuit that carries the blood from the heart to the cells of the body. The cells then transfer oxygen, nutrients, hormones, and waste with the blood before it is carried back to the heart. (26)
systems drawings: Engineering drawings that show the relationship of components in mechanical, fluidic (hydraulic and pneumatic), and electrical and electronic systems. (32)
systems thinking: The process of understanding and mentally exploring systems in multiple ways. (9)

T

table: Numerical data or textual information displayed in column format. (8)
tail assembly: A structure that provides steering capability for an aircraft. (23)
tankers: Large vessels used to move liquids. (23)
technical standards: Previously established norms and guidelines. (7)
technical writing: 1—A form of expository, informative, or explanatory writing used to communicate complex information to those who need it for a specific reason. (3); 2—Writing style that conveys complex information in a clear and concise manner. (8)
technicians: Skilled workers in laboratories and product-testing facilities. (10)
technological design: Open-ended, trial-and-error process of creating a problem solution to meet needs and desires. (2)
technological future: The type of future that examines the type of human-built world that is desired. (29)
technologically literate: Individuals who understand and are able to direct new technology. (1)
technological products: The artifacts that people build. (33)
technological services: The outputs that people use, but do not own. (33)
technological systems: A group of components designed by humans to function together to complete a desired task. (9)
technology: Humans using objects (tools, machines, systems, and materials) to change natural and human-made (built) environments. (1)
technology assessment: Evaluation of the impacts of technology on people, society, and the environment by groups of people. The task of controlling technology is placed on the government and the courts. (33)
technology transfer: Technology that migrates from one field to another to solve a different type of problem. (2)
telecommunications: The transmission of information over a distance for the purpose of communication. (11)
telecommunications technology: A communication process that uses electromagnetic waves to carry a message over distances. (11)
telemedicine: The provision of healthcare at a distance through telecommunication technology. (21)
tendons: Connective tissues that connect skeletal muscles to bones by connective tissues. (26)
tensile strength: The maximum pulling force, or tension, a material can withstand prior to failure. (16)
tension: Force that pulls material apart. (18)
terminals: 1—Vehicular support system that contains structures that house passenger and cargo storage, and loading facilities. (11); 2—A structure where transportation activities begin and end. They are used to gather, load, and unload passengers and goods. (23)
tests: Procedures designed to examine a particular technological device to ensure it works as designed and that it meets the desired specifications. (2)
therapeutic radiology: The treatment of diseases or disorders with radiation. (25)
thermal conditioning: Heat is used to change the internal properties of a material. (15)
thermal conductivity: The extent a material can conduct, or transfer, heat. (16)
thermal energy: Energy created by the internal movement of atoms in a substance. Atoms that move or vibrate rapidly emit heat. (19)
thermal fatigue: The amount of stress a material can withstand from repeated changes in heating and cooling. (16)

thermal processes: The use of heat to melt and reform a natural resource. (14)
thermal processing: Food processing method that uses heat as the primary energy to convert a food. (27)
thermal properties: The effect temperature has on the characteristics of a material. (16)
thermal sensors: Sensor used to determine changes in temperature. (12)
thermodynamics: The study of heat and temperature and the relation of these factors to work, energy, and power. (20)
thermoplastics: Polymers that do not share electrons between atoms, and thus have no covalent bonding. Thermoplastics can be softened when heated to be remolded. (16)
thermosets: Polymers that cannot be reheated to form a new mold once already cured. (16)
thermosphere: The region in the earth's atmosphere that lies just above the mesosphere. (23)
thinking skills: The ability to use mental processes to address problems and issues. (30)
third-class lever: A lever in which the effort is placed between the load and the fulcrum. (10)
thrust: A force that pushes outward at the base. (18)
tillage: The process of breaking and pulverizing soil to condition seedbeds. (27)
timber cruising: Teams of foresters measure the diameters and heights of the trees to find stands of trees that can be economically harvested. (14)
time: A measurement of how long an event lasts. (9)
tissue engineering: The practice of developing these biologically active materials. (26)
tolerance: The amount a dimension can vary and still be acceptable. (7)
tools: Artifacts that expand what humans are able to do. (9, 10)
top management: Level of management that is responsible for the entire company's operation. (31)
top plate: Double ribbon of 2 × 4s at top of a wall. (17)
torque: The measure of the twisting force on an object. (20)
torsion: An applied rotational force that causes an object to twist. (18)
trachea: Part of the body that functions as a filter for the air we breathe and channels the air into the two branches of the bronchi. (26)
trade-offs: Decisions that arise in which a choice must be made between two competing items. (5)
transaxle: A part in a front-wheel drive vehicle that combines the transmission and differential into a single unit. (23)
transducers: Devices that change energy from one form to another. (21)
transfemoral prosthesis: A prosthesis used to replace the knee joint as well as the lower part of the leg. (26)
transhumeral prosthesis: A prosthesis used when the elbow joint is missing from the arm. (26)
transistors: Semiconductor components that can amplify signals for components such as speakers or act as electronic switches. (22)
transition elements: Elements that can add additional electrons to the second to last orbital, as well as their valence orbital. The two outer shells can chemically bond with other elements to form chemical compounds. (28)
transmitting: Sending a coded message from a sender to a receiver. (11)
transportation: All activities that relocate people or their possessions. (23)
transportation engineering: Study of the design and optimization of transportation infrastructure. (18)
transportation technologies: Technologies that move people and products from one location to another through land, water, air, and space travel. (13)
transradial prosthesis: A prosthesis that is attached below the elbow. (26)
transtibial prosthesis: A prosthesis used to replace parts missing below the knee. (26)
trap: A water-filled *U*-shaped piece of pipe that prevents sewer gases from leaking into a building. (17)
treatment: Using knowledge, technological devices, and other means to fight disease, correct body malfunctions, or reduce the impact of a physical condition. (11)
Trombe wall: A black concrete or masonry wall that has glass panels in front of it. It is used in an indirect-gain solar system. (19)
troposphere: The lowest region of space above the earth in which spacecraft and satellites operate. (23)
truss: Triangle-shaped structure that contains both rafters and ceiling joists in one unit. They are manufactured in a factory and then shipped to the building site. (17)
truss bridge: A type of bridge that uses small triangular parts to support the deck. (17)
turbofan engine: An engine that is used in most commercial aircraft and that operates at lower speeds than a turbojet engine. (23)
turbojet engine: A type of jet engine developed during World War II. It operates at high speeds and is used in military aircraft. (23)
turboprop engine: A variation of a jet engine that operates efficiently at low speeds. (23)
turning machines: Machines in which a workpiece is held and rotated on an axis. (6)

twist drills: Shafts of steel with points on the end to produce a chip. (6)

U

ultrasonic sensor: Device that functions like a human eye to detect distances, shapes, sizes, speed, and direction. (11)

ultrasound: An imaging technique that uses high frequency sound waves and their echoes to develop an image of the body. (25)

underground mining: Mining method that requires shafts in the earth to reach coal deposits. (14)

undesirable outputs: Outputs that are not wanted or planned for. (12)

unintended outputs: Outputs that were not anticipated when a system was designed. (12)

unit: A physical quantity that is consistent to a standard form of measurement. (3)

unlimited liability: The proprietor of a business is responsible for all the debts the business incurs. Business income and liabilities are not separate from the owner's personal finances. (31)

US customary system: Measurement system, used in the United States, that includes units such as the inch, foot, mile, pound, pint, quart, and gallon. (6)

utilities: System of a structure that provides water, electricity, heating, cooling, or communications. (11)

V

valence electrons: Electrons in the outermost shell (valence orbit) of an atom. (16, 28)

valence orbital: The outermost orbital of an atom; it holds only eight electrons. (28)

value: A measure of the functional worth the customer sees in the product. (32)

vegetables: Crops with edible leaves, stems, roots, and seeds that provide important vitamins and minerals for the daily diet. (27)

vehicles: A technological artifact designed to carry people and cargo on a pathway. (23)

vehicular systems: Onboard technical systems that operate a vehicle. (11)

veins: The blood vessels that carry the blood back from the capillaries to the heart. (26)

Venturi effect: The reduction of fluid pressure when the fluid flows through a constricted section of pipe. (24)

vice presidents: Managers who report to the president or CEO. (31)

viscous fluid: Fluid that often sticks to a moving object's surface. (24)

voltage: In an electrical circuit, the pressure of the electrons waiting to move toward the positive charge. (22)

W

wage: A set rate paid for each hour worked. (32)

wage control: Monitoring the number of hours worked by employees who produce products. (12)

wastewater: Water that comes from sinks, showers, tubs, toilets, and washing machines. It is not safe to drink. (17)

water resource engineering: The study of ways to provide safe and reliable access to water for a variety of uses, to remove and treat wastewater, and to avoid the damages that can result from excess water. (18)

water-transportation systems: A system that uses water to support a vehicle. (23)

water turbine: A device used to power electric generators in hydroelectric power plants. It consists of a series of blades arranged around a shaft. As water passes through the turbine at a high speed, the blades spin the shaft. (19)

watt: A unit equal to 1 J of work per second. (20)

wedge: An inclined plane that is used to split and separate materials and to grip parts. (10)

wellness: The state of being in good health, both physically and mentally. (25)

what-if scenarios: Method for developing preliminary design solutions that start with a wild proposal, investigates the proposal's good and bad points, and then uses the good points to develop solutions. (5)

wheel and axle: A simple machine on which a shaft is attached to a disk. (10)

white blood cells: Blood cells that work to fight infections. (26)

windrows: Bands of hay. (27)

wire-frame model: Type of computer model developed by connecting all the edges of the object, producing a structure made up of straight and curved lines. (5)

work: Application of a force that moves a mass a distance in the direction of the applied force. (19)

World Wide Web (WWW): A computer-based network of information resources. (21)

x-rays: Electromagnetic waves that are short enough to pass through solid materials, such as paper and human tissue, but are absorbed by denser materials, such as metals and bones. (25)

yarding: The process of gathering logs in a central location. (14)

yield point: The point at which, when sufficient force is applied, a material does not return to its original shape. (15)

INDEX

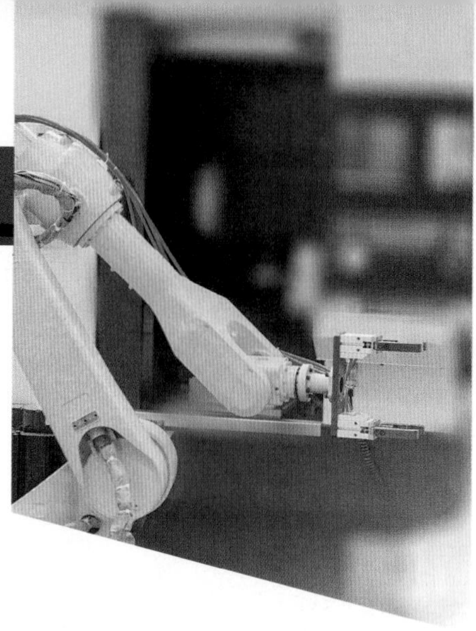

A

B

C

D

E

F

G

H

I

J

K

L

M

N

O

P

Q

R

S

T

U

V

W

Z